Springer Proceedings in Advanced Robotics 19

The Springer Proceedings in Advanced Robotics (SPAR) publishes new developments and advances in the fields of robotics research, rapidly and informally but with a high quality.

The intent is to cover all the technical contents, applications, and multidisciplinary aspects of robotics, embedded in the fields of Mechanical Engineering, Computer Science, Electrical Engineering, Mechatronics, Control, and Life Sciences, as well as the methodologies behind them.

The publications within the "Springer Proceedings in Advanced Robotics" are primarily proceedings and post-proceedings of important conferences, symposia and congresses. They cover significant recent developments in the field, both of a foundational and applicable character. Also considered for publication are edited monographs, contributed volumes and lecture notes of exceptionally high quality and interest.

An important characteristic feature of the series is the short publication time and world-wide distribution. This permits a rapid and broad dissemination of research results.

Indexed by WTI Frankfurt eG, zbMATH.

More information about this series at http://www.springer.com/series/15556

spar

Bruno Siciliano · Cecilia Laschi ·
Oussama Khatib
Editors

Experimental Robotics

The 17th International Symposium

Springer

Editors
Bruno Siciliano
Department of Electrical Engineering
and Information Technology
University of Naples Federico II
Naples, Italy

Cecilia Laschi
Department of Mechanical Engineering
National University of Singapore
Singapore, Singapore

Oussama Khatib
Department of Computer Science
Stanford University
Stanford, CA, USA

ISSN 2511-1256 ISSN 2511-1264 (electronic)
Springer Proceedings in Advanced Robotics
ISBN 978-3-030-71150-4 ISBN 978-3-030-71151-1 (eBook)
https://doi.org/10.1007/978-3-030-71151-1

This Springer imprint is published by the registered company Springer Nature Switzerland AG
The registered company address is: Gewerbestrasse 11, 6330 Cham, Switzerland

Foreword

At the dawn of the century's third decade, robotics is reaching an elevated level of maturity and continues to benefit from the advances and innovations in its enabling technologies. These all are contributing to an unprecedented effort in bringing robots to the human environment in hospitals and homes, factories and schools, in the field for robots fighting fires, making goods and products, picking fruits and watering the farmland, saving time and lives. Robots today hold the promise for making a considerable impact in a wide range of real-world applications from industrial manufacturing to health care, transportation and exploration of the deep space and sea. Tomorrow, robots will become pervasive and touch upon many aspects of modern life.

The *Springer Tracts in Advanced Robotics (STAR)* was launched in 2002 with the goal of bringing to the research community the latest advances in the robotics field based on their significance and quality. During the latest fifteen years, the STAR series has featured publication of both monographs and edited collections. Among the latter, the proceedings of thematic symposia devoted to excellence in robotics research, such as ISRR, ISER, FSR and WAFR, has been regularly included in STAR.

The expansion of our field as well as the emergence of new research areas has motivated us to enlarge the pool of proceedings in the STAR series in the past few years. This has ultimately led to launching in 2016 a sister series in parallel to STAR. The *Springer Proceedings in Advanced Robotics (SPAR)* is dedicated to the timely dissemination of the latest research results presented in selected symposia and workshops.

This volume of the SPAR series brings the proceedings of the seventeenth edition of the International Symposium on Experimental Robotics, ISER 2020, in Malta. The nine-part volume edited by Bruno Siciliano, Cecilia Laschi and Oussama Khatib is a collection of 54 contributions spanning a wide range of topics in robotics, including aerial robots, design and prototyping, field robotics, human–robot interaction, machine learning, mapping and localization, multi-robots, perception, planning and control. Experimental validation of algorithms, concepts or techniques is the common thread running through this large research collection.

The seventeenth edition of ISER brings this unique reference on the current developments and new directions of experimental robotics—a genuine tribute to its contributors and organizers!

January 2021

Bruno Siciliano
Oussama Khatib
SPAR Editors

Preface

The International Symposium on Experimental Robotics, ISER, is a series of biennial international symposia. Sponsored by the International Foundation of Robotics Research, IFRR, the goal of ISER symposia is to provide a single-track forum for research in robotics that focuses on the novelty of theoretical contributions validated by experimental results.

Experimental robotics is the collection of contributions for the 17th edition of ISER. A total of 54 papers were selected via peer review out of 108 submissions. These contributions cover a wide range of topics in robotics and are organized in nine chapters: aerial robots, design and prototyping, field robotics, human–robot interaction, machine learning, mapping and localization, multi-robots, perception and planning and control.

This edition should have taken place in Malta, November 9–12, 2020, but it has been postponed due to the pandemic. As we write this preface, we still hope to be able to safely gather in Malta later in 2021. We are grateful to the contributors of this volume for their work and enthusiasm. Some have been contributing to ISER over the years, and it is because of their continued participation that the symposium maintains its high quality. We also thank the colleagues in the steering committee for their important role in ensuring peer reviews of the papers for the symposium, as well as those in the advisory committee for their advice assisting us with their suggestions.

We are grateful to the Springer staff for their cooperation and assistance on this SPAR volume and all past volumes of the ISER series. Our special thanks and appreciation go to Fanny Ficuciello, Lourdes Cini and Daniela Passariello for their commitment, tireless work and great support to ISER 2020.

Bruno Siciliano
Cecilia Laschi
Oussama Khatib

Contents

Design and Prototyping

Field Robotics

Multi-robots

Perception

Aerial Robots

Radar-Inertial State Estimation and Obstacle Detection for Micro-Aerial Vehicles in Dense Fog

Andrew Kramer(✉) and Christoffer Heckman

University of Colorado Boulder, Boulder, CO 80309, USA
{andrew.kramer,christoffer.heckman}@colorado.edu

Abstract. Disaster response and search and rescue missions are among the most difficult missions in which an autonomous robot can be deployed. These require a robot to autonomously navigate chaotic, unstructured indoor and outdoor environments. However popular state estimation and mapping methods using vision and lidar are severely handicapped by fog, smoke, or other airborne particulates; conditions common in disaster scenarios. This work presents radar-based methods for state estimation and mapping that are not affected by smoke and fog. We demonstrate the performance of these methods are comparable to other popular methods in favorable conditions. We also demonstrate visual and lidar-based methods degrade quickly in fog, while our methods do not.

1 Introduction

Autonomous robots in disaster response and search and rescue missions are often required to autonomously navigate chaotic, unstructured environments. GPS and other external references are often unavailable, so they must perform state estimation and mapping with only their onboard sensors. Visual, lidar, and inertial sensors are often used for these tasks, but the quality of the estimates from visual and lidar methods quickly degrades in the presence of fog, smoke, or other airborne particulates that occlude visual information and scatter lidar beams. Further, such conditions, referred to as visually degraded conditions, are common in disaster response missions.

Reliable methods for perception in visually degraded environments (VDEs) have not yet been developed. Most state-of-the-art methods for perception and state estimation are significantly impaired in VDEs, e.g. cameras and lidar cannot capture useful information in the presence of heavy smoke or fog as shown in Fig. 1. It is clear that there exists a need for reliable methods for state estimation and perception which are capable of generalizing across diverse environments and operating conditions.

To this end we present novel methods for state estimation and mapping using millimeter wave radar. Our state estimation method uses low-power, lightweight sensors and provides estimates in three dimensions, while previous methods have

B. Siciliano et al. (Eds.): ISER 2020, SPAR 19, pp. 3–16, 2021.
https://doi.org/10.1007/978-3-030-71151-1_1

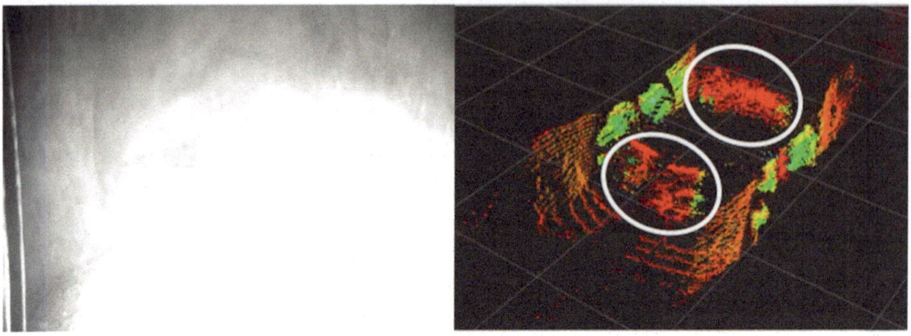

Fig. 1. In the presence of dust, measurements from a monocular camera (left) are saturated. VLP-16 lidar scans (right) are corrupted with unbearable noise (white circles).

been limited to 2D environments. Our mapping method uses a novel radar-specific sensor model and a unique learned noise filter based on the PointNet architecture [17]. These two advances could allow for the implementation of radar-based planning and control on lightweight, agile vehicles like micro aerial vehicles (MAVs).

2 Related Work:

Good results for ego-velocity estimation have been obtained using stereo camera configurations and RGB-D sensor systems [14,22,24]. The use of optical flow, an IMU, and a range sensor is proposed in [19] and [18]. Similarly, methods like [27] take advantage of a lidar or a combination of lidar and IMU measurements as in [8] and [6]. Although these approaches are able to efficiently estimate ego-velocity, they are ill-equipped for use in VDEs.

Several other methods have been proposed for perception in smoke and fog with applications to disaster response robotics. [16] proposes system using structured light for 3d perception under strong ambient light or low levels of fog using a custom built projector and sensor unit. [10] proposes a navigation system using thermal imaging and inertial measurements. These are promising areas of research but there is another sensing option that is well suited for perception in VDEs: millimeter wave radar.

Millimeter wave radar is well-established in the automotive industry and is commonplace in advanced driver assistance systems (ADAS) [5]. But very few methods for using radar in robotic state estimation have been presented in the literature. The most consequential of these have focused on 2 dimensional odometry and SLAM using only millimeter wave radar. [2] presents methods for radar landmark detection, scan matching, and odometry in diverse and challenging conditions; this work is extended in [3]. Additionally [21] and [20] have presented methods for radar-based SLAM using arrays of automotive-grade radar sensors. Lastly, [26] made use of Doppler velocity measurements for body-frame velocity estimation as part of their odometry system.

3 Technical Approach

3.1 State Estimation

Our approach estimates the global-frame velocity $\mathbf{v}_W$ and orientation $\mathbf{q}_{WS}$ of the sensor platform over a sliding window [23] of K successive radar measurements. These states are linked by integrated accelerometer and gyroscope measurements from the IMU. The full state vector is then $\mathbf{x} = [\mathbf{v}_W^T, \mathbf{q}_{WS}^T, \mathbf{b}_g^T, \mathbf{b}_a^T]^T$.

We formulate our radar-inertial ego-velocity estimation as an optimization over the cost function

$$J(\mathbf{x}) := \underbrace{\sum_{k=1}^{K} \sum_{d \in \mathcal{D}^k} \mathbf{e}_d^T \mathbf{e}_d}_{\text{Doppler term}} + \underbrace{\sum_{k=1}^{K-1} {\mathbf{e}_s^k}^T \mathbf{W}_s^k \mathbf{e}_s^k}_{\text{inertial term}} \tag{1}$$

where K is the number of past radar measurements for which states are estimated, $\mathcal{D}^k$ is the set of targets returned from the radar measurement at time k, $\mathbf{e}_d$ is the Doppler velocity error term, $\mathbf{e}_s$ is the IMU error. The IMU error terms are weighted by the information matrix $\mathbf{W}_s$. Doppler errors are not explicitly weighted, but a maximum likelihood sample consensus (MLESAC) [25] step is used to reject outliers and robust norms are used to further down-weight poor measurements.

Doppler Error Term. A radar measurement consists of a set of targets $\mathcal{D}$. Each $d \in \mathcal{D}$ consists of $[\tilde{r}_S, \tilde{v}_R, \tilde{\theta}_S, \tilde{\phi}_S]^T$, the range, Doppler (radial) velocity, azimuth, and elevation for target d. Theoretically, the Doppler velocity measurement v_R is equal to the projection of the relative velocity vector between the target and sensor $\mathbf{v}_S$ onto the ray between sensor origin and the target $\mathbf{r}_S$

$$v_R = \mathbf{C}_{SW} \mathbf{v}_W \left(\frac{\mathbf{r}_S}{\|\mathbf{r}_S\|_2} \right)^T \tag{2}$$

where $\mathbf{C}_{SW}$ is the rotation matrix representing rotation from the world frame to the sensor frame.

However, the actual operation of the radar sensor complicates this process significantly. Not only is there uncertainty in the Doppler measurement, $\tilde{v}_R$, there is also significant uncertainty in the other measurement components: $\tilde{r}_S$, $\tilde{\theta}_S$, and $\tilde{\phi}_S$. So the measured ray $\tilde{\mathbf{r}}_S = \mathbf{r}_S + \mathbf{e}_r$ where $\mathbf{r}_s$ is the true target location in the sensor frame and $\mathbf{e}_r$ is a random error. Similarly, the Doppler measurement $\tilde{v}_R = v_R + e_v$. So Doppler velocity measurements are modeled as

$$\tilde{v}_R - e_v = \mathbf{C}_{SW} \mathbf{v}_W \left(\frac{\tilde{\mathbf{r}}_S - \mathbf{e}_r}{\|\tilde{\mathbf{r}}_S - \mathbf{e}_r\|_2} \right)^T \tag{3}$$

If we assume the doppler targets represent stationary objects then each can serve as a constraint when regressing the sensor platform's velocity.

Because there is error on both the dependent and independent variables in the sensor model we formulate the doppler error term as

$$\mathbf{e}_d^k(\mathbf{x}_R^k, \mathcal{D}^k) = [e_v \; \mathbf{e}_r]^T \tag{4}$$

and the error is minimized by varying $\mathbf{v}_W$, $\mathbf{q}_{WS}$ and a ray correction factor $\mathbf{e}_r$ for each Doppler constraint in the optimization. The ray correction factors are treated both as states to be estimated and residuals to be minimized. This is similar to total least squares regression as presented in [4], and is a significant improvement over our previous work in [11].

Lastly, if any of the targets in the current radar scan represent false positives or if they represent objects that actually *are* moving in relation to the rest of the scene they will be outliers with respect to the sensor model. Radar is particularly prone to false positives so we must assume all radar scans will contain a significant number of outliers. In our testing we found around 5% of raw radar targets were outliers with respect to the sensor model in Eq. (3). Although we employ robust norms in our optimization to reduce the influence of such outliers, we also use maximum likelihood sample consensus (MLESAC) [25] as an explicit outlier rejection step. We have found the addition of explicit outlier rejection significantly improves the accuracy of our Doppler velocity estimates.

IMU Kinematics. In our system, IMU measurements are used to measure the system's change in global-frame velocity and orientation between radar measurements. Our IMU model is very similar to those used in OKVIS [12,13] and MSCKF [15], except we do not use the IMU to measure the change in the system's full pose, only its velocity, attitude, and heading. The states are propagated via the following differential equations:

$$\begin{aligned} \mathbf{q}_{WS} &= \frac{1}{2}\boldsymbol{\Omega}(\boldsymbol{\omega}_S)\mathbf{q}_{WS} \\ \mathbf{v}_W &= \mathbf{C}_{WS}\mathbf{a}_S + \mathbf{g}_W \\ \mathbf{b}_g &= \mathbf{w}_{b_g} \\ \mathbf{b}_a &= -\frac{1}{\tau}\mathbf{b}_a + \mathbf{w}_{b_a} \end{aligned} \tag{5}$$

The elements of $\mathbf{w} := [\mathbf{w}_g^T, \mathbf{w}_a^T, \mathbf{w}_{b_g}^T, \mathbf{w}_{b_a}^T]^T$ are zero-mean, uncorrelated Gaussian white noise, and $\mathbf{g}_W$ is the gravity vector in the world frame. The gyro and accelerometer measurements, $\tilde{\boldsymbol{\omega}}_S$ and $\tilde{\mathbf{a}}_S$ respectively, are defined as the true acceleration and angular rate with added bias and white noise

$$\begin{aligned} \tilde{\boldsymbol{\omega}}_S &= \boldsymbol{\omega}_S + \mathbf{b}_g + \mathbf{w}_{b_g} \\ \tilde{\mathbf{a}}_S &= \mathbf{a}_S + \mathbf{b}_a + \mathbf{w}_{b_a} \end{aligned} \tag{6}$$

and the matrix $\boldsymbol{\Omega}$ is formed from the estimated angular rate as

$$\boldsymbol{\Omega}(\boldsymbol{\omega}_S) := \begin{bmatrix} -\frac{1}{2}\boldsymbol{\omega}_S \\ 1 \end{bmatrix}^{\oplus} \tag{7}$$

The $\oplus$ operator is defined in [1].

In order to optimize over this model we need the linearized, discrete time version of the state equations in Eq. (5). First, the continuous time state transition matrix is

$$\mathbf{F_c}(\mathbf{x}) = \begin{bmatrix} \mathbf{0}_{3\times3} & \mathbf{0}_{3\times3} & \bar{\mathbf{C}}_{WS} & \mathbf{0}_{3\times3} \\ [\bar{\mathbf{C}}_{WS}\mathbf{a}_S]^\times & \mathbf{0}_{3\times3} & \mathbf{0}_{3\times3} & -\bar{\mathbf{C}}_{WS} \\ \mathbf{0}_{3\times3} & \mathbf{0}_{3\times3} & \mathbf{0}_{3\times3} & \mathbf{0}_{3\times3} \\ \mathbf{0}_{3\times3} & \mathbf{0}_{3\times3} & \mathbf{0}_{3\times3} & -\frac{1}{\tau}\mathbf{I}_{3\times3} \end{bmatrix} \tag{8}$$

where the $[.]^\times$ operator denotes the skew symmetric matrix associated with the cross product of the vector. Next, an approximate discretization of $\mathbf{F_c}$ is found via Euler's method

$$\mathbf{F_d}(\mathbf{x}, \Delta t) = \mathbf{I} + \mathbf{F_c}(\mathbf{x})\Delta t \tag{9}$$

and the covariance is propagated as

$$\mathbf{P}^{k+1} = \mathbf{F}_d(\hat{\mathbf{x}}^k, \Delta t)\mathbf{P}^k\mathbf{F}_d(\hat{\mathbf{x}}^k, \Delta t)^T + \mathbf{G}(\hat{\mathbf{x}}^k)\mathbf{Q}\mathbf{G}(\hat{\mathbf{x}}^k)^T \tag{10}$$

where $\mathbf{Q} := \text{diag}(\sigma_g^2, \sigma_a^2, \sigma_{b_g}^2, \sigma_{b_a}^2)$ and

$$\mathbf{G}(\mathbf{x}) = \begin{bmatrix} \bar{\mathbf{C}}_{WS} & \mathbf{0}_{3\times3} & \mathbf{0}_{3\times3} & \mathbf{0}_{3\times3} \\ \mathbf{0}_{3\times3} & \mathbf{I}_{3\times3} & \mathbf{0}_{3\times3} & \mathbf{0}_{3\times3} \\ \mathbf{0}_{3\times3} & \mathbf{0}_{3\times3} & \mathbf{I}_{3\times3} & \mathbf{0}_{3\times3} \\ \mathbf{0}_{3\times3} & \mathbf{0}_{3\times3} & \mathbf{0}_{3\times3} & \mathbf{I}_{3\times3} \end{bmatrix} \tag{11}$$

IMU Error. The IMU provides measurements at many times the rate of the radar sensor and the IMU and radar measurements are not synchronized. To address this we first interpolate between the IMU measurements immediately before and after the radar measurements to obtain estimated IMU readings that align temporally with the radar measurements at timesteps k and $k+1$. We then use the method for on-manifold IMU pre-integration presented in [7] to combine the set of IMU measurements between times k and $k+1$ into a single relative constraint. The IMU error is then defined a

$$\mathbf{e}^k(\mathbf{x}^k, \mathbf{x}^{k+1}, \mathbf{z}^k) = \begin{bmatrix} 2\left[\hat{\mathbf{q}}_{WS}^{k+1} \otimes \mathbf{q}_{WS}^{k+1^{-1}}\right]_{1:3} \\ \hat{\mathbf{v}}_W^{k+1} - \mathbf{v}_W^{k+1} \\ \hat{\mathbf{b}}_g - \mathbf{b}_g \\ \hat{\mathbf{b}}_a - \mathbf{b}_a \end{bmatrix} \tag{12}$$

where the $\otimes$ operator is as defined in [1]. The Jacobian of the error with respect to the state at $k+1$ is defined as

$$\frac{\partial \mathbf{e}^k}{\partial \hat{\boldsymbol{\chi}}^{k+1}} = \begin{bmatrix} \left[\hat{\mathbf{q}}_{WS}^{k+1} \otimes \mathbf{q}_{WS}^{k+1^{-1}}\right]^{\oplus}_{1:3,1:3} & \mathbf{0}_{3\times9} \\ \mathbf{0}_{9\times3} & \mathbf{I}_{9\times9} \end{bmatrix} \tag{13}$$

The Jacobian with respect to the state at time k is somewhat more difficult to calculate because the IMU error term is found by iteratively applying the IMU integration. Differentiating the error with respect to the state at time k requires use of the chain rule, leading to

$$\frac{\partial \mathbf{e}^k}{\partial \hat{\boldsymbol{\chi}}^k} = \left(\prod_{i=0}^{R} \mathbf{F}_d(\hat{\mathbf{x}}^i, \Delta t^i) \right) \frac{\partial \mathbf{e}^k}{\partial \hat{\boldsymbol{\chi}}^{k+1}} \tag{14}$$

Marginalization Strategy. When a radar measurement passes out of the optimization window, the states and constraints associated with that measurement are not simply dropped, as in our previous work [11]. Instead the states and constraints are marginalized through a process similar to that used in [12,13]. The Gauss-Newton system of equations is constructed from all the error terms, Jacobians and information: it takes the form $\mathbf{H}\delta\boldsymbol{\chi} = \mathbf{b}$. Let us consider a set of states to be marginalized out, $\mathbf{x}_\mu$, the set of all states related to those by error terms, $\mathbf{x}_\lambda$, and the set of remaining states, $\mathbf{x}_\rho$. Due to conditional independence, we can simplify the marginalization step and only apply it to a sub-problem:

$$\begin{bmatrix} \mathbf{H}_{\mu\mu} & \mathbf{H}_{\mu\lambda} \\ \mathbf{H}_{\lambda\mu} & \mathbf{H}_{\lambda\lambda} \end{bmatrix} \begin{bmatrix} \delta\boldsymbol{\chi}_\mu \\ \delta\boldsymbol{\chi}_\lambda \end{bmatrix} = \begin{bmatrix} \mathbf{b}_\mu \\ \mathbf{b}_\lambda \end{bmatrix} \tag{15}$$

Application of the Schur complement results in:

$$\mathbf{H}^*_{\lambda\lambda} := \mathbf{H}_{\lambda\lambda} - \mathbf{H}_{\lambda\mu}\mathbf{H}^{-1}_{\mu\mu}\mathbf{H}_{\mu\lambda} \tag{16}$$

$$\mathbf{b}^*_\lambda := \mathbf{b}_\lambda - \mathbf{H}_{\lambda\mu}\mathbf{H}^{-1}_{\mu\mu}\mathbf{b}_\mu \tag{17}$$

where $\mathbf{b}^*_\lambda$ and $\mathbf{H}^*_{\lambda\lambda}$ are nonlinear functions of $\mathbf{x}_\lambda$ and $\mathbf{x}_\mu$.

Equations 16 and 17 describe a single marginalization step which is applied repeatedly each time a frame passes out of the sliding window. We incorporate the resulting information from this step as a prior in the optimization. When marginalizing states we fix the linearization point around $\mathbf{x}_0$, the value of $\mathbf{x}$ at the time of marginalization. The perturbation $\Delta\boldsymbol{\chi} := \Phi^{-1}(\log(\bar{\mathbf{x}} \boxplus \mathbf{x}_0^{-1}))$ represents state updates after the marginalization step, where $\bar{\mathbf{x}}$ is our current estimate of $\mathbf{x}$ and the operator $\boxplus$ is defined in [12]. In other words, $\mathbf{x}$ is composed as

$$\mathbf{x} = \exp(\Phi(\delta\boldsymbol{\chi})) \boxplus \underbrace{\exp(\Phi(\Delta\boldsymbol{\chi})) \boxplus \mathbf{x}_0}_{=\bar{\mathbf{x}}} \tag{18}$$

This generic formulation allows us to apply prior information on the minimal representation of any of our state variables including the orientation quaternion. Introducing $\Delta\boldsymbol{\chi}$ allows us to make the following first order approximation to the right hand side:

$$\mathbf{b} + \left.\frac{\partial \mathbf{b}}{\partial \Delta\boldsymbol{\chi}}\right|_{\mathbf{x}_0} \Delta\boldsymbol{\chi} = \mathbf{b} - \mathbf{H}\Delta\boldsymbol{\chi} \tag{19}$$

Again using the partition into μ and λ, we can write Eq. (19) as the right hand side of Eq. (17):

$$\begin{bmatrix} \mathbf{b}_\mu \\ \mathbf{b}_\lambda \end{bmatrix} = \begin{bmatrix} \mathbf{b}_{\mu,0} \\ \mathbf{b}_{\lambda,0} \end{bmatrix} - \begin{bmatrix} \mathbf{H}_{\mu\mu} & \mathbf{H}_{\mu\lambda} \\ \mathbf{H}_{\lambda\mu} & \mathbf{H}_{\lambda\lambda} \end{bmatrix} \begin{bmatrix} \Delta\boldsymbol{\chi}_\mu \\ \Delta\boldsymbol{\chi}_\lambda \end{bmatrix} \tag{20}$$

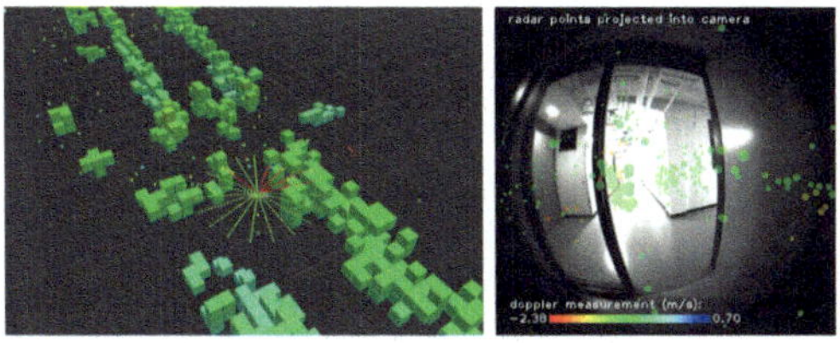

Fig. 2. Example of a radar map generated using our technique. The left image shows a robot-centric view of the radar map; the right image shows the view from the robot's camera with projected radar points for reference.

In this form Eq. (17) becomes

$$\mathbf{b}_\lambda^* = \underbrace{\mathbf{b}_{\lambda,0} - \mathbf{H}_{\lambda\mu}\mathbf{H}_{\mu\mu}^{-1}\mathbf{b}_{\mu,0}}_{\mathbf{b}_{\lambda,0}^*} - \mathbf{H}_{\lambda,\lambda}^* \Delta\boldsymbol{\chi}_\lambda \tag{21}$$

The marginalization procedure thus consists of iteratively applying Eq. (16) to update $\mathbf{H}$ and Eq. (21) to update $\mathbf{b}$. We can then remove the nonlinear terms consumed and add the marginalized $\mathbf{H}_{\lambda\lambda}^*$ and $\mathbf{b}_\lambda^*$ as summands in the overall Gauss-Newton system. Their contribution to the chi-square error may be written as $\boldsymbol{\chi}_\lambda^2 = \mathbf{b}_\lambda^{*T}\mathbf{H}_{\lambda\lambda}^{*+}\mathbf{b}_\lambda^*$.

3.2 Radar Mapping

Using radar scans with common occupancy mapping methods such as octomap [9] creates two problems: 1) the beam sensor model used in octomap does not accurately model the operation of the radar sensor, and 2) radar is particularly prone to false positives from noise, sidelobes, and multipath reflections. To address the first of these problems we employ a new, radar-specific sensor model. To combat the noise problem, we developed a learned radar noise filter. With these two developments we are able to use octomap to create radar maps as shown in Fig. 2.

Radar Sensor Model. The beam sensor model as employed in [9] uses ray casting to update the occupancy probability of a voxel. This process is pictured on the left of Fig. 3. The update rule for the beam model, given a sensor measurement z_t, is formulated as:

$$L(n|z_t) = \begin{cases} l_{\text{occ}} & \text{if beam ends within voxel,} \\ l_{\text{free}} & \text{if beam traversed voxel.} \end{cases} \tag{22}$$

However, a radar sensor does not cast a discrete set of rays. Also, radar can return multiple targets along the same ray. To address these problems we employ a new, radar-specific sensor model pictured on the right of Fig. 3. The update rule for this model is

$$L(n|z_t) = \begin{cases} l_{\text{occ}} & \text{if a target occurs within voxel,} \\ l_{\text{free}} & \text{if no target occurs within voxel.} \end{cases} \tag{23}$$

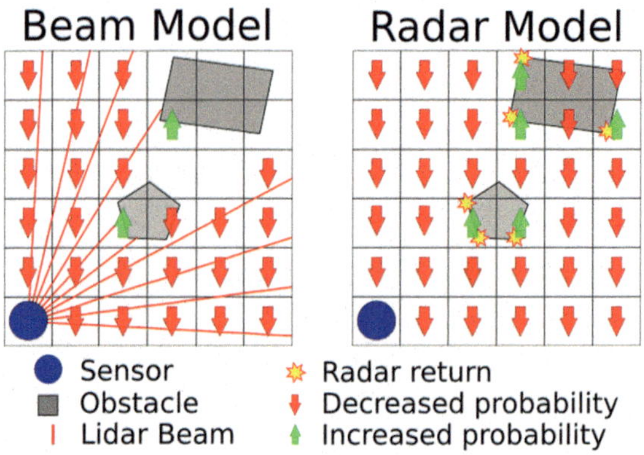

Fig. 3. The beam sensor model as implemented in [9] (left) contrasted with the radar model used in this work (right). Map cells whose occupancy probability is increased by the sensor measurement are marked with green up arrows. Decreased probability is marked with red down arrows.

Learned Noise Filter. This new sensor model solves one problem, but it does not address the high rate of false positives present in radar scans. For this we developed a learned noise filter using a modified version of the PointNet segmentation network [17]. In our approach, the model is trained to segment true detections from false positives. PointNet accepts tensors of unordered points identified only by their position in cartesian space. These tensors have size of $N \times 3$ where N is the number of points. An input transformation is performed on these points to align them to a canonical orientation. After this step, we concatenate the intensity and doppler readings for each point onto the transformed point tensor, resulting in a tensor of size $N \times 5$. After this step our PointNet implementation operates as described in [17].

We employ an automated method to create labeled training data for our learned noise filter. We gather data using a sensor rig that includes a TI AWR1843 radar sensor, an Intel Realsense T265 tracking camera, and an Ouster OS1 lidar sensor. For each run we spatially align the lidar and radar pointclouds using odometry from the Realsense T265. We label a radar point as "keep" if the distance from the radar point to the nearest lidar point is below a certain threshold and label it as "discard" otherwise. Using this method we generated around 30,000 labeled radar point clouds, of which 80% are used for training and 20% for testing. After training over 10 epochs our learned noise filter is able to classify "keep" and "discard" points with 89% accuracy in testing.

4 Experiments

We made quantitative comparisons of our state estimation and mapping methods against other popular methods. We also ran experiments in conditions with fog. Because of the difficulty of obtaining groundtruth data in fog these experiments are used for qualitative comparisons.

4.1 Quantitative Experiments

We compare our radar-inertial state estimator's performance to visual-inertial and lidar methods using a hand-carried sensor rig and a vicon motion capture system as groundtruth. For the visual-inertial benchmark we used the popular Intel Realsense T265 tracking camera, and for lidar we used a velodyne VLP-16 with the LAMP odometry system [6].

We also compare our radar mapping method with visual mapping using an Intel Realsense d435 depth camera. For this comparison lidar is used as groundtruth. To ensure the maps from all methods are in the same coordinate frame, motion capture is used to fuse new scans for all mapping methods. For every published map throughout the run, we calculate two values: first, the "false positive distance," which is the distance from each occupied voxel in the query map to the nearest occupied voxel in the corresponding groundtruth map. If the false positive distance for a voxel in the query map is zero it is a true positive, while increasing values indicate it is further from ground truth. Second, we calculate the reverse measurement, i.e. the distance from an unoccupied voxel in the ground truth map to the nearest unoccupied voxel in the query map; this is termed the "false negative distance."

4.2 Qualitative Experiments

In addition to our quantitative experiments, which were performed in ideal conditions, we also ran experiments in varying levels of fog. Fog density was quantified by visibility distance and varied from 12 m to less than 6 m. The test space is shown in no fog and heavy fog conditions in Fig. 4. Because groundtruth is not available in heavy fog these tests are compared qualitatively.

(a) No fog, unlimited visibility (b) High fog, 6m visibility

Fig. 4. The CoStar lab space used for evaluation of the state estimation and mapping methods with a handheld sensor rig. Shown with no fog on the left, and high fog on the right.

5 Results

5.1 State Estimator Tests

Table 1 shows the mean and standard deviation of the RMS error in velocity estimates for our radar-inertial state estimator as well as visual-inertial and lidar odometry methods in a brightly-lit indoor environment with no fog. These results show the accuracy of our radar-inertial state estimator is on par with the state of the art in visual-inertial and lidar-based methods in favorable conditions. The errors for a typical run with no fog are shown in Fig. 5.

Table 1. Mean (μ) and standard deviation (σ) in m/s of the RMS velocity estimate errors in x, y, and z (over 10 runs of approx 60 s each) for each estimator tested in no fog. Ground truth is provided by a motion capture system.

	Radar-Inertial		Visual-Inertial		Lidar	
	μ	σ	μ	σ	μ	σ
x	0.2576	0.0935	0.2019	0.0898	0.2270	0.1197
y	0.2272	0.0827	0.1830	0.0845	0.2271	0.1325
z	0.1125	0.0268	0.0809	0.0124	0.2054	0.0649

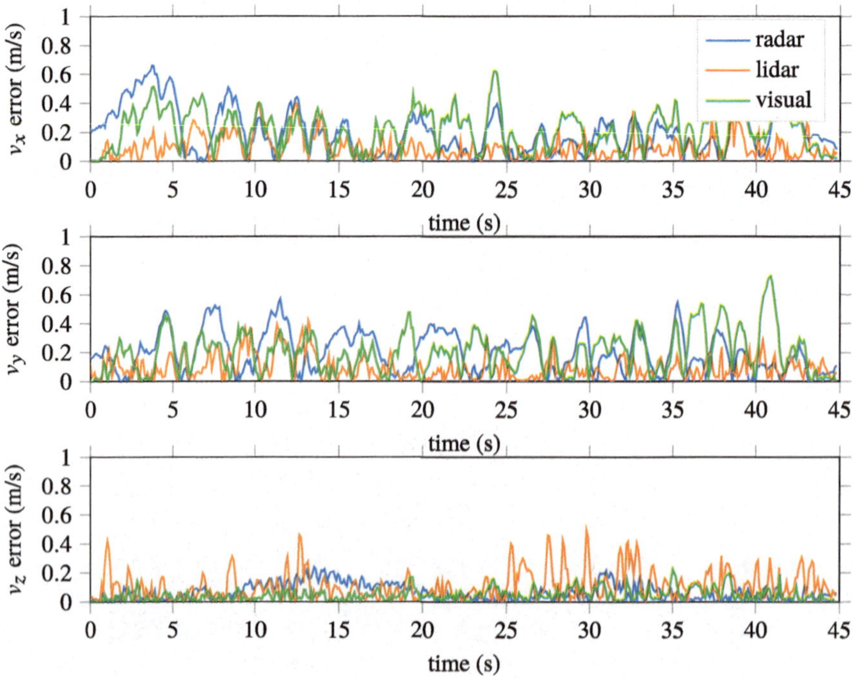

Fig. 5. Errors in global frame velocity estimates for each method for a prototypical run in no fog. The proposed radar-inertial method is plotted in blue, lidar is plotted in orange, and visual-inertial is plotted in green.

Figure 6 shows state estimates for all methods on a typical run in high fog. We can see the visual-inertial and radar-inertial methods tend to agree while lidar does not. This indicates lidar state estimation is more affected by fog than the visual and radar-based methods. We found lidar-based odometry failed even in low fog. VIO, on the other hand, is more robust to fog but it did fail when visibility dropped to around 6 m. For instance, the visual and radar estimates disagree at points in Fig. 6. It is not possible to determine whether the radar or visual method is failing in this case due to the lack of groundtruth, but the properties of the radar sensor would lead one to believe it should be more robust to fog than vision.

5.2 Occupancy Mapping Tests

Table 2 shows statistics on the false positive and false negative distances for radar and visual maps for a 60 s run in the JPL lab space. As expected, the mean number of occupied voxels per published map is smaller for radar. However, the percent of true positives per published map is higher for radar than for vision. Additionally, the mean, median, and variance of the false positive distance is lower for radar mapping than for visual mapping. This means we can be more

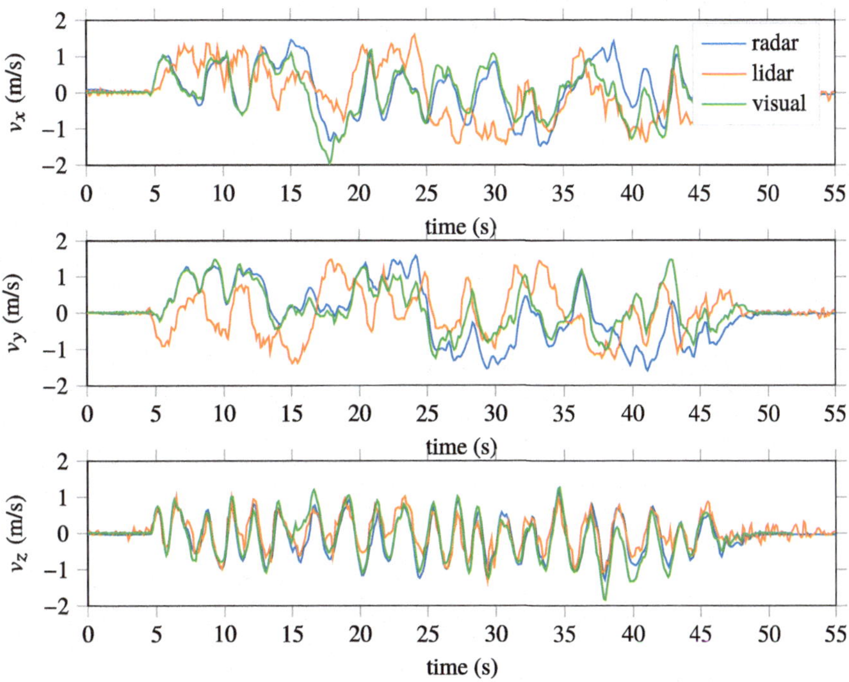

Fig. 6. Global frame velocity estimates for each method for a typical run with high fog, thereby disabling motion capture. The proposed radar-inertial method is plotted in blue, lidar is plotted in orange, and visual-inertial is plotted in green.

confident that occupied voxels in a radar map represent occupied space. On the other hand, radar maps are more prone to false negatives; this is supported by the data in Table 2, which shows the mean, median, and variance of the false negative distance are much greater for radar mapping than visual mapping. For closed-loop control applications, pairing radar with other short-range sensors is a potential mitigation strategy to overcome these false negatives.

Table 2. Comparison of the accuracy of radar and visual maps in no-fog conditions

	Radar	Visual
Mean number of occupied voxels per published map	180.87	12083
Mean percent true positives per published map	46.0	31.2
Max false positive distance (m)	1.80	26.5
Median false positive distance (m)	0.173	0.20
Mean false positive distance (m)	0.222	0.498
False positive distance variance (m^2)	0.101	1.59
Max false negative distance (m)	11.8	10.5
Median false negative distance (m)	3.41	0.283
Mean false negative distance (m)	3.70	1.09
False negative distance variance (m^2)	2.57	1.76

6 Experimental Insights

Our experiments demonstrate the accuracy of the proposed state estimation method is comparable to VIO and lidar-based methods for state estimation in conditions favorable to all methods. Further, the our proposed radar-inertial state estimation method's accuracy far exceeds that of VIO and lidar when dense fog is introduced.

We also demonstrate the false positive rate of our proposed radar-based mapping method is comparable to mapping with a depth camera in favorable conditions, though it has a far higher rate of false negatives. We also demonstrate that lidar and depth-camera-based mapping methods are incapacitated by fog while our method is completely unaffected. This is well demonstrated by the included video, which shows comparisons between each mapping method in conditions with and without fog. It also demonstrates our full state estimation and mapping pipeline in use with a hand-carried sensor rig.

All of this suggests our radar-based state estimation and mapping methods could be used to facilitate autonomous flight of an MAV in dense smoke and fog.

Acknowledgements. The authors would like to thank the members of the CoStar team for sharing both their lab space and technical expertise. We would also like to thank Shakeeb Ahmad of the CU Boulder Mechanical Engineering department for his help in demonstrating our radar state estimator's use in closed-loop control of a micro-aerial vehicle as shown in our video submission.

References

1. Barfoot, T., Forbes, R.J., Furgale, P.: Pose estimation using linearized rotations and quaternion algebra: Acta Astronaut. **68**, 101–112 (2011)
2. Cen, S.H., Newman, P.: Precise ego-motion estimation with millimeter-wave radar under diverse and challenging conditions. In: 2018 IEEE International Conference on Robotics and Automation (ICRA). IEEE, Brisbane, QLD (2018)
3. Cen, S.H., Newman, P.: Radar-only ego-motion estimation in difficult settings via graph matching. arXiv:1904.11476 [cs] (2019)
4. Deming, W.: Statistical Adjustment of Data. J. Wiley & Sons, Incorporated (1943). https://books.google.com/books?id=9awgAAAAMAAJ
5. Dickmann, J., Appenrodt, N., Bloecher, H., Brenk, C., Hackbarth, T., Hahn, M., Klappstein, J., Muntzinger, M., Sailer, A.: Radar contribution to highly automated driving. In: 2014 44th European Microwave Conference (2014)
6. Ebadi, K., Chang, Y., Palieri, M., Stephens, A., Hattel, A., Heiden, E., Thakur, A., Morrell, B., Carlone, L., Aghamohammadi, A.: Lamp: large-scale autonomous mapping and positioning for exploration of perceptually-degraded subterranean environments. In: 2020 IEEE International Conference on Robotics and Automation (ICRA). Paris, France (2020)
7. Forster, C., Carlone, L., Dellaert, F., Scaramuzza, D.: On-manifold preintegration for real-time visual-inertial odometry. Trans. Rob. **33**(1), 1–21 (2017)
8. Geneva, P., Eckenhoff, K., Yang, Y., Huang, G.: Lips: lidar-inertial 3D plane slam. In: 2018 IEEE/RSJ International Conference on Intelligent Robots and Systems (IROS), pp. 123–130 (2018)
9. Hornung, A., Wurm, K.M., Bennewitz, M., Stachniss, C., Burgard, W.: OctoMap: An efficient probabilistic 3D mapping framework based on octrees. Autonomous Robots (2013). Software available at http://octomap.github.com
10. Khattak, S., Papachristos, C., Alexis, K.: Keyframe-based direct thermal-inertial odometry. CoRR **abs/1903.00798** (2019). http://arxiv.org/abs/1903.00798
11. Kramer, A., Stahoviak, C., Santamaria-Navarro, A., Aghamohammadi, A., Heckman, C.: booktitle=2020 IEEE International Conference on Robotics and Automation (ICRA)
12. Leutenegger, S., Furgale, P., Rabaud, V., Chli, M., Konolige, K., Siegwart, R.: Keyframe-based visual-inertial slam using nonlinear optimization (2013)
13. Leutenegger, S., Lynen, S., Bosse, M., Siegwart, R., Furgale, P.: Keyframe-based visual-inertial odometry using nonlinear optimization. Int. J. Robot. Res. **34**(3), 314–334 (2015)
14. Michael, N., Shen, S., Mohta, K., Kumar, V., Nagatani, K., Okada, Y., Kiribayashi, S., Otake, K., Yoshida, K., Ohno, K., Takeuchi, E., Tadokoro, S.: Collaborative mapping of an earthquake-damaged building via ground and aerial robots. J. Field Robot. **29**(5), 832–841 (2012)
15. Mourikis, A.I., Roumeliotis, S.I.: A multi-state constraint kalman filter for vision-aided inertial navigation. In: Proceedings 2007 IEEE International Conference on Robotics and Automation, , Rome, Italy, pp. 3565–3572. IEEE (2007)
16. O'Toole, M., Achar, S., Narasimhan, S.G., Kutulakos, K.N.: Homogeneous codes for energy-efficient illumination and imaging. ACM Trans. Graph. **34**(4), 1–13 (2015)
17. Qi, C.R., Su, H., Mo, K., Guibas, L.J.: PointNet: Deep Learning on Point Sets for 3D Classification and Segmentation. arXiv:1612.00593 [cs] (2017). http://arxiv.org/abs/1612.00593. ArXiv: 1612.00593

18. Santamaria-Navarro, A., Loianno, G., Solà, J., Kumar, V., Andrade-Cetto, J.: Autonomous navigation of micro aerial vehicles using high-rate and low-cost sensors. Autonom. Robots **42**(6), 1263–1280 (2018)
19. Santamaria-Navarro, A., Solà, J., Andrade-Cetto, J.: High-frequency mav state estimation using low-cost inertial and optical flow measurement units. In: Intelligent Robots and Systems (IROS), 2015 IEEE/RSJ International Conference on, pp. 1864–1871 (2015)
20. Schuster, F., Keller, C.G., Rapp, M., Haueis, M., Curio, C.: Landmark based radar SLAM using graph optimization. In: 2016 IEEE 19th International Conference on Intelligent Transportation Systems (ITSC), pp. 2559–2564 (2016)
21. Schuster, F., Wörner, M., Keller, C.G., Haueis, M., Curio, C.: Robust localization based on radar signal clustering, pp. 839–844 (2016)
22. Shen, S., Mulgaonkar, Y., Michael, N., Kumar, V.S.A.: Vision-based state estimation and trajectory control towards high-speed flight with a quadrotor. In: Robotics: Science and Systems (2013)
23. Sibley, G.: A Sliding Window Filter for SLAM p. 17
24. Tomic, T., Schmid, K., Lutz, P., Domel, A., Kassecker, M., Mair, E., Grixa, I.L., Ruess, F., Suppa, M., Burschka, D.: Toward a fully autonomous UAV: research platform for indoor and outdoor urban search and rescue. IEEE Robot. Autom. Mag. **19**(3), 46–56 (2012). https://doi.org/10.1109/MRA.2012.2206473
25. Torr, P.H.S., Zisserman, A.: MLESAC: a new robust estimator with application to estimating image geometry. Comput. Vis. Image Underst. **78**, 138–156 (2000)
26. Vivet, D., Checchin, P., Chapuis, R.: Localization and Mapping Using Only a Rotating FMCW Radar Sensor. Sensors (Basel, Switzerland) **13**(4), 4527–4552 (2013). https://www.ncbi.nlm.nih.gov/pmc/articles/PMC3673098/
27. Zhang, J., Singh, S.: Loam: lidar odometry and mapping in real-time. In: Robotics: Science and Systems (2014)

Experimental Flights of Adaptive Patterns for Clouds Exploration with UAVs

Titouan Verdu[1,2](✉), Nicolas Maury[3], Pierre Narvor[2], Florian Seguin[2], Gregory Roberts[3], Fleur Couvreux[3], Grégoire Cayez[3], Murat Bronz[1], Gautier Hattenberger[1], and Simon Lacroix[2]

[1] ENAC, Université de Toulouse, Toulouse, France
[2] LAAS-CNRS, Université de Toulouse, CNRS, Toulouse, France
[3] METEO-FRANCE Toulouse, Toulouse Cedex 01, Toulouse, France

1 Introduction

The evolution of cumulus clouds is still no fully understood by atmospheric scientists [1]. The classic means to gather data for the development and validation of cloud models are either remote sensors, on ground or on-board satellites, or in-situ sensors on-board of planes or balloons. These acquisition systems do not provide enough data when it comes to model highly dynamic phenomena such as cumulus clouds, for which scientists need precise and dense measurements *within* clouds. The advent of UAVs has offered the possibility to fly sensors within clouds (e.g. [2,3]), but the gathered data with single systems remain too sparse to observe a dynamic 3D phenomenon.

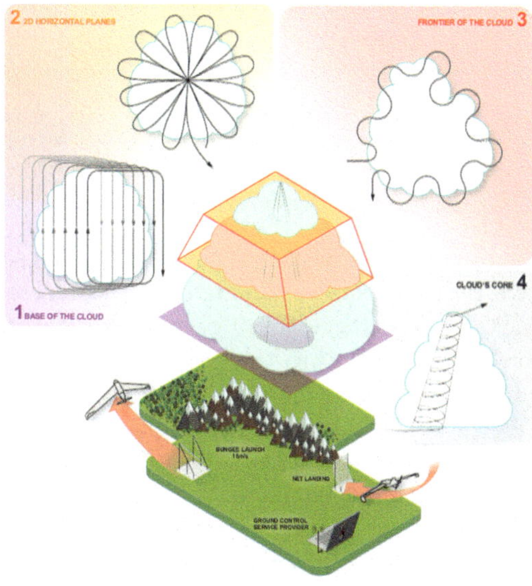

Fig. 1. Cloud interest zones (illustration and design by Sarah Gluschitz)

The Nephelae project is motivated by this need to collect dense data within clouds. Its goal to deploy a fleet of UAVs with the appropriate sensing and decisional capacities to follow the evolution of a cloud during its complete lifespan. The project gathers three French institutes: Météo France (atmosphere scientists, in charge of on-board sensor development and cloud modeling research),

B. Siciliano et al. (Eds.): ISER 2020, SPAR 19, pp. 17–27, 2021.
https://doi.org/10.1007/978-3-030-71151-1_2

LAAS-CNRS (roboticists, in charge of developing cloud mapping and exploration strategies), and ENAC (experts in UAV science, responsible of the UAV structure development and flight control with the open source Paparazzi autopilot [4]). Previous collaboration between these institutes have lead to researches on cloud exploration strategies and dedicated flight controllers [5,6], and this paper reports on a first flight campaign in Barbados in the beginning of 2020.

Clouds are complex 3D environments, of varying size, shape and position that quickly evolve over time. Due to this high dynamic, the geo-referenced flight patterns classically handled by autopilots (*.e.g.* straight lines, circles, swathing patterns) are not suited to efficiently gather cloud data. On the contrary, *adaptive* flight patterns that are servoed on real-time cloud data allow to fly the UAV in specific areas of a cloud. Discussions with the atmosphere scientists have lead to define a series of relevant cloud areas to collect data. Four such areas of interest are illustrated in Fig. 1: the cloud base (in purple), pairs of horizontal planes (orange) which mapping yield the estimate of liquid water and mass fluxes, the cloud core (dotted lines) and finally the cloud border (in red).

The remaining of the paper[1] summarizes the overall approach to explore clouds, and focuses on the development and experimental validation of the adaptive flight patterns.

2 Approach

2.1 Adaptive Flight Patterns

In order to explore the zones of interest, several sensor based flight patterns have been developed [6]. Figure 1 illustrates different such patterns, dedicated to explore specific zones. Two different patterns are dedicated to map the cloud border, by exploiting the Liquid Water Content sensor (Fig. 2, [7]) to assess whether the UAV is inside or outside of the cloud. The *Lace* pattern switches the UAV heading in a kind of bang-bang manner (the radius of the resulting circles being configurable) every time the cloud border is crossed (Fig. 3a). A second pattern dedicated to map the cloud border, called *Trinity* (Fig. 3b), is a little bit more subtle: the first time the UAV crosses the cloud border, a first circle center is generated in a direction normal to the trajectory (say leftward). The UAV follows this circle, and every time it crosses the cloud border, an other circle is generated in the other direction (rightward), which is then tracked *without changing the rotation direction.* This pattern generates longer trajectories, but has shown to be more robust to track small clouds.

The *Rosette* pattern (Fig. 4) is intended to cover a horizontal section of the cloud, sampling data more densely near the cloud center – which is where updrafts are stronger within cumulus clouds. It also exploits the LWC sensor to assess the cloud border crossing and triggers a turn back in the cloud, oriented

[1] This paper is a more complete and updated version of a former paper published in the IEEE/RSJ International Conference on Intelligent Robots and Systems, Las Vegas (USA), 2020.

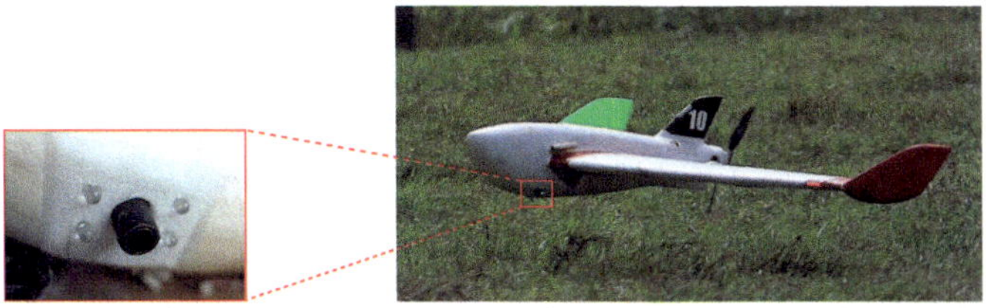

Fig. 2. The X6 UAV used during the experiments, close-up of the cloud sensor [7]

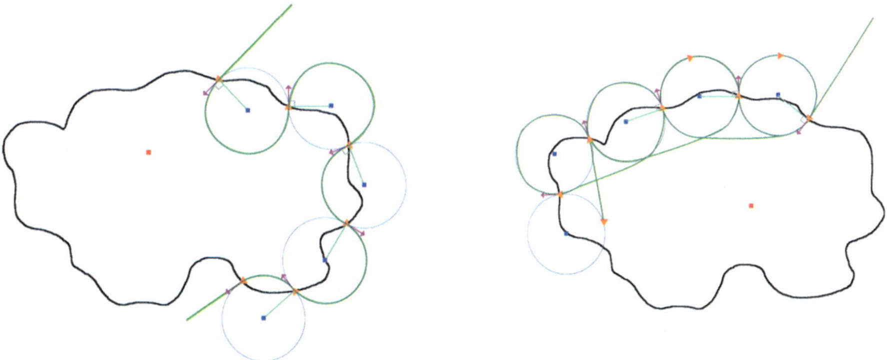

(a) (b)

Fig. 3. The two patterns Lace and Trinity dedicated to explore the cloud border

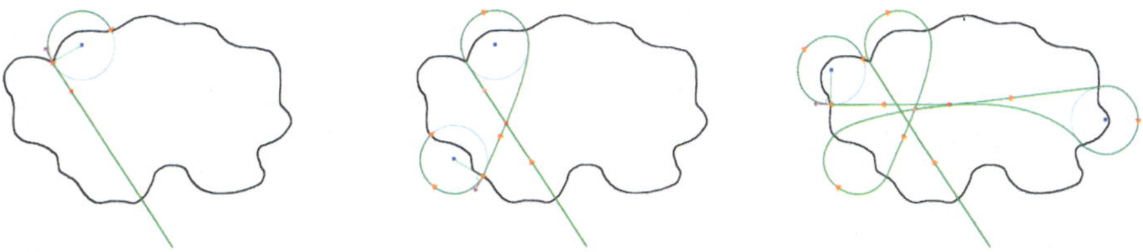

Fig. 4. Three stages of the execution of the Rosette pattern. The cloud LWC barycenter is denoted by a red square.

towards the cloud LWC barycenter which is updated in real time on the basis of the cloud map built from the gathered data (CAMS in Sect. 2.2).

2.2 System Architecture

The Fig. 5 sketches the complete architecture of the Nephelae system. The components of the Paparazzi autopilot (PPRZ, in blue) encompass the on-board flight control, the individual UAV Ground Control Station and the communications with the UAVs – established by a central modem that broadcasts data over the PPRZ Ivy middleware. The Cloud Adaptive Mapping System (CAMS,

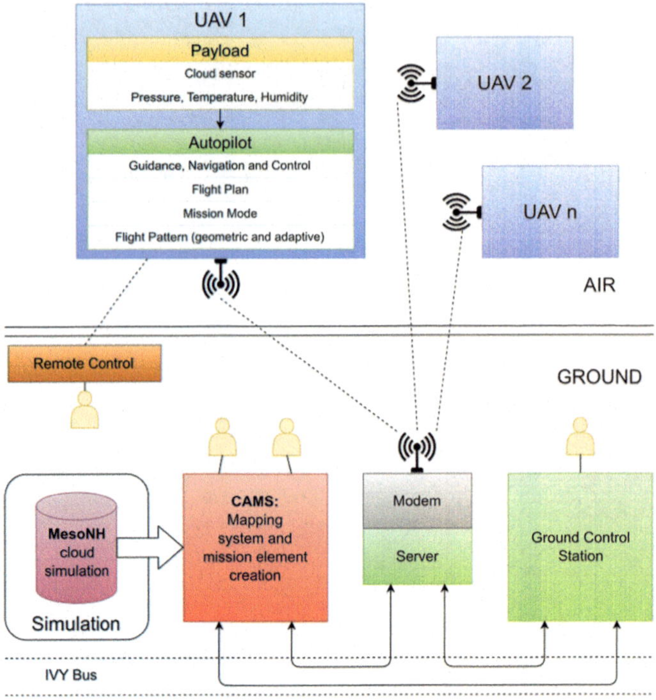

Fig. 5. Schematic of the whole system architecture

in red) is central component which builds in real-time cloud maps using the data gathered by the UAVs and that schedules new tasks (flight patterns) to be achieved by the UAVs. 3D wind, Liquid Water Content (LWC), Pressure, Temperature and Humidity and Aerosol concentration are all data of interest for the atmosphere scientists that can be mapped, however during our experiments we only exploited the LWC, which characterises the presence of a cloud. In its current version, CAMS is mostly an interface for the operators, who controls the fleet on the basis of the built maps.

3 Simulation

Deploying a fleet of UAVs calls for heavy logistics and is time consuming. Therefore, preliminary tests are required to validate the system behavior and ensure the flights safety. The overall Nephelae architecture has been designed to allow simulated flights, resorting to realistic cloud models computed beforehand[2], that

[2] Shown as a purple box in Fig. 5: such models implements the clouds microphysics MesoNH model, and require tens of hours of computation of dedicated clusters to be produced.

provides wind data to the flight simulator and cloud data to the simulated sensors, and that is also used as a reference map for the mapping software. The Nephelae architecture is used to simulate UAV flights on different cloud environments, so as to evaluate the flight patterns and tune their parameters.

First tests were conducted on a cloud simulation fixed in time: the cloud is neither growing nor moving. A series of trials on a variety of clouds could help to fine-tune the pattern parameters. Figure 6 shows screenshots of the CAMS interface that illustrates the result of the application of the Trinity and Rosette patterns on a mid-size cloud.

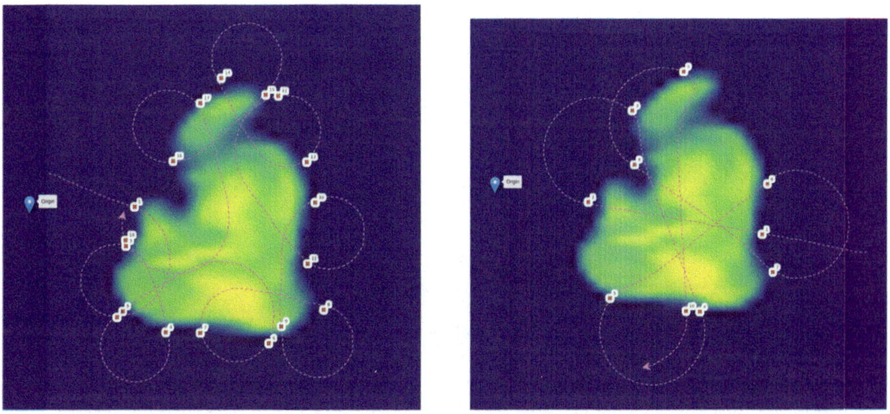

(a) (b)

Fig. 6. Trajectories resulting from the execution of the flight patterns Trinity and Lace on a simulated static cloud. The detected cloud border points are highlighted (cloud colors represent the LWC).

For more realistic simulations, the adaptive flight patterns have been evaluated on dynamic cloud simulations in which clouds evolve in shape over time, and move along with a constant advection wind. The Fig. 7 shows the resulting trajectory of the UAV while executing a Lace adaptive flight pattern while tracking an evolving and moving cloud in a dynamic simulation. The resulting trajectory may seem a bit strange, because it is displayed in a geo-referenced frame: when plotted in the frame associated to the constant wind, it takes the expected Lace shape.

Overall, the resulting trajectories are characteristic of the defined adaptive flight patterns. To evaluate the efficiency of each pattern, several assessment criteria have been defined, such as the time to perform one cloud lap, the number of border points detected in a giving time, the time spend inside the cloud... Overall, the *Lace* is well suited to quickly estimate the border of a cloud, the *Rosette* collects a lot of data inside the cloud while maintaining a correct cloud border estimation, and the *Trinity* allows to collect a lot of data both at the edge and inside the cloud.

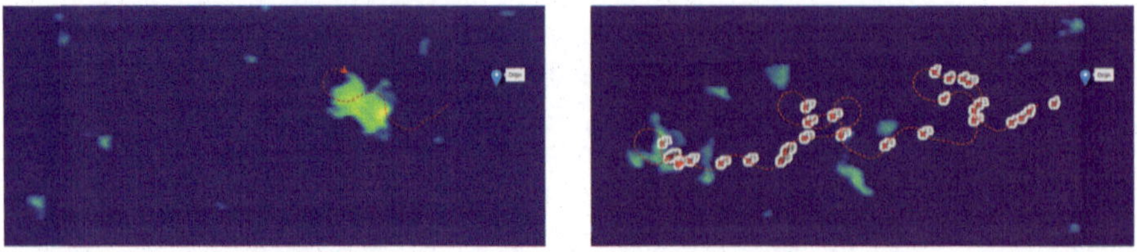

Fig. 7. Screenshots of the simulation on a dynamic cloud while executing a Lace adaptive flight pattern. Left: trajectory of the UAV at the beginning of the cloud evolution, right: complete trajectory of the UAV after chasing the cloud, with all detected borders highlighted.

Fig. 8. X6 UAV in the landing phase, Barbados

Finally, the Nephelae architecture also allows hybrid simulations, during which real UAVs fly in simulated clouds: the simulated cloud environment provides cloud data that are used to control the UAV flight. Such hybrid simulations were performed in Lannemezan (France) at the end of 2019, and gave a helpful experience to improve the operational procedures, and especially the to asses the proper integration of the flight patterns within the PPRZ autopilot and the overall Nephelae architecture.

4 Barbados Flight Campaign

4.1 Context

The flight campaign took place in the Barbados island in early 2020. This location was chosen because clouds start to develop at few kilometers from the east cost and are moved by a constant westwards wind. The initial plan was to deploy the fleet to collect data in clouds during 3 entire weeks. Due to unexpected logistic and administrative issues, the fleet was only deployed during two weeks. A total of ten people came on the field to work on the campaign fulling different roles and during different work periods. This campaign has been made possible thanks the invitation to the world atmosphere science campaign organized by EUREC4A.

4.2 Organization and Operation Sequence

The system requires at least five persons to be operated safely. A security pilot on the field controls the UAVs during the take off and landing phases. A remote operator is managing the UAVs during flight with the PPRZ Ground Control Station. CAMS requires two operators: an atmosphere scientist analyses the data and cloud maps in real time, decides which areas should be explored, and a fleet supervisor configures and sequences the flight patterns. Finally, an operator supervises all the operations and the interactions between each actor. The top picture on the Fig. 9 show the Ground Control Station and the CAMS operators.

The objective of the project was to collect enough data for atmosphere scientists to improve their cloud evolution models. To this end, several standard flight scenarios have been elaborated. A first group of scenarios was used to validate the behavior of the UAVs, to test the long range modems and to calibrate the sensors.

Then, two different scenarios were mostly used. The main one is the "cloud chase", which consist in the following actions:

- Take off ; waiting for the fleet ; spiral climb at desired altitude
- Go away to the coast to perform an hippodrome pattern ("research" mode)
- Tuning of the filter and the threshold on the LWC sensor to ensure precise cloud border detection
- Once a cloud is detected by the UAV and validated by the atmospheric operator:
  - The PPRZ operator turns the UAV into the "mission" mode
  - The CAMS fleet controller creates and configure a new mission element with an adaptive pattern
  - The UAV applies the flight pattern and follows the cloud until it crosses the border of the restricted flight area.

During this scenario, the UAVs could chase for one or two clouds before running out of battery. If the track of the cloud is lost during the execution of a pattern, the operators can decide to redirect it with an approximate target, send it back to "search" mode or land it.

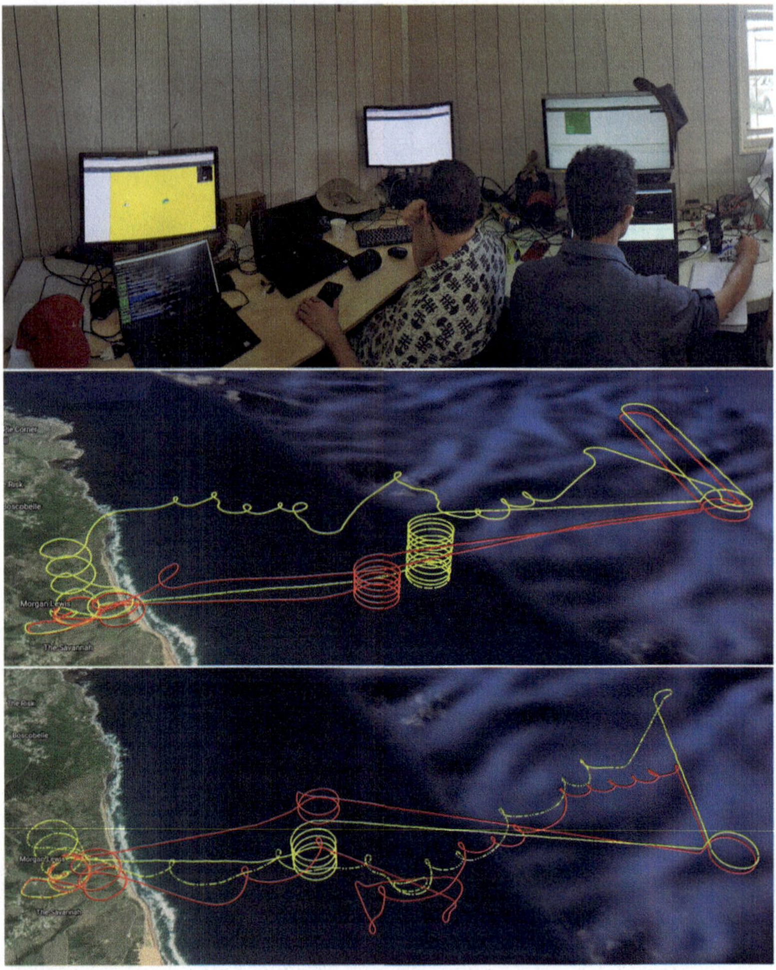

Fig. 9. Top: the ground segment of the overall architecture. Bottom: two trajectories involving 2 UAVs after a typical acquisition flight.

4.3 Logistic and Flight Security

According to the specific parameters of the island for clouds development, the campaign took place on the east cost, on an improvised airfield 500 m inland: three containers, one being power supplied, were deployed in a crop field of 500 m long and 100 m large. The site was chosen with a particular attention to the local neighborhood insuring that no houses were in the area within 500 m.

In flight, the teams operating in the same zone were responsible of the deconfliction of the different UAVs. The flight zone allowed for the Nephelae fleet was a rectangle of 15 km by 2.5 km with a maximum altitude of 2000 m. Each flight session was preceded by a call to the local control tower to ensure no plane was in the area.

The UAVs used during the campaign were Skywalker X6 equipped with PPRZ autopilot and MicroHard modem for the long range communications. These fixed wings UAVs of 1.5 m wingspan and 0.7 m fuselage were able to cruise at a speed of 16–20 m/s with a maximum autonomy of one hour. Their weights of 2 kg include the classic UAV electronics and a particular payload composed of Pressure, Temperature, Humidity sensors and the LWC sensor. Due to their load, the UAVs were launched with the help of a bungee fixed in the ground.

4.4 Flight Results

This section briefly exposes some results obtained during the Barbados flight campaign. Figure 10 presents the geo-referenced trajectory of one UAV applying the Lace pattern, colored according to whether the UAV is in or outside the cloud, which is assessed by thresholding the LWC sensor data.

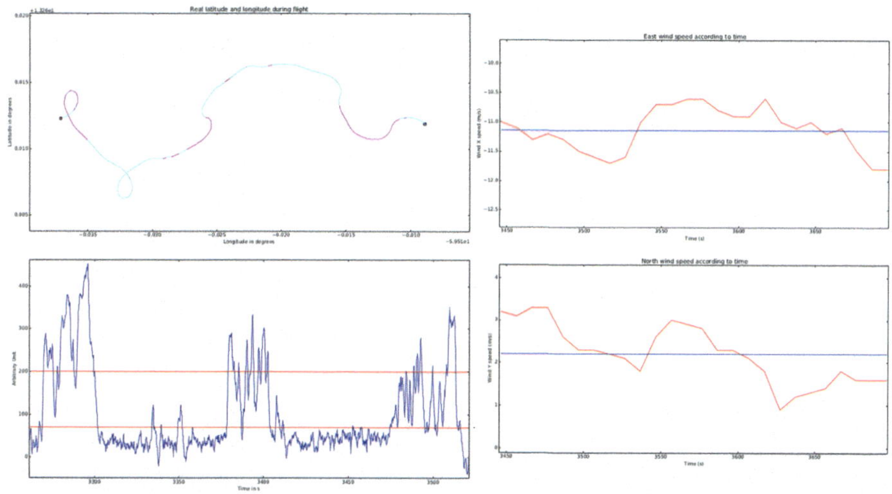

Fig. 10. Top left: geo-referenced trajectory of one UAV, colored according to the cloud sensor value (purple: inside; blue: outside); Bottom left: cloud sensor filtered value; Right: horizontal wind estimation (North and West, computed as in [8]). The blue line corresponds to the average wind estimate.

Figure 11 shows different plots of the achieved trajectories for the three different patterns, shown in a moving frame associated to the estimated mean advection wind. Once post processed, the shape and the location of the cloud are more distinguishable. This results are also used to analyse and validate the behavior of the pattern. During the flight campaign the fleet was confronted to real exploration conditions and several execution errors have been spotted. The improvement of the adaptive flight patterns will be a part of the future work to make the system more stable.

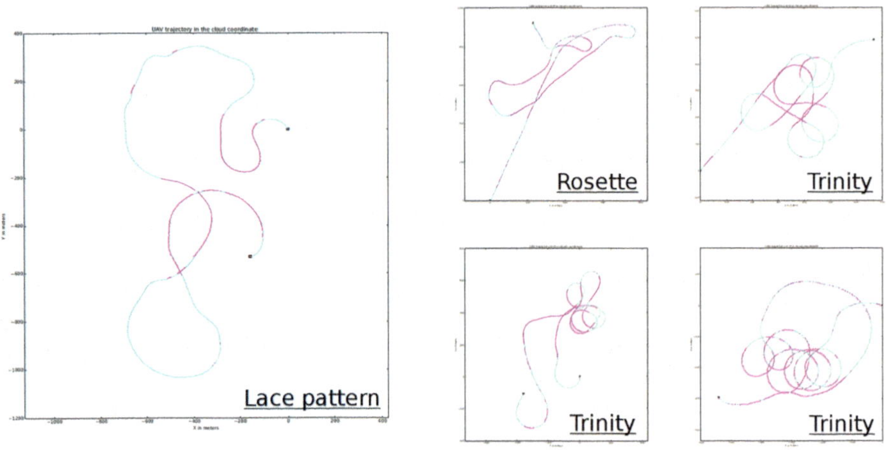

Fig. 11. Resulted trajectory of the UAV colored in the cloud coordinate. The left trajectory (Lace pattern) corresponds the geo-referenced trajectory shown Fig. 10.

5 Conclusion

During the flight campaign, 50 flights were achieved and 45 h of data was recorded. Half of the flight were realised with two UAVs and the average time of flight is around 53 min per flight. Only 14 flight gave good cloud tracking result. Detecting a cloud only based on the measurement of a cloud LWC sensor is quite complex and it will be essential for the future to add a vision system to detect the initial position of the cloud. Moreover, the errors on the patterns met during the operation proved that the adaptive patterns require a local supervisor that assesses the situation and repair execution faults.

The objective to deploy a fleet of UAVs to map clouds is partially achieved and the strategy of adaptive sampling has shown encouraging results. The system is fully operational and different axes of improvement are considered for future research. For example the flight patterns strategy can be improved to be more robust in case of disruption during the execution. The mapping process and the fleet management contained in CAMS will be upgrades allowing better map results and a more operable system. The number of UAVs in the fleet still are the most challenging part of the future work. Our original plans were to explore a cloud with at least 5 or 6 UAVs, but the real conditions of the campaign showed how complex it is to manage such a number of UAVs. For the future, the development of fleet management solutions has to be considered, that in particular allocate UAVs to specific cloud areas to explore.

References

1. de Rooy, W.C., Bechtold, P., Fröhlich, K., Hohenegger, C., Jonker, H., Mironov, D., Pier Siebesma, A., Teixeira, J., Yano, J.-I.: Entrainment and detrainment in cumulus convection: an overview. Q. J. Roy. Meteorol. Soc. **139**(670), 1–19 (2013). https://rmets.onlinelibrary.wiley.com/doi/abs/10.1002/qj.1959
2. Reineman, B.D., Lenain, L., Melville, W.K.: The use of ship-launched fixed-wing UAVs for measuring the marine atmospheric boundary layer and ocean surface processes. J. Atmos. Oceanic Technol **33**(9), 2029–2052 (2016). http://journals.ametsoc.org/doi/10.1175/JTECH-D-15-0019.1
3. Egorova, T., Gatsonis, N.A., Demetriou, M.A.: Estimation of gaseous plume concentration with an unmanned aerial vehicle. J. Guidance Control Dyn. **39**(6), 1314–1324 (2016). http://arc.aiaa.org/doi/10.2514/1.G001453
4. Hattenberger, G., Bronz, M., Gorraz, M.: Using the paparazzi UAV system for scientific research. In: IMAV 2014, International Micro Air Vehicle Conference and Competition 2014, Delft, Netherlands, pp. 247–252 (2014). https://hal-enac.archives-ouvertes.fr/hal-01059642
5. Reymann, C., Renzaglia, A., Lamraoui, F., Bronz, M., Lacroix, S.: Adaptive sampling of cumulus clouds with UAVs. Auton. Rob. **42**(2), 491–512 (2018). http://link.springer.com/10.1007/s10514-017-9625-1
6. Verdu, T., Hattenberger, G., Lacroix, S.: Flight patterns for clouds exploration with a fleet of uavs. In: 2019 International Conference on Unmanned Aircraft Systems (ICUAS), pp. 231–237 (2019)
7. Harrison, R.G., Nicoll, K.A.: Note: active optical detection of cloud from a balloon platform. Rev. Sci. Instrum. **85**(6), 066104 (2014). https://doi.org/10.1063/1.4882318
8. Mayer, S., Hattenberger, G., Brisset, P., Jonassen, M.O., Reuder, J.: A "no-flow-sensor" wind estimation algorithm for unmanned aerial systems. Int. J. Micro Air Veh. **4**(1), 15–30 (2012). https://hal-enac.archives-ouvertes.fr/hal-00934666

Design and Experimental Evaluation of Distributed Cooperative Transportation of Cable Suspended Payloads with Micro Aerial Vehicles

Guanrui Li(✉) and Giuseppe Loianno

Tandon School of Engineering, New York University, 5 MetroTech Center, Brooklyn, NY 11201, USA
{lguanrui,loiannog}@nyu.edu

Abstract. Micro Aerial Vehicles (MAVs) can play an important role in helping to accomplish tasks that are physically too dangerous or complex for humans. A team of aerial robots can cooperatively transport objects that are too large or heavy to be carried by a single MAV. This paper address the design and experimental evaluation of a distributed cooperative control method for cooperative transportation of cable-suspended payloads using MAVs. The key contributions of this paper are: i) a distributed control method for cooperative transportation of cable-suspended payloads with a team of quadrotors is experimentally designed and evaluated and allows to control both the load position and orientation, ii) our solution is light-weight and can run on-board small robots equipped with computationally limited units, iii) we study and analyze the system's performances and maneuverability based on different cable lengths.

Keywords: Aerial robotics · Aerial transportation

1 Introduction

Micro Aerial Vehicles (MAVs), equipped with on-board sensors have great potential to help humans in complex or dangerous tasks such as exploration [13], inspection [9], mapping [6], interaction with the environment [1,12], search and rescue [8], and transportation. Particularly, the aerial transportation offers a faster and more versatile solution compared to the ground transportation in congested urban environments. In addition, in post disaster scenarios, a team of physically interconnected aerial robots can deliver supplies and establish communication in areas where GPS signal is intermittent or unavailable. In order to increase the payload capacity, one can either use a large aerial vehicle or utilize a team of MAVs that cooperatively transport a payload. Although the complexity of the system increases with the number of robots, a team of MAVs can potentially provide additional resilience to the task compared to a single MAV in case of vehicle failures.

B. Siciliano et al. (Eds.): ISER 2020, SPAR 19, pp. 28–36, 2021.
https://doi.org/10.1007/978-3-030-71151-1_3

In the early research literature on cooperative transportation of a cable suspended rigid body with aerial vehicles, researchers mostly focus on quasi-static and centralized control solutions [7,10]. The same problem can also be approached in a distributed fashion with the advantage of not requiring a centralized coordinator. The vehicles share a common goal and each agent calculates its control actions based on its attach point on the payload. In [3] and [14], the authors propose a distributed approach to control the 6 Degrees of Freedom (DoF) of cable suspended rigid body payloads. However, both papers only provide a theoretical stability analysis and simulation results without dealing with the experimental design challenges on real robots equipped with embedded computationally limited units. Recently, leader-follower approaches have also been proposed [2,11]. However, such approaches are subject to the single point of failure since the system relies on the master agent as the central unit for navigation. In this work, we propose and experimentally demonstrate a distributed cable suspended payload control solution with multiple quadrotors.

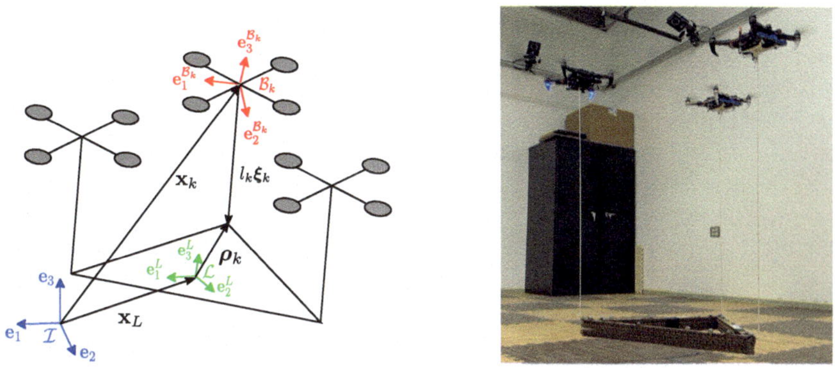

(a) (b)

Fig. 1. System convention definition $\mathcal{I}$, $\mathcal{L}$, $\mathcal{B}_k$ denote the inertial, payload, and the kth robot body frames respectively (left). The experimental setup deployed at ARPL (right).

This paper presents multiple contributions. First, we experimentally design and evaluate a distributed cooperative control method for the cable suspended rigid body payloads transportation with multiple quadrotors considering full nonlinear system dynamics. The experimental setup for the transported object and vehicles are depicted in Fig. 1(b). We show the ability to concurrently control the load position and orientation during the motion. Second, our solution is distributed and lightweight. It runs on each vehicle using a small on-board computational unit. Third, we analyze the effects of cable lengths on the system's response. This is the first time that a distributed control solution for multiple quadrotors with cable suspended payload is experimentally designed and validated. The proposed approach has the potential to be deployed in real world

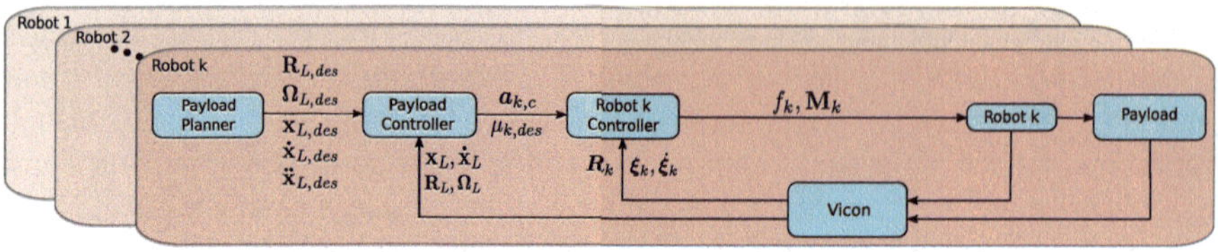

Fig. 2. System architecture of the proposed cooperative transportation system.

applications such as package delivery, search and rescue, construction, and communication in GPS-denied areas.

2 Technical Approach

We consider n quadrotors that cooperatively transport a rigid body payload through massless cables with respect to the world frame as depicted in Fig. 1(a).The position and velocity of the center of mass of the payload with respect to $\mathcal{I}$ are denoted by the vector $\mathbf{x}_L, \dot{\mathbf{x}}_L \in \mathbb{R}^3$. The orientation of the payload with respect to $\mathcal{I}$ is represented by rotation matrix $\boldsymbol{R}_L \in SO(3)$ and $\boldsymbol{\Omega}_L \in \mathbb{R}^3$ is the angular velocity of the payload expressed in the payload frame. The cable of the kth quadrotor is attached to the payload at point k. The location of the attach point k with respect to $\mathcal{L}$ is represented by $\boldsymbol{\rho}_k \in \mathbb{R}^3$, which is constant. The direction of the kth cable is represented by a unit vector $\boldsymbol{\xi}_k \in S^2$. The control force executed by the kth quadrotor is denoted by $\mathbf{u}_k \in \mathbb{R}^3$, which is decoupled into two components $\mathbf{u}_k^{\parallel}$ and $\mathbf{u}_k^{\perp}$. The first one is the projection of $\mathbf{u}_k$ on $\boldsymbol{\xi}_k$, and the second one is the orthogonal projection of $\mathbf{u}_k$ on the plane that is normal to $\boldsymbol{\xi}_k$. The force $\mathbf{u}_k$ will be mapped to the thrust f_k and moment $\mathbf{M}_k$ as input to the kth quadrotor system. The nonlinear system dynamics model is based on [3]. In the following, we introduce the proposed controller.

2.1 Payload Controller

The proposed control architecture is shown in Fig. 2. The desired forces and moments acting on the payload can be chosen as

$$\begin{aligned} \mathbf{F}_d &= m_L \left(\mathbf{K}_p \mathbf{e}_{\mathbf{x}_L} + \mathbf{K}_d \mathbf{e}_{\dot{\mathbf{x}}_L} + \mathbf{K}_i \int_0^t \mathbf{e}_{\mathbf{x}_L} d\tau + \ddot{\mathbf{x}}_{L,des} + \mathbf{g} \right), \\ \mathbf{M}_d &= \mathbf{K}_{\boldsymbol{R}_L} \mathbf{e}_{\boldsymbol{R}_L} + \mathbf{K}_{\boldsymbol{\Omega}_L} \mathbf{e}_{\boldsymbol{\Omega}_L}, \end{aligned} \tag{1}$$

where $\ddot{\mathbf{x}}_{L,des}$ is the desired valued given by the trajectory planner, $\mathbf{K}_p, \mathbf{K}_d, \mathbf{K}_i, \mathbf{K}_{\boldsymbol{R}_L}, \mathbf{K}_{\boldsymbol{\Omega}_L}$ are diagonal positive constant matrices and $\mathbf{e}_{\mathbf{x}_L}, \mathbf{e}_{\dot{\mathbf{x}}_L}, \mathbf{e}_{\boldsymbol{R}_L}, \mathbf{e}_{\boldsymbol{\Omega}_L}$ are the payload, velocity, orientation and angular velocity errors defined in [3]. The forces and moments are distributed to the tensions $\boldsymbol{\mu}_{k,des}$ along each cable as

$$\begin{bmatrix} \boldsymbol{\mu}_{1,des} \\ \vdots \\ \boldsymbol{\mu}_{n,des} \end{bmatrix} = \text{diag}\left(\boldsymbol{R}_L, \cdots, \boldsymbol{R}_L\right) \mathbf{P}^\top \left(\mathbf{P}\mathbf{P}^\top\right)^{-1} \begin{bmatrix} \boldsymbol{R}_L^\top \mathbf{F}_d \\ \mathbf{M}_d \end{bmatrix}, \ \mathbf{P} = \begin{bmatrix} \mathbf{I}_{3\times3} & \mathbf{I}_{3\times3} & \cdots & \mathbf{I}_{3\times3} \\ \hat{\rho}_1 & \hat{\rho}_2 & \cdots & \hat{\rho}_n \end{bmatrix}, \quad (2)$$

where the hat map $\hat{\cdot} : \mathbb{R}^3 \to SO(3)$ is defined such that $\hat{\boldsymbol{a}}\mathbf{b} = \boldsymbol{a} \times \mathbf{b}, \forall \boldsymbol{a}, \mathbf{b} \in \mathbb{R}^3$ and $\mathbf{I}_{3\times3}$ is the identity matrix. The desired acceleration at each payload attach point is

$$\boldsymbol{a}_{k,c} = \mathbf{K}_p \mathbf{e}_{\mathbf{x}_L} + \mathbf{K}_d \mathbf{e}_{\dot{\mathbf{x}}_L} + \mathbf{K}_i \int_0^t \mathbf{e}_{\mathbf{x}_L} d\tau + \ddot{\mathbf{x}}_{L,des} + \mathbf{g} + \boldsymbol{R}_L \hat{\boldsymbol{\Omega}}_L^2 \rho_k,$$

2.2 Robot Controller

Once the desired forces and moments acting on the payload are selected, we map them to the thrust and moments acting on each robot. The desired direction $\boldsymbol{\xi}_{k,des}$ and the desired angular velocity $\boldsymbol{\omega}_{k,des}$ of the kth cable link can then be obtained as

$$\boldsymbol{\xi}_{k,des} = -\frac{\boldsymbol{\mu}_{k,des}}{\left\|\boldsymbol{\mu}_{k,des}\right\|}, \ \boldsymbol{\omega}_{k,des} = \boldsymbol{\xi}_{k,des} \times \dot{\boldsymbol{\xi}}_{k,des},$$

where $\dot{\boldsymbol{\xi}}_{k,des}$ is the derivative of $\boldsymbol{\xi}_{k,des}$. The thrust f_k and moments $\mathbf{M}_k$ acting on the kth quadrotor are

$$f_k = \mathbf{u}_k \cdot \boldsymbol{R}_k \mathbf{e}_3 = \left(\mathbf{u}_k^{\parallel} + \mathbf{u}_k^{\perp}\right) \cdot \boldsymbol{R}_k \mathbf{e}_3, \ \mathbf{M}_k = -\mathbf{K}_R \mathbf{e}_{R_k} - \mathbf{K}_{\boldsymbol{\Omega}} \mathbf{e}_{\boldsymbol{\Omega}_k} + \boldsymbol{\Omega}_k \times \mathbf{J}_k \boldsymbol{\Omega}_k, \quad (3)$$

where, $\mathbf{K}_R$ and $\mathbf{K}_{\boldsymbol{\Omega}}$ are diagonal positive constant matrices, $\mathbf{e}_{R_k}$ and $\mathbf{e}_{\boldsymbol{\Omega}_k}$ are the orientation and angular velocity errors respectively defined in [3], $\boldsymbol{R}_k$ is the rotation matrix of the kth robot with respect to $\mathcal{I}$, $\boldsymbol{\Omega}_k$ is the angular velocity of the kth robot in its body frame, $\mathbf{J}_k$ is its inertia, and $\mathbf{u}_k^{\perp}$ and $\mathbf{u}_k^{\parallel}$ are designed as

$$\begin{aligned} \mathbf{u}_k^{\parallel} &= \boldsymbol{\xi}_k \boldsymbol{\xi}_k^\top \boldsymbol{\mu}_{k,des} + m_k l_k \left\|\boldsymbol{\omega}_k\right\|^2 \boldsymbol{\xi}_k + m_k \boldsymbol{\xi}_k \boldsymbol{\xi}_k^\top \boldsymbol{a}_{k,c}, \\ \mathbf{u}_k^{\perp} &= m_k l_k \boldsymbol{\xi}_k \times \left[-\mathbf{K}_{\xi_k} \mathbf{e}_{\xi_k} - \mathbf{K}_{\omega_k} \mathbf{e}_{\omega_k} - (\boldsymbol{\xi}_k \cdot \boldsymbol{\omega}_{k,des}) \dot{\boldsymbol{\xi}}_{k,des} - \boldsymbol{\xi}_k \times \boldsymbol{\xi}_k \times \boldsymbol{\omega}_{k,des}\right] \\ &\quad - m_k \boldsymbol{\xi}_k \times \boldsymbol{\xi}_k \times \boldsymbol{a}_{k,c}, \end{aligned}$$

where $\mathbf{K}_{\xi_k}$, $\mathbf{K}_{\omega_k}$ are diagonal positive constant matrices, and $\mathbf{e}_{\xi_k}, \mathbf{e}_{\omega_k}$ are the cable direction and cable angular velocity errors defined in [3].

3 Experimental Results

The experiments are conducted in an indoor testbed with a flying space of $7 \times 5 \times 4$ m^3 of the ARPL lab at New York University. We conduct the experiments with a triangular payload suspended from 3 quadrotors. Each quadrotor is connected to the payload with a cable and we denote l_k as its cable length. The quadrotor platform used in the experiments is equipped with a Qualcomm®Snapdragon™

board for on-board computing [4]. The framework has been developed in ROS[1]. The mass of the payload is 288 g, which exceeds the payload capacity of each single vehicle as well as of our previous system [5]. The odometry of the payload, quadrotors, the position and velocity of attachment points are estimated using a Vicon[2] motion capture system at a frequency of 100 Hz. The unit vector of each cable direction $\boldsymbol{\xi}_k$ and the corresponding velocity $\dot{\boldsymbol{\xi}}_k$ are estimated by

$$\boldsymbol{\xi}_k = \frac{\mathbf{x}_{pk} - \mathbf{x}_k}{\|\mathbf{x}_{pk} - \mathbf{x}_k\|}, \quad \dot{\boldsymbol{\xi}}_k = \frac{\dot{\mathbf{x}}_{pk} - \dot{\mathbf{x}}_k}{l_k}, \tag{4}$$

where $\mathbf{x}_{pk}, \dot{\mathbf{x}}_{pk}$ are position and velocity of the kth attach point in $\mathcal{I}$ and $\mathbf{x}_k, \dot{\mathbf{x}}_k$ are position and velocity of the kth robot in $\mathcal{I}$, all of which are estimated by the motion capture system.

3.1 Step Response

We first conduct several experiments with a step input position in each direction and different cable lengths l_k to evaluate the system response and its maneuverability with respect to variations in the mechanical setup. The results are shown in Fig. 3.

We use two different cable lengths, 1 m and 0.5 m, for all the 3 cables and same control gains. As we can observe, changing l_k does not affect the step response in z direction since the $\mathbf{u}_k^{\|}$ will not be affected by l_k if $\|\boldsymbol{\omega}_k\| = 0$. The response in x, y direction reveals that a reduction in cable length causes a quicker system response and shorter settling time. Intuitively, this is due to the fact that reduced cable lengths induce reduced cable swing. The system settles after an average of 1 s with 0.5 m cable length, while it settles in over 2 s with 1 m cable length.

3.2 Trajectory Tracking

We then evaluate the tracking performance of the proposed system with 1 m cables and the same controller gains used in the step response experiments. The results are shown in Fig. 4. Two types of payload trajectories are executed, a circular trajectory and a $\infty-$shape Lissajous trajectory. The circular trajectory is designed as

$$\mathbf{x}_{L,des}^{c}(t) = \begin{bmatrix} r\cos\frac{2\pi t}{T_c} & r\sin\frac{2\pi t}{T_c} & h \end{bmatrix}^\top$$

and the Lissajous trajectory as,

$$\mathbf{x}_{L,des}^{l}(t) = \begin{bmatrix} A_x(1-\cos\frac{2\pi t}{T_l}) & A_y\sin\frac{4\pi t}{T_l} & h \end{bmatrix}^\top$$

with period $T_c = 6$ s, $T_l = 12$ s, radius $r = 0.6$ m, amplitude $A_x = 1.0$ m, $A_y = 0.5$ m and a constant height $h = 0.3$ m. Two different desired pitch angles

[1] www.ros.org.
[2] www.vicon.com.

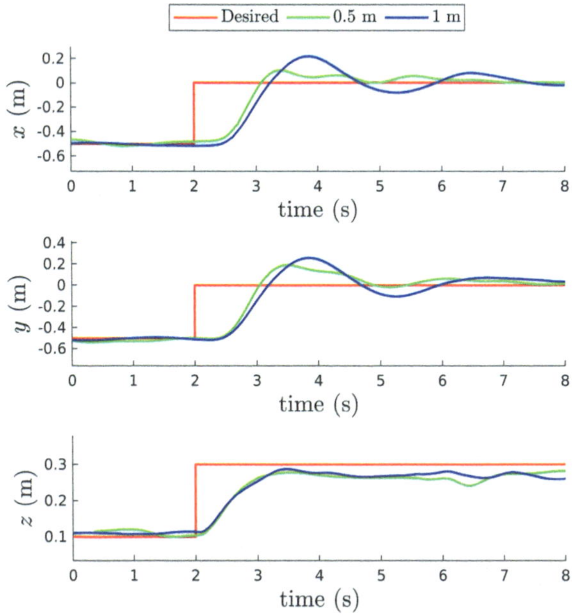

Fig. 3. Step response of the payload position along the three Cartesian x, y, z axis of $\mathcal{I}$ frame.

(0° and 15°) are set for the payload during the trajectory tracking experiments. In Fig. 4, we can observe that the position and velocity of the center of mass of the payload closely follows the commanded trajectory with different pitch angles. The RMSE of position and pitch angle of the payload compared with the desired trajectory are reported in Table. 1. From Table. 1 we can observe that the position tracking errors are within 0.1 m range and the pitch angle tracking errors are within 0.05 rad level, equivalently around 2.5°. The payload maximum speed reaches around 1 m/s. We noticed similar results during multiple tests with speeds up to 1–1.5 m/s.

Table 1. Position and pitch angle RMSE of the payload compared with the desired trajectory

RMSE	Circular, Pitch = 0°	Circular, Pitch = 15°	Lissajous, Pitch = 0°	Lissajous, Pitch = 15°
x (m)	0.0948	0.0805	0.0371	0.0317
y (m)	0.131	0.110	0.0836	0.0777
z (m)	0.00847	0.0137	0.0164	0.0158
Pitch (rad)	0.0702	0.0508	0.0650	0.0381

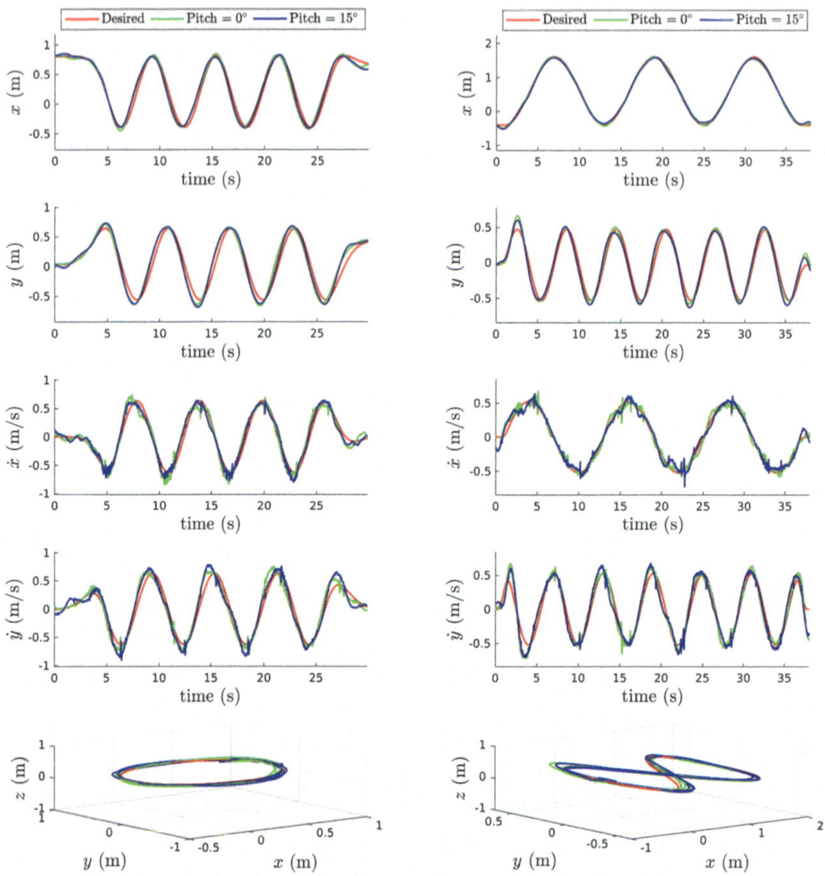

(a) (b)

Fig. 4. Tracking results of the payload following a circular trajectory and a Lissajous trajectory with different pitch angles. (a) Cartesian payload position, velocity, and 3D path for the circular case. (b) Cartesian payload position, velocity, and 3D path for the Lissajous case.

3.3 Discussion

The experimental results reveal several insights. First, we showed the possibility to transport a rigid body payload to follow a given trajectory cooperatively by a team of micro quadrotors in a distributed fashion. Second, we proposed a control approach, which is robust to different cable lengths and enables payload to follow desired trajectories with small tracking errors. Third, we observed that the cable length will affect the maneuverability of the system. By decreasing the suspended cable length, the system reacts faster. This is obtained at a price of a reduced payload-vehicles relative mobility. These properties can be exploited to improve sensing and to increase flexibility and maneuverability in different transportation/manipulation tasks.

4 Conclusion

In this paper, we presented the design and experimental evaluation of a distributed cooperative control method for cooperative transportation of a suspended rigid body payload using a team of MAVs. The proposed approach runs in real time on-board small vehicles that carry computationally constrained units. The experimental results show that the presented system can enable the rigid body payload to closely follow given desired trajectories with variations in both position and orientation and speeds up to 1–1.5 m/s. We have also analyzed and discussed the system response with different cable lengths and observed that shorter cables induce faster system response and this comes at the price of a reduced payload-vehicles relative mobility. We believe that the proposed approach can impact several real world applications.

Future work will involve bridging the gap between the proposed system and deployment in real world settings by closing the loop using on-board sensing and real-time estimation. We will also investigate the best cable length choice considering its effects and the trade-off in terms of payload-vehicle relative mobility, payload maneuverability, and system's sensing capabilities and response time. Furthermore, we will study the optimal tension distribution among the vehicles to guarantee a safety distance among the agents and consider the sensor and actuator constraints at the planning and control levels. Finally, we will be also interested in exploring the scalability of the proposed control approach varying the number of vehicles. We will relax the a-priori knowledge on the mechanical and geometric properties and estimate them with on-board sensors.

Acknowledgments. This work was supported by Qualcomm Research, the ARL grant DCIST CRA W911NF-17-2-0181, the Technology Innovation Institute, Nokia, NYU Wireless, and the young researchers program "Rita Levi di Montalcini" 2017 grant PGR17W9W4N.

References

1. Forte, F., Naldi, R., Marconi, L.: Impedance control of an aerial manipulator. In: American Control Conference (ACC), Montreal, Canada, pp. 3839–3844 (2012)
2. Gassner, M., Cieslewski, T., Scaramuzza, D.: Dynamic collaboration without communication: Vision-based cable-suspended load transport with two quadrotors. In: 2017 IEEE International Conference on Robotics and Automation (ICRA), pp. 5196–5202 (2017)
3. Lee, T.: Geometric control of quadrotor uavs transporting a cable-suspended rigid body. In: 53rd IEEE Conference on Decision and Control. IEEE (2014)
4. Loianno, G., Brunner, C., McGrath, G., Kumar, V.: Estimation, control, and planning for aggressive flight with a small quadrotor with a single camera and IMU. IEEE Robot. Autom. Lett. **2**(2), 404–411 (2017)

5. Loianno, G., Kumar, V.: Cooperative transportation using small quadrotors using monocular vision and inertial sensing. IEEE Robot. Autom. Lett. **3**(2), 680–687 (2018)
6. Loianno, G., Thomas, J., Kumar, V.: Cooperative localization and mapping of MAVS using RGB-D sensors. In: IEEE International Conference on Robotics and Automation (ICRA), pp. 4021–4028 (2015)
7. Michael, N., Fink, J., Kumar, V.: Cooperative manipulation and transportation with aerial robots. Autonom. Robots **30**(1), 73–86 (2011). https://doi.org/10.1007/s10514-010-9205-0
8. Michael, N., Shen, S., Mohta, K., Kumar, V., Nagatani, K., Okada, Y., Kiribayashi, S., Otake, K., Yoshida, K., Ohno, K., Takeuchi, E., Tadokoro, S.: Collaborative mapping of an earthquake-damaged building via ground and aerial robots. J. Field Robot. **29**(5), 832–841 (2012)
9. Ozaslan, T., Loianno, G., Keller, J., Taylor, C.J., Kumar, V., Wozencraft, J.M., Hood, T.: Autonomous navigation and mapping for inspection of penstocks and tunnels with mavs. IEEE Robot. Autom. Lett. **2**(3), 1740–1747 (2017). https://doi.org/10.1109/LRA.2017.2699790
10. Sreenath, K., Kumar, V.: Dynamics, control and planning for cooperative manipulation of payloads suspended by cables from multiple quadrotor robots. Robot. Sci. Syst. (RSS) (2013)
11. Tagliabue, A., Kamel, M., Verling, S., Siegwart, R., Nieto, J.: Collaborative transportation using MAVS via passive force control. In: 2017 IEEE International Conference on Robotics and Automation (ICRA), pp. 5766–5773 (2017)
12. Thomas, J., Loianno, G., Daniilidis, K., Kumar, V.: Visual servoing of quadrotors for perching by hanging from cylindrical objects. IEEE Robot. Autom. Lett. **1**(1), 57–64 (2016)
13. Tomic, T., Schmid, K., Lutz, P., Domel, A., Kassecker, M., Mair, E., Grixa, I., Ruess, F., Suppa, M., Burschka, D.: Toward a fully autonomous UAV: research platform for indoor and outdoor urban search and rescue. IEEE Robot. Autom. Mag. **19**(3), 46–56 (2012)
14. Wu, G., Sreenath, K.: Geometric control of multiple quadrotors transporting a rigid-body load. In: 53rd IEEE Conference on Decision and Control, pp. 6141–6148. IEEE (2014)

CMPCC: Corridor-Based Model Predictive Contouring Control for Aggressive Drone Flight

Jialin Ji, Xin Zhou, Chao Xu, and Fei Gao(✉)

College of Control Science and Engineering, Zhejiang University, Hangzhou, China
{jlji,iszhouxin,cxu,fgaoaa}@zju.edu.cn

1 Introduction

Among the criteria of designing autonomous quadrotors, generating optimized trajectories and tracking the flight paths precisely are two critical components in the action aspect. As shown in our recent work *Teach-Repeat-Replan* [2], a cascaded planning framework with global trajectory generation and local collision avoidance support agile flights under the user preferable routines. In work [2], though the first criteria is met by the global and local planners, the controller has no guarantees on tracking the generated motion precisely. Also, in industrial applications, the planner and controller of a quadrotor are mostly independently designed, making it hard to tune the joint performance in different applications.

Some works [4,7] attempt to compensate uncertainties introduced by disturbances, by designing error-tolerated trajectory planning methods based on Hamilton-Jacobi Reachability Analysis [1]. These works set handcrafted disturbance bounds, making it too conservative to find a feasible solution among dense obstacles. *Tal et al.,* [9] propose control systems for accurate trajectory tracking that improves tracking accuracy. Nevertheless, they still try to track the unreachable trajectory when facing violent disturbance instead of adjusting the primary trajectory. If un-negligible disturbance occurs, local replanners such as [10,11] can plan motions to rejoin the reference quickly, but they are inferior to give a proper temporal distribution. The closest work to this paper [5] applies Model predictive contouring control (MPCC) [3] as the planner for miniature car racing, where safety constraints are established by modeling linear functions from the boundary of the racing track. However, such constraints are not directly available for a quadrotor in unstructured environments. What's more, due to the limited planning horizon, MPCC cannot guarantee feasibility and heavily relies on proper parameter tuning.

To bridge this gap, we propose an efficient, receding horizon, local adaptive low-level planner as the middle layer between our original planner and controller. Our method is named as corridor-based model predictive contouring control (CMPCC) since it builds upon on MPCC [3] and utilizes the flight corridor as hard safety constraints. It optimizes the flight aggressiveness and tracking accuracy simultaneously, thus

J. Ji and X. Zhou—Equal contributors.

This work was supported by the Fundamental Research Funds for the Central Universities underGrant 2020QNA5013.

B. Siciliano et al. (Eds.): ISER 2020, SPAR 19, pp. 37–46, 2021.
https://doi.org/10.1007/978-3-030-71151-1_4

improving our system's robustness by overcoming unmeasured disturbances. Our method features its online flight speed optimization, strict safety and feasibility, and real-time performance and it is released as a low-level plugin for a large variety of quadrotor systems. We summarize our contribution as follows:

1. We propose an efficient and disturbance-adaptive receding horizon low-level planner, which generates a collision-free and dynamic feasible trajectory with adjusted temporal allocation in real time.
2. We propose a method of building appropriate constraints by constructing a forward spanning polygon tube from corresponding polyhedron and setting a terminal velocity from reference trajectory.
3. We integrate the proposed methods into a fully autonomous quadrotor system, and release our software for the reference of the community[1].

2 Methodology

We get the global optimized reference trajectory and the flight corridor from our previous work *Teach-Repeat-Replan* [2], as shown in Fig. 1.

Fig. 1. Global reference trajectory and Flight corridors

The global trajectory $p(t)$ is an optimized smooth curve in the space, parameterized by its original time t, given by

$$p = [x_p, y_p, z_p]^T, p(t) : [t_0, t_m] \rightarrow \mathbb{R}^3. \tag{1}$$

[1] https://github.com/ZJU-FAST-Lab/CMPCC.

We design a re-timing function $\{t(\tau) : \tau \to t\}$ to map the original time variable t to a new time variable τ, shown in Fig. 2(a). And we also design the adjusted local trajectory $s(\tau)$, given by

$$s = [x, y, z]^T, s(\tau) : [0, \tau_m] \to \mathbb{R}^3, \quad (2)$$

where τ_m is the duration of predictive horizon. The optimization objective is shown in Fig. 2(b), where $v_p = \dot{p}(t(\tau))$ indicates the speed of the point on global reference trajectory after re-timing, and the tracking error $e(\tau) = \|s(\tau) - p(t(\tau))\|$. The objective trades off the minimization of $\{\int_0^{\tau_m} e(\tau)d\tau\}$ and the maximization of $\{\int_0^{\tau_m} v_p d\tau\}$ by optimizing $t(\tau)$ and $s(\tau)$.

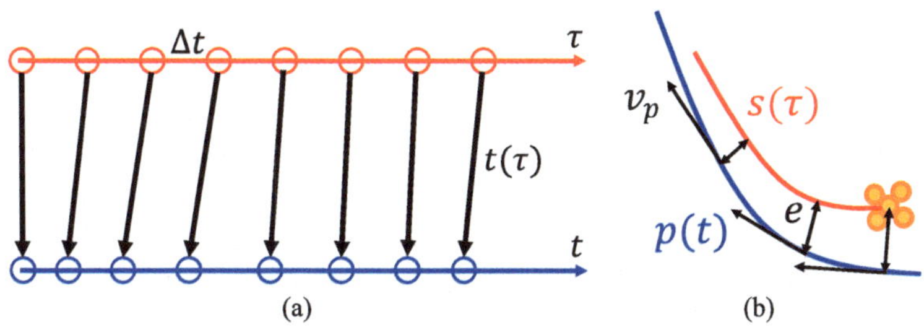

Fig. 2. (a) Re-timing function (b) Components of the objective of CMPCC

Note that $v_p = \dot{p}(t(\tau)) = p'(t) \cdot \dot{t}(\tau)$, where $p'(t)$ is definite according to the global trajectory $p(t)$ but $\dot{t}(\tau)$ truly indicates the traveling progress, i.e. $\dot{t}(\tau) > 1$ means the drone travels faster than and $\dot{t}(\tau) < 1$ means the opposing situation. Thus the objective indicating the tradeoff of tracking error $\|s(\tau) - p(t(\tau))\|$ and traveling progress $\dot{t}(\tau)$ is given by

$$J = \int_0^{\tau_m} \left(\|s - p\|^2 - \rho \dot{t} \right) d\tau, \quad (3)$$

where ρ is a weight of the aggressiveness. We model the system as a 3rd-order integral model and mark $v_\mu := \dot{\mu}(\tau)$, $a_\mu := \ddot{\mu}(\tau)$, $j_\mu := \dddot{\mu}(\tau), \mu = x, y, z, t$. Then the states and inputs of the system is given by

$$\mathbf{x} = [x, v_x, a_x, y, v_y, a_y, z, v_z, a_z, t, v_t, a_t]^T, \quad (4)$$

$$\mathbf{u} = [j_x, j_y, j_z, j_t]^T. \quad (5)$$

They are discretized by Δt with the length of predictive horizon N, and the optimization problem turns into a receding horizon MPC with linear state-transfer equations given by

$$\mathbf{x}^{(k+1)} = \mathbf{A_d}\mathbf{x}^{(k)} + \mathbf{B_d}\mathbf{u}^{(k)} \quad (6)$$

at the k-th time-step, where $\mathbf{A_d}$ and $\mathbf{B_d}$ are given by

$$\mathbf{A_d} = \bigoplus_{i=1}^{4} \begin{pmatrix} 1 & \Delta t & \frac{1}{2}\Delta t^2 \\ 0 & 1 & \Delta t \\ 0 & 0 & 1 \end{pmatrix}, \mathbf{B_d} = \bigoplus_{i=1}^{4} \begin{pmatrix} 0 \\ 0 \\ \Delta t \end{pmatrix}, \tag{7}$$

and the objective (3) is discretized as

$$J = \sum_{k=1}^{N} \left\{ \sum_{\mu=x,y,z} \left(\mu^{(k)} - \mu_p(t^{(k)}) \right)^2 - \rho \cdot v_t^{(k)} \right\}, \tag{8}$$

which is a nonlinear function of $\mathbf{x}$ and $\mathbf{u}$. Thanks to the framework of receding horizon, we linearize $\mu_p(t^{(k)})$ by

$$\mu_p(t^{(k)}) = \mu_p(\theta^{(k)}) + \mu_p'(\theta^{(k)}) \cdot \left(t^{(k)} - \theta^{(k)} \right), \tag{9}$$

where $\theta^{(k)}$ is the optimized result of $t^{(k+1)}$ in the last horizon. Thus the objective (8) can be represented as quadratic, given by

$$J = \sum_{k=1}^{N} \left\{ \mathbf{x}^{(k)T} S_1^T Q_k S_1 \mathbf{x}^{(k)} + q_k^T S_2 \mathbf{x}^{(k)} \right\}, \tag{10}$$

where S_1 and S_2 are selection matrices such that

$$\left(x^{(k)}, y^{(k)}, z^{(k)}, t^{(k)} \right)^T = S_1 \mathbf{x}^{(k)}, \tag{11}$$

$$\left(x^{(k)}, y^{(k)}, z^{(k)}, t^{(k)}, v_t^{(k)} \right)^T = S_2 \mathbf{x}^{(k)}, \tag{12}$$

and

$$Q_k = \begin{pmatrix} 1 & 0 & 0 & -x_p'(\theta^{(k)}) \\ 0 & 1 & 0 & -y_p'(\theta^{(k)}) \\ 0 & 0 & 1 & -z_p'(\theta^{(k)}) \\ -x_p'(\theta^{(k)}) & -y_p'(\theta^{(k)}) & -z_p'(\theta^{(k)}) & \sum_{\mu=x,y,z} c_\mu^{(k)2} \end{pmatrix}, \tag{13}$$

$$q_k = \left(2c_x^{(k)}, 2c_y^{(k)}, 2c_z^{(k)}, \sum_{\mu=x,y,z} \left\{ -2\mu_p'(\theta^{(k)}) c_\mu^{(k)} \right\}, -\rho \right)^T, \tag{14}$$

$$c_\mu^{(k)} = \mu_p'(\theta^{(k)}) \cdot \theta^{(k)} - \mu_p(\theta^{(k)}), \mu = x, y, z. \tag{15}$$

Then we construct linear inequality constraints. For a given reference point $p(\theta^{(k)})$ on the global trajectory, we define Ω as the intersection of v_p's normal plane Φ with the corresponding polyhedron. As shown in Fig. 3(a), the resulting Ω is a convex polygon. Then each edge of Ω expands a plane sweeping along the direction of $v_p^{(k)}$, which gives a polygon tube, as shown in Fig. 3(b). The inner side of this tube is considered as the safe space near $p(\theta^{(k)})$ and will be modeled as inequality constraints:

$$\mathbf{C}^{(k)} \cdot [x^{(k)}, y^{(k)}, z^{(k)}]^T \le \mathbf{b}^{(k)}, k = 1, 2, 3, \ldots, N. \tag{16}$$

The sequence of the polygon tubes is shown in Fig. 4.

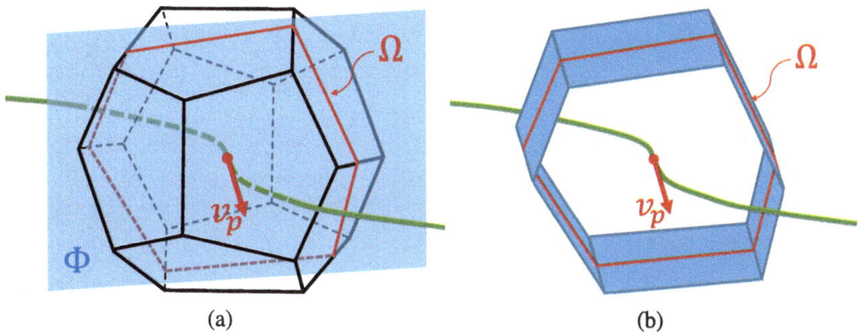

Fig. 3. Generation of polygon tube

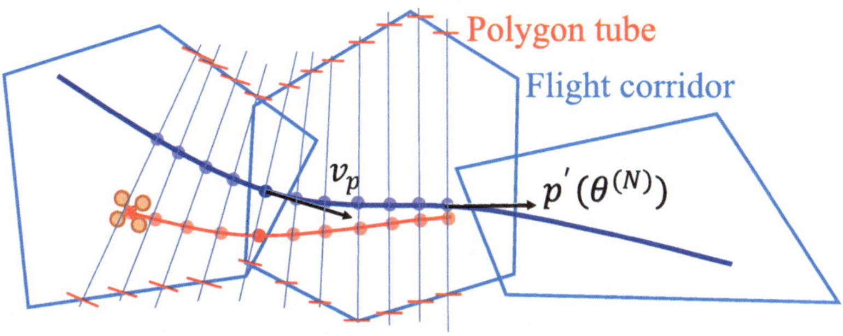

Fig. 4. Sequence of linear inequality constraints for safety

In order to guarantee the dynamic feasibility, physical limits for each state are set as inequality constraints:

$$l_x \leq S_3\mathbf{x}^{(k)} \leq u_x, \tag{17}$$

$$l_u \leq \mathbf{u}^{(k)} \leq u_u, \tag{18}$$

where l_x, l_u, u_x, u_u are lower and upper bounds of velocity, acceleration and jerk at each time step. S_3 is another selection matrix such that

$$[v_x^{(k)}, a_x^{(k)}, v_y^{(k)}, a_y^{(k)}, v_z^{(k)}, a_z^{(k)}]^T = S_3\mathbf{x}^{(k)} \tag{19}$$

In addition to physical limits for each state, a terminal velocity constraint is added such that the terminal speed should be less than $p'(\theta^{(N)})$ in the predictive horizon. Thus, feasibility can be guaranteed since the reference trajectory is globally optimal.

The optimization problem is formulated as a standard quadratic program (QP):

$$\min_{\bar{\mathbf{x}}} \bar{\mathbf{x}}^T Q\bar{\mathbf{x}} + q^T\bar{\mathbf{x}} \tag{20}$$

$$\textbf{s.t.} \quad P\bar{\mathbf{x}} = h \tag{21}$$

$$C\bar{\mathbf{x}} \leq b \tag{22}$$

where

$$\bar{\mathbf{x}} = [\mathbf{x}^{(1)};\mathbf{u}^{(1)};\mathbf{x}^{(2)};\mathbf{u}^{(2)};...;\mathbf{x}^{(N)};\mathbf{u}^{(N)}] \tag{23}$$

and it is solved by OSQP [8] with warm start speed-up. In practice, we choose a $1s$ predictive horizon and the sampling interval $\Delta t = 0.05s$, which means $N = 20$. The performance of our algorithm is tested on an Intel i7-6700 CPU, with the average solving time around $5ms$.

3 Experiments

3.1 Experiments Configuration

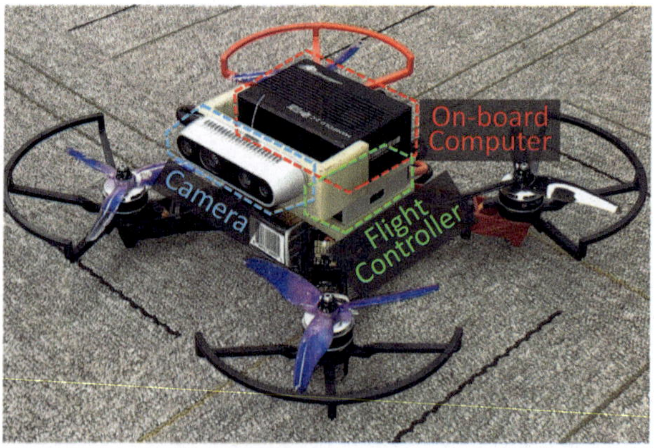

Fig. 5. The hardware of the UAV

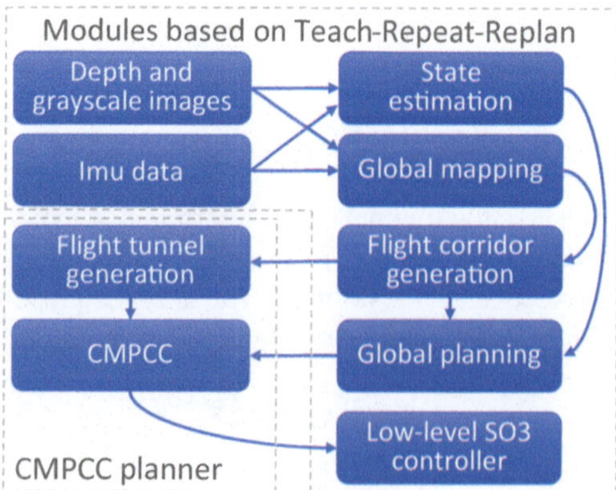

Fig. 6. The Software architecture of the UAV

We use a self-developed quadrotor with an Intel Realsense D435i stereo camera[2] and a DJI N3 flight controller[3] for state estimation. All modules run solely on a DJI Manifold 2-C onboard computer[4]. Our system inherits the localization, mapping and global planning from the *Teach-Repeat-Replan* system [2], where readers can check these modules in detail. The overall hardware and software architecture of our system is shown in Fig. 5 and Fig. 6.

3.2 Autonomous Flight with Contact Disturbance

(a) (b)

Fig. 7. Circumstance of instant force disturbance

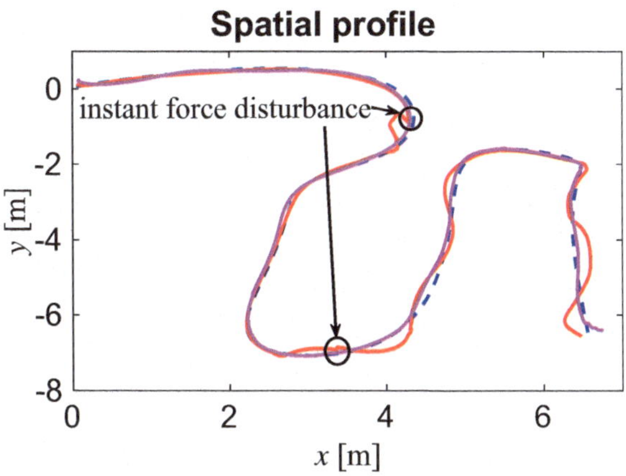

Fig. 8. Spatial profile facing instant force disturbance

[2] https://www.intelrealsense.com/depth-camera-d435i/.
[3] https://www.dji.com/n3.
[4] https://www.dji.com/manifold-2.

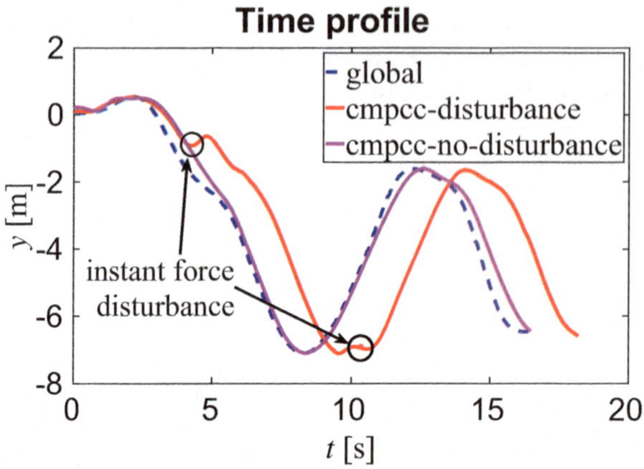

Fig. 9. Time profile facing instant force disturbance

We apply challenging force disturbance to the drone to test the performance of the proposed CMPCC, as shown in Fig. 7(a). The global trajectory (blue), our locally re-planned trajectory (red), and the geometry constraints (magenta) after the hit are visualized in Fig. 7(b). As seen in the top-down view of the experiment in Fig. 8, oscillation occurs near the hit position, but the local trajectory soon converges to the global one. Also, as shown in Fig. 9, the instant force disturbance changes the temporal distribution of the optimal trajectory, resulting in the delaying of the trajectory of cmpcc (red) relative to which without disturbance (magenta) and the global trajectory (blue).

3.3 Autonomous Flight with Wind Disturbance

We also test our method with wind disturbance by a fan, as shown in Fig. 10(a). Without the proposed CMPCC, the quadrotor tracks the global trajectory generated by *Teach-Repeat-Replan* with only a feedback controller, and collides with the nearby obstacle soon. However, thanks to the safety guarantee, the proposed CMPCC re-plans a safe trajectory and rejoins the global reference quickly under the wind disturbance, as shown in Fig. 10(b).

The video of the experiments is available[5].

[5] https://www.youtube.com/watch?v=_7CzBh-0wQ0.

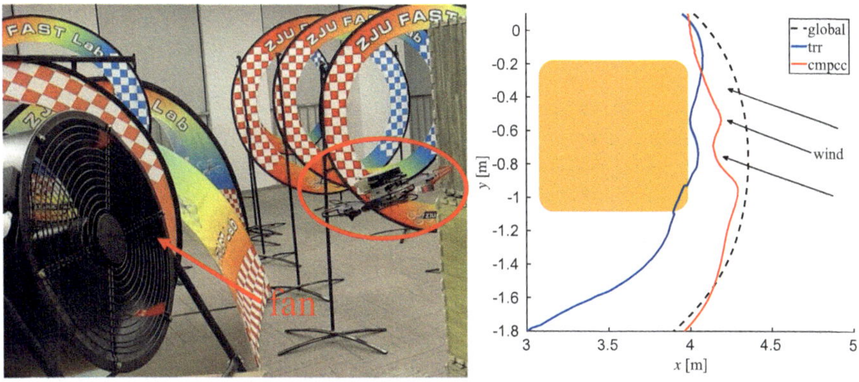

(a) (b)

Fig. 10. The circumstance of wind disturbance

4 Main Experimental Insights

In practice, the flight performance of a quadrotor can be affected by many factors. Among all issues, the unexpected and unmeasurable disturbance is always an essential one for quadrotor autonomous navigation, especially for fast and aggressive flight. Recently, most autonomous quadrotor systems [2,6] are developed with several independent modules include controller, planner, and perception, with the assumption that a properly designed, smooth, derivative bounded trajectory can be tracked by a controller within high bandwidth. However, this assumption does not always hold. No matter how robust the feedback controller is, it's noted that it may fail when encountering drastic disturbance, such as immediate contact and a gust of wind, which are demonstrated in our experiments. Traditionally, people have to spend tons of time tuning the parameters of the feedback controller until a satisfactory performance. In this work, as validated by our challenging experiments, the proposed intermediate low-level replanner successfully compensates disturbances by planning local safe trajectories and automatically adjusting the flight aggressiveness. Therefore, the robustness of fast autonomous flight is improved significantly. Moreover, thanks to the convex formulation, the proposed CMPCC is solved within 5 *ms*, which suits onboard usage well.

In experiments, we also observe that the polygon tube now we use heavily depends on the static corridor. Therefore it cannot handle the variation of the environment or dynamic obstacles. In the future, we plan to investigate the way to generate safety constraints for CMPCC online.

References

1. Bansal, S., Chen, M., Herbert, S., Tomlin, C.J.: Hamilton-jacobi reachability: a brief overview and recent advances. In: 2017 IEEE 56th Annual Conference on Decision and Control (CDC), pp. 2242–2253. IEEE (2017)

2. Gao, F., Wang, L., Zhou, B., Zhou, X., Pan, J., Shen, S.: Teach-repeat-replan: a complete and robust system for aggressive flight in complex environments. IEEE Trans. Robot. **36**(5), 1526–1545 (2020)
3. Lam, D., Manzie, C., Good, M.: Model predictive contouring control. In: 49th IEEE Conference on Decision and Control (CDC), pp. 6137–6142. IEEE (2010)
4. Li, Z., Arslan, O., Atanasov, N.: Fast and safe path-following control using a state-dependent directional metric. *arXiv preprint* arXiv:2002.02038 (2020)
5. Liniger, A., Domahidi, A., Morari, M.: Optimization-based autonomous racing of 1: 43 scale rc cars. Optimal Control Appl. Method **36**(5), 628–647 (2015)
6. Oleynikova, H., Lanegger, C., Taylor, Z., Pantic, M., Millane, A., Siegwart, R., Nieto, J.: An open-source system for vision-based micro-aerial vehicle mapping, planning, and flight in cluttered environments. J. Field Robot. **37**(4), 642–666 (2020)
7. Seo, H., Lee, D., Son, C.Y., Tomlin, C.J., Kim, H.J.: Robust trajectory planning for a multi-rotor against disturbance based on hamilton-jacobi reachability analysis. In: 2019 IEEE/RSJ International Conference on Intelligent Robots and Systems (IROS), pp. 3150–3157. IEEE (2019)
8. Stellato, B., Banjac, G., Goulart, P., Bemporad, A., Boyd, S.: OSQP: an operator splitting solver for quadratic programs. Math. Program. Comput. 1–36 (2020)
9. Tal, E., Karaman, S.: Accurate tracking of aggressive quadrotor trajectories using incremental nonlinear dynamic inversion and differential flatness. In: 2018 IEEE Conference on Decision and Control (CDC), pp. 4282–4288. IEEE (2018)
10. Usenko, V., Von Stumberg, L., Pangercic, A., Cremers, D.: Real-time trajectory replanning for mavs using uniform b-splines and a 3d circular buffer. In: 2017 IEEE/RSJ International Conference on Intelligent Robots and Systems (IROS), pp. 215–222. IEEE (2017)
11. Zhou, B., Gao, F., Pan, J., Shen, S.: Robust real-time uav replanning using guided gradient-based optimization and topological paths. *arXiv preprint* arXiv:1912.12644 (2019)

Multirotor Docking with an Airborne Platform

Ajay Shankar[1(✉)], Sebastian Elbaum[2], and Carrick Detweiler[1]

[1] Department of Computer Science and Engineering, University of Nebraska-Lincoln, Lincoln, USA
ashankar@cse.unl.edu

[2] Department of Computer Science, University of Virginia, Charlottesville, USA

Abstract. Multirotor systems have traditionally been employed for missions that ensure minimal contact with the objects in their vicinity. However, their agile flight dynamics lets them sense, plan and react rapidly, and therefore perform highly dynamic missions. In this work, we push their operational envelope further by developing a complete framework that allows a multirotor to dock with a moving platform. Our approach builds on state-of-the-art and optimal methods for estimating and predicting the state of the moving platform, as well as for generating interception trajectories for the docking multirotor. Through a total of 25 field tests outdoors, we demonstrate the capabilities of our system in docking with a platform moving at different speeds and in various operating conditions. We also evaluate the quality of our system's trajectory following at speeds over 2 m/s to effect docking within 10 s.

1 Motivation and Background

Recent advances in multirotor unmanned aerial systems' (UASs) precision in sensing and control have allowed them to be used more efficiently in a wide array of aerial manipulation tasks. These involve picking up objects, inspecting surfaces, and applying contact forces on other objects in the world [1]. Performing such aerial manipulation of stationary objects in outdoor environments still remains challenging due to the sensing and kino-dynamic constraints of the UAS. The problem becomes even more difficult when considering *non-stationary* target objects outdoors, since (a) errors and ambiguities in the target's relative motion estimates can undermine the advantages of precise maneuvering and control, and, (b) manipulating a moving target often imposes other mission contraints such as the time horizon or the number of attempts to engage successfully.

In this paper, we introduce a framework that allows a follower multirotor to dock with a non-stationary docking platform moving on a predictable (non-evasive) path, akin to that of a recovery [leader] aircraft. Illustrated in Fig. 1, we mimic a towed platform by suspending it over a moving "zipline" system, and effect motions at different speeds outdoors. Our approach minimally equips the docking platform with a fiducial marker and a real-time kinematic (RTK) GPS

B. Siciliano et al. (Eds.): ISER 2020, SPAR 19, pp. 47–59, 2021.
https://doi.org/10.1007/978-3-030-71151-1_5

Fig. 1. Snapshots depicting a multirotor docking with a suspended platform outdoors. Our outdoor setup comprises of pulleys interconnected by an actuated "zipline" mechanism to regulate the translational velocity of the platform suspended from it.

unit. The follower multirotor fuses the high-rate visual feedback with sporadic GPS measurements to estimate the platform's trajectory over a time horizon, h. A smooth, energy-optimal, end-to-end polynomial trajectory is then planned and refined over time with new measurements to effect docking within this constrained horizon. Our evaluations are conducted outdoors under realistic constraints and operational conditions.

We consider docking with a moving platform as a variant of the classic problem of landing on a moving platform, coupled with the mechanical and state estimation challenges of an aerial manipulation task. UASs landings on moving platforms have been demonstrated for slow [2] and fast moving targets [3], as well as using entirely onboard estimation [4]. Aerial manipulation problems have been studied within certain applications such as payload transportation and applying contact forces on objects [1].

Unlike landing, in-flight docking necessitates a specialized mechanism to engage with the moving platform. This is akin to robotic manipulators and end-effectors that extend outward from the UAS's center of mass to safely grasp another object [5,6]. However, for docking, the target object (the docking platform) will typically have limited surface area as it is either a part of the vehicle or an extension towed by an aerial vehicle. Consequently, the positioning tolerances in forward-flight are considerably lower for docking (compared to landing, for instance). We note that while guidance principles have been derived and studied for docking fixed-wing aircraft [7,8], limited work has been shown for docking stationary platforms [9]. Furthermore, the leader-follower approaches mentioned prior typically employ reactive control strategies that do not accommodate trajectory plans. To the best of our knowledge, there is no previous work demonstrating planned docking for multirotors and moving platforms.

Problem Statement. In a local (fixed) world (Fig. 2), $\mathcal{W}$, given a docking platform $\mathcal{D}$ at position $D(t_i)^{\mathcal{W}}$ with a true velocity vector $\mathbf{D}$, a follower multirotor $\mathcal{F}$ at position $F(t_i)^{\mathcal{W}}$, and a time horizon h, the problem is to generate and execute a valid trajectory, $\mathbb{T}(t)$ such that $dock(\mathcal{D}, \mathcal{F}, t) = true$ and $t \in [t_i, t_i + h]$.

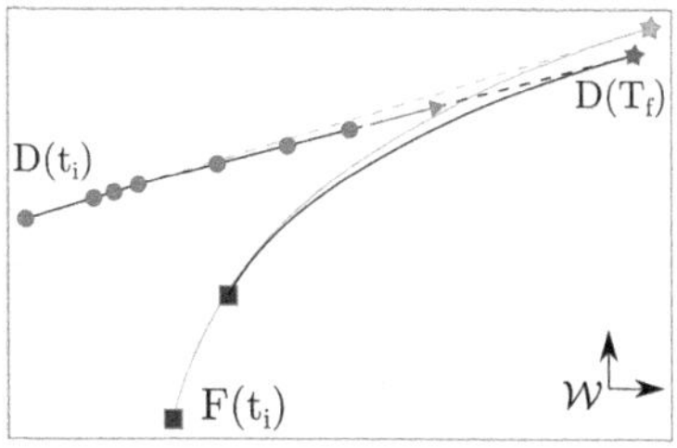

Fig. 2. A top-down graphical illustration of docking: follower UAS at F plans an interception trajectory towards $D(T_f)$, the projected final location for the leader.

Solution Space. We aim to develop a complete framework that enables the follower multirotor UAS to precisely estimate the trajectory of a moving docking platform, and continually refine safe trajectories to dock with it. We also seek cost-effective solutions by integrating off-the-shelf subsystems that can be mounted on a small multirotor UAS.

2 Technical Approach

We first describe the sensing, planning, and control components needed to meet our goal of state estimation and generating interception trajectories, and then provide a description of the physical mechanism that enables the docking to complete the mission.

A mission is successful when the UAS autonomously docks with the moving platform and disengages its flight controller within the specified time horizon. In the following, let $\mathcal{W}$ denote the origin of the world-fixed (local map) frame in a North-East-Down (NED) convention, and let $F^{\mathcal{W}}, D^{\mathcal{W}} \in \mathbb{R}^3$ respectively denote the origins of the NED frames of the follower UAS ($\mathcal{F}$) and the docking platform ($\mathcal{D}$) expressed in the fixed frame. The first objective is to accurately estimate the pose of $\mathcal{D}$ in the fixed frame for some final time, T_f.

2.1 Sensing and Estimation: Leader

We equip the docking platform with two complementary sources - a real-time kinematic (RTK) GPS unit and a passive fiducial marker. Using the constant velocity assumption, a straight line trajectory is estimated for the platform using the observed positions such that, starting with an initial measurement of the platform's position, $D(t_i)$, we can express its position at some time t as

$$D(t) = D(t_i) + t\mathbf{D}, \tag{1}$$

where $\mathbf{D} \in \mathbb{R}^3$ is the NED velocity of the platform expressed in world frame. To estimate $\mathbf{D}$, we stack k such observations, so that,

$$\begin{pmatrix} D(t_1) - D(t_i) \\ D(t_2) - D(t_i) \\ \vdots \\ D(t_k) - D(t_i) \end{pmatrix} = \begin{pmatrix} t_1\mathbb{I}_{3x3} \\ t_2\mathbb{I}_{3x3} \\ \vdots \\ t_k\mathbb{I}_{3x3} \end{pmatrix} * \mathbf{D} \tag{2}$$

where $\mathbb{I}$ is the identity matrix. The equations have been rearranged in the form of $b = Ax$, and a closed-form solution can be obtained as $\mathbf{D} \equiv x = (A^\top A)^{-1}A^\top b$. The location of the leader at time T_f can now be calculated from Eq. (1). This form of linear regression is resilient to outliers that are often encountered in GPS measurements, and yields robust estimates of the constant $\mathbf{D}$.

In some cases, the velocity of the leader may not be constant, but the resultant path may still be predictable (e.g., parabolic). This is true for an emulated leader setup suspended on a gantry/zipline system, which exhibits a longitudinal sag due to its own weight. For a leader system moving along this path, the resultant trajectory in the vertical axis can be approximated as a parabola in time. We can extend the same approach to define a polynomial regression over the observed path points $D_{\hat{\imath}}$ along any axis $\hat{\imath}$. As before, k such observations can be stacked, and the set of parabolic equations can be rewritten as

$$\begin{pmatrix} D_{\hat{\imath}}(t_1) \\ D_{\hat{\imath}}(t_2) \\ \vdots \\ D_{\hat{\imath}}(t_k) \end{pmatrix} = \begin{pmatrix} 1 & t_1 & t_1^2 \\ 1 & t_2 & t_2^2 \\ & \vdots & \\ 1 & t_k & t_k^2 \end{pmatrix} * \begin{pmatrix} c_1 \\ c_2 \\ c_3 \end{pmatrix}, \tag{3}$$

where $[c_1, c_2, c_3]^\top$ are the coefficients of a parabola. Once again, the system of equations yield a closed-form solution for $[c_1, c_2, c_3]^\top$ by recognizing the form to be $b = Ax$ and solving $(A^\top A)^{-1}A^\top b$ in this case. The location and velocity in the $\hat{\imath}$ axis at the final time, T_f, can then be expressed as

$$D_{\hat{\imath}}(T_f) = c_1 + c_2 T_f + c_3 T_f^2, \text{ and, } \dot{D}_{\hat{\imath}}(T_f) = c_2 + 2c_3 T_f. \tag{4}$$

In this case, the $\hat{\imath}$ component of $\mathbf{D}$ is simply substituted with $\dot{D}_{\hat{\imath}}(T_f)$ instead.

For dynamically adjusting to the changes in the platform's state, especially upon closer approach, we require fast and accurate observations from the calibrated monocular camera. Using a fiducial marker, the camera-frame position of the platform, $z_{\text{cam}} \equiv D(t)^{\mathcal{C}} \in \mathbb{R}^3$, is obtained directly through a fast and open-source localization framework [10]. We can then express the position in world-fixed frame as $D(t)^{\mathcal{W}} = R_{\mathcal{F}}^{\mathcal{W}} R_{\mathcal{C}}^{\mathcal{F}} D(t)^{\mathcal{C}}$, where the general notation R_B^A denotes the transform from a frame B to frame A. The final location of the platform at time T_f is computed using Eq. (1).

These complementary sources of information are fused using a first-order Kalman filter, with two measurement update cycles for z_{cam} and z_{gps}. We treat the low-rate RTK-GPS measurements as ground-truth (with respect to $\mathcal{W}$ frame), and thus associate a lower measurement uncertainty to it.

2.2 Sensing and Estimation: Follower

The follower UAS is modeled as a six degree of freedom (DOF) object in a feedback-linearized system of equations. The UAS state vector, denoted as $\boldsymbol{x} \equiv [p_n, p_e, p_d, v_n, v_e, v_d, \psi]^\top$, is comprised of position ($p_{(\cdot)}$) and velocity ($v_{(\cdot)}$)

elements in North, East and Down axes, and the heading (ψ). The 2nd-order dynamic system of equations can then be represented as

$$\begin{aligned}\dot{\boldsymbol{x}} &= A_F\boldsymbol{x} + B_F\boldsymbol{u} \\ y &= C_F\boldsymbol{x}.\end{aligned} \quad \Bigg| \quad A_F = \begin{pmatrix} 0_{3x3} & \mathbb{I}_{3x3} & 0_{3x1} \\ 0_{3x3} & 0_{3x3} & 0_{3x1} \\ 0_{1x3} & 0_{1x3} & 0 \end{pmatrix} B_F = \begin{pmatrix} 0_{3x3} & 0_{3x1} \\ \mathbb{I}_{3x3} & 0_{3x1} \\ 0_{1x3} & 1 \end{pmatrix} C_F = \mathbb{I}_{7x7}. \tag{5}$$

An extended Kalman filter (EKF) is used to produce the best estimate of the system state, $\hat{\boldsymbol{x}}$, by fusing low rate RTK-GPS and high rate attitude measurements from the onboard inertial measurement unit (IMU). The controller uses this $\hat{\boldsymbol{x}}$ to regulate $\boldsymbol{x}$ along the trajectory for the docking mission.

2.3 Trajectory Planning and Control

A trajectory for the UAS is generated at time t_i, such that it begins at $F(t_i)^{\mathcal{W}}$ and terminates at $D(T_f)$, and is smooth up to the 3rd order (to minimize drastic changes in acceleration). For a smooth terminal contact, we also require that the trajectory is parameterized in time to account for velocity constraints, such that $\dot{F}(T_f) = \mathbf{D} + \mathbf{v}_\epsilon$, where $\mathbf{v}_\epsilon \in \mathbb{R}^3$ is a small constant. The choice of $\mathbf{v}_\epsilon$ is a design parameter that allows the follower to use the transfer of momentum from the contact to trigger a grasping mechanism.

To meet these criteria, we use closed-form representations of energy optimal trajectories that can be computed algebraically [11]. Thus, given a starting state $F(t_i)$ and $\dot{F}(t_i)$, a trajectory $\mathbb{T}(t)$ is defined as a ordered collection of 3D points in time that minimize the total jerk over time,

$$\mathbb{T}(t) = \{F(t) \mid \min \int_{t_i}^{T} \|\frac{\mathrm{d}^3\mathbb{T}}{\mathrm{d}t^3}\|^2 \mathrm{d}t\}, \qquad t_i < t \le T_f, \tag{6}$$

such that $\mathbb{T}(T_f) = D(T_f)$ and $\frac{\mathrm{d}\mathbb{T}(T_f)}{\mathrm{d}t} = \mathbf{D} + \mathbf{v}_\epsilon$. Since the translational end-states are known (estimated in Sect. 2.1), the resultant trajectory is fully defined and can be computed through closed-form expressions.

A linear quadratic regulator (LQR) outer-loop controller governs the execution of the generated trajectory. The resultant path from Eq. (6) is a 4th-order polynomial in time. We use the path points, and their first derivative (velocity) as reference states for the controller, while the second derivative (acceleration) is used as a feed-forward element. The trajectory is undefined for $t > T_f$, and thus, the controller maintains a stable hover if docking has failed, or is disengaged if successful.

Denoting $\boldsymbol{x}_r = [\mathbb{T}(t), \dot{\mathbb{T}}(t)]^\top$ as the reference state for the controller, we compute the control input to the system as the sum of feedback and feedforward elements as,

$$\boldsymbol{u} = -K_{\mathrm{lqr}}(\hat{\boldsymbol{x}} - \boldsymbol{x}_r) + \ddot{\mathbb{T}}(t), \tag{7}$$

where K_{lqr} is the LQR feedback gain matrix. The control input, $\boldsymbol{u}$, in this case consists of acceleration commands in the three translational axes. These are converted to angle and thrust commands by a non-linear inversion map, such that, $[\phi_d, \theta_d, \psi_d, T_d] = g_{\text{kin}}(\boldsymbol{u}, m)$, with m denoting the total mass of the system, and $(\cdot)_d$ denoting a desired target.

2.4 Docking Subsystem

A distinctive element in the docking problem (compared to landing, for instance) is the mechanical subsystem that ensures a physical lock. The mechanism must sense the contact with the docking platform and react rapidly to ensure a successful "grasp" of the docking platform. Since the relative velocity is non-zero at the time of contact (due to $\mathbf{v}_\epsilon$), the mechanism can utilize the transfer of momentum as a sensory input.

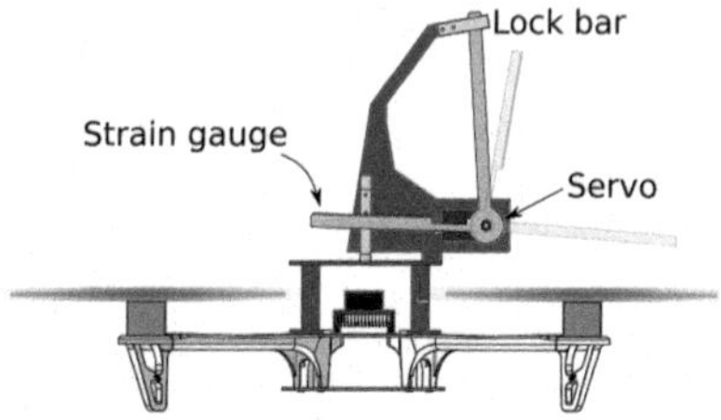

Fig. 3. A CAD schematic for the docking assembly atop the UAS.

Figure 3 shows a side-view illustration of the docking mechanism. The servo-actuated rotary "gate" is mounted on top of the UAS such that a strain gauge affixed to one of the gate arms is exposed (shown as faded). A microcontroller senses the calibrated output of the strain gauge, and actuates the gate to a "closed" position when a threshold for contact force is met. Using a fast-actuation servo, we minimize the reaction time to an external contact down to 0.2 s.

3 Field Studies

We begin by first outlining the details of our implementation setup, including the various off-the-shelf components utilized for a fully characterizable set of tests. We then present our results from outdoor field studies in two parts: first, for a stationary leader, and then, for a leader moving in linear trajectory at different velocities. Our evaluations focus on the ability of the system to (a) correctly estimate the location of the docking platform and (re)plan trajectories towards it, (b) regulate the distance to the final predicted location in a smooth and predictable fashion within the specified time horizon, and finally, (c) dock with the platform.

Implementation Details

Leader System. As mentioned earlier, we utilize a zipline-like mechanism to emulate a leader aircraft to aid consistent and repeatable missions. The zipline consists of two pulleys mounted rigidly on poles approximately 17 m apart. A light-weight string loops around the pulleys and runs through a DC motor system (driver, battery and a micro-controller). The motor can affect different translational speeds for the zipline system through a serial command interface. A 3D-printed clip and frame are affixed on the string to suspend the docking platform on it. The docking platform is a metal bar held horizontal by two strings in an 'A'-frame fashion. The bar is a hollow 20 mm square extruded to a length of 0.5 m. The total available contact surface spans approximately 0.4 m. A fiducial marker is placed separated 0.2 m from the docking platform. The RTK GPS receiver (uBlox ZED-F9P) is also mounted on this A-frame, exposed to the sky.

The platform exhibits translational velocities due to the motion of the zipline. These are constant in time in the lateral (north-east) plane, resulting in a linear trajectory. Since the strings are flexible, the weight of the platform causes the zipline to sag in the vertical (down) axis, resulting in a curved path that can be approximated as parabolic. The platform also exhibits 3-axis rotational velocities due to wind and noise from the standing waves induced in the zipline's strings. These are only naturally damped by the material properties of the strings.

Follower System. We use a DJI Flamewheel frame with the open-source Pixhawk autopilot as the follower vehicle. The docking platform described in Sect. 2.4 is mounted on top of the frame as shown in Fig. 3. All of state estimation, prediction, trajectory generation and flight control is implemented entirely onboard on an Odroid XU4 single-board computer. No external motion-capture or ground controllers are utilized.

3.1 Stationary Leader

We first assess the quality of our state estimation, trajectory control and the docking subsystem by keeping the leader system stationary. This primarily allows us to isolate the follower's performance by removing the prediction errors that might arise when the target is in motion. It also lets us validate that the electro-mechanical elements of the docking system function as expected upon contact with the docking platform. For these tests, the zipline system is commanded a zero velocity, while the leader (docking platform) simply suspends from it.

Figure 4 shows the distribution of this stationary leader's estimated velocities as a probability density function. Due to ambient wind, the leader can exhibit a 'bobbing' motion with a small velocity (usually $\leq$0.01 m/s), which can cause erroneous predictions over a long time horizon. It is possible to use Eq. (1) with the assumption that the leader remains stationary. However, to avoid over-simplifying the missions, we make no such assumptions, and carry out state estimation, projection, and trajectory (re-)planning without any simplifications. For context, an estimated velocity of 0.01 m/s over a horizon of 8 s produces an error of 0.08 m which is ~70% the size of our docking mechanism's gate.

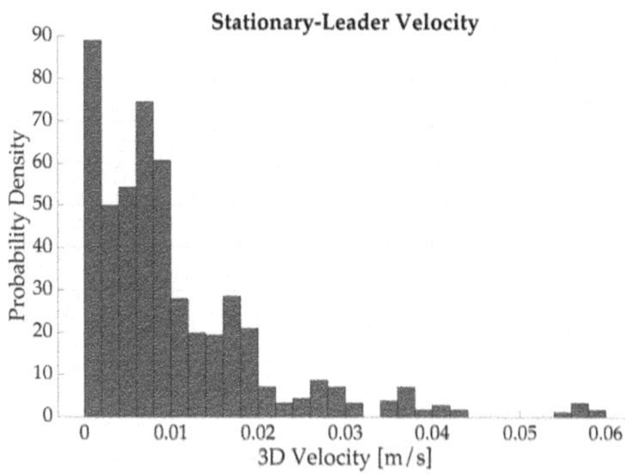

Fig. 4. Probability function of the leader's velocities even when stationary.

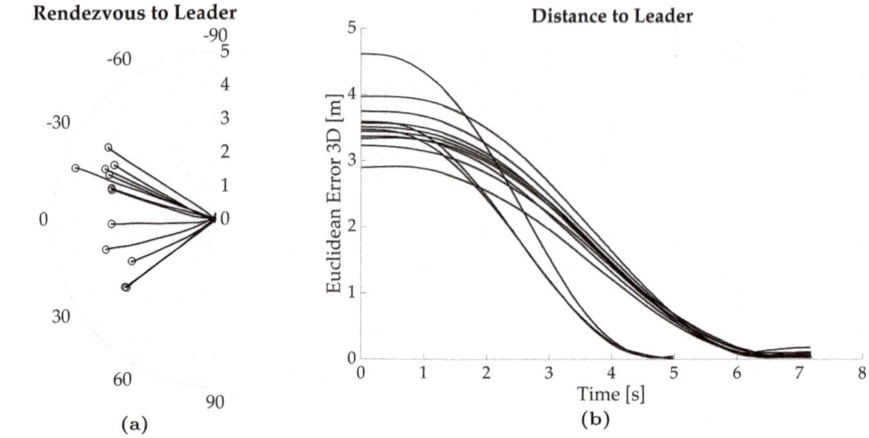

Fig. 5. Docking with a stationary leader, shown as, (a) top-down polar view of all the docking tests, framed to keep the leader at the origin; and (b) temporal evolution of the 3D Euclidean distance to the leader. Each test lead to a successful docking.

Figure 5 graphically summarizes the results from the stationary tests, conducted under ambient wind speeds of 0.75–1.8 m/s. We conduct a total of 12 trials starting at different relative positions between the leader and the follower. These are graphically shown on a polar plot of the lateral axes in Fig. 5(a), rotated and re-centered to keep the leader at the origin. For 9 of these, the fixed time-horizon is set to 7 s. To stress the system further, for the final 3 tests, we shrink this horizon to 5 s. In all cases, the controller is able to regulate the UAS's position exactly within these selected time windows and effect a successful docking. In addition, one of the 5 s missions also begins at the largest separation to

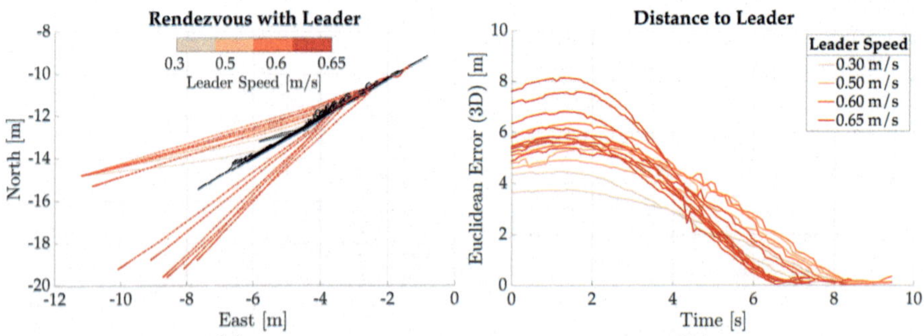

Fig. 6. Docking with a moving leader. (a) The top-down (North-East) view of trajectories (leader moves on the same path at different speeds). The initial locations are on the lower-left of the image, and time evolves from left to right. The follower's trajectories are color-coded for the 4 different speeds of the leader. (b) Temporal evolutions of the 3D Euclidean distance to the leader. Docking occurs between 7–9 s.

the leader, thus creating a challenging mission. This particular trial exhibits a peak velocity of almost 2 m/s, and still leads to a successful docking.

3.2 Moving Leader

We now present our evaluations from 15 tests with the leader system in motion: 8 successive tests at 4 different translational speeds, and then 7 additional tests with different initial conditions. For the speed tests, we select a common starting location for the follower. It maintains a stable hover at this location as the zipline begins to move at the commanded speed. The four speeds we test at are 0.3, 0.5, 0.6 and 0.65 m/s. We set $t_{\text{obs}} = 10$ s for all of these tests, and choose either 8.5 s or 9.5 s as the mission horizon. These numbers are chosen empirically based on the operational size of our test arena and the speed of the zipline. For instance, approximately 20 s of motion at 0.65 m/s would move the zipline from one extremity to another.

In Fig. 6(a), we show a top-down (North-East) view of all these trials. The black trajectory is that of the leader/zipline, which moves on the same path (starting at lower-left) but at the different speeds we listed above. The follower UAS begins its mission at the common location at the lower-left, and the trajectories for the four leader speeds are shown with different colors. We see all the planned paths meet the leader's path correctly. As expected, longer paths are generated for a faster moving leader, since its predicted location at T_f will be farther away.

Figure 6(b) also shows the trajectory of the follower's distance errors to the leader's location. These are noisier than the corresponding error trajectories for docking with a stationary leader (in Fig. 5). We note that the distance to the leader is bridged in a consistent and repeatable fashion in spite of the higher noise. The peak velocity attained by the follower in these tests can exceed 2 m/s for the fastest moving leader.

Table 1 summarizes the key statistics from these tests. The leader's actual velocity, as well as the average and the peak velocity of the follower UAS during each mission are listed therein. Additionally, we also list the 3D error between the leader's final predicted location at $t = T_f$ and its actual location at $t = T_f$. We see that this error is typically less than 0.05 m. However, this can be occasionally much higher (for instance, 0.24 m in Exp.4). The mission still results in a successful docking since the majority of the error is situated in the lateral axes, which have a larger margin because of the width of the docking platform.

We continue at the highest speed of the leader ($\sim$0.65 m/s) and assess the repeatability the docking missions through 7 more tests. For these, the follower is initialized at different locations and the leader moves at a constant speed over the same path. The time horizon is further reduced to 7.5 s to expose the system to a more challenging operational constraint. The resultant tracks are shown in Fig. 6(a) corresponding the leader velocity of 0.65 m/s. In Fig. 6(b), the corresponding evolutions of the 3D Euclidean distances to the leader can be visualized – these are the tracks that complete docking around the 7 s mark. The docking is successful in 6 out of these 7 tests. In the one failed run, the final state error is approximately 4 cm outside the tolerance, and the docking mechanism never makes contact with the platform.

Table 1. Summary of statistics from outdoor (moving) tests.

Exp.	Leader Vel (m/s)	Obsv (s)	Mission (s)	Pred. Err 3D (m)	UAS Err Final (m)	UAS Vel (m/s) Avg, Max	*dock?*
1	0.31	10	8.50	0.013	0.24	0.78, 1.25	*true*
2	0.30	10	9.50	0.024	0.08	0.60, 1.00	*true*
3	0.50	10	9.50	0.057	0.15	0.93, 1.67	*true*
4	0.48	10	9.50	0.049	0.24	0.90, 1.42	*true*
5	0.58	10	9.50	0.050	0.08	1.15, 1.97	*true*
6	0.57	10	9.50	0.121	0.29	1.02, 1.72	*true*
7	0.65	10	8.50	0.244	0.18	1.14, 2.06	*true*
8	0.64	10	8.50	0.050	0.15	1.17, 2.13	*true*
9–15	0.65	10	7.50	0.099 (median)	0.13 (median)	1.31, 2.35 (median)	6**true* 1**false*

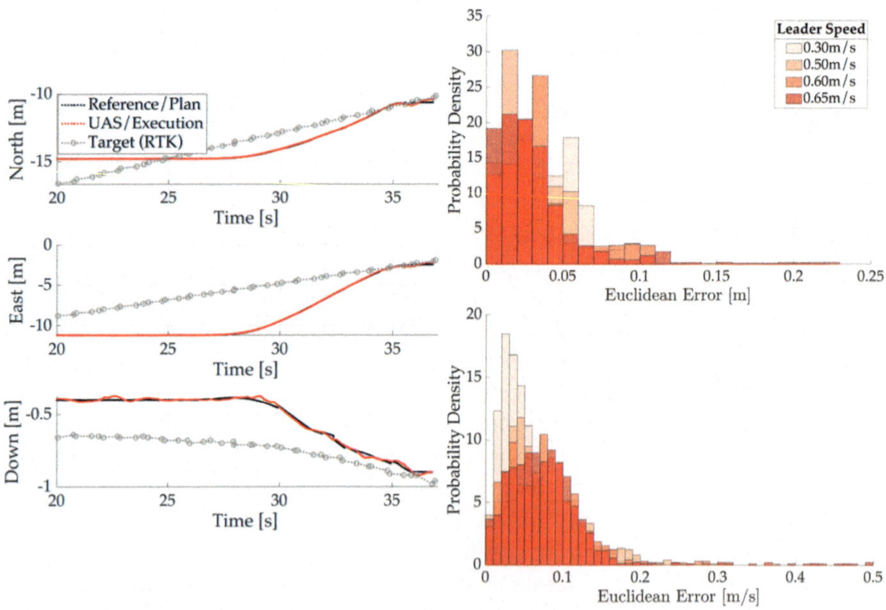

Fig. 7. Trajectory following performance represented as [left] temporal graphs for each axis, and [right] probability histograms of position and velocity tracking errors for different leader speeds. The errors are measured as $||\boldsymbol{x} - \boldsymbol{x}_r||_2$.

Figure 7 also shows key evaluations of our system's trajectory following performance in docking with a moving target. On the left, we show the commanded and the executed trajectories in each axes for one sample mission. We see that the trajectory of the leader (target) exhibits a parabolic curvature in the vertical plane (Down), while it is linear in the lateral plane (North and East). The follower UAS begins its mission at around the 28 s mark, and successfully docks around 36.5 s. The planned and the executed trajectories are in close agreement with each other (note that the magnification on Down axis is nearly 20x).

The tracking errors in position and velocity references for all missions are also assimilated into two histograms on the right. We see that the majority of the position errors lie within 0.05 m, while the velocity errors are usually under 0.1 m/s. We also note that velocity errors are higher when the leader is moving at a higher speed. The peak for the velocity error for a slower mission is around 0.05 m/s. These statistics collectively indicate that our controller is able to regulate the position of the follower UAS sufficiently for a docking mission.

To dock with a moving leader, an accurate estimate of its position and velocity at the specified time horizon is necessary. This is prone to measurement noise, since the velocity must be inferred by the follower UAS using its own observations. Figure 8 shows this error in the leader's estimated position at $t = T_f$ compared to its true position at $t = T_f$ for one sample mission (Exp. 6 in Table 1). The estimates are obtained using Eqs. (1)–(3) for a time horizon of 9.5 s and $k = 5$. The figure shows that the error rapidly diminishes within 4–5 s, and then increases dramatically as noisier observations are introduced. Towards the end of the mission (at closer proximity), the error diminishes again, such that the final 3D error, as reported in Table 1 is approximately 0.12 m. Notice that since the North and East components contribute most to this final error, the system can still dock successfully.

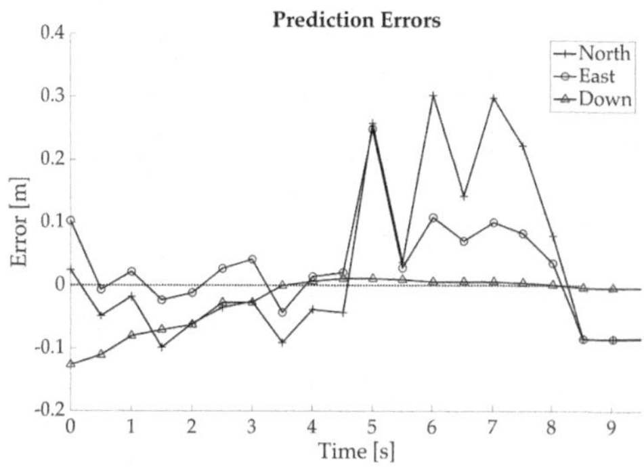

Fig. 8. The error in a 10 s-horizon prediction for the final location of a platform at 0.5 m/s.

4 Discussion and Future Work

Our outdoor tests with a stationary and moving leader have shown consistent and repeatable docking with an emulated leader system. These results indicate that we are able to achieve the accuracy in estimation and control necessary for such missions even under realistic operating constraints (such as the relative sizes of the docking platform and docking subsystem).

We observe in Fig. 6(a) that a significant amount of noise can corrupt the leader's position estimate during certain intervals. These are due to the combined effect of ambient wind, and disagreements between camera and RTK-GPS measurements. The disagreements can manifest themselves due to various factors, such as latency in GPS measurements, calibration errors on the camera, and the different motions exhibited by the fiducial marker and the RTK antenna mounted on the trapeze.

In our current implementation, we require the observation and prediction horizons be specified at the start of a mission. In practice, however, these values must be automatically inferred. For instance, the system can continue to collect observations until the residual of new data (or alternatively, the covariance) falls within a certain threshold. Similarly, one strategy to pick the mission duration automatically is to choose the shortest duration that leads to feasible trajectories. Our future work will investigate these methods. We also plan to expand the operational range of our setup to allow higher speeds of the platform, carried by another unmanned system.

Acknowledgments. This work was supported in part by NSF-IIS-1925052 -1924777, IIS-1638099, IIS-1925368, and USDA-NIFA 2017-67021-25924. Thanks to the members of the Nimbus Lab (Paul Fletcher, Ji Young Lee and Daniel Rico) for assisting with the field tests, and Jacob Hogberg (Research Engineer) for his design contributions to the docking subsystem.

References

1. Ruggiero, F., Lippiello, V., Ollero, A.: Aerial manipulation: a literature review. IEEE Robot. Autom. Lett. **3**(3), 1957–1964 (2018)
2. Saripalli, S., Sukhatme, G.: Landing on a moving target using an autonomous helicopter. In: Field and Service Robotics, pp. 277–286. Springer, Heidelberg (2006)
3. Borowczyk, A., Nguyen, D.T., Nguyen, A.P.V., Nguyen, D.Q., Saussié, D., Ny, J.L.: Autonomous landing of a multirotor micro air vehicle on a high velocity ground vehicle. CoRR abs/1611.07329 (2016). http://arxiv.org/abs/1611.07329
4. Falanga, D., Zanchettin, A., Simovic, A., Delmerico, J., Scaramuzza, D.: Vision-based autonomous quadrotor landing on a moving platform. In: 2017 IEEE International Symposium on Safety, Security and Rescue Robotics (SSRR), pp. 200–207 (2017)
5. Kim, S., Choi, S., Kim, H.J.: Aerial manipulation using a quadrotor with a two DOF robotic arm. In: 2013 IEEE/RSJ International Conference on Intelligent Robots and Systems, November 2013
6. Jimenez-Cano, A., Martin, J., Heredia, G., Ollero, A., Cano, R.: Control of an aerial robot with multi-link arm for assembly tasks. In: 2013 IEEE International Conference on Robotics and Automation (ICRA), pp. 4916–4921. IEEE (2013)
7. Tandale, M.D., Bowers, R., Valasek, J.: Trajectory tracking controller for vision-based probe and drogue autonomous aerial refueling. J. Guidance Control Dyn. **29**(4), 846–857 (2006). https://doi.org/10.2514/1.19694
8. Wilson, D.B., Göktogan, A., Sukkarieh, S.: Guidance and navigation for UAV airborne docking. In: Robotics: Science and Systems (2015)
9. Miyazaki, R., Jiang, R., Paul, H., Ono, K., Shimonomura, K.: Airborne docking for multi-rotor aerial manipulations. In: 2018 IEEE/RSJ International Conference on Intelligent Robots and Systems (IROS), pp. 4708–4714, October 2018

10. Krajník, T., et al.: A practical multirobot localization system. J. Intell. Robot. Syst. **76**(3), 539–562 (2014). https://doi.org/10.1007/s10846-014-0041-x
11. Mueller, M.W., Hehn, M., D'Andrea, R.: A computationally efficient motion primitive for quadrocopter trajectory generation. IEEE Trans. Rob. **31**(6), 1294–1310 (2015)

BAXTER: Bi-Modal Aerial-Terrestrial Hybrid Vehicle for Long-Endurance Versatile Mobility

Hyungho Chris Choi[1(✉)], Inhwan Wee[1], Micah Corah[2], Sahand Sabet[3], Taeyeon Kim[1], Thomas Touma[2], David Hyunchul Shim[1], and Ali-akbar Agha-mohammadi[2]

[1] Korea Advanced Institute of Science and Technology, Daejeon, Republic of Korea
[2] NASA Jet Propulsion Laboratory, Caltech, Pasadena, CA, USA
[3] Univesity of Arizona, Tucson, AZ, USA

Abstract. Unmanned aerial vehicles are rapidly evolving within the field of robotics. However, their performance is often limited by payload capacity, operational time, and robustness to impact and collision. These limitations of aerial vehicles become more acute for missions in challenging environments such as subterranean structures which may require extended autonomous operation in confined spaces. While software solutions for aerial robots are developing rapidly, improvements to hardware are critical to applying advanced planners and algorithms in large and dangerous environments where the short range and high susceptibility to collisions of most modern aerial robots make applications in realistic subterranean missions infeasible. To provide such hardware capabilities, one needs to design and implement a hardware solution that takes into the account the Size, Weight, and Power (SWaP) constraints. This work focuses on providing a robust and versatile hybrid platform that improves payload capacity, operation time, endurance, and versatility. The Bi-modal Aerial and Terrestrial hybrid vehicle (BAXTER) is a solution that provides two modes of operation, aerial and terrestrial. BAXTER employs two novel hardware mechanisms: the *M-Suspension* and the *Decoupled Transmission* which together provide resilience during landing and crashes and efficient terrestrial operation. Extensive flight tests were conducted to characterize the vehicle's capabilities, including robustness and endurance. Additionally, we propose Agile Mode Transfer (AMT), a transition from aerial to terrestrial operation that seeks to minimize impulses during impact to the ground which is a quick and simple transition process that exploits BAXTER's resilience to impact.

1 Introduction

Problem Statement and Challenges: Consider an autonomous robot traversing through a large and complicated subterranean environment with narrow corridors, prone to impact, and disconnected stretches of ground-traversable terrain. In such an environment, both conventional drones and unmanned ground

B. Siciliano et al. (Eds.): ISER 2020, SPAR 19, pp. 60–72, 2021.
https://doi.org/10.1007/978-3-030-71151-1_6

Fig. 1. Agile Mode Transfer. BAXTER exhibits a simple transfer from aerial to terrestrial operation via a short free fall with body inclined to reduce impulses to wheels.

vehicles (UGV's) would have limited scope of operation due to limited operational time and lack of traversability, respectively. The problem mentioned above is a widespread challenge for traversing subterranean environments—including mines, caves, and urban sprawl—which require extensive endurance to impact and versatile solutions that can traverse across rough and steep terrain [1]. Although potential functions of drones continue to expand by various algorithmic advances [2,3], the development of hardware platforms that simultaneously achieve both the versatility of aerial vehicles and the endurance of ground vehicles is an important topic for hardware engineering in robotics. The challenges in developing such platforms include delicate management of payload and hardware features in limited size, weight, and power (SWaP) constraints, maintaining structural and aerial stability, and all-around robustness of co-dependent components (e.g. coupled drive and transmission) to mitigate risk of system failure.

Proposed Work and Contribution: This publication outlines the development of the **Bi**-modal **A**erial-**TER**restrial Hybrid Vehicle (BAXTER), shown in Fig. 1, which aims to provide a robust hardware-based solution for the aforementioned challenges. The highlights of the contributions are:

- **Improved Operating Scope and Range** via the efficient use and transition between two modes of operation: *Terrestrial* (low-versatility and high-efficiency) and *Aerial* (high-versatility and low-efficiency).
- **Improved Resilience** through introduction of wheel assemblies around the aerial propulsion units as well as the suspension system, emphasizing resilience to impact while reasonably considering SWaP constraints.
- **Novel Mechanical Concepts** (the *M-Suspension* and *Decoupled transmission*) were introduced to realize the contributions above.
- **Agile Mode Transfer** demonstrates the versatility and resilience of the platform through a simple transition from aerial to terrestrial operation.

Outline: In this paper, novel mechanical concepts are proposed, realized, and tested to provide the above-mentioned contributions. Section 2 summarizes notable developments in hybrid vehicles, morphing drone concepts, and general-purpose aerial vehicles. Section 3 states the goals and context of this paper.

Section 4 explains the design concept and dynamics, and Sects. 5 and 6 summarize the realization and testing of the first prototype. Sections 7 and 8 provide insight into the development process, propose future works, and conclude.

2 Related Work

This section outlines previous hardware-based efforts on hybrid platforms, morphing drone concepts, and multi-functional aerial vehicles. Additionally, the end of this section highlights the design iterations toward developing the BAXTER robot platform.

Hybrid Platforms: Hybrid Platforms are a class of platforms that include more than one mode of operation [2,4,5]. In the scope of this publication, unmanned hybrid platforms incorporating aerial and terrestrial operation are the main focus. The current Hybrid platforms can be categorized in two ways:

- **Separate Articulation in Terrestrial Operation:** Platforms utilizing separate means of propulsion in terrestrial mode are classified to have *Active Terrestrial Mode* of operation. Vehicles with Active Terrestrial Mode [6,7], compared to vehicles with Passive Terrestrial Mode [2,4,5], provide a more efficient and mutually independent operation, while their payload capacity is partially reduced due to the addition of the driving system.
- **Morphing Mechanism:** Platforms utilizing separate built-in mechanisms for changing the internal form factor are *Morphing* platforms. Platforms can be *Active Morphing* [7], *Passive Morphing*, or *Non-Morphing* [2,4,5]. Morphing capabilities in hybrid vehicles provide vehicles with specialized forms for each operation, but may limit payload capacity or scale.

Morphing Drone Concepts. Apart from hybrid platforms, this class of novel aerial vehicle design focuses on morphing or controlled dynamic reconfiguration of the aerial structure to increase efficiency or versatility. Single-entity approaches manipulate the form of individual drones to achieve favorable characteristics. Examples include actuation of thrust vectors to achieve additional freedom in motion [8–10], translation of the thrusters to assume narrow profiles for traversing tight spaces [11–13], along with other unique morphing concepts [13–15]. Multi-entity approaches arrange or integrate many drone units with similar characteristics to achieve different forms and characteristics to produce multi-link chains [16,17] or other topologies.

2.1 Design Iteration Toward BAXTER

The development of BAXTER followed the design, manufacture, and evaluation of two predecessor models: MODEL 1 "TUMBLER" and MODEL 2 "HEXWING"[1] (illustrated in Fig. 2). Notable contributions of each design include:

[1] One iteration, MODEL 3, was designed and not produced.

(a) MODEL 1 "TUMBLER" (b) MODEL 2 "HEXWING" (c) MODEL 4 "BAXTER"

Fig. 2. Summary of Design Iterations. (a) The initial working prototype of protected hybrid operation. (b) The initial working prototype for spherical wheel-rotor subassembly. (c) Incorporation of novel mechanical concepts to emphasize resilience and versatility.

- **MODEL 1 "TUMBLER"** An initial working prototype for a robot with a spherical protective cage and active terrestrial operation. Limited in payload capacity and versatility (Fig. 2a).
- **MODEL 2 "HEXWING"** A proof of concept with spherical subassemblies around each arm. Increased versatility in terrestrial operation. Limited in resilience, scalability, and payload capacity. (Fig. 2b)
- **MODEL 4 "BAXTER"** Incorporation of novel mechanical concepts (the *M-Suspension* and the *Decoupled Transmission*) to emphasize resilience and versatility. Increased payload capacity. (Fig. 2c)

3 Problem Statement

This section describes the objectives for the development of BAXTER with appropriate context.

Goal Statement: The development of BAXTER aims to provide a hardware-based solution for the following capabilities: (1) A hybrid vehicle that significantly increases operational scope and range by utilizing two modes of travel: *Aerial* and *Terrestrial.* (2) A collision resistant hardware platform to maintain each mode and transition between them. (3) A robust, non-specific hardware platform for use in general, third party applications. (4) A controlled demonstration of the transition from aerial to terrestrial mode. Due to energy expenses in flight, the main mode of operation for BAXTER is Terrestrial, and its aerial operation is intended to be briefly used to traverse through environments that do not permit terrestrial operation. This addresses navigation through challenging subterranean environments such as tunnels, subway stations, and caves, which include tight and narrow spaces where many existing platforms would be prone to collision. Further, the design directives described above are in line with the challenging environments from the DARPA Subterranean Challenge, which inspired the development of BAXTER [1].

4 Technical Approach

This section describes the BAXTER hardware in detail along with the simple dynamic basis for the demonstrative landing procedure.

Base Design: Toward realizing the capabilities discussed in Sect. 3, BAXTER's design falls into the class of robots with an *active terrestrial mode* and *passive morphing*, according to Sect. 2. The design is centered on a concept of rotating, separately powered cages around each unit of propulsion (Fig. 3). For aerial operation, an X8 configuration[2] is used with 7-inch rotors mounted on 2500KV class motors. For terrestrial operation, a single DC motor powers each pair of wheels via the novel transmission system. The rendering and overall dimensions of BAXTER's mechanical design are illustrated in Fig. 3 (a)-(b).

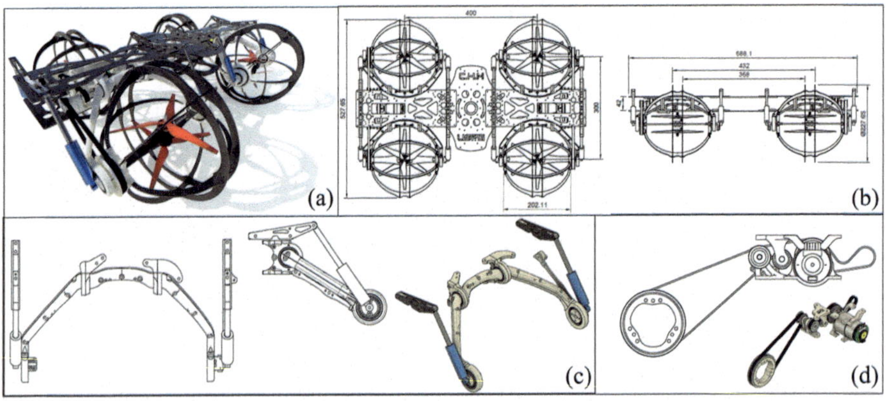

Fig. 3. (a) Rendering of BAXTER's Full Body Without Sensors/Processors (b) Overall dimensions of the platform (c) The M-Suspension design provides resilience during landing, crashes, and terrestrial operation (d) The Decoupled Transmission delivers motor-power to each wheel through a moving shock absorbing system.

Novel Concepts: BAXTER features novel mechanical suspension and motor-power transmission systems. The need for reliable shock resistance for both drop impact and terrestrial operation motivated the incorporation of a suspension system into the aerial platform. The *M-Suspension* (Fig. 3(c)) is a novel approach for vertical shock reduction in a hybrid aerial vehicle. This new type of passive suspension design focuses on ease of manufacturing based on 3D printing and minimizing the amount of additional weight. However, this suspension introduces a new obstacle: transmitting power through the moving joint during active terrestrial operation. The *Decoupled Transmission* (Fig. 3(d)) is a power transmission that has a timing pulley on an axis coinciding with each suspension

[2] The X8 configuration is a planar configuration of 8 rotors, arranged in 4 coaxial pairs.

joint. This design keeps the total pitch line length of the timing belt constant at all angles of the suspension joint (so that the belt tension remains constant at all times). In conjunction with the suspension design, this completes the mechanical concept to deliver motor power to each wheel through a moving, shock absorbing system.

4.1 Agile Mode Transfer

A simplified kinematic model of a quick and simple procedure for transfer from aerial to terrestrial mode referred to as *Agile Mode Transfer (AMT)* is discussed here (see Fig. 1). AMT includes a controlled impact with the ground at an angle which provides a fast transition from aerial to terrestrial operation.

The basis of the simplified kinematics of AMT is outlined below. Figure 4 illustrates the four phases of the process (Free Fall, Initial Impact, Roll, and Final Impact). The subscripts used in Fig. 4 indicate the number of the phase associated with given variable. In the following, $\vec{v}_n$ and $\boldsymbol{\omega}_n$ respectively represent the translational velocity and the angular velocity of the center of mass (CoM) of the chassis at the end of the respective phase. It is assumed that the CoM of the system is located at the center of the chassis. The term $\boldsymbol{h}_n$ represents the final height of the CoM, vector $\boldsymbol{d}_2$ connects the contact point of the *Back Wheel* to the CoM, and vector $\boldsymbol{d}_4$ connects the contact point of the *Front Wheel* to the CoM. The four phases shown in Fig. 4 are discussed below:

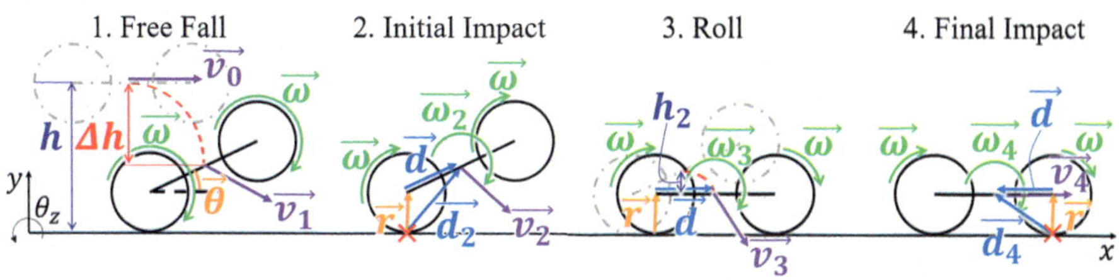

Fig. 4. Simplified Kinematics Model for BAXTER's Agile Mode Transfer (AMT). AMT models the landing procedure as four phases: 1. Free Fall, 2. Initial Impact, 3. Roll, 4. Final Impact.

1. **Free Fall:** Vehicle keeps a positive landing angle (θ) and is subjected to free-fall motion until making contact with the ground. Governing principle: parabolic motion.

$$v_{1x} = v_{0x} \qquad\qquad v_{1y} = -\sqrt{2g\Delta h} \tag{1}$$

Here Δh is the magnitude of the vertical displacement of the CoM during free fall, v_{0x} is the desired entrance velocity in x-direction, and g is the gravitational acceleration.

2. **Initial Impact:** The first wheel (i.e. *Back Wheel*) makes contact with the ground. Governing principles: Conservation of angular momentum and no-slip condition.

$$\vec{d}_2 \times \vec{v}_1 = \vec{d}_2 \times \vec{v}_2 + I\vec{\omega}_2 \qquad \vec{\omega}_2 \times \vec{d} + \vec{\omega} \times \vec{r} = \vec{v}_2 \tag{2}$$

   Here $\vec{d}$ is the displacement vector connecting the *Back Wheel* center to the CoM and $\boldsymbol{r}$ connects the contact point to the center of the *Back Wheel.*

3. **Roll:** The first wheel rolls while the second wheel is in motion toward the ground. Governing principles: conservation of energy and no-slip condition.

$$\frac{1}{2}mv_2^2 + \frac{1}{2}Iw_2^2 + mgh_2 = \frac{1}{2}mv_3^2 + \frac{1}{2}Iw_3^2 \tag{3}$$

$$\vec{\omega_3} \times \vec{d} + \vec{\omega} \times \vec{r} = \vec{v_3} \tag{4}$$

   Here the terms $\vec{\omega}$, m, and I present the angular velocity, mass, and moment of inertia of the chassis, respectively.

4. **Final Impact:** The second wheel (i.e. *Front Wheel*) makes contact with the ground. Governing principles: Conservation of angular momentum, no-slip condition for both wheels.

$$\vec{d}_4 \times \vec{v}_3 + I\vec{\omega}_3 = \vec{d}_4 \times \vec{v}_4 + I\vec{\omega}_4 \qquad \vec{\omega} \times \vec{r} + \vec{\omega}_4 \times \vec{d} = \vec{v}_4 \qquad \boldsymbol{v_4} = \boldsymbol{v_f} \tag{5}$$

Given the desired entrance and exit velocity (v_0, v_f), this model is used to determine the optimal landing angle (θ) that minimizes the maximum impact with the ground. To achieve this angle, the impact impulses at phase 2 $(\vec{I}_2)$ and phase 4 $(\vec{I}_4)$ are defined as:

$$\vec{I}_2 = m\vec{v}_2 - m\vec{v}_1 \qquad \vec{I}_4 = m\vec{v}_4 - m\vec{v}_3 \tag{6}$$

Then, the maximum impact $I_{\max}$ is defined as:

$$I_{\max} = \max(|\vec{I}_2|, |\vec{I}_4|) \tag{7}$$

Finally, the landing angle (θ) that minimizes $I_{\max}$ is calculated numerically.

5 Experiment Design

Preliminary flight tests with the prototype verified BAXTER's viability. Now, in this section, additional tests (outlined below) verify the flight characteristics, versatility, and resilience of the platform.

Basic Flight and Endurance Tests: Basic maneuvers (hover, rectangle, etc.) validate the flight characteristics of the prototype. The results from these tests provide a quantitative evaluation (*Flight and Mission Time, Maximum Speed, and Control Limits*) of the vehicle along with qualitative assessments.

Drop Test: BAXTER's suspension system is unique since it is designed to absorb step-like impulses from impact with the ground on landing or crashing.

So, to measure the suspension endurance, a drop test was performed with the prototype in the all-up configuration[3] (4.2 kg total) and in an upright position from incrementally increasing heights to obtain the following metrics (Drop Limits): (1) *Maximum Intact Height*, the maximum height where the prototype takes no irreversible damage, and (2) *Maximum Flyable Height*, the maximum height where the prototype can still maintain aerial operation.

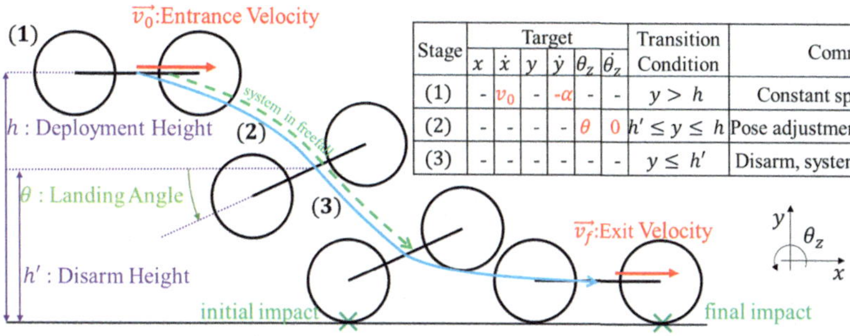

Stage	Target						Transition Condition	Comments
	x	$\dot{x}$	y	$\dot{y}$	θ_z	$\dot{\theta}_z$		
(1)	-	v_0	-	$-\alpha$	-	-	$y > h$	Constant speed descent
(2)	-	-	-	-	θ	0	$h' \leq y \leq h$	Pose adjustment in zero thrust
(3)	-	-	-	-	-	-	$y \leq h'$	Disarm, system uncontrolled

Fig. 5. Agile Mode Transfer Mission Diagram, Mission Controller Stages (which are distinct from the phases of the landing process in Sect. 4) and Parameters. This figure and the attached target table (controlled set-points for each phase are highlighted in red) illustrates and describes targets for the AMP in the mission perspective with 3 Stages: (1): Before Deployment, (2): Deployment (vehicle adjusts pose in free fall), and (3): Disarm (flight controller switches off)

Agile Mode Transfer Test: The agile mode transfer experiments exploit and demonstrate the effectiveness of the hardware design. From the kinematics model described in Sect. 4, a mobile landing procedure will be devised by providing the parameters in Fig. 5, and calculating the optimum landing angle (θ) that minimizes the load at impact in landing (I_{max}). The implementation of this maneuver is accomplished via ROS and the attitude and velocity control capability of the flight controller [18], with input from onboard sensors (Fig. 6).

Autonomous Flight System: To operate a mission autonomously in terrestrial and aerial mode, BAXTER uses odometry for velocity and attitude. Attitude estimation is done through the 9-axis IMU (Inertial Measurement Unit) within the flight controller. For velocity and position estimation, an Intel RealSense T265 tracking camera was linked to a PX4 via ROS (Robot Operating System) running in the companion computer through VIO (Vision Inertial Odometry). The mission controller for AMT runs on the companion computer (Nvidia Jetson TX2), which implements the 3 mission phases outlined in Fig. 5.

Summary of Prototype: Finally, the first prototype of BAXTER presents a tangible proof of concept for the mechanical design described in Sect. 4. The details of the circuit-level hardware are listed in Fig. 6-(a).

[3] All-up refers to the maximum payload configuration which uses additional weights as a proxy for sensing and computation payloads.

6 Results

This section provides the qualitative and quantitative outcomes of the design and testing of BAXTER, described in Sect. 5.

Basic Flight Parameters: Table 1 presents the results of the *Basic Flight and Endurance Tests* and *Drop Test* outlined in Sect. 5. These results were obtained and verified over 3 trials. Inertial measurements were conducted with the bifilar pendulum method [19].

Agile Mode Transfer Test: With the verified maximum intact drop height of around 1.3 m and control limits drawn from the previous experimental results, agile mode transfer solutions (varying Entrance and Exit Velocity) that were

Table 1. Obtained Flight Parameters for BAXTER. (a) Obtained results from measurements (b) Obtained results from experiments

(a)

Category	Sub-category	Metric
Mass	Chassis	1.2 kg
	Propulsion System	1.2 kg
	Maximum Payload	1.8 kg
Inertia (All-Up)[6]	I_{xx}(roll axis)	$0.351\ kg \cdot m^2$
	I_{zz}(pitch axis)	$0.125\ kg \cdot m^2$
	I_{yy}(yaw axis)	$0.254\ kg \cdot m^2$

(b)

Category	Sub-category	Metric
Time Limit	Flight (Manual)	5 min
	Flight (Autonomous)	4 min
	Drive (Drive Only)	30 min
Control Limit	Aerial Speed	2 m/s
	Terrestrial Speed	1 m/s
	Roll Angle	30 deg
	Pitch Angle	25 deg
Drop Limit	Max. Intact	1.3 m
	Max. Flyable	1.4 m

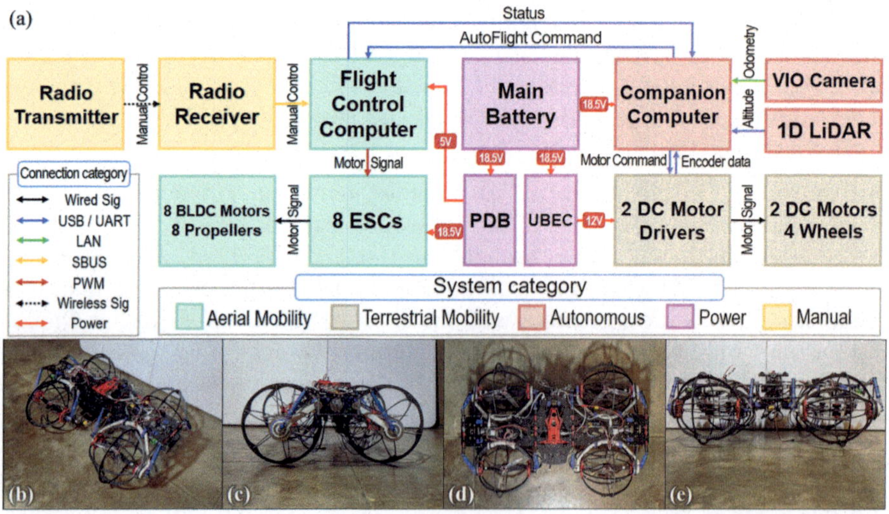

Fig. 6. Overview of the First Prototype of BAXTER. (a) Electrical System Diagram of BAXTER (b)-(e) Aspect Photos of BAXTER Prototype

within the control limits were tested and verified to be repeatable. Figure 7 depicts the case of: $v_i = 1.3, v_f = 0.5, h = 0.65$ with respect to potential landing angles (θ). It can be deduced that the landing angle of around 20 deg produces the least maximum impulse on impact. Such optimal landing angles are calculated internally to the mission controller and are implemented as a sequence of controller stages described in Fig. 5. A representative frame-on-frame picture of this landing procedure is shown in Fig. 1.

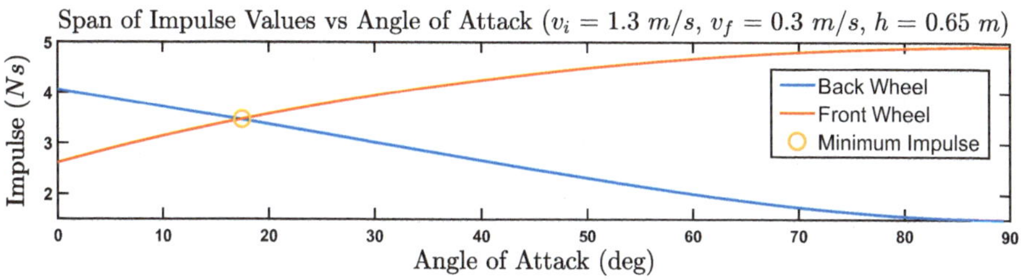

Fig. 7. Plot of expected front (I_2) and back (I_1) wheel impulse values, varying the landing angle ($v_i = 1.3$ m/s, $v_f = 0.5$ m/s, h = 0.65 m) using the kinematics model described in Sect. 4.1. The optimum landing angle, yielding the lowest maximum impulse value, is highlighted with a yellow circle.

7 Experimental Insights

This section lists notable insights obtained regarding the overall flight characteristics and test results of BAXTER.

Overall Insight: Much insight into suspension systems for hybrid platforms was gained through development and experimentation on BAXTER. To the best of our knowledge, BAXTER is the first platform to incorporate a passive suspension system into a hybrid aerial vehicle. Notably, BAXTER's flight characteristics are affected by deflection of the suspension joints, but fine adjustment of the flight parameters allowed for reasonable performance in autonomous control.

Vibration Due to the Suspension System: This was evident in low-frequency vibration of the vehicle (magnitude of $1Hz$) during preliminary flight tests. Although these vibrations can be mitigated by careful tuning of the flight controller, doing so restricts the control limits.

Motor Translation due to Body Distortion: As a side effect of deflecting passive suspension joints, passive morphing occurs, and the thrusters translate along with the wheel assemblies in flight. This introduces irregularities in flight behavior due to distortions of the body that are not present in conventional aerial vehicles. A variant of this phenomenon has been studied with respect to a folding drone where a motor-actuated morphing mechanism involving translating thrusters introduced undesired displacement in flight behaviour [11]. A potential research topic can be to dynamically change parameters in the control system to mitigate the deflection of the suspension system in order to expand the control limits for BAXTER.

Insight on Resilience: As intended, the results from the drop tests (see Sect. 5) demonstrate the resilience of BAXTER in contact-laden environments and crash-prone operation. A notable takeaway from the drop tests is that for many times that the prototype took damage its onboard electrical components remained intact while primarily easily-replaceable 3D-printed parts took concentrated damage. A potential future design objective cloud be to concentrate structural weak spots within the platform to easily accessible parts (Commercially available parts, 3D printed parts, etc.).

Insight on Agile Mode Transfer: Agile Mode Transfer tests were a successful demonstration of BAXTER's potential for improved operational scope and resilience. Overall, BAXTER was able to reliably perform Agile Mode Transfer (AMT) without irreversible damage to the hardware.

8 Conclusion and Future Works

Summary of BAXTER: BAXTER presents a novel hardware-based approach to aerial-terrestrial hybrid platforms. It also offers a solution for multiple fronts in challenging environments where both resilience and extensive operational scope and range are required. This publication has detailed the design philosophy and prototype of the BAXTER concept and successfully showed viability of its hybrid capabilities by evaluation of conventional flight characteristics and testing the novel Agile Mode Transfer.

Future Works: BAXTER provides a reliable platform to operate in large and contact-prone environments where operation of conventional aerial robots would prove to be infeasible. BAXTER also opens avenues for applications (such as subterranean exploration) that exploit its resilience which is demonstrated with the Agile Mode Transfer. Introducing a contact-flexible platform also opens potential for new software features that exploit contacts such as for computation of odometry [3].

Further iteration on this design will continue to push the boundaries of hybrid robot platforms. For BAXTER, the introduction of the *M-Suspension* and the *Decoupled Transmission* provided notable improvements. But, as described in Sect. 7, challenges in aerial mobility arise as a trade-off between including a passive suspension system for resilience and adverse changes to the vehicle's flight characteristics. One hardware-based solution to this challenge is the introduction of a simple active suspension, specifically a suspension system that can be switched on or off. This change to the suspension system will be a main focus for the design of the next iteration (MODEL 5).

Acknowledgements. This work was supported by the Institute for Information & communications Technology Promotion (IITP), funded by the Korean government

(MSIP) (Development of AI-powered Autonomous Drone for Complex Indoor Environment) under the guidance of Unmanned System Research Group of Korea Advanced Institute of Science and Technology (KAIST USRG). Further guidance from the Jet Propulsion Laboratory (JPL) of the National Aeronautics and Space Administration (NASA). We want to thank Hyunjee Ryu, Brian Kim, and Hanseob Lee of KAIST USRG, along with Brett Lopez and Team CoSTAR of JPL for various technical support and fruitful discussions.

References

1. DARPA: Unearthing the subterranean environment (2020). Accessed 3 Jun 2020
2. Fan, D.D., Thakker, R., Bartlett, T., Miled, M.B., Kim, L., Theodorou, E., Agha-mohammadi, A.: Autonomous hybrid ground/aerial mobility in unknown environments. In: Proceedings of the IEEE/RSJ International Conference on Intelligent Robots and Systems (IROS), pp. 3070–3077 (2019)
3. Lew, T., Fan, D., Tara, B., Santamaria-Navarro, A., Thakker, R., Agha-mohammadi, A.: Contact inertial odometry: collisions are your friends. In: Proceedings of the International Symposium of Robotics Research, Hanoi, Vienam (2019)
4. Sabet, S., Agha-mohammadi, A.A., Tagliabue, A., Elliott, D.S., Nikravesh, P.E.: Rollocopter: an energy-aware hybrid aerial-ground mobility for extreme terrains. In: IEEE Aerospace Conference, Big Sky, Montana (2019)
5. Kalantari, A., Spenko, M.: Design and experimental validation of HyTAQ, a Hybrid Terrestrial and Aerial Quadrotor. In: IEEE International Conference on Robotics, and Automation, Karlsruhe, Germany (2013)
6. Kalantari, A., Touma, T., Kim, L., Jitosho, R., Strickland, K., Lopez, B.T., Agha-mohammadi, A.: Drivocopter: a concept hybrid aerial/ground vehicle for long-endurance mobility. In: IEEE Aerospace Conference, Big Sky, Montana (2020)
7. Meiri, N., Zarrouk, D.: Flying STAR, a hybrid crawling and flying sprawl tuned robot. In: IEEE International Conference on Robotics and Automation, Montreal, Canada (2019)
8. Zheng, P., Tan, X., Kocer, B.B., Yang, E., Kovac, M.: TiltDrone: a fully-actuated tilting quadrotor platform. IEEE Robot. Autom. Lett. **5**(4), 6845–6852 (2020)
9. Kamel, M., Verling, S., Elkhatib, O., Sprecher, C., Wulkop, P., Taylor, Z., Siegwart, R., Gilitschenski, I.: The Voliro omniorientational hexacopter: an agile and maneuverable tiltable-rotor aerial vehicle. IEEE Robot. Autom. Mag. **25**(4), 34–44 (2018)
10. Ryll, M., Bülthoff, H.H., Giordano, P.R.: A novel overactuated quadrotor unmanned aerial vehicle: Modeling, control, and experimental validation. IEEE Trans. Control Syst. Technol. **23**(2), 540–556 (2015)
11. Falanga, D., Kleber, K., Mintchev, S., Floreano, D., Scaramuzza, D.: The foldable drone: a morphing quadrotor that can squeeze and fly. IEEE Robot. Autom. Lett. **4**(2), 209–216 (2018)
12. Riviere, V., Manecy, A., Viollet, S.: Agile robotic fliers: a morphing-based approach. Soft Robot. **5**(5), 541–553 (2018)
13. Bucki, N., Mueller, M.W.: Design and control of a passively morphing quadcopter. In: IEEE International Conference on Robotics and Automation, Montreal, Canada (2019)
14. Hedayati, H., Suzuki, R., Leithinger, D., Szafir, D.: PufferBot: actuated expandable structures for aerial robots. arXiv preprint arXiv:2008.07615 (2020)

15. Zhao, N., Luo, Y., Deng, H., Shen, Y.: The deformable quad-rotor: design, kinematics and dynamics characterization, and flight performance validation. In: Proceedings of the IEEE/RSJ International Conference on Intelligent Robots and Systems (IROS), Vancouver, Canada, pp. 2391–2396 (2017)
16. Nguyen, H., Dang, T., Alexis, K.: The reconfigurable aerial robotic chain: modeling and control. In: IEEE International Conference on Robotics and Automation (ICRA), Paris, France, pp. 5328–5334 (2020)
17. Zhao, M., Kawasaki, K., Chen, X., Noda, S., Okada, K., Inaba, M.: Whole-body aerial manipulation by transformable multirotor with two-dimensional multilinks. In: IEEE International Conference on Robotics and Automation (ICRA), Singapore, pp. 5175–5182 (2017)
18. Meier, L.: MAVLINK common message set for offboard control (2020). Accessed 23 Jun 2020
19. Then, J.W., Chiang, K.: Experimental determination of moments of inertia by the bifilar pendulum method. Am. J. Phys. **38**, 537 (1970)

Design and Prototyping

Characterization of Compliant Parallelogram Links for 3D-Printed Delta Manipulators

Pragna Mannam(✉), Oliver Kroemer, and F. Zeynep Temel

Carnegie Mellon University, Pittsburgh, PA 15213, USA
{pmannam,okroemer,ztemel}@andrew.cmu.edu

Abstract. We design a compliant delta manipulator using 3D-printing and soft materials. Our design is different from the traditionally rigid delta robots as it is more accessible through low-cost 3D-printing, and can interact safely with its surroundings due to compliance. This work focuses on parallelogram links which are a key component of the delta robot design. We characterize these links over twelve dimensional parameters, such as beam and hinge thickness, and two material stiffness settings by displacing them, and observing the resulting forces and rotation angles. The parallelogram links are then integrated into a delta robot structure to test for delta mechanism behavior, which keeps the end-effector parallel to the base of the robot. We observed that using compliant hinges resulted in near-delta behavior, laying the groundwork for fabricating and utilizing 3D-printed compliant delta manipulators.

1 Introduction

Personal care robots operating in homes require dexterous and compliant manipulators to ensure that interactions with human and objects are performed reliably and safely. Delta robots, Fig. 1(a), are ideal for these applications due to their low-inertia and high-precision [1]. The dexterity and precision of these robots enable them to manipulate objects in unstructured home environments. However, most delta robots are made from rigid materials [2] which precludes them from interacting safely with humans. The compliance of soft materials, on the other hand, can allow them to conform to objects, exhibit physical robustness, and execute safe human interactions [4]. Furthermore, 3D-printing delta robots using soft materials reduces fabrication costs and makes them accessible to a wider community. This paper presents the process of designing the first 3D-printable compliant delta robot using soft materials. We characterize parallelograms, an integral component of the delta robot, with varying dimensional parameters and materials to study their behavior as part of a delta mechanism. More specifically, we focus on parallelogram hinge and beam thicknesses, and material stiffness. These design choices can then be leveraged to achieve desired end-effector behaviors of 3D-printed delta robots composed of compliant parallelogram links.

B. Siciliano et al. (Eds.): ISER 2020, SPAR 19, pp. 75–84, 2021.
https://doi.org/10.1007/978-3-030-71151-1_7

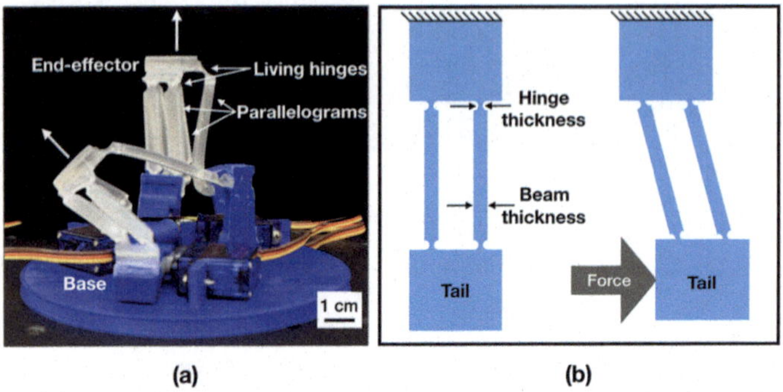

Fig. 1. (a) A delta robot with each arm consisting of a rigid portion and a flexible parallelogram with living hinges that transfers motion from motors at base to end-effector. (b) Ideal behavior of displaced tail of the parallelogram in characterization experiments testing different hinge and beam thicknesses and materials.

Printing compliant delta robots requires a departure from the classical delta design due to the use of soft materials, especially with respect to the joints. Rigid universal joints are replaced by two types of living hinges[1], one in the compliant parallelogram links of the delta robot and the other between the end-effector and the parallelogram links, as seen in Fig. 1(a). Properties inherent to soft materials like bending and hysteresis can alter the delta mechanism behavior. As shown in Fig. 1(a), we expect the end-effector (or platform) of the delta robot to always stay parallel to its base as seen in the upright position; however, this orientation is not guaranteed in all configurations when using compliant materials. This deviation from the ideal kinematic behavior can be measured by the end-effector rotation with respect to the base across the robot configuration space. In this work, we study the characteristics of centimeter-scaled compliant parallelograms with 12 varied combinations of hinge and beam thicknesses, and 2 materials by displacing them incrementally and observing the force required and resulting rotation of the parallelogram tail, as shown in Fig. 1(b). Subsequently, we evaluate these parallelograms by integrating them into the delta robot structure and measuring the end-effector rotation with respect to the base in specific configurations of the robot. These experiments allow us to infer the parameters of the compliant parallelogram link that achieve the ideal kinematic behavior, bringing us one step closer to the goal of creating a 3D-printed compliant delta manipulator.

2 Related Work

Compliant robots have been used in biomedical applications and are built on micro or milli scale using flexure-based mechanisms. Manufacturing compliant

[1] A **living hinge** is an articulated joint created by locally reducing the width of the material to make it more compliant in that region.

robots on such a small scale is difficult using traditional techniques and are typically built using either Micro Electro Mechanical Systems (MEMS) [5,7,8], or Electro-Discharge Machining (EDM) [6,9–11]. These manufacturing techniques require sophisticated tools that are not accessible to the wider community. An alternative approach is to utilize origami-inspired self-folding robots [14–16] or laminate robot mechanisms [5,13,17]. Our work is closest to the millimeter-scaled Delta robot using laminate structures [5]. Unlike the manual assembly required for laminate robot mechanisms, we look to successful prototyping of compliant mechanisms using 3D-printing [12]. Our goal is to create a low-cost compliant delta manipulator without using complex manufacturing techniques by relying on 3D-printing.

A popular alternative parallel manipulator is the 6-DOF Stewart platform [3] that can translate and tilt the end-effector unlike the delta robot which only allows the end-effector to translate parallel to the base. Both of these manipulators can move at high speeds and accelerations with low-inertia because their relatively heavy motors are stationary and at the base. Delta robots, unlike Stewart platforms, have simple closed-form kinematic solutions [2]. The characteristics improve as the mechanisms are scaled down, which is why delta robots like the mm-scaled delta (milliDelta) can move with precision down to 5 micrometers in a 7mm^3 workspace and lend themselves to micromanipulation [5]. Previous mm-scaled delta robots required multi-step processes or manual assembly which introduced irregularities in the final platforms [5,13]. In contrast, we use 3D-printing to make repeatable and accessible delta robots.

3 Approach

The parallelogram structure plays an integral role in the delta robot's functionality [1]. The key components of the parallelogram are the beam and hinge thicknesses, as labeled in Fig. 1(b). Hence, we start with characterizing the parallelograms with varying beam and hinge thicknesses, as well as two different materials, by recording the effect on force and position when displacing the tail of the parallelogram as shown in Fig. 1(b). Subsequently, we study the compounding effects of three parallelogram links in a delta robot structure on the resulting end-effector orientation. This enables us to model the mechanical properties of different materials and dimensional parameters of the compliant parallelogram that result in motions close to the idealized delta mechanism.

We characterized and evaluated the parallelogram links over twelve combinations of dimensional parameters: 3 beam thickness values of 2.5 mm, 3.5 mm and 4.5 mm, and hinge thickness values of 0.375 mm, 0.5 mm, 0.75 mm, and 1 mm to study deviations from the ideal delta kinematic model caused by different hinge and beam thicknesses. For our experiments, we used Ultimaker S5 3D-printer and filaments. We chose the two most compliant Ultimaker filament materials which are best suited to living hinge mechanisms, polypropelene (PP) and thermoplastic polyurethane (TPU) 95A. For comparison, the tensile modulus is 220 MPa (using ISO 527) and 26 MPa (using ASTM D638) for PP and TPU respectively. Additionally, the flexural strength is 13 MPa and 4.3 MPa (both measured using ISO 178) for PP and TPU respectively. These values indicate that TPU is more compliant than PP, which is beneficial for the living hinges while the stiffness of PP is useful to maintain rigidity of the parallelogram beams.

3.1 Characterization of Parallelogram

The two parameters for characterization experiments were the parallelogram beam and hinge thicknesses (Fig. 1(b)). The remaining dimensions were fixed, such as the length and depth of the beams which were 27 mm and 4 mm respectively, and the length of the hinge was always 2 mm. Compliant parallelograms were made from PP and TPU while supporting and mounting materials were 3D-printed using rigid polylactic acid (PLA) material. Characterization experiments involved mounting these parallelograms vertically and using a linear actuator to push the parallelogram tails in 5mm increments to measure the exerted force, and resulting displacement and orientation change of the tails. We vertically mounted parallelograms to reduce the effects of gravity in pulling the parallelogram out of plane. As shown in Fig. 2(a), an Actuonix L12-R Micro Linear Servo is mounted with a Transducer Techniques GS0-150 load cell to measure the force required to displace the parallelogram tail. For PP parallelograms with beam thickness of 4.5 mm, we used the GS0-500 load cell instead to measure the higher force values that were required to displace these parallelograms. Three grey markers on the parallelogram are tracked as a rigid body by an OptiTrack V120:Trio camera with submillimeter accuracy to record the position and orientation of the parallelogram tail. The load stem was lubricated to overcome frictional effects when pushing the parallelogram. Force measurements were recorded after a six minute delay to allow for stabilization of any creep (deformation) effect. Additionally, rotation data from the markers on the parallelogram indicate if the tail of the parallelogram rotated out-of-plane or twisted in plane.

Using the same reference axes as shown in Fig. 2(a), we observe in-plane and out-of-plane rotations of the parallelogram tail. The force applied by the actuator causes the tail of the parallelogram to move in an arc in the plane of the parallelogram. Under ideal conditions, the tail would remain parallel to the base. However, due to non-ideal hinge and beam behaviors, the tail exhibits additional rotations within the plane of movement, i.e., about the Z axis. Additionally, the restorative force of the parallelogram (caused by gravity and joint stiffness) will push back against the actuator. Misalignment between this restorative force and the actuator's pushing force can result in the parallelogram twisting (Y axis rotation) or bending out of plane (X axis rotation). Thus, we only focus on the in-plane rotations to test whether the parallelogram tail stays parallel to the mounted end of the parallelogram.

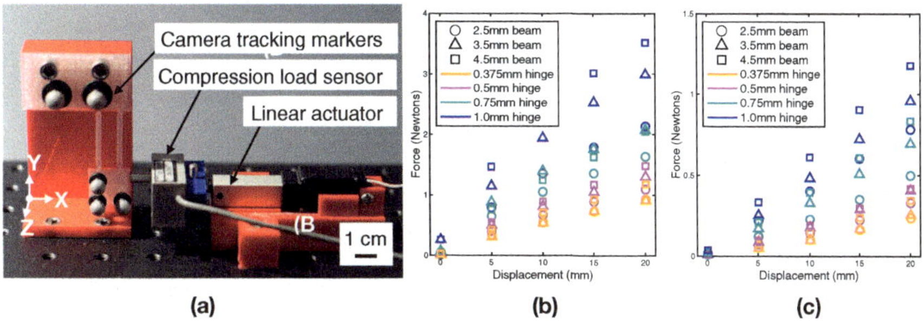

Fig. 2. (a) Parallelogram characterization experiment setup: linear actuator mounted with compression load sensor to displace the tail of a PP parallelogram with beam thickness = 2.5 mm and hinge thickness = 0.375 mm. (b) Force-displacement plots for PP and (c) TPU for varying beam thicknesses (shown with shapes) and hinge thicknesses (shown with colors). Each marker represents the mean across five repetitions of the experiment.

3.2 Evaluation of Delta Mechanism

Rotations of the compliant parallelogram links directly relate to the rotations of the delta robot end-effector. Hence, using the same parallelograms tested in the characterization experiments, we validated the results by integrating the parallelograms as links in an actuated delta robot as shown in Fig. 4. The parallelogram links have an added living hinge on each side, oriented 90° apart from the compliant links' revolute joint, to mimic one universal joint at each end of the beams as shown in Fig. 1(a). Three of these parallelograms are attached to a triangular end-effector platform, all created with either PP or TPU throughout. Then, the compliant delta tops are secured to 3D-printed rigid links made with Ultimaker Tough PLA (shown in red in Fig. 4) which are attached to servo motors.

To measure the end-effector rotation with respect to the base at different configurations of the robot, we define two configurations for each of the arms, namely up and down, that are separated by a 50° offset. We consider two configurations of the robot, one in which two arms are down and the other arm is up, and the other in which two arms are up and the other arm is down. The three marker points on the end-effector are used to calculate the end-effector plane and normal axis. The normal axis is computed at each position of the delta robot, such that the angle between the normal vector and a reference base normal is used to compute the rotation of the end-effector with respect to the base.

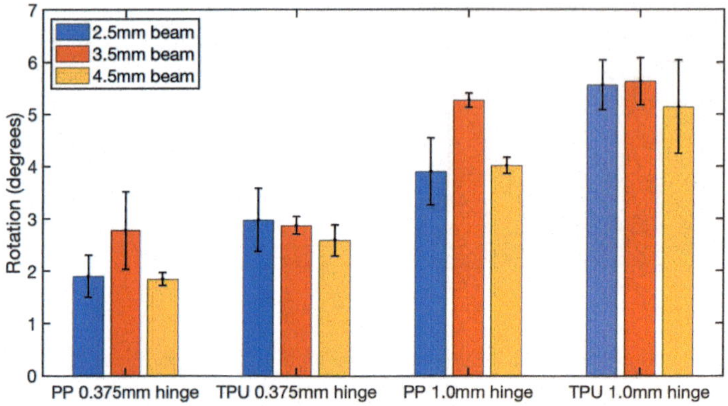

Fig. 3. Characterization experiment results for Z rotation angles (in degrees) of PP and TPU parallelograms at the largest displacement (20 mm). The mean and standard deviation shown as error bars are taken over five repetitions of the experiment. The most compliant hinges (0.375 mm) and least compliant hinges (1.0 mm) are grouped together, and the colors denote different beam thicknesses of 2.5, 3.5, and 4.5 mm.

4 Experimental Results

4.1 Characterization Experiments

We tested parallelograms of varying hinge thickness (0.375, 0.5, 0.75, and 1 mm) with beam thickness of 2.5 mm, 3.5 mm, and 4.5 mm. The force values at the end of each 5 mm increment push until 20 mm for both PP and TPU, averaged over five repetitions, are shown in Fig. 2(b) and (c) respectively. The circle, triangle, and square markers denote 2.5 mm, 3.5 mm, and 4.5 mm respectively, and each color denotes a difference hinge thickness between 0.375 and 1 mm. Five repetitions were executed for a given set of dimensional parameters and material on a newly printed parallelogram, resulting in a total of 120 characterization experiments. There is a clear trend for both materials where displacing the parallelogram by a larger distance takes more force. Overall, PP parallelograms require more force to displace than ones made from TPU, needing almost 4N versus close to 1.5N for the stiffest (thickest) hinges at 20 mm displacement.

Using OptiTrack markers, Z rotation angles of the parallelogram tail were recorded after displacements of 5 mm increments up to 20 mm. In Fig. 3, the mean and standard deviation for Z rotation angles are shown for both PP and TPU at the largest (20 mm) displacement. For the same beam thickness, increasing the hinge thickness tends to result in larger Z rotations across both materials implying that in-plane rotations are smaller for thinner hinges. We can also observe that for a given beam and hinge thickness, TPU results in larger in-plane rotations when compared to PP. This can be attributed to higher compliance of TPU when compared to PP. Interestingly, for a fixed hinge thickness varying the beam thickness does not show consistent behavior across materials and hinge thicknesses. We leave exploring the reasons to future work.

Table 1. Characterization experiment results for X, Y, and Z rotation of parallelograms at 20 mm displacement. The mean (μ) and standard deviation (σ) is taken over five repetitions of the experiment. The last two columns consist of data from evaluation experiments integrating parallelograms of varying beam and hinge thicknesses and calculating the rotation of end effector (in degrees). The mean and standard deviation of three repetitions raising 2 arms and lowering one arm (2 up, 1 down) as well as lowering 2 arms and raising 1 arm (2 down, 1 up) by 50° is listed in fourth and fifth columns respectively.

(Beam th., Hinge th.)	X-rotation		Y-rotation		Z-rotation		2 up, 1 down		2 down, 1 up	
	μ	σ	μ	σ	μ	σ	μ	σ	μ	σ
PP (2.5, 0.375)	2.31	2.11	2.28	0.85	**1.90**	0.41	**6.27**	1.68	**5.65**	1.66
PP (3.5, 0.375)	1.03	0.38	2.35	1.85	2.78	0.74	8.12	3.80	5.87	1.49
PP (4.5, 0.375)	2.08	0.84	4.68	1.56	**1.84**	0.12	8.75	2.61	6.15	3.83
PP (2.5, 1.0)	1.20	1.11	3.10	2.65	3.90	0.64	20.08	5.91	17.01	3.31
PP (3.5, 1.0)	1.04	0.38	1.20	0.57	5.26	0.14	15.23	0.86	13.22	1.69
PP (4.5, 1.0)	3.87	1.62	9.32	3.37	4.01	0.16	15.01	0.72	11.57	4.04
TPU (2.5, 0.375)	2.26	1.39	2.06	1.70	**2.97**	0.61	7.84	3.51	7.69	2.23
TPU (3.5, 0.375)	1.21	1.03	1.41	0.72	**2.87**	0.17	5.5	0.24	5.44	1.48
TPU (4.5, 0.375)	1.06	1.42	1.60	0.95	**2.58**	0.30	**4.72**	1.42	**5.00**	2.09
TPU (2.5, 1.0)	0.62	0.24	5.86	3.82	5.56	0.48	17.18	4.83	13.55	2.00
TPU (3.5, 1.0)	0.73	0.59	2.03	1.83	5.63	0.45	18.54	5.95	15.35	4.36
TPU (4.5, 1.0)	0.93	0.26	2.50	1.96	5.14	0.90	20.49	9.99	14.67	2.08

The X, Y, and Z rotation angles with mean and standard deviation at 20mm displacement of the parallelogram can be found in the first three columns of Table 1. As expected from our discussion in Sect. 3.1, X and Y rotations, corresponding to out-of-plane rotations, have high standard deviations due to misaligned restorative and actuator pushing forces. We have included these rotation values for completeness in Table 1. We can conclude that for both PP and TPU materials, a hinge thickness of 0.375 mm results in the least in-plane rotation. For beam thickness, the table indicates that both 2.5 mm and 4.5 mm for PP material have small in-plane rotations. Similar to Fig. 3, we can infer that PP material achieves smaller mean in-plane rotation when compared to TPU material across all settings which can be again attributed to the difference in stiffness between the materials.

4.2 Evaluation Experiments

We evaluated the effects of integrating parallelograms from the characterization experiments, TPU and PP parallelograms with the most and least compliant hinges (0.375 mm and 1.0 mm), into full delta mechanisms. Specifically, we studied the effect on the end-effector orientation, relative to the base, when it is moved towards the edge of the configuration space, by lowering or raising each arm by 50°, as shown in Fig. 4. Both configurations were repeated three times with each of the three arms. We present the mean and standard deviation of

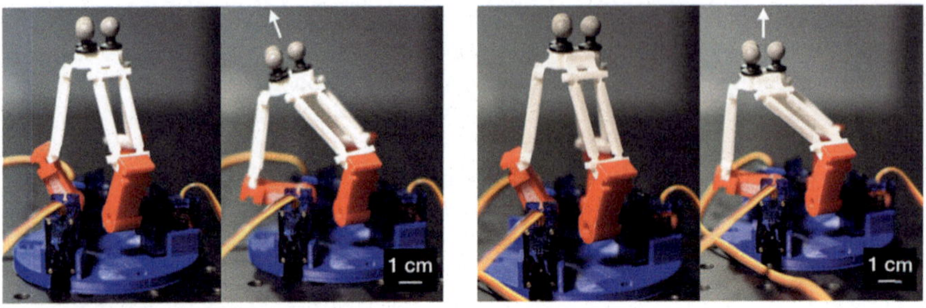

(a) TPU hinge thickness of 1 mm

(b) TPU hinge thickness of 0.375 mm

Fig. 4. Evaluation experiment parallelograms with beam thickness 2.5 mm and hinge thickness of (a) 1 mm and (b) 0.375 mm are shown before and after lowering one arm (2 arms up, 1 down) by 50°.

the end-effector rotation in Table 1 (last two columns.) For both PP and TPU materials, hinge thickness of 0.375 mm results in a smaller end-effector rotation when compared to a hinge thickness of 1.0 mm across all settings. This highlights the importance of compliance in living hinges to achieve delta behavior that is closer to ideal. With a thick hinge, parallelogram links are stiff and unable to maintain the end-effector orientation to be parallel to the base. This is similar to the conclusion drawn from our characterization experiments where we observed that smaller hinge thicknesses are better for minimal in-plane rotations.

Among parallelograms with hinge thickness of 0.375 mm, a beam thickness of 2.5 mm works best for PP while a beam thickness of 4.5 mm works best for TPU. This can be attributed to the stiffness of PP which requires thinner beam to be flexible enough to maintain the end-effector orientation across different configurations. On the other hand, TPU is more compliant and requires thicker beams to ensure that the beams stay parallel to each other which is important to achieve an end-effector orientation that is parallel to the base.

The amount of permissible rotation is dependent on the application, but for manipulating large objects (centimeter scale or larger) 7° of rotation or less is manageable. From a visual perspective as well, about 7° of rotation in Fig. 4(b) still demonstrates near-delta robot behavior.

5 Conclusion and Future Work

We tested 3D-printed parallelograms made with two materials of significantly varying stiffness, PP and TPU. The characterization experiments presented in Sect. 4.1 studied the effects of displacing the parallelogram tail. Due to the large variation in amount of force required for each of the parallelogram dimensions and materials, the force profile can be chosen for a specific application. For example, a delta robot that is interacting with humans for applications like assisted feeding is expected to be more compliant and therefore require less force to be displaced. Thus, we can tailor the parameters and material of the parallelogram links to the desired application using the force profile. Our characterization experiments have also indicated that more compliant hinges and PP material are better for achieving minimal in-plane rotations. Our evaluation experiments in Sect. 4.2 also show that more compliant hinges result in end-effector orientations that are closer to ideal delta behavior. In addition, our evaluation experiments have also provided insight into the ideal beam thickness for PP and TPU materials that result in minimal end-effector rotation.

Working with soft materials is challenging as their properties change over time and across repetitions. In order to control the 3D-printed delta for manipulation, we will have to learn how to accommodate for the material properties and adapt to changes. Our work is an initial step in this direction and the experiments show the potential of a 3D-printed delta using soft materials. Future extensions of this work involve manipulation tasks using opposing delta manipulators, arrays of delta robots for distributed actuation platforms, and modeling compliant delta robots. Modeling the delta robots also involves studying the error from rigid delta kinematic models and how to compensate for the error using robust control and state estimation techniques.

Acknowledgements. We thank Avi Rudich, Anirudh Vemula, Ankit Bhatia, Tess Hellebrekers, and Keene Chin for their assistance and support. The research presented in this work was funded by National Science Foundation under Grant No. CMMI-2024794 as well as an Amazon Research Award.

References

1. Rey, L., Clavel, R.: The delta parallel robot. In: Parallel Kinematic Machines, pp. 401-417. Springer (1999)
2. Tsai, L.: Robot Analysis: The Mechanics of Serial and Parallel Manipulators. John Wiley & Sons (1999)
3. Stewart, D.: A platform with six degrees of freedom. Proc. Inst. Mech. Eng. **180**(1), 371–386 (1965)
4. Laschi, C., Mazzolai, B., Cianchetti, M.: Soft robotics: technologies and systems pushing the boundaries of robot abilities. Sci. Robot. **1**(1) (2016)
5. McClintock, H., Temel, F.Z., Doshi, N., Koh, J.S., Wood, R.J.: The millidelta: a high-bandwidth, high-precision, millimeter-scale delta robot. Sci. Robot. **3**(14) (2018)
6. Tian, Y., et al.: Design and forward kinematics of the compliant micro-manipulator with lever mechanisms. Precis. Eng. **33**(4), 466–475 (2009)
7. Belfiore, N.P., Simeone, P.: Inverse kinetostatic analysis of compliant four-bar linkages. Mech. Mach. Theory **69**, 350–372 (2013)
8. Balucani, M., et al.: The development of a MEMS/NEMS-based 3 DOF compliant micro robot. In: 19th International Workshop on Robotics in Alpe-Adria-Danube Region (RAAD 2010), pp. 173–179. IEEE (2010)
9. Bhagat, U., et al.: Design and analysis of a novel flexure-based 3-DOF mechanism. Mech. Mach. Theory **74**, 173–187 (2014)
10. Culpepper, M.L., Anderson, G.: Design of a low-cost nano-manipulator which utilizes a monolithic, spatial compliant mechanism. Precis. Eng. **28**(4), 469–482 (2004)

11. Yi, B.J., et al.: Design and experiment of a 3-DOF parallel micromechanism utilizing flexure hinges. IEEE Trans. Robot. Autom. **19**(4), 604–612 (2003)
12. Pierre, R.S., Paul, N., Bergbreiter, S.: 3DFlex: a rapid prototyping approach for multi-material compliant mechanisms in millirobots. In: IEEE International Conference on Robotics and Automation (ICRA), pp. 3068–3073. IEEE (2017)
13. Correa, J.E., et al.: Laminated micro-machine: design and fabrication of a flexure-based Delta robot. J. Manufact. Process. **24**, 370–375 (2016)
14. Mehta, A., DelPreto, J., Rus, D.: Integrated codesign of printable robots. J. Mech. Robot. **7**(2) (2015)
15. Kim, S.J., et al.: An origami-inspired, self-locking robotic arm that can be folded flat. Sci. Robot. **3**(16) (2018)
16. Shin, B., et al.: Self-assembling sensors for printable machines. In: IEEE International Conference on Robotics and Automation (ICRA), pp. 4417–4422. IEEE (2014)
17. Koh, J., et al.: A modular folded laminate robot capable of multi modal locomotion. In: International Symposium on Experimental Robotics, Springer, Cham, pp. 59–70 (2016)

L.U.N.A. - A Laser-Mapping Unidirectional Navigation Actuator

Jasper Zevering, Anton Bredenbeck, Fabian Arzberger, Dorit Borrmann(✉), and Andreas Nüchter

Informatics VII – Robotics and Telematics, University of Würzburg, Am Hubland, 97074 Würzburg, Germany
borrmann@informatik.uni-wuerzburg.de

Abstract. This paper proposes an autonomous approach to 3D mapping using the concept of impulse by conservation of angular momentum (IBCOAM) as a unidirectional drive to roll a 2D laser scanner in an IMU equipped, pose-tracked spherical robot system. An experimental prototype of the robot is introduced, giving details about the hardware. The laser scanning results as well as the IBCOAM drive data that have been gathered using the prototype are analyzed, revealing technical challenges.

1 Introduction

In today's world, autonomous robots have found their way into everyday life in a variety of ways. This includes, but is not limited to, the vacuum cleaner that independently navigates one's living-room or mobile robots employed for exploration of areas that are too dangerous for humans. To foster new advances in the latter, specifically for underground environments, the Defense Advanced Research Projects Agency (DARPA) of the US Defense Department established the yearly "SubT" Challenge in 2017. In this challenge, teams are tasked to "Drive novel approaches and technologies to allow warfighters and first-responders to rapidly map, navigate, and search dynamic underground environments." [1] proving the demand for further research in this domain. One subtask of this challenge is building an accurate 3D model of the environment, i.e., mapping the surroundings. This paper shows a proof of concept of a such a novel approach and validates it with experiments.

One such approach using a 2D laser scanner to scan 3D indoor environments has been proposed in [11]. The authors mount a 2D laser scanner on a cylindrical structure. An operator then initiates a rolling motion by manually pushing the contraption. This enables the scanner to sense the 3D environment successfully. However, manually pushing the scanner is not practical, especially for long scans.

Previous work includes our RADLER (RADial LasER scanning device), which consists of a 2D laser scanner attached to the axle of a unicycle [5]. An operator pushes the unicycle along a requested path. The inherent rotation of the wheel creates a radial 3D laser scanning pattern. However, this approach still requires an operator, therefore does not fulfill the autonomy requirements.

B. Siciliano et al. (Eds.): ISER 2020, SPAR 19, pp. 85–94, 2021.
https://doi.org/10.1007/978-3-030-71151-1_8

A more autonomous approach was taken by Fang et al. [7]. The authors mounted a rotating 2D laser-scanner on top of a turtle-bot thus removing the need of an operator. In contrast to the RADLER however, the turtle-bot does not provide an inherent rotation. Therefore an additional actuator is required to create the radial 3D scanning-pattern.

This paper builds upon the results of the RADLER and has a specific application of mapping lunar craters autonomously in mind. We propose a novel approach to low-cost 3D laser scanning using a 2D laser scanner inside a spherical robot based on impulse by conversation of angular momentum (IBCOAM): the L.U.N.A. - sphere (Laser-mapping Unidirectional Navigation Actuator). The 2D laser scanner is fixed to the spherical structure, hence a similar situation as with the RADLER is given: the inherent rotation of the sphere creates a radial 3D scanning pattern. Using the format of a spherical robot permits the system to be designed more compact. Additionally, the spherical shell doubles as a protective layer for the actuators, sensors and electronics. This is especially valuable for applications in rough terrain or scenarios in which non-minimal impact is expected, such as space applications. During a launch, withstanding large G-forces is a necessary requirement, which can be better implemented using the spherical format. Furthermore, an operator is no longer required given a drive implemented in the robot.

2 State of the Art

As evolution of RADLER, a hand-driven radial scanner device, a self-driving spherical approach was chosen to ensure robustness and autonomy. The following presents different approaches to spherical robots.

An early idea of an self-driving sphere has been introduced by J.L. Tate in 1893 [15] for a sphere, driven by an inner moving counterweight, which got its torque from an spring. This idea of an actuator attached to an counterweight pointing to the bottom and therefore the torque being transferred to the sphere and moving it, is still a widespread approach for spherical robots.

In [6] a basic motion control system for the BYQ-III is introduced. The BYQ-III has a mass of 25 kg and a diameter of 600 mm and its driving mechanism has been proposed in [3] by Hanxu et al. It contains a counterweight pendulum, four gyro actuators, providing movement for two axes and one IMU mounted on the case of the gyro. There is no extra payload or sensor, nor would there be space for a centered measurement unit due to the centered counterweight. Therefore the counterweight leads to a steady movement, not relying on acceleration but on velocity of the actuators and therefore providing continuous speed.

A second spherical robot with its driving mechanism relaying on inner counterweight is presented in [17]. This robot was designed for movement on a water surface and therefore has fins mounted to the shell orthogonal to the movement. Two actuators attached to the shell and the inner counterweight provide movement around one axis. In contrast to the BYQ-III a middle-centered sensor would be possible, but this would have no movement relative to the surface, as it would be part of the relatively non moving inner counterweight. It also has steady, well controllable movement. The spheres presented in [12] and [4] provide a solution for a driving system which does not rely on a moving inner counterweight but uses internal reaction wheels to provide torque. This leads to theoretically having middle-centered space available, which would be rotating with respect to the surface. However, the prototype provided by Muralidharan et al. [12] shows less controllability then counterweight driven spheres. Furthermore, this robot is driven by acceleration and not velocity which leads to limited movement capabilities.

A third approach for spherical robots relies on an internal unit which drives inside the sphere. A design and control approach is provided in [2] where a four-wheeled vehicle moves in the sphere to initiate rolling by moving the center of mass in the desired direction. This technical solution is capable of a nearly maximum size of possible payload in relation to the overall-size, but also does not provide a rotation of the sensor is needed for 3D laser scanning. Additionally, this provides good controllability, similar to the counterweight-driven approach. It is not stable regarding external perturbations or forces, as there is no fixed connection to the shell. This makes it not suitable for missions with extreme forces and unknown starting conditions like space applications involving a rocket launch or a hard landing to get to the starting point. In a worst case scenario this leads to a start with the unit being rotated by 180° and therefore not being able to bring torque to the sphere. If the sphere is stuck a similar situation occurs. Then the inner unit tries to perform a whole revolution ("looping") inside the sphere, which might result in a supine position.

Halme et al. [9] overcome this shortage of the inner unit with just relying on gravity to ensure contact to the shell. They introduce the approach of a rod, expanded by a spring to the maximum size and having a wheel on one side of the rod and the original unit on the opposite site. The wheel generates the motion by moving the center of mass, so the non-reversible supine position does not exist anymore. Even if the center of mass is at the top, the vehicle is still pressed to the shell by the spring and therefore is capable of maintaining its movement. This approach does not rotate a sensor placed in the center without further contraptions.

Inspired by the seminal Cubli [8], we started designing a spherical robot based on a impulse by conservation of momentum (IBCOAM) drive.

3 Technical Approach

As result of the impulse-momentum theorem, for a change of the momentum of the flywheels, an impulse, rooted in a force, is needed. The motors provide this force, and according to Newtons' third, a force is applied back, resulting in the opposite change of momentum of the sphere. So the rotation of the sphere is not a direct consequence of the angular momentum of the flywheels, but rather of the impulse needed to spin them. The well known Cubli [8] also uses this phenomenon, utilizing the impulse generated by breaking spinning wheels, to generate a rotation of the whole system. The force does not come from the momentum due to rotation of the flywheels but through the impulse of the braking. Unlike Cubli, a sphere does not loose energy to gravity compensation due to its form. A sphere starts rolling by itself and only loses energy to friction.

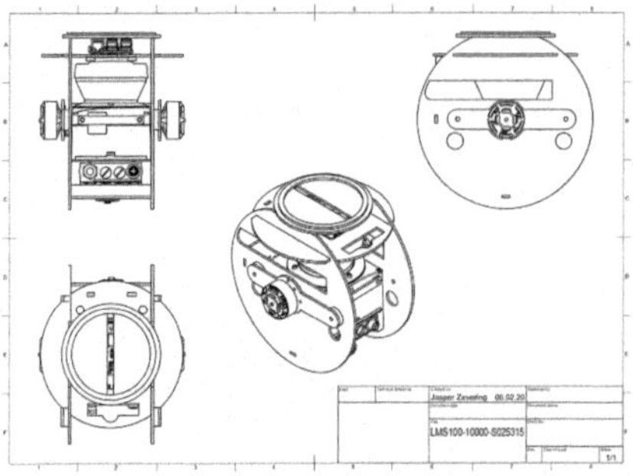

Fig. 1. Blueprint of the mechanical structure of the spherical robot.

Figure 1 shows a CAD blueprint of the overall interior layout of the mechanical structure of the L.U.N.A sphere, ignoring the outside sphere, flywheels and wiring. The LMS-100-10000 laser scanner is mounted to the supporting structural components which are made of acrylic glass. It is an afforable laser scanner with high precision and a work-range of 0.5 m to 20 m. The Raspberry Pi 3B boardcomputer is placed on top of the laser. It was chosen because of its high availability and its USB and Ethernet port, as is simplifies the connection of the laser scanner (Ethernet) and the 3 IMUs (USB). This avoids further adapters or hubs in the already limited space. Furthermore, it has enough computing power for the required tasks. Above the Pi, another supporting structure holds additional counterweights to correct for inhomogeneous weight distribution. The battery finds its place in front of the laser scanner on another supporting structure. It is an 1000 mAh, 30C and 3s lithium battery. It is a compromise of the acceptable run-time at operation of approximately 10 min and its dimensions. It powers the laser scanner and the ESCs directly, as well as an 12 V to 5 V step down converter for the Pi, mounted under the same plate as itself. Due to the experimental state, the connections use standard connection-clamps, instead of a power-board, which ensures high flexibility while building. The two brushless motors were each placed on one side of the supporting structure. The chosen Turnigy Park480 Brushless Outrunner 1320 kV [16] provide high rotational speed but also the required height of the body (without shaft) of less than 34 mm. The motors are mounted with spacers, that leave room for the side IMU underneath one of the motors. The PhidgetsSpacial 10441B IMUs [14] are highly affordable IMUs with acceptable precision. Two other IMUs are placed in front of and beneath the laser to ensure coverage of all axes. Finally a battery checker is mounted on the battery-holding plate, as a small battery combined with high power consumption would lead to deep discharge in a short period of time.

Figure 2 shows the final hardware setup of the robot. In order to reduce complexity with respect to the 3D-transformation calculations, the laser scanner was placed at the center of a spherical acrylic glass shell as precisely as possible. This limits the laser scanners movement to rotational movement and removes translational movement completely. Strong epoxy glue attaches the weights to the motor shafts and shells. As the flywheels start spinning with respect to the structural components of the sphere, the sphere itself starts spinning with respect to the ground.

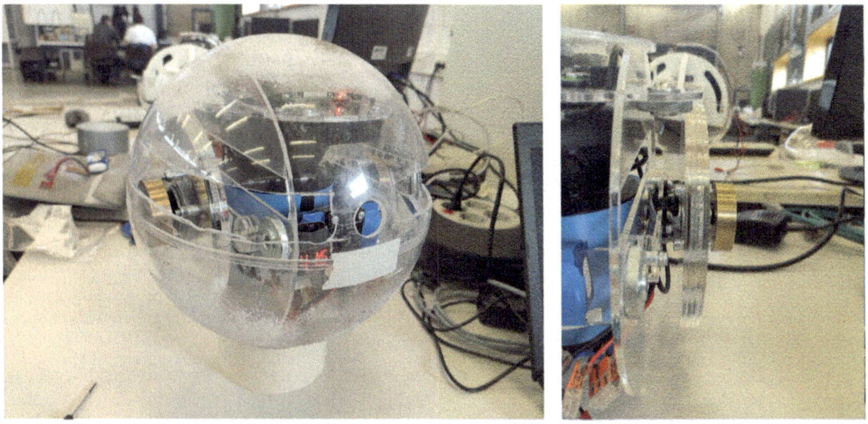

Fig. 2. Hardware setup of the L.U.N.A sphere prototype, including notches in the shell and friction granule (left). IMU (beneath supporting structure) and brushless motor including flywheel mass (above supporting structure) of the L.U.N.A sphere (right).

Figure 2 also shows that the top and the bottom of the shell are covered in table salt, which made a good granule to increase friction to the ground in the early testing phase. Unfortunately, the laser scanner measurements are affected by blockades due to components of the sphere. Specifically, the outside shell is an inhibiting factor as an object with the distance of the radius is measured at all times, which is why measurement slits in the front side of the shell exist, to increase permeability for the laser.

3.1 Sensor Integration

The sensor integration is fully implemented with the Robot Operating System (ROS) using the Ubuntu distribution ROSberryPi which contains a pre-installed ROS version.

Overall, three seperate IMUs keep track of the pose of the sphere. Each IMU is placed in such a way that the IMUs Z-axis corresponds to one possible rotation axis of the sphere. Therefore, each IMU is perpendicular to the other two. Combining the axes measurements leads to a "virtual" IMU, which emulates being an IMU positioned at the center of the sphere. Hence, isolating the measurements of the resulting virtual IMU to only the rotation in the given axis. Figure 3 illustrates this principal. Previous prototypes have shown that transforming the data of only one non-centered IMU leads to lower quality measurements. However, combining the

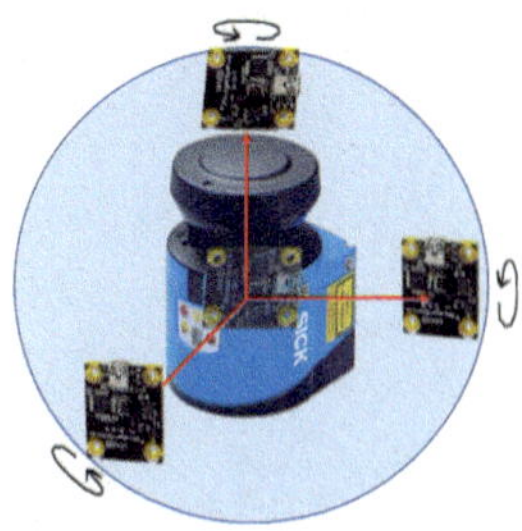

Fig. 3. Illustration of the combination of 3 IMUs into 1 virtual IMU that simulates being at the center of the sphere.

measurements of three IMUs, where each measures only the static rotation around one of their rotational axes (which also represents a rotation axis of the sphere), leads to less noise, as seen in Figs. 4a–4d.

The IMUs also ship with accelerometers that are used to determine the full pose of the sphere. Each IMU calculates their pose separately, using a quaternion extended Kalman filter (QEKF). However, combining those poses into one does not have any positive effect, but only makes the software more resource demanding and slow. Thus only the pose of the bottom IMU's accelerometer is used to keep track of the pose.

The motors are controlled using the piGPIO library. The GPIO signals are forwarded by the pins to two ESCs that drive the motors.

Unfortunately, the brass weights are not drilled in the very center, causing an unbalance when rotating. The resulting vibrations inhibit the movement of the sphere. Thus a controller was implemented that measures the extend of the vibrations using standard deviations of the non-rotating axes of the IMU and adjusts the throttle of the motors accordingly. This was done with a two-point controller with hysteresis. Considering the translational velocity of the sphere in a controller is not possible. The speed of the sphere is calculated by the rotational speed, which is why slippage of the sphere causes such a controller not to produce the desired motion.

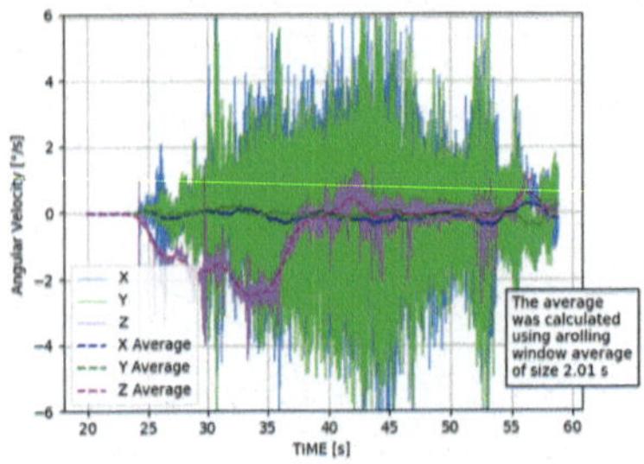

(a) IMU 1: Z-Axis corresponds to Sphere X-Axis

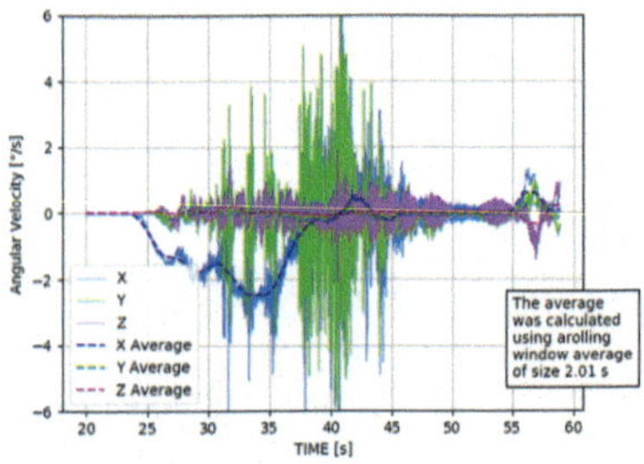

(b) IMU 2: Z-Axis corresponds to Sphere Z-Axis

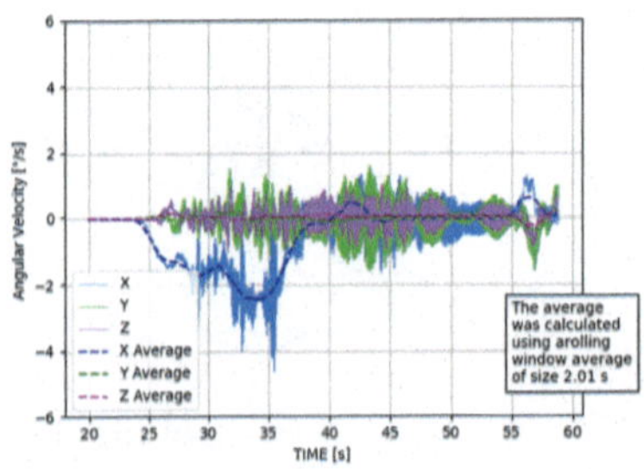

(c) IMU 3: Z-Axis corresponds to Sphere Y-Axis

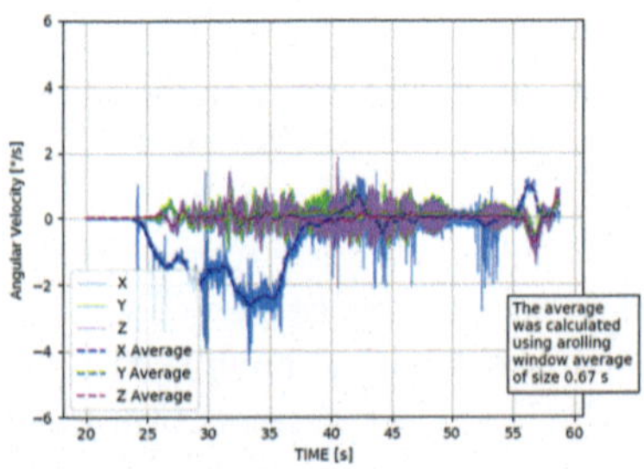

(d) Merged virtual IMU. Maps the Z-Axis of all other IMUs to the rotational axes.

Fig. 4. Angular velocity measurements of singular IMUs and the combined IMU.

3.2 Point Cloud Processing

For the processing of the point cloud the 3D Toolkit (3DTK) was used. This provides multiple methods and algorithms for processing 3D point clouds, especially the 6D continuous time Simultaneous Localization And Mapping (SLAM) algorithm (see [10,13]) as post-processing. Therefore only the time-stamped raw data of the IMUs and laser scanner is transferred and the estimation of the pose and the SLAM algorithm itself is performed externally. For this prototype the transfer is realized using the host-function of ROS, giving the external PC the possibility to subscribe to the topics and process them. 3DTK itself takes pairs of files, one representing the pose, one the laser scanner data, and each file named by the time-stamp with an identifier if it is scan-data or pose. Also, the use of USB-connected IMUs and ROS as transfer-mechanism of data leaves potential for enhancement and therefore reducing the load of the internal controller.

4 Experimental Results

4.1 3D Laser Scanning

The test of the 3D laser scanning consists of 3 phases. First a semi stationary scan is performed by moving the robot manually in a limited manner (tilting it by approximately 45°) without its exterior shell. This results in the 3D point cloud in Fig. 5a. For the second test the same movement is used. However, this time the outside shell of the robot is present. This results in Fig. 5b. Finally,

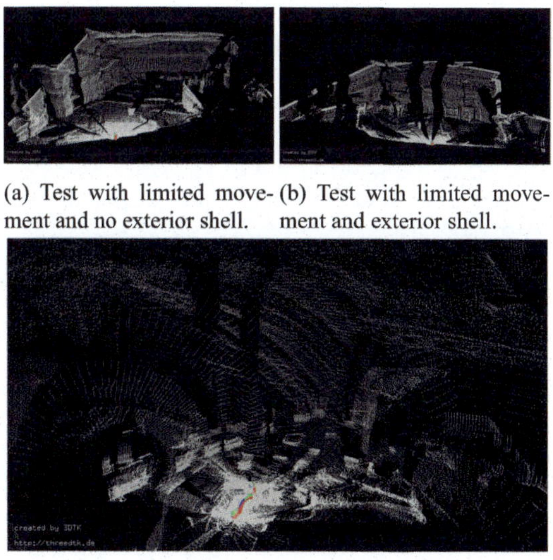

(a) Test with limited movement and no exterior shell. (b) Test with limited movement and exterior shell.

(c) Test with exterior shell and full movement.

Fig. 5. 3D Point cloud results of different test scenarios of the system.

the IBCOAM-drive enabled test of the overall system leads to a 3D point cloud consisting of multiple revolutions. Figure 5c shows results.

The first test yields the clearest 3D point cloud (Fig. 5a). This is contributed to the fact, that in this test the exterior shell is absent. Since the shell partially reflects the beams of the laser scanner, the noise level increases. In this case the laser scanner measures the shell instead of the environment around it. In Fig. 5b lines are blocked that are measured in the previous test further showing the reflection of the laser beams by the shell. The rolling motion of the robot increases this effect. Specifically, a large number of measurement points along the path of the robot manifests this (Fig. 5c). We assume that the more powerfull laser scanner LMS141 reduces this effect.

Furthermore, the time asynchrony of the pose determination sub-system and the laser scanner introduces further systematic errors to the 3D point cloud. Specifically, Fig. 5c shows that the laser scanner detects points underneath the surface the robot was rolling on.

4.2 IBCOAM Drive

The conversation of angular momentum drive accelerates the L.U.N.A. sphere reliably. Figure 4 shows the angular acceleration of the whole sphere measured by the IMU system in one test run. Furthermore, it shows that the acceleration along the rotational axis of the flywheels rises while the accelerations along the other axes remain lower, albeit are noisy. However, it also shows the decrease in noise due to the combination of the IMU measurements. The vibrations and tilt of the robot contribute to the velocities along the other axes. The vibrations are results of inexact drilling of the flywheels such that there is an unbalance. At the main test site the ground is a hard, clean and low friction concrete floor. In such a scenario the vibrations add up and lead to slippage. However, a rubber surface (a running track) absorb the vibrations, such that the acceleration process happens reliably.

5 Conclusions

This paper has presented a new cost efficient approach to 3D laser scanning: The L.U.N.A. sphere. It uses a 2D laser scanner mounted inside a spherical robot and the inherent rotational movement to form a radial scanning pattern, hence creating a 3D point cloud. The spherical robot is based on conversation of angular momentum and uses flywheels to drive the robot forward.

The prototype developed for the tests in this paper was able to move in one direction reliably on soft surfaces (such as rubber), however had difficulties with slippage on hard and low friction surfaces.

With regards to 3D laser scanning this paper has delivered a proof of concept, even though the results leave room for improvement. Needless to say, a lot of work still has to be done. In future work, we plan to overcome the biggest issues, i.e., the reflection of the laser beams by the exterior shell and synchronization issues between the IMU system and the laser scanner.

This includes improving the field of view of the laser scanner and extending the robot to two dimensional movement control. This enables autonomous mapping of environments using the L.U.N.A. sphere.

Acknowledgements. The authors thank Dieter Ziegler and Sergio Montenegro for supporting our work. We acknowledge funding from the ESA Contract No. 4000130925/20/NL/GLC for the "DAEDALUS – Descent And Exploration in Deep Autonomy of Lava Underground Structures" Open Space Innovation Platform (OSIP) lunar caves-system study and the Elite Network Bavaria (ENB) for providing funds for the academic program "Satellite Technology".

Authors Note. In an attempt to abide by the Fair-Principles of open science the authors provided all code developed and further information at their GitHub page.

References

1. Booz Allen: Unearthing the subterranean environment. https://www.subtchallenge.com. Accessed 06 Apr 2020
2. Alves, J., Dias, J.: Design and control of a spherical mobile robot. Proc. Inst. Mech. Eng. Part I J. Syst. Control Eng. **217**(6), 457–467 (2003)
3. Hanxu, S., Aiping, X., Qingxuan, J., Liangqing, W.: Omnidirectional kinematics analysis on bi-driver spherical robot. J. Beijing Univ. Aeronaut. Astronaut. **31**, 735–739 (2005)
4. Bhattacharya, S., Agrawal, S.K.: Spherical rolling robot: a design and motion planning studies. IEEE Trans. Robot. Autom. **16**(6), 835–839 (2000)
5. Borrmann, D., Jörissen, S., Nüchter, A.: RADLER - A RADial LasER scanning device. In: Proceedings of the 16th International Symposium of Experimental Robotics (ISER '18), Springer Tracts in Advanced Robotics, pp. 655–664, Buenos Aires, Argentina (2018)
6. Liu, D., Sun, H., Jia, Q., Wang, L.: Motion control of a spherical mobile robot by feedback linearization. In: 2008 7th World Congress on Intelligent Control and Automation, pp. 965–970 (2008)
7. Fang, Z., Zhao, S., Wen, S., Zhang, Yu.: A real-time 3D perception and reconstruction system based on a 2D laser scanner. J. Sens. **1–14**(05), 2018 (2018)
8. Gajamohan, M., Merz, M., Thommen, I., D'Andrea, R.: The cubli: a cube that can jump up and balance. In: International Conference on Intelligent Robots and Systems, pp. 3722–3727 (2012)
9. Halme, A., Suomela, J., Schönberg, T., Wang, Y.: A spherical mobile micro-robot for scientific applications. ASTRA, 96 (1996)
10. Lauterbach, H.A., Borrmann, D., Nüchter, A., Rossi, A.P., Unnithan, V., Torrese, P., Pozzobon, R.: Mobile mapping of the la corona lavatube on lanzarote. In: Proceedings of the ISPRS Geospatial Week 2019, Laserscanning 2019, ISPRS Annals Photogrammetry and Remote Sensing, Spatial Information Science, IV-2/W5, pp. 381–387, Enschede, Netherlands (2019)
11. Lehtola, V., Virtanen, J.-P., Kukko, A., Kaartinen, H., Hyyppä, H.: Localization of mobile laser scanner using classical mechanics. ISPRS J. Photogramm. Remote Sens. **99**, 25–29 (2015)

12. Muralidharan, V., Mahindrakar, A.D.: Geometric controllability and stabilization of spherical robot dynamics. IEEE Trans. Autom. Control **60**(10), 2762–2767 (2015)
13. Nüchter, A., Bleier, M., Schauer, J., Janotta, P.: Improving Google's cartographer 3D mapping by continuous-time SLAM. In: Proceedings of the 7th ISPRS International Workshop 3D-ARCH 2017: "3D Virtual Reconstruction and Visualization of Complex Architectures", ISPRS Archives Photogrammetry and Remote Sensing Spatial Information Science, vol. XLII/W3, pp. 543–549, Nafplio, Greece (2017)
14. Phidgets Imu Website. https://www.phidgets.com. Accessed 15 Apr 2020
15. Tate, J.L.: Method and system for measurement of road profile. US Patent 508,558, 14 Nov 1893
16. Turnigy Park480 Brushless Outrunner. https://hobbyking.com/en_us/turnigy-park480-brushless-outrunner-1320kv.html. Accessed 17 Apr 2020
17. Vahid Alizadeh, H., Mahjoob, M.J.: Quadratic damping model for a spherical mobile robot moving on the free surface of the water. In: 2011 IEEE International Symposium on Robotic and Sensors Environments (ROSE), pp. 125–130 (2011)

Hybrid Wheel-Leg Locomotion in Rough Terrain

Bhanuka Silva[1], Uvindu Perera[1], Thilina Thilakarathna[1], Isuru Athapattu[1], Peshala Jayasekara[1], Thilina Lalitharatne[2], Navinda Kottege[3], and Tirthankar Bandyopadhyay[3(✉)]

[1] Department of Electronic and Telecommunication Engineering, University of Moratuwa, Moratuwa, Sri Lanka
bhanuka.silva@outlook.com, peshala@uom.lk

[2] Department of Mechanical Engineering, University of Moratuwa, Moratuwa, Sri Lanka
thilina@mech.mrt.ac.lk

[3] Robotics and Autonomous Systems Group, Commonwealth Scientific and Industrial Research Organisation, Pullenvale, QLD 4069, Australia
{navinda.kottege,tirtha.bandy}@csiro.au

Abstract. This paper introduces a novel locomotion strategy of combining wheeled and legged actuation in a parallel structure. This enables the robot to traverse rough terrains using the articulated limbs and use wheeled locomotion for more efficient traversal over smooth terrains. Authors present a prototype robot architecture to embody the concept and show experimentally the robot traversing a variety of indoor and outdoor terrains. Additionally, a novel butterfly gait is introduced for the robot that enables stable and efficient traversal of challenging terrains.

Keywords: Hybrid wheel-leg locomotion · Reconfigurable robots · Terrain characterisation

1 Introduction

Real world applications require ground mobile robots to operate in complex environments consisting of structured/unstructured and even/uneven terrains where manoeuvrability, urgency and efficiency are crucial factors [1–3]. Among the three most commonly used locomotion mechanisms for ground mobile robots, wheel locomotion can achieve high speeds with low power consumption, yet its ability to overcome obstacles, stability and ease of navigation in unstructured environments are generally limited [4]. Biologically inspired legged robots are suitable for applications in both structured and unstructured environments and they can overcome uneven terrains by adapting their gaits accordingly [5]. However, legged robots are relatively slow, inefficient in energy consumption and require more actuators and complex control systems. Tracked robots are better suited to uneven and soft terrains due to the maximised ground contact surface

B. Siciliano et al. (Eds.): ISER 2020, SPAR 19, pp. 95–104, 2021.
https://doi.org/10.1007/978-3-030-71151-1_9

area of the tracks but they are slow, consume more energy and create a great degree of disturbance to the external environment [4].

It is important for a ground mobile robot to achieve manoeuvrability in a wide range of environments and maximise average energy efficiency by selecting a suitable locomotion mechanism. Wheel-leg hybrid locomotion could provide the best of both worlds [4]: i.e. it can combine the energy efficiency and low cost of transportation of wheels while the legs can help navigate uneven terrains by overcoming obstacles. Placing the wheels at the end or at a suitable position of the leg links can be seen as the most common technique to achieve wheel-leg hybrid locomotion [6,7]. Another technique of having wheel modules that can transform into legs [8] suffers from the disadvantage of mechanical complexity. A third approach of incorporating additional legs to a wheeled robot is rarely adopted due to the increased mass of the robot by two separate locomotion systems [4]. However, locomotion system independence is an advantage.

Reconfigurability allows adaptive navigation by changing the locomotive interaction and kinematic configuration of a mobile robot to optimise in a navigational task [9]. Optimum gait pattern selection for a reconfigurable mobile robots with a legged subsystem [10], maintaining a stability margin by regulating the center of mass of a robot [10], reduction of the cost of transportation of leg-wheel hybrid robot over counterpart legged robots [6] and ability to overcome obstacles with the aid of reconfigurability [11] are state-of-the-art results that inherit advantages from the reconfigurability.

Humans use bipedal locomotion for regular movement while they additionally make use of arms when difficult environments are encountered. For instance, a skier would use poles for balance and propulsion on snow; a swimmer would use his hands for propulsion; a soldier would use arms for crawling. In this work, a novel hybrid wheel-leg locomotion method inspired by human locomotion in extreme conditions is proposed. Figure 1 portrays the final prototype design, which utilises wheeled locomotion in smooth terrain and wheel-leg hybrid locomotion with suitable gait patterns for adaptive navigation in rough terrains.

Fig. 1. Prototype robot design performing an outdoor navigational task

2 Technical Approach

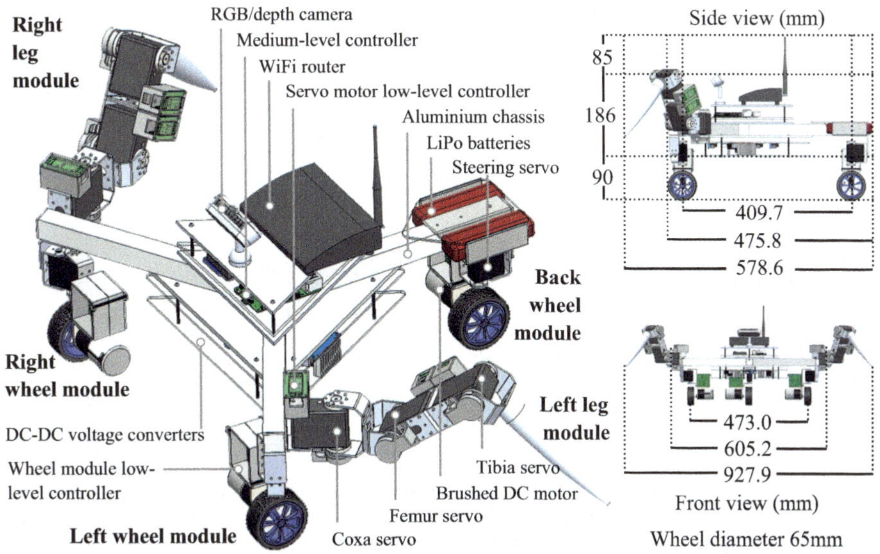

Fig. 2. Hardware architecture

The hardware architecture of the proposed robot is depicted in Fig. 2. The wheeled subsystem contains three wheel modules with independent steering and propulsion. The legged subsystem contains two leg modules with three degrees of freedom for each, and will be supported by the back wheel module for balancing and propulsion requirements in a unique wheel-leg crawling locomotion configuration. Every module comprises sufficient low-level processing power to perform module specific commanding, sensing and information feeding-back tasks of the actuators.

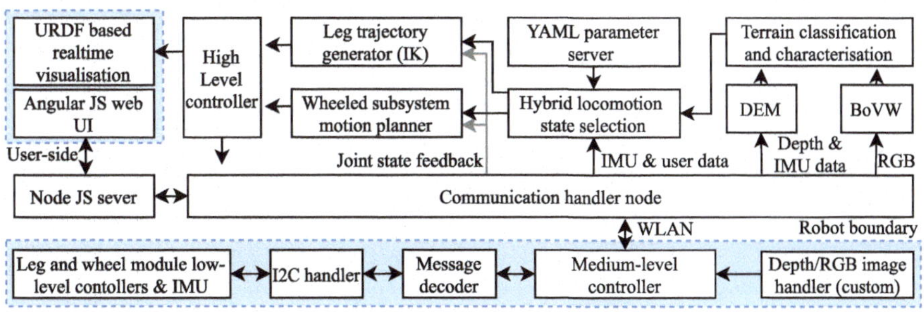

Fig. 3. Communication architecture

The communication architecture of the robot is depicted in Fig. 3. A wireless local area network (WLAN) is used to communicate between the user interface (UI), the high-level controller computer and the robot. A Raspberry Pi medium level controller with Robot Operating System (ROS) is used to decode and distribute the commands and feedbacks from low level controllers. Each actuator is attached to a low level controller that allows dedicated feedback control, networking with the medium level controller (e.g. daisy chaining) and extreme condition handling (e.g. overloading). Medium-level controller acts as a bridge in transmitting visual and point cloud feedbacks obtained from an Intel RealSense D435 depth camera to the high-level controller.

2.1 Wheel-Leg Hybrid Locomotion

Inspired by the butterfly stroke of swimming, a novel gait pattern called "butterfly gait" is developed as the main contribution of the proposed hybrid wheel-leg crawling locomotion mechanism. Figure 4 depicts the correspondence between the butterfly stroke and the butterfly gait as a left to right sequence.

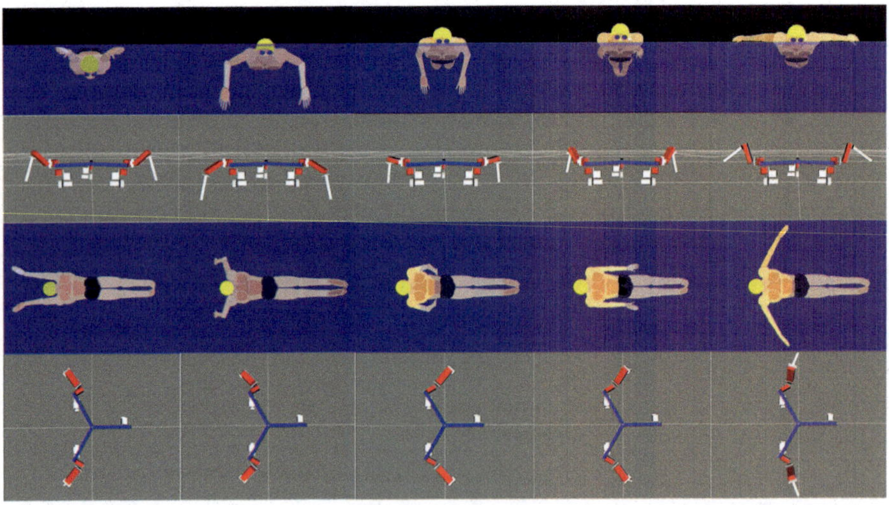

Fig. 4. Correspondence between butterfly stroke and butterfly gait

Both leg tips follow identical trajectories in the designed gait patterns and the kinematics of the right leg is shown in Fig. 5(a). A "short butterfly gait" is inherited from the butterfly gait with a shorter stride length allowing more stability margin for the movement of the centre of gravity (CoG) of the robot in highly unstructured and uneven environments. An "extended butterfly gait" is derived with 10% additional stride height to navigate in more unstructured terrains, where the leg tip can sink between loose objects. The original butterfly

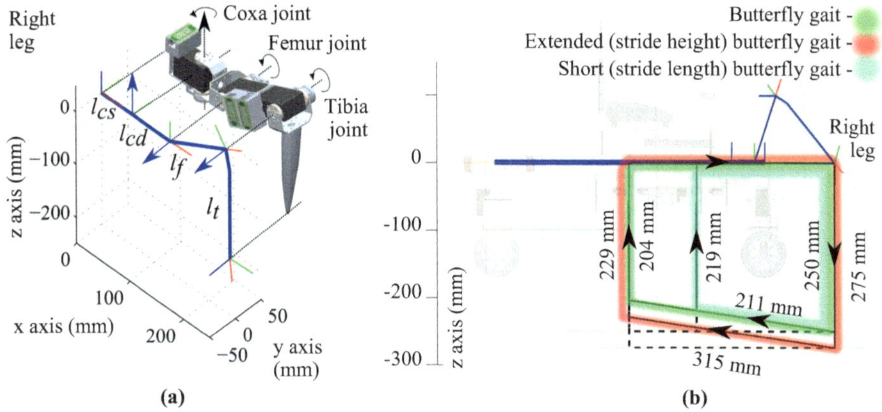

Fig. 5. (a): Right leg configuration. (b): Right leg tip trajectory for the three variants of the butterfly gait. Coordinate system is identical in both sub-figures. Coxa-static (l_{cs}), Coxa-dynamic (l_{cd}), Femur (l_f) and Tibia joint to tip (l_t) link lengths are 59.5, 70, 119 and 202.1 mm respectively (Fig. 5a).

gait and the variants are shown in Fig. 5(b). "Back-lift gait" is incorporated to pivot the robot from front two wheel-ground contact points and lift the back wheel module up to overcome any obstacles for the back wheel. The "skiing gait" is derived to aid the wheeled subsystem with additional points of ground contact and propulsion for terrains with less traction such as gravel or sand.

The trajectories for the right leg tip are visualised in 3D in Fig. 6. The trajectory way points are placed considering the workspace feasible margins and joint torque values to achieve maximum stride height and stride length for a particular gait variant.

2.2 Gait Selection from Terrain Feedback

The robot dynamically reconfigures itself and chooses the suitable locomotion technique based on the immediate terrain. Classification of the immediate terrain is performed using bag of visual words (BoVW) method from the real-time RGB video feed of the camera. Support vector machine (SVM) algorithm is used as the classifier for clustered speeded up robust features (SURF) of each RGB frame where the four classes of categorisation are shown in Fig. 7.

Characterisation of the immediate terrain is performed based on a digital elevation model (DEM) created using the point cloud data feed from the camera and orientation information feed from inertial measurement unit (IMU). Root mean square height (RMSH), as given in Eq. (1), is the most commonly used measure of surface roughness, especially for scattering elevation data [12],

$$RMSH = \sqrt{\frac{\sum_{i=1}^{n}(z_i - \bar{z})^2}{n-1}} \tag{1}$$

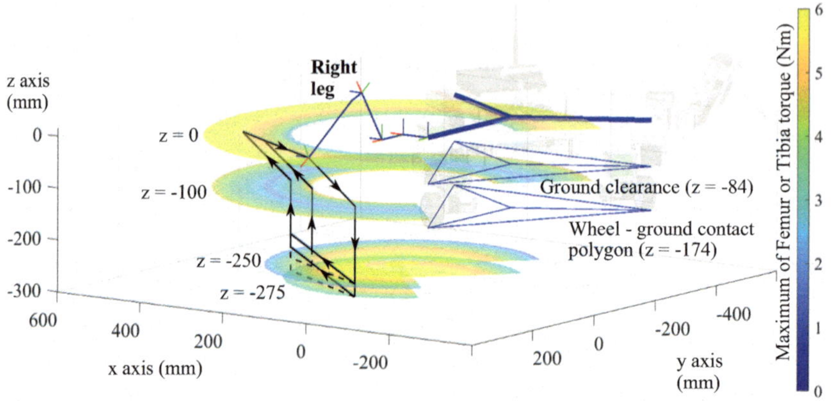

Fig. 6. Feasible workspace for right leg tip. Static torque values for leg joints are calculated assuming a constant payload of 20N for the right leg at each tip position. Each leg joint servo actuator's stall torque is 6Nm.

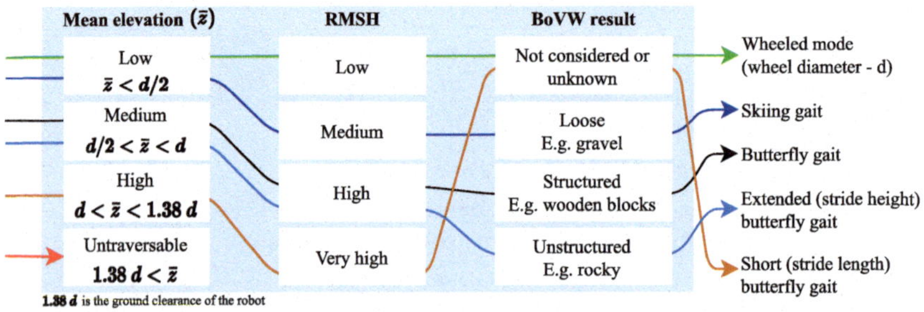

Fig. 7. Primary states of hybrid locomotion state selector node

where n represents the number of data points, z_i is the elevation of the ith data point and $\bar{z}$ is the mean elevation of all n data points.

3 Experiments Completed

Three types of navigational experiments–overcoming obstacles, complex terrain navigation, and inclined surface navigation, are conducted to evaluate the performance of the robot. Figure 8 illustrates the four stages of an obstacle overcoming sequence. Figure 8 (a), (b) and (c) are frames when the short butterfly gait is used to lift and clear the front of the robot from the obstacle. In frame Fig. 8(d), the robot employs the back-lift gait to clear the back wheel from the obstacle.

An indoor multi-terrain test bed is constructed to assess the robot's navigational capabilities in a complex terrain that comprises four terrain types. Figure 9 depicts snapshots for each type of terrain. Inclination tests were conducted to identify the maximum ramp angle the robot can navigate.

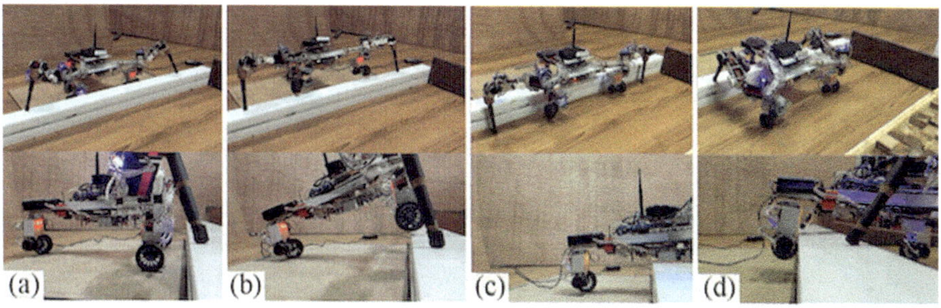

Fig. 8. Four stages the robot executes when overcoming an obstacle.

Fig. 9. Test-bed: (a) structured-even, (b) unstructured-even, (c) unstructured-uneven, and (d) structured-uneven

For evaluation of motion efficiency, the cost of transport (CoT) is defined as,

$$CoT = \frac{P}{mgv} \tag{2}$$

where P is the overall power consumption of the robot. Each Atmega328 based custom low-level controller provides feedback on power consumption of each actuator at 5 Hz frequency. Power consumption of other electronics are measured using separate sensors at the same frequency. In Eq. (2), m is the mass of the robot, which is 5.72 kg, g is the gravitational acceleration, and v is the average velocity of the robot calculated by measuring the distance travelled using a particular gait pattern over a measured duration. The robot stays statically in the traversing direction for approximately 70% of the butterfly gait period, and therefore, substituting an average velocity instead of instantaneous velocity is preferred to calculate CoT.

4 Results and Analysis

The robot is capable of surmounting an obstacle of height 1.38 times the diameter of a wheel, which is also the ground clearance of the developed hardware. It was

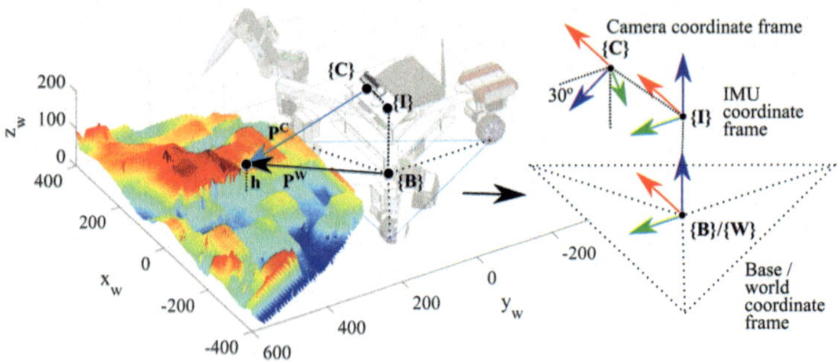

Fig. 10. Coordinate systems for generating the DEM.

Fig. 11. RGB image feed (top) and generated DEM result (bottom) for four instances while the robot navigates in the test bed.

observed that the crawling gait patterns do not increase the navigating angle (35.8°) when compared to the wheeled mode on a flat ramp. Instead, the leg modules can be oriented to maintain the CoG such that all three wheels can have the maximum ground reaction force, therefore, the maximum grip. The results confirm that the crawling locomotion gait patterns have assisted navigation in unstructured and uneven ramps, where leg tips maintain a stronger grip with the ramp than wheels. Figure 10 assumes that the robot base coordinate frame is coincident with world cooridnate frame at $t = 0$. Point cloud information is obtained with respect to the camera coordinate frame $\{C\}$ and is transformed to the world coordinate frame to obtain the DEM. Field of view of the camera can be visualised from the same figure.

Structured-even terrain (RMSH $<$ 3; e.g. Fig. 11(a)) can be navigated using omni-directional wheeled mode. Unstructured-even terrain such as gravel (RMSH range 3–7.5; e.g. Fig. 11(b)) requires more traction, therefore, skiing gait is used where the wheels and legs collaboratively provide five points of contacts and propulsion. Stride height is maintained such that front wheels as well as leg tips maintain a positive ground reaction force. Unstructured-uneven terrain (RMSH range 7.5–15; e.g. Fig. 11(c)) is navigated using the extended butterfly gait. Having an extended stride length is beneficial as the leg tip is to sink between the loose rocks. Robot maintains three to five ground contact points in this gait pattern. Structured-uneven terrain (RMSH range 15–25; e.g. Fig. 11(d)) is navigated using the butterfly gait and the robot maintains three points of contact at any given time.

The average cost of transport (CoT) measure can be calculated for a navigation task of one meter distance using wheeled mode, butterfly gait and short butterfly gait on a structured-even surface.

Table 1. Experimental outcomes related to energy efficiency

Parameter	Wheeled mode	Butterfly gait	Short butterfly gait
Average CoT	4.949	41.767	71.280
Total energy consumption per 1 m (kJ/m)	0.202	1.873	3.159
Range for 80% battery capacity (m)	695.6	75.1	44.52

According to Table 1, wheeled locomotion improves CoT by 93.05% than short butterfly gait, and by 88.15% than butterfly gait. That is, the robot can revert back to wheeled locomotion whenever it encounters structured-even terrain during a navigation task for higher energy efficiency, which is not possible with a pure legged robot. Increasing the stride length more than that value of the butterfly-gait decreases the CoT. Increased stride length however brings two out of three ground contact points' vertical projection plane closer to the CoG during the rear-most way point of the leg swing trajectory, and therefore it is suitable more towards lesser uneven surfaces. Inclination test results emphasise that the inclusion of two leg modules is effective and actively contribute in maintaining the CoG of the robot to provide stability as well as the maximum traction for the wheels during structured ramps. Moreover, the legs are capable of active propulsion in unstructured ramps.

5 Conclusions

The experimental results demonstrate that the robot is capable of achieving manoeuvrability in a wide range of environments while maximising the average energy efficiency in compatible environments as well as minimising disturbances to the external environment. The research has further enlightened potential capabilities for future research in leg-wheel hybrid collaborative and modular locomotion. Removing one of the leg modules can reduce the curb weight of the fully assembled robot by 14.93% when operating in a structured-even terrain while increasing the payload limits by the same margin. Modularity also allows easy diagnosis and replacement capacity for an operator in case of individual component failures in operative environments. Since the wheeled subsystem can alone maintain the static stability in any type of environment, the legged subsystem can be used for certain manipulation tasks such as actuation or interaction with objects within the feasible workspace of the legs when the robot is stationary. Furthermore, wheeled subsystem motion planner and legged subsystem gait engines can collaboratively achieve improvised motions together with slip detection and avoidance mechanisms. Since the autonomous dynamic-reconfigurability between wheeled locomotion and leg-wheel hybrid locomotion upon terrain classification and characterisation is already achieved, state-of-the-art autonomous navigation nodes can easily be implemented in the ROS nodal architecture.

References

1. Seo, K., Park, J.W., Oh, S., Hahm, J., Suh, J.: Requirement analysis of simulator-based integration for disaster response robots. In: International Conference on Control, Automation and Systems, Jeju, pp. 1295–1299 (2017)
2. Takada, R., Oishi, N.: Priority issues of infrastructure inspection and maintenance robot: a part of COCN 2012 project "disaster response robot and its operational system". In: IEEE Region 10 Humanitarian Technology Conference, Sendai, pp. 166–171 (2013)
3. Tadokoro, S.: Challenge of disaster robotics. In: Chinese Control Conference, Hangzhou, pp. 21–22 (2015)
4. Bruzzone, L., Quaglia, G.: Locomotion systems for ground mobile robots in unstructured environments. Mech. Sci. **3**, 49–62 (2012)
5. Bjelonic, M., Kottege, N., Beckerle, P.: Proprioceptive control of an over-actuated hexapod robot in unstructured terrain. In: IEEE/RSJ International Conference on Intelligent Robots and Systems, Daejeon, pp. 2042–2049 (2016)
6. Bjelonic, M., et al.: Keep rollin'-whole-body motion control and planning for wheeled quadrupedal robots. IEEE Robot. Autom. Lett. **4**(2), 2116–2123 (2019)
7. Lauria, M., Piguet, Y., Siegwart, R.: Octopus - an autonomous wheeled climbing robot (2002)
8. Tadakuma, K., et al.: Mechanical design of the wheel-leg hybrid mobile robot to realize a large wheel diameter. In: IEEE/RSJ International Conference on Intelligent Robots and Systems, Taipei, pp. 3358–3365 (2010)
9. Heredia, M.V., Martínez-García, E.A., Torres-Córdoba, R.: Unstructured terrain adaptive navigation of self-reconfigurable quadruped robot. IFAC Symp. Robot Control SYROCO **48**(19), 183–188 (2015)
10. Wang, J., Ramirez-Serrano, A.: Stair-climbing and energy consumption evaluation of a leg-tracked quadruped robot. In: IEEE International Conference on Advanced Intelligent Mechatronics, Banff, AB, pp. 1448–1453 (2016)
11. Ben-Tzvi, P.: Experimental validation of a hybrid mobile robot mechanism with interchangeable locomotion and manipulation. In: IEEE/RSJ International Conference on Intelligent Robots and Systems, St. Louis, MO, pp. 420–421 (2009)
12. Fan, L., Atkinson, P.: A new multi-resolution based method for estimating local surface roughness from point clouds. ISPRS J. Photogramm. Remote Sens. **144**, 369–378 (2018)

Digger Finger: GelSight Tactile Sensor for Object Identification Inside Granular Media

Radhen Patel(✉), Rui Ouyang, Branden Romero, and Edward Adelson

Massachusetts Institute of Technology, Cambridge, USA
{radhen,brromero}@mit.edu, nouyang@alum.mit.edu, adelson@csail.mit.edu

1 Introduction

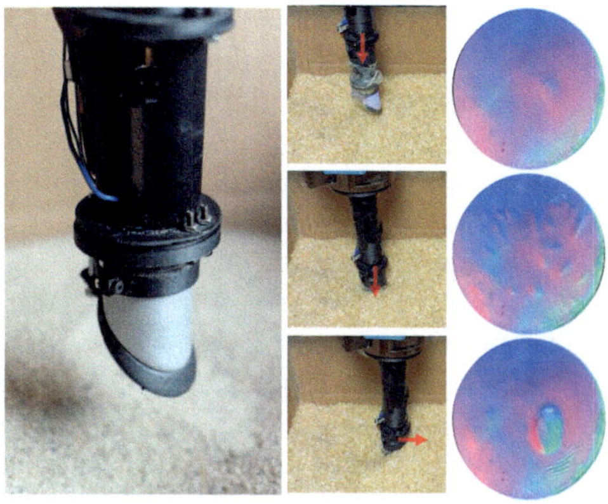

Fig. 1. *Left:* Digger Finger. *Middle:* Penetration motion. *Right:* Tactile data showing zero contact, granular media (rice), and object contact

Imagine you are at the beach with a metal detector which goes off, and you stick your hands in the sand to find the metal object. Even though the granular media (sand) is constantly affecting your sense of touch on your fingers and palm, its acuity combined with your cognition enables you to easily find buried objects. Manipulating granular media, just like manipulating rigid or deformable objects, comes very naturally to us. For robots, however, it remains a challenging task. Research in robotic manipulation has focused on rigid objects more than deformable objects and granular media. One reason is due to the difficulty of modelling the complex dynamics of the latter two. Another reason is that the perceptual understanding of the latter two via tactile based hardware devices and algorithms, compared to vision-based methods, is poor. Yet, when dealing with physical interactions, tactile sensation can be more critical than visual information. Due to these compounding limitations, robotic manipulation of deformable objects and granular media remains poorly explored. With this motivation, we take on the challenge of using touch feedback to search for objects buried in granular media. A robot with such capabilities can prove useful in areas such as deep sea exploration [1], mining, excavation, decommissioning explosive ordinates [2,3], agricultural robotics and other areas where task dependant information can be occluded. In this paper we present an early prototype of the Digger Finger (Fig. 1) that is designed to easily penetrate granular media and is equipped with the GelSight sensor [4,5].

B. Siciliano et al. (Eds.): ISER 2020, SPAR 19, pp. 105–115, 2021.
https://doi.org/10.1007/978-3-030-71151-1_10

We begin by providing a brief overview of the related work in the domain of robotic manipulation for granular media. We then present our technical approach in Sect. 2 starting with the design of the sensor in Sect. 2.1 and then the manufacturing process in Sect. 2.2. Section 3 describes the two experimental procedures we used to evaluate the performance of the Digger Finger in searching object buried inside granular media. Section 3.1 illustrates the ability of the Digger Finger to fluidize granular media during penetration. Section 3.2 illustrates the ability of the Digger Finger to identify objects that are buried inside granular media. Finally, in Sect. 4 we provide our concluding remarks.

Robotics research related to granular media has been fall primarily within the scope of automated operation of construction equipment such as scooping [6], legged locomotion [7,8], gripper design [9], manipulators [10], haptic displays [11,12] and in robotic pouring tasks [13,14], to name a few. In contrast, work on robotic manipulation of and within granular media has only recently begun receiving attention from the research community. In [15], the authors teach a robot how to scoop and dig into a pile of beans by learning the dynamics of the media from visual data. Building on [15], the authors in [16] and [17] learn to use tactile, visual and auditory feedback to estimate the flow and amount of granular materials during scooping and pouring tasks respectively. Regarding object identification in granular media in particular, two non-destructive methods that are popular for finding buried objects are Ground Penetrating Radars (GPR) and ultrasonic vibrations [18,19]. These methods, though reliable, are only good at estimating rough geometry and approximate location of the buried objects. Closest to our work is [20] and [21]. In [20] the authors use the BioTac sensors from SynTouch Inc. on a three fingered Barrett hand to detect contact with a cylinder fixed inside a bed of granular media. They take advantage of multimodal sensory data to classify contact and no-contact events. Similarly in [21] the authors again only classify the presence or absence of an object inside granular media. Unlike [20,21], we use the high resolution data from the Digger Finger to identify different objects inside granular media. Also, our design enables deeper penetration in granular media with the help of mechanical vibrations.

2 Technical Approach

2.1 Sensor Design

The three main goals of our design were, (*i*) enable the sensor to easily penetrate the granular media, (*ii*) provide rich tactile sensing to identify objects that are buried inside granular media. (*iii*) achieve human finger like form factor so that the sensor can easily be fitted on existing robot hands. The GelSight [4,5] is one type of vision based tactile sensor [22,23]. It consists of a camera that observes a directionally illuminated clear elastomer gel coated in a Lambertian or semi-specular membrane. Objects pressed into the gel deform the surface and the illumination allows estimation of surface normal along the deformation. The 3D geometry of the deformation can then be recovered using photometric stereo. Previous GelSight sensors had flat sensing surfaces [24,25], which restricted their integration with manipulators to parallel grippers for planar contact interactions. The four major modifications that we make to the previous GelSight sensors to make it suitable for searching for objects in granular media are (*i*) sensor shape, (*ii*) illumination source, (*iii*) addition of mechanical vibration and (*iv*) gel. Following are the details on the same.

First, our design adds a pointed tip to the prior work on creating curved GelSight sensors [26]. Second, we replace two color LEDs (green and red) with fluorescent acrylic paint from Liquitex. This allows for a simple and compact design especially at the tip of the Digger Finger which has to face the brunt of digging. We use six blue color LEDs (3528) from Chazon to excite the fluorescent paints from the top of the Digger Finger (ref. Fig. 2). To excite the red and green color paint to the required intensity we shine an excessive amount of blue light into the Digger Finger. This causes the image to be dominated by the blue spectrum compared to the red and green parts. To compensate this we use a small piece of yellow filter that we put on top of the image sensor of the camera by placing the filter inside the lens assembly. We also experimentally set the white balance setting of the camera to suite our needs. The camera that we use is from Arducam (SKU: B006603) which we interface with using Raspberry Pi 4. Third, this new design aids digging by fluidizing granular media using vibrations. For this we mount a high speed micro vibration motor (6–12 V, 18000 rpm) on to the Digger Finger. Fourth, we replace the silicone gel used in the previous GelSight sensors with a 3 mm wide and 1.5 mm thick double sided, transparent polyurethane tape (ZSHK Happy Cover HC06). We also keep two things from previous GelSight designs. First, we use a mirror as did [25] to have a camera view that is perpendicular to the gel sensing surface, rather than looking at the surface at an angle. This allows us to have a slim device without worrying about self-occlusions of the sensing surface when contact is made. Second, we use the concept of light piping illustrated in [26] to have light rays pass across a curved surface with negligible loss in light intensity.

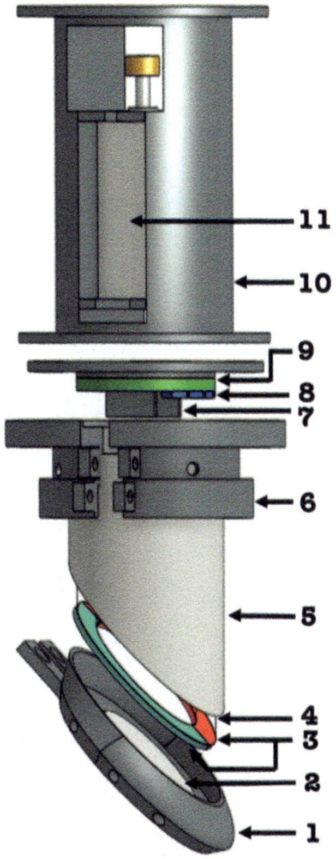

Fig. 2. Exploded view of the Digger Finger. Numbered arrows are 1. Bottom housing, 2. Mirror 3, Fluorescent paint, 4. Clear acrylic tube, 5. Gel, 6. Top housing, 7. Camera housing, 8. Blue LEDs, 9. PCB, 10. Vibrator housing, 11. Vibrator motor

2.2 Manufacturing Process

The manufacturing process of the Digger Finger consists of several simple rapid prototyping techniques including 3D printing, laser cutting and spray painting, to name a few. The Digger Finger consists of a sensing module and a vibration module. The Digger Finger's sensing module's core is shaped like a cylindrical wedge. We use a piece of optically clear acrylic tube with inner diameter of 16 mm and outer diameter of 22 mm (see Fig. 2). The tube is first diagonally cut using a band saw. The top and bottom surfaces of the cut tube are then sanded and polished until optically clear. The bottom surface of the tube is then painted with red and green fluorescent paint. Afterwards, we design and 3D print two custom housing (top and bottom) to house the necessary components of the Digger Finger (cameras, LEDs, mirror, gel boundaries). A thin piece of mirror is cut in shape of an ellipse using a laser cutter and attached to the center of the bottom housing using VHB double sided tape. Blue colored LEDs are soldered onto a custom designed printed circuit board (PCB) that is milled from a copper plate. This PCB is friction fit to the top housing along with the camera. The polyurethane double-sided tape that we use as the gel is brushed with aluminum flake powder, and then over coated with a thin layer of thermoplastic polyurethane (TPU) dissolved in a solvent. The gel is left to rest until the coating is dry, after which the uncoated side is applied across the clear acrylic tube. With the fluorescent paint and gel on the acrylic tube, we then assemble the top and bottom housing with the acrylic tube. This finishes the complete assembly of the sensing module of the Digger Finger. The vibration module is a 3D printed housing for the vibrator motor which is also cylindrical in shape. This module gets attached on top of the sensing module using screws and nuts.

3 Experiments and Results

3.1 Fluidizing Granular Media

The goal of this experiment is to study the effect of mechanical vibrations on the jamming of granular media. An immersed intruder moving in a granular media will experience strong forces because of the particle jamming effect. These forces are a function of several geometrical and material properties of the intruder and the granular media particles. It is well studied that the force chains formed among the granular particles during jamming can be weakened and the granular media fluidized by blowing air [27,28] or by mechanical vibrations [29,30]. Following [30] we use a vibrator motor to study the relation between the force acting on the Digger Finger and its penetration depth in granular media, in the presence and absence of mechanical vibrations.

The experimental setup as shown in Fig. 3 (left) consists of a robot arm (UR5) from Universal Robots Inc. The Digger Finger is coupled to the robot wrist using a 3D printed part. We use two types of granular media in our experiments, sand and rice. We choose these two particular media because of two very distinct effects they have on the object identification task, explained in the next section.

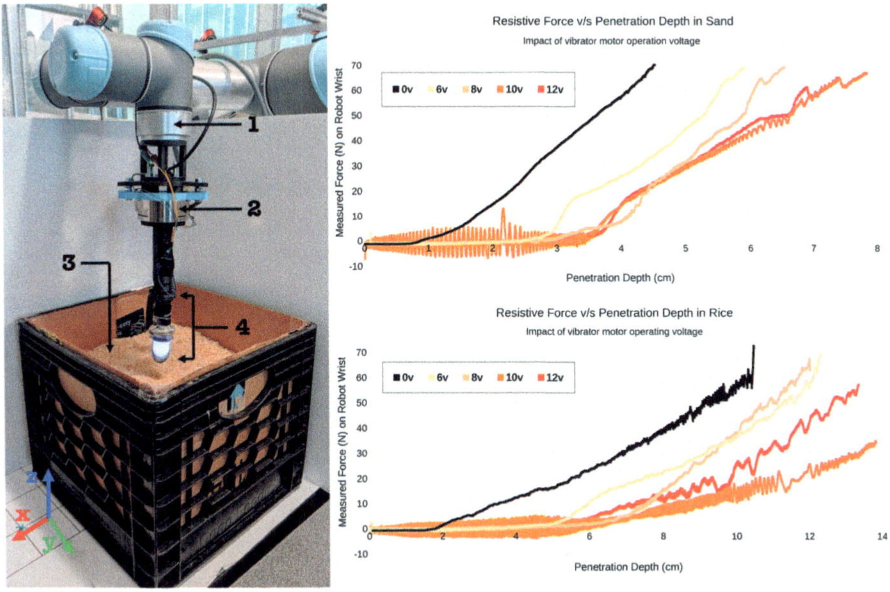

Fig. 3. *Left:* Experimental setup. Arrows show 1. UR5 robot arm, 2. F/T sensor, 3. Granular media (rice), 4. Digger Finger *Right:* As vibrator motor operating voltage increases, the force (measured by F/T sensor) required to move a given vertical distance in the granular media decreases.

For the sake of this experiment it is more important to note that both these media have bulk densities equal to 1578.56 kg/m^3 and 941.48 kg/m^3 respectively, resulting in different force versus distance relationships. Both the granular media are placed in a plastic container (12 × 12 × 11 in.) up to 6 in. deep. We choose these container dimensions to reduce the *edge effect* during penetration [31]. We attach a 6-axis force torque sensor (ATI Gamma) at the wrist of the robot arm to measure the ground truth forces acting on the Digger Finger during penetration. For each experimental trial the robot arm is first positioned such that the tip of the Digger Finger touches the surface of the granular media. The robot arm is then commanded to vertically penetrate the granular media at a speed of 2 mm/s. We keep the velocity of penetration fixed for all trials. The robot arm is moved until it reaches the safety limit of its joints and stalls. We experimentally find that this distance, both in the case of sand and rice, is less than the depth of the container. In the first trial the vibrator motor is off. For each of the next trials we increase the operating voltage of the vibrator motor by 2 V, starting from 6 V and going up to 12 V, thereby increasing the vibrating frequency and amplitude of oscillation of the motor. Prior to each trial, the state of the granular media is reset by vigorously shaking the container and shoveling the granular media with a tool. For each trial the 3D position of the robot wrist and force from the ATI Gamma sensor are synchronously recorded. The force in z-axis is first smoothed using an exponential filter ($\alpha = 0.1$) and is then plotted against the z position vector of the robot arm for each trial as shown in Fig. 3 (right). It is evident from the figure that when the vibrator motor is off i.e. at 0 V, the force on the Digger Finger begins to increase earlier in contrast to when the vibrations are on. This early rise in the force is approximately 2 cm in the sand case and 4 cm in the rice case. Moreover, in the cases when the motor is on, all the curves are slightly less steep resulting in almost double the penetration at stall as compared to the case when the motor is off.

We quantify the vibrations from the vibrator motor at 6 V, 8 V, 10 V & 12 V voltages in terms of frequency and acceleration amplitude using a 3-axis accelerometer. As the motor is rated to run at 18000 rpm at 12 V, theoretically the motor cannot vibrate more than 300 Hz. So we use an accelerometer present in LSM9DS1 sensor chip[1] that we can sample at a maximum frequency of 500 Hz. The board is attached to the tip of Digger Finger using a double sided tape. The vibrator motor is run at the above specified four operating voltages and data from the accelerometer is recorded for five seconds. The fundamental frequency of the vibrations is found by running fast Fourier transform on the data. The average of the fundamental frequency over two trials for the four operating voltage is shown in Table 1. The drop in the frequency and acceleration amplitude going from 10 V to 12 V is attributed to the resonance in the mechanical vibrations that we observe at 10 V. Due to this resonance the Digger Finger vibrates with the largest amplitude at 10 V. This is clearly visible in Fig. 3. While all the others curves are smoothed out with the same smoothing constant (i.e. $\alpha = 0.1$), the curve at 10 V still shows periodic oscillation at certain sections of the curve, both for sand and rice, indicating the need for an $\alpha < 0.1$ to obtain a smooth curve like the rest. We further verify the vibration frequency using the accelerometer on a smart phone using two Android software applications. One (iDynamics[2]) analyzes mechanical vibration and the other (Spectroid[3]) audio. We find that the fundamental frequency values for the four operating voltages range between 160–210 Hz. The above observations clearly indicate that the Digger Finger is able to fluidize densely packed granular media with vibrations of approximately 150–200 Hz and 10–24 m/s^2 acceleration amplitude, resulting in deeper penetration.

Table 1. Vibrating frequency and acceleration amplitude of the Digger Finger tip as a function of vibrator motor voltage.

Motor operating voltage (V)	Frequency (Hz)	Acceleration amplitude (m/s^2)
6	156	9.6
8	189	19.8
10	213	23.6
12	172	14.7

[1] https://www.sparkfun.com/products/13944.
[2] https://www.bauing.uni-kl.de/en/sdt/idynamics/.
[3] https://download.cnet.com/Spectroid/3000-20432_4-77833231.html.

3.2 Object Identification

This experiment aims to study the performance of the Digger Finger in identifying objects in different granular media. For simplicity we narrow down the problem to classify four simple shapes i.e. triangle, square, hexagon and circle as shown in Fig. 4, in two different types of granular media i.e. sand and rice. We choose rice and sand because both media have exhibit very different behavior in the way their grains interact with the Digger Finger and the object shapes we are interested in identifying. Through ad hoc experiments with different granular media (washed sand, chia seeds, lentils, and mung beans) we observed that there are three distinct phenomenona that occur when a Digger Finger immersed in granular media approaches a stationary object, depending upon the geometry and material properties of the grains as well as the Digger Finger and the object. The first occurs when very small grains such sand get permanently stuck between the Digger Finger and the object. The second occurs with grains (in our case rice) that are similar in size to the buried object, in which case the grains can completely block the object from coming in contact with the Digger Finger. The only way to come in contact with the object is to vibrate or twist the Digger Finger to push the grains to the side (refer to supplementary video[4]). We found twisting to be the most efficient way of pushing rice grains to the sides.

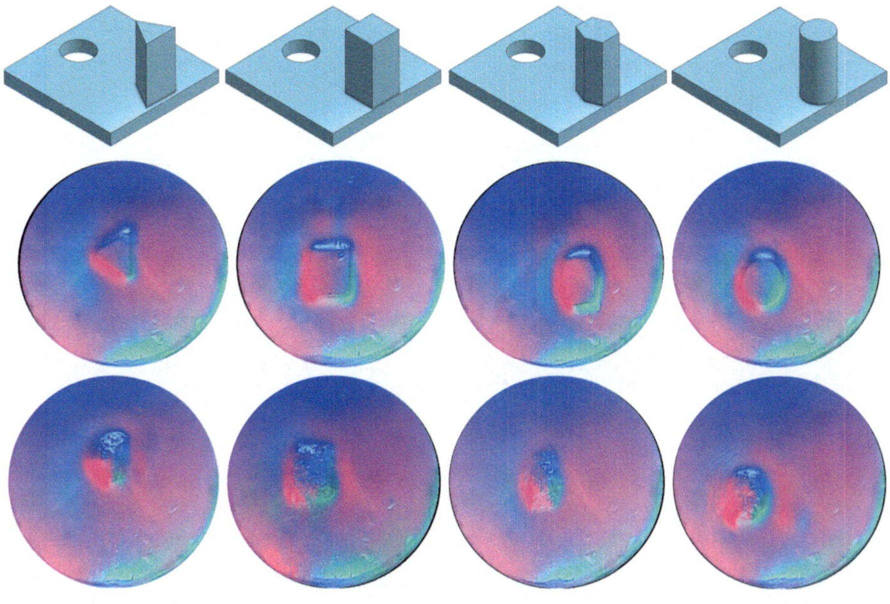

Fig. 4. *Top row:* Design of buried 3D printed objects. *Middle row:* Digger Finger image showing imprint of object shapes in absence of any granular media. *Bottom row* Imprints of shapes in the presence of granular media (sand).

[4] https://sites.google.com/view/diggerfinger.

Third case is when the material properties of the grains e.g. mung beans and lentils are such that they become slippery and do not get stuck between the Digger Finger and object. Based on this ad hoc experiment we narrow down the choice of granular medias to sand and rice. Figure 4 shows the different 3D printed objects that we make use of to collect data for this experiment and their corresponding imprints on the Digger Finger as seen from the camera in the form of RGB images.

For data collection we manually press the 3D printed objects on the Digger Finger and collect around 3000 images for each object shape. We repeat this procedure in a container filled with sand to collect images for cases when the sand grains get stuck between the Digger Finger. As shown in Fig. 4, the sand grains distort the boundaries of the object shapes, potentially making them ambiguous. These two rounds of data collection result in a total of eight classes: four classes with just the object shapes and four with sand obscuring the objects. We add a ninth class for the zero contact case. The dataset for the first eight classes are cleaned by manually removing images where the object is making no or partial contact with the Digger Finger. This data cleaning process leaves us with around 1500 images for each of the first eight classes.

For classification we rely on convolution neural networks because of their growing popularity for image analysis. In particular we train a residual neural network [32] (ResNet50) on our data set by performing a popular technique in machine learning called transfer learning. Transfer learning focuses on storing knowledge gained while solving one problem and applying it to a different but related problem. For example, ResNet50 network is often first trained to classify real world images from ImageNet[5] data set. We use this pre-trained network as the starting point and re-train it on our own object shape data set. We do this by only training the last convolutional block of the network and an uninitialized fully connected layers (classifier block) of size 128 at the end of the network. We split the 1500 images of each class into training (1200), validation (200) and testing (100) data sets. We augment the training data by randomly cropping and rotating the images. We also add Gaussian noise to each color channel. The network is trained for 10 epochs with a learning rate scheduler

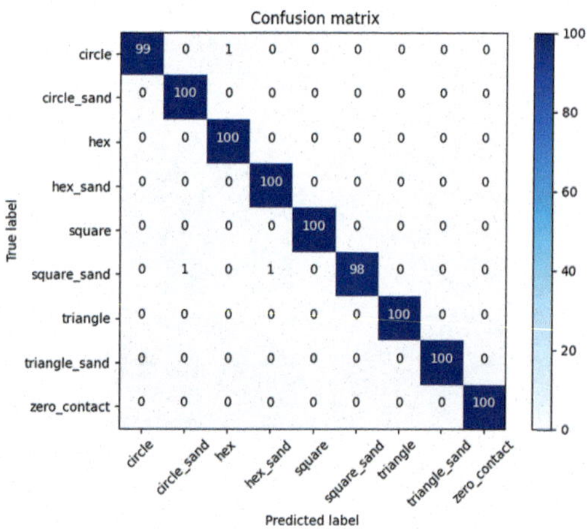

Fig. 5. Confusion matrix for object identification task

[5] http://www.image-net.org/.

and batch size of 64, at the end of which it attains 99% training 98% validation accuracy. The confusion matrix calculated on the test images is shown in Fig. 5.

4 Conclusion and Future Works

Identifying objects buried in granular media using tactile sensors is a challenging task. In our experiments we identify several difficulties. First, it is difficult to actually reach the object because the granular media will start to jam and prevent downward movement. Second, even when the tactile sensor can reach the object, the granular media particles tend to get stuck between the sensor and object, distorting the actual shape of the object. To tackle these challenges we present a novel tactile sensor that we call Digger Finger. We build on previous GelSight tactile sensor designs and introduce several innovations, including the use of red and green fluorescent paint and polyurethane tape as the gel, to design a compact wedge-shaped sensor. We design the Digger Finger to fluidize granular media during penetration using mechanical vibrations. We use the high resolution tactile sensing provided by the Digger Finger to successfully identify different object shapes even when distorted by granular media particles.

For fluidizing granular media, we have only presented results for vertical penetration, but moving horizontally in granular media is also increasingly challenging at greater depths. Further experiments are required to understand the nature of vibration and robot arm motion needed to fluidize granular media while moving horizontally. For object identification a major issue is that over time the paint on the gel incurs wear causing noticeable artifacts in the RGB sensor image data. The tactile data after wear looks different than the data used to train the neural network for object identification. This causes the network to make false positive predictions, in particular for the zero contact class. The network may also make false predictions when the granular media touches the Digger Finger during free motion or when the gel wrinkles during penetration. Ideally, the network should be able predict the type of the granular media and also robustly predict zero contact. We therefore plan to train the network on additional classes of granular media and also experiment with better data augmentation techniques to make the network more robust to gel artifacts. We also plan to calibrate the Digger Finger in order to construct 3D geometry data from the raw RGB image data. In particular this would let us use algorithms trained on simulated 3D geometry data, as opposed to trying to simulate raw RGB sensor image which requires modelling the exact illumination of the Digger Finger. Using 3D geometry data would also allow object detection algorithms to better transfer to real world tactile data from multiple Digger Finger units.

We believe that such a tactile sensor paves the way for the robotic manipulation within and of granular media, whether for everyday tasks such as scooping rice or litter, or for more industrial applications such as finding and inspecting buried cables and other objects.

Acknowledgements. This research was supported by the Toyota Research Institute, the Office of Naval Research (ONR) [N00014-18-1-2815], and the GentleMAN project of the Norwegian Research Council. We would like to thank Achu Wilson, Shaoxiong Wang, Sandra Liu, Yu She, Filipe Veiga, and Megha Tippur for insightful discussions. We also thank Siyuan Dong, Daolin Ma and Alberto Rodriguez for lending us the force-torque sensor.

References

1. Nadeau, P., Abbott, M., Melville, D., Stuart, H.S.: IEEE International Conference on Robotics and Automation (2020)
2. Burtness, A.J.: Evaluation of haptic feedback methods for teleoperated explosive ordnance disposal robots. Master's thesis, Johns Hopkins University (2011)
3. T4^{TM} Robotic System (2019). https://www.harris.com/solution/t4-robotic-system. Library Catalog: www.harris.com
4. Johnson, M.K., Adelson, E.H.: 2009 IEEE Conference on Computer Vision and Pattern Recognition, pp. 1070–1077 (2009)
5. Johnson, M.K., Cole, F., Raj, A., Adelson, E.H.: ACM Transactions on Graphics (TOG) **30**(4), 1 (2011)
6. Sarata, S., Osumi, H., Kawai, Y., Tomita, F.: IEEE International Conference on Robotics and Automation, 2004. In: Proceedings. ICRA 2004, vol. 4, pp. 3488–3493. IEEE (2004)
7. Li, C., Zhang, T., Goldman, D.I.: Science **339**(6126), 1408 (2013)
8. Hauser, S., Eckert, P., Tuleu, A., Ijspeert, A.: 2016 6th IEEE International Conference on Biomedical Robotics and Biomechatronics (BioRob) (2016)
9. Brown, E., Rodenberg, N., Amend, J., Mozeika, A., Steltz, E., Zakin, M.R., Lipson, H., Jaeger, H.M.: Proceedings of the National Academy of Sciences, vol. 107(44), p. 18809 (2010)
10. Cheng, N.G., Lobovsky, M.B., Keating, S.J., Setapen, A.M., Gero, K.I., Hosoi, A.E., Iagnemma, K.D.: 2012 IEEE International Conference on Robotics and Automation, pp. 4328–4333. IEEE (2012)
11. Brown, J.P., Farkhatdinov, I.: 2020 IEEE Haptics Symposium (HAPTICS) (2020)
12. Stanley, A.A., Gwilliam, J.C., Okamura, A.M.: 2013 World Haptics Conference (WHC), pp. 25–30. IEEE (2013)
13. Yamaguchi, A., Atkeson, C.G., Ogasawara, T.: Int. J. Humanoid Rob. **12**(03), 1550030 (2015)
14. Matl, C., Narang, Y., Bajcsy, R., Ramos, F., Fox, D.: arXiv preprint arXiv:2003.08032 (2020)
15. Schenck, C., Tompson, J., Fox, D., Levine, S.: arXiv preprint arXiv:1709.02833 (2017)
16. Clarke, S.: Robot learning for manipulation of granular materials using vision and sound. Master's thesis, Carnegie Mellon University Pittsburgh (2019)
17. Clarke, S., Rhodes, T., Atkeson, C.G., Kroemer, O.: CoRL, pp. 529–550 (2018)
18. Travassos, X.L., Avila, S.L., Ida, N.: Applied Computing and Informatics (2020)
19. Martin, J.S., Fenneman, D.J., Codron, F.T., Rogers, P.H., Scott, W.R., Larson, G.D., McCall II, G.S.: Detection and Remediation Technologies for Mines and Minelike Targets VII, vol. 4742 (International Society for Optics and Photonics, vol. 4742, pp. 606–616 (2002)
20. Jia, S., Santos, V.J.: Proceedings of the RSS Workshop Tactile Sensing for Manipulation: Hardware, Modeling, and Learning, Boston, MA, USA, vol. 15 (2017)

21. Syrymova, T., Massalim, Y., Khassanov, Y., Kappassov, Z.: 2020 IEEE/ASME International Conference on Advanced Intelligent Mechatronics (AIM), pp. 175–180. IEEE (2020)
22. Abad, A.C., Ranasinghe, A.: IEEE Sens. J. (2020)
23. Shimonomura, K.: Sensors **19**(18), 3933 (2019)
24. Yuan, W., Dong, S., Adelson, E.H.: Sensors (2017)
25. Donlon, E., Dong, S., Liu, M., Li, J., Adelson, E., Rodríguez, A.: 2018 IEEE/RSJ International Conference on Intelligent Robots and Systems (IROS), pp. 1927–1934 (2018)
26. Romero, B., Veiga, F., Adelson, E.: ArXiv abs/2005.09068 (2020)
27. Brzinski III, T.A., Durian, D.J.: Soft Matter **6**(13), 3038 (2010)
28. Naclerio, N.D., Hubicki, C.M., Aydin, Y.O., Goldman, D.I., Hawkes, E.W.: 2018 IEEE/RSJ International Conference on Intelligent Robots and Systems (IROS), pp. 5918–5923. IEEE (2018)
29. Firstbrook, D., Worrall, K., Timoney, R., Suñol, F., Gao, Y., Harkness, P.: Proceedings of the royal society a: mathematical. Phys. Eng. Sci. **473**(2198), 20160673 (2017)
30. Texier, B.D., Ibarra, A., Melo, F.: PLoS ONE **12**, 4 (2017)
31. Seguin, A., Bertho, Y., Gondret, P.: Phys. Rev. E **78**(1), 010301 (2008)
32. He, K., Zhang, X., Ren, S., Sun, J.: Proceedings of the IEEE Conference on Computer Vision and Pattern Recognition, pp. 770–778 (2016)

Toward High Power-to-Weight Ratio Electro-hydrostatic Actuators for Robots

Takahiro Nakanishi(✉), Mitsuo Komagata, Ko Yamamoto, and Yoshihiko Nakamura

University of Tokyo, 7-3-1 Hongo, Bunkyo-ku, Tokyo 113-8656, Japan
{nakanishi,komagata,yamamoto.ko,nakamura}@ynl.t.u-tokyo.ac.jp

Abstract. Electro-hydrostatic actuators (EHAs) exhibit excellent backdrivability and therefore mechanical robustness. However, it has been difficult to design an EHA which satisfies the requirements on the same level of power-to-weight ratio as widely used actuators with gear reducers. To address this challenge, we developed a revolute EHA module realizing a high power-to-weight with a frameless motor and integrated modular design. Its output power was preliminary estimated by formulating its losses prior to the future experiment. In addition, its backdrive performance was experimentally evaluated to demonstrate the superiority of our EHA over gear reducers. The design method, achieving a high power-to-weight ratio and high backdrivability, offers the possibility of the EHA as an alternative to existing robotic actuators.

1 Introduction

In spite of their notable advantages, electro-hydrostatic actuators (EHAs) [1] have problems with their weight and size. The hydraulic circuit of a revolute EHA is generally characterized by a one-to-one connection of a pump and a vane motor and does not have any valves connected in series with them. This means that one joint always needs one set of a pump, a vane motor and auxiliary components, resulting its weight increase. This is the main shortcome of an EHA.

The gear reducers such as *Harmonic Drives* [2] with high reduction ratio, have been used in many robots, and they seemingly meet the demands of compactness and precision. Nevertheless improving weight and size of EHAs would be beneficial because EHAs realize higher backdrivability than gear reducers. These are preferable characteristics of a transmission, especially for robots that require sensitive force control to cooperate with humans. In order to grant these advantages to robots, small-sized high power EHAs have been developed. An example is the EHA-driven robotic manipulator Hydracer [3], which consists of the involute gear pump [4], the linear cylinder [5] and the single vane motor [3]. In spite of the remarkable improvements, the power-to-weight ratio (PWR) of robotic EHAs is not yet as high as that of the actuators with gear reducers like *Harmonic Drives*. In fact, while the target output power of the revolute EHA on Hydracer [3] is 98 W, the total mass of the electric motor, the pump, the vane

B. Siciliano et al. (Eds.): ISER 2020, SPAR 19, pp. 116–125, 2021.
https://doi.org/10.1007/978-3-030-71151-1_11

motor, the pipeline, etc. is approximately 2.3 kg, namely 43 W/kg of PWR. It is still inferior to other reported revolute actuators with gear reducers as shown in Table 5.

The goal of this study is to design an electro-hydrostatic revolute actuator whose PWR is comparable to that of typical gear-reduced actuators. Robotic revolute actuators that combine frameless motors and gear reducers have been developed [6–10]. In this paper, we also use a frameless motor and use its design freedom to integrate with a hydraulic pump and a vane motor as well as the other components. In particular, we adopt a flat-type frameless motor with inner hollow space because (1) we can use the space for housing a pump, and (2) the specification of such motors tends high-torque and low-speed, which is suitable for the input of hydraulic pumps, as explained in the next section.

2 Specifications and the Flow of Design for High PWR EHAs

In the case of revolute joints of Hydracer [3] the unit of an electric motor and a pump is connected by a flexible tube to the single vane motor as shown in Fig. 1(a). The separate design has an advantage that the placement of the unit and the vane motor is the matter of the designer's choice. On the other hand, the translational joints of Hydracer [3] adopt the module design which means that an electric motor, a pump and a hydraulic cylinder make a single unit, which leads to another kind of advantage that it reduces the complexity and improves the reliability. It is also noteworthy that the module design tends to have higher PWR since the two housings are now one. We paid a careful attention to the fact and further explored the advantage of module design.

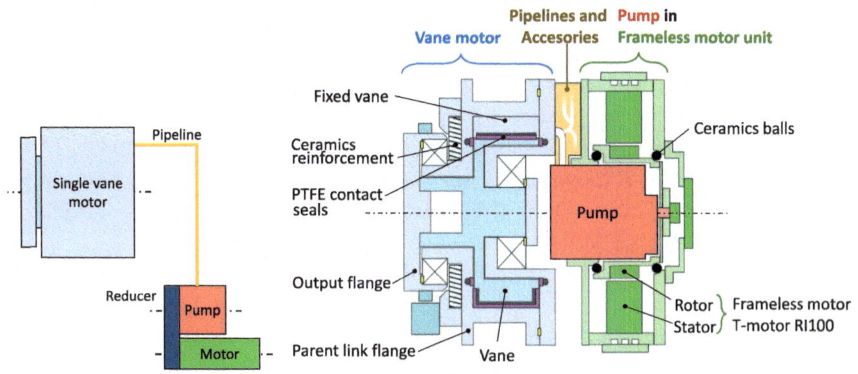

(a) Conventional EHA [3]. (b) Simplified section of the proposed EHA.

Fig. 1. Comparison of the configurations between the conventional EHA on Hydracer [3] and the proposed EHA module. The proposed design method aligns the frameless motor, the pump and the vane motor compactly and coaxially.

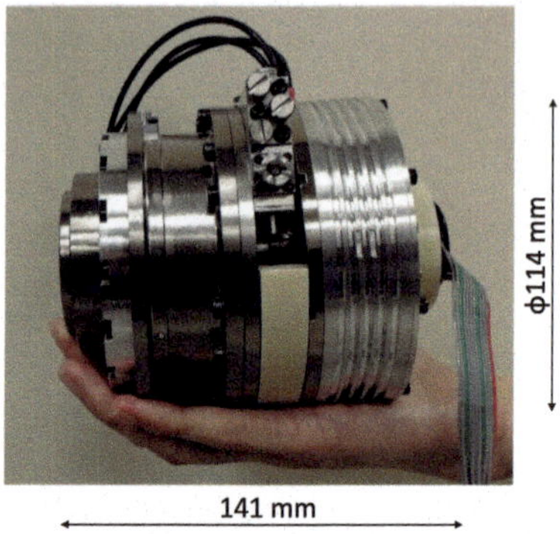

Fig. 2. Developed revolute EHA module with high power-to-weight ratio. The weight is approximately 2.8 kg and the maximum output is about 300 W.

Figure 1(b) shows the proposed design of the EHA. Using a flat-type frameless motor and setting a pump in the hollow space, we can pursue short axial-length and small-volume. The prototype shown in Fig. 2 demonstrated the possibility of manufacturing a palm-sized module containing all the component of a revolute EHA.

Komagata [4] investigated the power efficiency of EHAs and pointed out that the maximum angular velocity of pump should stay low. This is why a mechanical gear train was adopted between the motor and the pump in Fig. 1(a) since the rated angular velocity of high-power motors like maxon EC series goes beyond 10000 rpm. However Wensing et al. [11] presented that using a motor with a large gap radius r_g between the rotor and the stator leads to an optimal actuator design. Given a constant mass budget and output torque requirement, they showed the relationship between required reduction ratio N and motor gap radius r_g, $N \propto 1/r_g$. Therefore we should select a motor with a large r_g to reduce N, and such design achieves a less gear mass, less friction and higher torque density with no tradeoffs. Based on this idea, the pump on our proposed EHA was designed to be powered by a flat-type frameless motor with a larger diameter and shorter axial length than the motors on a conventional EHAs [3]. Accordingly, we can remove the gear train and directly connect the rotor of the electric motor to the input shaft of the pump (see Fig. 1(b)). This is also advantageous for high PWR design.

As described in Fig. 1(b), The frameless motor directly drives a pump, and houses the pump in the hollow space. Because of the low radial load realized by this direct drive method, the motor bearings, which are usually made of steel bearing, are made of aluminum alloy and ceramic balls, reducing the weight and axial-length. Traditional ball bearings are made of bearing steel (e.g. SUJ2) which has high hardness and wear resistance, whereas aluminum alloys generally do not have those qualities and can be expected to have an insufficient life. Therefore we plated aluminum bearing race with electroless nickel-PTFE plating.

The surface treatment was selected based on the following criteria: (1) low friction, (2) high hardness, and (3) flaking resistance. Table 1 shows the comparison of hardness of bearing materials and surface treatments. The hydraulic pump by Komagata [4] has thrust bearings with high-hardness ceramic balls and diamond-like carbon (DLC) coated steel gears. Actually flaking of DLC coating was occured in the pump after long hours of driving. The reason is speculated that the hardness of DLC coating and gear material (NAK55) was very different (see Table 1) and therefore the contact stress from the ceramic balls deformed the surface of the steel gears greater than the coating, making the coating easy to peel off. Taking advantage of this know-how, we selected A7075 (HV ~170) for bearing races and electroless nickel-PTFE plating (HV ~750) so that difference in hardness of the base material and plating is less than that of the pump by Komagata [4].

Table 1. The hardness of bearing materials.

Bearing	Bearing race material	Hardness (HV)
Common steel bearings	SUJ2	~700
Thrust bearing between a pump gear and a casing by Komagata [4]	NAK55	~400
	DLC coating	~3000
Proposed light-weight and small-sized motor bearing	A7075	~170
	Electroless nickel-PTFE plating	~750

The vane motor has ball bearings made of silicon nitride Si_3N_4. The total mass of the two Si_3N_4 bearings is 111 g, which is approximately 150 g less than the weight of steel bearings of the same dimension.

A hydraulic equipment has a loss known as internal leakage, which means the hydraulic oil flow from the high pressure side to the low pressure side. In a high-pressure hydraulic system, the internal pressure deforms the casing, which increases internal gaps, resulting in increased internal leakage. One method to suppress this effect is to increase the rigidity of the casing by using ceramic reinforcements [3,4]. The proposed EHA also has ceramic reinforcements (see Fig. 1) to reduce the axial deformation of the pump and the vane motor casing. In addition, every sliding part of the vane motor has PTFE contact seals pushed by the elasticity of a compressed NBR rope or O-ring. This design of contact seals achieves low internal leakage and low friction at the same time.

3 Preliminary Calculation of the Output Power of Proposed EHA

We preliminarily analyzed the losses in the proposed EHA module prior to the experimental evaluation of its output power. The losses in an EHA are generally divided into internal leakage loss, viscous friction loss and Coulomb friction loss. Table 2 summarizes the parameters of a revolute EHA.

Table 2. Parameters of a revolute EHA

Symbol	Description	Unit
τ_p	Input torque of a pump	Nm
τ_v	Output torque of a vane motor	Nm
$\dot{\theta}_p$	Rotational speed of a pump	rad/s
$\dot{\theta}_v$	Rotational speed of a vane motor	rad/s
d_p	Discharge of a pump	m^3/rad
d_v	Discharge of a vane motor	m^3/rad
k_p	Internal leakage resistance of a pump	Pa·s/m^3
k_v	Internal leakage resistance of a vane motor	Pa·s/m^3
Δp	Differential pressure	Pa
p	Internal pressure	Pa
$\tau_{p,\mathrm{viscous}}(\dot{\theta}_p)$	Viscous friction torque at a pump	Nm
$\tau_{p,\mathrm{coulomb}}(\Delta p)$	Coulomb friction torque at a pump	Nm
$\tau_{v,\mathrm{coulomb}}$	Coulomb friction torque at a vane motor	Nm

Table 3. Loss factors of a revolute EHA. In the table, p denotes the internal pressure, and Δp denotes the differential pressure between high pressure side and low pressure side.

Component	Taken in consideration	Not taken in consideration
Pump	1. Internal leakage through axial gaps between a gear and a casing 2. Viscous friction at axial gaps between a gear and a casing 3. Coulomb friction between a gear and a casing 4. Coulomb friction due to oil seals (depending on p)	1. Increase of internal leakage flow rate due to the high internal pressure p
Vane motor	1. Internal leakage through all gaps (Leakage reduction effect due to PTFE contact seals) 2. Coulomb friction due to PTFE contact seals 3. Coulomb friction due to oil seals (depending on p)	1. Increase of internal leakage flow rate due to the high internal pressure p 2. Increase of coulomb friction loss depending on the differential pressure Δp

The relationships between the input $\tau_p, \dot{\theta}_p$ and the output $\tau_v, \dot{\theta}_v$ are represented by the following formula. Equation (1) shows that the input torque of the EHA is the sum of the torque due to the differential pressure and the Coulomb friction torque of the pump. Equation (2) states that the flow rate produced by the pump is equal to the sum of the flow rate corresponding to the rotational displacement of the vane motor and the internal leakage flow rate. Equation (3) shows that the output torque of the EHA is the torque due to the differential pressure minus the friction torque at the vane motor.

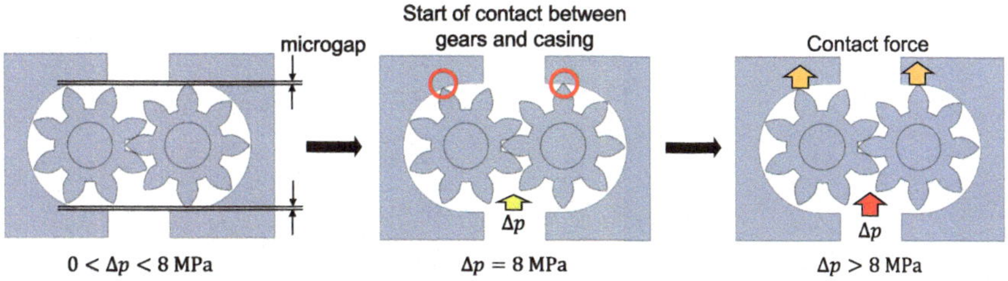

Fig. 3. Pump gears get eccentric by the differential pressure Δp. We assumed the contact force between the gears and the casing appears when $\Delta p > 8$ MPa.

$$\tau_p = d_p \Delta p + \tau_{p,\text{viscous}}(\dot{\theta}_p) + \tau_{p,\text{coulomb}}(\Delta p) \tag{1}$$

$$d_p \dot{\theta}_p = d_v \dot{\theta}_v + \left(\frac{1}{k_p} + \frac{1}{k_v} \right) \Delta p \tag{2}$$

$$\tau_v = d_v \Delta p - \tau_{v,\text{coulomb}} \tag{3}$$

Internal leakage loss can be estimated by fluid dynamics. Komagata et al. [4] assumed the plane Poiseuille flow (described in Fig. 4) at the internal gap in a gear pump to estimate the internal leakage resistance k_p. By extending the method, we estimated the internal leakage resistance of our vane motor k_v as well as that of our pump k_p in Eq. (2). All the gap sizes (x in Fig. 4) were determined based on the median value of machining tolerance, and the internal leakage resistances $k_\bullet$ are obtained by the equation:

$$1/k_\bullet = \int_0^w \frac{x^3 \mathrm{d}w}{12\mu l(w)}. \tag{4}$$

Viscous friction loss is also determined by fluid dynamics. Komagata et al. formulated the viscous friction loss by considering the shear of hydraulic oil at the internal gap in a gear pump. This method enabled us to estimate the viscous friction torque at our pump $\tau_{p,\text{viscous}}(\dot{\theta}_p)$ in Eq. (1).

Coulomb friction loss was estimated by assuming the internal pressure p, differential pressure Δp and friction coefficients. As Fig. 3 shows, the pump gears get eccentric by the differential pressure, and the contact force with the casing is not negligible. We estimated the friction loss by this phenomenon using the relationship obtained by FEM between the differential pressure Δp and the amount of eccentricity, and also by the assumption of no contact between the gear and the casing under the condition of $\Delta p < 8$ MPa. In addition, contact forces between rotating parts and seals were taken in consideration as shown in Table 3.

Considering above-mentioned loss factors, the output power and efficiency were calculated prior to the experiments. Figure 5 shows the estimated (a) output power, (b) required input torque and (c) required input speed. The horizontal and vertical axes indicate the output speed and the output torque, respectively. Figure 5(a) indicates that the maximum output torque is about 120 Nm, and it

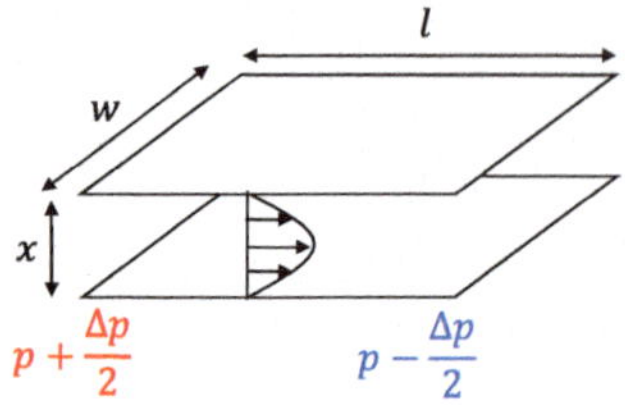

Fig. 4. Plane Poiseuille flow. The flow rate is obtained by $q = \Delta p w x^3 / 12\,\mu l$, where μ denotes the kinematic viscosity.

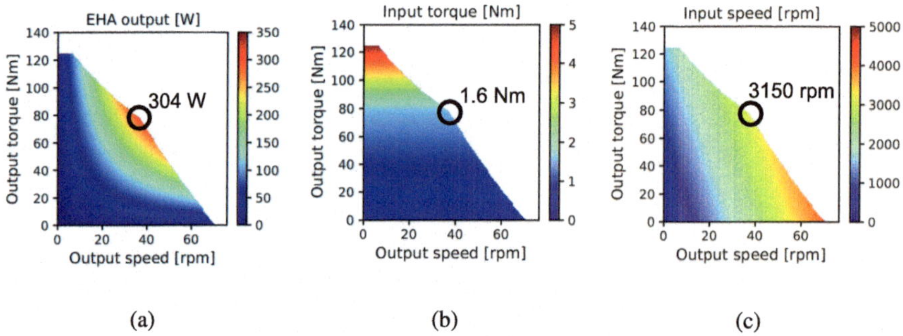

Fig. 5. The preliminary calculation of (a) output power, (b) required input torque and (c) required input speed of the proposed EHA. The horizontal and vertical axes indicate the output speed and the output torque, respectively. This result estimates that the maximum output power of 300 W is obtained when the output speed and the output torque are 36 rpm and 80 Nm, respectively. In this condition, input torque of about 1.6 Nm and input speed of 3000 rpm are estimated to be required, respectively. Discontinuous slope change seen in the graph (b) where the output torque is 80 Nm shows the effect of the gear contact within the pump described in Fig. 3.

decreases as the output speed increases. The maximum output power is estimated to be 300 W at the output torque of 80 Nm and the output speed of 36 rpm, where required input torque (b) is 1.6 Nm and required input speed (c) is 3150 rpm.

This results of the calculation are not very accurate because some factors are not taken into account as Table 3 shows, and because the internal leakage loss (Eq. 4) is sensitive to the size x of the microgaps. However, this estimation helps us to find the operating condition that produces the maximum output power in the future experiment.

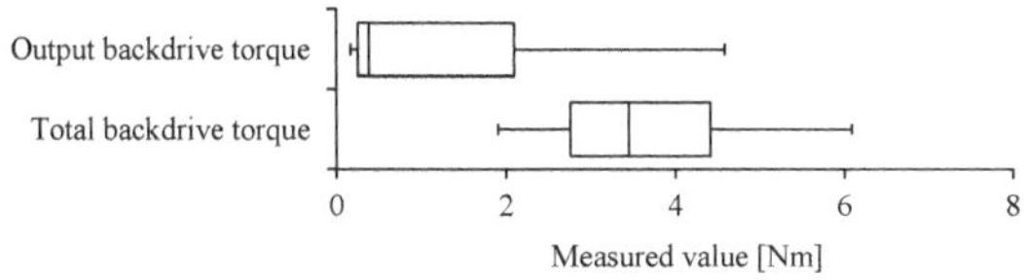

Fig. 6. Distributions of experimentally measured output backdrive torque and total backdrive torque. The maximum value of total backdrive torque is 6.0 Nm.

4 Experimental Evaluation of the Level of Backdrivability

Since backdrivability is one of the performance indicators of a transmission itself, the frameless motor is disconnected from the EHA, making the input shaft of the pump free.

There are two types of backdrive in an EHA: output backdrive and total backdrive [12]. If the output torque of the vane motor is τ_m, the angular velocity of the vane motor is τ_m, and the angular velocity of the pump is $\dot{\theta}_p$, backdrive conditions are defined as follows:

1. Output backdrive: $\mathrm{sgn}(\tau_m \cdot \dot{\theta}_m) < 0$,
2. Total backdrive: $\mathrm{sgn}(\tau_m \cdot \dot{\theta}_p) < 0$,

where $\mathrm{sgn}(\bullet)$ is the signum function.

In particular, total backdrivability is the ease that the torque applied to the vane motor (output side) rotates the pump (input side). The higher total backdrivability means the higher mechanical robustness of the EHA.

We measured output backdrive torque and total backdrive torque of the developed EHA, and compared it to gear reducers made by Harmonic Drive Systems. The backdrivability indicator of a Harmonic Drive is the overdrive starting torque, which is defined as the torque that the input side starts rotation when a torque is applied on the output side. By the definition, overdrive starting torque of a Harmonic Drive is equivalent to total backdrive torque of an EHA. For comparison with the EHA, a Harmonic Drive CSF-32-50, which has a similar size and torque capacity as Table 4 shows, was chosen.

Figure 6 shows the distribution of the experimental data of output backdrive torque and total backdrive torque of the developed EHA. The maximum value of the total backdrive torque was 6.0 Nm, which is still smaller than the overdrive starting torque of a Harmonic Drive CSF-32-50 [2], 16 Nm, as shown in Table 4. This result claims that the developed EHA has more than 2.7 times the backdrivability of the Harmonic Drive.

Table 4. Comparison of backdrive torque between our EHA and Harmonic Drive CSF-32-50. Our EHA realizes more than 2.7 times the backdrivability of Harmonic Drives.

	Our EHA	CSF-32-50
Outer diameter [mm]	114	110
Maximum permissible average load torque [Nm]	120	108
Backdrive torque [Nm]	< 6.0	16

Table 5. Comparison of the proposed EHA to the existing revolute actuators with gear reducers.

Model	PWR [W/kg]	Peak output [W]	Weight [kg]
Our EHA (estimated)	107	300	2.8
Revolute EHA of Hydracer [3]	43	98	2.3
ARMAR-6 [6] Large	110	243	2.2
WALK-MAN [7,8] A	1144	2288	2.0
CHIMP [9] (NGT-200)	95	493	5.2
WAREC-1 [10]	64	372	5.8

5 Conclusions

In Table 5, we compare the estimated output power and PWR of the proposed EHA to the existing actuators. Four gear-reduced actuators were selected based on the following criteria: (1) driven by a frameless motor, (2) high output power and (3) a revolute module. It should be noted that the output power of each actuator listed depends on not only its mechanical structure but also motor driving condition. For example, WALK-MAN [7,8] achieves very high PWR by employing a motor with the ingenious coil insulator that can be used in high temperature. Although a careful comparison is necessary, estimated PWR of the proposed EHA seems to be almost the same level as the listed actuators, except for those of WALK-MAN.

WALK-MAN actuators have excellent output power and PWR, but a Harmonic Drive CPL-25 (reduction ratio = 80) used in the Large-type actuator has the backdrive torque of 7.7 Nm [2]. On the other hand, the backdrive torque of our developed EHA is less than 6.0 Nm, so our EHA has an advantage in backdrivability. We can namely say that the proposed design method realizes high PWR comparable to many existing gear-reduced actuators, as well as higher backdrivability.

The potential applications of our proposed EHA can be demonstrated by existing actuators and their applications. For example, ARMAR-6 [6] was designed to perform various tasks in the industrial field with a target payload of 10 kg. The maximum power of its elbow joints is 283 W and its peak torque is 123 Nm. On the other hand, our proposed EHA is expected to achieve the maximum power of 300 W and maximum torque of approximately 120 Nm. These values are similar to those of ARMAR-6, and therefore, our revolute EHA module can be used for the tasks with almost the same payload.

We can design the output speed and torque with some design freedom depending on the application scenario. Namely, reducing the flow rate per revolution of the pump will increase the output torque, while reducing the flow rate per revolution of the vane motor will increase the output speed. Besides this fact, since the proposed design method can achieve a PWR as high as existing gear-reduced actuators, our EHA can be optimal in many applications.

Acknowledgements. This work was supported by Joint Research between Nachi-Fujikoshi Corporation and The University of Tokyo (2020–2022).

References

1. Kaminaga, H., et al.: Mechanism and control of whole-body electro-hydrostatic actuator driven humanoid robot Hydra. In: International Symposium on Experimental Robotics (2016)
2. Harmonic Drive Systems. https://www.hds.co.jp/. Accessed 28 Nov 2020
3. Suzuki, R., et al.: A 6-DOF electro-hydrostatically driven robot manipulator with a 3-DOF wrist complex consisting of three vane motors and manifolds. In: Robotics Symposia, p. 5C1 (2019)
4. Komagata, M., et al.: Design and development of compact ceramics reinforced pump with low internal leakage for electro-hydrostatic actuated robots. In: IFToMM World Congress on Mechanism and Machine Science, pp. 2439–2448. Springer (2019)
5. Komagata, M., et al.: 6-DOF manipulator with EHA drive module and energy efficiency analysis of a pump/module system. In: Robotics Symposia (2020)
6. Rader, S., et al.: Highly integrated sensor-actuator-controller units for modular robot design. In: IEEE International Conference on Advanced Intelligent Mechatronics (AIM), pp. 1160–1166. IEEE (2017)
7. Negrello, F., et al.: A modular compliant actuator for emerging high performance and fall-resilient humanoids. In: IEEE-RAS 15th International Conference on Humanoid Robots (Humanoids), pp. 414–420. IEEE (2015)
8. Tsagarakis, N.G., et al.: WALK-MAN: a high-performance humanoid platform for realistic environments. J. Field Robot. **34**(7), 1225–1259 (2017)
9. Stentz, A., et al.: CHIMP, the CMU highly intelligent mobile platform. J. Field Robot. **32**(2), 209–228 (2015)
10. Hashimoto, K., et al.: WAREC-1 - a four-limbed robot having high locomotion ability with versatility in locomotion styles. In: IEEE International Symposium on Safety, Security and Rescue Robotics (SSRR), pp. 172–178. IEEE (2017)
11. Wensing, P.M., et al.: Proprioceptive actuator design in the MIT Cheetah: impact mitigation and high-bandwidth physical interaction for dynamic legged robots. IEEE Trans. Robot. **33**(3), 509–522 (2017)
12. Kaminaga, H., et al.: Development of backdrivable hydraulic joint mechanism for knee joint of humanoid robots. In: IEEE International Conference on Robotics and Automation, pp. 1577–1582. IEEE (2009)

Ori-Vent: Design and Prototyping of Accessible and Portable Origami-Inspired Ventilators

Abudula Aihaitijiang, Shou-Shan Chiang(✉), Erik H. Skorina, and Cagdas D. Onal

Worcester Polytechnic Institute, 100 Institute Road, Worcester, MA 01609, USA
cdonal@wpi.edu
http://softrobotics.wpi.edu/

Abstract. The COVID-19 pandemic is responsible for hundreds of thousands of deaths worldwide and remains a challenge that humanity must still overcome. As hospitals continue to admit critical patients, they face medical equipment shortages that may hinder life-saving treatments. We seek to combat this equipment shortage by presenting a proof-of-concept portable automated ventilator that is simple, lightweight, and easy-to-fabricate based on an origami-inspired robotic approach. We discuss the design, fabrication, and initial testing of three such devices, two folded from a thin plastic film into a flexible bellows shape, and the other a 3D-printed version of the same design. We show that these initial designs satisfy most of the ventilator performance requirements for emergency portable automated use, with the 3D printed bellows being more flexible and capable of generating higher pressures. The next step is to expand our design and perform further testing to ensure our device is ready to save lives.

Keywords: Portable ventilator · Origami-inspired robotics · Pneumatics

1 Motivation, Problem Statement, Related Work

1.1 Motivation

There are more than 70 million confirmed cases of COVID-19 worldwide, and the death toll has surpassed 1.5 million [1]. As of Mid-December 2020, the US was reporting an additional hundred thousand cases per day, and hospitals remained in danger of being overwhelmed with patients who may require critical care and relatively long-term ventilation to stay alive. A spike in cases could lead to a dangerous shortage of ventilators, resulting in more deaths. It is clear that there is a growing need for inexpensive and easy-to-manufacture medical equipment to supplement hospitals' existing supplies and enable portable automated operation.

B. Siciliano et al. (Eds.): ISER 2020, SPAR 19, pp. 126–136, 2021.
https://doi.org/10.1007/978-3-030-71151-1_12

1.2 Problem Statement

The main requirements for our novel ventilator design are affordability, accessibility, and functionality. Our design must use inexpensive materials and simple manufacturing methods. It must not depend on any medical parts that may also suffer a shortage or take away from their primary uses, such as bag valve masks (BVMs). Additionally, the design must be small and portable, without depending on a hospital's infrastructure or large rigid machines. In order to design a functional ventilator, we aim to meet the following criterion [2]:

* Capable of functioning for at least 48 h
* Have a connection method to allow an adjustable O_2 source to be connected
* Capable of providing a respiratory rate of 12–40 breaths per minute (BPM)
* Capable of providing 5–24 cm-H_2O of positive end expiratory pressure (PEEP)
* Capable of delivering tidal volumes range 160–1500 cc with every breath
* Capable of limiting the peak inspiratory pressure to 18–40 cm-H_2O
* Powered using 120 V or lower (ideally with a 12 V car battery)

1.3 Related Work

Because of the COVID-19 pandemic, the field of alternative ventilator design has recently become an urgent matter that has caught the attention of many organizations and labs. Some organizations have rapidly published open-source ventilator projects with the hope of quickly saving lives. For example, the team from University of Florida [3] is working on a bellows-based ventilator. MIT's E-VENT [4] project uses a linkage mechanism to compress BVM as the pressure source. Another low-cost open-source ventilator [5] is using vacuum to generate a pressure differential. These designs depend on relatively large machinery, a vacuum source, or a BVM to function, while our goal is to utilize inexpensive materials and create a portable design that can be added to the current stockpile of manual and automated ventilation machines and that can be built from scratch with off-the-shelf commonly available materials.

Origami structures, two-dimensional patterns that are folded into complex structures, are popular actuator designs because they are lightweight, controllable, and inexpensive to manufacture. One existing approach focuses around the proprieties [6,7] and applications [8] of origami springs and bellows.

2 Technical Approach

Our proposed ventilator design leverages our existing work with origami continuum manipulators, where we have developed cable-driven origami-inspired soft robotic bending and axially compressing modules with integrated electrical motors, sensing, control, and communication systems [9]. The ventilator is based on this module crease pattern and consists of a flexible, air-tight bellows, either made from folded plastic or NinjaFlex, a flexible 3-D printed material. A single

L-Bend geared DC Motor (224:1 gear ratio) mounted inside the structure drives a cable attached to the other end. When the motor wraps the cable around a spool, the bellows contracts and the system performs its intended function to push air as a ventilator (Cables are pulled with a clockwise rotation of the motor and are released with a counterclockwise rotation). The prototypes of the proposed design are shown in Fig. 1.

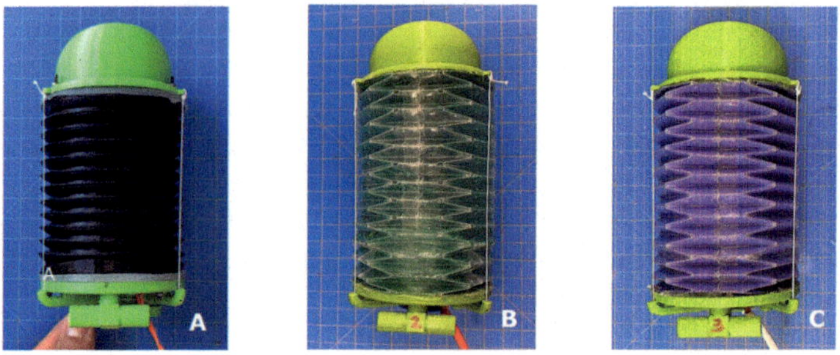

Fig. 1. Prototypes of a low-cost ventilator. (A) 3D Printed NinjaFlex Bellows. (B) Origami Bellows sealed with Stretchlon200 bagging film.(C) Origami Bellows sealed with Nitrile Latex.

2.1 Design and Prototyping

Origami Bellows Ventilator Design. Origami mechanisms have proved to be effective actuation modules because they are highly controllable and have a very simple design, making them easy to fabricate. The main body of our origami ventilators are based on the Yoshimura origami crease pattern, which results in a bellows that is easy to folds given the CAD model (shown in Fig. 2(B)). We use a CO_2 laser cutter to create perforation lines that create creases, but a simple plotting machine or manual folding are also possible. The material can be a thin film of PET (polyethylene terephthalate), or any other low-cost and accessible plastic or even cardboard.

We control the origami structure with a single DC motor, which sits on a 3D-printed mount and turns a spool, which pulls three cables firmly attached to the three corners of the other end of the bellows. We attach the motor mount to the origami module, then route the three cables outside the module.

Table 1. Major parameters of the prototypes

	3D printed bellows ventilator	OBV (Stretchlon200)	OBV (Nitrile)
Volume (ideally)	≈350 ml	≈350 ml	≈ 350 ml
Bellows diameter	8 cm	8 cm	8 cm
Spool diameter	2.5 cm	2.5 cm	2.5 cm
Full length	17 cm	17 cm	17 cm
Gear motor	224:1	224:1	224:1
Voltage	6–8 v	6–8 v	6–8 v
Weight	158 g	126 g	124 g

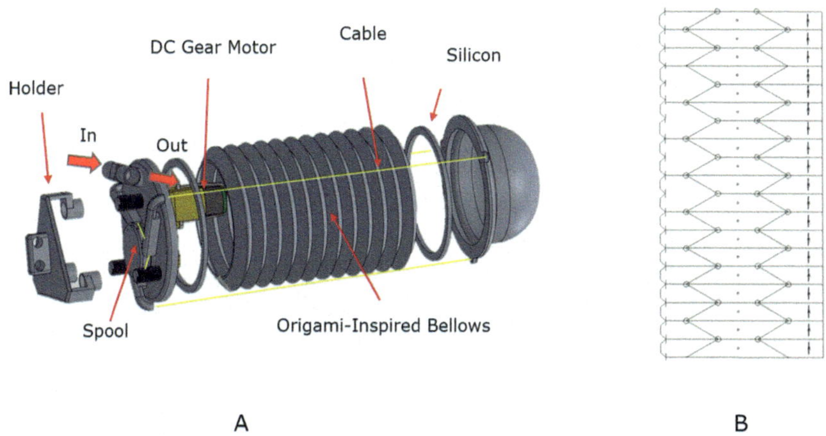

Fig. 2. CAD Design of Ventilators. (A) CAD model of ventilator. The rigid parts per printed out of PLA (B) Origami two-dimensional pattern cut using a laser cutter.

To seal the holes within the origami structure, we tested two different materials and sealing processes. We selected sealing materials based on material's flexibility and safety data sheet. Nitrile exam glove material is perfectly suited to most medical environments, being exceptionally puncture-resistant. We glued multiple pieces of Nitrile exam glove onto the inner surface of the origami module using E6000 flexible, paintable glue. The other material we used was Stretchlon200 vacuum bagging film, which was chosen because it is suited for stretching over complex shapes and has a high elongation, which reduces bridging in corners. We glued the film to the inner surface of the partially-assembled origami module before folding completely. Both of these methods results in an air-tight and lightweight bellows, as shown in Fig. 1. As the motor pulls the three cables, the origami structure compresses and the ventilator delivers air.

3-D Printed Bellows Ventilator Design. As an alternative to origami, we also printed a bellows out of Ninjaflex, a flexible TPU filament by Ninjatek. This design requires less assembly than the origami design, as a user can print a complete bellows without folding the structure. The 3-D printed ventilator compresses the same way as the origami design, with the same overall cable mechanism where a DC motor spools cables routed outside the bellows to push air.

Ventilator and Parameters Setting. The three ventilators presented above are initial proof-of-concept designs and are easily adjustable to change size and increase air volume output as required. For testing, we used the parameters shown in Table 1.

3 Experimental Results

Force-Displacement Tests. To characterize the origami and 3D-printed bellows, we first performed a force-displacement test to relate the linear force applied to the actuator and the corresponding length change of the bellows. We placed each bellows on a scale, then manually compressed it while reading the scale's output. We compressed the bellows by 6.8 cm, and found that they each exhibited similar force-displacement curves as shown in Fig. 3. The Stretchlon200 origami bellows was slightly stiffer (with 6.8 cm displacement occurring after 23N of force instead of 20N for the other two bellows), likely a result it requiring additional glue in order to properly secure the film. However, this difference is small and allows each bellows to function properly with the same motor and gearbox.

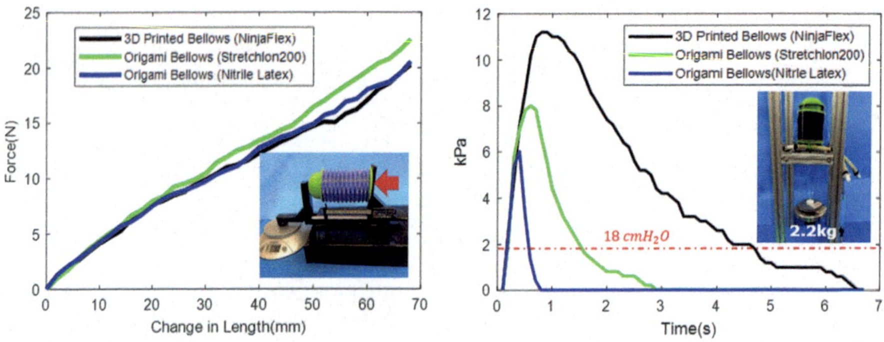

Fig. 3. Force-Displacement (left) and Sealing Test (right) Results: Force-displacement response of the proposed origami-inspired ventilator prototypes as they were compressed. Prototypes have a similar maximum force output when fully compressed.

Sealing Test Results. To validate and characterize the proof-of-concept ventilator system, we performed initial tests to show that the bellows have a sufficient seal and can maintain a pressure when compressed and released. We sealed the inlet of each bellows, and applied a constant 22N force, measuring the pressure

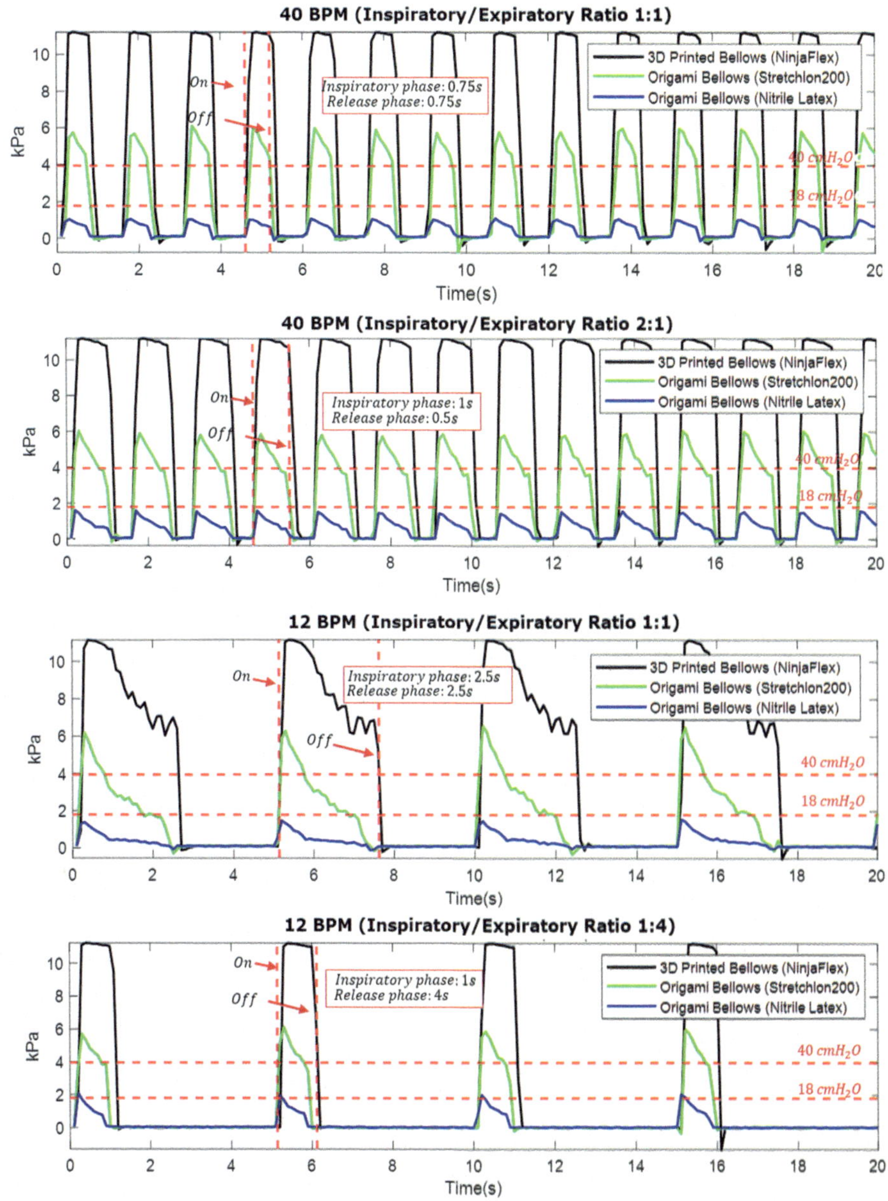

Fig. 4. Pressure generated within the proposed ventilators during cyclic operation with the outlet sealed. 40 Breaths Per Minute (BPM), Inspiratory/Expiratory ratio [1:1] and [2:1]. 12 Breaths Per Minute (BPM), Inspiratory/Expiratory ratio [1:1] and [1:4]. This highlights the range of rations our system can achieve.

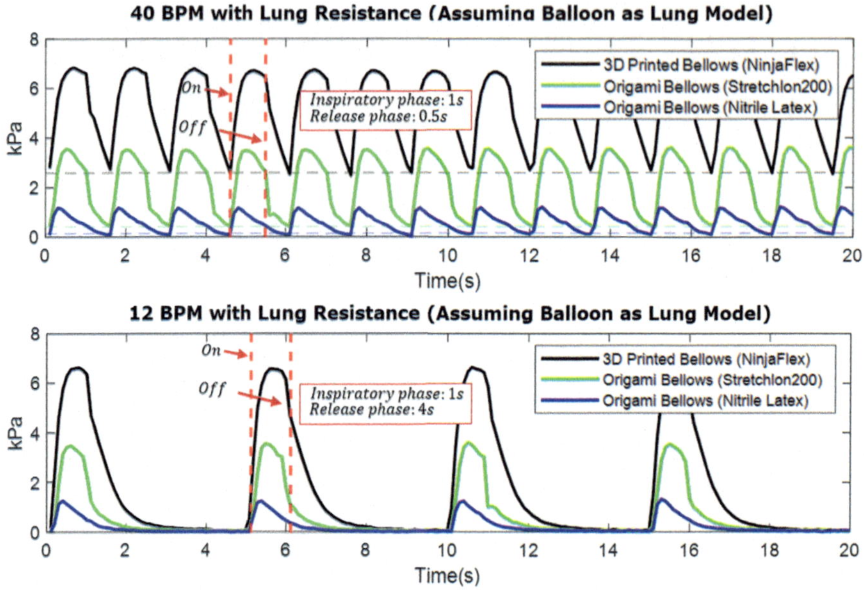

Fig. 5. Pressure readings from the ventilators with at 40 BPM and 12 BPM with a pierced balloon mimicking the resistance of a lung (balloon as lung model).

inside as air leaked out The experimental results (Fig. 3 right) show that proposed origami-inspired ventilator prototypes have a sufficiently good seal to generate the required 18 cmH_2O. Of the three bellows, the 3D printed one achieved the highest pressure and returned to atmospheric the slowest, likely the result of its being formed out of a single continuous piece of material, with leaks only possible at the interfaces on either end. The Nitrite-sealed bellows leaked the most, probably because it was sealed with a larger quantity of smaller sheets of the material.

System Functionality Results. We performed experiments to test ventilator performance repeatedly contracting the bellows and measuring the pressure output. The Ventilator prototypes and the sensor are run by an Arduino Uno control board. We had the ventilators contract and relax themselves (instead of applying an external force) at 40 PBM (with two different I:E (inspiratory/expiratory) ratio [1:1], [2:1]) and 12 PBM (with I:E ratio [1:1], [1:4]) while sealing the exit and measured the resulting pressure waveform. The results of these experiments can be seen in Fig. 4, with the 3-D printed ventilator and Stretchlon200 and Nitrile ventilators generating peak pressures of 11 kPa, 6 kPa and 1.8–2 kPa (112, 61 and 18–20 cm-H_2O) respectively. This is in agreement with our previous results, with the 3D printed bellows maintaining the best seal and this achieving higher pressures. Additionally, we performed 40 BPM and 12 BPM pressure experiment under Non-Zero airflow rate. We attached balloon with little hole to the outlet tube of the ventilator, which we used to represent a lung. We tested both 40 and 12 BPM for each bellows, the experimental results of which are shown in Fig. 5. We can see that the 3D printed bellows still can output more than the maximum pressure required, while the Stretchlon200 bellows can output pressure in the middle of the required range.

Airflow Rate and Volume Test. We also performed experiments to verify the flow rate the bellows could produce. We connected a balloon with a whole in it to the outlet of the ventilators to mimic them being connected to a patient's lungs, and mounted a custom differential pressure sensor to the tube to measure the flow rate. We run each of the origami ventilators at 40 PBM and 12 PBM for 60 s. The raw data is 100 Hz sampling frequency and filtered by simple moving average. The results of these experiments can be seen in Fig. 6. The average tidal volumes from the 40 BPM experiments were 213 cc for the NinjaFlex, 120 cc for the Stretchlon200, and 62 cc for the Nitrile, while for the 12 BPM experiments they were 362 cc for the NinjaFlex, 257 cc for the Stretchlon200, and 78 cc for the Nitrile.

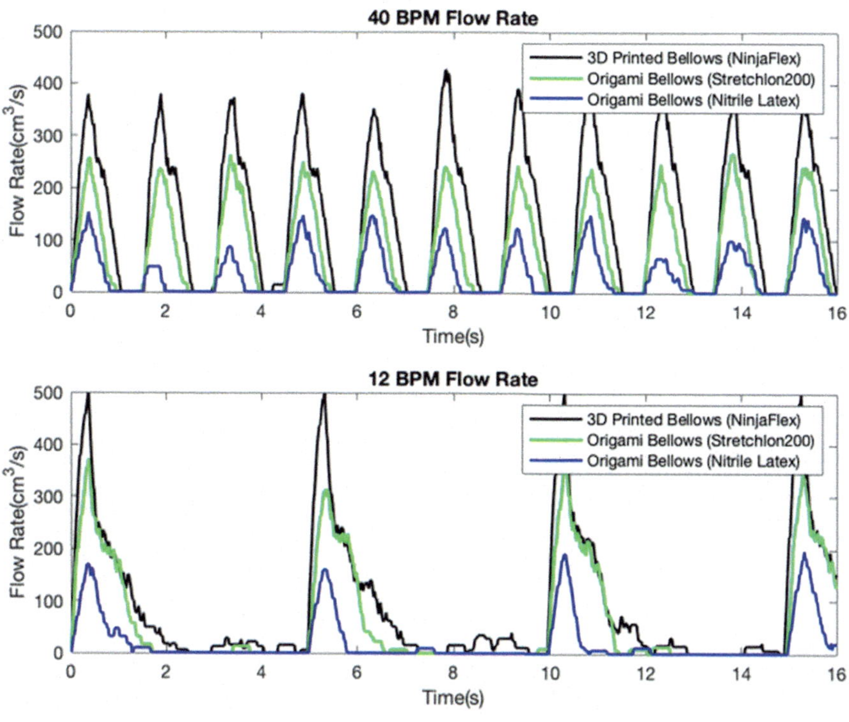

Fig. 6. The experimental results The Inspiratory Airflow Rate, where a custom flow rate sensor was to measure the unrestricted flow rate that the three ventilator prototypes could output.

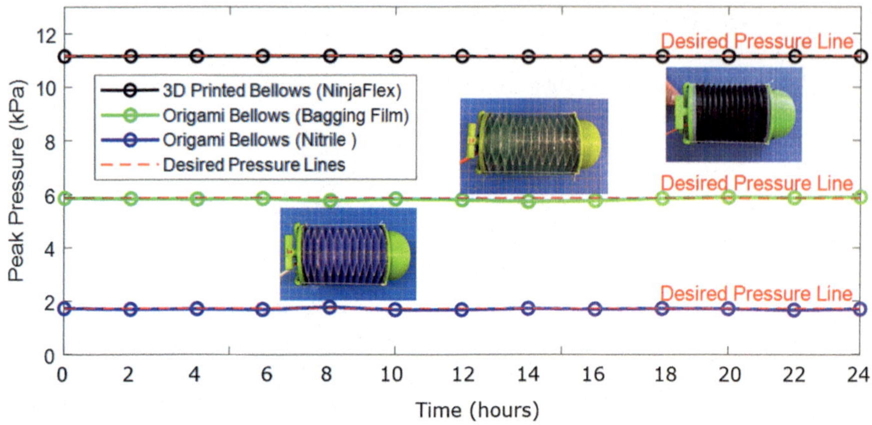

Fig. 7. Results from the robustness test, where we ran the ventilator prototypes at 0.5 Hz for 24 h straight. The peak pressure was recorded for 10 consecutive cycles once every 2 h, with the averages reported here. All three ventilators showed no change in behavior.

Robustness Test. To assess the robustness of our bellows, we ran them continuously for 24 h at a constant rate of 30 cycles per minute (frequency 0.5 Hz). We recorded the peak pressure data of 10 cycles every 2 h and averaged them, the results of which can be seen in Fig. 7. During the 43200 cycles of this experiment, all three ventilators continued to generate pressure without any changes in performance. This, combined with the fact that these same bellows did not require repair in between the other experiments, makes us confident that they could perform more than 48 h of continuous operation.

The performed experiments have provided the proof-of-concept validation to proceed towards the later stages of this project. Planned experiments include the ability to control the inhalation/exhalation frequency, maintaining desirable pressure and flow waveforms, as well as the overall robustness of the proposed devices including studies of variations in user requirements and fatigue.

4 Conclusion and Future Work

We have presented three different designs for a novel origami-inspired, lightweight, low-cost (less than $20), portable, and automated ventilator design as well as initial validation of their functionality as ventilator systems. Our experiments show that all three designs can generate the pressure required for use, and are robust enough to maintain this pressure over 24 h of continuous use. Additionally, we have shown that the 3-D printed bellows prototype has a better seal. This may point to the 3-D printed bellows as the superior design, though more testing is required to confirm. On the other hand, the ability to 3-D print using flexible filaments may not be commonly available, while folding a thin plastic film into a bellows shape may be more accessible.

Our requirements, as stated in the Problem Statement, were affordability, accessibility, and functionality. The initial design uses inexpensive materials and is easy to manufacture, and initial testing has proven basic functionality. Additionally, the system is highly portable and does require external infrastructure or heavy, rigid components. We have met four of the seven ventilator requirements we discussed earlier: the ventilator allows for an adjustable O_2 source, can provide a respiratory rate and PEEP above the required ranges while runing on a 6 V power supply.

Though the volumes of air displaced by the prototypes discussed in this paper are small, would be relatively easy to scale the system up. This could be done either by running several of these ventilators in parallel (which would also add increased robustness in the event of a catastrophic failure of one) or by a scaling up the bellows and increasing the size of the motor. This latter would represent a simple engineering task, as both the origami and 3d printed bellows designs can be easily changed to increase their height and diameter.

One requirement that we have not addressed was to limit the peak inspiratory pressure to 18–40 cm-H_2O. This is just a matter of performing closed-loop control with pressure sensors and the motor controlling the contraction of the bellows. Nevertheless, it is an import component of our future work to make the system discussed fully suitable for use in the medical field.

Another aspect of future work is to improve the seal of all the bellows, but in particular the nitrile oragimi bellows. This could be done via a more secure seal around the ends of the bellows (especially for the 3D printed bellows), or by improving the seams between the material on the bellows.

We believe that our design, which emphasizes accessibility, portability, and ease of manufacturing, has the potential to seamlessly support a hospital's existing medical equipment, allowing doctors and nurses to work unhindered and save more lives.

Acknowledgements. This material is based upon work supported by the National Science Foundation under Grant Nos. CMMI-1752195 and DGE-1922761. Any opinions, findings, conclusions or recommendations expressed in this material are those of the authors and do not necessarily reflect the views of the NSF. The authors would also like to acknowledge Raagini Rameshwar for her contributions to the manuscript.

References

1. Who coronavirus disease (covid-19) dashboard (2020). http://covid19.who.int/
2. DIY Hack-a-Vent Innovation Challenge: North Carolina Military Business Center. http://www.ncmbc.us/announcements/diy-hack-a-vent-innovation-challenge/
3. Open-Source Ventilator Project. https://simulation.health.ufl.edu/technology-development/open-source-ventilator-project/
4. MIT E-VENT—Emergency ventilator design toolbox. https://e-vent.mit.edu/
5. Lee, J.: Jcl5m1/ventilator. https://github.com/jcl5m1/ventilator

6. Morgan, J., Magleby, S.P., Howell, L.L.: An approach to designing origami-adapted aerospace mechanisms. J. Mech. Des. **138**(5), 052031 (2016). https://doi.org/10.1115/1.4032973. https://asmedigitalcollection.asme.org/mechanicaldesign/article/138/5/052301/473012/An-Approach-to-Designing-Origami-Adapted-Aerospace
7. Butler, J., Morgan, J., Pehrson, N., Tolman, K., Bateman, T., Magleby, S.P., Howell, L.L.: Am. Soc. Mech. Eng. Digit. Collect. https://doi.org/10.1115/DETC2016-59060. https://asmedigitalcollection.asme.org/IDETC-CIE/proceedings/IDETC-CIE2016/50169/V05BT07A001/259263
8. Santoso, J., Skorina, E.H., Luo, M., Yan, R., Onal, C.D.: 2017 IEEE/RSJ International Conference on Intelligent Robots and Systems (IROS), pp. 2098–2104 (2017)
9. Luo, M., Yan, R., Wan, Z., Qin, Y., Santoso, J., Skorina, E.H., Onal, C.D.: OriSnake: design, fabrication, and experimental analysis of a 3-D origami snake robot. IEEE Robot. Autom. Lett. **3**(3), 1993 (2018)

Field Robotics

Towards a Reliable Heterogeneous Robotic Water Quality Monitoring System: An Experimental Analysis

Monika Roznere[1(✉)], Mingi Jeong[1], Lily Maechling[1], Nicole K. Ward[3], Jennifer A. Brentrup[4], Bethel Steele[5,8], Denise A. Bruesewitz[7], Holly A. Ewing[5], Kathleen C. Weathers[6], Kathryn L. Cottingham[2], and Alberto Quattrini Li[1]

[1] Department of Computer Science, Dartmouth College, Hanover, NH, USA
[2] Department of Biological Sciences, Dartmouth College, Hanover, NH, USA
monika.roznere.gr@dartmouth.edu
[3] Department of Biological Sciences, Virginia Tech, Blacksburg, VA, USA
[4] Department of Biology and Environmental Studies, St. Olaf College, Northfield, MN, USA
[5] Environmental Studies Program, Bates College, Lewiston, ME, USA
[6] Cary Institute of Ecosystem Studies, Colby College, Waterville, ME, USA
[7] Environmental Studies, Colby College, Waterville, ME, USA
[8] Cary Institute of Ecosystem Studies, Millbrook, NY, USA

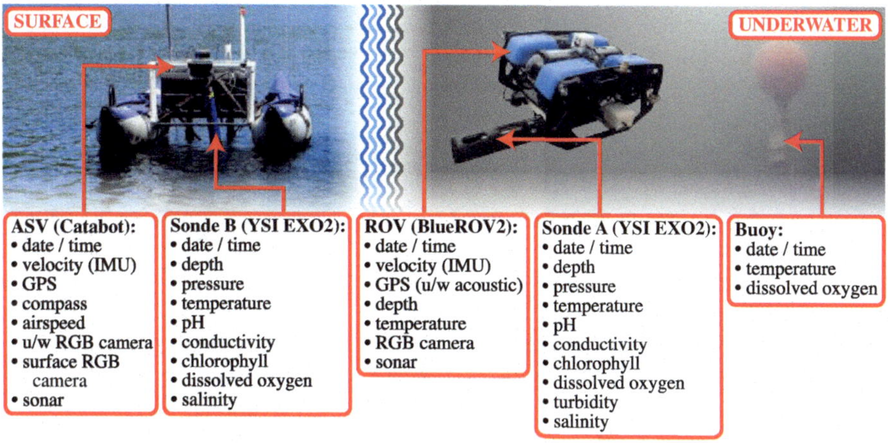

Fig. 1. The instrumentation and parameters that make up the sensor system come from three sources: an ROV, an ASV, and a sensored underwater buoy.

1 Problem Statement, Motivation, and Related Work

This paper describes experiments that tested the effect of robotic movement on the reliability of aquatic sensor readings. It also demonstrates the utility of a

M. Roznere and M. Jeong–These authors contributed equally to the paper.

B. Siciliano et al. (Eds.): ISER 2020, SPAR 19, pp. 139–150, 2021.
https://doi.org/10.1007/978-3-030-71151-1_13

heterogeneous system of robots to advance limnological monitoring and research. An Autonomous Surface Vehicle (ASV) and an underwater Remotely Operated Vehicle (ROV), both equipped with multiparameter water quality sondes, were deployed weekly in Lake Sunapee, NH, to collect routine measurements horizontally over the water surface and vertically in the water column, respectively. We then compared the robot-collected data with data from fixed underwater instrument platforms (buoys) outfitted with a complementary suite of sensors as well as manually collected samples. The ASV was also deployed on one date in China Lake, ME, to test robotic procedures and evaluate potential environmental effects, including heterogeneity in water quality. We found that sensor response time and robotic movement (e.g., turns, stops, collisions) produced small discrepancies between the robot-derived and other datasets. Further, robotic coverage patterns impacted water quality parameter measurements, affecting our understanding of horizontal heterogeneity in biological and chemical data across the lake surface.

A principal goal of aquatic monitoring is to provide data which can be used to inform protection of and management of good water quality for drinking or recreation. However, the limited spatial and temporal coverage of traditional limnological monitoring may constrain inferences about water quality. For example, monitoring typically involves measuring water clarity, temperature, dissolved oxygen, nutrients, and biological variables such as chlorophyll-*a*, at weekly to monthly intervals, usually during the ice-free season, at a single site. Although continuous sensor measurements on fixed-location buoys have increased temporal sampling frequency of many variables [15], the number of buoys in a lake and type/number of sensors are typically limited by cost constraints, thus limiting spatial coverage. Moreover, buoy locations and depth of sensors are often selected based on discretized zones, such as pelagic (open water) areas or above and below the thermocline [12]. However, heterogeneity within commonly accepted lake zones is increasingly being recognized as critical for understanding lake ecosystem functioning [12]. We therefore seek to monitor the key factors that drive water quality by collecting and integrating higher spatial and temporal resolution data using autonomous vehicles.

A limnology-computer science interdisciplinary approach of designing sampling strategies for ASVs and underwater ROVs within robot-based limitations is a promising method to monitor waterbodies more effectively [3,9,10,13]. Two types of robotic approaches have been used: *complete coverage* and *adaptive sampling* [2]. *Coverage approaches* are based on area decomposition to identify unit cells covered with a specific motion (e.g., lawnmower pattern) and on optimization techniques to order the cells to visit, which minimizes a cost function (e.g., traveled distance) [1,6,11]. For example, Garneau *et al.* [7] used an ASV to collect high resolution water quality data, but due to the slow response time of the dissolved oxygen probe, the coverage was limited only to a 2D transect line across a lake. As an alternative, *adaptive sampling techniques* process real-time sensor data strategically visiting spots of high interest [4,5]. Hitz *et al.* [8] extended their earlier work [7,9] with a 3D real-time planner for an ASV that continuously collects data at different sensor depths while traveling to pre-established waypoints. They did not use a dissolved oxygen sensor, removing the limitations enforced on robotic speed. These prior studies highlight that research is needed to better understand how robot movement may interact with sensor characteristics to impact data quality and bridge the gap for implementing effective and reliable sampling methods in real-world environments.

We address the following questions:

- How do robotic motions affect water quality sensor response time?
- How do water quality measurements compare between robots and fixed-location buoys?
- How can robotic coverage paths best complement and extend monitoring data from discrete water samples (i.e., chlorophyll-a concentrations)?

The main contribution of this paper is to provide insights that will improve *robotic sampling* techniques for collecting reliable sensor data, thereby providing higher resolution temporal and spatial information about water quality.

2 Experimental Methodology and Setup

Our integrated system for water quality data collection is illustrated in Fig. 1. The system includes the following robots:

- BlueROV2: An underwater ROV with downward-looking sonar, front-facing RGB camera, acoustic transducer for short baseline localization, and YSI EXO2 sonde (Sonde A) recording water temperature, depth, pH, conductivity, chlorophyll fluorescence, dissolved oxygen (DO), salinity, pressure, and turbidity at 1 s intervals;
- Catabot [10]: An ASV with sonar, surface and underwater RGB cameras, GPS, anemometer, and Sonde B (similar to Sonde A, but no turbidity sensor).

The weekly robotics experiments reported here occurred at Lake Sunapee, NH, from ice-out in April until September 2020. We were particularly interested in comparing robotic water quality measurements to existing limnological data:

- Winter Buoy: A fixed, instrumented, subsurface buoy (44 m from the home location in Fig. 6) with sensors along its vertical line (1–8 m). Dissolved oxygen (DO) was measured at two depths (1 and 8 m) at 10 min intervals using PME Mini-DOT loggers. Water temperature was measured every 0.5–1 m along the full vertical profile at 2 h intervals using HOBO pendant temperature loggers;
- Summer Buoy: A fixed, instrumented, surface buoy (313 m from the home location in Fig. 6) with the same sensors along its vertical line as the Winter Buoy. Here, we focus on the PME Mini-DOT sensor at 1 m depth that measured water temperature and DO at 10 min intervals;
- Hand-collected water samples (grab samples) that were subsequently analyzed for chlorophyll-a [14].

In this analysis, we compare vertical data on temperature and DO from the ROV to the Winter Buoy, and surface measurements of temperature, DO, and chlorophyll-a made by the ASV near the Summer Buoy and home location.

We also evaluated the effect of sensor response time and field conditions on data quality at Lake Sunapee, NH and China Lake, ME. In the laboratory, we measured sensor response time defined as the time to equilibrium for temperature and DO when moving between buckets of hot (40 °C) vs. cool water (20 °C) or still vs. bubbled water. These times were then compared to field measurements in which the ROV traveled vertically along the Winter Buoy line, stopping at individual sensor locations for set amounts of time. For the ASV, we analyzed sensor response time and instability by internal factors – when it moved (with linear and angular motion) and stopped in one place or continued to move to waypoints during a path coverage – and external factors – when it encountered adversarial environmental conditions. We also quantified the impact of the ROV hitting bottom and stirring up the sediment on sensor readings.

Finally, we determined how different ASV coverage paths inform the adequacy of sampling spatial resolution relative to spatial variability in water quality variables within the environment at both Lake Sunapee and China Lake. At Lake Sunapee, the ASV took three lawnmower path patterns (see Fig. 6) with overlap to gain better spatial distribution: 1) *Black* path – vertical lawnmower, 2) *Blue* path – diagonal lawnmower, and 3) *Red* path – quadrilateral. We determined the principal direction of the paths by avoiding dominant adversarial external conditions, i.e., wind (from *NW*) and current (towards *SE*) to attain motion efficiency and sensor reliability from our previous study [10]. The three paths were executed in the following order: 1) *Black*, 2) *Blue*, and 3) *Red*.

3 Experimental Results and Analysis

Our experiments revealed (1) effects of sensor characteristics and robot motions on sonde readings; (2) small differences between data collected with the robots and those collected from stationary buoys or via grab samples; and (3) a measure of spatial heterogeneity of water quality measurements as compared to grab samples.

3.1 Sensor Response Times and Robot Motions Have a Modest Impact on Sensor Reliability

Sensor response times under controlled scenarios (i.e., hot-cold, still-bubble) averaged 26 s (3.4 s standard deviation [SD]) for temperature (°C) and 27 s (4.3 s SD) for DO (mg/L) to achieve a stable reading. Although these two times are very similar, the proportional temperature differential was greater than that for DO, suggesting that DO readings take slightly longer time to stabilize for a given differential. This is expected, as optical-based sensors have slower response times than electrical sensors. For reference, temperatures are unlikely to change more than 5 °C over 0.5 m, whereas DO can change drastically from 4–6 mg/L to < 2 mg/L over the same water column.

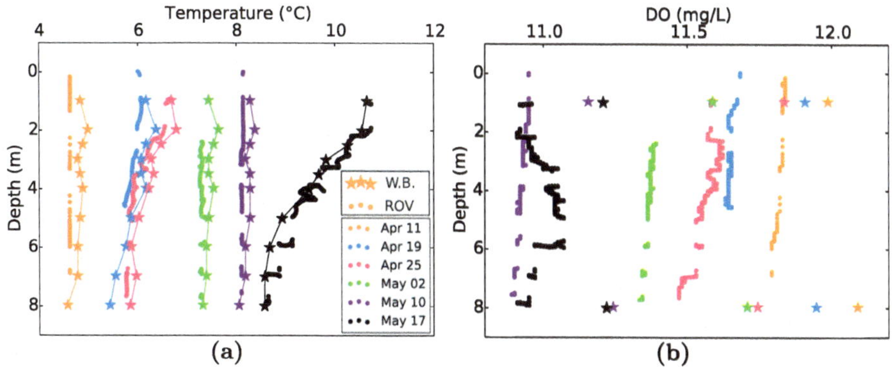

Fig. 2. ROV with Sonde A (points) vertical profiles compared to fixed-location Winter Buoy sensors (star) at Lake Sunapee, NH. **(a)** Temperature (°C) measurements match well. On May 17 (black), the ROV captured the beginning of thermal stratification. **(b)** DO (mg/L) measurements also agree – values are within known measurement error. Note, the DO range is very small. The results reinforce the need for higher resolution data throughout the water column. (Color figure online)

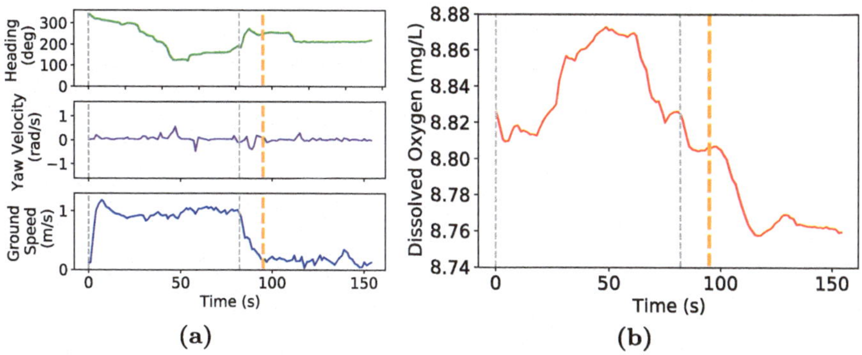

Fig. 3. ASV deployment at Lake Sunapee, NH on June 17. **(a)** Proprioceptive data of ASV. **(b)** Measurements of surface water DO (mg/L). After the ASV initiated and finished motion (between two gray dashed vertical lines; 0 s and 82 s), it completely stopped in one place starting at 95 s (orange dashed vertical line). DO measurements took about 30 s to stabilize after the ASV ceased movement.

To compare these response times to those in real environmental conditions (see Fig. 2), the ROV collected data as it moved vertically along the buoy line and stayed at each depth associated with a sensor on the buoy for 1 min. In these profiles, temperature and DO measurements from the ROV sensors stabilized after about 10–15 s, as indicated by SD below 0.01. Each sensor's response time is not only based on the *type of sensor* it is, but also the current *variability of the environment* (i.e., change in the variable being measured over area or depth). For example, the onset of thermal stratification by May 17 meant that

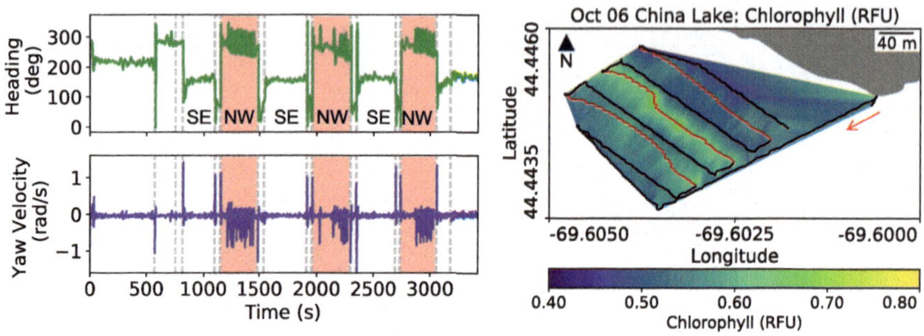

(a) (b)

Fig. 4. ASV deployment at China Lake, ME on October 6. **(a)** Proprioceptive data of ASV. The red shaded areas indicate when ASV adjusted its heading to counteract external forces while following the NW direction. **(b)** Coverage paths and measurements of surface water chlorophyll (RFU) linearly interpolated across an area of 0.04 km^2. The red tracks indicate the same time intervals as the shaded area in (a). The red arrow is the starting direction.

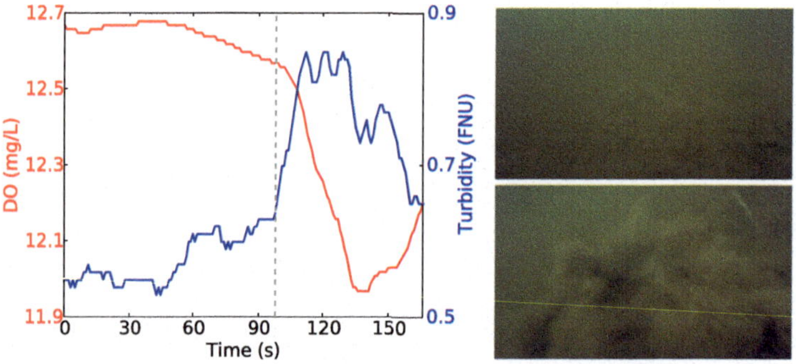

(a) (b)

Fig. 5. ROV deployment at Lake Sunapee, NH. **(a)** The ROV hit the bottom of the lake and stirred up sediment around 100 s. Sonde A's turbidity sensor (FNU) detected an increase in sediment particles, and a few seconds later the DO (mg/L) readings dropped. **(b)** Before and after underwater images from the ROV.

both temperature and DO now changed with depth. In comparison with the nearly isothermal conditions in previous weeks, the ROV control would need to actively be aware of how readings change with a given depth and pause longer at specific depths when needed. Furthermore, water movement (e.g., due to wind or currents) and required ROV motion (e.g., depth hold) caused some fluctuation in sonde readings, even after readings had stabilized.

For the ASV, two main factors affected sensor stability during a path coverage: *motion attributes* (internal) and *environmental conditions* (external). First, rotational and translational movement at each waypoint may delay sensor responses. For example to monitor the sensor reading stability, the ASV arrived at home location at 82 s and stayed in one place at 95 s to monitor the sensor reading stability (Fig. 3). The DO sensor readings stabilized after approximately 30 s, indicating that the previous motion of the ASV impacted sensor readings. Second, environmental conditions independent of ASV movement can alter the ASV path (e.g., primarily when following the NW direction in Fig. 4), which may also affect the sensor reliability. When the ASV adjusted its heading to counteract external forces (Fig. 4 (a) – red shaded area), the motions caused higher variability in DO measurements, in particular between 2736-3051 s. However, these fluctuations were relatively small and likely not biologically meaningful (SD: 0.06 mg/L). A similar event occurred when a powerboat circled the ASV, causing water disturbances from wake currents to heavily affect the ASV's heading and yaw, but variability in the DO readings was also small enough to not impact biological interpretations (SD: 0.01 mg/L; data not shown). However, high variability in sonde readings for future deployments may be a useful indicator of potential sensitivity to robot behavior.

In other situations, sensor data reliability was more prominently impacted by *unintentional interactions* with the environment. For example, the ROV hit the bottom of the lake and stirred up sediment (Fig. 5). This led to misleading water quality readings, including an increase in turbidity followed by a decrease in DO that was simply a result of the unintentional interaction with the sediment. In this case, it was important to have a turbidity sensor paired with the DO sensor, so the cause of the inadvertent sudden drop in DO could be easily identified. The faster response time of the turbidity sensor compared to the DO sensor could predict incoming reading changes for DO or other water quality measurements.

3.2 Robots vs. Fixed Buoys: Robotic Sampling Extends DO Vertical Coverage

The robot-driven sensors and fixed buoy sensors provided complementary information about Lake Sunapee following ice-out (Fig. 2). Traditionally, the period following ice-out is dangerous for manual grab sampling, as the water is so cold and floating ice can be dangerous to navigate around, while robot sampling avoids these issues. ROV sampling along the Winter Buoy line improved the vertical underwater depth resolution, while the higher-frequency data from Sonde A increased the temporal resolution beyond that of the 10 min recording interval of the fixed buoy. ASV sampling extended the horizontal sampling coverage of the lake surface with higher temporal resolution, which will be discussed in Sect. 3.3.

ROV and Winter Buoy temperature measurements generally agreed (Fig. 2 (a)). During the first 5 weeks, Sonde A temperatures were slightly cooler than the buoy's, while the last week, May 17, shows the opposite pattern. From April 25 to May 17, the ROV consistently stayed at each depth for approximately 1 min. Yet, on May 17 Sonde A produced more variable readings at most depths, possibly due to the greater difference between surface and bottom temperatures requiring a longer response time for the sensor. A comparison of DO concentrations between Sonde A on the ROV and at the top and bottom of the buoy line were also quite similar (Fig. 2 (b)) – the observed discrepancies are within the measurement error of the PME Mini-DOT (±5% of measurement or ±0.3 mg/L, whichever is larger). The offsets may also be due to drift in the sensor calibrations, since neither type of DO sensor was recalibrated during the course of the study. While differences in sensor response times need to be resolved, ROV with Sonde A *greatly increased the vertical resolution of DO data.* In fact, the higher resolution ROV data revealed the development of an oxygen bulge in the middle of the water column on May 17 (Fig. 2 (b)).

A similar correspondence between sensors was observed between data from the ASV with Sonde B when it paused near the Summer Buoy during each *Black* path coverage (Fig. 6 (a)) for a total of 12 expeditions from May 31 to September 11. Sonde B water *temperatures were strongly positively correlated with buoy temperatures* ($r = 0.97$, $N = 12$, p-value $p < 0.01$), as were DO measurements ($r = 0.9$, $N = 12$, $p < 0.01$).

3.3 Robots vs. Grab Samples: Robotic Sampling Captures Horizontal Patterns of Chlorophyll Data

ASV deployments improved general coverage area and provided real-time data on spatial heterogeneity in the two lakes (Lake Sunapee, NH and China Lake, ME) we sampled. The lakes strongly differ in trophic status, i.e., a classification of waterbodies based on the biological productivity. Lake Sunapee is oligotrophic with low chlorophyll values and high water clarity, while China Lake has higher chlorophyll and lower clarity and would be classified as meso-eutrophic. Our ASV collected chlorophyll fluorescence readings (RFU) each week at three strategic locations in Lake Sunapee (near the home location and the Summer Buoy in Fig. 6). These readings were positively correlated with lab-based measurements of chlorophyll made from grab samples collected during the ASV sampling at these same locations ($r = 0.47, N = 18, p = 0.05$). Measured concentrations in Lake Sunapee were relatively low (3–5 µg/L) and are likely near the low end for reliable readings from the chlorophyll sensor. In comparison, measured concentrations in China Lake on October 6 were higher (8.42 µg/L, SD: 0.07 µg/L). Interpolated maps of chlorophyll (in RFU) at both Lake Sunapee and China Lake suggested that the continuous data obtained over a wide spatial area from the ASV provided information about spatial heterogeneity in chlorophyll patterns, particularly in China Lake (Fig. 6 and Fig. 4 (b)).

Different spatial coverage paths performed in a single waterbody (i.e., Lake Sunapee) demonstrated horizontal variability in chlorophyll during short-term deployments (Fig. 6, Table 1). Note that the intended proceeding speed was 0.5 m/s for consistent operation, but the actual proceeding speed may have varied depending on environmental factors (Table 1). Although a cyanobacterial bloom did not occur in Lake Sunapee during the expeditions, the coverage methods can help understand the patterns of phytoplankton distribution. A spatial pattern in chlorophyll fluorescence was more visible from the first path coverage scenario (e.g., the line artifacts in Fig. 6 (a) *Black* path rather than (b) *Blue* path, (c) *Red* path), possibly because that was the first path attempted on this

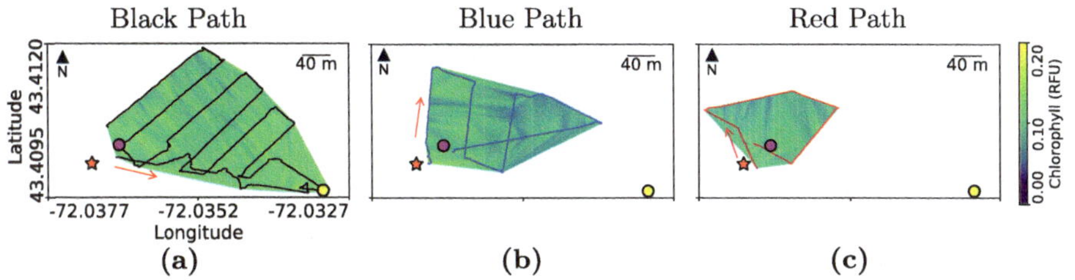

Fig. 6. Three ASV coverage patterns for Lake Sunapee, NH and measurements of surface water chlorophyll fluorescence (RFU, linearly interpolated) on July 15. **(a)** Black path – route to the Summer Buoy and vertical lawnmower. **(b)** Blue path – diagonal lawnmower. **(c)** Red path – quadrilateral. Yellow and magenta circles are the Summer and Winter Buoy's locations, respectively. The red star is the home location and the red arrow is the starting direction.

Table 1. Lake Sunapee (July 15) and China Lake (October 6) coverage path operation data for the ASV, which includes total area, distance and time traveled, and the area and distance coverage rate. Summary statistics of the chlorophyll fluorescence (RFU) measurements in different scales are also included.

Region	Path	Area (m^2)	Dist. (m)	Time (s)	Covg. (m^2/s)	Speed (m/s)	Mean Chl (RFU ± SD)	Range Chl (RFU)
Sunapee	Black	47215.22	1921	3016	15.65	0.63	0.11 ± 0.01	0.07 – 0.16
	Blue	24232.06	793	1423	17.03	0.56	0.095 ± 0.02	0.05 – 0.16
	Red	14702.29	530	854	17.21	0.62	0.096 ± 0.02	0.06 – 0.13
China		43267.35	2108	3418	12.66	0.62	0.54 ± 0.05	0.44 – 0.71

date. Data combined from all paths on the same day may slightly obscure this observation, potentially due to water mixing from the ASV movements. However, mean chlorophyll fluorescence was very similar across all three coverage paths (Table 1). Since all paths had a similar speed, the operator/robot can decide on a path pattern and scale to best sample a region of interest, which could be useful for calculating the sparsity of samples for *complete coverage* or updating real-time models in *adaptive sampling*.

ASV deployments may also enhance understanding of the relationship between daily environmental conditions, such as wind and water currents, and within-lake chlorophyll patterns. The line artifacts perpendicular to the long trajectories of the paths (i.e., Fig. 6 (a) *Black* path) could have been from wind or water currents, which usually follow the SE direction, development of Langmuir circulation (counter-rotating vortices on the water surface aligned with the wind; also referred to as windrows), or from interpolation noise within the small range of chlorophyll values. In addition, at China Lake, patterns observed along the NE direction (Fig. 4 (b)) could have been from similar environmental effects on the sampling date. The highest chlorophyll concentration occurred in the middle of the coverage area with residual high concentrations in the SW region, so an additional iteration of the coverage path may have shown a movement of the phytoplankton in the NE direction.

4 Discussion and Experimental Insights

Our experiments confirmed that robotic water quality measurements are fairly robust to sensor response times and robotic motions, though there are a few situations that may impact biological interpretations. In particular, varied sensor response times, unintentional robot-environment interactions, and ground truthing need to be considered when designing sampling strategies and in post-sampling data quality control measures. Additionally, future path planning methodologies will improve adaptive sampling to target specific locations and data reliability.

Based on our results to date, we offer the following guidelines for operating robotic water-quality sampling strategies: (1) A robot should stop or reduce speed to provide sufficient time for different sensors to collect accurate readings, especially optical sensors. While speed is commonly recognized as a strategy constraint, we note that multi-directional motions by ROV and horizontal rotation by ASV may impact data quality in some situations. As shown in Fig. 2 and Fig. 3, response times for sensors in lakes can be different than baseline response times, depending on environmental factors. (2) Allow the robot to self-adjust waiting times for sensor equilibration, similar to adaptive sampling routines; the robot decides when the sensor readings are stable enough to proceed to the next sampling point. (3) The robot should consider factors such as environmental interactions and external forces and minimize their effects when possible. For instance, downward-facing sonar can act as an obstacle avoidance mechanism and help prevent the ROV from stirring up the bottom. For an ASV, studying proper maneuvers can prevent too close approaches from passing boats and counteract leeway caused by wind and currents. Although the ASV did not encounter meaningful instability of the sensor readings during the expeditions, there could be instances where such motions have a significant effect. (4) Robot-collected data should be validated with other data streams when possible, especially when working in new systems so that robot-collected data can be used to scale in both space and time. (5) When robotic movements that may affect sensor reliability can not be avoided, a post-sampling data quality control protocol should be developed specific to the robot instrumentation and environmental conditions. To achieve the future possibility of robots collecting data over larger areas than the traditional limnological sampling that occurs at a few widely separated static places, the robot should plan a path considering (1)–(5), regardless of robotic approaches aimed at *complete coverage* or *adaptive sampling.*

5 Conclusion

The results show that ROV and ASV sampling, when performed with robust validation and quality assurance/quality control measures, can lead to the collection of high quality, spatially and temporally resolved, water quality data. These data are necessary for characterization of whole-lake water quality, identification of hot spots of concern, and ultimately, for prediction of water quality.

Our next step is to conduct extensive tests on different ASV coverage paths (Fig. 6) during a period of decreased water quality, such as a cyanobacterial bloom, to have a thorough understanding of the balance between high resolution of spatial heterogeneity and deployment time. In addition, we will expand our investigation on how to determine, with and without a turbidity sensor, if potential disturbances in the environment cause significant fluctuations in the sonde readings. If such events do affect the reliability of sensors, how should the robot respond? One option, as mentioned in the guidelines, is to let the robot stop and determine when readings are sufficiently stabilized before continuing the path. A more complex solution is to first determine if the location where the unreliable measurements were taken will provide substantial information if resampled, then either send the robot back to resample or let the robot continue on its path, but include adjustments in control to minimize the chance of the event repeating itself.

We aim to extend the system to multiple ROVs and ASVs with a strategic motion planning methodology that is self-aware of sensor characteristics (e.g., response time) and of the impacts from motion attributes (e.g., speed, turning, stopping, collisions). A well-informed system of robots, combined with data collected from existing monitoring efforts, can help create a more holistic picture of spatial and temporal variability within a lake and provide limnologists and lake managers with better data and tools for understanding and predicting changes in water quality, such as cyanobacterial blooms.

Acknowledgment. This work was supported in part by NSF CNS-1919647, OIA-1923004, ICER-1517823, DEB-1753639. We thank the Eliassen family who provided access to sites; the Lake Sunapee Protective Association (LSPA) for boat and buoy assistance; and all the students in the ES 494 Senior Research Capstone class of Fall 2020 in Colby College, in particular, Grace Neumiller, Taryn Waite, and Grace Andrews, for their support in chlorophyll sampling at China Lake, ME.

References

1. Arzamendia, M., Gregor, D., Reina, D.G., Toral, S.L.: An evolutionary approach to constrained path planning of an autonomous surface vehicle for maximizing the covered area of ypacarai lake. Soft Comput. **23**, 1723–1734 (2019)
2. Dunbabin, M., Marques, L.: Robots for environmental monitoring: Significant advancements and applications. IEEE Robot. Autom. Mag. **19**(1), 24–39 (2012)
3. Ferri, G., Manzi, A., Fornai, F., Ciuchi, F., Laschi, C.: The hydronet ASV, a small-sized autonomous catamaran for real-time monitoring of water quality: from design to missions at sea. IEEE J. Oceanic Eng. **40**, 710–726 (2015)
4. Flaspohler, G., Preston, V., Michel, A.P., Girdhar, Y., Roy, N.: Information-guided robotic maximum seek-and-sample in partially observable continuous environments. IEEE Robot. Autom. Lett. **4**(4), 3782–3789 (2019)
5. Fossum, T.O., Fragoso, G.M., Davies, E.J., Ullgren, J.E., Mendes, R., Johnsen, G., Ellingsen, I., Eidsvik, J., Ludvigsen, M., Rajan, K.: Toward adaptive robotic sampling of phytoplankton in the coastal ocean. Sci. Robot. **4**(27), (2019)

6. Galceran, E., Carreras, M.: A survey on coverage path planning for robotics. Robot. Autom. Syst. **61**(12), 1258–1276 (2013)
7. Garneau, M.E., Posch, T., Hitz, G., Pomerleau, F., Pradalier, C., Siegwart, R., Pernthaler, J.: Short-term displacement of planktothrix rubescens (cyanobacteria) in a pre-alpine lake observed using an autonomous sampling platform. Limnol. Oceanogr. **58**(5), 1892–1906 (2013)
8. Hitz, G., Galceran, E., Garneau, M.È., Pomerleau, F., Siegwart, R.: Adaptive continuous-space informative path planning for online environmental monitoring. J. Field Robot. **34**(8), 1427–1449 (2017)
9. Hitz, G., Pomerleau, F., Garneau, M.È., Pradalier, C., Posch, T., Pernthaler, J., Siegwart, R.Y.: Autonomous inland water monitoring: design and application of a surface vessel. IEEE Robot. Autom. Mag. **19**, 62–72 (2012)
10. Jeong, M., Roznere, M., Lensgraf, S., Sniffen, A., Balkcom, D., Quattrini Li, A.: Catabot: Autonomous surface vehicle with an optimized design for environmental monitoring. In: Proceedings OCEANS (2020)
11. Karapetyan, N., Moulton, J., Lewis, J.S., Quattrini Li, A., O'Kane, J.M., Rekleitis, I.: Multi-robot dubins coverage with autonomous surface vehicles. In: Proceedings ICRA, pp. 2373–2379 (2018)
12. Kraemer, B.: Rethinking discretization to advance limnology amid the ongoing information explosion. Water Res. **178** (2020)
13. Teece, M.A.: An inexpensive remotely operated vehicle for underwater studies. Limnol. Ocean.: Methods **7**(3), 206–215 (2019)
14. Welschmeyer, N.A.: Fluorometric analysis of chlorophyll-a in the presence of chlorophyll-b and pheopigments. Limnol. Oceanogr. **39**, 1985–1992 (1994)
15. Wilkinson, A., Hondzo, M., Guala, M.: Vertical heterogeneities of cyanobacteria and microcystin concentrations in lakes using a seasonal in situ monitoring station. Global Ecol. Conserv. **21**, e00838 (2020)

Experimental Evaluation of a Hierarchical Operating Framework for Ground Robots in Agriculture

Stuart Eiffert, Nathan D. Wallace, He Kong(✉), Navid Pirmarzdashti, and Salah Sukkarieh

Australian Centre for Field Robotics, University of Sydney, Sydney, Australia
{s.eiffert,n.wallace,h.kong,navidp,salah}@acfr.usyd.edu.au

Abstract. For mobile robots to be effectively applied to real world unstructured environments—such as large scale farming—they require the ability to generate adaptive plans that account both for limited onboard resources, and the presence of dynamic changes, including nearby moving individuals. This work provides a real world empirical evaluation of our proposed hierarchical framework for long-term autonomy of field robots, conducted on University of Sydney's Swagbot agricultural robot platform. We demonstrate the ability of the framework to navigate an unstructured and dynamic environment in an effective manner, validating its use for long-term deployment in large scale farming, for tasks such as autonomous weeding in the presence of moving individuals.

Keywords: Field robotics · Navigation · Dynamic path planning

1 Introduction

Mobile robots are seeing increasing application in real world environments. However, for longer term deployments the planning of the robot's operation must take its resource constraints into account, which, for applications such as large scale farming, may include energy and other materials necessary to perform tasks around the farm—such as herbicide for weed spraying. Consideration of these constraints is especially important due to the often broad scale of the areas the robot is expected to operate over.

Additionally, to achieve robust autonomy, the robot must be able to adapt any mission level plan in response to the presence of both unknown and dynamic elements in these unstructured environments. Currently, autonomous field robots are limited to use in more structured environments, including row crops [1] or

S. Eiffert and N.D. Wallace—Contributed equally to this work.

B. Siciliano et al. (Eds.): ISER 2020, SPAR 19, pp. 151–160, 2021.
https://doi.org/10.1007/978-3-030-71151-1_14

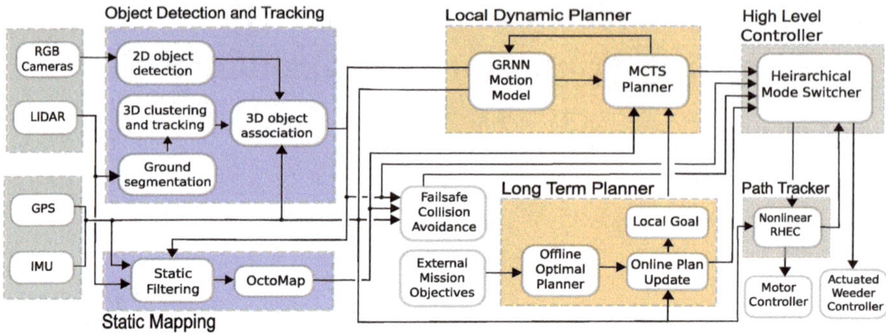

Fig. 1. System overview of the hierarchical framework, illustrating the communication between each planner, the failsafe collision module, and the high level controller. Mission objectives are provided externally to the long-term planner.

orchards [2], where dynamic elements such as moving individuals are not a critical consideration. These approaches generally make use of reactive planners, which can adapt to unforeseen static obstacles. Planning in the presence of moving individuals, however, requires an understanding of how they might respond to the planned motion of the robot, in order to avoid the 'frozen-robot problem' [3] which would otherwise occur in the presence of crowds and herds.

To address the above challenges, in our prior work [4] we have proposed a hierarchical framework for long-term and robust deployment of field ground robots in large-scale farming. In this paper, we present a real world evaluation of the former framework which integrates a local, interaction-aware dynamic path planner with a longer term resource aware mission planner to achieve long-term autonomy in unstructured and dynamic environments. Through trials conducted on the University of Sydney's Swagbot agricultural robot platform, we demonstrate that our framework is able to navigate an unstructured environment in an energy-efficient manner in the presence of moving individuals, enabling the completion of tasks such as weed spraying in farm environments.

2 Technical Approach

In our previous work [4], we developed a hierarchical framework for off-road, multi-objective robot navigation in the presence of both moving individuals and static obstacles. The performance of this framework was validated using simulated trials, in which moving agents were modelled using the ORCA reciprocal collision avoidance model of motion [5]. In the present work, we further extend this hierarchical framework—shown in Fig. 1—to include the operation of an actuated weeding arm, and evaluate its use in a real world agricultural environment, in the context of spraying a set of pre-identified weeds in the presence of moving individuals.

The hierarchical framework is composed of a few key components: an object detection and tracking module, a long-term planner, a local dynamic planner, a

Fig. 2. Swagbot Robotic Platform used for testing. (a) Photo of Swagbot with actuated weeder extended. (b) Top-down illustration of sensor FOV, and the blind spot arising due to the front-mounted sensor installation.

high level switching controller, and a path tracker. The multimodal perception solution used in this work utilises both LIDAR and RGB for object detection, as per our simulated work in [4]. However, due to limited sensor field of view (FOV), as shown in Fig. 2, detection is not possible directly behind the robot. The tracking module used has been updated to be aware of the sensor field of view, ensuring that updating of tracked object confidence based on missed observations is performed at a slower rate when it enters the blind spot, in order to maintain object permanence.

The long-term planner is based on [6–9], in which an energy efficient path is generated by searching over a Probabilistic Roadmap of the environment to find the lowest cost paths between all the goal waypoints—determined using a learnt energy cost of motion metric—and by then solving an asymmetric Travelling Salesman problem over this emergent graph to yield the optimal plan (alternatively, the long-term plan can also be generated using our recent work in [10] on the orienteering problem with replenishment, accounting for when, where, and for how long to recharge resources). This plan takes into consideration the topography of the area, the slip conditions of the terrain, and any previously known non-traversable areas. The local dynamic planner used in this work is adapted from our prior work in [11,12], using generative Recurrent Neural Networks (RNNs) and Monte Carlo Tree Search (MCTS). A high level controller switches between planning modes based on proximity to detected moving individuals, static obstacles, and waypoints.

3 Experimental Evaluation

Real world empirical evaluation of our approach has been conducted on the University of Sydney's Swagbot agricultural robot platform, a four-wheeled independently driven and steered electric ground vehicle capable of omnidirectional motion, which was designed for use in environments with uneven terrain. Testing

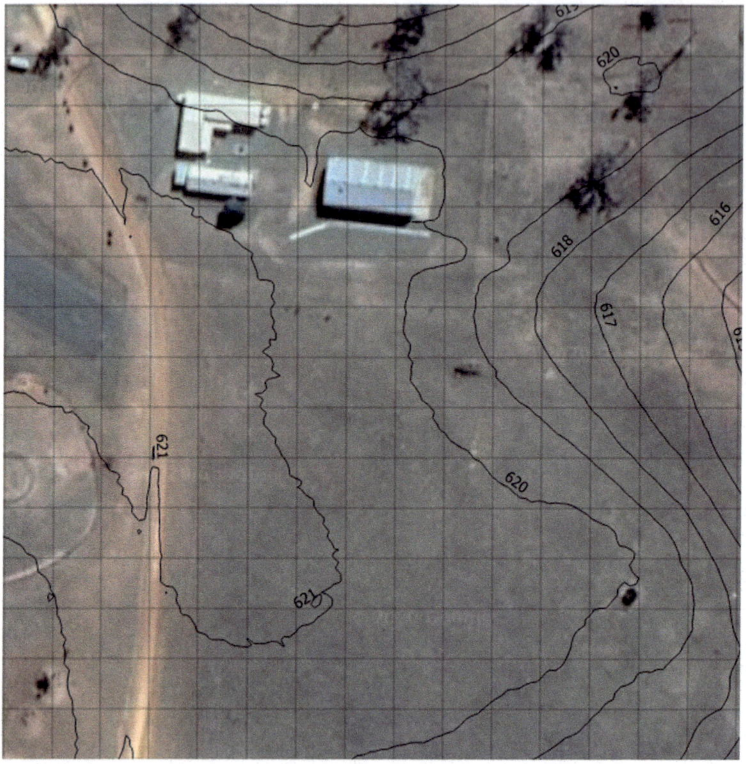

Fig. 3. Satellite imagery of the Arthursleigh Farm test site, where the experiments were conducted. Terrain topography indicated by the contour lines, and spatial extent indicated by grid lines spaced at 20 m intervals.

was performed at the University's Arthursleigh Farm—pictured in Fig. 3—and involved the continuous navigation between updated sets of externally provided mission waypoints across an unstructured field 2 ha in size. At each waypoint the robot was required to accurately spray a pre-located weed, operating in the presence of both moving individuals and unknown obstacles. Each iteration of the trial began at a home waypoint, where a set of 5–8 objective waypoints were supplied to the robot.

As per Sect. 2, an optimal energy-efficient path was then computed and used as the reference path for online local planning during navigation to each objective. Upon returning to base, a new set of waypoints were supplied and the trial repeated. A total of 3 sets of waypoints were reused, with testing continuing until the robot exhausted its energy resources. An overview map of the testing area showing the repeated route of the robot is illustrated in Fig. 4. Total time of the trial was 2 h 44 min, covering a distance of 5.49 km, including 37 distinct interactions with groups of moving agents. For the purposes of this trial, people—rather than livestock—were used as agents, both for safety, and due to the availability

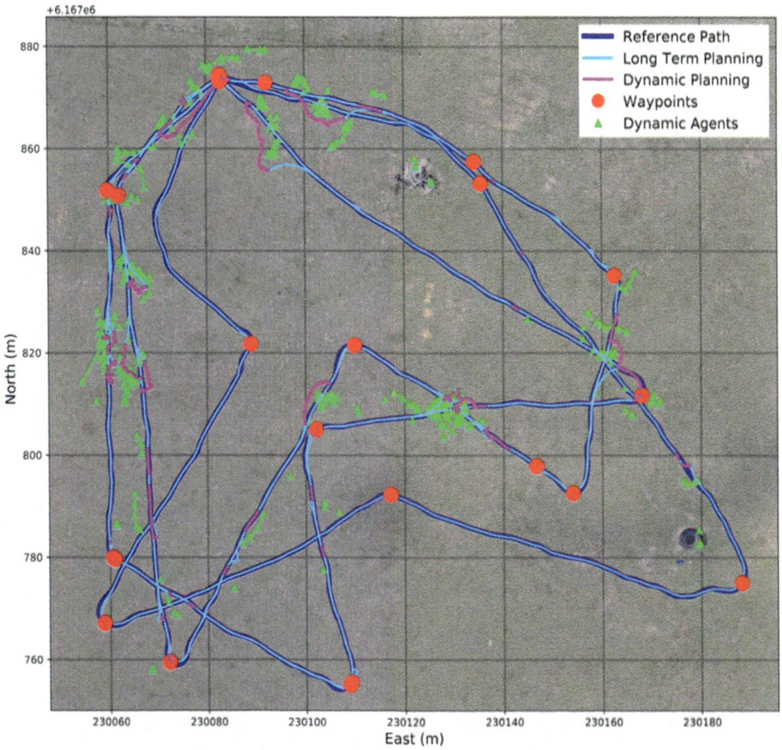

Fig. 4. Complete course, showing repeated route of the robot during the trial, all waypoints, and detected agents. The reference path (dark blue) is overlayed with the actual taken path, differentiating between when the long-term planner module (light blue) and dynamic planning module (purple) were each in use.

of a response prediction model of person-robot interactions from our prior work. Performance was determined using metrics of: (1) number of near-collisions with moving individuals; (2) average speed between waypoints; (3) deviation from the reference path.

4 Results

4.1 Dynamic Collision Avoidance

Throughout the trial 9.42% of time was spent in dynamic planning mode (a total of 15.5 min). Switching between dynamic and long-term planning modes is handled by the high level controller as shown in Fig. 1, dependent on proximity of detected agents both to the robot and the prior output of the path tracker module. As shown in Fig. 5, this allows the robot to continue in long-term mode even when in close proximity to nearby agents if the current planned path will not result in a collision. Figure 5 also demonstrates the ability of the dynamic planning mode to successfully avoid collisions and navigate around dynamic

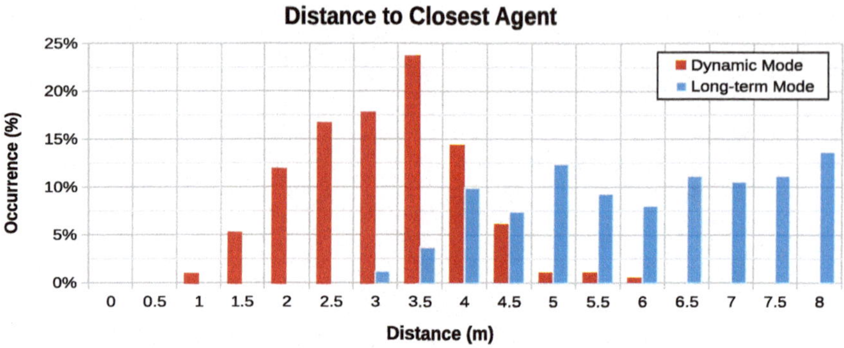

Fig. 5. Histogram of distances to closest agent by planner mode, shown as a percentage of occurrence of times when an agent was within 8 m. Dynamic planning mode is able to successfully take over from the long-term planner when in the presence of other agents and avoid collisions. Mode handover is dependent both on proximity of agents to the robot and to the prior planned path, resulting in the overlap shown.

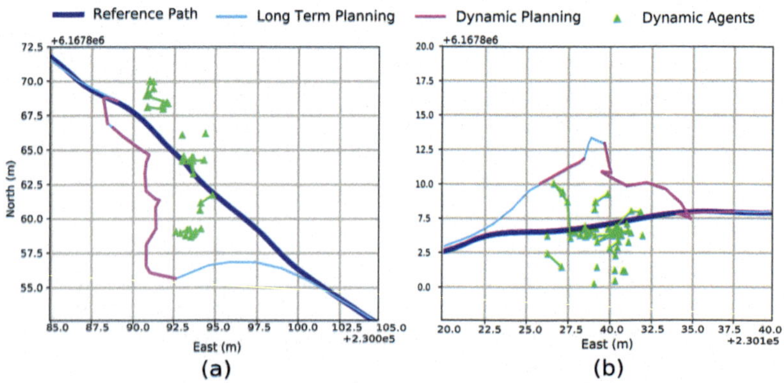

Fig. 6. Navigation around moving individuals, highlighting when the dynamic planning module was in use (purple). (a) the robot initially attempts to pass between agents, but instead takes wider route as the agents continue their path down. (b) less desirable behaviour, where tracking of nearby agents failed when the robot was facing away. Both examples show the robot returning to the reference path when clear of agents.

agents. The robot was able to maintain an average distance of 3.36 m from all agents whilst in dynamic mode. Figure 6 (a) shows an example of this desired behaviour, whilst example (b) shows the single occasion in which an agent came within the fail-safe distance of 1.5 m, reaching a minimum distance of 1.30 m before the robot stopped. During this interaction the robot turned on the spot without realising that a person was present in its 'blind spot'. This occurred both as a result of limited sensor field of view and inconsistent implementations of motor controller between prior simulated work in [4] and this experiment. In [4], planning was done using a non-holonomic simulated robot configured for

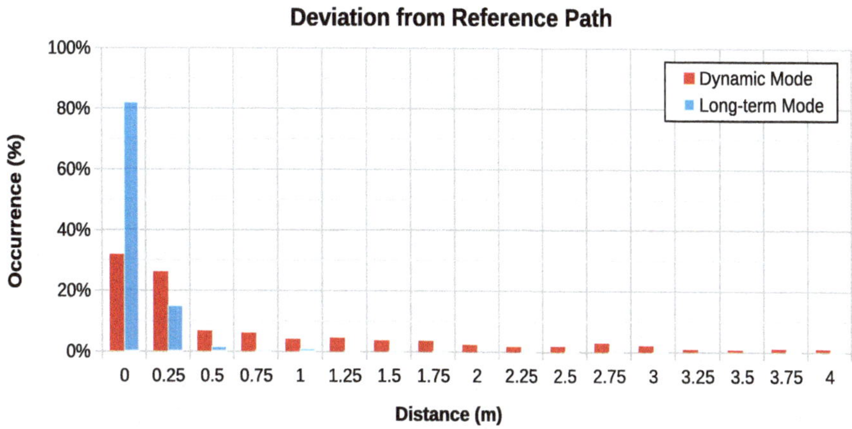

Fig. 7. Distances from the offline computed energy optimal reference path throughout the trial, by planner mode. Both the long-term and dynamic mode tend to remain close to the reference, however the dynamic planner can deviate significantly in order to navigate around nearby moving individuals, as shown by the protracted tail.

Ackermann motion, limiting motion to forward direction only. However, the real world robot is capable of omnidirectional motion, and while the planner constrained the robot to drive forwards, preventing it from travelling into its blind spot, it could nevertheless still turn on the spot whilst tracking both the dynamic and long-term paths.

4.2 Navigation Efficiency

Figure 7 illustrates how significant deviations from the prior optimal path can occur whilst navigating through groups of individuals, although for the most part the dynamic planner tends to remain close to the reference path when possible. This is desirable since keeping close to the long-term reference path will retain the energy efficiency merits and minimise resource usage. Similarly, a comparison of the velocity of the robot along the optimal path, compared to its overall velocity, allows an analysis of how much energy is required navigating around dynamic obstacles. Figure 8 highlights the impact that the presence of moving individuals has on the efficiency of the robot's navigation. As shown in Fig. 8(a), when in long-term planning mode, almost all of the robot's velocity is directed along the optimal path, with average velocity towards goal of 1.32 m/s nearly matching average overall velocity of 1.37 m/s. However when dynamically navigating around moving agents the robot achieves an overall velocity of only 0.60 m/s. Additionally, a significant amount of energy is directed away from the optimal path, as shown in Fig. 8(b). Over 35% of motion is directed away from the local goal resulting in an average velocity towards goal of 0.44 m/s, which is 33.3% of that achieved in long-term mode.

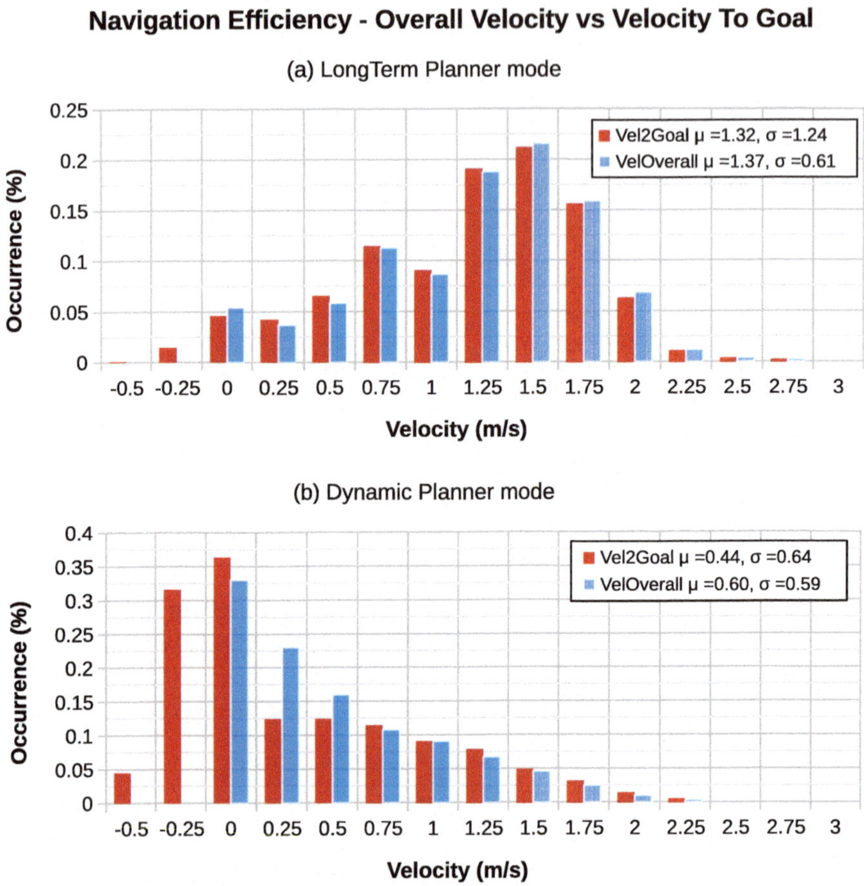

Fig. 8. Comparison of overall velocity (VelOverall) and velocity along the reference path (Vel2Goal) when in long-term planning mode (a) versus dynamic planning (b) illustrating the impact of nearby moving individuals on the robot's ability to follow the reference path. The average VelOverall in dynamic mode is less than half of that achieved in long-term mode. Additionally, the ratio of Vel2Goal to VelOverall is significantly less in dynamic mode, with over 35% of time spent travelling away from the optimal direction.

5 Discussion

This work has shown that the proposed system allows for safe operation of a mobile robot around moving individuals during the completion of tasks such as weeding an agricultural field. Overall, the presented results have successfully demonstrated the implementation of our hierarchical framework in the real world. These results validate the hierarchical framework for use in large scale farming, allowing the safe and efficient operation in dynamic and unstructured environments.

However, a number of outstanding challenges remain in order to extend the proposed system for long-term autonomy. Improved perception around moving individuals is required, including the ability to better detect and track agents on a robotic platform with a limited field of view, and the proper integration of the planner's dynamics and the real world platform dynamics. Additionally, real applications of long-term autonomy require inclusion of a recharging station, rather than simply designating a base waypoint as such as was done in this work. Whilst other works [13] have shown how this can be applied to mobile robots in agriculture, it will need to be included in the proposed hierarchical framework in future. Based on the results discussed in Sect. 4.2, an understanding of how dynamic adjustments to the offline computed optimal path influence energy usage and navigation efficiency should also be considered. This suggests the need for consideration of crowd or herd density during extended operation, to ensure mobile robots are able to effectively reach a recharging station when required.

References

1. Bechar, A., Vigneault, C.: Agricultural robots for field operations. Part 2: operations and systems. Biosyst. Eng. **153**, 110–128 (2017)
2. Carpio, R.F., et al.: A navigation architecture for Ackermann vehicles in precision farming. IEEE Robot. Autom. Lett. **5**(2), 1103–1110 (2020)
3. Trautman, P., Krause, A.: Unfreezing the robot: navigation in dense, interacting crowds. In: IEEE/RSJ International Conference on Intelligent Robots and Systems (IROS), pp. 797–803 (2010)
4. Eiffert, S., Wallace, N., Kong, H., Pirmarzdashti, N., Sukkarieh, S.: A hierarchical framework for long-term and robust deployment of field ground robots in large-Scale farming. In: Proceedings of IEEE CASE, pp. 948–954, Hong Kong (2020)
5. Van Den Berg, J., Guy, S.J., Lin, M., Manocha, D.: Reciprocal n-body collision avoidance. In: Proceedings of the 14th International Symposium ISRR. Springer Tracts in Advanced Robotics, vol. 70, pp. 3–19 (2011)
6. Wallace, N., Kong, H., Hill, A., Sukkarieh, S.: Receding horizon estimation and control with structured noise blocking for mobile robot slip compensation. In: Proceedings of IEEE ICRA, pp. 1169–1175, Montreal, Canada (2019)
7. Wallace, N., Kong, H., Hill, A., Sukkarieh, S.: Structured noise blocking strategies for receding horizon estimation and control of mobile robots with slip. In: Proceedings of Australasian Conference on Robotics and Automation (ACRA), pp. 1–10 (2018)
8. Wallace, N., Kong, H., Hill, A., Sukkarieh, S.: Experimental validation of structured receding horizon estimation and control for mobile ground robot slip Compensation. In: Proceedings of the 12th Conference on Field and Service Robotics, pp. 1–16 (2019)
9. Wallace, N., Kong, H., Hill, A., Sukkarieh, S.: Motion cost characterisation of an omnidirectional WMR on uneven terrains. In: Proceedings of the 12th IFAC CAMS and 1st IFAC WROCO, vol. 52, No. 22, pp. 31–36, Daejeon, Korea (2019)
10. Wallace, N., Kong, H., Hill, A., Sukkarieh, S.: The orienteering problem with replenishment. In: Proceedings of the IEEE CASE, pp. 973–978, Hong Kong (2020)

11. Eiffert, S., Kong, H., Pirmarzdashti, N., Sukkarieh, S.: Path planning in dynamic environments using Generative RNNs and Monte Carlo tree search. In: Proceedings of IEEE ICRA, pp. 10263–10269, Paris, France (2020)
12. Eiffert, S., Sukkarieh, S.: Predicting responses to a robot's future motion using generative recurrent neural networks. In: Proceedings of Australasian Conference on Robotics and Automation (ACRA) (2019)
13. Varandas, L., Gaspar, P., Aguiar, M.: Standalone docking station with combined charging methods for agricultural mobile robots, world academy of science, engineering and technology, open science index 145. Int. J. Mech. Mechatron. Eng. **13**(1), 38–42 (2019)

Autonomous Off-Road Navigation over Extreme Terrains with Perceptually-Challenging Conditions

Rohan Thakker[1], Nikhilesh Alatur[2(✉)], David D. Fan[1], Jesus Tordesillas[3], Michael Paton[1], Kyohei Otsu[1], Olivier Toupet[1], and Ali-akbar Agha-mohammadi[1]

[1] Jet Propulsion Laboratory (JPL), California Institute of Technology (Caltech), 4800 Oak Grove Drive, Pasadena, CA, USA
{rohan.a.thakker,david.d.fan,michael.paton,kyohei.otsu,olivier.toupet, aliakbar.aghamohammadi}@jpl.nasa.gov

[2] Swiss Federal Institute of Technology, Zurich, Switzerland
alaturn@ethz.ch

[3] Massachusetts Institute of Technology, Cambridge, USA
jtorde@mit.edu

Abstract. We propose a framework for resilient autonomous navigation in perceptually challenging unknown environments with mobility-stressing elements such as uneven surfaces with rocks and boulders, steep slopes, negative obstacles like cliffs and holes, and narrow passages. Environments are GPS-denied and perceptually-degraded with variable lighting from dark to lit and obscurants (dust, fog, smoke). Lack of prior maps and degraded communication eliminates the possibility of prior or off-board computation or operator intervention. This necessitates real-time on-board computation using noisy sensor data. To address these challenges, we propose a resilient architecture that exploits redundancy and heterogeneity in sensing modalities. Further resilience is achieved by triggering recovery behaviors upon failure. We propose a fast settling algorithm to generate robust multi-fidelity traversability estimates in real-time. The proposed approach was deployed on multiple physical systems including skid-steer and tracked robots, a high-speed RC car and legged robots, as a part of Team CoSTAR's effort to the DARPA Subterranean Challenge, where the team won 2nd and 1st place in the Tunnel and Urban Circuits, respectively.

R. Thakker, N. Alatur, D.D. Fan and J. Tordesillas—Equally contributed.
N. Alatur and J. Tordesillas—Work done while at the JPL, Caltech.

B. Siciliano et al. (Eds.): ISER 2020, SPAR 19, pp. 161–173, 2021.
https://doi.org/10.1007/978-3-030-71151-1_15

(a) Rough Terrain

(b) Stairs (Legged) (c) Cluttered environment (d) Sloped terrain

(e) Negative Obstacles (f) Mud (g) Stairs (Tracked)

Fig. 1. Mobility-stressing components at test sites: a) Arch Coal Mine, WV, b,c,e) Powerplant in Satsop Business Park, WA, d) Mars Yard, NASA-JPL, CA,and f) Beckley Exhibition Mine, WV. Robots demonstrated long-range autonomous navigation capability, such as in Arch Mine (1100 m), Beckley Exhibition Mine (1000 m).

Supplementary material: A video is available at https://youtu.be/cjxN2Tu-qVY

1 Introduction

Autonomous off-road navigation over extreme terrains and perceptually-degraded environments has a wide range of applications from space exploration [1] to terrestrial search and rescue, including subterranean (tunnels, mines, caves) exploration [2]. This poses a set of **key challenges**:

[C1] *Negotiating mobility-stressing elements* such as rough, high-sloped, and cluttered terrains with stairs and narrow passages while avoiding hazards such as rocks, overhangs, and negative obstacles including pits, cliffs, etc. (Fig. 1).

[C2] *Perceptually degraded GPS-denied environments* require the robots to cope with large uncertainties and degraded sensing due to presence of obscurants (dust, fog, smoke), low-light conditions, lack of features/landmarks, motion blur and issues with dynamic range (Fig. 4).

[C3] *Resilient self-reliant operations in comms-denied environment* without any operator intervention are required due to degraded communication. The robot is also required to return back to communication range if the mission is infeasible or violates the desired risk posture.
[C4] *On-board operations in unknown environments* at high-speeds are required since lack of prior maps and degraded communications eliminate any possibility of prior or off-board processing and finite mission time constraints make operations at maximum speed supported by hardware, desired.
[C5] *Scalability to different environments and robots* such as tracked, wheeled and legged which have disparate speed and traversability abilities (Table 1).

Related Work: Papadakis [3] summarises the prior work on traversability estimation by classifying them in proprioceptive methods and exteroceptive methods. [4] performs fast, approximate settling by computing upper and lower bounds for the settled pose. In [5], the authors use principal component analysis on the local surface to estimate the settled pose. The settling algorithms can operate on instantaneous scans [6] which are robust to localization noise or by fusing multiple scans [7] that provide a longer range.

The work in [8] runs an iterative algorithm to estimate the contact points between an articulated, tracked robot and the terrain mesh [9] for settling. [10] proposes a computationally fast method to run a physics simulation that drops the robot under the influence of gravity onto the surface, which is represented as B-Patch [11]. [12] provides a probabilistic formulation by extending a discrete elevation map [13] to account for uncertainty in localization. The objective of this work is to develop a holistic solution that can address all challenges [C1]–[C5], simultaneously.

Contributions:

1. To address these challenges, we propose a Resilient Multi-fidelity Architecture (RMF) design which shows how these different prior works can be combined to form a resilient end-to-end system. RMF leverages a multi-fidelity traversability estimator which generates accurate estimates in short-range by using instantaneous scans and approximate estimates in mid-range and long-range by using a temporal map. Furthermore, it is designed to eliminate single-point failures and achieve further resiliency by exploiting redundancy and heterogeneity of different sensing modalities [C2].
2. We propose a fast settling algorithm that estimates traversability in extreme terrains [C1] for multiple robot types [C5] in real time [C4].
3. We propose a sequence of field-tested autonomous recovery behaviors to eliminate human interventions [C3].
4. We push the boundaries of state-of-practice and demonstrate the proposed solution through deployment on a variety of ground vehicles including wheeled (Ackermann and skid-steered), tracked, and legged systems, by conducting intense field testing in various subterranean environments (Fig. 1, Table 1).

The proposed architecture was successfully deployed by Team CoSTAR[1] as our local navigation solution for large-scale, subterranean exploration and mapping missions during the DARPA Subterranean (SubT) Challenge[2]. Together with other autonomy modules for large-scale localization and mapping [14], resilient state estimation [15] and perception-aware global planning [16], it formed the overall autonomy solution called *NeBula* (Networked Belief-aware Perceptual Autonomy) [17] that led to Team CoSTAR winning 2nd place in the Tunnel Circuit and 1st place in the Urban Circuit of the SubT Challenge.

2 Methodology

We push the boundaries of the state-of-practice by enabling on-board navigation in complex rough terrain under severe perceptual, computational and real-time constraints. In this section, we discuss an architecture and supporting techniques of the proposed system.

2.1 Resilient Multi-Fidelity Architecture

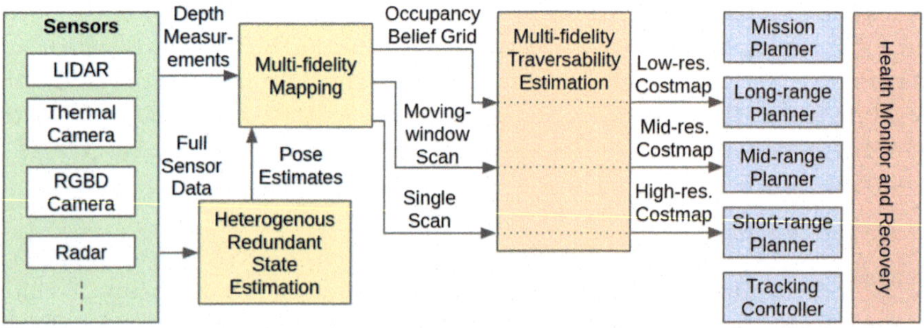

Fig. 2. Resilient Multi-Fidelity (RMF) architecture.

Heterogeneity and Redundancy of Sensors to Eliminate Single-Point Failures: Figure 2 shows our pipeline to generate the multi-fidelity traversability cost estimates for motion planning. To design a resilient system capable of extreme terrain navigation, we start by building a stack of sensors with heterogeneous modalities such as lidar, radar, thermal and visible cameras, etc. Further redundancy is achieved by adding multiple sensors of the same modality, pointing in different directions to achieve larger coverage and eliminate single point failures.

Heterogeneous Resilient State Estimation: These sensors signals are process by our estimation module HeRO (Heterogeneous Redundant Odometry) [15] that further adds additional analytical redundancy by running multiple

[1] https://costar.jpl.nasa.gov/.
[2] https://www.subtchallenge.com/.

odometry algorithms [18–20], performs confidence checks to quantify uncertainty and finally produces robust pose estimates by multiplexing to the most robust odometry source. This provides resiliency to state estimation failures which are usually triggered by many factors like dust, low-light conditions, lack of features/landmarks, motion blur and issues with dynamic range.

Robust Multi-Fidelity Traversability Estimation in Real-Time: To achieve further robustness to noise in perception, we perform traversability estimation at three different levels. First, a high-resolution short-range traversability costmap is constructed using the most recent depth scan which does not suffer from localization noise. This allows robust high-fidelity traversability estimation by preventing the propagation of localization noise into the planning and control layers. Second, mid-resolution mid-range planner uses a moving window of last N depth scans for traversability estimation which allows coverage over a larger region.

While these depth estimates are sensitive to localization noise, the traversability estimation is less susceptible to this noise since it is performed at a lower fidelity than the high-resolution costmap. Specifically, features which are within the magnitude of perception noise are ignored (e.g. rocks smaller than 30 cm). Third, the low-resolution long-range costmap uses an occupancy belief grid that maintains a probability distribution of occupancy which is a sufficient statistic of all past depth measurements [21,22]. This allows for incorporating highly noisy yet informative measurements at long ranges. The low-resolution costmap achieves robustness to this noise by estimating traversability at the least fidelity (e.g. only detecting large obstacles such as walls). This multi-fidelity approach is an efficient and practical way of handling perception noise without adding too much computational burden. In the next subsection, we describe the details of the settling algorithm used to estimate traversability.

Real-Time On-Board Planning and Control: To achieve real-time performance with limited on-board computation, the planning and controls problems is decoupled into several hierarchical layers. The mission planner keeps track of high-level states such as battery status, time remaining, coverage goals, etc. [23] and generates a desired goal for the long-range planner. The long-range planner finds a path on a long-range roadmap (0–10 km range) and generates a goal for the mid-range planner. The mid-range planner performs search on the mid-resolution costmap (0–50 m range) to obtain a geometric path to this goal which is tracked by the local kinodynamic planner by generating an optimal trajectory from a set of motion primitives prioritized by using the high-resolution costmap (0–10 m range). Finally, the low-level tracking controller tracks this kinodynamic trajectory. [17] describes the details of how this was implemented for a legged robot.

Resilience at System-Level: Note that while the long-range and mid-range planner can produce intermittent unsafe goals due to presence of localization errors, our system is still resilient to these failures since the short-range planner closes the loop using only instantaneous or near-instantaneous scans (past <1 s), thus eliminating past errors in localization. The health monitor asynchronously monitors the system state to trigger recovery behaviors when failures are detected.

2.2 Settling-Based Traversability Analysis

This section describes how traversability is assessed over a point cloud for a single query pose. The traversability assessment for a query pose consists of three steps. First, the input surface point cloud is segmented into ground and obstacle points. Second, the robot is virtually *settled* on a ground point cloud at a given query pose. Third, various traversability metrics are computed from the settled pose, surface point cloud, and its interaction.

The algorithm begins by first segmenting the surface point cloud into ground and obstacle points. In rough terrain (which is a target of this paper), the boundary between ground and obstacles is vague. To have a consistent method across different environments, we set a following definition based on the mechanical capability of the robot platform: ground points are surface points of an area where the robot can mechanically drive, and obstacle points are remainder points that locate above the ground points. Multiple implementations are possible for this segmentation task. We employed a simple geometric approach based on line-fitting [24].

After extracting the ground point cloud, a robot model is placed on arbitrary locations on the point cloud to evaluate traversability. The process is depicted in Fig. 3. A settling algorithm is used to solve for an $SE(3)$ pose of the robot given an $SE(2)$ query pose. Different settling methods should be considered based on the platform types. One approximation method that is applicable to many ground vehicles is to fit a plane using points under the robot footprint.

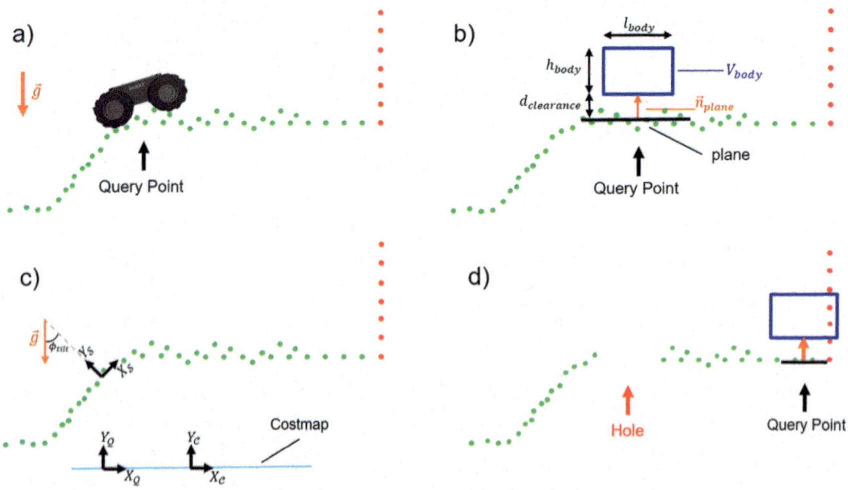

Fig. 3. Settling-based traversability: a) settling, b) body collision model, c) stability check, d) collision detection and negative obstacle detection using low point density.

The normal of fitted plane is used to approximately estimate the orientation of settled pose and derive the placement of the body collision model.

Once a full pose is determined for a query pose via settling, the following metrics are computed using the pose and point cloud. The final traversability is evaluated based on the combination of these metrics:

- *Tip-over stability*: The robot must avoid steep slopes, where it could tip over. In theory, the robot's footprint and orientation highly influence the stability of the robot. In [8], the authors list several stability metrics that can be used for stability checking. However, for robots whose footprint have an aspect ratio close to one, one can simplify the stability analysis by applying a threshold on the angle between the surface normal and the gravity vector. This stability check is depicted in Fig. 3c.
- *Positive obstacles*: Positive obstacles, such as rocks below the robot's belly, overhangs, walls, or larger obstacles, may collide with robot's body. The surface points that interfere with the robot's collision model are counted, and if the number of colliding points is larger than a threshold, that location is marked as untraversable, see Fig. 3d.
- *Negative obstacles*: Negative obstacles include cliffs and holes, which can severely damage the vehicle if they are not avoided. We create two categories based on sensor coverage. If the shape of negative obstacle is fully visible, the previous slope and collision checks are applied to evaluate if it is safe to drive. If the negative space is occluded or not visible, we check the density of points around the negative space. If too few ground points are available, we assume the worst case and mark the area as untraversable.

By changing the size of the robot bounding box used in the settling algorithm, we obtain different fidelities of traversability estimates. A larger bounding box will generate higher fidelity estimates which can detect hazards such as very small rocks, while a smaller bounding box will result in lower fidelity estimates to detect large hazards such as walls. A larger size will be more sensitive to localization noise and hence is only applied to near-instantaneous measurements, whereas a smaller size is robust to localization noise and can be applied to temporally fused depth maps.

2.3 Recovery Behaviors

In the real world, failures are unavoidable. Some failures (e.g., mechanical failure) are fatal, while others (e.g., algorithmic failure) can potentially be resolved autonomously, allowing the robot to continue the mission. Hence, as a last line-of-defense, we design behaviors to recover the system from non-fatal failures. The following recovery behaviors are implemented and triggered when relevant failures are detected:

- Clear the map and start building it from scratch.
- Backtrack the traversed path.
- Move the robot in an open loop to the direction of maximum clearance.
- Trigger wall following (which does not require global localization).

3 Experiments

The proposed system has been extensively tested and validated in many different natural and man-made environments, including coal/gold mines, power plants, school/office buildings, and the JPL Mars Yard; as shown in Fig. 1. The navigation system was successfully used by Team CoSTAR at the DARPA SubT Challenge Circuit events (Tunnel and Urban), where the robots autonomously explored more than a kilometer in each run.

3.1 Hardware and Software

The experimental platforms are shown at the top of Table 1. The platforms were heavily customized with a heterogeneous sensor suite, computing units, batteries and speed controllers. The odometry sources consist of wheel-inertial odometry (WIO), lidar-inertial odometry (LIO), visual-inertial-odometry (VIO), thermal-inertial-odometry (TIO) and kinematic-inertial-odometry (KIO). HeRO [15] monitors the different inputs and selects the most reliable source to produce a continuous state estimate under perceptually degraded conditions. The proposed architecture is implemented with the ROS navigation framework[3].

Table 1. Specification of the different robots. Superscript on the robot type indicates location/source of data used for computing statistics: [1]DARPA SubT Urban Competition, [2]DAPRA SubT Tunnel Competition, [3]Beckley Exhibition Coal Mine, WV.

Specification	Skid-steer[1]	Tracked[3]	Ackermann[2]	Quadruped[1]
Model	Husky A200	Telemax Pro	X-Maxx	Spot
Max. speed	1.0 m/s	1.1 m/s	22 m/s	1.6 m/s
Local planner	DWA [25]	DWA [25]	TEB [26]	Proprietary
Controller	PID	PID	PID	Proprietary
Distance traveled	5.98 km	0.76 km	0.39 km	2.85 km
Avg. speed during traverse	0.93 m/s	0.57 m/s	0.83 m/s	0.65 m/s
Autonomous recoveries / km	6.9	0.16	0.0	13.0
Critical failures / km	0.2	0.0	0.0	1.1

[3] http://wiki.ros.org/navigation.

3.2 Results

The internals of the proposed algorithm are visualized in Fig. 4, for a mission in the Arch Coal Mine in Beckley, WV. The costmaps for different traversability challenges, including positive obstacles, negative obstacles and rubble, are shown in Fig. 5 for the wheeled platform A. Figure 6 visualizes the costmap for the Ackerman platform in a gold mine, for the quadruped in the power plant and for the tracked platform on stairs. With a modified settling algorithm [4], we were able to demonstrate autonomous stair climbing with the tracked platform. The difference boils down to a more conservative stability assessment that considers the attainable worst case attitudes instead of average attitudes that result from simple plane fitting. The addition of conservatism is motivated by the higher slope and greatly reduced contact area, which makes stairs a very challenging terrain to drive on.

Figure 7 shows ≈1 h of autonomous exploration and mapping in various extreme environments. These environments include the locales for the DARPA Subterranean Challenge. Utilizing our resilient navigation methods, both wheeled and legged robots were able to successfully navigate autonomously and in comm-denied environments over kilometer-range distances.

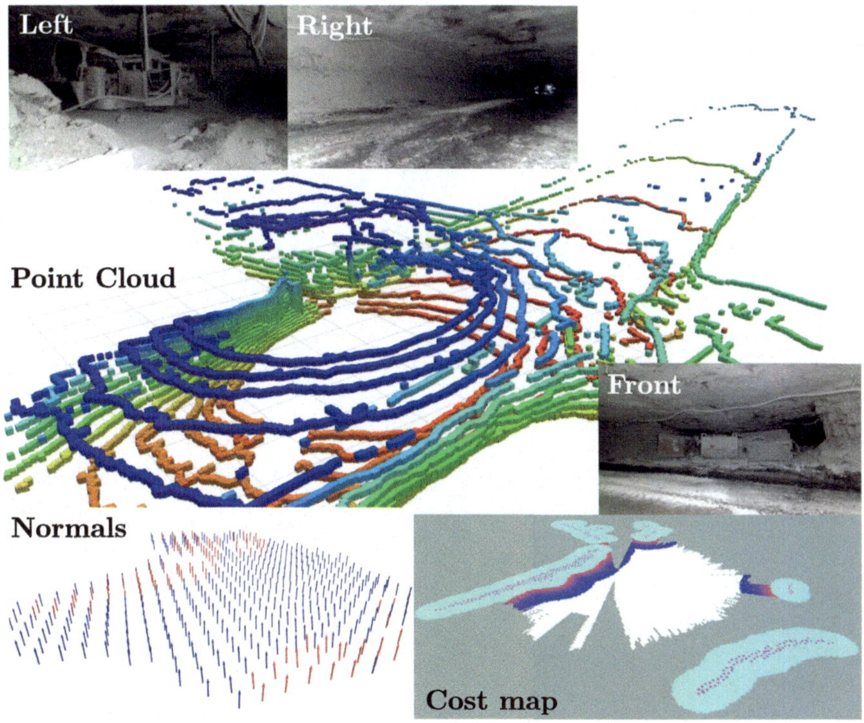

Fig. 4. Traversability estimation pipeline visualization showing point cloud from sensors, normal vector of settled robot pose (hazards marked in red) and costmap at Arch coal mine.

3.3 Lessons Learned and Future Work

The following is a summary of the lessons learnt from our intense test campaign; to stem future research:

- **Blind Spots in Sensing:** Our architecture uses instantaneous depth measurements for generating optimal local kinodynamic trajectories in the robot's ego frame. While this prevents any localization noise from percolating into the costmap, it requires large number of sensors to ensure sufficient coverage and density. Violation of this requirement can result in false negative obstacles especially in the blind-spots of the sensors. This problem can be mitigated with better sensor coverage or perception-aware planning.
- **Very Small Obstacles and Narrow Passages:** It is challenging to detect obstacles whose size is comparable to the noise levels of the depth sensors. Similar challenge exists while navigating through passages whose size is very close to the robot width. This issues can be potentially resolved with better ground clearance and smaller width of the robot.

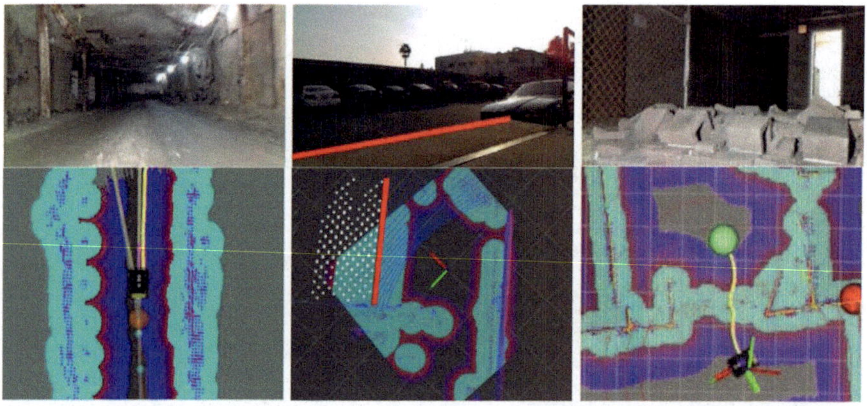

Fig. 5. Left to right: positive obstacles, negative obstacles, rubble

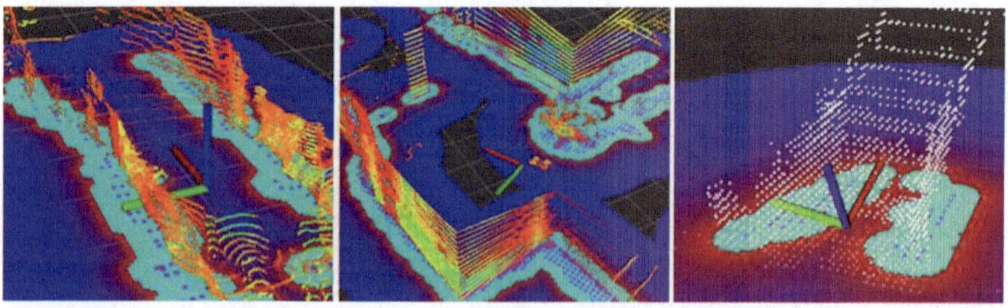

Fig. 6. Left to right: skid-Steer in mine, quadruped in power plant, tracked on stairs

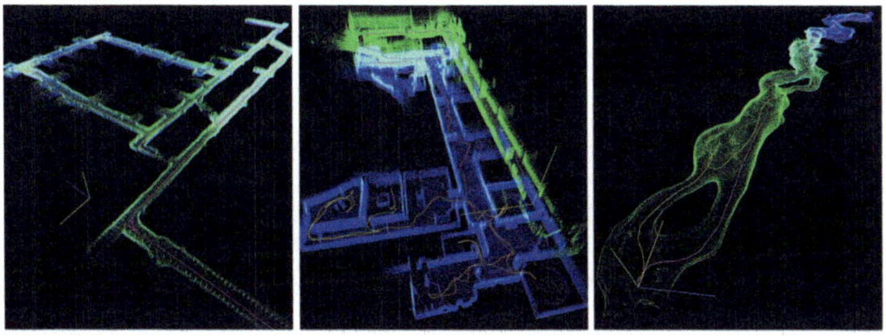

Fig. 7. Extreme environments successfully explored and mapped using various robotic platforms. From left to right: Husky in a coal mine environment, Spot in a power plant environment, Spot in a lava-tube cave environment. Colors of points (blue to green) indicate z height. Purple to yellow line indicates path traveled by robot. Blue/Yellow/Green axes indicate 10 m scale.

- **Non-Geometric Hazards:** There were a few cases where a robot was immobilized due to non-geometric hazards such as mud and water. These hazards mainly triggered slips on the ground contact points and made the wheeled robots get stuck and the legged robots fall. One potential solution to this problem is the use of machine learning techniques to detect these non-geometric hazards.

4 Conclusion

A resilient solution to navigation in environments with extreme traversability and perceptual conditions (Fig. 1) was proposed by leveraging a multi-fidelity traversability estimation using a combination of instantaneous depth scans and occupancy belief maps constructed using localization that exploited redundancy and heterogeneity in sensing. The proposed solution was adaptive, real-time, on-board and works without any prior maps or GPS. Single-point failures were further eliminated by using health monitoring and recovery behaviors. Through intense field test campaigns, we demonstrated that the proposed approach scales to different environments and four different types of robots. Future work includes addressing non-geometric hazards such as mud, dirt, etc. and testing on platforms with higher speed.

Acknowledgements. The authors would like to thank Anushri Dixit and Joel Burdick for support with the Ackermann robot and members of Team CoSTAR for their hardware and testing support. The work is partially supported by the Jet Propulsion Laboratory, California Institute of Technology, under a contract with the National Aeronautics and Space Administration (80NM0018D0004), and Defense Advanced Research Projects Agency (DARPA).

References

1. Haruyama, J., Morota, T., Kobayashi, S., Sawai, S., Lucey, P.G., Shirao, M., Nishino, M.N.: Lunar holes and lava tubes as resources for lunar science and exploration. In: Moon, pp. 139–163. Springer, Heidelberg (2012)
2. Murphy, R.R.: Disaster Robotics. MIT Press, Cambridge (2014)
3. Papadakis, P.: Terrain traversability analysis methods for unmanned ground vehicles: a survey. Eng. Appl. Artif. Intell. **26**(4), 1373–1385 (2013)
4. Otsu, K., Matheron, G., Ghosh, S., Toupet, O., Ono, M.: Fast approximate clearance evaluation for rovers with articulated suspension systems. J. Field Robot. **37**(5), 768–785 (2019)
5. Krüsi, P., Furgale, P., Bosse, M., Siegwart, R.: Driving on point clouds: motion planning, trajectory optimization, and terrain assessment in generic nonplanar environments. J. Field Robot. **34**(5), 940–984 (2017)
6. Florence, P., Carter, J., Tedrake, R.: Integrated perception and control at high speed: evaluating collision avoidance maneuvers without maps. In: Workshop on the Algorithmic Foundations of Robotics (WAFR) (2016)
7. Florence, P.R., Carter, J., Ware, J., Tedrake, R.: NanoMap: fast, uncertainty-aware proximity queries with lazy search over local 3D data. In: IEEE International Conference on Robotics and Automation (ICRA)
8. Brunner, M., Fiolka, T., Schulz, D., Schlick, C.M.: Design and comparative evaluation of an iterative contact point estimation method for static stability estimation of mobile actively reconfigurable robots. Robot. Auton. Syst. **63**(P1), 89–107 (2015)
9. Brunner, M., Brüggemann, B., Schulz, D.: Autonomously traversing obstacles: metrics for path planning of reconfigurable robots on rough terrain. In: International Conference on Informatics in Control, Automation and Robotics (ICINCO), vol. 2, pp. 58–69 (2012)
10. Ma, Y., Shiller, Z.: Pose Estimation of Vehicles Over Uneven Terrain (2019)
11. Iagnemma, K., Shimoda, S., Shiller, Z.: Near-optimal navigation of high speed mobile robots on uneven terrain. In: IEEE/RSJ International Conference on Intelligent Robots and Systems, pp. 4098–4103 (2008)
12. Fankhauser, P., Bloesch, M., Hutter, M.: Probabilistic terrain mapping for mobile robots with uncertain localization. IEEE Robot. Autom. Lett. **3**(4), 3019–3026 (2018)
13. Goldberg, S.B., Maimone, M.W., Matthies, L.: Stereo vision and rover navigation software for planetary exploration. IEEE Aerosp. Conf. **5**, 2025–2036 (2002)
14. Ebadi, K., Chang, Y., Palieri, M., Stephens, A., Hatteland, A., Heiden, E., Thakur, A., Morrell, B., Carlone, L., Agha-Mohammadi, A.: LAMP: Large-scale autonomous mapping and positioning for exploration of perceptually-degraded subterranean environments. In: IEEE International Conference on Robotics and Automation, pp. 80–86 (2020)
15. Santamaria, A., Thakker, R., Fan, D.D., Morrell, B., Agha, A.: Towards resilient autonomous navigation of drones. In: International Symposium on Robotics Research (2019)
16. Agha-Mohammadi, A., Chakravorty, S., Amato, N.: FIRM: sampling-based feedback motion planning under motion uncertainty and imperfect measurements. Int. J. Robot. Res. **33**(2), 268–304 (2014)

17. Bouman, A., Ginting, M.F., Alatur, N., Palieri, M., Fan, D.D., Touma, T., Pailevanian, T., Kim, S.-K., Otsu, K., Burdick, J., Agha-Mohammadi, A.: Autonomous spot: long-range autonomous exploration of extreme environments with legged locomotion. In: IEEE/RSJ International Conference on Intelligent Robots and Systems, pp. 1057–1064 (2020)
18. Ebadi, K., Chang, Y., Palieri, M., Stephens, A., Hatteland, A., Heiden, E., Thakur, A., Funabiki, N., Morrell, B., Wood, S., et al.: LAMP: large-scale autonomous mapping and positioning for exploration of perceptually-degraded subterranean environments. In: 2020 IEEE International Conference on Robotics and Automation (ICRA), pp. 80–86. IEEE (2020)
19. Kramer, A., Stahoviak, C., Santamaria-Navarro, A., Agha-Mohammadi, A.-A., Heckman, C.: Radar-inertial ego-velocity estimation for visually degraded environments. In: 2020 IEEE International Conference on Robotics and Automation (ICRA), pp. 5739–5746. IEEE (2020)
20. Tagliabue, A., Tordesillas, J., Cai, X., Angel Santamaria-Navarro, A., How, J.P., Carlone, L., Agha-Mohammadi, A.-A.: Lion: Lidar-inertial observability-aware navigator for vision-denied environments. In: International Symposium on Experimental Robotics (2020)
21. Agha-Mohammadi, A.-A., Heiden, E., Hausman, K., Sukhatme, G.: Confidence-rich grid mapping. Int. J. Robot. Res. **38**(12–13), 1352–1374 (2017)
22. Hornung, A., Wurm, K.M., Bennewitz, M., Stachniss, C., Burgard, W.: OctoMap: an efficient probabilistic 3D mapping framework based on octrees. Auton. Robots **34**(3), 189–206 (2013)
23. Otsu, K., Tepsuporn, S., Thakker, R., Vaquero, T.S., Edlund, J.A., Walsh, W., Miles, G., Heywood, T., Wolf, M.T., Agha-Mohammadi, A.-A.: Supervised autonomy for communication-degraded subterranean exploration by a robot team. In: IEEE Aerospace Conference, pp. 1–9 (2020)
24. Himmelsbach, M., Hundelshausen, F.V., Wuensche, H.-J.: Fast segmentation of 3D point clouds for ground vehicles. In: Intelligent Vehicles Symposium (IV), 2010 IEEE, pp. 560–565. IEEE (2010)
25. Fox, D., Burgard, W., Thrun, S.: The dynamic window approach to collision avoidance. IEEE Robot. Autom. Mag. **4**(1), 23–33 (1997)
26. Rosmann, C., Hoffmann, F., Bertram, T.: Kinodynamic trajectory optimization and control for car-like robots, pp. 5681–5686 (2017)

Online Soil Classification Using a UAS Sensor Emplacement System

Adam Plowcha(✉), Jacob Hogberg, Carrick Detweiler, and Justin Bradley

Department of Computer Science and Engineering, University of Nebraska-Lincoln, Lincoln, NE 68588, USA
adam.plowcha@huskers.unl.edu, {jhogberg,jbradley}@unl.edu, carrick@cse.unl.edu

Abstract. Deployment of sensors in hard-to-access locations can improve data gathering for scientific studies. We have developed a sensor emplacement system that can be mounted to unmanned aircraft systems with vertical takeoff and landing capabilities to autonomously auger a sensor into the ground. Various techniques can be chosen to enhance the augering process when certain characteristics of the soil are known. Moisture content and compressive strength are the soil characteristics that most impact the augering process, yet directly measuring them would require additional sensors to an already-burdened airframe. We address this through a novel means of predicting these soil characteristics within the first 30 s of an average 85 s augering evolution using onboard sensors and a Gaussian process regression scheme that predicts the soil moisture content and compressive strength with accuracy of 86.53% and 90.53% of the respective measured values.

Keywords: Soil classification · Field robotics · Machine learning

1 Motivation and Problem Statement

Remote deployment of sensors in hard-to-access locations can enable improved data gathering for scientific study. Some sensors, such as seismic or soil moisture sensors, function best when placed into the soil. We have developed an in-ground sensor emplacement system for an unmanned aircraft system (UAS) capable of remotely augering these types of sensors into the soil (Fig. 1).

This material is based upon work supported by US Strategic Command through the 55th Contracting Squadron under Contract No. FA4600-12-D-9000 and was also funded, in part, by NSF IIS-1925368, IIS-1638099, IIS-1925052 and USDA-NIFA 2017-67021-25924.

B. Siciliano et al. (Eds.): ISER 2020, SPAR 19, pp. 174–184, 2021.
https://doi.org/10.1007/978-3-030-71151-1_16

The general concept of operations for our system is as follows. The UAS arrives in the area where the sensor is to be placed and lands. It then uses a custom augering mechanism to drill a sensor into the ground [8, 13]. If the system cannot successfully emplace the sensor into the ground, then it relocates to a new location and tries again. If sensor emplacement is successful, then the UAS departs the area.

Fig. 1. Sensor emplacement system. a) Augering b) Emplaced sensor

The sensor is housed inside the body of the auger which is left behind at the completion of the emplacement sequence. The sensor can be any sort of generic device that can fit inside the 100 mm long by 35 mm diameter hollow section of the auger body (see Fig. 2). The emplacement system is mechanically robust and allows for fine control of the downforce delivered on bit as well as the revolutionary speed of the auger itself. The auger's vertical position in the soil column and the downforce on bit is controlled by an advanced elevator platform that allows for rapid up and down movement enabling us to employ a technique known as "pecking" [4,6]. The upward movement of the pecking motion allows soil that has been broken up and potentially clogging the lower portions of the auger flutes to be transported up and out of the hole. This creates space for the soil in the bottom of the hole to move into the newly vacant flute areas when the auger is pushed back down into the hole. Soil parameters determine the choice of an effective pecking profile, i.e. the speed and distance of the peck.

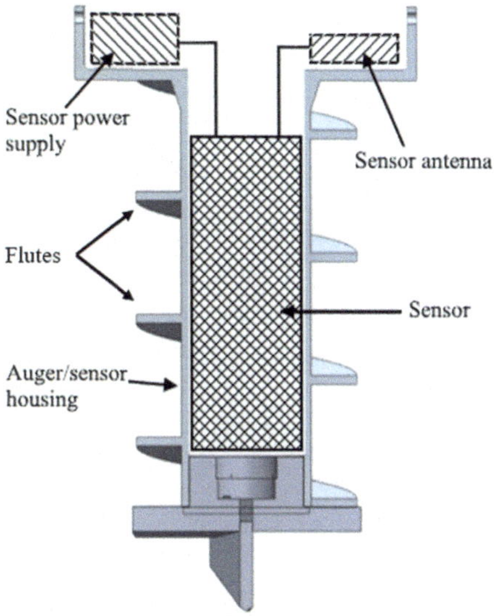

Fig. 2. Auger with internal sensor

Augering and, if necessary, relocating to a new location place considerable demand on the available energy stored in the system's batteries. Therefore, efficiently emplacing the sensor or rapidly determining that a new a location must be tried are two key factors in the overall success of our system. Emplacing the sensor as fast as possible and with the greatest chance of success requires continuous adjustment of the auger's rotational speed, downward force, and pecking motions but is highly dependent upon key soil parameters (e.g., water content).

As a result, knowledge of soil parameters, especially during augering activities, greatly increases the chance of a fast and successful sensor emplacement or determination of imminent failure.

In this paper, we describe a novel, online soil classification strategy that takes information from the on-board sensor suite and determines key soil parameters, enabling us to adjust our augering strategy in real time or quickly determine that a new location must be tried. Previous work has estimated the relative hardness of a surface with a UAS [2,3,11], although these methods did not leverage an in-ground emplacement augering system that makes direct contact with lower layers of the soil. This capability, including direct interaction with the soil, provides a rich dataset from which an online classifier can be trained.

Soil classification covers a wide range of parameters [1], however, for the purposes of in-ground sensor emplacement to 150 mm, we have found water content and soil compressive strength are the characteristics that most greatly impact the chance of successful emplacement. These key characteristics help us determine how much downforce, torque, and speed to apply, whether or not to engage higher level augering strategies (e.g., pecking), and predict whether or not the current digging effort will be successful. However, direct measurement of these parameters is difficult and would require additional equipment to be mounted to the UAS, which cannot be done with current size, weight, and power restrictions. As a result, we leverage auger RPM, motor current use, downward force on the auger, and system vibration levels on the UAS, alongside a custom classifier to determine the water content and compressive strength of the soil.

Rocks, tree roots, or other impediments can prevent emplacement of the sensor. However, the highly stochastic nature of their locations in soil make prediction especially difficult. Here, we focus on regularly predictable soil parameters that impact emplacement in the absence of significant halting impediments. This paper makes the following contributions:

- A novel classifier for soil water content and compressive strength through indirect means
- A comparison of various machine learning technique in their application to classifying soil water content and compressive strength
- A large, expansive data set of 2.8 million points of data over 150 augering evolutions

The on-board sensors used to classify a soil in terms of its water content and compressive strength are not the respective purpose-built moisture sensors and penetrometers, but rather the sensors used for monitoring the system performance of the auger mechanism: auger motor RPM, auger motor current use, weight on auger bit, system vibration (via accelerometers), and time. The data from these sensors is analyzed using machine learning techniques such as decision trees, linear discriminant analysis, naive Bayesian analysis, k-nearest neighbor (knn) analysis, and Gaussian process regression. We also examine the effectiveness of each technique in assessing the soil composition within the first 30 s of an emplacement operation. Barring any stoppages of the emplacement process, it takes a minimum of 60 s to emplace a sensor. This minimum time is increases with an increase in sensor (and subsequent auger) size. We show that Gaussian process regression outperforms the other methods at the 30 s mark with an overall average predictive accuracy of 86.53% when determining moisture content and 90.53% when determining soil compressive strength.

2 Related Work

Machine learning can be used to classify or predict new data based on previously observed values. Our work examines the effectiveness of decision trees, linear discriminant analysis, naive Bayes prediction, k-nearest neighbor, and Gaussian process regression to predict the current values of moisture and soil compressive strength based on the sensor data available to our sensor emplacement system while engaged in an augering evolution. These particular algorithms have been previously employed in the analysis of soil composition.

Pekel examined the use of decision trees to predict soil moisture using atmospheric measurements obtained from stationary HOBO U30 weather data loggers as predictors [7]. The loggers are left in the field and gather data over several days. Suthar used eight soil-specific predictors, to include moisture content and the amoung of lime sludge present, to predict the compressive strength of stabilized pond ash [14]. Gathering these parameters from a given soil sample required transport of the sample to a lab and upwards of seven days of curing. Valaee et al. have used linear discriminant analysis and the magnetic properties of soil measured by external instruments such as Kappameters to predict its moisture content [15]. Yamaç et al. predicted moisture content of soil using k-nearest neighbor analysis with lime content, organic matter, soil particle size, and bulk density as predictors [16]. Rajeswari and Arunesh used naive Bayesian analysis to classify soil in terms of iron content versus organic content [9]. For these last two, organic content was analyzed using a LECO CN-2000 combustion oven that was not located at the sample site.

In all of the aforementioned works, measuring the parameters to be used in the prediction schemes required either equipment or facilities not available for a UAS-sized platform. Additionally, these methods often require the sampling and removal of material from the environment. The material must be analyzed external to the device using the predictive policy in order to provide inputs for that policy [9,14,16]. This is where our approach differs from the above. While, like the other approaches, we generate our policy offline, our system is able to gather the required data in situ and use it in our predictive policy as it is being gathered. Our approach to analyzing soil moisture and compressive strength is unique in that we use the sensors internal to our emplacement system during the physical act of drilling into the soil to determine our predictions.

3 Technical Approach

3.1 Description of the Emplacement System

The sensor emplacement system is a modular design that can be mounted to the underside of unmanned aircraft systems that have vertical takeoff and landing capability. The system is housed in an aluminum chassis that can be adapted to fit on any applicable air frame capable of operating with a 2.7 kg payload and supplying the system with 24 V DC.

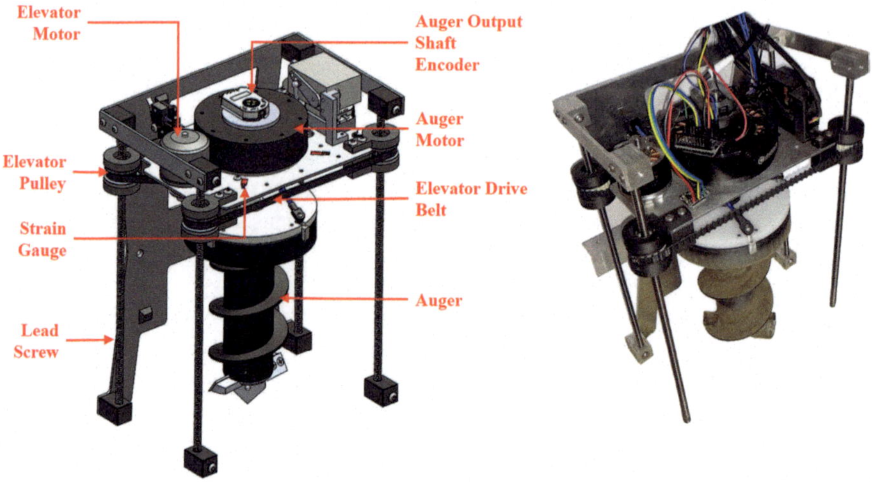

Fig. 3. Emplacement mechanism: model (left) and system (right)

The emplacement system (Fig. 3) consists of an auger with a 150 mm long shaft and diameter of 75 mm. It is attached to a T-Motor A80-6 24 V brushless motor with an integral planetary gear transmission and rotary encoder used for measuring auger RPM. During augering operations, a proportional control law is used to maintain auger rotational speed at 200 RPM. In difficult soil conditions where maintaining 200 RPM causes excessive current draw, auger RPM is allowed to decrease in order to maintain safe operating conditions for the auger motor. The auger motor is capable of outputting continuous 6 Nm of torque under a 12 A load. The motor/auger combination is mounted to an aluminum plate that advances downward at a rate of 0.1375 cm/s during augering operations. Strain gauges mounted to the elevator plate measure vertical force applied during augering. An inertial measurement unit (IMU) mounted to the aluminum plate provides data for vibration analysis. The elevator consists of a smaller T-Motor MN3520 brushless motor, driving a belt connected to pulleys on each corner of the aluminum plate to raise or lower the aluminum platform on four lead screws. A rotary encoder is calibrated to measure the vertical distance the platform travels. An Odrive Robotics motor controller is used to control the auger and elevator motors, while an ATMega-based microcontroller is used as the primary computing device that manages communications with the Odrive controller and outputs the various measured parameters via serial connection.

For this work, the emplacement system is mounted to an aluminum frame for ease of testing, although normally the system is attached to an unmanned aircraft as shown in Fig. 1.

3.2 Measured Parameters

The following parameters are continuously monitored by the emplacement system and output to a serial communications line at a rate of 10 Hz. Their values provide quantitative insight into the augering process and are aggregated to classify the type of soil.

- Revolutions per minute of the auger motor - measured in RPM
- Current draw of the auger motor - measured in Amps
- Weight on auger bit (WOB) - measured by the strain gauges in kg
- Elevator position relative to top - measured in cm
- Acceleration in the X, Y, and Z axes - measured in m/s^2
- Time - each line of logged output is timestamped in seconds

3.3 Soil Parameters of Interest

Soil can be described by various parameters ranging from its particle sizes to its organic material content [1]. In our case, we are concerned with the physical properties that have the most impact on successfully augering into the soil. In our previous work, we have determined that higher moisture content and/or higher compressive strength coincide with a reduced chance of success for an augering evolution [8,13]. An increase in moisture leads to an increase in friction between the soil and the auger surfaces. Higher compressive strength means the soil is more compact and requires that more force be applied in order to loosen the soil for transport up the auger's flutes.

4 Experimental Setup

We conducted 150 trials of our emplacement system in order to gather the required data for our analysis. 110 trials were conducted in the silty clay soil commonly found in eastern Nebraska with an additional 40 trials conducted between north central Pennsylvania, western Virginia, and north central Kentucky. The soils in Nebraska and Pennsylvania exhibited similar characteristics to each other, with each having a higher clay content than the soils in Virginia and Kentucky. Figure 4 shows the relative differences in the soils.

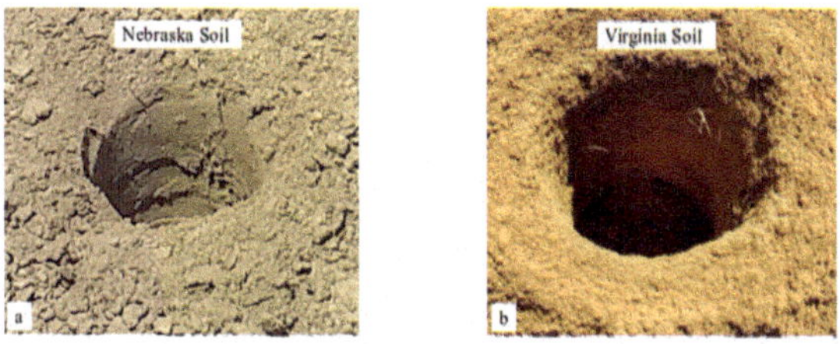

Fig. 4. Representative soil types - a) shows the higher clay content of the Nebraska testing area (with Pennsylvania being similar), and b) shows the relatively more sandy soil of the Virginia testing area (with Kentucky being similar)

To obtain truth data, we manually measured the soil moisture content and soil compressive strength with tools and techniques accepted in the pedology community: a capacitive moisture sensor, volumetric analysis, and pocket penetrometer [1,10,12]. Specifically, the percentage of soil moisture was measured with a capacitive device for each trial, with every tenth trial verifying the moisture content by volumetric means (i.e., weighing the soil before and after baking the moisture out in an oven). The moisture values over the 150 trials ranged from 5% to 80%. Measurements were taken in the upper, middle, and lower thirds of the soil column at the completion each augering operation (see Fig. 5).

The unconfined compressive strength of soil is defined as the amount of force required to crush or displace the soil within a given area [10,12] and is measured in kg/cm^2. The soil in our trials was measured with a pocket penetrometer by probing the side wall of the resultant hole left by the auger. Measurements were taken near the surface, in the middle third of hole, and at the bottom (Fig. 5). The compressive strength values over the 150 trials ranged from 0.5 kg/cm^2 to 4.6 kg/cm^2. Figure 6 shows the distribution of these measurements.

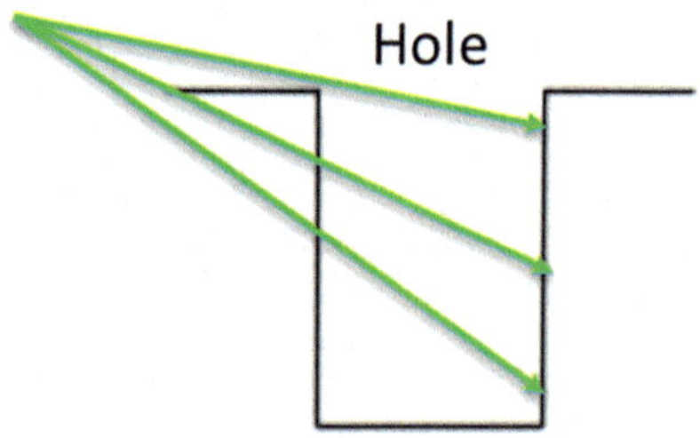

Fig. 5. Soil measurement locations in the upper, middle, and lower thirds of an excavated soil column

The initial 4 trials were conducted indoors using an 18 liter bucket filled with soil that was compressed to the desired soil strength using a hydraulic press. The remaining 146 trials were conducted outdoors in areas where soil moisture content and compressive strength varied in order to gather data over a range of soil conditions. Figure 7 shows the indoor testing area and a representative outdoor testing area.

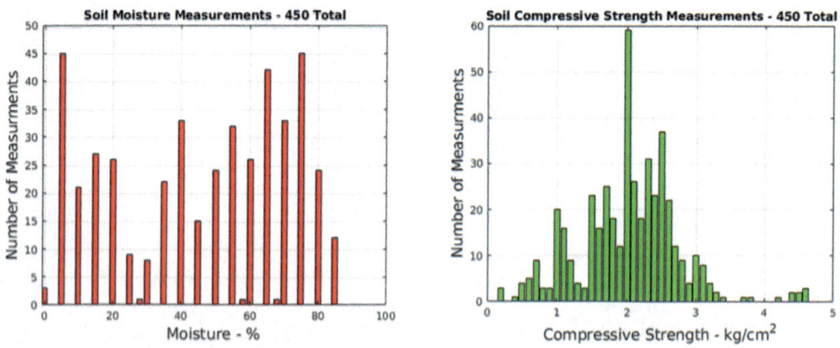

Fig. 6. Distribution of soil measurements taken over the 150 trials

Auger RPM, auger current, auger depth, weight on bit, and acceleration values were logged at a rate 10 Hz during each augering evolution. These parameters were then used as predictors in the following machine learning algorithms:

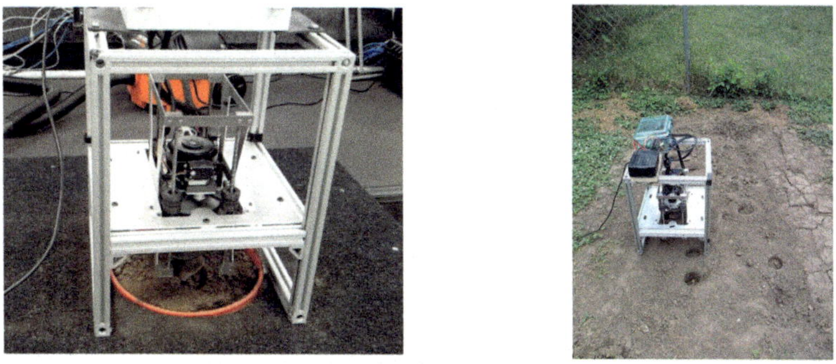

(a) Indoor testing (b) Outdoor testing

Fig. 7. Testing areas

- Decision tree
- Linear discriminant
- Naive Bayes
- K-Nearest neighbor
- Gaussian process

5 Experimental Results

Figure 8 shows a representative plot of the parameters we monitored during each trial of our system: auger RPM, auger motor current, weight on bit, depth, acceleration, and time. These parameters are used as the predictor variables in following survey of classification/regression schemes: decision tree, linear discriminant, naive Bayes, k-nearest neighbor, and Gaussian process regression. The responses in these schemes are our desired soil moisture content and soil compressive strength. We use Matlab®'s "fitctree()," "fitcdiscr()," "fitcnb()," "fitcknn()," and "fitrgp()" methods to generate our predictive models [5]. We randomly chose 50 of our trials to provide the training data for our classification schemes. We then simulated our predictive models against the data from our remaining 100 trials.

As one of our main goals is to determine the soil moisture content and compressive strength as quickly as possible, we examine the predictive accuracy of our models at 5, 10, 20, and 30 s into an augering evolution. We define predictive accuracy as how closely a model calculates the soil moisture content or compressive strength compared to the actual measured value for that trial at the auger depth for the given point in time. Table 1 shows the results of our simulations. Gaussian process regression was the most accurate prediction method during all phases of an augering evolution.

Table 1. Prediction scheme accuracy

Accuracy		Decision tree	Linear discriminant	Naive bayes	k-Nearest neighbor	Gaussian process
5 s	Moisture content	82.29%	83.54%	84.66%	84.85%	**86.86%**
	Soil strength	86.34%	89.59%	85.25%	87.44%	**89.92%**
10 s	Moisture content	86.34%	84.44%	86.40%	85.81%	**88.86%**
	Soil strength	88.03%	86.57%	84.58%	87.40%	**90.02%**
20 s	Moisture content	85.96%	82.07%	84.98%	85.83%	**87.90%**
	Soil strength	88.63%	85.11%	88.00%	87.36%	**90.65%**
30 s	Moisture content	76.36%	70.99%	80.59%	76.85%	**82.51%**
	Soil strength	88.18%	90.02%	90.70%	87.38%	**91.51%**
Average	Moisture content	82.74%	80.26%	84.16%	83.34%	**86.53%**
	Soil strength	87.79%	87.82%	87.15%	87.39%	**90.53%**

Table 2. Prediction accuracy with original system parameters (3rd column), compared with new system parameters (4th column).

Accuracy		RPM, Current, Depth, Time	RPM, Current, Depth, Time, WOB, Acceleration
5 s	Moisture content	84.29%	**86.86%**
	Soil strength	89.11%	**89.92%**
10 s	Moisture content	86.07%	**88.86%**
	Soil strength	89.60%	**90.02%**
20 s	Moisture content	87.34%	**87.90%**
	Soil strength	**90.96%**	90.65%
30 s	Moisture content	82.20%	**82.51%**
	Soil strength	91.31%	**91.51%**
Average	Moisture content	84.98%	**86.53%**
	Soil strength	90.24%	**90.53%**

Additionally, we examine how the choice of predictors can influence the accuracy of the chosen classification/regression schemes. In our previous work we used auger RPM, auger motor current, auger depth, and time to predict whether an augering evolution would succeed or not [8]. For that work, we were limited to those four parameters as a function of the system design. Our current emplacement system is a complete redesign of our initial system and allows for the addition of recording weight on bit and acceleration values. Table 2 shows an increase in predictive accuracy for our Gaussian process regression when incorporating weight on bit and x, y, and z axis acceleration values.

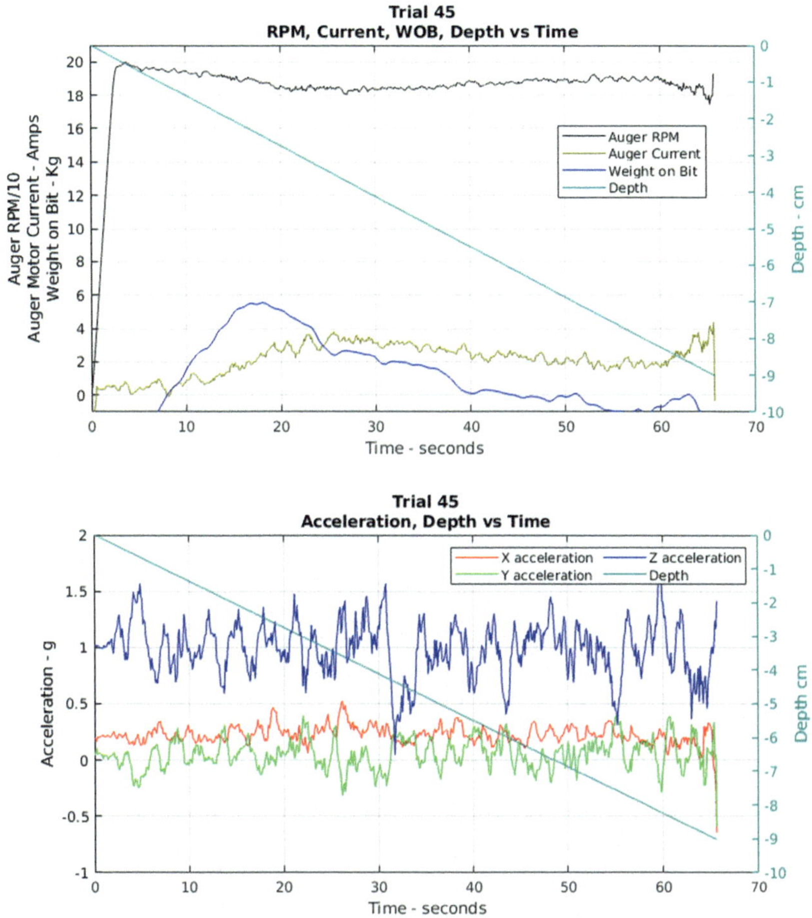

Fig. 8. Example of parameters gathered during system trials. The top graph shows auger RPM, motor current, weight on bit, and depth over time. The bottom graph show accelerations in the x, y, and z axes and depth over time

6 Conclusion and Future Work

We show that it is possible to predict soil moisture content and compressive strength using the available sensor that provide auger RPM, auger motor current, auger depth, weight on bit, and acceleration data. Gaussian process regression generates the most accurate policy of the schemes that were tested. It can predict within the first 30 s of an average 85 s augering evolution the soil moisture content within 86.53% of the actual value and soil compressive strength within 90.53% of the actual value.

Encoding this policy on the hardware that controls the augering mechanism is the next step in our research. Additionally, this will allow us to use the real-time predicted soil composition to adjust our augering technique in order to optimally drill our sensor into the soil.

References

1. American Society for Testing and Materials International (ed.): ASTM D2487-17e1, Standard Practice for Classification of Soils for Engineering Purposes (Unified Soil Classification System). American Society for Testing and Materials International (2017)
2. Anthony, D., Basha, E., Ostdiek, J., Ore, J.P., Detweiler, C.: Surface classification for sensor deployment from UAV landings. In: 2015 IEEE International Conference on Robotics and Automation (ICRA), pp. 3464–3470 (2015)
3. Fox, C.W., Mitchinson, B., Pearson, M.J., Pipe, A.G., Prescott, T.J.: Contact type dependency of texture classification in a whiskered mobile robot. Auton. Robots **26**(4), 223–239 (2009)
4. Fujishima, M., Nishiura, I., Kakino, Y., Matsubara, A.: Integration of adaptive control functions for drilling in intelligent machine tools. In: Proceedings of the Sixth International Conference on Automation Technology, vol. 1, pp. 531–535 (2000)
5. The Mathworks, Inc., Natick, Massachusetts: MATLAB version 9.7.0.1190202 (R2019b) (2019)
6. Mellinger, J.C., Burak Ozdoganlar, O., DeVor, R.E., Kapoor, S.G.: Modeling chip-evacuation forces and prediction of chip-clogging in drilling. J. Manuf. Sci. Eng. **124**(3), 605–614 (2002)
7. Pekel, E.: Estimation of soil moisture using decision tree regression. Theoret. Appl. Climatol. **139**(3), 1111–1119 (2020)
8. Plowcha, A., Sun, Y., Detweiler, C., Bradley, J.: Predicting digging success for unmanned aircraft system sensor emplacement. In: Proceedings of the 2018 International Conference on Experimental Robotics (ISER 2018). Buenos Aires, Argentina (2018)
9. Rajeswari, V., Arunesh, K.: Analysing soil data using data mining classification techniques. Indian J. Sci. Technol. **9**(19), 1–4 (2016)
10. Robertson, P.: Cone penetration test (CPT)-based soil behaviour type (SBT) classification system–an update. Can. Geotech. J. **53**(12), 1910–1927 (2016)
11. Romano, J.M., Kuchenbecker, K.J.: Methods for robotic tool-mediated haptic surface recognition. In: 2014 IEEE Haptics Symposium (HAPTICS), pp. 49–56. IEEE (2014)
12. Sanglerat, G.: The Penetrometer and Soil Exploration. Elsevier, Amsterdam (2012)
13. Sun, Y., Plowcha, A., Nail, M., Elbaum, S., Terry, B., Detweiler, C.: Unmanned aerial auger for underground sensor installation. In: 2018 IEEE/RSJ International Conference on Intelligent Robots and Systems (IROS 2018) (2018)
14. Suthar, M.: Applying several machine learning approaches for prediction of unconfined compressive strength of stabilized pond ashes. Neural Comput. Appl. **32**, 1–10 (2019)
15. Valaee, M., Ayoubi, S., Khormali, F., Lu, S.G., Karimzadeh, H.R.: Using magnetic susceptibility to discriminate between soil moisture regimes in selected loess and loess-like soils in northern Iran. J. Appl. Geophys. **127**, 23–30 (2016)
16. Yamaç, S.S., Şeker, C., Negiş, H.: Evaluation of machine learning methods to predict soil moisture constants with different combinations of soil input data for calcareous soils in a semi arid area. Agric. Water Manag. **234**, 106121 (2020)

Simulation, Learning and Control Methods to Improve Robotic Vegetable Harvesting

Simon Birrell(✉) and Fumiya Iida

Bio-Inspired Robotics Lab, Department Engineering, Cambridge University, Cambridge, UK
{sab233,fi224}@cam.ac.uk

Abstract. Agricultural robots are subject to a much harsher environment than those in the factory or lab and control strategies need to take this into account while maintaining a low cycle time. Three control strategies were tested on Vegebot, a lettuce-picking robot, in both simulation and on the real robot. Between a fast open loop that was vulnerable to environmental noise and a slow but robust visual servoing technique, a Learned Open Loop strategy was tested where the robot learned from successful picks to pick at an intermediate speed. This reduced the projected cycle time from 31 s to 17.2 s, a 45% reduction.

1 Introduction

Agriculture has historically been labour-intensive and subject to the vicissitudes of the weather, but automation has increased efficiency and had a downward effect on prices. However, automation has not been achieved for many crops and a series of challenges for agricultural robots have been exposed. Farmers' fields are not as sheltered as factories and robots will encounter vibrations, hard-to-navigate terrain and rough weather conditions, affecting every aspect of how they are designed and built. The present paper describes different approaches to tackling one particular problem for harvesting robots, that of reaching for and grasping the target vegetable under environmental noise.

This paper builds on the previously published work on the Vegebot [1], a lettuce-picking robot (Fig. 1). Lettuce harvesting is still a difficult manual task for farm labourers; the goal for Vegebot is to autonomously and accurately harvest lettuces without damaging them, in the shortest possible cycle time. The first iteration of Vegebot automated harvesting, achieving a cycle time of 31 s (compared to a human cycle time of 6 s), a harvest success rate of 88% and a damage rate of 38% [1]. All of these metrics need to be improved, but the most glaring failing is a cycle time that is 5 times slower than a human. Speed is critical to make the device economically viable, and new techniques are required for this.

Vegebot needs to swiftly move the end effector to the visually identified target lettuce. One problem that complicates this seemingly simple goal is the unpredictable and rough environment in which agritech robots find themselves. Robots

B. Siciliano et al. (Eds.): ISER 2020, SPAR 19, pp. 185–197, 2021.
https://doi.org/10.1007/978-3-030-71151-1_17

Fig. 1. The Vegebot: (a) Vegebot deployed for field trials (b) Vegebot in the lab with a "virtual" lettuce.

with a less than perfect self-model and subject to environmental noise and wear and tear have difficulties in transforming a visual cue into a precise location in robot task space. Early experiences in the field with Vegebot suggested that dealing with these conditions will be a central challenge for many different types of harvesting devices.

The purpose of this paper is threefold: (1) to develop a realistic simulation platform for Vegebot, (2) to test new harvesting algorithms, including visual servoing and learned behaviours and (3) to optimise the operation of Vegebot in the real world in terms of speed and accuracy under environmental perturbations. With a robust perception and reaching strategy, harvesting robots could produce a greater yield with a shorter cycle time.

Extensive detection and manipulation work has been performed on other individual crops such as broccoli [2], strawberries [3] and apples [4]. A good overall review with important metrics definitions can be found in [5]. Many of the crops that have been successfully automated such as tomatoes and sweet peppers have done so in semi-sheltered environments [6], where they are less subject to the environmental wear and tear tackled here.

Visual servoing has been extensively studied; see [7] for a review and [8] for an example application in harvesting. The use of multiple cameras for visual servoing along with authority switching algorithms has also been explored, for example in [9], which uses distance derived from an RGBD sensor to determine the change of authority. RGBD sensors are problematic outdoors, so Vegebot simply uses the presence or absence of the target in the end effector camera to switch authority.

2 Technical Approach

Vegebot has now been deployed in three environments: in physical field trials (see Fig. 1a), in the lab (see Fig. 1b) and in simulation (see Fig. 2). Field trials have been described in the previous paper [1]; for this paper experiments were performed in the lab and in simulation.

2.1 The Real Vegebot

The Vegebot platform is two metres wide, 55 cm high and 140 cm long; the wheels straddle a standard sized lane of lettuces in the field. A Universal Robotics UR10 arm is mounted on top of the platform and carries a custom-designed end effector, which contains a soft gripper (for holding the lettuce) and a blade (for cutting the stalk) driven by pneumatic actuators via a circular belt. In its current form, the end effector weighs almost 10 kg, at the limit of what the arm can support.

Vegebot has a two-camera vision system. One camera (640 × 480 pixels) is mounted on an overhead support and provides a broad view of the workspace. A second camera (320 × 240 pixels) is mounted inside the end effector and allows for fine adjustment in positioning as the end effector descends over the lettuce. Both cameras can localize lettuces, while the overhead camera can also estimate the relative pose of the platform and end effector from their attached Aruco boards (see Fig. 2). The robot arm will generally occlude much of the platform and end effector in the overhead camera's view, but the partially visible Aruco boards provide consistent readings nevertheless at an update rate 6 Hz.

A two-stage localisation/classification pipeline detects the lettuces visible in each camera. First, a YOLO network localises the lettuces, passing the cropped images of each one to the second stage. This latter stage classifies them as harvest-ready, immature or diseased. YOLO is a fast object detection neural network architecture [11]. The two stages are necessary to compensate for the unbalanced datasets available and could be collapsed into one on a full production system [1]. The output of the full pipeline is a set of lettuce hypotheses, updated at 5.5 Hz on current hardware: each hypothesis with a bounding box in normalized u, v pixel coordinates and a class label, as shown in Fig. 2.

Finally, the robot contains an IMU (Adafruit IMU BNO055) for measuring the absolute orientation of the platform and a laptop with a gaming GPU for managing the whole system. The IMU is connected to an Arduino which streams quaternions to the laptop via a USB connection. The Vegebot software was built on ROS (Robot Operating System) in Python, PyTorch and JavaScript. Inverse kinematics and dynamics are calculated by the UR10's built-in controller.

2.2 The Simulated Vegebot

The purpose of building a simulated Vegebot (see Fig. 2) was to provide a platform for experimentation of different control and learning strategies. CoppeliaSim 4.0 was selected, which provides a ROS-compatible environment with a ready-made model of the UR10 and pluggable physics engines. The Newton physics engine provided the best stability, and a model of the Vegebot platform was created using the real robot's dimensions and weights. The simulation ran at around 10 frames per second.

Virtual cameras were used for overhead and end effector cameras. Photographs of earth from the lettuce fields were mapped onto the ground textures. Wrapping photographs of lettuces onto spheres did not work with the real

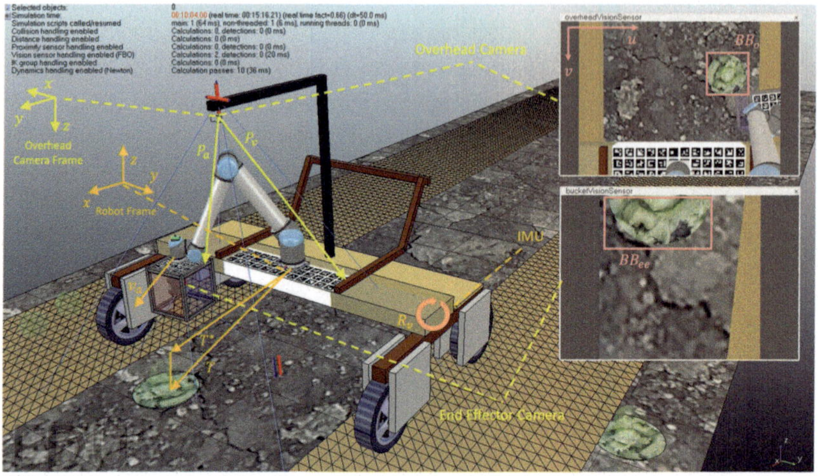

Fig. 2. The Vegebot simulated using CoppeliaSim. Virtual camera fields are inset and a "virtual" lettuce is shown.

robot's vision system; presumably the features picked up by YOLO were distorted and no longer recognisable. Instead, an overhead photograph of a lettuce was mapped onto a flat circle, and this was topped with an almost transparent sphere, to give the simulated end effector something to grip onto (see Fig. 2. With that, the vision system worked from the virtual cameras in the same way as on the physical robot, providing bounding boxes (see camera insets on Fig. 2).

There remained the problem of reproducing the arm's movements from inverse kinematics in simulation. Previous work found the onboard UR10 controller to be more reliable than the ROS kinematics package. Fortunately, Universal Robotics provide a simulation of the controller called URSim which accepts position or velocity commands and streams the position of a virtual UR10 arm. This stream is translated into ROS messages and fed to the arm model in CoppeliaSim which then tracks the controller's virtual arm model. The software arrangement can be seen in Fig. 3. The bulk of the Vegebot software can be used unaltered in the two configurations.

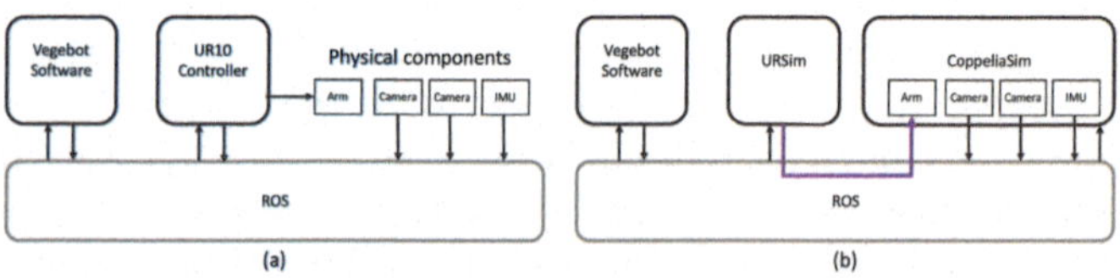

Fig. 3. The Vegebot Software Architecture: (a) on the real Vegebot (b) in simulation, with simulated cameras and IMU, plus a simulated arm tracking the output of the URSim simulated controller.

2.3 The Three Control Approaches

Once the vision system has detected the bounding box of a target lettuce, the end effector must be moved to a position just above the target and then brought down to the ground to envelope, grip and cut it. Three new control approaches were tested, first in simulation and then on the real Vegebot. Each method has two sequential stages (Fig. 4) and begins with the **Overhead Camera** feeding a stream of 640 × 480 images to the **Lettuce Detector**, which generates one or more bounding boxes BB_o.

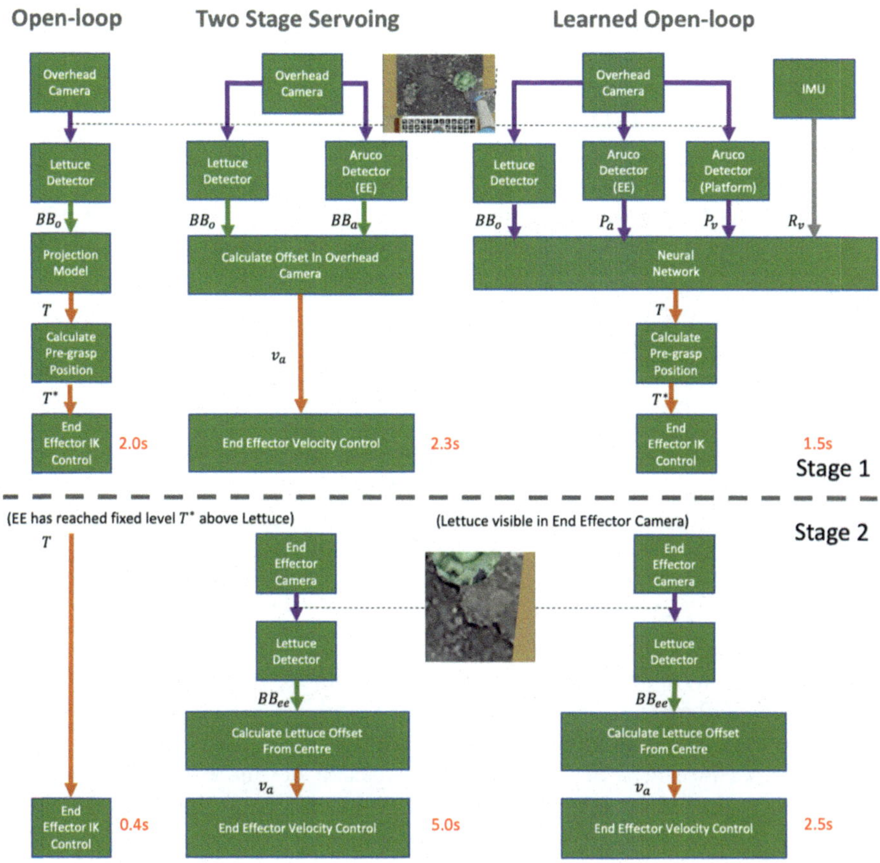

Fig. 4. Control system information flow for the three harvesting methods. The sources of delay are indicated in red.

In the first of the new approaches, **Open Loop** control, a static **Projection Model** transforms the bounding box coordinates BB_o into an estimated 3D location in the robot arm space T. This model derives T by using knowledge of the camera's geometry to project a vector from the camera through the centre of the bounding box to where it intersects with an estimated ground plane.

The system then calculates the **Pre-Grasp Position** T^*, which is some 20 cm vertically above T.

In Stage 1 of the movement, the end effector is moved using inverse kinematics to the **Pre-grasp position** T^*, and then in Stage 2 drops down to T so that the lettuce is left in the centre of the end effector cage. This **End Effector Inverse Kinematics Control** is executed by the UR10's controller (or URSim in simulation). The resulting movements are fast, but not robust to noise or to inaccuracies in the described projection model. It is a variant of the method used on the original Vegebot.

In the second approach, **Two Stage Servoing**, visual servoing is used to continually adjust the velocity v_{xyz} in robot arm space of the end effector to approach the visual target. The **Lettuce Detector** provides a bounding box BB_o for the lettuce as before, but the overhead camera image stream is also passed to an **Aruco Board Detector (EE)** which returns a Bounding Box BB_a for the top of the end effector.

An offset d_{uv} between the centres of the two bounding boxes BB_o and BB_a is calculated in normalized pixel coordinates, and this is used to derive the overall desired end effector velocity v_{xyz} in the module **Calculate Offset in Overhead Camera**:

$$v_{xy} = -\alpha_{cam} d_{uv} \tag{1}$$

$$v_z = -\beta_{cam}[(z - z_{target}) + \frac{\gamma_{cam}}{|d_{uv}| + 0.1}] \tag{2}$$

where z is the z position of the end effector in robot arm space, z_{target} is the estimated z position of the ground and α_{cam}, β_{ca} and γ_{cam} are constants that are defined separately for each camera. The resulting velocity commands v_{xyz} are passed to the UR10 Controller (or URSim) in the module **End Effector Velocity Control**. The end effector will accelerate towards the lettuce in the XY plane while slowly descending.

The overhead camera retains authority until the lettuce becomes visible in the end effector camera, at which point Stage 2 commences. Authority is switched to the **End Effector Camera** the output of which is used by the **Lettuce Detector** to derive a new, close-up bounding box BB_{ee}. The offset d_{uv} is now the offset of the centre of BB_{ee} from the centre of the visual field and the module **Calculate Lettuce Offset from Centre** derives the desired velocity v_{xyz} using the same Eqs. (1) and (2) but with different constants for α_{cam}, β_{cam} and γ_{cam}. The end effector continues to approach the centre of the lettuce while now descending faster; given that the lettuce is close in the XY place, there is less chance of missing the target by moving down too quickly.

The Two Stage Servoing approach is limited in velocity by the speed of updates from the perceptual system (5 Hz on the current hardware) but is more robust to environmental noise and changes in body shape: as the Vegebot receives blows, the trajectory self-corrects (within limits).

The third approach, **Learned Open-Loop**, attempts to combine the speed advantages of the first approach with the robustness of the second. As well as the

Table 1. Input and output variables to the Learned Open Loop neural network.

Var.	Description	Units	Frame of reference	Dim.
P_a	Pose of end effector Aruco board	Position (x, y, z) Orientation (x, y, z, w)	Overhead camera	7
P_v	Pose of platform Aruco board	Position (x, y, z) Orientation (x, y, z, w)	Overhead camera	7
R_v	IMU orientation	Orientation (x, y, z, w)	World	4
BB_o	Bounding box of lettuce in overhead camera	Centre, width, height (u,v,w,h) in normalized pixel coordinates	Overhead camera image	4
T	Pose of end effector that places it over the lettuce	Position (x, y, z) Orientation (x, y, z, w)	Robot arm	7

Lettuce Detector, the **Aruco Detector (EE)** now generates an estimated pose P_a for the end effector, and an additional **Aruco Detector (Platform)** derives an estimated pose P_v for the platform using the second Aruco Board, both poses being estimated in overhead camera space. The IMU measures the absolute orientation R_v of the platform as a quaternion in world space.

Next, a **Neural Network** that accepts BB_o, P_a, P_v and R_v as inputs and generates position T as an output, estimating the lettuce pose in robot arm space using the available sensory data (Eq. 3). T^* is derived in the usual way, begin 20 cm above T. These variables are listed in Table 1 and are shown in Fig. 2.

$$T = f(P_a, P_v, R_v, BB_o) \tag{3}$$

The input variables are mapped to T using 4 fully-connected layers, tanh activation functions, a mean squared error loss function and the Adam optimizer. The input and output layer size is determined by the dimensions of the variables in Table 1. The size and number of the hidden layers was arrived at by iterative experimentation, trading off training speed against accuracy. The neural network was trained for 2000 iterations (Fig. 5), using data taken from successful Two-Stage Servoing picks. Separate datasets were gathered in simulation and reality: in simulation 2000 samples were gathered, while on the real robot a smaller dataset of 50 successful picks was duplicated to form 2000 samples and found to work quite acceptably.

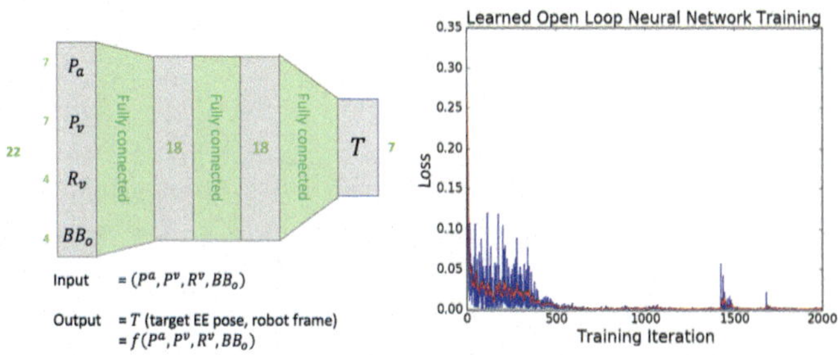

Fig. 5. (a) The inputs and outputs to the Learned Open Loop neural network. (b) Training the neural network over 2000 iterations.

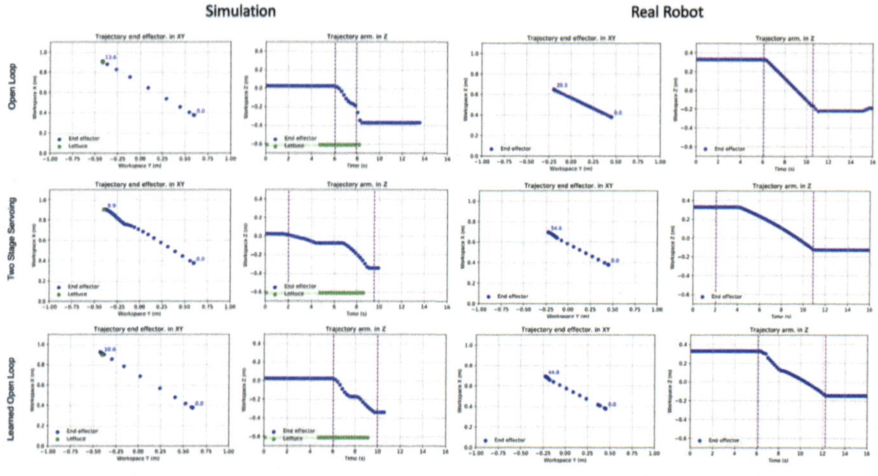

Fig. 6. Experiment One: End effector trajectories under the different control systems, (a) in simulation and (b) on the real robot. The graphs on the left show the end effector's trajectory in the X-Y plane with the start and end points labelled with time t. The graphs on the right show the end effector's trajectory along the Z axis, with time t along the bottom. The vertical purple lines indicate the start and end of the picking sequence.

Stage 1 of the movement is open loop and fast, with the end effector moving swiftly to T^*. Stage 2 is identical to stage 2 of Two-Stage Servoing, with the end effector descending over the lettuce, self-correcting as it goes. As a result, Learned Open Loop is faster than Two Stage Servoing, while retaining much of its robustness.

3 Experiments and Insights

3.1 Experiment One: Three Different Control Systems

Experiment: The three different control methods were tested for speed in simulation and then on the real Vegebot robot in the lab environment. A lettuce (either a paper photograph for the lab environment or a sphere and texture mapped disc in simulation) was placed on the ground and each control method was tested 5 times.

Results in Simulation: The **Open Loop** method is the fastest method (2.4s duration) as it can ignore any further sensory information once started (see Fig. 6). The end effector accelerates and moves rapidly in a straight line towards T^* as quickly as the hardware permits. On reaching T^* it accelerates downwards towards the ground. The changeover between the two stages can be seen as a kink in the time series graph of Fig. 6 (a) (Open loop, Simulation), at $t = 8.0$ s. Note that the end point of the trajectory does not coincide perfectly with the lettuce centre, due to noise in the vision system and an imperfect static model.

The **Two Stage Servoing** method is the slowest (7.3 s duration). While the arm could travel faster, the bounding box updates will then lag behind and the end effector may miss the lettuce. In the first stage the descent in Z is slower (between $t = 2.0–4.5$ s) than the first method (Fig. 6a, centre right). Descending too rapidly runs the risk of the lettuce falling outside the field of view of the end effector camera, which then misses the target. In the second stage (between $t = 4.5 - 9,3$ s) the descent is faster, but still limited by the need to arrive cleanly at the XY position of the lettuce before touching ground.

The **Learned Open Loop** is a compromise between the first two methods (4.0 s duration). The first stage is open loop (between $t = 6.0 - 7.4$ s) with rapid descent. The second stage uses the same algorithm and parameters as the Two Stage Servoing, using the bounding box BB_{ee} to drive the velocity control (between $t = 7.4 - 10.0$ s). This second stage could potentially be further optimised for speed, sacrificing robustness to errors in the learned value of T. This would steepen the second stage curve (Fig. 6, bottom right) reducing overall travel time to close to the Open Loop value, but missing the lettuce in the face of extreme environmental noise.

Results on Real Robot: The trajectories on the real robot are slower, but follow the same pattern as in simulation. Open Loop is the fastest (4.8 s duration), Two Stage Servoing is the slowest (8.9 s duration) and Learned Open Loop is a compromise (6.2 s duration).

The velocities of the different stages of the trajectories had to be adjusted downwards on the real robot for two main reasons. Firstly, more protective stops were triggered by the UR10 controller on the real robot than on the URSim software. This suggests that the URSim's simulation of the behaviour of the arm controller under real conditions is not exact. Secondly, the geometry of the real USB cameras differed from the simulated ones, meaning that the successful handover from overhead to end effector cameras was more sensitive to end effector velocity. The simulated cameras will be altered in future work.

Overall though, the three different control methods that were prototyped in simulation worked in reality, with minimal adjustments.

3.2 Experiment Two: Robustness to Perturbations and Distance

Experiment: The second experiment was designed to test the control methods' robustness to perturbations in the system and to the initial distance of the lettuce from the parked end effector.

Perturbations were added to the simulation by moving the UR10 arm off its usual alignment. In the simulator, a virtual ball joint was introduced under the arm and the yaw, pitch and roll modified. On the real robot, the platform itself was inclined while keeping the overhead camera support vertical. The magnitude of the perturbations is given by the IMU and measured in radians, referring to the yaw pitch and roll angles introduced.

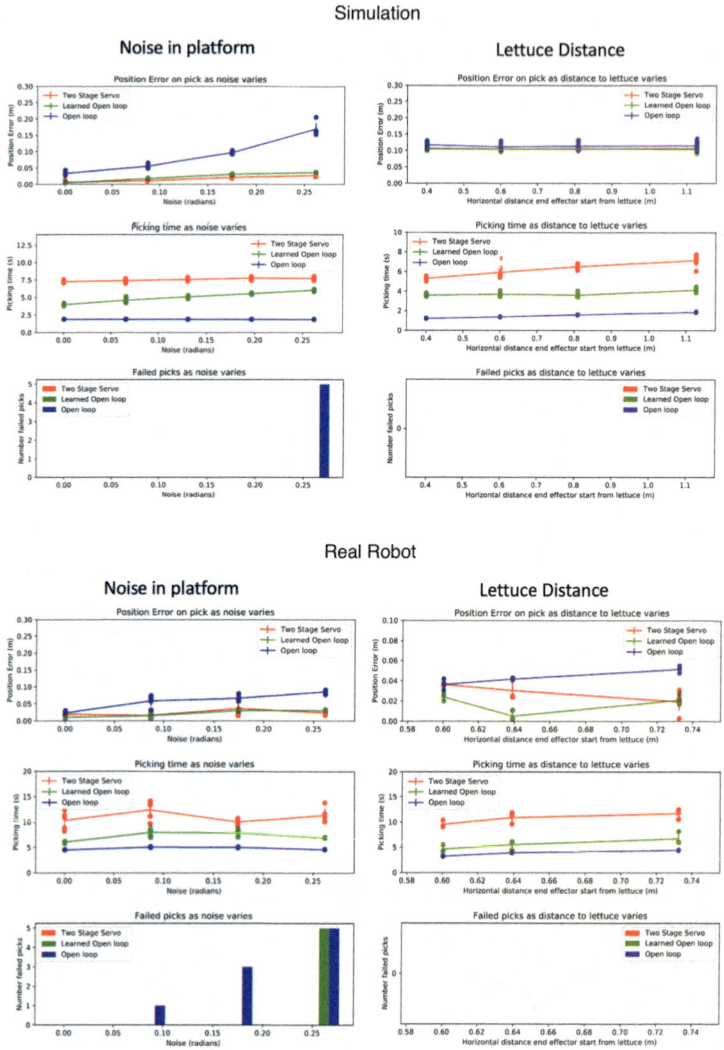

Fig. 7. Experiment Two: The three control systems in simulation and reality as perturbations are added (left) and the distance of the parked end effector to the lettuce is increased (right). The top graphs in each block show the final distance of the end effector centre from the lettuce. The middle graphs show the trajectory time. The lower graphs show failed picks from a total of 5 attempts.

On the real Vegebot platform, for practical reasons, the perturbations were modelled slightly differently. The end of the platform was jacked up to incline it to a similar angle to the ball joint method. The overhead camera was then adjusted back to a vertical orientation. This is similar, but not geometrically identical to the method used in simulation. 5 sample picks were made for each combination of perturbation size and control method.

In separate tests, the distance of the lettuce from the parked en effector was varied to ensure that the methods worked over the range of the workspace. 5 sample picks were made for each combination of distance and control method.

Results in Simulation: The results in simulation can be seen on the top half of Fig. 7). The position error of Open Loop degrades as the noise is increased, to the point where 100% of the picks fail when the noise reaches 0.25 rad. Learned Open Loop and Two Stage servoing are robust to the noise, as the visual servoing component brings the end effector back to the correct position. In terms of picking time though, Open Loop is fastest as once the initital lettuce localisation is performed, no further vision system updates are required. Two Stage Servoing is slowest, as it needs to move slow enough not to outpace the localisation update rate. In some ways, it is an over-cautious strategy: it would work even if the lettuce moved. Learned Open Loop, a combination of a fast open loop approach for stage 1 and slower servoing in stage 2, resulting in a picking time between the two extremes.

The results for varying lettuce distance are as expected: there is no variation in picking error with distance from the parked end effector, suggesting that the projection model is accurate.

Results on Real Robot: Results on the real robot follow the trends seen in simulation. The increase in position error on Open Loop (from 3 to 9 cm) is not as extreme as in simulation: practical considerations prevented noise being applied to all three rotational exes. Learned Open Loop and Two Stage Servoing are robust to low noise, as in simulation, and the differences in picking time follow the same ordering. In general, it proved necessary to run the real robot at a lower velocity than the simulated version to prevent misses and protective stops.

Failures in Open Loop start at a lower noise level (0.10 rad), than in simulation (0.25 rad). In addition, Learned Open Loop fails as much as Open Loop does at the highest noise level (0.25 rad): the targeted positions are too far from the real ones to give the servoing time to compensate. Picking error distance increases with lettuce distance, suggesting inaccuracies in the projection model that grow towards the edge of the visual field.

4 Conclusion

All the tested methods show the expected trade-offs in simulation and in reality. **Open loop** harvesting is fast but vulnerable to pertubations and to model imperfections. **Two Stage Servoing** is more accurate and robust, but slower where the frequency of vision updates is constrained by the vision system hardware. The speed of the **Learned Open Loop** method lies between the other two, trading off some robustness to improve speed. There is an analogy here to human reaching, where an open-loop initial ballistic reach moves the wrist to the vicinity of the target object, followed by a more measured feedback-driven grasping process [10].

The Vegebot therefore has a number of picking strategies to choose from. Under normal circumstances, it should choose the fastest strategy that works. To the extent that it can measure its own picking error (or receive feedback from a human operator), it could adaptively switch between methods. This suggests a combined strategy that would be resistant to slight perturbations and changes in body shape caused by wear and tear in the field.

Each time Vegebot successfully picks a lettuce, the final position could be added to a "successful pick buffer". Once this buffer reaches a certain size, the samples could be augmented through cloning and the neural network retrained. While 2000 training iterations were required to train the network (Fig. 6 (b)), only 50 successful pick samples were actually used on the real robot. This suggest that Vegebot could adapt quickly to wear and tear by falling back to a robust but slow method like Two Stage Servoing as soon as it starts detecting picking errors, gather a set of successful picks and then retrain itself. By using Learned Open Loop, the approach time of 20 s can be reduced to 6.2 s and so the overall cycle time decreased from 31 s to 17.2 s, a 45% improvement and closer to the target human value of 6 s.

Acknowledgments. This project was possible thanks to EPSRC Grant EP/L01-5889/1, the Royal Society ERA Foundation Translation Award (TA160113), EPSRC Doctoral Training Program ICASE AwardRG84492 (cofunded by G's Growers), EPSRC Small Partnership AwardRG86264 (in collaboration with G's Growers), and the BBSRC Small Partnership GrantRG81275. Special thanks to G Growers, George Walker and Josie Hughes for their invaluable assistance.

References

1. Birrell, S., Hughes, J., Cai, J., Iida, F.: A field-tested robotic harvesting system for iceberg lettuce. J. Field Robot. (2019). https://doi.org/10.1002/rob.21888
2. Kusumam, K., Krajnik, T., Pearson, S., Cielniak, G., Duckett, T., et al.: Can you pick a broccoli? 3D-vision based detection and localisation of broccoli heads in the field. In: IEEE (2016)
3. Hayashi, S., et al.: Evaluation of a strawberry-harvesting robot in a field test. Biosyst. Eng. **v105**(2), 160–171 (2010)
4. Silwal, A., et al.: Design, integration, and field evaluation of a robotic apple harvester. J. Field Robot. **v34**(6), 1140–1159 (2017)
5. Bac, C.W., et al.: Harvesting robots for high-value crops: State-of-the-art review and challenges ahead. J. Field Robot. **v31**(6), 888–911 (2014)
6. Hua, Y.: Recent advances in intelligent automated fruit harvesting robots. Open Agric. J. (2020) https://doi.org/10.2174/1874331501913010101
7. Chaumette, F., Hutchinson, S., Corke, P.: Visual servoing. In: Springer Handbook of Robotics, 2nd Ed. Springer, Cham (2016)
8. Mehta, S.S., MacKunis, W., Burks, T.F.: Robust visual servo control in the presence of fruit motion for robotic citrus harvesting. Comput. Electron. Agric. **123**, 362–375 (2016)

9. Cuevas-Velasquez, H., Li, N., Tylecek, R., Saval-Calvo, M., Fisher, R.B.: Hybrid Multi-camera Visual Servoing to Moving Target. arXiv preprint arXiv:1803.02285 (2018)
10. MacKenzie, C., Iberall, T.: The Grasping Hand, Elsevier, Amsterdam (1994)
11. Redmon, J., Farhadi, A.: Yolov3: an incremental improvement. Retrieved from (2018) http://arxiv.org/abs/1804.02767

Incorporating Noise into Adaptive Sampling

Christopher E. Denniston(✉), Aravind Kumaraguru, David A. Caron, and Gaurav S. Sukhatme

University of Southern California, Los Angeles, CA, USA
{cdennist,akumarag,dcaron,gaurav}@usc.edu

Abstract. Adaptive sampling is important in robotic environmental monitoring, allowing a robot to intelligently select sampling locations to build an informative model of a phenomenon of interest. Most adaptive sampling techniques assume the localization noise does not vary with location, or that this variation is negligible, and thus do not model this behavior. In practice, the noise will vary greatly depending on the robot's trajectory and location. Additionally, prior surveys collected by other means, e.g., satellite or drone imagery, may use different state estimators or parameters. If these are used to drive sampling, this dependence may be significant. We provide a unified framework for adaptively collecting and modeling samples when heteroskedastic noise is present. Our framework is agnostic to the distribution of the noise. Our method outperforms others which do not take into account localization noise, validated by simulated trials and noise from a real state estimator.

1 Introduction

Adaptive sampling is the process of using robots to intelligently gather information about an environment. Typically the robot does this by creating an internal model and selecting sampling positions that improve the environment model over time [1].

Using adaptive sampling, a robot can create a high quality survey of an area of interest more efficiently (e.g., in less time or by using less energy) than a pre-planned coverage pattern when certain key properties of the environment, like smoothness or isotropy, can be exploited. This has far reaching impacts in various fields such as the study of algae blooms, oil spill cleanup and search and rescue where the survey area may be expensive to survey or too large for a single vehicle to completely cover. Adaptive sampling systems generally assume that the samples are taken at precise locations. In practice, this assumption is violated because there is always some uncertainty in the robot's pose. This is especially true in the case of autonomous underwater vehicles (AUVs) when

This work was supported by the Southern California Coastal Water Research Project Authority under prime funding from the California State Water Resources Control Board on agreement number 19-003-150.

B. Siciliano et al. (Eds.): ISER 2020, SPAR 19, pp. 198–208, 2021.
https://doi.org/10.1007/978-3-030-71151-1_18

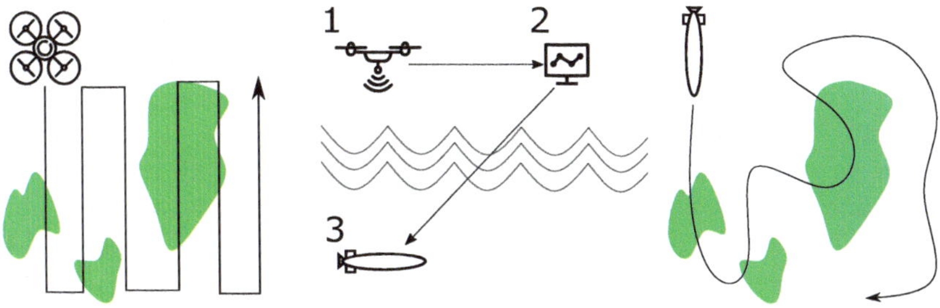

(a) (b) (c)

Fig. 1. A proposed mission. (a) A UAV performs a lawnmower pilot survey. (b) A base station compiles imagery and sends it to an AUV. (c) The AUV plans the sampling mission from this prior data and newly collected data.

the robot does not have GPS and relies integrating inertial measurements. To model this uncertainty, a state estimator is used to derive a belief distribution over the current pose of the robot. Here, we utilize this distribution to more accurately model the phenomenon of interest instead of using the expectation of the robot's location as a point estimate. When the noise is heteroskedastic, the robot should pay more attention to areas of higher positional noise since the model in these regions can be improved with additional noisy samples [2]. We are motivated by the design of aquatic sampling missions (Fig. 1) that exploit a previously generated map of a phenomenon of interest from overhead imagery from an unmanned aerial vehicle (UAV) as a prior for an autonomous underwater vehicle (AUV). The adaptive sampling method must take into account both the uncertainty on the prior data from the UAV as well as the uncertainty of the sampling robot while it is progressing in the mission.

In this work we aim to extend prior methods for long horizon adaptive sampling and adaptive sampling with uncertain inputs to produce a method which can perform long term planning with localization uncertainty that is agnostic to the belief distribution about the robot state. Adaptive sampling with localization uncertainty and a finite budget can be formulated as follows

$$P^* = \underset{P \in \Phi}{\operatorname{argmax}} \mathbb{E}[F(P)] | c(P) \leq B \tag{1}$$

where Φ is the space of all robot trajectories, B is some budget, F is some objective function and c is some cost function, P^* is the optimal trajectory [3]. In this work we maximize the expectation the agent has for the objective over the trajectory. This differs from the statement in [3] which does not use the expected objective, but the objective directly. Our work makes the same assumptions about the form of the cost function and objective function (*i.e.* the cost function is monotonically increasing and the objective function is submodular) as those in [3].

2 Related Work

The adaptive sampling survey [1] notes the usage of filters to remove sensor and localization noise, but does not cover methods explicitly incorporating this noise. Prior work [4] using UAV data to determine sampling locations, does not explicitly incorporate either the UAV or ground robot's localization noise. A framework for Bayesian optimization when a robot's pose is uncertain is found in [5]. The authors probabilistically bound the regret and show applications to robots, using a greedy method for objective maximization. A framework for multiple step prediction with uncertainty in input location with a Monte Carlo approximation is given in [6]. An approach [3] incorporating Rapidly Exploring Random Trees (RRT) into the adaptive sampling setting shows that a modified RRT algorithm (RIG-Tree) can maximize submodular functions. This assumes that there is certainty in the pose of the robot and that the pose can be fully specified. In [2] the effectiveness of informative path planning with uncertainty on the input is shown, with the key insight that modeling localization uncertainty can improve the overall mapping quality. This is achieved by augmenting the acquisition function with an explicit term for uncertainty. The modeling assumption in [2] assumes that the mean position is a good description of the state when the sample is taken, which may not be true in multimodal localization scenarios or when localization is particularly bad in some areas. Previous works have investigated using noisy samples from the environment to decrease the uncertainty in the robot pose using techniques similar to adaptive sampling [7]. Unlike our work, [7] does not try to use the characteristics of these noisy samples to approximate the underlying concentration. Our work extends the framework [2,5] by utilizing the RIG-Tree algorithm to perform longer term planning, using a different modeling technique to incorporate noise, and allowing non-Gaussian sampling distributions.

3 Technical Approach

Gaussian processes are widely used modeling tools for adaptive sampling because they provide a non-parametric, continuous representation with uncertainty quantification. Gaussian processes approximate an unknown function from its known outputs by utilizing the similarity between points from a kernel function ($k(\cdot,\cdot)$) [8]. Function values y_* at any input location $\mathbf{x}_*$ are approximated by a Gaussian distribution:

$$y_*|y \sim \mathcal{N}\left(K_*K^{-1}y,\ K_{**} - K_*K^{-1}K_*^T\right) \tag{2}$$

where y is training output, y_* is test output, $\mathbf{x}$ is training input, and $\mathbf{x}_*$ is test input. K is $k(\mathbf{x},\mathbf{x})$, K_* is $k(\mathbf{x},\mathbf{x}_*)$, K_{**} is $k(\mathbf{x}_*,\mathbf{x}_*)$,

Gaussian Processes With Input Uncertainty: To incorporate localization uncertainty into the environmental model, we extend the framework in [8] for querying a Gaussian process with uncertainty on the input points. The Gaussian process posterior with uncertain inputs [6] is calculated as $p(y|D) =$

$\int p(y|x)p(x)dx$ where y^* is the estimated sample value, x is a sampling location drawn from D (the belief distribution over states at that time from the state estimator). When it is possible to sample from D, Monte Carlo sampling can be used, resulting in $p(y|D) \simeq \frac{1}{N}\Sigma_{n=1}^{N} p(y|x^n)$ where x^n are independent samples from $p(D)$. Approximating $p(y|D)$ as normal by taking the mean and variance of the resulting samples yields $\bar{\mu}(D)$ and $\bar{\sigma}(D)$.

$$\bar{\mu}(D) = \mathbb{E}_x[\mu(x)] \qquad \bar{\sigma}^2(D) = \mathbb{E}_x[\sigma^2(x)] + var_x(\mu(x)) \tag{3}$$

We extend this formulation in two ways. We calculate the likelihood of the samples according to the Gaussian Process distribution and draw the sampled points from their distribution at every step. Critically, our formulation only depends on the ability to sample the robot localizer's belief state at past and future locations and is agnostic to the parameterization of the belief state, which allows this method to support localizers such as pose graphs, Kalman filters and particle filters.

Algorithm 1. Gaussian Process With Sample Uncertainty

procedure MONTE CARLO QUERY UNCERTAIN GP(θ, D_q)
  for $i \in iterations$ **do**
    $x_t^i \leftarrow sample(D_t)$
    $l^i \leftarrow P(y_t|GP(x_t^i, y_t, \theta))P(x_t^i|D_t)$
    $x_q^i \leftarrow sample(D_q)$
    $\mu_q^i, \sigma_q^i = GP(x_t^i, y_t, \theta).query(x_q^i)$
  $\mu \leftarrow \mathbb{E}_q[\mu_q]$
  $\sigma^2 \leftarrow \mathbb{E}_q[\sigma_q^2] + var_q(\mu_q)$

D_q is the belief distribution over the query points, D_t is the belief distribution over the previously collected points, l^i is the ith likelihood which is used for weighting and sub-sampling and θ are pretrained hyper-parameters.

Path Planning for Adaptive Sampling with Input Uncertainty: RIG-Tree [3] is an algorithm for long-horizon adaptive sampling planning. Each node in the tree keeps track of the total objective and cost along a path to that node. The objective functions for adaptive sampling generally use μ and σ from the Gaussian process model which can be extended to the stochastic case using Algorithm 1. The objective used for adaptive sampling is conditional entropy (which has also been used in prior work as the objective for sensor placement [9]).

$$f(x) = \frac{1}{2}\log(2\pi e\bar{\sigma}^2(D)) \tag{4}$$

Given the ability to calculate the stochastic objective function, we present Uncertain Input RIG-Tree, an extension of RIG-Tree to the stochastic case using Algorithm 1 to compute the expected information. Uncertain Input RIG-Tree must also extend the formulation of RIG nodes to keep track of the state of the filter at each proposed location along a trajectory. This is done in order to calculate the objective at proposed sampling location.

4 Experiments

4.1 Simulation Experiments

To test the performance of Algorithm 1 when applied to adaptive sampling, we test both the ability of the Monte-Carlo approximation to approximate the underlying distribution and the ability to approximate the objective function. Testing was done in environments designed to test interpolation methods (Eqs. (5) and (6)). x, y and z are the robot's global coordinates.

$$f(x, y, z) = 4(x - 2 + 8y - 8z^2)^2 + (3 - 4y)^2 + 16\sqrt{z+1}(2z - 1)^2 \tag{5}$$

$$f(x, y, z) = 100(e^{-2/x^{1.75}} + e^{-2/y^{1.5}} e^{-2/z^{1.25}})^2 \tag{6}$$

Comparison of Monte-Carlo Methods. In order to test the predictive power of Algorithm 1, we sample the environment at 1000 evenly spaced samples and add Gaussian noise. Permuted subsets of the samples are used as training data for the different methods to be tested. This training data are used to predict the ground truth and compare the error between the two. The data span a 3D cube with all dimensions in $[0, 1]$. The noisy data were generated by adding Gaussian noise with $\sigma = 0.01$. The following are tested: A noiseless GP representing the ground truth, a traditional GP representing a model on noisy data without any corrections, subsampling the data based on likelihood, and weighing the data based on likelihood. Figure 2 shows the outcome of this experiment, illustrating that likelihood weighting and keeping all samples produces the best estimate. When subsampling, *All* keeps all iterations, *100* keeps only the 100 most likely iterations and *Single* keeps only the most likely iteration.

Noisy Observations. We compare our approach with RIG-Tree without doing noise modeling. Both methods use Eq. (4) as the objective, but the baseline does not use Algorithm 1 for computing the objective. The environments are the same as Fig. 2. The robot runs for a fixed time budget of 3000 m in a 30 × 30 × 30 m cube taking samples every meter. The internal model is compared to a ground truth model over 30^3 evenly spaced points. Prior data are given to the robot in the form of 10 by 10 samples (modeled after pictures from a UAV) taken as noisy samples from the function, with $\sigma = 5$. For the noise characteristics, we use a simplified noise model. The noise characteristics for the underwater vehicle are varied linearly from $\sigma = 1$ at minimum depth to $\sigma = 3$ at maximum depth in all spatial directions. Figure 3 illustrates the results of this experiment. Using Monte Carlo sampling reduces the variance between seeds and, in the first environment, lowers the average error. In environment two, the average error is comparable, but the variance between runs is smaller.

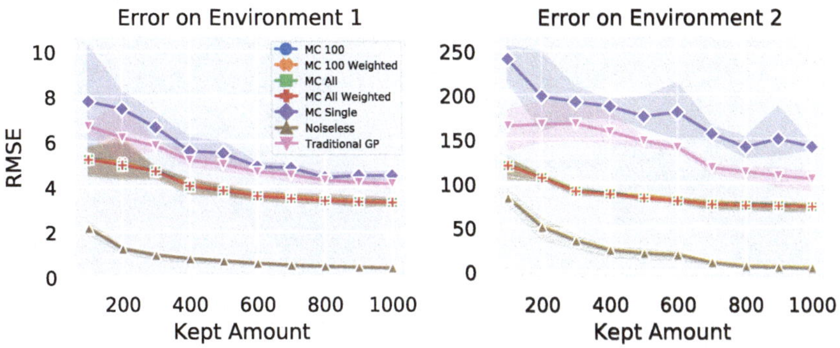

Fig. 2. Error when predicting from noisy data using Algorithm 1. Environment 1 is sampling Eq. (5), Environment 2 is Eq. (6). All MC methods were run for 1000 iterations. Six seeds each.

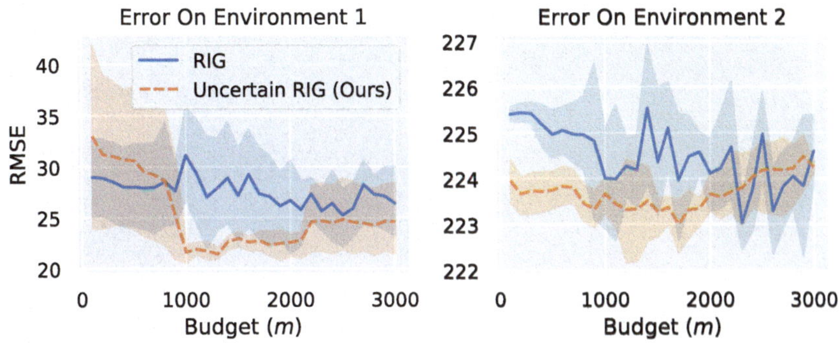

Fig. 3. Error when predicting from noisy data using Algorithm 1. Environment 1 is sampling Eq. (5), Environment 2 is Eq. (6). The robot is simulated for a fixed budget and compared against a noiseless evenly spaced ground truth. Each method runs with three seeds.

4.2 Experiments with Data Collected from Real Environment

In order to evaluate the usability of our algorithm on real world data we evaluate the error of a robot taking samples from orthomosaics created by flying a drone over water. These orthomosaics were created from data taken in Clear lake, California and created using structure from motion software. The UAV used to collect these images carries a hyperspectral imaging camera with a 710 nm band. We use the reflectivity of this band as the desired concentration to be modeled.

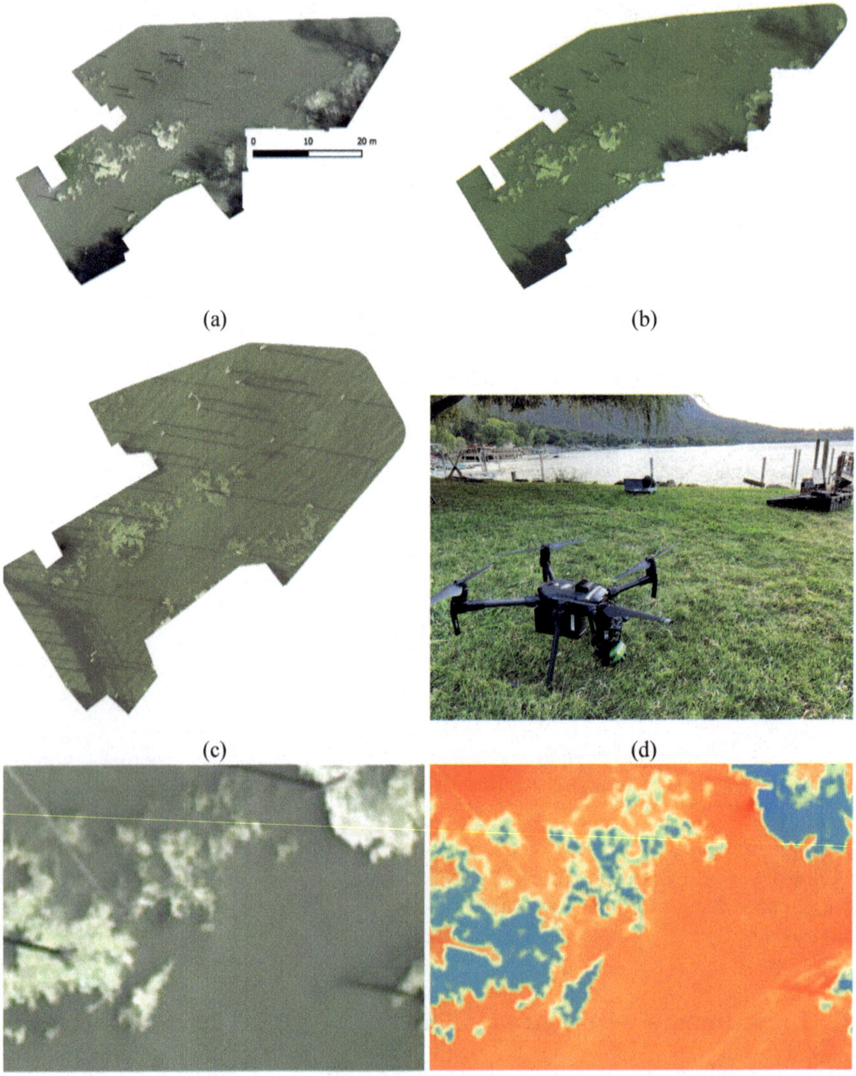

Fig. 4. Environments created from data gathered in Clear lake, California from data collected by a UAV with a hyperspectral camera (Fig. 4d). Figure 4e shows an algal mass in true color while Fig. 4f shows the hyperspectral reflectivity at the target band.

Pose Graph-Based Localization. To model the covariance of the agent at different locations, we take the GPS/IMU readings from the UAV and construct and solve a pose graph using GTSAM [10]. This is to model the error the robot would have if it were to plan using the raw imagery provided by the UAV, in a system like the one described in Fig. 1. The robot is run for 2000m, re-planning every 20m. We compare with a baseline method which simply integrates the noisy samples and attempts to predict from those. This uses the RIG-Tree planner method similar to the last experiment. This is compared against two other methods. The first method does not use Algorithm 1 in calculating the variation for the objective equation, Eq. (4), but does use Monte Carlo estimation for computing the underlying concentration to compare for computing the error. The final method uses Algorithm 1 for calculating both the objective function and the estimate of the underlying concentration.

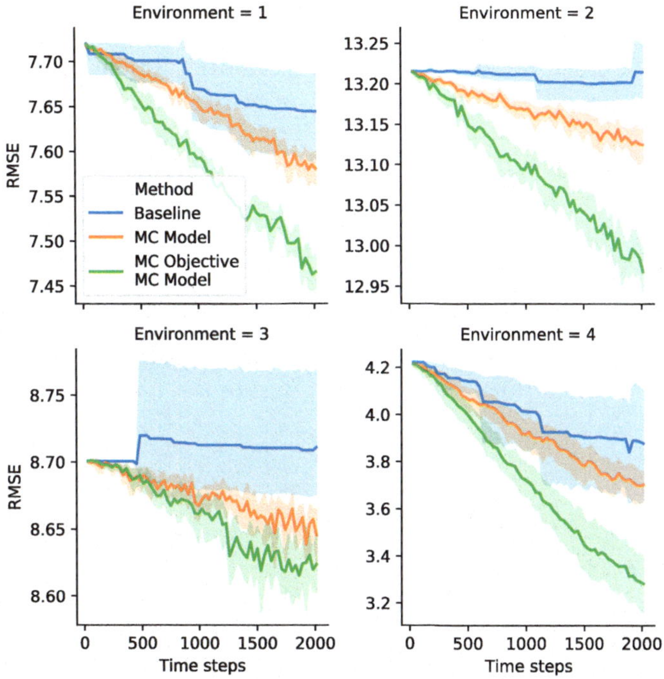

Fig. 5. Error between noisy samples and ground truth on four environments represented in Sect. 4.2. Noise is generated from a pose-graph based localization using the UAV's GPS/IMU. 10 seeds used.

The results of the experiment are seen in Fig. 5. The full method outperforms both the baseline method and the method which does not plan with the added variance from the uncertainty in the robots pose. Both improvements work together to provide a boost in accuracy from both a better estimate of the underlying value and better planning to take more samples from higher noise areas. In environment two the baseline method on average performs better but has a very high variance in the final error, while the Monte Carlo methods tend to have a low variance in the end error. In environment four the robot cannot make a better model by taking noisy samples because there is too much noise on the samples to accurately reconstruct without doing a Monte Carlo estimation.

Particle Filter-Based Localization. To test the ability of the proposed method to work with non-parameteric posterior distributions from state estimators, we simulate a surface vehicle which takes noisy ranges measurements from 10 landmarks and localizes itself using a particle filter. The robot is run for 2000m, re-planning every 20m. To update the localizer, the robot receives noisy measurements to 10 randomly placed landmarks with $\sigma = 3$. The filter is initialized with 100 particles normally distributed around the robot's true starting location. The baseline method uses the weighted mean of the particle filter as the sampling location while the Monte-Carlo methods sample the distribution of particles according to their weight to sample from D. To simulate the filter when planning, the particles and weights are added to each node. When extending samples, the particle filter is simulated forward with the parent's particles and weights in order to calculate the objective for the sampling location. As seen in Fig. 6 the Monte-Carlo based methods outperform the baseline method with a non-Gaussian posterior.

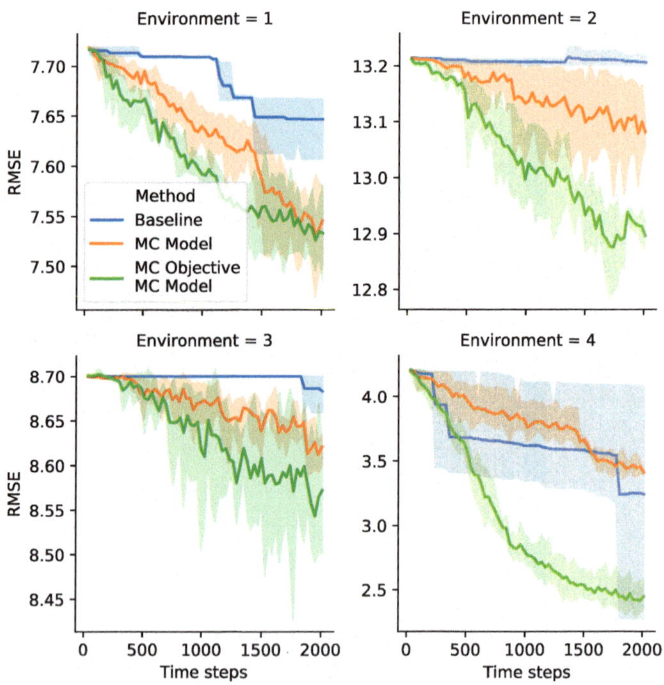

Fig. 6. Error between noisy samples and ground truth on four environments collected from the environments represented in Sect. 4.2. A particle filter with noisy measurement to 10 landmarks is used. 3 seeds run.

In environment four, the Monte Carlo run is outperformed by the baseline method. This environment has the lowest data range of all the environments and may not benefit from using a Monte Carlo estimation as much. Both Monte Carlo methods do suffer from high variance which could potentially be reduced with more iterations of the model estimator.

5 Conclusion

We propose a novel way to incorporate localization noise into adaptive sampling. When there is sufficient and varying noise on the state observations the robot's state estimator will not be accurate. This is common in field scenarios where the pose observations contain noise. We find that frameworks which use the state estimator's maximum likelihood estimate of the robot's pose may fail to properly model the true underlying field being sampled. By doing Monte-Carlo estimation from the belief distribution over robot states generated by the localizer, we are able to better estimate the concentration of an external field from sparse samples compared to previous methods. We show that the field estimation may be poor if the uncertainty in the robot's state estimator is not taken into account. Our framework is a novel approach for adaptive sampling under different types of robot pose uncertainty, such as those generated by a state of the art pose estimation framework. We show that doing a Monte-Carlo estimate of both the objective function during planning and the field estimation improves the planning and model used in adaptive sampling. This modification of an existing adaptive sampling algorithm improves performance when modeling from noisy pose observations and reduces the variance in the error between the robot's model and ground truth. We validate our findings on synthetic functions showing that the plan and model can be improved by our method. We also validate our findings on four real world data-sets from aerial imagery over a lake using complex state estimators from real world dynamics.

References

1. Hwang, J., Bose, N., Fan, S.: AUV adaptive sampling methods: a review. Appl. Sci. **9**, 3145 (2019)
2. Popovic, M., Vidal-Calleja, T.A., Chung, J.J., Nieto, J., Siegwart, R.: Informative path planning for active mapping under localization uncertainty. In: Robotics & Automation Letters (2019)
3. Hollinger, G.A., Sukhatme, G.S.: Sampling-based robotic information gathering algorithms. IJRR **33**(9), 1271–1287 (2014)
4. Manderson, T., Manjanna, S., Dudek, G.: Heterogeneous Robot Teams for Informative Sampling. Workshop on Informative Path Planning and Adaptive Sampling at Robotics Science and Systems, June (2019)
5. Oliveira, R., Ott, L., Ramos, F.: Bayesian Optimisation Under Uncertain Inputs. In: 2nd International Conference on Artificial Intelligence and Statistics (AISTATS 2019)], February 2019
6. Girard, A., Rasmussen, C.E., Candela, J.Q., Murray-Smith, R.: Gaussian process priors with uncertain inputs application to multiple-step ahead time series forecasting. In: Advances in Neural Information Processing Systems, vol. 15, pp. 545–552, MIT Press (2003)

7. Xu, N., Low, K.H., Chen, J., Lim, K.K., Ozgul, E.B.: GP-Localize: Persistent Mobile Robot Localization using Online Sparse Gaussian Process Observation Model. arXiv:1404.5165 [cs, stat], April 2014
8. Rasmussen, C.E., Williams, C.K.I.: Gaussian Processes for Machine Learning. MIT press, Cambridge (2006). ISSN: 0129-0657
9. Guestrin, C., Krause, A., Singh, A.P.: Near-optimal sensor placements in gaussian processes. In: Proceedings of the 22Nd International Conference on Machine Learning, ICML 2005, New York, NY, USA, pp. 265–272. ACM (2005)
10. Forster, C., Carlone, L., Dellaert, F., Scaramuzza, D.: IMU Preintegration on Manifold for Efficient Visual-Inertial Maximum-a-Posteriori Estimation. in Robotics: Science and Systems (2015)

Human-Robot Interaction

A New Approach to Estimate the Apparent Mass of Collaborative Robot Manipulators

Sebastian Herbster(✉), Roland Behrens, and Norbert Elkmann

Fraunhofer IFF, Magdeburg, Germany
{sebastian.herbster,roland.behrens,norbert.elkmann}@iff.fhg.de

Abstract. Collaborative robots have finally found their way into the factories as companions to human workers for releasing them from repetitive and exhausting tasks. Internal torque sensors or tactile skins make these robots sensitive to external forces, which is a critical requirement for acting safely, as their direct force feedback allows them to react instantaneously on dangerous collisions with humans. Biomechanical limit values give robot users a reference to determine the injury potential of their robots. In some cases, those limits are available as energy limits. A valuable advantage of such limits is that any user can easily convert them into velocity constraints if the apparent mass of the robot is known. The apparent mass is a virtual quantity that projects the inertia properties of the robot to a directional point mass. In our paper, we show why the currently-available method to determine the apparent mass needs to be extended. Our findings reveal that the impact conditions determine whether the drive inertia must be taken into account or the apparent mass may only be calculated based on the dynamic mass parameters of the link. The verification relies on data from collision tests with two real collaborative robots. Our results fit well to the results we obtained from the simulation models, why our proposed modification seems to be necessary in order to achieve reliable velocity constraints for operating collaborative robots safely.

1 Introduction

Collaborative robots are ideally suited to improve the working conditions in factories with a high share of manual work. They can work safely near humans, for instance to relieve them from repetitive and exhausting tasks. However, the design of collaborative workplaces must guarantee that the collaborative robot installed for a particular application does not expose an injury risk to the human. This is especially relevant, if physical contact between both partners is possible or even necessary to run the process [6]. Biomechanical limit values, given as energy- or force- and pressure-based limit values are a frequently used measure to mitigate the injury potential of robots [3,5]. If certain parameters are specified, different

The original version of this chapter was revised: The abstract to this chapter has been added. The correction to this chapter is available at
https://doi.org/10.1007/978-3-030-71151-1_55

B. Siciliano et al. (Eds.): ISER 2020, SPAR 19, pp. 211–221, 2021.
https://doi.org/10.1007/978-3-030-71151-1_19

methods exist to estimate safe velocities that will prevent the robot from exceeding the limit values. Most methods have in common that the estimation is only possible, if the apparent mass of the robot is known.

Khatib developed an analytical method to calculate the apparent mass of a robot manipulator [16,17]. The combination of the apparent mass and energy-based limit values not only allows the estimation of safe velocities [11,12,18–20] but also to derive inherently safe robot designs [18] and trajectories [21,24,26]. Besides energy-based methods, there is another approach incorporating a three mass oscillator that takes into account the distinct influence of the joint elasticity and motor inertia with the goal of evaluating collision forces. Simulations with such impact models reveal that low joint elasticity decouples the motor inertia and reduces the impact force [7,10,13,14].

This article presents an experimental study to determine the apparent mass of any collaborative robot by impact tests in order to eliminate the contradictions between the different models. The test setup is based on theoretical investigations and findings of the analytical solution of a three mass oscillator. The study scheme was tested in multiple test series with two commercially available collaborative robots (KUKA iiwa14 and UR10). Based on the test results, the quality and precision of the models mentioned above are briefly evaluated and compared.

2 Technical Approach

Energy-based limits specify a maximum allowable amount of energy, which the robot must not transfer to the human under any circumstances. In general, the validity of such values is subject to a specific part of the human body and shape of the energy transferring surface. Let $\hat{E}$ be an energy threshold for a certain shape and body part. Then the robot must not transfer more energy during an impact, meaning that its kinetic energy T_I, which is virtually stored in the impact point on the robot surface, must be

$$\hat{E} \leq T_I \qquad T_I = \frac{1}{2} m_I v_I^2 \,, \tag{1}$$

while m_I is the impact mass and v_I the impact velocity. If m_I is known, the maximal allowable velocity of the robot can be easily determined. A simplified impact model considers robot and human as two point masses m_R and m_H, connected by the contact stiffness c_H which represents the compliance of the stressed human tissue [9,27]. Key of the model is to reduce the degrees of freedom by one, making the system a single mass oscillator with a "reduced" impact mass

$$m_I = \left(\frac{1}{m_H} + \frac{1}{m_R} \right)^{-1} \,. \tag{2}$$

However, a three mass oscillator is better suited to represent the impact behavior, especially for a lightweight robot with elastic joints [7,13]. As shown in Fig. 1, the three masses originate from the inertia of the joint drives m_D, the link masses m_L, and the mass of the colliding object m_H. The elasticity between the drives

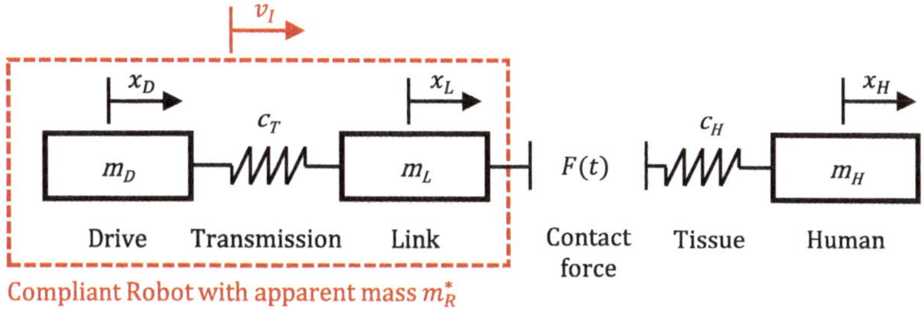

Fig. 1. Three mass oscillator as an impact model for human-robot interactions

and joints is c_T, and the one at the affected human body location is c_H. Based on the illustration in Fig. 1, the following system of differential equations can be obtained

$$\begin{aligned} \ddot{y}_1 + y_1 \underbrace{\left(\frac{c_H}{m_H} + \frac{c_H}{m_L}\right)}_{a} - y_2 \underbrace{\frac{c_T}{m_L}}_{b} &= 0 \\ \ddot{y}_2 - y_1 \underbrace{\frac{c_H}{m_L}}_{c} + y_2 \underbrace{\left(\frac{c_T}{m_L} + \frac{c_T}{m_D}\right)}_{d} &= 0 \end{aligned} \tag{3}$$

with

$$y_1 = x_H - x_L \quad \text{and} \quad y_2 = x_L - x_D \; . \tag{4}$$

To solve Eq. (3) we introduce the variables

$$y_3 = \dot{y}_1 \quad \text{and} \quad y_4 = \dot{y}_2 \; , \tag{5}$$

which brings us to a differential equation of first order in the matrix form

$$\begin{bmatrix} \dot{y}_1 \\ \dot{y}_2 \\ \dot{y}_3 \\ \dot{y}_4 \end{bmatrix} + \begin{bmatrix} 0 & 0 & -1 & 0 \\ 0 & 0 & 0 & -1 \\ a & -b & 0 & 0 \\ -c & d & 0 & 0 \end{bmatrix} \begin{bmatrix} y_1 \\ y_2 \\ y_3 \\ y_4 \end{bmatrix} = \dot{\mathbf{y}}(t) + \mathbf{A}\mathbf{y}(t) = \mathbf{0}^{4\times 1} \; . \tag{6}$$

The solution of the differential equation is given by

$$\mathbf{y}(t) = e^{\mathbf{A}t}\mathbf{y}_0 \; , \tag{7}$$

where $e^{\mathbf{A}t}$ is the matrix exponential function and $\mathbf{y}_0$ represents the start conditions. For our considerations, we assume that at t_0 all masses are at their neutral position, where no spring forces occur and the masses m_D and m_L are moving at the same and constant impact velocity v_I towards the non-moving mass m_H, which leads to

$$\mathbf{y}_0 = \begin{bmatrix} x_H - x_L \\ x_L - x_D \\ \dot{x}_H - \dot{x}_L \\ \dot{x}_L - \dot{x}_D \end{bmatrix} = \begin{bmatrix} y_1 \\ y_2 \\ y_3 \\ y_4 \end{bmatrix} = \begin{bmatrix} 0 \\ 0 \\ -v_I \\ 0 \end{bmatrix} \; . \tag{8}$$

To determine the deformation at the contact point between the robot and the human only $y_1(t)$ is of interest. With the solution of $e^{\mathbf{A}t}$ and $\mathbf{y}_0$ we receive

$$y_1(t) = v_I \underbrace{(\alpha \sin(\omega_1 t) + \beta \sin(\omega_2 t))}_{P(t)} \tag{9}$$

with the magnitudes α and β and the eigenfrequencies ω_1 and ω_2 of the system. Note that α, β, ω_1 and ω_2 consist only of the mass and stiffness parameters of the system. The oscillator function $P(t)$ is introduced to summarize the two superimposed sinusoidal oscillations. For an unconstrained contact between human and robot only the first impact is of interest, since c_H can only transmit compression forces and appropriately configured collaborative robots usually initiate an emergency stop after a collision has been detected. With $\hat{y}_1$ describing the maximum deformation between human and link and $\hat{P}$ being the corresponding maximum of the oscillation function, the maximum impact force $\hat{F}$ is given by

$$\hat{F} = c_H \hat{y}_1 = c_H v_I \hat{P} \,. \tag{10}$$

3 Results

For the analysis of the apparent mass of the robot, we first investigate the oscillation behavior of the three mass oscillator. Equation (9) reveals that the impact velocity v_I has a linear effect on $y_1(t)$ and therefore on $F(t)$. Consequently, the qualitative collision behavior does not depend on v_I and it is sufficient to further analyze $P(t)$.

The graphs on the left in Fig. 2 indicate the course of $P(t)$ over time for different transmission stiffness c_T compared to a pure sinusoidal shape. The pure sinus is fitted from $t = 0$ to the first maximum (blue vertical line). The solid line in green illustrates the error between the fitted sinus and $P(t)$. The graphs on the right depict the magnitudes α and β and the eigenfrequencies ω_1 and ω_2 over c_T. Although $P(t)$ is composed of two superimposed oscillations, it corresponds to the pure sinus over its entire course for the case of $c_T \to \infty$ and $c_T \to 0$. This behavior is the result of the magnitudes, which show that $\alpha \to 0$ for $c_T \to \infty$ and $\beta \to 0$ for $c_T \to 0$. For $c_T \approx c_H$ (in this example at $10\,\mathrm{N\,mm^{-1}}$) the oscillation differs from a pure sinus, since both magnitudes and eigenfrequencies are unequal to zero. Up to the first maximum, however, the error is negligibly small. Based on these findings, an external observer who measures the course up to the first maximum would conclude, that the compliant robot behaves similar to a single object with the apparent mass m_R^*. Assuming a fully elastic linear contact, the apparent impact mass becomes

$$m_I^* = \frac{\hat{F}^2}{v_I^2 c_H} \,, \tag{11}$$

by equating the kinetic with the spring energy. The apparent mass of the robot m_R^* can then be calculated by solving Eq. (2) with Eq. (11) after substituting $\hat{F}$

with Eq. (10)

$$m_R^* = \frac{c_H m_H \hat{P}^2}{m_H - c_H \hat{P}^2} . \tag{12}$$

Next, the model is parameterized to one of the examined robots. For this purpose, the effective masses m_D and m_L, and the effective stiffness c_T need to be calculated. Khatib presented an approach to project the mass matrix of a robot manipulator $\mathbf{M}(\mathbf{q}) \in \mathbb{R}^{n \times n}$ to a directional point mass

$$m_R = [\mathbf{u}^T \mathbf{\Lambda}_t^{-1} \mathbf{u}]^{-1} \tag{13}$$

with $\mathbf{u}$ being the impact direction and $\mathbf{\Lambda}_t$ the translational part of the robot's mass matrix $\mathbf{\Lambda}(\mathbf{x})$ given by [16,17]

$$\mathbf{\Lambda}^{-1} = \mathbf{J}(\mathbf{q})\mathbf{M}(\mathbf{q})^{-1}\mathbf{J}(\mathbf{q})^T . \tag{14}$$

Assuming that $\mathbf{M}$ includes only the inertia parameters of the robot links, then m_R reduces to $m_R^- = m_L$. If $\mathbf{M}^+$ is the robots mass matrix including the inertia of the drive $\mathbf{B}$, m_R is given by

$$m_R^+ = m_L + m_D . \tag{15}$$

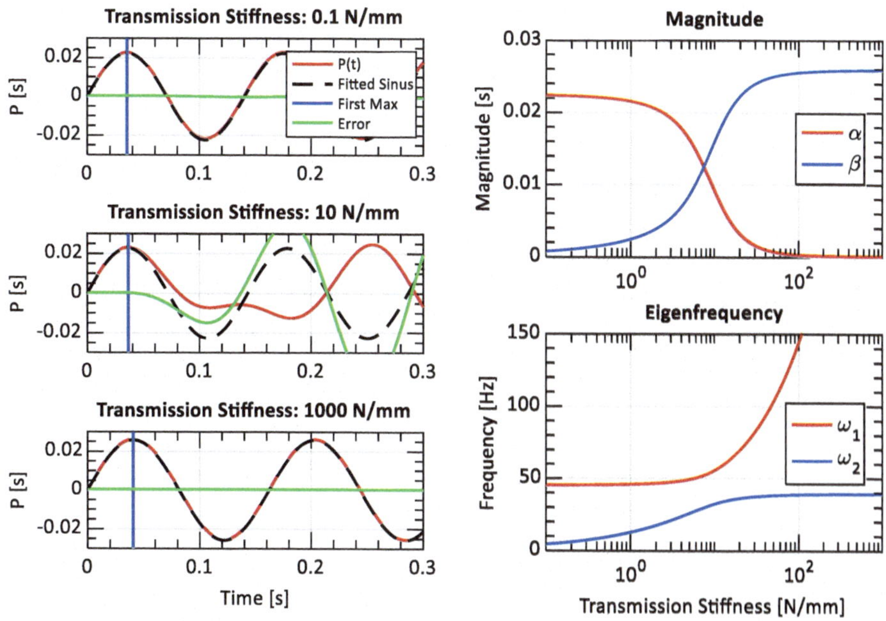

Fig. 2. Oscillator function $P(t)$, Magnitudes α, β and Eigenfrequencies ω_1, ω_2 for different Transmission Stiffness c_T with fixed values for the contact stiffness $c_H = 10\,\mathrm{N\,mm^{-1}}$ and the masses $m_D = m_L = m_H = 10\,\mathrm{kg}$

The robot's effective stiffness c_T can be obtained in the same way, but based on the stiffness matrix $\mathbf{K}_q$ instead of $\mathbf{M(q)}$ [4,25]. Finally, the values are inserted into the analytical solution of Eq. (9) and evaluated for m_R^* according to Eq. (12). Figure 3 represents a sample of results for different body parts. It shows m_R^* over c_T for multiple results with different c_H and m_H as listed in ISO/TS 15066. The diagram shows that m_R^* saturates for $c_T \to \infty$ to m_R^+ and for $c_T \to 0$ to m_R^-. The latter case shows a full decoupling of m_D from m_L. Between both cases there is a transition area whose position depends on the parameters of the colliding object.

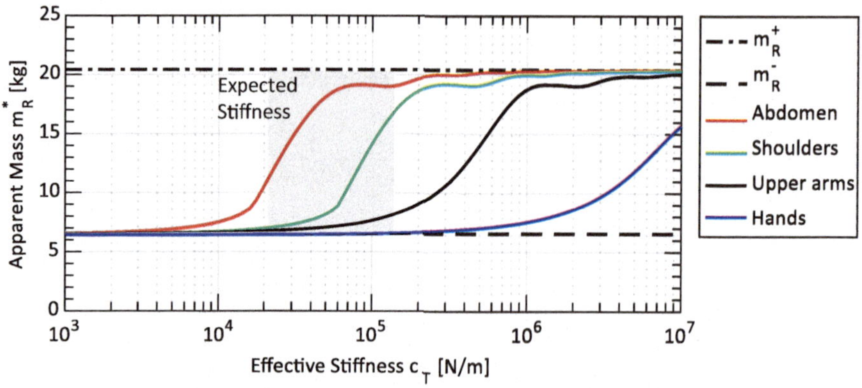

Fig. 3. Apparent mass simulated for a KUKA iiwa 14 in joint configuration 1 (see Table 1) for different body parts: Abdomen (m_H = 40 kg, c_H = 10 N mm^{-1}), Shoulders (m_H = 40 kg, c_H = 35 N mm^{-1}), Upper Arms (m_H = 3 kg, c_H = 30 N mm^{-1}) and Hands ($m_H = 0.6kg$, c_H = 75 N mm^{-1}) [3]

The colored box indicates the effective stiffness c_T of a KUKA iiwa 14 for the joint configurations listed in Table 1 that have been obtained from $\mathbf{K}_q$ [8]. The simulation indicates that for body parts with high mass and low stiffness the influence of the motor inertia cannot be neglected. For body parts with a low total mass and high stiffness, however, only the link mass m_R^- may be taken into account.

4 Experiments

The goal of the experiments was to measure the apparent mass of a robot. As the impact model revealed, the apparent mass depends on the mass and stiffness properties of the colliding object, as well as the joint configurations of the robot. Figure 4 depicts the setup created for the experiments. It consists of a collaborative robot on a welding table and a free-hanging pendulum that simulates the mass of the human body part m_H. To analyze the influence of m_H on the apparent mass m_R^*, two masses of 4 kg and 15 kg were used. For the variation of c_H, the lightweight impact tool was mounted on a changeable spring. The spring was also attached to a load-cell that acquires the maximum contact force $\hat{F}$. This force sensor measures with a sample rate of 10 kHz and 16 bit in a range of 0 ... 500 N. Different springs were applied that lie within the stiffness range of the human body

Table 1. Tested configurations with the vector of joint positions $\mathbf{q} \in \mathbb{R}^n$ in [°]

Config	Robot	q_1	q_2	q_3	q_4	q_5	q_6	q_7	Type	Velocity
1		0	90	0	0	90	–90	0	PTP	10, 20, 30, 40, 50 deg/s
2	iiwa14	0	35	0	–115	90	–90	0	PTP	20, 32.5, 45 deg/s
3		0	50	0	–83	0	–43	0	LIN	150, 275, 500 $\mathrm{mm\,s^{-1}}$
4		0	–180	0	180	10	–90	-	PTP	10, 20, 30, 40 deg/s
5	UR10	0	–119	–106	–133	20	–90	-	PTP	15, 30, 45, 60 deg/s
6		0	–135	–80	–145	–90	–90	-	LIN	100, 300, 500, 700 $\mathrm{mm\,s^{-1}}$

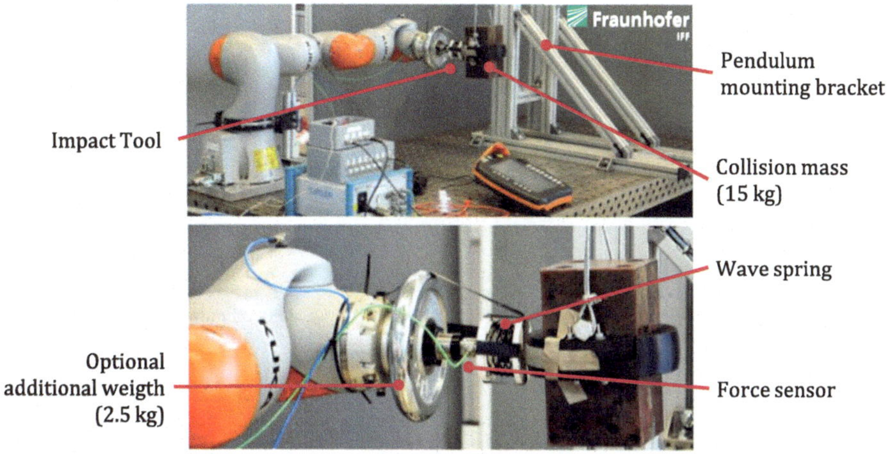

Fig. 4. Experimental setup to measure the effective mass of a robot in configuration 1 (see Table 1) at the impact position

(c_H = 10 ... 120 $\mathrm{N\,mm^{-1}}$) [3]. For each combination, the robots collided at different velocities according to Table 1 with the free-hanging pendulum. The programmed trajectory ensured that the robot reached the given velocity at initial contact with the pendulum mass. Every combination and velocity level has been measured five times. The Point-to-Point (PTP) movement was executed by the first axis only. In case of the linear trajectories (LIN), the robot moves away from its base in the horizontal plane.

5 Main Experimental Insights

The apparent robot mass m_R was determined from the recorded maximum contact force $\hat{F}$, the known pendulum mass m_H and spring stiffness c_H according to Eq. (2) and Eq. (11). For c_H the exact spring rate was measured and used for the evaluation. The collision velocity v_I was calculated from the angular joint positions, joint velocities and a kinematic model of the robot under test, which also includes the end-effector. The blue bars in Fig. 5 illustrate the experimental result as mean values with corresponding standard deviation (SD). The red bars represent the result of the analytically solved three mass oscillator. The horizontal lines m_R^+ and m_R^- were calculated based on Eq. (13) with and without joint inertia, respectively. The evaluated data for m_R^* did not depend on v_I, but on c_H, m_H, and the joint configuration. This observation is consistent with the predictions made with the three mass oscillator in Fig. 3. Higher m_H and lower c_H increase the apparent robot mass. The comparison of the experimental data with the three mass oscillator provides consistent results for most configurations. In some constellations the measured apparent mass is even smaller than m_R^-. This effect is probably caused by the link stiffness that decouples a part of the links' mass itself [15]. Some results indicate an increased standard deviation (e.g. Configuration 3, $c_H = 10\,\mathrm{N\,mm^{-1}}$, $m_H = 10\,\mathrm{kg}$). Since none of the input parameters indicated extraordinary variations, a sensitivity analysis was performed. It revealed that the sensitivity with respect to the impact force

$$\frac{\partial m_R^*}{\partial \hat{F}} = \frac{2c_H m_H^2 v_I^2 \hat{F}}{(\hat{F}^2 - c_H m_H v_I^2)^2} \tag{16}$$

was up to 40 kg N^{-1} in some test configurations.

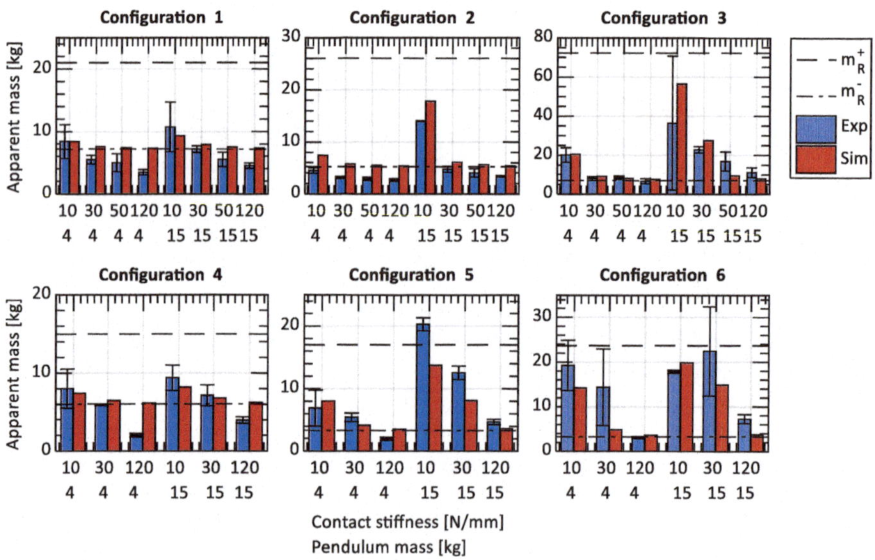

Fig. 5. Experimental results of the apparent mass for different robot configurations, contact stiffnesses and pendulum masses (mean ± SD)

6 Conclusion and Outlook

The article presents an experimental and simulation-based procedure to determine the apparent mass of a robot with compliant joints. The results of the experiments and simulation highlight that established methods for calculating the apparent mass based on m_R^-, which neglect the effect of the drive inertia, underestimate the apparent mass of robots. This finally leads to velocities that are too high and therefore unsafe. If, however, the drives are always considered (m_R^+) extreme values which provide too conservative estimates for the safe velocity of the robot are the result.

The analysis of the three mass oscillator with respect to the apparent mass showed, that especially the stiffness ratio between transmission c_T and soft tissue c_H as well as the mass of the colliding human body part m_H determines the collision behavior. For the examined robots, the conservative assessment (m_R^+) is required for body parts, such as the Abdomen, Back, Shoulders, and Lower Extremities, with $m_H \geq 40\,\mathrm{kg}$ or $c_H \leq 30\,\mathrm{N\,mm^{-1}}$. For body parts of $m_H \leq 4.4\,\mathrm{kg}$, including Hands, Upper and Lower Arms, it is sufficient to consider m_R^- only. The findings are not only relevant for end-users of collaborative robots but also of particular importance for robot manufacturers. They can design robot joints such that, regardless of the joint configuration the effective stiffness c_T is lower than that of the human soft tissue c_H. This allows to consider m_R^- in any contact scenario and increases the safe velocity.

Based on our findings, we propose to skip the explicit calculation of the apparent mass for robots with compliant joints, but to use the analytical solution of the three mass oscillator to estimate safe velocities. Since the impact velocity v_I has a linear influence on the oscillating function $P(t)$, the straight forward estimation

$$v_I = \frac{\hat{F}}{c_H \hat{P}} = \sqrt{\frac{2\hat{E}}{c_H \hat{P}^2}} \tag{17}$$

based on the admissible maximum impact force $\hat{F}$ or maximum amount of transferred energy $\hat{E}$ is possible. The approach allows to consider all relevant parameters without the use of numerical solvers. It is therefore ideally suited for planning tools and real-time applications.

Future studies will examine the legitimacy of neglecting the controller. It is presumed that this simplification depends likewise on the stiffness ratio. Consequently, a validity range for the model will be defined. The definition of the range is complex, since the controller architecture of each robot is different and in most cases unknown. Furthermore, the presented three mass oscillator only considers a linear stiffness for c_H. However, in vivo measurements have shown that human soft tissue behaves nonlinearly [23] and follows a hysteresis shape [22,28]. Further simulations and experiments have to investigate this effect on the impact behavior. In addition, the mass and stiffness parameters of the robot are often not publicly accessible, but are essential for the calculation of accurate estimates. Chawda and Niemeyer performed a system identification in the frequency domain for the first joint of a KUKA iiwa to quantify its stiffness, link and controlled motor inertia [8]. It is planned to use a similar approach and define a test procedure that can determine all relevant parameters for the model presented in this article.

References

1. 2012 IEEE/RSJ International Conference on Intelligent Robots and Systems (IROS), Vilamoura, Portugal, October 2012
2. 2017 IEEE/RSJ International Conference on Intelligent Robots and Systems (IROS), Vancouver, Canada, September 2017
3. ISO/TS 15066. Robots and robotic devices - collaborative robots. Standard, International Organization for Standardization (2016)
4. Albu-Schaffer, A., Fischer, M., Schreiber, G., Schoeppe, F., Hirzinger, G.: Soft robotics: what cartesian stiffness can obtain with passively compliant, uncoupled joints? In: Proceedings of the 2004 IEEE/RSJ International Conference on Intelligent Robots and Systems (IROS), pp. 3295–3301, vol. 4, Sendai, Japan, February 2004. https://doi.org/10.1109/IROS.2004.1389925
5. Behrens, R., Elkmann, N.: Study on meaningful and verified thresholds for minimizing the consequences of human-robot collisions. In: Proceedings of the 2014 IEEE International Conference on Robotics & Automation (ICRA), pp. 3378–3383, Hong Kong, China, May 2014. https://doi.org/10.1109/icra.2014.6907345
6. Behrens, R., Saenz, J., Vogel, C., Elkmann, N.: Upcoming technologies and fundamentals for safeguarding all forms of human-robot collaboration. In: Proceedings of the 8th International Conference Safety of Industrial Automated Systems, pp. 18–23, Königswinter, Germany, November 2015
7. Bicchi, A., Tonietti, G.: Fast and "soft-arm" tactics.IEEE Robot. Autom. Mag. **11**(2), 22–33 (2004). ISSN 1070-9932. https://doi.org/10.1109/MRA.2004.1310939
8. Chawda, V., Niemeyer, G.: Toward torque control of a kuka lbr iiwa for physical human-robot interaction. In: Proceedings of [2], pp. 6387–6392. https://doi.org/10.1109/iros.2017.8206543
9. Haddadin, S., Albu-Schäffer, A., Hirzinger, G.: Requirements for safe robots: measurements, analysis and new insights. Int. J. Robot. Res. **28**(11–12), 1507–1527 (2009)
10. Haddadin, S., Albu-Schäffer, A., Eiberger, O., Hirzinger, G.: New insights concerning intrinsic joint elasticity for safety. In: Proceedings of the 2004 IEEE/RSJ International Conference on Intelligent Robots and Systems (IROS), pp. 2181–2187, Taipei, Taiwan, October 2010. https://doi.org/10.1109/IROS.2010.5652037
11. Haddadin, S., Haddadin, S., Khourym, A., Rokahr, T., Parusel, S., Burgkart, R., Bicchi, A., Albu-Schäffer, A.: On making robots understand safety: Embedding injury knowledge into control. Int. J. Robot. Res. **31**(13), 1578–1602 (2012). ISSN 0278-3649. https://doi.org/10.1177/0278364912462256
12. Haddadin, S., Haddadin, S., Khoury, A., Rokahr, T., Parusel, S., Burgkart, R., Bicchi, A., Albu-Schäffer, A.: A truly safely moving robot has to know what injury it may cause. In: Proceedings of [1], pp. 5406–5413. https://doi.org/10.1109/iros.2012.6386163
13. Haddadin, S., Krieger, K., Mansfeld, N., Albu-Schäffer, A.: On impact decoupling properties of elastic robots and time optimal velocity maximization on joint level. In: Proceedings of [1], pp. 5089–5096. https://doi.org/10.1109/IROS.2012.6385913
14. Haninger, K., Surdilovic, D.: Bounded collision force by the sobolev norm. In: Proceedings of the 2019 IEEE International Conference on Robotics & Automation (ICRA), pp. 8529–8535, Montreal, Canada, May 2019. https://doi.org/10.1109/ICRA.2019.8793711
15. Hu, M., Wang, H., Pan, X., Tian, Y.: Research on elastic deformation modeling of collaborative robots. In: Proceedings of the 2017 IEEE International Conference

on Robotics and Biomimetics (ROBIO), pp. 2462–2467, Macau, China, December 2017. https://doi.org/10.1109/ROBIO.2017.8324789
16. Khatib, O.: A unified approach for motion and force control of robot manipulators: the operational space formulation. IEEE J. Robot. and Autom. **3**(1), 43–53 (1987). https://doi.org/10.1109/JRA.1987.1087068
17. Khatib, O.: Inertial properties in robotic manipulation: An object-level framework. Int. J. Robot. Res. **14**(1), 19–36 (1995). ISSN 0278-3649. https://doi.org/10.1177/027836499501400103
18. Lee, S.-D., Kim, B.-S., Song, J.-B.: Human-robot collision model with effective mass and manipulability for design of a spatial manipulator. Adv. Robot. **27**(3), 189–198 (2013). https://doi.org/10.1080/01691864.2012.754076
19. Liang, J., Meng, D., Wang, X., Liang, B., She, Y.: Modeling and optimization of head-collision of a flexible joint robot. In: Proceedings of the 36th Chinese Control Conference (CCC), pp. 6298 – 6933, Dalian, China, June 2017. https://doi.org/10.23919/ChiCC.2017.8028450
20. Lucci, N., Lacevic, B., Zanchettin, A.M., Rocco, P.: Combining speed and separation monitoring with power and force limiting for safe collaborative robotics applications. IEEE Robot. Autom. Lett. **5**(4), 6121–6128 (2020). https://doi.org/10.1109/LRA.2020.3010211
21. Mansfeld, N., Djellab, B., Veuthey, J.R., Beck, F., Ott, C., Haddadin, S.: Improving the performance of biomechanically safe velocity control for redundant robots through reflected mass minimization. In: Proceedings of [2], pp. 5390–5397. https://doi.org/10.1109/IROS.2017.8206435
22. Moorhouse, K.: An improved normalization methodology for developing mean human response curves. In: Proceedings of the 23rd International Technical Conference on the Enhanced Safety of Vehicles (ESV), pp. Moorhouse 1–11, Seoul, South Korea, May 2013
23. Pailler-Mattei, C., Bec, S., Zahouani, H.: In vivo measurements of the elastic mechanical properties of human skin by indentation tests. Med. Eng. Phys. pp. 599–606 (2008). https://doi.org/10.1016/j.medengphy.2007.06.011
24. Rossi, R., Polverini, M.P., Zanchettin, A.M., Rocco, P.: A pre-collision control strategy for human-robot interaction based on dissipated energy in potential inelastic impacts. In: 2015 IEEE/RSJ International Conference on Intelligent Robots and Systems (IROS), pp. 26–31 (2015). https://doi.org/10.1109/IROS.2015.7353110
25. Salisbury, J.-K.: Active stiffness control of a manipulator in cartesian coordinates. In: Proceedings of the Conference on Decision and Control including the Symposium on Adaptive Processes, pp. 95–100 (1980). https://doi.org/10.1109/CDC.1980.272026
26. Sloth, S., Petersen, H.G.: Computation of safe path velocity for collaborative robots. In: Proceedings of the 2018 IEEE/RSJ International Conference on Intelligent Robots and Systems (IROS), pp. 6142–6148, Madrid, Spain, October 2018. https://doi.org/10.1109/IROS.2018.8594217
27. Vemula, B.R., Ramteen, M., Spampinato, G., Fagerström, B.: Human-robot impact model: For safety assessment of collaborative robot design. In: Proceedings of the 2017 IEEE 5th International Symposium on Robotics and Intelligent Sensors (IRIS), pp. 236–242, Ottawa, Canada, October 2017. https://doi.org/10.1109/IRIS.2017.8250128
28. Wang, Z., Behrens, R., Elkmann, N.: Fe-simulation of impact loads on the human body: Methodology for the development of tissue models. In: Proceedings of the Dynamore Forum 15, Bamberg, Germany, October 2018

A New Conversion Method to Evaluate the Hazard Potential of Collaborative Robots in Free Collisions

Sebastian Herbster(✉), Roland Behrens, and Norbert Elkmann

Fraunhofer IFF, Magdeburg, Germany
{sebastian.herbster,roland.behrens,norbert.elkmann}@iff.fhg.de

Abstract. With the increasing importance of collaborative robots in industrial manufacturing, their economic efficiency is becoming more and more important. Today, one approach to prevent collaborative robots from injuring humans is to assure that the robot cannot exceed biomechanical limits in the event of an accidental collision or clamping. The ability of the robot to avoid collision forces beyond the limits must be validated with a biofidelic measurement device that mimics the biomechanical behavior of the human. For reliable use, the measurement devices must be attached to a rigid frame. Consequently, the test setup is solely able to simulate the contact dynamics and biomechanical consequences of a clamping contact in which the human body part cannot move. Free collisions that allow the human to move freely reduce the collision forces, but can only be evaluated with such a measurement device. This technical limitation leads to slower robots and thus a loss of productivity. The study presents a method that increases the efficiency of safe collaborative robots by adding a new validation procedure. The presented method incorporates a model-based conversion of measurements that enables safety experts to validate robots in free collisions. Data from experimental tests with a collaborative robot and a biofidelic measurement device show a good fit of the model-based prediction and thus confirm our approach. This new approach has great potential to increase the productivity of collaborative robots, since our method will allow them to move at faster but still safe speeds.

1 Introduction

In the closest form of human-robot interaction, robots and humans complete common tasks next to each other at the same time [3]. At such workplaces, accidental contacts such as clamping or unconstrained collisions constitute a risk of human injury that technical measures cannot eliminate entirely, since there is always the possibility of a technical failure or foreseeable misuse. As a metric and reference for acceptable risks, ISO/TS 15066 defines biomechanical

The original version of this chapter was revised: The abstract to this chapter has been added. The correction to this chapter is available at
https://doi.org/10.1007/978-3-030-71151-1_55

B. Siciliano et al. (Eds.): ISER 2020, SPAR 19, pp. 222–232, 2021.
https://doi.org/10.1007/978-3-030-71151-1_20

force and pressure limits for multiple body parts. The technical risk mitigation measures of the robot must reliably prevent these limits from being exceeded [7].

During the validation measurement, the velocity of the robot must be reduced until the measured contact forces and pressures do not exceed the applicable limit values [12]. The proper use of the PFMD requires mounting it on a rigid structure that avoids any undesired displacements of the PFMD. This requirement, however, does not reflect the conditions of an unconstrained contact in which the human body part hit by the robot can move freely.

Falco et al. and Oberer-Treitz presented moveable PFMDs that are capable of simulating the dynamics of free collisions [5,10]. They attached a PFMD to a linear guide rail and adhered additional weight, so that the total mass of the movable parts reflect exactly the mass of the human body part under test. The US American de-facto standard RIA T R15.806-2018 incorporated the concept of a free-moving PFMD and specifies that it should be used to evaluate free collisions instead of a fixed PFMD. However, a movable PFMD has the major limitation that it can only measure collisions within the horizontal plane.

This article presents a method that allows to convert the results recorded with a fixed PFMD into values that reflect the results taken with a free-moving PFMD. Validation experiments with three collaborative robots (UR3e, UR10e, Doosan M0607) and two different PFMDs confirm that our conversion method provides correct and sufficiently accurate results.

2 Technical Approach

The development of the proposed conversion method can be demonstrated by the analysis of an accurate and simplified impact model. The accurate model in Fig. 1 (top) reproduces the impact behavior of a compliant robot that collides with a biofidelic PFMD. The collaborative robot is modeled as a two mass oscillator [4,6]. The first mass represents the contribution of the robot's drive inertia m_D and the second the mass of its links m_L , while c_T describes the joints elasticity. The other part of the model covers the PFMD. The parameters c_D and d_D represent the damping behavior and elasticity of the damping material of the PFMD. The damper is attached to a bumper of mass m_B. A mechanical spring of stiffness c_S absorbs the impact energy from the bumper and transmits the generated force to a load cell that records the impact force $F_I(t)$. To simulate a properly fixed PFMD the housing is considered to be immovable. To simulate an unconstrained collision the load cell is attached to a body of mass m_M with static friction F_R.

To convert the maximum contact force $\hat{F}_C$ from a fixed PFMD into the maximum contact force $\hat{F}_F$ of a movable PFMD, the complexity of the simulation model must be reduced. This ultimately leads to the model shown in Fig. 1 (bottom). This model assumes that m_D, m_L, and c_T are included in an apparent robot mass m_R^*. Due to its simplified structure, the comparison of the energies stored in elasticities and masses of both devices (fixed and free) lead to the

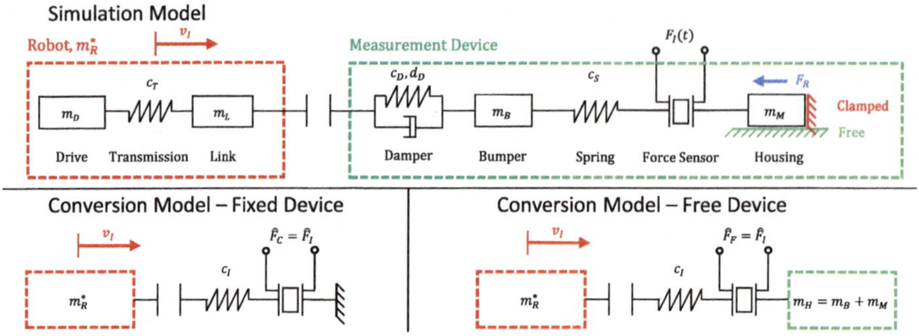

Fig. 1. Simulation and conversion model of a fixed and free measurement device

following conversion

$$\hat{F}_F = \hat{F}_C \sqrt{\frac{m_H}{m_H + m_R^*}} \; . \tag{1}$$

The expression reflects the relative difference between the maximum force of a free $\hat{F}_F$ and a clamped collision $\hat{F}_C$. In contrast to m_H, the apparent robot mass m_R^* is unknown, but can be estimated with the momentum of the impact

$$p = m_R^* v_I = \int_{t_0}^{t_{max}} F_C(t) dt \; , \tag{2}$$

where v_I is the impact velocity, t_0 the time of initial contact and t_{max} the time at which the clamped measured impact force $F_C(t)$ reaches $\hat{F}_C$.

3 Model-Based Results

A simulation with the accurate model (Fig. 1, on the top) was performed to study the conversion method. Therefore, the applied model parameters were precisely adjusted to the robot and PFMD, that we later used in the experimental tests for model validation. The robot masses m_L and m_D have been calculated with the method presented by Khatib [8,9]. This method projects the mass matrix of the robot manipulator $\mathbf{M}^+$ (incl. the contribution of drives and links) to a directional point mass m_R that equals the sum of m_L and m_D

$$m_R = m_L + m_D \; . \tag{3}$$

In this context, m_L denotes the directional point mass that can be calculated with the mass matrix $\mathbf{M}$, which does not contain contributions from the joint inertias. Then, m_D can easily be calculated by subtracting m_L from m_R. The robot's effective stiffness c_T was obtained in a similar way, but based on the stiffness matrix $\mathbf{K}_q$ instead of the $\mathbf{M}$ [2,11]. The compliant configuration of the PFMD used for the simulation corresponded to the Back of the Hand (ID 25 in

ISO/TS 15066). A spring of 75 N mm^{-1} combined with a damping material of shore hardness 70 (SH70) simulates the biomechanical characteristics of this specific body part [12]. Both stiffness parameters, c_D for the damping material and c_S for the spring, were previously measured with a testing system and the same impactors that were used in the experiments. The damping parameter d_D was estimated in impact experiments at various collision speeds. Also, the friction F_R of the free PFMD was determined by measurements. Figure 2 shows the impact force over impact velocity obtained from the simulation with a PFMD for fixed and free conditions. The solid line in yellow indicates the results for the fixed PFMD. The results for the free PFMD (solid lines with red diamonds and blue boxes for 6 kg and 22 kg) clearly show that the impact force decreases in case m_H can move freely. The dashed lines in red and blue are the results obtained with the conversion method according to Eq. (1). For both masses (m_H = 6 kg and 22 kg) the curves confirm a good match with the simulation. The intersection points of all lines with the black horizontal line indicate the maximum allowable robot velocity when assuming a force limit of 280 N (here: Back of the Hand). For the given scenario, the maximum allowable robot velocity for the fixed PFMD is 271 mm s^{-1}. In contrast, the simulation of a free collision with a body part of 22 kg (which is far beyond the expected weight of an outreached hand) enables an increase of the safe velocity by 62 %. For the same configuration, the conversion method estimates an increase of 60 %. The considerable low difference between both results demonstrates that the proposed conversion method obtains sufficiently accurate results.

4 Experiments

As the analysis of the simulation data has shown, the mass of the free moving Pressure and Force Measurement Device (PFMD) significantly affects the maximum impact force. Therefore, it was necessary to gather results from collision experiments with a freely moving PFMD of different total masses. Figure 3 shows the test setup that was developed for the experiment. It is based on the concept of a freely movable PFMD as illustrated in RIA T R15.806-2018.

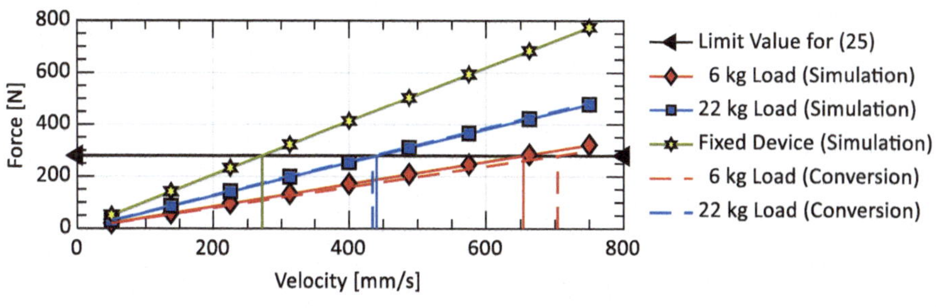

Fig. 2. Result of the simulated impact force compared to the predicted force of the conversion method for an UR10e robot with a payload of 66 %

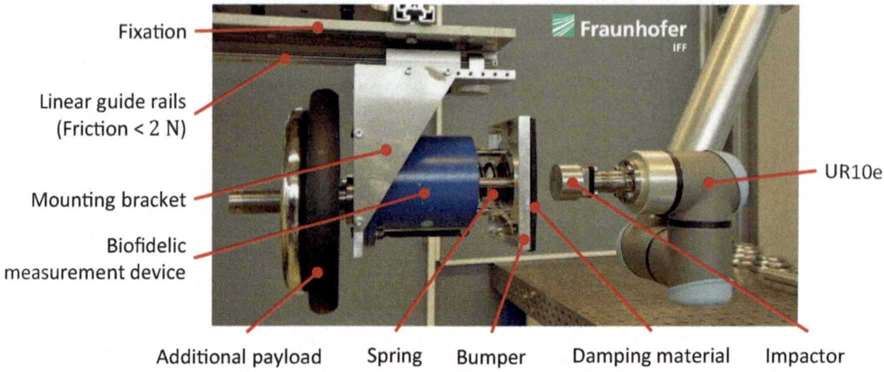

Fig. 3. Experimental test-setup to measure the impact force of a collaborative robot with a movable biofidelic measurement device

The PFMD was positioned on a movable carrier that was placed on low-friction linear guides with a static friction force of approximately 2 N. Tests with the simulation model revealed that a friction force within this range has a negligible influence on the measurement result. A large industrial robot was used to hold the assembly consisting of guides and the PFMD. Additional weights were mounted on the moveable PFMD to represent different body masses. For the measurement of the impact force, two different PFMDs with a similar design were used as listed in Table 1. The collisions were executed by three collaborative robots (UR3e, UR10e, Doosan M0607) that were equipped with a cylindrical impactor of 50 mm in diameter as a tool simulant. The robots were tested in the joint configuration listed in Table 2. In the test with free collisions, the total mass of the freely moving PFMD ranged from 4 to 22 kg (see Table 3.). For the tests with a fixed PFMD the movable carrier and the system's ground plate were connected. For a precise validation of the conversion method, the experiments aimed to obtain accurate results for any given combination of damping material and spring. Therefore, every test was performed with combinations that correspond to three different body parts, namely the Abdomen (10), the Sternum (8) and the Back of the Hand (25) (IDs according to ISO/TS 15066). Furthermore, three different payloads were tested with the UR10e in order to exclude possible negative influences of the robot's payload on the accuracy of the conversion method.

During the experiments the collaborative robot executes a linear movement at different velocities (starting from 50 and increasing to 650 mm s^{-1}) perpendicular towards the PFMD. The robot's starting position was chosen so that the desired velocity was ensured to be reached before colliding with the PFMD. The safety

Table 1. Characteristics of the biofidelic PFMDs used for the experimental validation

Name	Manufacturer	Force range [N]	Sampling frequency [kHz]
PRMS	PILZ	0 ... 500	2
KOLROBOT	IFA	0 ... 1000	10

Table 2. Joint configurations of the robots at the impact position

Robot	q_1[deg]	q_2[deg]	q_3[deg]	q_4[deg]	q_5[deg]	q_6[deg]
UR3e	0	−120	−126	247	−90	0
UR10e	0	−120	−126	67	90	0
Doosan M0607	0	−45	−90	0	45	0

Table 3. Test configurations of the different robots

	UR3e	UR10e	Doosan M0607
Payload robot [kg]	2.1	3.6; 6.9; 10.2	0.6
Payload robot [%]	70	36; 69; 102	10
Force threshold for safety stop [N]	50	100	50
Impact velocity [mm s^{-1}]	50 ... 450	50 ... 650	50 ... 650
Velocity steps	9	13	13
Repetitions per velocity step	1	5 at Payload 6.9 kg	1
PFMD	KOLROBOT	KOLROBOT, PRMS	KOLROBOT
Mass of free PFMD [kg]	KOLROBOT: 6; 12; 17; 22		
	PRMS: 4; 10; 15; 20		
Biofidelic configuration	Abdomen (10): Spring: 10 N mm^{-1}; Damper: SH10		
	Sternum (8): Spring: 25 N mm^{-1}; Damper: SH70		
	Back of the Hand (25): Spr.: 75 N mm^{-1}; Damp.: SH70		

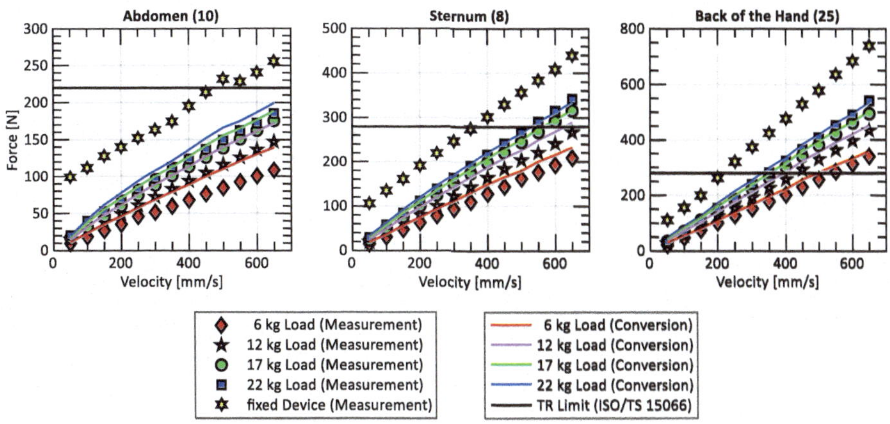

Fig. 4. Comparison of maximum forces from tests with a fixed and moveable PFMD with those obtained with the conversion method (UR10e with Payload 6.9 kg, Measurement Device: KOLROBOT)

functions of the robot monitored the external forces acting on the robot. They triggered an emergency stop, if the magnitude of these forces exceeded the configured threshold (see Table 3).

5 Experimental Results

The analysis of the experimental results compares the maximum impact forces of the moveable PFMD with those that were obtained with the conversion method. For the conversion method, the impact force recorded by the fixed PFMD was used. Figure 4 shows one sample of the obtained experimental results. Each of the three diagrams illustrates the results measured with different configurations of the examined PFMD. The markers indicate mean values of the measurement results from tests with the fixed and moveable PFMD. The corresponding mean standard deviation over all measurements shown in Fig. 4 was $\pm 3.4\,\text{N}$ for the fixed and $\pm 2.8\,\text{N}$ for the moveable measurement device. The solid lines represent the results obtained with the conversion method. The results confirm a linear relationship between the impact force and the impact velocity. Moreover, the results demonstrate that the impact forces of a moveable PFMD are significantly lower than those of a fixed PFMD. This effect is enhanced by decreasing the total mass of the human body m_H. Furthermore, the forces obtained with the conversion method are slightly higher than those measured with the movable PFMD. In conclusion, the conversion method achieves almost the same accuracy in the results from experimental tests as from simulations.

6 Discussion

A sensitivity analysis regarding the input parameters of the conversion method is performed to study the influences of uncertainties on the overall accuracy. Of particular interest is the following part of Eq. (1)

$$c = \sqrt{\frac{m_H}{m_H + m_R^*}}\,. \tag{4}$$

that calculates a factor ≤ 1 to convert $\hat{F}_C$ into $\hat{F}_F$. To determine the sensitivity with respect to the input parameters the partial derivatives for m_H and m_R^* must be calculated

$$\begin{aligned}\frac{\partial c}{\partial m_H} &= \frac{m_R^*}{2(m_H + m_R^*)^2\sqrt{\frac{m_H}{m_H+m_R^*}}}\\ \frac{\partial c}{\partial m_R^*} &= -\frac{m_H}{2(m_H + m_R^*)^2\sqrt{\frac{m_H}{m_H+m_R^*}}}\,.\end{aligned} \tag{5}$$

The summation of the partial derivatives gives the total error

$$\Delta c = \frac{\partial c}{\partial m_H}\Delta m_H + \frac{\partial c}{\partial m_R^*}\Delta m_R^*\,. \tag{6}$$

Figure 5 shows the courses of sensitivity for each parameter. It indicates that the collision masses m_H and m_R^* have a non-linear effect on the conversion result. The absolute magnitude of the sensitivity for m_H and m_R^* decreases with increasing mass. However, the sensitivity of m_R^* with respect to c is negative denoting that an underestimation of m_R^* will result in a higher scaling factor or higher impact force ($\Delta m_R^* < 0$ leads to $\Delta c > 0$ and then to $\Delta \hat{F}_F > 0$). Since m_H is assumed to be given, the estimation of m_R^* must be analyzed in more depth. The partial derivatives of Eq. (2) with respect to the input parameters momentum p and impact velocity v are

$$\begin{aligned} \frac{\partial m_R^*}{\partial v} &= -\frac{p}{v^2} \\ \frac{\partial m_R^*}{\partial p} &= \frac{1}{v}, \end{aligned} \tag{7}$$

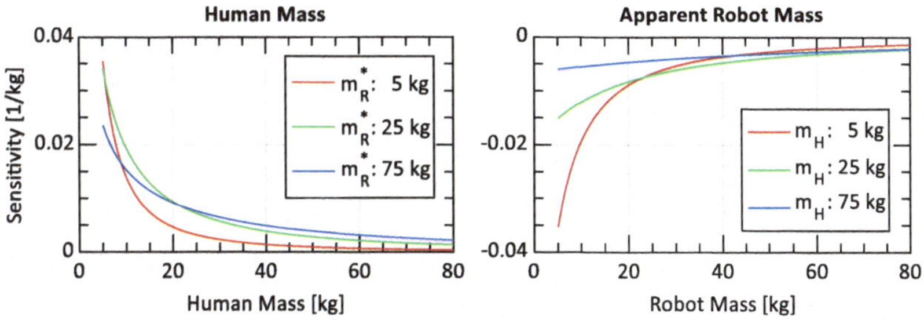

Fig. 5. Sensitivity analysis of the conversion with respect to the input parameters human mass m_H and apparent mass of the robot m_R^*

that further give the following expression for the total error

$$\Delta m_R^* = \frac{\partial m_R^*}{\partial v}\Delta v + \frac{\partial m_R^*}{\partial p}\Delta p\,. \tag{8}$$

Equation (7) indicates a proportional influence of p and a negative non-linear one of v. Since $\frac{\partial m_R^*}{\partial v}$ describes a hyperbolic function, the absolute magnitude of the sensitivity of m_R^* decreases with increasing v. This behavior indicates that the higher the impact velocity, the better the estimation accuracy. To study whether the sensitivity has a significant influence on the conversion method, the following relative error

$$\Delta x_{rel} = \frac{\hat{F}_{conv}}{\hat{F}_{meas}} - 1 \tag{9}$$

and absolute error

$$\Delta x_{abs} = \hat{F}_{conv} - \hat{F}_{meas} \tag{10}$$

is introduced. Both can be calculated from the converted force $\hat{F}_{conv}$, obtained with the conversion method, and the impact force $\hat{F}_{meas}$ measured with a movable PFMD. The green bars in Fig. 6 display the distribution of the errors in a histogram. It includes all data from the experimental tests, which incorporate more than 1000 individual results. The error distribution shows that 91 % of the impact forces obtained with the conversion method exceed the forces measured in tests

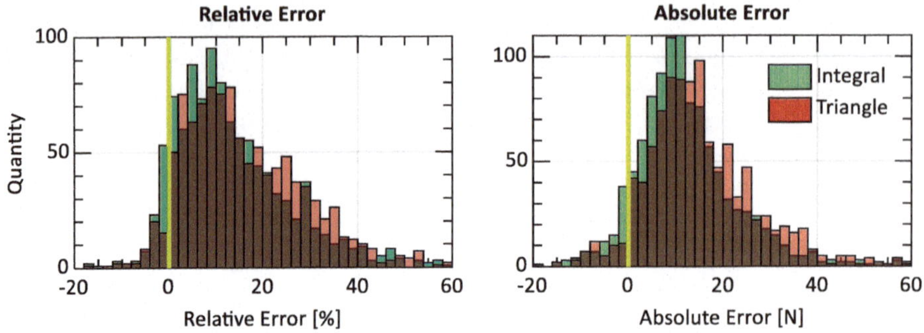

Fig. 6. Distribution of the relative and absolute error of the conversion method for different approaches to calculate the apparent mass m_R^* (the green bars use Eq. (2), the red bars Eq. (11) to calculate m_R^*; the yellow line indicates 100 % accuracy; each error on the right hand-side of the yellow line indicates that the conversion method yields to conservative and, thus, safer results)

with the movable PFMD. The mean errors are $\Delta x_{rel} = 13.4\,\%$ and $\Delta x_{abs} = 12.2\,\text{N}$. Considering this tendency, it can be concluded that the conversion method yields to slightly increased impact forces. Consequently, the conversion method can be considered as conservative as higher forces ultimately result in lower but safer robot velocities. The observations confirm that the parameters with positive influence on Δc dominate. Since $\hat{F}_C$ and m_H can be considered as accurately known or given, the conservative tendency of the converted values must be caused by a systematic underestimation of m_R^*. Consequently, the technique to estimate m_R^* results in too small values, which derive from $\Delta v > 0$ or $\Delta p < 0$ originating from Eq. (8). We assume that the calculation of p tends to yield smaller values, since the exact time of initial contact t_0 is difficult to detect. Considering this result, it must be highlighted that the conversion method can underestimate the impact forces even though the probability of this effect is very low with 10 %. An evaluation of the dataset shows that a safety factor of $S = 1.1$ multiplied with $\hat{F}_{conv}$ is sufficient to reduce the probability of underestimations to less than 1 %. This safety factor additionally leads to lower velocities of about 9 % for the conversion method.

7 Simplification of the Conversion Method

The error analysis has shown that underestimations of p lead to conservative estimates and lower robot velocities. This finding can be used to further simplify the estimation of the apparent robot mass m_R^*. Especially for end-users who do not have direct access to the measurement data or the knowledge to apply Eq. (2), it might be difficult to obtain p. An approximation of the integral can be obtained when considering p_{app} as the area of a triangle

$$p_{app} = \frac{1}{2}(t_{max} - t_0)\hat{F}_C \ , \tag{11}$$

where $\hat{F}_C$ is the maximum contact force at time t_{max}. Figure 7 compares the triangle approximation to the exact integral. The area of the triangle always cor-

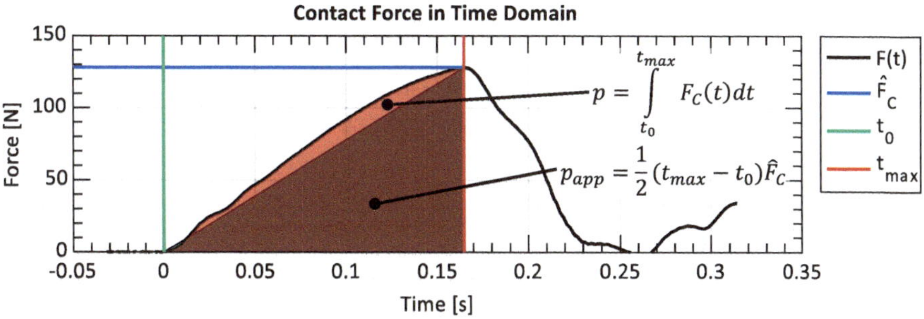

Fig. 7. Approximation of the momentum using the area of a triangle (Force signal of a UR10e robot with a payload of 66 % at an impact velocity of 150 mm s^{-1} colliding with a fixed measurement device with Spring 10 N mm^{-1} and Damper SH10)

responds to a slightly smaller momentum Δp. According to Eq. (7), a negative momentum error $\Delta p < 0$ leads to an underestimation of m_R^*, consequently to higher collision forces $\hat{F}_F$ (see Eq. (5)) and ultimatley to lower, but safe robot velocities. The simplified conversion method that considers p_{app} was applied to all experimental results to calculate m_R^*. The errors compared to the measurements with the movable PFMDs are shown as red bars in Fig. 6. As expected, the simplified conversion method provides a higher relative $\Delta x_{rel} = 15.9\,\%$ and absolute mean error $\Delta x_{abs} = 15.1\,\text{N}$, whereas the shape of the error distribution remains almost identical.

8 Conclusion and Outlook

This article presents a new method to convert impact forces recorded with a fixed Pressure and Force Measurement Device (PFMD) into such that would occur in a free collision. The technique has been validated with three different collaborative robots and two different PFMDs. The results obtained from the simulation and experiments show a significant increase of the allowable robot velocities. The findings of this study have a decisive contribution to increase the overall productivity of collaborative robots without compromising their ability to reduce the risk of human injuries. Another contribution of the conversion method to the robot community is the opportunity to evaluate unconstrained human-robot collisions of any direction even in confined spaces. Both advantages outweigh the slightly lower, but more conservative allowable robot velocity that the conversion method provides.

The moveable PFMD intends to represent a free collision between a robot and a part of the human body. One limitation is that it cannot reproduce the exact human behavior since it only has one Degree of Freedom (DoF), while the human body is a multi-body system with several DoFs. However, the short impact duration of free collisions (30 ... 200 ms) and the elasticity of the human joints suggest that the affected human body part initially behaves like a one DoF system of specific apparent mass.

Further work will focus on the estimation of the robots velocity, which is necessary to calculate the apparent mass m_R^*. In fact, the translational velocity of the robot must be assumed as unknown, especially if the robot executes Point-to-Point movements or if the collision point is not at the end-effector. However, it is always possible to use the velocity threshold as it is configured in the safety controller, since a safety-rated function will always ensure that the robot is never exceeding this value.

Acknowledgement. Parts of this work have been funded by the ROSSINI-Project, which has received funding from the European Union's Horizon 2020 research and innovation programme under grant agreement No 818087. The biofidelic measurement device PRMS (PILZ Robot Measurement System) used for the experiments was provided by courtesy of PILZ.

References

1. 7th International Conference Safety of Industrial Automated Systems, Montreal, Canada, October 2012
2. Albu-Schäffer, A., Fischer, M., Schreiber, G., Schoeppe, F., Hirzinger, G.: Soft robotics: what Cartesian stiffness can obtain with passively compliant, uncoupled joints? In: Proceedings of the 2004 IEEE/RSJ International Conference on Intelligent Robots and Systems (IROS), Sendai, Japan, February 2004, vol. 4, pp. 3295–3301 (2004). https://doi.org/10.1109/IROS.2004.1389925
3. Behrens, R., Saenz, J., Vogel, C., Elkmann, N.: Upcoming technologies and fundamentals for safeguarding all forms of human-robot collaboration. In: Proceedings of the 8th International Conference on the Safety of Industrial Automated Systems, pp. 18–23, Königswinter, Germany, November 2015
4. Bicchi, A., Tonietti, G.: Fast and soft-arm tactics. IEEE Rob. Autom. Mag. **11**(2), 22–33 (2004). ISSN 1070-9932. https://doi.org/10.1109/MRA.2004.1310939
5. Falco, J., Marvel, J., Norcross, R.: Collaborative robotics: measuring blunt force impacts on humans. In: Proceedings of the 7th International Conference on the Safety of Industrial Automated Systems (2012)
6. Haddadin, S., Krieger, K., Mansfeld, N., Albu-Schäffer, A..: On impact decoupling properties of elastic robots and time optimal velocity maximization on joint level. In: Proceedings, pp. 5089–5096, Vilamoura, Portugal, October 2012. https://doi.org/10.1109/IROS.2012.6385913
7. Huelke, H.-J., Ottersbach, M.: How to approve collaborating robots - the IFA force pressure measurement system. In: Proceedings of the 7th International Conference on the Safety of Industrial Automated Systems SIA, pp. 5089–5096 (2012)
8. Khatib, O.: A unified approach for motion and force control of robot manipulators: the operational space formulation. IEEE J. Rob. Autom. **3**(1), 43–53 (1987). https://doi.org/10.1109/JRA.1987.1087068
9. Khatib, O.: Inertial properties in robotic manipulation: an object-level framework. Int. J. Rob. Res. **14**(1), 19–36 (1995). ISSN: 0278-3649. https://doi.org/10.1177/027836499501400103
10. Oberer-Treitz, S.: Abschaetzung der Kollisionsfolgen von Robotern zur Bewertung des sicheren Einsatzes in der Mensch-Roboter-Kooperation. Dissertation, Universitaet Stuttgart (2017)
11. Salisbury, J.-K.: Active stiffness control of a manipulator in Cartesian coordinates (1980)
12. Fachbereich Holz und Metall der DGUV. Dguv-information - collaborative robot systems: design of systems with power and force limiting function (2017)

Towards a Robotically Steerable System for High Dose Rate Brachytherapy

Nancy J. Deaton[1(✉)], Yash Chitalia[1], Pretesh Patel[2], and Jaydev P. Desai[1]

[1] Medical Robotics and Automation (RoboMed) Laboratory, Wallace H. Coulter Department of Biomedical Engineering, Georgia Institute of Technology, Atlanta, GA, USA
njdeaton@gatech.edu
[2] Winship Cancer Institute, Emory University, Atlanta, GA, USA

Abstract. Brachytherapy (BT) is a commonly employed percutaneous approach for cancer treatment. In this work, we propose a new tendon-driven omnidirectional bending joint for a steerable stylet that can be inserted into a commercial BT needle. This joint is made by laser cutting a diamond pattern with or without vertical supporting members into a nitinol tube. Simulations are used to compare the effect of different geometric pattern parameters and determine the most optimal parameter values. Joint simulation results are used in a model for bending the joint-needle system and compared to experimental results. We demonstrate the ability of the joint to achieve the desired bending angle in the needle.

Keywords: Steerable needle · Continuum joints · Brachytherapy

1 Motivation, Problem Statement, Related Work

Brachytherapy (BT) is the seminal form of radiotherapy that was first used to treat cancer in the early 1900s. BT is the implantation of radioactive sources into malignant tumors leading to unparalleled concentration of radiation dose and excellent clinical outcomes. For easily implantable cancers, such as cervical cancer [1] and prostate cancer [2], brachytherapy has been the standard of care for decades and remains so today. In fact, when BT is omitted from the modern treatment of cervical cancer [3] or aggressive prostate cancer [4], cancer-related patient outcomes are worse.

High dose rate (HDR) BT is a modern form of BT where hollow, often plastic, needles are implanted into cancerous tissues and a highly active radiation source, usually Iridium-192, is after-loaded into the needles based on an image-guided radiation treatment plan. The implantation procedure requires manual placement of BT needles into the cancerous tissues while avoiding normal tissue injury and yielding optimal radiation dose exposure to the target. Highly customized radiation dose clouds are possible using HDR BT. HDR BT is commonly used to treat cervical and prostate cancer but there is growing interest to use it for many other forms of cancer as well. The literature describes many

B. Siciliano et al. (Eds.): ISER 2020, SPAR 19, pp. 233–244, 2021.
https://doi.org/10.1007/978-3-030-71151-1_21

experiences where BT has been used to treat head and neck cancer, breast cancer, liver cancer, kidney cancer, pancreatic cancer, brain cancer and sarcoma. The use of HDR BT is particularly attractive in cancers where there is clear evidence of benefit of radiation dose escalation, such as pancreatic cancer [5,6], or avoidance of normal tissue injury such as liver cancer in patients with liver cirrhosis. However, the widespread adoption of HDR BT for treatment of many cancers has been hampered by difficulties in the implantation of BT needles into the target tissues.

Perioperative HDR BT has shown some promise for the treatment of pancreatic cancer, especially for the possibility of increased local control [7]. A feasibility study of interstitial BT has proven the potential for this approach in pigs and emphasized the need for a reliable and accurate delivery system for percutaneous BT to become a treatment option for pancreatic cancer [8]. Steerable needles have the potential to correct path errors, avoid obstacles, reduce the number of punctures, and provide other advantages for targeting lesions in difficult to reach areas [9].

Prior work has been done on both passive and active steerable needles. Steering mechanisms include base manipulation, bevel tip, programmable bevel, precurved inner stylet, active cannula, and tendon actuation [10]. Bevel tip needles rely on uneven forces from inserting the needle to steer and rotate the needle to control the trajectory [11,12]. Programmable bevel-tip needles rely on the same principles but extend or retract sections with differently oriented bevels instead of rotating to steer [13,14]. Both the pre-curved inner stylet and active cannula approaches control the rotation and relative insertion length of pre-curved inner tube(s) with different stiffnesses to steer [15–17]. Tendon actuated needles involve creating a flexible joint in the needle and either pulling a tendon or contracting a nitinol wire to bend the joint [18–20].

We aim to develop an actively steerable robotic system that will enable percutaneous HDR BT for pancreatic cancer. In this paper, we focus on the design of a tendon driven steerable stylet (i.e. part of the overall robotic system for HDR BT) that works with existing hollow BT needles. The steerable stylet will be able to correct path errors from puncturing the pancreas as the stylet and needle are driven into the tumor. It must be able to bend the needle tip in all directions up to 20° (specified by clinician collaborator) to enable precise positioning and be removed once the needle is in place to facilitate loading of a radioactive seed in the needle. In this paper, we propose potential designs for the joint of the steerable stylet (Fig. 1(a,b)). The joint will be tested both alone and inside the needle (Fig. 1(c,d)).

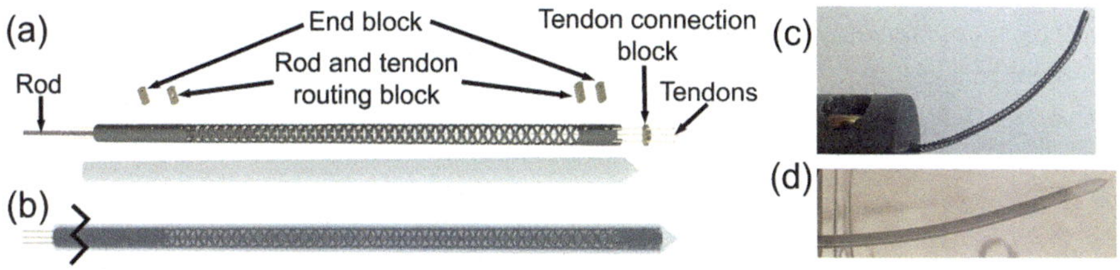

Fig. 1. (a) Exploded diagram of steerable stylet assembly (top) and needle (bottom), (b) Assembled needle and stylet, (c) Stylet joint, (d) Stylet joint in needle.

2 Technical Approach

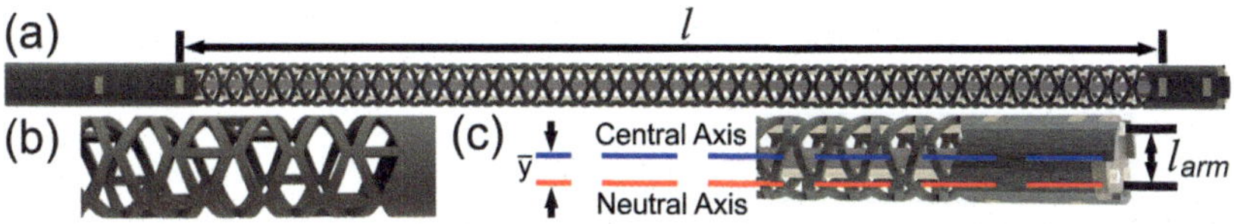

Fig. 2. (a) Length, l, of the bending segment of the stylet assembly, (b) Joint with vertical members, (c) Moment arm of the active tendon indicated by l_{arm}.

One of the following two patterns is cut into the nitinol tube to create a continuum joint for the steerable stylet: a diamond pattern (Fig. 2(a)) or a diamond pattern with vertical members for axial stiffness (Fig. 2(b)). A cylindrical nitinol rod is placed in the center of the stylet joint to prevent axial compression beyond 2 mm. The steerable stylet is placed in the hollow BT needle. This joint can be modeled to approximate a spring for the desired bending range of 20° for a 3 cm long joint. The rod can be modeled using an Euler beam model to relate the applied bending moment (M_{rod}) to the angle of deflection (θ) at the tip of the joint, with θ given as follows:

$$\theta_{rod} = \frac{M_{rod} l}{E_{rod} I_{rod}} \tag{1}$$

where θ_{rod} is the bending angle of the rod, E_{rod} is the Young's modulus, $I_{rod} = \pi d_{rod}^4/64$ is the area moment of inertia, and l is the length of the bending segment (Fig. 2(a)). The variable d_{rod} is the diameter of the cylindrical rod. Due to tolerance between the holes cut for the rod in the routing blocks and the diameter of the rod, the bending angle of the rod may be offset compared to the bending angle of the joint as follows:

$$\theta_{rod} = \theta_{joint} - \theta_{tolerance} \tag{2}$$

where θ_{joint} is the bending angle of the joint, and $\theta_{tolerance}$ is the offset of the bending angle of the rod due to the tolerance in the routing blocks.

The moment required to bend the joint can be found using Finite Element Analysis (FEA). Since the joint is modeled as a spring, according to Hooke's law the axial deflection (Δl) and θ are proportional to the force (F) and moment (M_{joint}), respectively, applied at the joint tip as follows:

$$F = k\Delta l \quad \text{and} \quad M_{joint} = k_\theta \theta_{joint} \tag{3}$$

where k is the spring constant for axial deflection and k_θ is the spring constant for angular deflection. This relationship can be used to calculate the force to compress the joint ($F_{compression}$) at a given Δl. Both k and k_θ are proportional to the length of the spring. Therefore, the results of a FEA simulation run using a joint of free length l_1 can be applied to a joint of free length l_2 by multiplying a scaling factor by the axial ($\Delta l_2 = \Delta l_1 \frac{l_2}{l_1}$) and angular ($\theta_2 = \theta_1 \frac{l_2}{l_1}$) deflections. The total moment ($M_{rod} + M_{joint}$) can be found by multiplying the force of the tendon by the moment arm (l_{arm}) (see Fig. 2(c)), which is measured from the tendon to the neutral axis of the assembly and is given as follows:

$$l_{arm} = \frac{ID_{joint}}{2} - \frac{d_{tendon}}{2} - c + \overline{y} \tag{4}$$

where ID_{joint} is the inner diameter of the joint, d_{tendon} is the diameter of the tendon, c is the distance between the tendon and the inner wall of the joint, and $\overline{y}$ is the location of the composite neutral axis from the stylet assembly central axis (see Fig. 2(c)). When the joint is uncompressed, the rod and joint bend about their central axes; therefore initially, $\overline{y} = 0$. However, as the joint bends, the central rod is deflected and runs along the wall opposite to the tendon, and the joint is compressed until it engages the rod which then prevents further compression. The new location of the composite neutral axis for the bending portion of the stylet assembly is given by $\overline{y} = \sum_j \frac{E_j A_j \overline{y}_j}{E_j A_j}$, where $j \in \{joint, rod\}$. The moment applied by the tendon must equal the sum of the moments required to bend the rod (M_{rod}) and the joint (M_{joint}). Therefore, the total force pulling the tendon (F_{total}) bending the stylet assembly must be the sum of these moments divided by the moment arm of the tendon. F_{total} and the force contribution from each component ($j \in \{joint, rod\}$) are given by:

$$F_{total} = \frac{M_{rod} + M_{joint}}{l_{arm}} \quad \text{and} \quad F_j = \frac{M_j}{l_{arm}} \tag{5}$$

From Eqs. (2), (3), and (5), we get the following relationship for the joint and rod stylet:

$$\begin{aligned} F_{total} l_{arm} l &= (E_{rod} I_{rod} + k_\theta l)\theta \\ &= (E_{stylet} I_{stylet})\theta \end{aligned} \tag{6}$$

where, the term $E_{stylet} I_{stylet}$ is the experimentally derived equivalent of the flexural rigidity of the stylet (since E_{rod} is derived experimentally and k_θ is derived through FEM). When the stylet is inserted into the needle and the tendon actuated, it exerts a uniform distributed load on the needle causing the needle to deflect. Therefore, incorporating the reaction forces from the needle on the stylet in Eq. (6), we get the following F_{total} vs. θ relationship:

$$F_{total} = \frac{(E_{stylet}I_{stylet} + E_{needle}I_{needle})\theta}{l_{arm}l} \tag{7}$$

where E_{needle} is the experimentally derived Young's modulus of the needle and $I_{needle} = \pi(OD^4 - ID^4)/64$ is the area moment of inertia of the needle. In this equation, OD and ID are the outer diameter and inner diameter of the needle. Both Eqs. (6) and (7) are validated in Sect. 4.

3 Experiments Completed

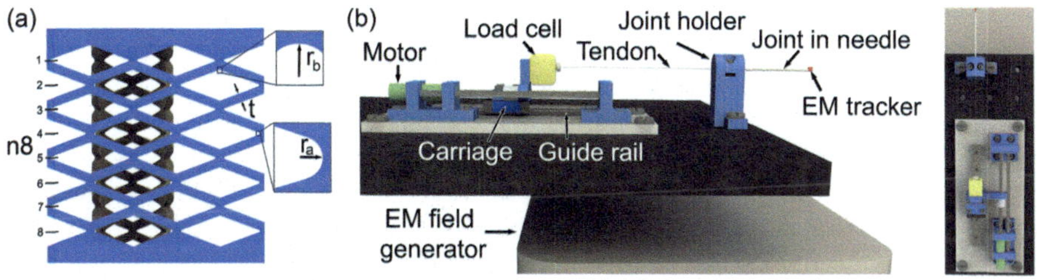

Fig. 3. (a) Joint sample parameters, (b) Experimental setup.

This paper primarily presents our work on the simulation and modeling of the proposed joint. All FEA simulations were conducted in ANSYS 2019 R2 or R3 (ANSYS Inc., Canonsburg, PA, United States). The material properties used for nitinol in these simulations were taken from literature [21]. Tetrahedral elements (TET10) were used for the mesh of these joints. A convergence study was performed on sample S2^{n8} (Table 1) until maximum stress and strain varied less than 3% and force, angle, and change in joint length varied less than 1% with further mesh refinement. This resulted in a mesh element face size of 0.043 mm along the walls of the diamonds and an element face size of 0.012 mm at the corners of the diamonds with one automatic refinement step which reduced the characteristic length of elements in this area to between 0.0014 mm and 0.0044 mm. The average aspect ratio of the mesh elements was 2.08 ± 0.64, and the average Jacobian Ratio for the Gauss points was $0.995 \pm 1.76e-2$. The same mesh size and refinement was then applied to the other joint samples. Multiple simulations were performed, varying the thickness of the bending member (t), radii of curvature of the diamonds (r_a and r_b), and length of the joint sample tested which is represented by the number of rows of diamonds cut into the joint (n) (Table 1, parameters defined in Fig. 3(a)). The diagonal lengths of the diamonds are constant with a 1.19 mm width and 0.45 mm height before the corners are

Table 1. FEA simulated joint samples.

Parameters					
Sample	t mm	r_a mm	r_b mm	n mm	Vertical members
S1*	0.12	0.02	0.02	8	No
S2	0.12	0.03	0.03	8, 10	No
$S2^{*}_{vert}$	0.12	0.03	0.03	8, 10	Yes
S3	0.12	0.04	0.04	8	No
S4	0.12	0.02	0.04	8	No
S5	0.12	0.04	0.02	8	No
S6*	0.11	0.03	0.03	8	No
S7	0.13	0.03	0.03	8	No

$*$ Samples experimentally validated using 3 cm (n = 78) long joints.

rounded. Simulations were run to test the performance of joint samples with vertical members, $S2^{n8}_{vert}$ and $S2^{n10}_{vert}$, added in a helix to increase the stiffness of the joint and potentially eliminate the need for a rod to prevent compression when applied in the needle. These simulations were used to compare the different joint samples. FEA simulations were also run on joints of different lengths to validate the scaling assumptions in the theoretical model. The joint samples which have been experimentally validated are marked in Table 1.

To fabricate the steerable stylet, a diamond pattern was cut into a nitinol tube of 1.17 mm outer and 0.97 mm inner diameter (Edgetech Industries LLC, FL, United States) using a femtosecond laser (Optec Laser S.A., Frameries, Belgium), and a nitinol rod of diameter 0.3 mm was placed inside the joint. The steerable stylet was tested in a commercial 5F needle (Nucletron B.V., Veenendaal, Netherlands). To verify the model above and test the performance of the steerable stylet, the testing setup depicted in Fig. 3(b) was used. This setup uses a DC motor with a lead screw (Maxon Precision Motors, MA, United States) to linearly translate a stage with a load cell with a maximum load capacity of 22 N (Transducer Techniques, CA, United States) mounted to it. One tendon of the steerable stylet joint is attached to the load cell and is thus pulled when the stage translates. An electromagnetic (EM) field generator (Northern Digital Inc., Waterloo, Ontario, Canada) is placed beneath the joint, and an EM tracker is attached to the tip of the joint to obtain the true joint deflection. The Young's modulus E_j of the needle and rod were experimentally determined by applying a perpendicular force to the tip of a segment of known length and measuring the force and tip displacement.

4 Results

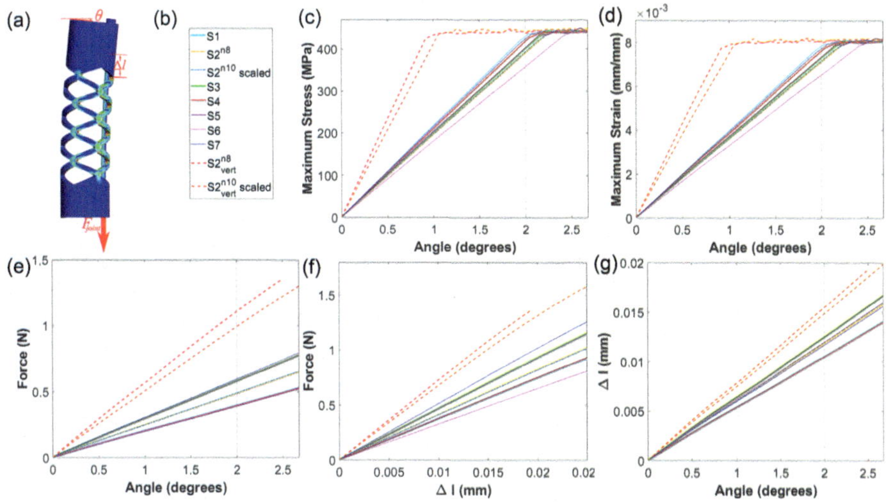

Fig. 4. (a) ANSYS sample S6 strain results showing applied force (F_{joint}), joint compression (Δl), and angle (θ), (b) legend for (c–g). Results of ANSYS simulations for the joint samples listed in Table 1: (c) Maximum stress, (d) Maximum strain, (e) Total tendon force vs. angle, (f) Total tendon force vs. joint compression, and (g) Joint compression vs. angle. Vertical lines signify a 2° bend in $n = 8$ joints and thus a 22° bend in 3 cm long ($n = 78$) joints.

Figure 4 shows the results of the ANSYS simulations for the joint samples listed in Table 1. A sensitivity analysis compares the bending angle (θ), compression (Δl), tendon force (F_{joint}), stress, and strain of joint samples of different parameters (Fig. 4(c–g)). The results are reported for a 2° bending angle as this corresponds to a 22° bending angle for the full 3 cm long joint. The parameter r_b has the least effect on the joint performance. At a 2° bending angle, the largest difference between samples S3 and S5 is a 0.67% difference in maximum stress and a 0.14% difference in tendon force. Similarly, samples S1 and S4 show a 3.56% difference in maximum stress and a 0.16% difference in tendon force. Both sets show less than 0.45% difference in compression. Samples S3 and S4, where S3 has a larger r_a value, show a 4.53% difference in maximum stress with S4 having the larger value. The tendon force for S3 is 47.9% higher than that of S4, and the compression is 19.3% higher than that of S4. The thickness of the bending member (t) also impacts the simulation results. sample S7 shows 19.0% higher maximum stress and a 53.3% higher tendon force as compared to sample S6 at a 2° bending angle. These two samples have the same values for r_a and r_b, and the compression of the two samples is similar at the given bending angle with only a 2.6% difference. The compression for sample S2, which also shares the same values for r_a and r_b, is in between the compression for samples S6 and S7. For all comparisons, the percent difference in strain values were approximately the same as the percent difference in stress values.

The results of the ANSYS simulations also show that the assumption that the spring constants for axial (k) and angular (k_θ) deflection of the joint are linearly scalable by length holds true for the joint samples without the vertical members but not for the joint samples with the vertical members. In Fig. 4(c–g), the results for samples $S2^{n10}$ and $S2^{n10}_{vert}$ are scaled as described in Sect. 2 to the length of the n=8 joint samples. Visual inspection of the results shows that $S2^{n10}_{vert} scaled$ and $S2^{n8}_{vert}$ do not overlap The percent differences between these two samples are 11.8% for force and 4.4% for compression at 2° and 17.9% for maximum stress at 0.91°. The maximum stress is compared at 0.91° because this is the end of the linear region for the maximum stress vs. angle for $S2^{n8}_{vert}$, before the nitinol undergoes a phase change and the slope becomes approximately zero. However, the results for $S2^{n10} scaled$ successfully match the results for $S2^{n8}$. At 2°, the percent differences between $S2^{n10} scaled$ and $S2^{n8}$ are 0.80% for maximum stress, 0.61% for tendon force, and 1.25% for compression.

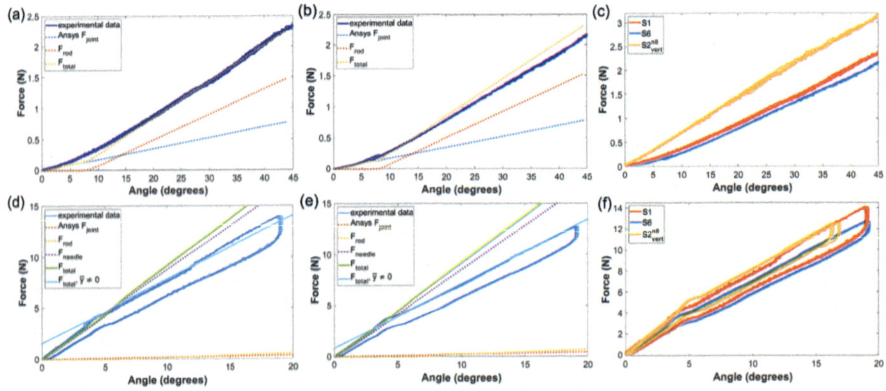

Fig. 5. (a) Comparison of the theoretical model to experimental data for the sample S1 stylet assembly (joint and rod), (b) Comparison of the theoretical model to experimental data for the sample S6 stylet assembly, (c) Experimental data for the sample S1, S6, and $S2_{vert}$ stylet assemblies, (d) Comparison of the theoretical model to experimental data for the sample S1 joint, rod, and needle system, (e) Comparison of the theoretical model to experimental data for the sample S6 joint, rod, and needle system, (f) Experimental data for the sample S1, S6, and $S2_{vert}$ joint, rod, and needle systems.

The Young's modulus of the nitinol rod (E_{rod}) and polyoxymethylene (POM) thermoplastic needle (E_{needle}) were experimentally determined to be 65.1 GPa and 3.27 GPa respectively. Figure 5(a–c) show the theoretical and experimental results for the stylet assemblies (joint and rod) for the full-length (n = 78) joint samples S1, S6, and $S2_{vert}$. Because the ANSYS results for the joint samples with vertical members were shown not to scale linearly with length, the experimental results for sample $S2_{vert}$ are not compared to the theoretical model which relies on the scaling assumption. The tolerance between the holes cut for the rod in the routing blocks and the diameter of the rod was used to calculate the maximum difference in bending angle between the joint and rod ($\theta_{tolerance}$) to be 8°. Therefore, in the theoretical model used, the rod only contributes to the tendon force once the joint reaches an 8° bend, and the bending angle of the rod is 8° less than that of the joint thereafter. This can be seen by the sudden change in the slope of F_{total} at this angle in Fig. 5(a,b).

Figure 5(d–f) shows the experimental data and application of the theoretical model for the bending of the joint, rod, and needle system for full-length (n = 78) joint samples S1, S6, and $S2_{vert}$. Similarly, the theoretical model is not compared to the experimental results for sample $S2_{vert}$. Hysteresis is seen in the straightening of the joint and is not modeled in this paper. Once the rod is engaged, the reduction of force required to bend the joint and needle is greater than can be accounted for solely by the reaction force of the rod preventing further compression. The rod is not constrained to the center axis along the length of the joint and was observed to contact the wall of the joint while bent (Fig. 1(c)). Once the joint has reached the maximum allowed compression such that the rod is engaged and preventing further compression, the rod and joint behave as a single body, bending about a combined neutral axis. The deflection of the rod to the wall opposite the tendon causes an offset to the neutral axis of the stylet assembly ($\overline{y}$) which was calculated to be 0.328 mm for both Samples S1 and S6. For this calculation, the central axis of the rod was considered to be at an offset of 0.333 mm from the central axis of the joint. This model using the offset neutral axis is therefore compared to the second section of the experimental data in Fig. 5(d,e).

5 Main Experimental Insights

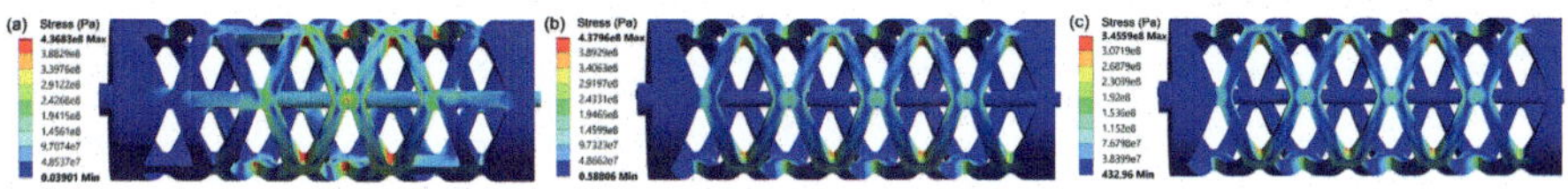

Fig. 6. Stress of joint samples at 2° bend [out of the plane of the page] (a) $S2^{n8}_{vert}$, (b) $S2^{n8}$, (c) S6

For this joint, the ideal would be pure bending (zero compression), in which case the rod would not be necessary. Based on the objective of maximizing bending while minimizing compression, joint samples S4 and S1 show the highest bending-to-compression ratio. The addition of vertical members increases both the axial and bending stiffness. However, at least for relatively short joint samples, the stiffness of the joints with vertical members does not scale linearly with length as it does for the joints without vertical members. If the stiffness scaled linearly with length, the total force necessary to bend the needle, rod, and $S2_{vert}^{n8}$ joint scaled to full length ($n = 78$) to 20° would cause an acceptable compression of 2.9 mm. However, additional stiffness from the vertical members decreases with length, and experimental results confirmed that the added stiffness is not sufficient to prevent a 3 cm long joint from primarily undergoing compression without bending when actuated within the needle which dominates the stiffness of the system. It is observed that the addition of the vertical members, which show low stress, increases the stress of the non-restricted bending members as compared to the same joint without vertical members. This causes the highest values of stress and strain of the simulated joint samples (Fig. 6(a,b)).

If the central rod is necessary to cause all proposed joint samples to bend rather than compress when actuated within the needle, then the joint that minimizes the bending stiffness is desirable, sample S6 of those simulated in this paper. Experimental results confirm that sample S6 requires the least force to bend, both with and without the needle. Sample S6 also shows the lowest maximum stress and strain which will reduce fatigue over extended use (Fig. 4(c,d) and Fig. 6(c)). Both the central rod and the needle contribute to the straightening the system when the tendon is not actuated, and the tendons can be used to actively straighten the system. Therefore, a stiff joint may not be necessary to assist in passive straightening. Further tuning of the joint can be done using the results of this study. The parameter r_b was found to have negligible effect on the performance of the joint while the thickness of the bending member can be kept low to reduce stiffness. The value of r_a can also be reduced in order to decrease the tendon force necessary for bending, but this will increase the maximum stress and strain and thus the fatigue experienced by the joint.

The experimental results are seen to validate the theoretical model for the S1 and S6 joint samples tested. This is seen to be the case both with and without the needle. Furthermore, the model of the joint, rod, and needle system shows that the needle dominates the stiffness of the system and contributes most to the tendon force required to reach a desired angle. This is also shown by the comparatively low forces necessary to bend the stylet assembly alone. Joint sample S6 is shown to require the least force of the joints tested to bend to the desired angle both with and without the needle.

In future application studies, characterization of the external forces that will act on the system in practice will determine if a stiffer joint is necessary to increase its resistance to these external forces and ensure that unwanted deflection does not occur.

Acknowledgement. This work is supported in part by Emory University and the National Science Foundation Graduate Research Fellowship Program under Grant No. DGE-1650044. Any opinions, findings, and conclusions or recommendations expressed in this material are those of the authors and do not necessarily reflect the views of Emory University or the National Science Foundation.

References

1. Koh, W.-J., Abu-Rustum, N.R., Bean, S., Bradley, K., Campos, S.M., Cho, K.R., Chon, H.S., Chu, C., Clark, R., Cohn, D., et al.: Cervical cancer, version 3.2019, NCCN clinical practice guidelines in oncology. J. Natl. Compr. Cancer Netw **17**(1), 64–84 (2019)
2. Mohler, J.L., Antonarakis, E.S., Armstrong, A.J., D'Amico, A.V., Davis, B.J., Dorff, T., Eastham, J.A., Enke, C.A., Farrington, T.A., Higano, C.S., et al.: Prostate cancer, version 2.2019, NCCN clinical practice guidelines in oncology. J. Natl. Compr. Cancer Netw **17**(5), 479–505 (2019)
3. Han, K., Milosevic, M., Fyles, A., Pintilie, M., Viswanathan, A.N.: Trends in the utilization of brachytherapy in cervical cancer in the United States. Int. J. Radiat. Oncol.* Biol.* Phys. **87**(1), 111–119 (2013)
4. Kishan, A.U., Cook, R.R., Ciezki, J.P., Ross, A.E., Pomerantz, M.M., Nguyen, P.L., Shaikh, T., Tran, P.T., Sandler, K.A., Stock, R.G., et al.: Radical prostatectomy, external beam radiotherapy, or external beam radiotherapy with brachytherapy boost and disease progression and mortality in patients with gleason score 9–10 prostate cancer. Jama **319**(9), 896–905 (2018)
5. Krishnan, S., Chadha, A.S., Suh, Y., Chen, H.C., Rao, A., Das, P., Minsky, B.D., Mahmood, U., Delclos, M.E., Sawakuchi, G.O., et al.: Focal radiation therapy dose escalation improves overall survival in locally advanced pancreatic cancer patients receiving induction chemotherapy and consolidative chemoradiation. Int. J. Radiat. Oncol.* Biol.* Phys. **94**(4), 755–765 (2016)
6. Han, Q., Deng, M., Lv, Y., Dai, G.: Survival of patients with advanced pancreatic cancer after iodine125 seeds implantation brachytherapy: a meta-analysis. Medicine **96**(5), e5719 (2017)
7. Waniczek, D., Piecuch, J., Rudzki, M., Mikusek, W., Arendt, J., Białas, B.: Perioperative high dose rate (HDR) brachytherapy in unresectable locally advanced pancreatic tumors. J. Contemp. Brachytherapy **3**(2), 84 (2011)
8. Sun, S., Qingjie, L., Qiyong, G., Mengchun, W., Bo, Q., Hong, X.: EUS-guided interstitial brachytherapy of the pancreas: a feasibility study. Gastrointest. Endosc **62**(5), 775–779 (2005)
9. De Jong, T.L., van de Berg, N.J., Tas, L., Moelker, A., Dankelman, J., van den Dobbelsteen, J.J.: Needle placement errors: do we need steerable needles in interventional radiology? Med. Dev. (Auckland, NZ) **11**, 259 (2018)
10. van de Berg, N.J., van Gerwen, D.J., Dankelman, J., van den Dobbelsteen, J.J.: Design choices in needle steering—a review. IEEE/ASME Trans. Mech. **20**(5), 2172–2183 (2014)
11. Kallem, V., Cowan, N.J.: Image guidance of flexible tip-steerable needles. IEEE Trans. Rob. **25**(1), 191–196 (2009)
12. Khadem, M., Rossa, C., Usmani, N., Sloboda, R.S., Tavakoli, M.: Semi-automated needle steering in biological tissue using an ultrasound-based deflection predictor. Ann. Biomed. Eng. **45**(4), 924–938 (2017)
13. Ko, S.Y., Davies, B.L., y Baena, F.R.: Two-dimensional needle steering with a "programmable bevel" inspired by nature: modeling preliminaries. In: 2010 IEEE/RSJ International Conference on Intelligent Robots and Systems, pp. 2319–2324. IEEE (2010)
14. Secoli, R., y Baena, F.R.: Closed-loop 3D motion modeling and control of a steerable needle for soft tissue surgery. In: 2013 IEEE International Conference on Robotics and Automation, pp. 5831–5836. IEEE (2013)

15. Okazawa, S., Ebrahimi, R., Chuang, J., Salcudean, S.E., Rohling, R.: Hand-held steerable needle device. IEEE/ASME Trans. Mech. **10**(3), 285–296 (2005)
16. Webster III, R.J., Romano, J.M., Cowan, N.J.: Mechanics of precurved-tube continuum robots. IEEE Trans. Rob. **25**(1), 67–78 (2008)
17. Sears, P., Dupont, P.: A steerable needle technology using curved concentric tubes. In: IEEE/RSJ International Conference on Intelligent Robots and Systems, pp. 2850–2856. IEEE (2006)
18. Henken, K.R., Seevinck, P.R., Dankelman, J., van den Dobbelsteen, J.J.: Manually controlled steerable needle for MRI-guided percutaneous interventions. Med. Biol. Eng. Comput. **55**(2), 235–244 (2017)
19. Datla, N.V., Hutapea, P.: Flexure-based active needle for enhanced steering within soft tissue. J. Med. Dev. **9**(4) (2015)
20. van de Berg, N.J., Dankelman, J., van den Dobbelsteen, J.J.: Design of an actively controlled steerable needle with tendon actuation and FBG-based shape sensing. Med. Eng. Phys. **37**(6), 617–622 (2015)
21. Eastwood, K.W., Francis, P., Azimian, H., Swarup, A., Looi, T., Drake, J.M., Naguib, H.E.: Design of a contact-aided compliant notched-tube joint for surgical manipulation in confined workspaces. J. Mech. Rob. **10**(1), 015001 (2018)

Kinesthetic Curiosity: Towards Personalized Embodied Learning with a Robot Tutor Teaching Programming in Mixed Reality

Thomas Groechel(✉), Roxanna Pakkar, Roddur Dasgupta, Chloe Kuo, Haemin Lee, Julia Cordero, Kartik Mahajan, and Maja J. Matarić

University of Southern California, Los Angeles, CA 90089, USA
{groechel,pakkar,udasgupt,cmkuo,haeminle,jrcorder, kmahajan,mataric}@usc.edu

Abstract. Personalizing interactions in socially assistive robot (SAR) tutoring has shown promise with a wide variety of learners, especially when using multiple interaction modalities. Many of those interactions, however, focus on seated learning contexts, creating a need for multimodal personalization measures in kinesthetic (i.e., embodied) learning contexts. This paper proposes a multimodal measure of **student kinesthetic curiosity** (KC^S) that combines a student's movement and curiosity measures into a single, personalized measure. This work evaluates the efficacy of KC^S in a SAR tutor interaction by conducting a within-subjects ($n = 9$) pilot study where participants completed kinesthetic mixed reality coding exercises alongside a curious robot tutor whose actions were determined by KC^S. The study results indicate that the stationarity assumptions needed for KC^S were met and that the robot tutor was able to successfully use KC^S to personalize its action policy, thereby positively affecting short term KC^S. However, no significant results were found for longer state changes for each student. The mixed reality visual programming language (MoveToCode) created for this work has been made open-source. This work aims to inform future online features and measures for mixed reality human-robot interactions.

Keywords: Socially assistive robotics · Augmented and mixed reality · Embodied learning · Visual programming languages

1 Motivation, Problem Statement, Related Work

A recent surveys of Socially Assistive Robot (SAR) tutors reports that SAR tutoring has shown great promise with various learners, especially when using multiple interaction modalities, but largely consists of one-on-one learning companions for children with an emphasis on personalizing the learning interaction [2]. This is often characterized by Bloom's two sigma problem [1] where students

B. Siciliano et al. (Eds.): ISER 2020, SPAR 19, pp. 245–252, 2021.
https://doi.org/10.1007/978-3-030-71151-1_22

performed two standard deviations better when tutored one-on-one compared to traditional one-to-many lecture contexts. To personalize toward individual student needs, SAR tutor interactions have adopted interfaces (e.g., tablets) that increase the observability of student actions. The data from those interfaces are used to pursue multimodal reasoning about hidden student state such as a student's knowledge [11], affect [12], or engagement levels [6]. The emphasis on seated learning interactions creates a need to explore kinesthetic learning contexts (i.e., embodied learning [8]) that include the well-documented benefits of the physical embodiment of the SAR tutor [3].

Embodied learning can be effectively explored in virtual, augmented, and mixed reality human-robot interaction (VAM-HRI) settings. Consequently, this area of research has grown in recent years [15], focusing on design [14], teleoperation [7], and signalling challenges [5,13]. Many of the studied interactions employ augmented reality head-mounted displays (ARHMD) that generate rich multimodal data minimizing or removing the need for external sensing.

This work explores using such rich, multimodal data for personalizing student interactions in an embodied learning context. We create synergies between SAR tutors, VAM-HRI, and embodied learning through the design and implementation of MoveToCode (Fig. 1), an open-source, mixed reality programming platform that interfaces with a robot tutor [4]. We introduce a real-time, multimodal measure of **student kinesthetic curiosity** (KC^S) and analyze how a curious robot tutor's actions impact KC^S during an interaction.

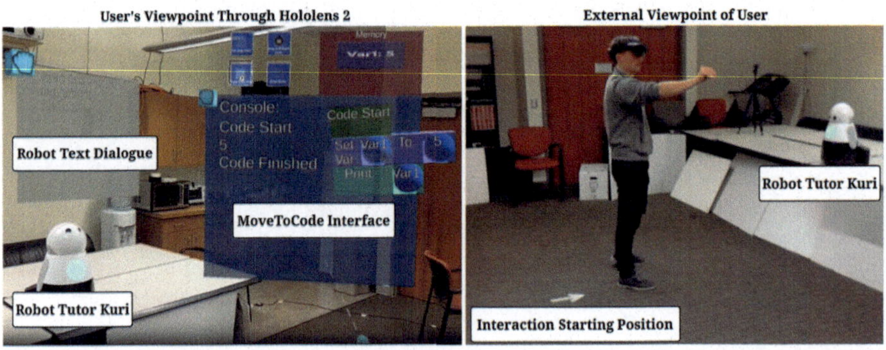

Fig. 1. MoveToCode Interaction; participants attempted to solve coding exercises involving 3D code blocks alongside the robot tutor, Kuri

2 Technical Approach

The key insight of this work is to combine the components of embodied learning (e.g., movement) and student curiosity (e.g., seeking new information) into a single measure of *kinesthetic curiosity* (KC^S) that is personalized for each student. To better understand how the design of KC^S can be leveraged for learning experiences with a robot, we designed a robot action policy that uses KC_t^S. With the hypotheses given in Sect. 2.3, participants completed mixed reality coding exercises with a curious SAR tutor described further in Sect. 4.

2.1 Measuring and Personalizing Kinesthetic Curiosity

To inform a robot's action policy, our real-time measures used a sliding-window approach to measure KC_t^S, a student's kinesthetic curiosity at a given time:

$$movement_t^S = \sum_{n=t-tw+1}^{t} dist(head\_pose_n, head\_pose_{n-1}) \tag{1}$$

$$curiosity_t^S = \sum_{n=t-tw}^{t} [ISA_n^S \neq NULL] \tag{2}$$

$$KC_t^S = w_0 * \frac{movement_t^S - \overline{movement^S}}{\sigma_{movement^S}} + w_1 * \frac{curiosity_t^S - \overline{curiosity^S}}{\sigma_{curiosity^S}} \tag{3}$$

where $movement_t^S$ (1) is measured with accumulated head pose change over a sliding time window tw, and $curiosity_t^S$ (2) is measured as the sum of information seeking actions (ISAs) over tw. ISAs are defined relative to the domain and action space of the learner. Specifically for this work, ISAs included snapping code blocks (Fig. 2), unsnapping code blocks, pressing interaction menu buttons (Fig. 1), and creating new code blocks. KC_t^S (3) assumes an underlying Gaussian distribution for $movement^S$ and $curiosity^S$ for all instances of time from 0 to t. This measure is a weighted combination of $movement_t^S$ deviation and $curiosity_t^S$ deviation from their respective mean (i.e., z-normalization [9]). The resulting normalized scores are therefore personalized to each student at time t.

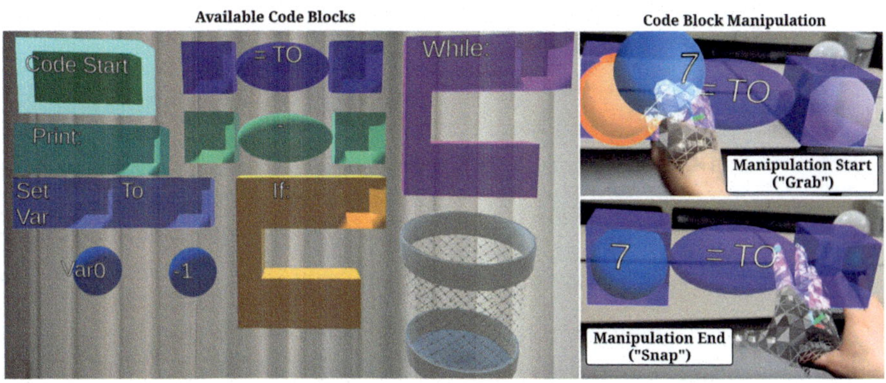

Fig. 2. Available MoveToCode code blocks (left) as seen by the participant through the Hololens 2. Code block manipulation (right) with a participant grabbing the block and letting it go to snap code blocks together.

For post-interaction analysis, a student's KC^S and a robot's KC^R values for an entire interaction consisted of the following:

$$KC^S = (movement^S, curiosity^S) \tag{4}$$

$$KC^R = (movement^R, curiosity^R) \tag{5}$$

where $movement^S$ is the student's total movement throughout the interaction, and $curiosity^S$ is the total number of ISAs throughout the interaction, allowing for normalization across interactions of varying lengths. Using the same set of actions to measure KC^S, a robot tutor's KC^R can be evaluated for both the real-time measure KC_t^R and the post-interaction analysis of KC^R.

2.2 *KC* Robot Tutor Action Policy

The robot's adjustable rule-based policy, shown in Table 1, was designed using a changeable threshold T^{KC} and time-window tw that triggered robot actions based on the time since last action $tsla^R$. For example, a lower T^{KC} causes the robot to take more information-seeking actions in order to motivate the user to do the same. The rule-based policy was designed to support a data collection for informing future data-driven policies.

Table 1. Robot action policy

Action	Activation state
Exercise goal dialogue	New exercise start
Virtual ISA	$KC_t^S < T^{KC}$ and $tsla^R \geq tw$
Positive physical affect	$KC_t^S \geq T^{KC}$ and $tsla^R \geq tw$ or dialogue
Congratulatory dialogue	Correct answer
Scaffolding dialogue	Incorrect answer and Scaffolding dialogue left
Encouraging dialogue	Incorrect answer and ¬Scaffolding dialogue left

2.3 User Study Hypotheses

We performed a user study (described in Sect. 4) to evaluate the following hypotheses regarding KC^S with equal weights ($w_0 = w_1 = 0.5$):

H1: KC^S data fulfill the stationarity assumption needed for z-normalization.
H2: Robot virtual information seeking actions (i.e., curious actions) will positively affect student KC_t^S.
H3: A more curios robot (i.e., lower T^{KC}) will encourage a higher KC^S when compared to a less curious robot (i.e., higher T^{KC}).

For **H1**, KC_t^S assumes an underlying Gaussian distribution over the time series which implies that the time series data are stationary. For **H2**, we examined if robot virtual information seeking action (i.e., curious actions) could positively affect KC_t^S to better inform future robot action selection policies. For **H3**, we tested the differences in conditions to examine longer interaction effects of a more or less curious robot.

3 Results

KC_t^S depends on the time series for $movement_t^S$ and $curiosity_t^S$ to be stationary as they are modeled with the underlying assumption of a constant mean and variance needed for z-normalization. We performed an Augmented Dickey-Fuller test on each participants' $movement_t^S$ and $curiosity_t^S$ measures over the interaction to test for stationarity. With the exceptions of $movement_t^{P1}(p = .017, DF_\tau = -3.248)$ and $curiosity_t^{P2}(p = .012, DF_\tau = -3.378)$, all tests reported a significance of $p < .01$, supporting **H1**.

To analyze the effect of robot virtual information seeking actions (ISAs), we calculated the difference of measure (M) from time t to time $t + tw$ ($\Delta M_{t,t+tw}$) for all robot ISAs at time t. The robot totaled 170 ISAs with $\Delta M_{t,t+tw}$ distributions tested for normality (Fig. 3). A two-sided, single sample t-test was performed against a mean of 0. Significant results were found for all measures: $\Delta movement_{t,t+tw}^S(m)$ ($t = 4.51, p < .001, \bar{x} = 0.495, d = 0.35$); $\Delta curiosity_{t,t+tw}^S(ISA)$ ($t = 4.637, p < .001, \bar{x} = 0.776, d = 0.36$); $\Delta z\left(movement_{t,t+tw}^S\right)$ ($t = 4.623, p < .001, \bar{x} = 0.477, d = 0.36$); $\Delta z\left(curiosity_{t,t+tw}^S\right)$ ($t = 5.087, p < .001, \bar{x} = 0.452, d = 0.39$); $\Delta KC_{t,t+tw}^S$ ($t = 6.764, p < .001, \bar{x} = 0.464, d = 0.52$). These findings demonstrate a positive short term effect of robot virtual ISAs on KC_t^S, supporting **H2**.

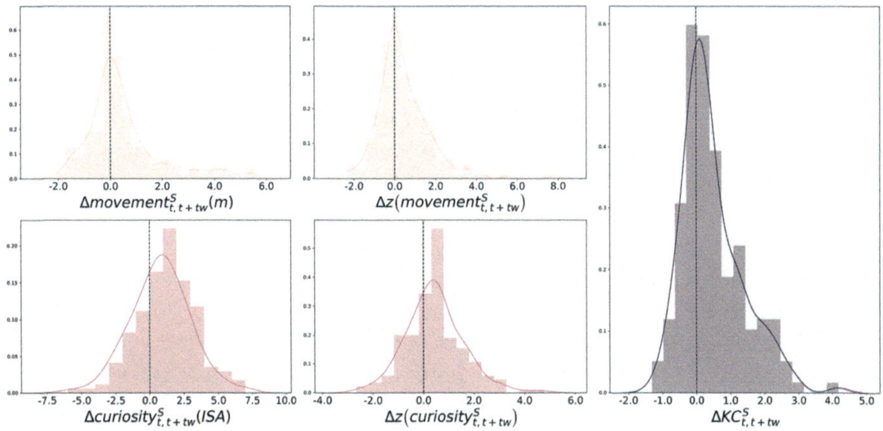

Fig. 3. Action distributions for all differences in measures where the robot took an ISA at time t to the score at time $t + tw$ ($\Delta M_{t,t+tw}$). Zero line is plotted to show distribution shifts.

We analyzed differences in total $movement^S$ and $curiosity^{S,R}$ measures between the robot action policy threshold conditions of $T_{high}^{KC} = 0.5$ and $T_{low}^{KC} = -0.5$ shown in Fig. 4. A Wilcoxon signed-rank test indicated a significant effect for $curiosity^R(ISA)$ ($\tilde{x}_{high} = 15, \tilde{x}_{low} = 3, W = 0, p = .008$).

No significant effect was found for $movement^S(m)$ ($\tilde{x}_{high} = 52.2, \tilde{x}_{low} = 39.93, W = 21, p = .859$) or $curiosity^S(ISA)$ ($\tilde{x}_{high} = 90, \tilde{x}_{low} = 130, W = 11, p = .172$). These findings support the T^{KC} thresholds chosen but do not support a difference in KC^S between conditions posited by **H3**.

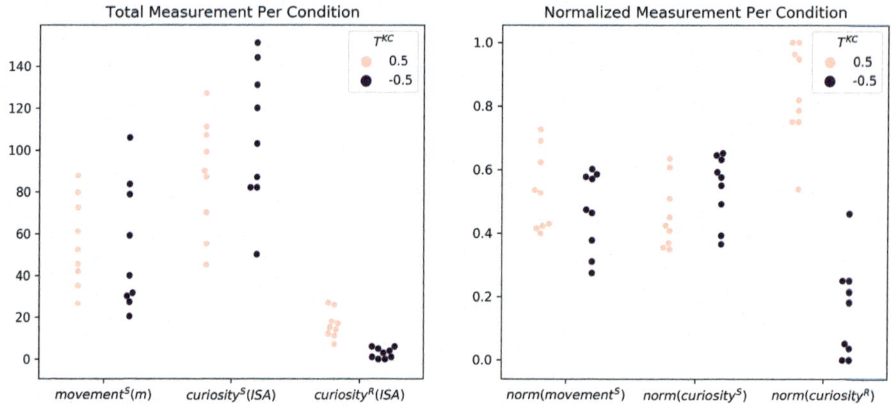

Fig. 4. Measures between the more curious ($T^{KC} = -0.5$) and less curious ($T^{KC} = 0.5$) robot conditions. The left depicts the total score for each measure for all participants. The right depicts normalized measures for each participant between conditions.

4 Completed Experiment

A single-session within-subjects experiment was conducted with approval by our university IRB (UP-17-00226) to evaluate a curious robot tutor policy (Table 1) with differing thresholds (T^{KC}). Ten participants (3F,7M) were recruited from the University of Southern California student population, with an age range of 19–27 ($\bar{x} = 22.8, \sigma = 2.9$). P8 experienced two separate operating system crashes (at 285.92 s and 341.44 s); having not experienced both experimental conditions, P8 was therefore removed from the behavioral data analysis.

Participants wore the ARHMD (Microsoft Hololens 2) and attempted to complete programming exercises with the help of a robot tutor, a Mayfield Robotics Kuri (Fig. 1). Kuri used the action policy described in Table 1 with tw empirically set to 20 s. The independent variable in the study was the robot KC threshold level using $T^{KC}_{high} = 0.5$ and $T^{KC}_{low} = -0.5$. The experiment lasted 20 min with T^{KC} substituted at 10 min. ISAs (see Eq. 2) in this experiment included snapping code blocks, unsnapping code blocks, pressing interaction menu buttons, and creating new code blocks.

5 Main Experiment Insights

This work developed a multimodal measure of kinesthetic curiosity (KC) and conducted and evaluated it in a pilot study. The study validated the efficacy of using multimodal data from an ARHMD to personalize student interactions via kinesthetic curiosity. MoveToCode, the programming language developed for this work, has been made fully open-source [4] and is extensible to any robot supporting the Robot Operating System (ROS) [10].

The study results demonstrate that z-normalization across modalities shows promise for providing a unitless comparison across measure modalities. Given time-synchronized data from the ARHMD, there is a significant benefit to further exploring additional interaction modalities the headsets contain, such as articulated hand tracking and eye gaze tracking. The logging system developed as part of MoveToCode has allowed multiple followup projects to use the collected behavioral data from this experiment with minimal processing, as each modality is logged at the same rate 50 Hz.

Our results indicate short-term effects of robot actions, but full interaction effects were not found across conditions. Based on the high post-interaction interview reports of novelty (7/10) within the interaction (e.g., users referring to the interaction as "cool"), future work will explore longer-term interactions with students. This work demonstrates that KC_t^S lends itself to reinforcement learning approaches for long-term personalization being explored in the rare long-term SAR tutoring interactions [2]. We plan to expand the virtual action space of the robot tutor and eventually study long-term personalization in school and home settings.

Acknowledgement. This research was supported by National Science Foundation National Robotics Initiative 2.0 grant for "Communicate, Share, Adapt: A Mixed Reality Framework for Facilitating Robot Integration and Customization" (NSF IIS-1925083).

References

1. Bloom, B.S.: The 2 sigma problem: the search for methods of group instruction as effective as one-to-one tutoring. Educ. Researcher **13**(6), 4–16 (1984)
2. Clabaugh, C., Matarić, M.: Escaping Oz: autonomy in socially assistive robotics. Ann. Rev. Control Rob. Auton. Syst. **2**, 33–61 (2019)
3. Deng, E., Mutlu, B., Mataric, M.J., et al.: Embodiment in socially interactive robots. Found. Trends® Rob **7**(4), 251–356 (2019)
4. Groechel, T., Kuo, C., Dasgupta, R.: Interaction-lab/movetocode: Doi release (2020). https://doi.org/10.5281/zenodo.3924514
5. Groechel, T., Shi, Z., Pakkar, R., Mataric, M.J.: Using socially expressive mixed reality arms for enhancing low-expressivity robots. In: 2019 28th IEEE International Symposium on Robot and Human Interactive Communication (RO-MAN). IEEE (2019)
6. Jain, S., Thiagarajan, B., Shi, Z., Clabaugh, C., Matarić, M.J.: Modeling engagement in long-term, in-home socially assistive robot interventions for children with autism spectrum disorders. Sci. Rob. **5**(39), (2020)
7. Lipton, J.I., Fay, A.J., Rus, D.: Baxter's homunculus: virtual reality spaces for teleoperation in manufacturing. IEEE Rob. Autom. Lett. **3**(1), 179–186 (2018)
8. Macedonia, M.: Embodied learning: why at school the mind needs the body. Front. Psychol. **10**, 2098 (2019)

9. Mariéthoz, J., Bengio, S.: A unified framework for score normalization techniques applied to text-independent speaker verification. IEEE Signal Process. Lett. **12**(7), 532–535 (2005)
10. Quigley, M., Conley, K., Gerkey, B., Faust, J., Foote, T., Leibs, J., Wheeler, R., Ng, A.Y.: Ros: an open-source robot operating system. In: ICRA Workshop on Open Source Software, Kobe, Japan, vol. 3, p. 5 (2009)
11. Schodde, T., Bergmann, K., Kopp, S.: Adaptive robot language tutoring based on bayesian knowledge tracing and predictive decision-making. In: Proceedings of the 2017 ACM/IEEE International Conference on Human-Robot Interaction, pp. 128–136 (2017)
12. Spaulding, S., Breazeal, C.: Frustratingly easy personalization for real-time affect interpretation of facial expression. In: 2019 8th International Conference on Affective Computing and Intelligent Interaction (ACII), pp. 531–537. IEEE (2019)
13. Walker, M., Hedayati, H., Lee, J., Szafir, D.: Communicating robot motion intent with augmented reality. In: Proceedings of the 2018 ACM/IEEE International Conference on Human-Robot Interaction, pp. 316–324. ACM (2018)
14. Williams, T., Hirshfield, L., Tran, N., Grant, T., Woodward, N.: Using augmented reality to better study human-robot interaction. In: HCII Conference on Virtual, Augmented, and Mixed Reality (2020)
15. Williams, T., Szafir, D., Chakraborti, T., Soh Khim, O., Rosen, E., Booth, S., Groechel, T.: Virtual, augmented, and mixed reality for human-robot interaction (vam-hri). In: Companion of the 2020 ACM/IEEE International Conference on Human-Robot Interaction, HRI 2020, pp. 663–664 (2020)

A Soft Robotic Cover with Dual Thermal Display and Sensing Capabilities

Yukiko Osawa(✉) and Abderrahmane Kheddar

CNRS-University of Montpellier, LIRMM, Interactive Digital Humans group, Montpellier, France
yukiko.akiyama@lirmm.fr

Abstract. We propose a new robotic cover prototype that achieves thermal display while also being soft. We focus on the thermal cue because previous human studies have identified it as part of the touch pleasantness. The robotic cover surface can be regulated to the desired temperature by circulating water through a thermally conductive pipe embedded in the cover, of which temperature is controlled. Besides, an observer for estimating heat from human contact is implemented; it can detect human interaction while displaying the desired temperature without temperature sensing on the surface directly. We assessed the validity of the prototype in experiments of temperature control and contact detection by human hand.

1 Introduction

Physical interaction between human and robot enables various robotic tasks: helping a frail person in daily life, programming a robot in manufacturing through touch guidance and demonstration, etc. In order to achieve such tasks, robots can be conceived in order to display features such as soft texture and warmth, in order to be pleasant to human touch. We claim that the interaction between humans and robots shall also be considered from an aesthetic and pleasantness (to human touch) hardware design viewpoint. In our view, combining functionalities with aesthetics and pleasantness is an important dimension in human-robot physical interaction and worth investigating in human-centric robotics. Moreover, robot cover's softness is at the intersection of both functionality (casting the environment irregularities, absorbing impacts, safety through inherent compliance, etc.) and pleasantness to touch when compared to current rigid design. In the other hand, warmth in physical interactions has beneficial effects [8,11]. Therefore, combining thermal exchanges and softness properties to design robotic covers potentially increases embodiment, acceptability, and human friendliness.

The challenge is that (i) most of the soft and low-density materials are not thermally conductive and hence cannot display changes in temperature on demand; (ii) most of the thermally conductive material are also electric conductive and have high-density mass which is problematic to cover all the robot

B. Siciliano et al. (Eds.): ISER 2020, SPAR 19, pp. 253–262, 2021.
https://doi.org/10.1007/978-3-030-71151-1_23

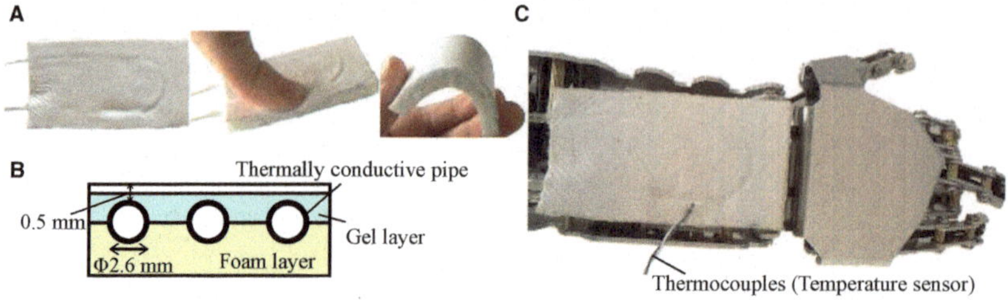

Fig. 1. The prototype of the robotic cover.

limbs (e.g. the humanoid case); (iii) if any cover material exist, one still need to provide desired temperature display using a heat source. But integrating a heat source–as in haptic displays technologies [2], for the entire cover surface is neither a realistic nor a plausible perspective. We solve this issue by using flowing liquid (water) with the Peltier pump that can heat or cool it.

Our aim is to achieve robotic covers that are both soft and capable of thermal display. Another side idea is to exploit the thermal cover in a dual thermal cues display and sensing. This would allow the robot to also detect human touch through monitoring the thermal exchange during the process of robotic cover thermal display regulation. Thermal display techniques are found in haptics research. Namely in portable haptic displays. Indeed, there are many kinds of thermal displays [2–4,6,10,13] based on a Peltier pump heat source. The latter can either heat or cool the surface (e.g. human fingertips when touched) based on the Peltier effect [12]. Yet, to our best knowledge, thermal display technology has not yet been considered to be part of the robotic cover: this is totally novel. More challenging, is to embed thermal display to a soft robot skin that appear to be a useful feature in multi-contact [1] and also to cast human touch.

2 Technical Approach

2.1 Design of the Robotic Cover and Its Control System

Figure 1A shows the prototype of our designed composite soft and thermal robotic cover. The small prototype, of which size is $3.0 \times 6.0\,\text{cm}^2$, is structured into two-layered materials, as shown in Fig. 1B. The upper (gel) part would conduct and hence diffuse the temperature. The foam part, being not thermally conductive, would isolate the inner robot body. Moreover, the chosen foam material has a very low mass density and good compliance properties, which allows designing an overall light whole-body cover that is soft with thermal rendering capabilities. These two kinds of materials consisting of the cover with high heat conductivity and low density made it possible to be soft, well thermally conductive, and light. Thanks to the structure, the weight of the prototype is only 4.6 g. For circulating water, a thermally conductive pipe is embedded between the two layers. A thin layer of gel is used to transfer temperature from the water

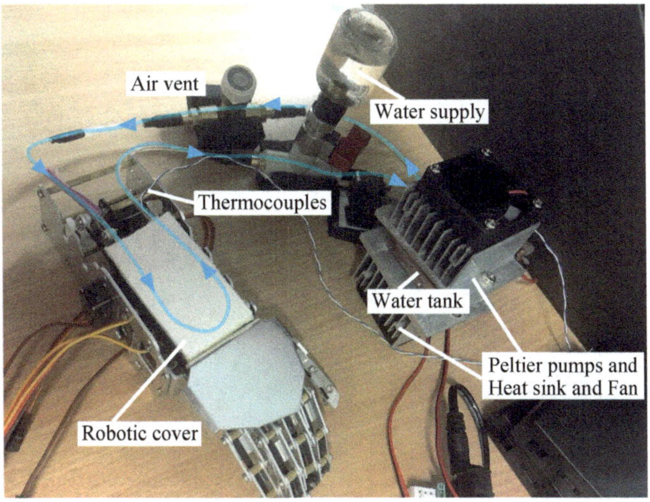

Fig. 2. Overall robotic cover system.

flowing inside the pipe. The thickness of the gel and foam layer is 1 mm and 5 mm, respectively. The overall cover thickness is thus 6 mm. Ideally, we would choose the thickness to meed desired compliance of the cover while being as thin as possible. To this end, our proposed cover achieves two opposing properties: softness and high thermal conductivity at the same time. Some experiments were conducted by attaching the cover to the small robot hand shown in Fig. 1C.

The overall system is shown in Fig. 2. It consists of a small water pump; a copper tank – on top of which is mounted a Peltier pump acting as a heat source; and finally, the soft robotic cover explained previously. The silicone pipe with a diameter of 4.5 mm, connects the water pump, the tank, and the robotic cover. The water stored in the tank is heated and cooled by the Peltier pump, regulating the circulating water temperature in a closed circuit throughout our proposed cover. Thermocouples that measure temperature are attached to the surface of the heat source, the copper water tank, and others to the soft robotic cover.

2.2 Heat Flow Observer for Human Interaction

To detect altered thermal response of the robotic cover subsequent to human touch, heat flow is used. Indeed, heat flow behavior depends on the intrinsic (thermal) parameters of each material, and hence can be monitored to detect that a contact has been established. Moreover, the heat flow dynamic can (in theory) reveal if the contact is actually a human touch. Heat flow sensors exist, but the material by which they are composed biases heat conduction properties when it lies between two other materials (i.e. the robot cover and the human skin). In this paper, heat flow induced from a human touch on the cover is

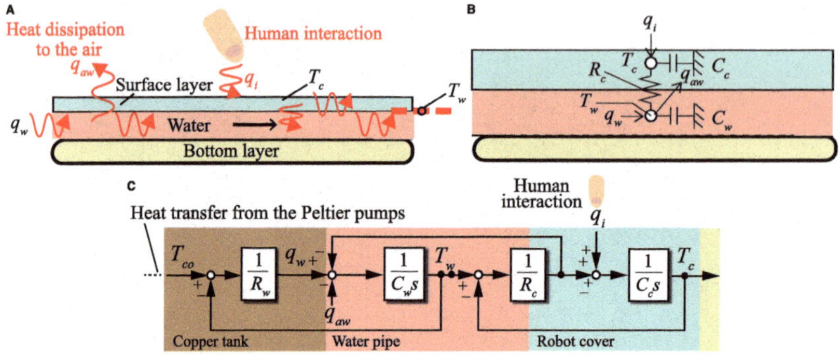

Fig. 3. Modeling of the circulating water system.

estimated from temperature rate information. The latter is obtained solely from thermocouples attached to the thermally conductive pipe (i.e. we do not use temperature measurement on the soft cover surface).

To distinguish human touch, a heat flow observer is implemented. Our algorithm is based on the disturbance observer developed in [7,9], already used in many control applications. The estimation algorithm is simple; by monitoring the temperature of the copper box (T_{co}) and the water pipe (T_w) as observer inputs, heat flow from human contact q_i is calculated based on the thermal model explained hereafter.

The thermal model of the circulating water system is shown in Fig. 3. It is derived from heat transfer modeling in [5]. Here, T_w, T_c, R_w, R_c, C_w, and C_c stand for temperature, thermal resistance, thermal capacitance of the water pipe (index $_w$) and the soft cover (index $_c$). T_{co} is the temperature of the copper tank, attached to the Peltier pumps. Heat flow q_w convects through the circulating water, which is biased due to human-touch subsequent heat flow q_i and heat

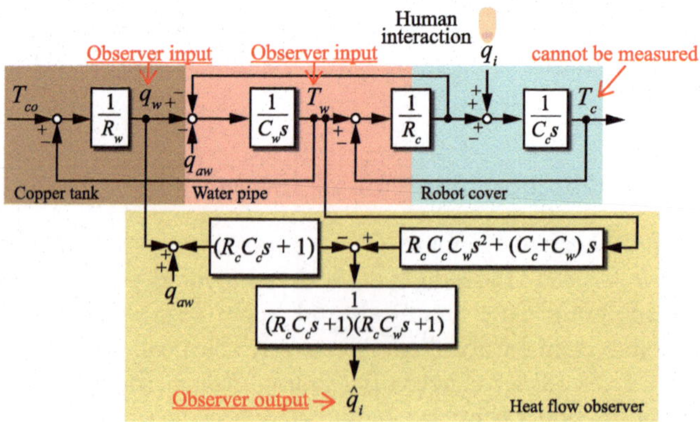

Fig. 4. The block diagram of the heat flow observer.

loss q_{aw}. Based on the block diagram in Fig. 3C, the transfer function of the circulating water system is

$$\frac{T_w}{q_w + q_{aw}} = \frac{R_c C_c s + 1}{R_c C_w C_c s^2 + (C_w + C_c)\, s}, \quad (1)$$

where s is the Laplace operator. Heat loss q_{aw} is confirmed and estimated in preliminary experiments. We compensate for q_{aw} to the transfer function input q_w in (1). The transfer function from q_i to T_w write as

$$\frac{T_w}{q_i} = \frac{(R_c C_w s + 1)(R_c C_c s + 1)}{R_c C_w C_c s^2 + (C_w + C_c)\, s}. \quad (2)$$

Using (1) and (2), T_w is expressed as

$$T_w = \frac{(R_c C_c s + 1)}{R_c C_w C_c s^2 + (C_w + C_c)\, s}(q_w + q_{aw}) + \frac{(R_c C_w s + 1)(R_c C_c s + 1)}{R_c C_w C_c s^2 + (C_w + C_c)\, s} q_i. \quad (3)$$

Based on Eq. (3), the observer for estimating q_i is derived as

$$\hat{q}_i = \frac{\left(R_c C_w C_c s^2 + (C_w + C_c)\, s\right) T_w - (R_c C_c s + 1)(q_w + q_{aw})}{(R_c C_w s + 1)(R_c C_c s + 1)}. \quad (4)$$

where $\hat{q}_i$ is the estimated value. By using Eq. (4), a structure of the observer is expressed as shown in Fig. 4. The observer output is directly affected by the parameter variation of (4). We conducted some identification tests by comparing to the model in Fig. 3C with experimental results, deriving the exact parameters shown in Table 1. As a result, the observer can extract only the heat flow from the contact object, regardless of the temperature that is displayed at the soft cover, as far as it is different from that of the object.

Table 1. Identified parameters of the robotic cover system.

Parameter	Value	For
$R_{com}C_{com}$	500 (heat), 450 (cool) seconds	Thermal display using T_c (Experiment 1)
L_d	45 (heat), 30 (cool) seconds	Thermal display using T_c (Experiment 1)
R_a	0.2 (heat), 0.3 (cool) K/W	Thermal display using T_c (Experiment 1)
$R_{com}C_{com}$	500 (heat), 410 (cool) seconds	Thermal display using T_w (Experiment 2)
L_d	45 (heat), 30 (cool) seconds	Thermal display using T_w (Experiment 2)
R_a	0.3 (heat), 0.7 (cool) K/W	Thermal display using T_w (Experiment 2)
C_{co}	1152.57 J/K	Thermal exchange observer (Experiment 2)
C_w	197.41 (heat), 182.79 (cool) J/K	Thermal exchange observer (Experiment 2)
C_c	0.40 (heat), 0.10 J/K	Thermal exchange observer (Experiment 2)
R_{co}	0.09 K/W	Thermal exchange observer (Experiment 2)
R_w	6.00 (heat), 5.56 (cool) K/W	Thermal exchange observer (Experiment 2)
R_c	120.12 (heat), 30.03 (cool) K/W	Thermal exchange observer (Experiment 2)
R_{aw}	2.1 K/W	Thermal exchange observer (Experiment 2)

2.3 Control Algorithm for Thermal Display

The closed-circuit water-circulation system has two controllable inputs: (i) the temperature of the Peltier pumps that are attached to the water tank; and (ii) the water flow rate, which can be changed by the voltage command of the water pump. A constant voltage applied to the water pump leads to a constant water flow rate and a linear model. In fact, given the size of the system and the slow dynamic of thermal heating and cooling, intermittent bang-bang type (max water flow, or no flow) command is sufficient. The heat transfer model of Fig. 3C can be approximated by summarizing multiple transfer layers (cooper water-tank, water-pipe...) as

$$T_c(s) = \frac{\exp(-L_d s)}{1 + R_{\text{com}} C_{\text{com}} s} \left(T_p(s) - q_a\right), \tag{5}$$

where s, T_p, T_c, R_{com}, C_{com}, q_a, and L_d stand for the Laplace operator, temperature of the Peltier pump and robotic cover, combined thermal resistance, combined thermal capacitance, heat loss at the surface of the robotic cover, and dead-time, respectively. A model preview controller (MPC) is used to regulate the temperature of the robotic cover. The cost function with additional constraints are chosen as

$$\min \sum_{i=1}^{H} \|\hat{T}_c(k+i|k) - T_c^{\text{cmd}}(k+i|k)\|_{W_1}^2 + \sum_{i=0}^{H-1} \|\hat{T}_p(k+i|k)\|_{W_2}^2 \tag{6}$$

subject to

$$T_c(k) = \frac{R_{\text{com}} C_{\text{com}}}{R_{\text{com}} C_{\text{com}} + t_s} T_c(k-1) + \frac{t_s}{R_{\text{com}} C_{\text{com}} + t_s} T_p(k - 1 - L_d/t_s) \tag{7}$$

$$T_{\text{min}}^{\text{th}} \leq T_p(k) \leq T_{\text{max}}^{\text{th}}, \tag{8}$$

where k, H, W_1, W_2, t_s, $\hat{T}_c$, T_c^{cmd}, $T_{\text{min}}^{\text{th}}$, and $T_{\text{max}}^{\text{th}}$ stand for the discrete time, prediction horizon, weight values of the cost function, sampling time, estimated value of T_c, temperature command of T_c, and minimum and maximum value of threshold of the calculated input, respectively. The first term of (6) works for tracking the temperature of the cover T_c to its temperature command T_c^{cmd}, and the second one is for suppressing rapid change of the Peltier pump input. Here, the continuous model shown in (5) is converted to the discrete model. The same model is used for controlling T_w with the observer explained in Sect. 2.2.

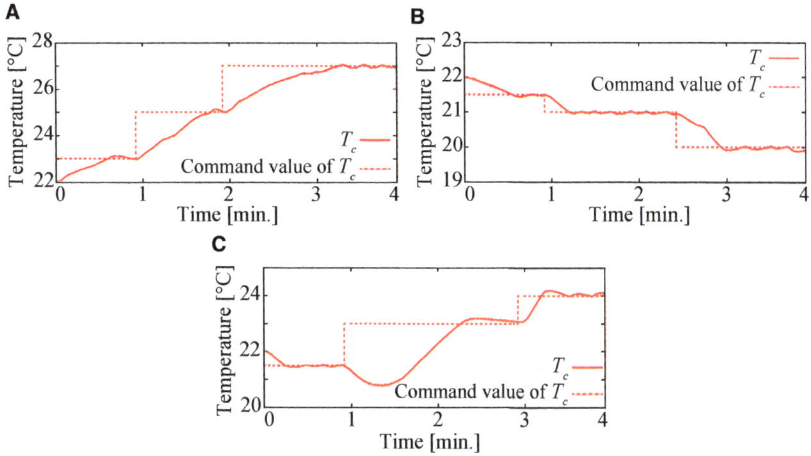

Fig. 5. Surface temperature control of the prototype cover (Experiment 1).

3 Experimental Results

We achieved two kinds of experiments: (i) temperature control of the cover (Experiment 1: controlling T_c); and (ii) heat flow estimation without sensing the cover's surface temperature (Experiment 2: controlling T_w and heat flow estimation). The identified parameters for both experiments are summarized in Table 1. R_{aw} is the thermal resistance between the water pipe temperature and ambient temperature, including q_{aw} in Fig. 4. These values were derived from experimental responses to step inputs and the model mentioned in Sects. 2.2 and 2.3.

3.1 Temperature Control of the Prototype Cover (Experiment 1)

In Experiment 1, the surface temperature of the prototype soft cover is controlled to be heated (with successive desired temperatures 23 °C, 25 °C, and 27 °C) shown in Fig. 5A, and then cooled (with successive desired temperature 21.5 °C, 21 °C, and 20 °C) shown in Fig. 5B. Figure 5C shows the system responses heated to 23 °C and 24 °C after cooling to 21.5 °C. The dotted lines plot desired temperatures, whereas plain lines plot the cover response when our control algorithm is applied.

The Peltier pump's temperature command is generated so that the temperature of the robotic cover T_c meets its desired T_c^{cmd} temperature, and the response of the Peltier pump achieved subsequent to the controller. The circulating water is stopped whenever the response reaches its commanded desired value. These experimental results show that the control system of the robotic cover is functioning correctly. When heating after cooling (or *vice-versa*), it takes more time to meet the desired temperature because of the slow dynamics of thermal transfers (see Fig. 5C). The main reason is related to the water's thermal capacitance,

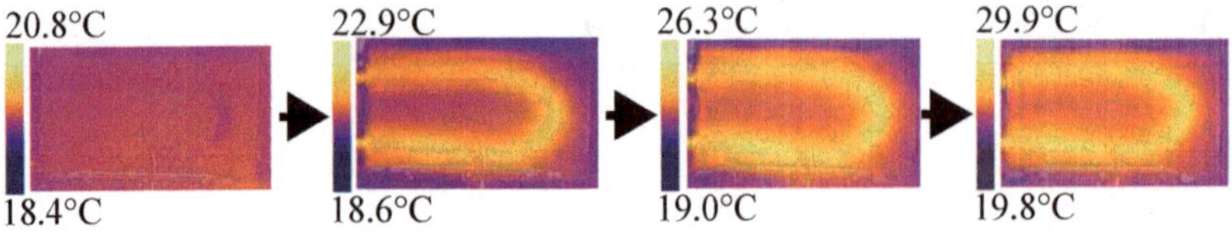

Fig. 6. Experimental results of Fig. 5A monitored by an infrared camera (Experiment 1).

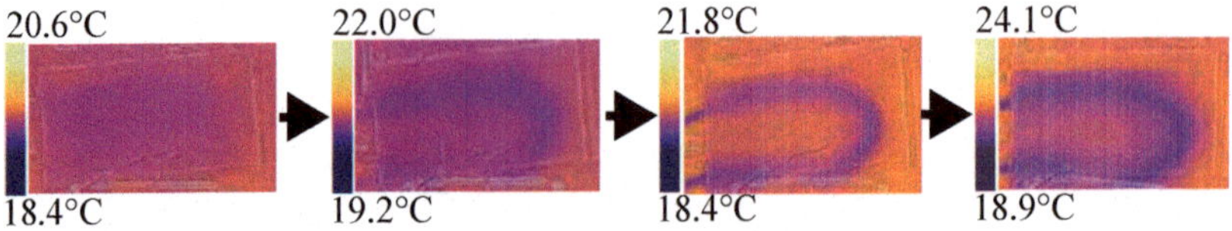

Fig. 7. Experimental results of Fig. 5B monitored by an infrared camera (Experiment 1).

taking about 30 s to heat or cool for the whole circulating water. As additional evidence, Figs. 6 and 7 show the surface of the prototype soft cover for the cases of Fig. 5A and B monitored by an infrared camera.

3.2 Heat Flow Estimation with Temperature Control (Experiment 2)

In Experiment 2, we achieve heat flow estimation with temperature control. The estimated heat flow is made with the thermal exchange observer in Sect. 2.2. We monitor two types of human contact with the soft cover: (i) grasp and (ii) soft touch, each lasting 5 s. The conductive pipe temperature is controlled with successive steps 23, 24 and 25°C for about 1 mn 30 s each. Resulting temperature and estimated heat flow, together with infra-red camera monitoring of the soft cover, are shown in Figs. 8 , 9, 10 and 11, respectively. The heat flow observer is operating independently of the controlled water-pipe's temperature. Recall that the observer is based on temperature change of the water pipe due to a contact at the cover. The observer output depends on the amount of heat exchange when the human skin contacts the cover. For instance, the contact area for a grasp is wider relatively to a soft-touch (see Figs. 8 and 9). Yet, a limitation of the observer remains for extending it to contact localization and even to robust contact detection. Theoretically, the observed heat flow is zero when there is no contact, and increases or decreases only when contact occurs. However, the temperature response of the water pipe exhibit sometimes chattering around the desired temperature. This is because the water pump repeatedly switches between on and off when the response settles to its desired value. Subsequently, the observer output is not always zero when there is no contact. In practice, we can of course anticipate this chattering as we know when the pump switches on and off. However, the frequency of the switches is relatively high around

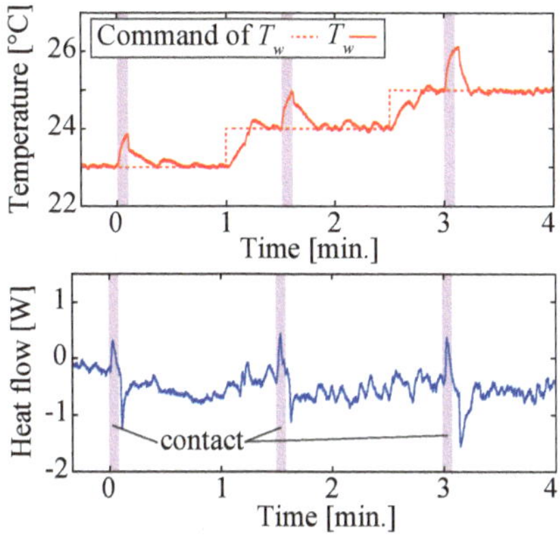

Fig. 8. Temperature and heat flow responses in grasping action (Experiment 2).

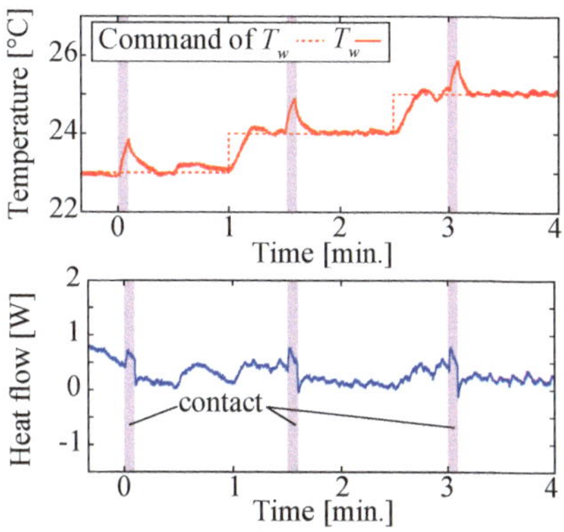

Fig. 9. Temperature and heat flow responses in soft touch (Experiment 2).

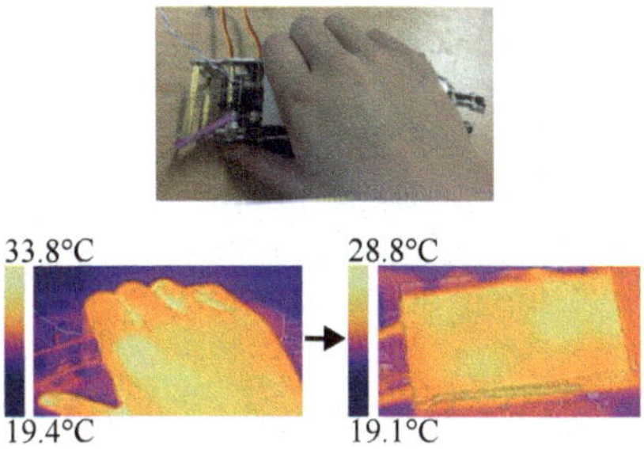

Fig. 10. Experimental results of Fig. 8 monitored by an infrared camera (Experiment 2).

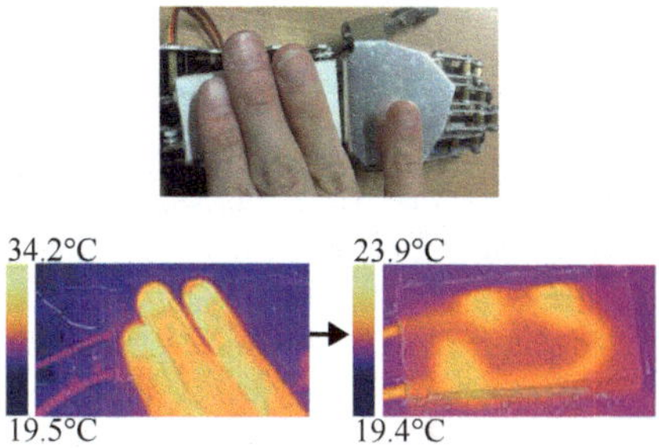

Fig. 11. Experimental results of Fig. 9 monitored by an infrared camera (Experiment 2).

the desired temperature and the contact can occur exactly at the moment of switch. The observer fails to detect touches or grasps lasting less than 5 s as it has relatively low bandwidth. Indeed, the observer's output is too small to be distinguished from noise. We keep alive this problem and test other contact taxonomies such as holding and stroking with an integrated whole-scale system in future work.

4 Conclusion and Future Work

We designed a new robotic cover that gathers both softness and thermal display (i.e. rendering) capabilities. On the bottom layer of the cover, we assembled thermally conductive pipes in which, water that can be heated or cooled, flows in a closed-circuit. We devised an integrated mechatronic system controller by means of a MPC that regulate the cover's surface temperature to any desired one. Moreover, we also noted that cover can also detect human contact through thermal features by implementing a thermal exchange observer. We hence obtain a dual soft thermal and sensing cover that can potentially be mounted on complex robots such as humanoids.

Acknowledgement. This work is supported in part by JSPS Overseas Research Fellowships No. 201960463.

References

1. Bouyarmane, K., Caron, S., Escande, A., Kheddar, A.: Multi-contact planning and control. In: Goswami, A., Vadakkepat, P. (eds.) Humanoid Robotics: A Reference, pp. 1763–1804. Springer, Dordrecht (2019)
2. Drif, A., Citerin, J., Kheddar, A.: Thermal bilateral coupling in teleoperators. In: IEEE/RSJ International Conference on Intelligent Robots and Systems, pp. 1301–1306 (2005). https://doi.org/10.1109/IROS.2005.1545403
3. Goetz, D.T., Owusu-Antwi, D.K., Culbertson, H.: Patch: pump-actuated thermal compression haptics. In: IEEE Haptics Symposium, pp. 643–649 (2020)
4. Guiatni, M., Benallegue, A., Kheddar, A.: Learning-based thermal rendering in telepresence. In: Ferre, M. (ed.) Haptics: Perception, Devices and Scenarios, pp. 820–825. Springer, Heidelberg (2008)
5. Holman, J.P.: Heat Transfer, vol. 1. McGraw-Hill Inc., New York (1990)
6. Jones, L.A., Ho, H.: Warm or cool, large or small? the challenge of thermal displays. IEEE Trans. Haptics **1**(1), 53–70 (2008)
7. Morimitsu, H., Katsura, S.: Sensorless control of heat inflow to a thermal display using a heat inflow observer. IEEE Trans. Ind. Electron. **62**(7), 4288–4297 (2015)
8. Nie, J., Park, M., Marin, A.L., Sundar, S.S.: Can you hold my hand? physical warmth in human-robot interaction. In: ACM/IEEE International Conference on Human-Robot Interaction, pp. 201–202 (2012)
9. Ohnishi, K., Shibata, M., Murakami, T.: Motion control for advanced mechatronics. IEEE/ASME Trans. Mechatron **1**(1), 56–67 (1996)
10. Osawa, Y., Katsura, S.: Thermal propagation control using a thermal diffusion equation. IEEE Trans. Ind. Electron. **65**(11), 8809–8817 (2018)
11. Park, E., Lee, J.: I am a warm robot: the effects of temperature in physical human-robot interaction. Robotica **32**(1), 133–142 (2014)
12. Rowe, D.M.: Thermoelectrics Handbook: Macro to Nano. CRC Press, Boca Raton (2018)
13. Yamamoto, A., Cros, B., Hashimoto, H., Higuchi, T.: Control of thermal tactile display based on prediction of contact temperature. In: IEEE International Conference on Robotics and Automation, vol. 2, pp. 1536–1541 (2004)

Ensuring Stable and Transparent High Stiffness Haptic Interaction Using Successive Force Augmention with Time Domain Passivity Approach

Harsimran Singh[1(✉)], Michael Rothammer[1], Chan-Il Lee[2], Thomas Hulin[1], Jee-Hwan Ryu[2], and Christian Ott[1]

[1] Institute of Robotics and Mechatronics, German Aerospace Center (DLR), 82234 Wessling, Germany
harsimran.singh@dlr.de

[2] IRiS Lab, Advanced Institute of Science and Technology (KAIST), 34141 Daejeon, Republic of Korea

Abstract. Even though a lot of work has been done to improve the stability of haptic interfaces, increasing the impedance range and enhancing the rate-hardness still remains a challenging task. The Successive Force Augmentation approach was proposed to improve the impedance range and rate-hardness of haptic interfaces by introducing an adaptive feed-forward offset to the feedback force, while maintaining the desired stiffness value. In this paper, we modify the Successive Force Augmentation approach to eliminate the push backs resulting from high stiffness haptic contacts and guarantee stability by implementing the Time Domain Passivity Approach in series with the modified Successive Force Augmentation approach. In order to validate the proposed controller, high stiffness haptic interaction and teleoperation with high stiffness virtual fixture experiments on two different devices are carried out.

Keywords: Telerobotics and teleoperation · Haptics and haptic interfaces · Physical human-robot interaction · Robust/adaptive control of robotic systems

1 Introduction

Kinaesthetic feedback is an important branch of haptics that can display forces and torques, which can improve the conveyance of information considerably. Especially for telepresence, it makes use of an additional human sense and can thereby enhance immersion. However, properly realizing kinaesthetic feedback can be very challenging, especially for impedance-type haptic devices. Any haptic device comes with an inherent impedance range that it can adequately display [1,2].

Kinaesthetic feedback along with virtual fixtures (VF) shows better performance for tasks that require human-in-the-loop intelligence along with accuracy

B. Siciliano et al. (Eds.): ISER 2020, SPAR 19, pp. 263–273, 2021.
https://doi.org/10.1007/978-3-030-71151-1_24

and precision of better-than-human level. Virtual Fixtures, through a haptic device, cue force information to the human operator that assists in the task by limiting or encouraging movement along the desired path. They are increasingly being used in medical simulators [3], aerial robots [4] and teleoperation [5].

It would be beneficial if the VFs could be rendered in a stable manner with an increased stiffness range, allowing the human operator to discriminate between different VFs or between a VF and real environment, in case of teleoperation. Compared to the amount of research carried out to stabilize haptic interfaces [6–10], not much work has been focused on enhancing its transparency. Transparency can be improved either by hardware modifications [11,12] or through inventing control schemes [13–16]. The challenges with the previously mentioned control schemes are either they introduce additional damping, which distorts the stiffness perceived by a human operator [17] or they are represented in frequency domain and Z-domain, which may not always be available online. Additionally none of those methods can enhance the rate-hardness, which is a transparency index for haptic interaction [18].

This paper presents a modification to the Successive Force Augment (SFA) approach [19], which can increase the impedance range and rate-hardness, thereby enabling the possibility to render VFs of varying stiffnesses. The SFA approach uses low stiffness feedback force to maintain stability while taking advantage of feedforward force offset to increase the impedance range, which is independent of stability. The modified version of SFA is combined with Time Domain Passivity Approach (TDPA) [6] to ensure that the 1-port haptic interface is passive. The proposed control scheme can be implemented to any impedance-type haptic device to stably render VF of any stiffness. This is experimentally validated by implementation on two different devices, (i) Phantom Premium 1.5, a 3-DoF force feedback device, and (ii) Franka Emika Panda, a 6-DoF force feedback robotic arm.

2 Review of Successive Force Augmentation Approach

The effect of quantization in haptic interfaces is one of the major sources of non-passive behavior. This active interface injects energy into the system, which might induce instabilities. If the actual stiffness of the virtual environment (VE), k_{de}, is lower than the critical stiffness, k_p, then the generated energy will be fully dissipated by the physical damping of the haptic device. On the other hand, if k_{de} is greater than k_p, then the generated energy may not be fully dissipated by the physical damping of the device, and therefore the haptic interaction may potentially become unstable. The critical stiffness as defined in [1] is as follows:

$$k_p \leq \frac{2b_m}{\Delta T} \tag{1}$$

where, b_m is the physical damping of the haptic device and ΔT the sampling time. Thus, if the stiffness of the rendered VE is less than k_p, then the haptic interface would be passive.

The SFA approach [19] was proposed to circumvent the trade-off between stability and impedance range of a haptic interface. It can, in successive successions, increase the displayed stiffness as close as possible to the desired stiffness by using low enough stiffness, k_v, so as to maintain stability, and by introducing a feedforward force offset, $\Delta 0$, which is independent of stability. An observer is implemented which detects whether the haptic interaction point (HIP) is being pushed in (pressing) or pulled out (releasing) of the VE by the user.

$$f(n) = k_v(x(n) - x_e(n)) + \Delta O \tag{2}$$

where, $f(n)$ is the feedback force, k_v is the value for stiffness chosen such that $k_v < k_p$, $x_e(n)$ is the position of the VE and $x(n)$ is the position of the HIP inside the VE. The feedforward force offset is updated only at the beginning of every in-contact interaction cycle, depending on whether the displayed stiffness (k_d) is less or greater than the desired stiffness (k_{de}). Each in-contact interaction cycle constitute of one pressing and its subsequent releasing path.

$$\Delta O = \begin{cases} \Delta O + \alpha, & \text{for} \quad k_d \leq k_{de} \\ \Delta O - \alpha, & \text{for} \quad k_d > k_{de} \end{cases} \tag{3}$$

α is chosen such that the gradient at the beginning of each pressing path is equal to k_p rather than k_v.

$$\alpha = [x(n) - x_e(n) - x(n-1) + x_e(n-1)](k_p - k_v) \tag{4}$$

where, $x(n-1)$ and $x_e(n-1)$ are the last sampled values of position of HIP and VE, respectively.

To increase the rate-hardness, the feedback force is computed using k_{de} during the first pressing path of the initial contact of HIP with VE, Fig. 1a. Following this, the feedback force is calculated using (2) with a finite non-zero ΔO, Fig. 1b.

3 Successive Force Augmentation Approach with TDPA

Rate-hardness is closely correlated to perceptual hardness of the VE and is given by $H_R = \frac{\dot{f}(n)}{\dot{x}(n)}$ [18], where, $\dot{f}(n)$ and $\dot{x}(n)$ are the rate of change of force and velocity, respectively, during the first contact with the VE. As explained in Sect. 2, the SFA enhances rate-hardness by following k_{de} only during the first pressing path and k_v with a finite non-zero ΔO thereafter. This is represented by the schematic of Fig. 1, where it can be seen that the generated energy after the end of the first interaction cycle might push back the user if it is not dissipated by the physical damping of the haptic device. A thorough user study was conducted, which showed that this initial high stiffness force computation was enough for the user to accurately discriminate between four VEs of different stiffness.

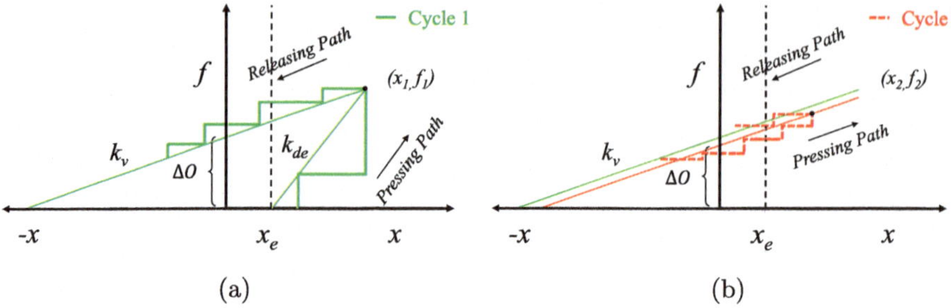

Fig. 1. Schematic diagram showing the increase of rate-hardness using SFA approach (a) First contact cycle, (b) Second in-contact cycle

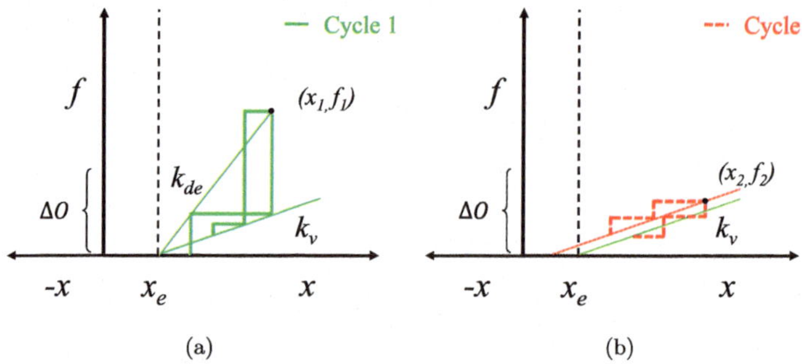

Fig. 2. Schematic diagram showing the increase of rate-hardness without generating as much energy (a) First contact cycle, (b) Second in-contact cycle

Therefore, to enrich the perception of hardness while generating less energy, we propose that the force rendering follows the desired high stiffness of the VE, k_{de}, during the initial contact's pressing path, and during the initial releasing path, the force is rendered based on k_v with $\Delta O = 0$. This does not produce a continuous high force to push the user out of the VE when k_{de} is high. Thereafter, from the second in-contact interaction cycle, the feedback forces are computed based on (2), as depicted in Fig. 2. This modification of the SFA helped retain the rate-hardness displaying capabilities during the transient response without generating as much energy. The algorithm for the proposed modified SFA is as follows:

```
if In Contact then
    if Pressing Path then
        if Initial Pressing Path then
            f_sfa(n) = k_de(x(n) − x_e(n))
        else
            if First Pressing Sample then
                α = [x(n) − x_e(n) − x(n − 1) + x_e(n − 1)](k_p − k_v)
                ΔO = { ΔO + α,  for k_d ≤ k_de
                     { ΔO − α,  for k_d > k_de
            f_sfa(n) = k_v(x(n) − x_e(n)) + ΔO
    else
        f_sfa(n) = k_v(x(n) − x_e(n)) + ΔO
else
    f_sfa(n) = 0
```

However, this modification only reduces the generated energy which might still destabilize the haptic interface, in case of high k_{de}.

To tackle this issue, a 1-port TDPA is implemented between the SFA controller and the haptic device (see Fig. 3), which makes sure that the 1-port haptic interface is passive and therefore the system remains stable. The power flow direction is defined by the discrete-time conjugate variables, and its summation results in energy. The passivity observer (PO) calculates the generated energy:

$$E_{obsv}(n) = \Delta T \sum_{j=0}^{n} f_{sfa}(j)\dot{x}(j) + \beta(j-1)\dot{x}(j-1)^2 \tag{5}$$

where, $f_{sfa}(n)$ is the force from the SFA controller, and $\dot{x}(n)$ is the velocity of the haptic device. If $E_{obsv}(n) \geq 0$ for every n, then the system dissipates energy. If at any instance $E_{obsv}(n) < 0$, then the system generates energy and the amount of generated energy is $-E_{obsv}(n)$. This makes the 1-port active which in turn may contribute to instability. The passivity controller (PC) takes the form of a time-varying dissipative element, $\beta(n)$, in series configuration due to impedance causality, to dissipate only the required amount of energy.

$$\beta(n) = \begin{cases} \frac{-E_{obsv}(n)}{\Delta T \dot{x}(n)^2}, & \text{if } E_{obsv}(n) < 0 \\ 0, & \text{if } E_{obsv}(n) \geq 0 \end{cases} \tag{6}$$

$$f_{pc}(n) = f_{sfa}(n) + \beta(n)\dot{x}(n) \tag{7}$$

where, $f_{pc}(n)$ is the force output from the PC.

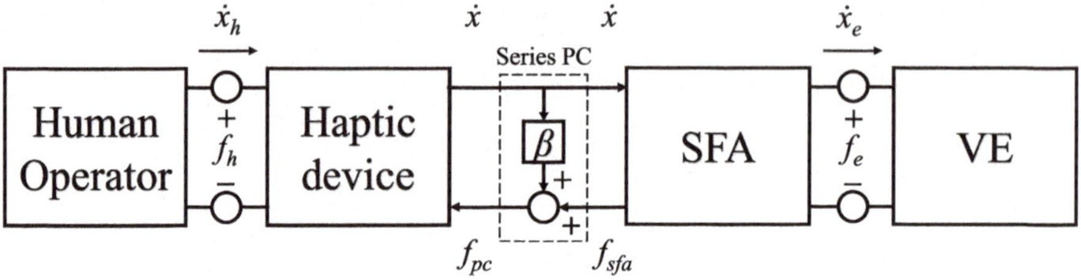

Fig. 3. Block diagram of the modified SFA with TDPA controller to improve the impedance range while guaranteeing stability.

4 Experimental Evaluation

4.1 Experimental Setup

To validate the stability and transparency of the proposed approach, it was implemented on two different devices (see Fig. 4), (i) Phantom Premium 1.5, and (ii) Franka Emika Panda. An F/T sensor, ATI 9105 Axia80-M8, was used at the end-effector of Frank Emika Panda. VEs of different stiffness (k_{de}) were rendered on both the devices. The value of k_v was experimentally selected to an optimal value, as shown in Table 1.

Table 1. Stiffness parameters

Device name	$k_v(N/mm)$	$k_p(N/mm)$	$k_{de}(N/mm)$
Phantom premium 1.5	0.8	1	8
Franka Emika Panda	2.6	3	30

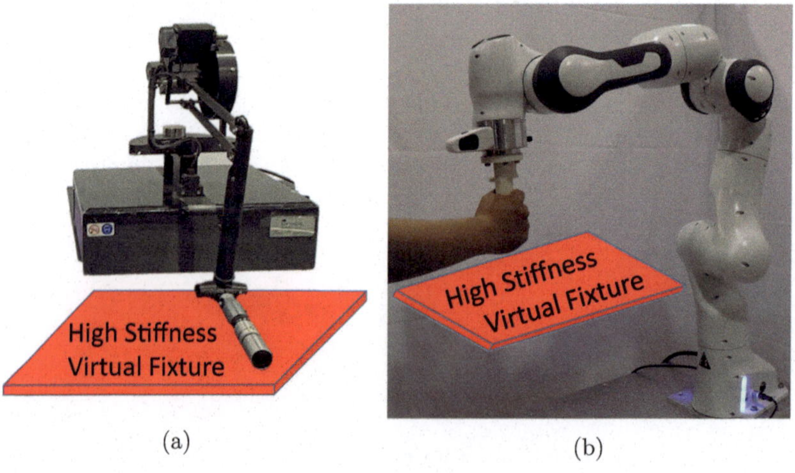

Fig. 4. Experimental setup (a) Phantom Premium 1.5, (b) Franka Emika Panda robotic arm, interacting with high stiffness virtual fixture.

4.2 High Stiffness Haptic Interaction

Figures 5 and 6 exhibit the overall results of implementing the modified SFA with TDPA on different haptic devices, where a human user interacts with a flat VF. It could be seen that for all the experiments, the interaction with the VF was stable, which could be observed from the position and force graphs. The spike in force at the beginning of each contact is due to the rate-hardness functionality of the SFA, which can be seen from the displayed stiffness graphs, where k_d during the initial contact is equal to the desired stiffness of the VF. The displayed stiffness (k_d) reaches up to the desired stiffness (k_{de}) of the VF within a few in-contact interaction cycles. In the case of Panda, k_{de} was set to 10 times the value of k_p and yet it managed to display that high stiffness stably. TDPA assures that $E_{obsv}(n) \geq 0$ during the entirety of wall contact. At times when $E_{obsv}(n) < 0$, the PC was activated which dissipated the excess energy, thereby maintaining a passive system. It should be noted that in all the experiments, the PC got activated only when the human user was moving out of contact with the VF, and never while maintaining contact. The user did not report any unwanted push or force jumps, other than the force jittering due to the PC.

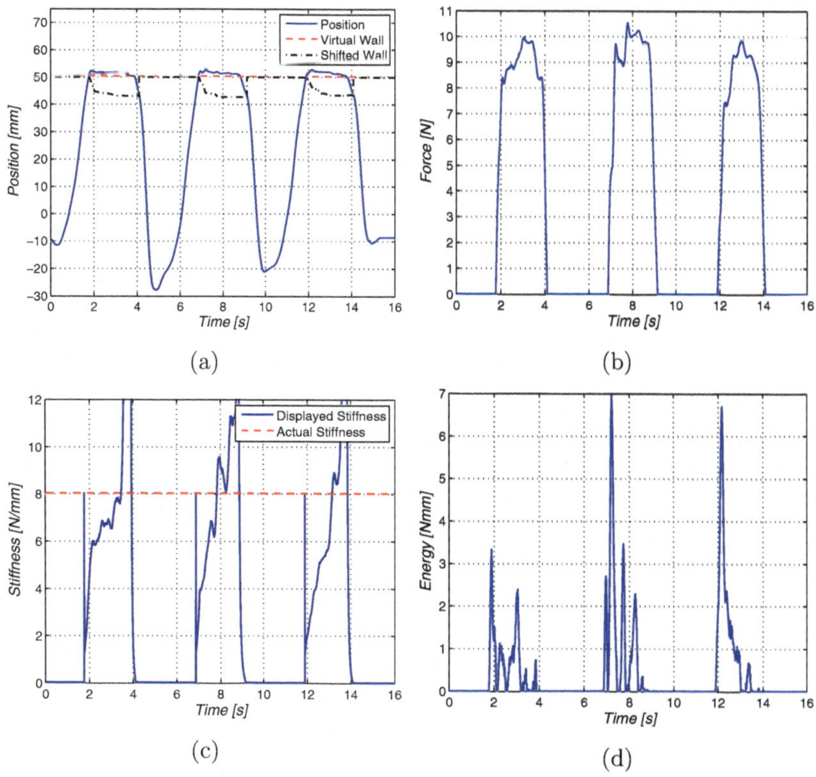

Fig. 5. Proposed SFA with TDPA controller implemented on Phantom Premium 1.5 with high stiffness VF, $k_{de} = 8\,\text{N/mm}$ (a) Position, (b) Force feedback, (c) Displayed stiffness, (d) Energy.

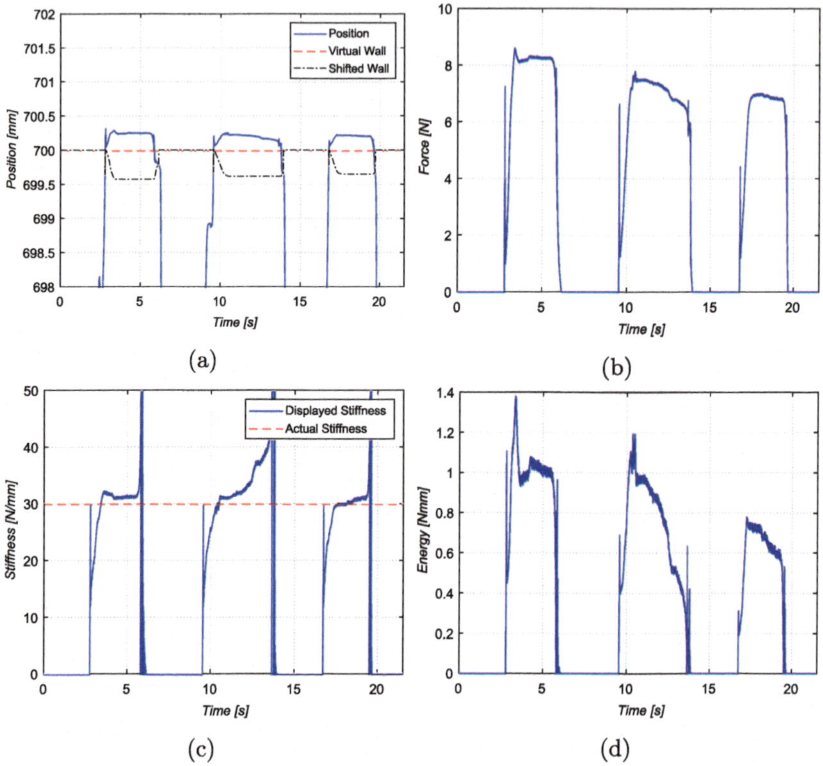

Fig. 6. Proposed SFA with TDPA controller implemented on Franka Emika Panda maintains stability even with high stiffness virtual fixture, $k_{de} = 30\,\text{N/mm}$ (a) Position, (b) Force feedback, (c) Displayed stiffness, (d) Energy.

4.3 Teleoperation with High Stiffness Virtual Fixtures

The Phantom premium (haptic device) and Panda (remote robot) were used to establish a position-force measured teleoperation architecture. A cube with a square hole cut through it, was used as a local VF on the haptic device (HD) side. The proposed SFA with TDPA was used to render its high stiffness VF, $k_{de} = 5\,\text{N/mm}$. For this experiment, the human user moved the HD to make contact with the VF and then moved the HD to make contact with the remote environment via the remote robot (RR). Please note that when the HD made contact with the VF, the RR was not in contact with its environment, and vice versa.

It could be observed from the results in Fig. 8 that the displayed stiffness was higher when making contact with the VF, compared to when making contact with the remote environment. This goes on to show that the user can discriminate between a VF and a real object, when the stiffness of the VF is set at a higher value. This could be a useful feature when performing complex manipulation tasks, where the user should know when he/she is interacting and operating

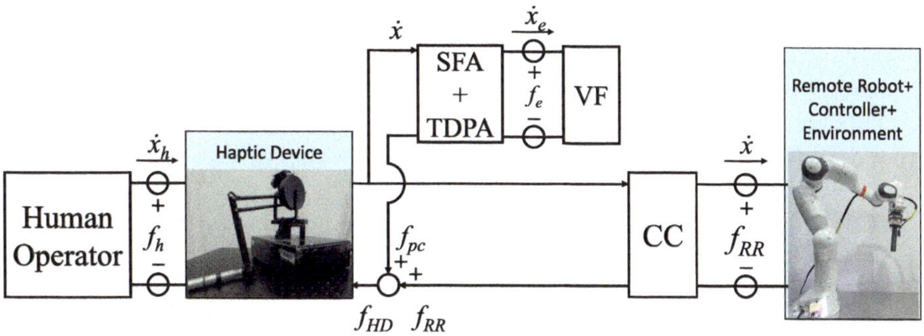

Fig. 7. Position-force measured teleoperation architecture with local VF on the HD side rendered using the modified SFA with TDPA.

with a real environment and when he/she has touched the boundary limitation described by the VF.

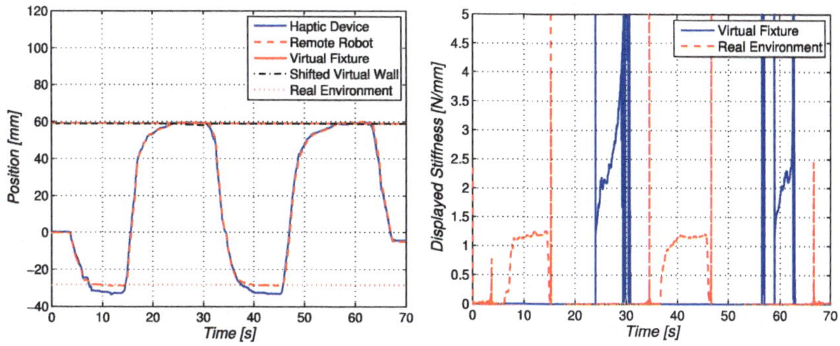

(a) (b)

Fig. 8. Position-force measured teleoperation architecture with local VF of stiffness 5 N/mm at the HD side (a) Position, (b) Displayed stiffness.

5 Conclusion

This paper proposes a modification to the SFA approach, which limits the generated energy and prevents a push back to the human user during contact with a high stiffness VE. Furthermore, to passivate the haptic interface, a 1-port TDPA is implemented in-between the haptic device and SFA controller. In order to validate the amalgamated approach, high stiffness haptic interaction experiments were carried out on a Phantom premium 1.5 and a Franka Emika Panda, with VE stiffness of 8 N/mm (8 times the critical stiffness) and 30 N/mm (10 times the critical stiffness), respectively. The experimental evaluation showed that the proposed control scheme is capable of displaying a wide range of stiffnesses in a stable manner. Another experiment conducted showed that the human user could discriminate a VF from a real environment, rather than being confused by it. This is because the controller allows the VF to have a much higher stiffness compared to the stiffness information transmitted through teleoperation, which is perceived through a F/T sensor. Therefore, the experiments demonstrate that the proposed control scheme can increase the impedance range and rate-hardness perception without sacrificing stability, which can assist the haptic information to be conveyed more comprehensively and in an informative manner.

Future work will include conducting a thorough user study to highlight the advantages of the proposed controller over the state-of-the-art. Additionally, a virtual mass spring passive filter will be considered to filter out the high frequency force jittering resulting from the impedance-type PC.

References

1. Colgate, J.E., Stanley, M.C., Brown, J.M.: Issues in the haptic display of tool use. In: Proceedings. 1995 IEEE/RSJ International Conference on Intelligent Robots and Systems 95.'Human Robot Interaction and Cooperative Robots, vol. 3, pp. 140–145. IEEE (1995)
2. Diolaiti, N., Niemeyer, G., Barbagli, F., Salisbury, J.K.: Stability of haptic rendering: discretization, quantization, time delay, and coulomb effects. IEEE Trans Robot. **22**(2), 256–268 (2006)
3. Hager, G.D., Okamura, A.M., Kazanzides, P., Whitcomb, L.L., Fichtinger, G., Taylor, R.H.: Surgical and interventional robotics: part iii [tutorial]. IEEE Robot. Autom. Mag. **15**(4), 84–93 (2008)
4. Cacace, J., Finzi, A., Lippiello, V.: A mixed-initiative control system for an aerial service vehicle supported by force feedback. In: 2014 IEEE/RSJ International Conference on Intelligent Robots and Systems, pp. 1230–1235. IEEE (2014)
5. Abbott, J.J., Marayong, P., Okamura, A.M.: Haptic virtual fixtures for robot-assisted manipulation. In: Robotics Research, pp. 49–64. Springer (2007)
6. Hannaford, B., Ryu, J.-H.: Time-domain passivity control of haptic interfaces. Robot. Autom. IEEE Trans. **18**(1), 1–10 (2002)
7. Niemeyer, G., Slotine, J.-J.: Stable adaptive teleoperation. IEEE J. Oceanic Eng. **16**(1), 152–162 (1991)
8. Anderson, R.J., Spong, M.W.: Bilateral control of teleoperators with time delay. IEEE Trans. Autom. control **34**(5), 494–501 (1989)
9. Kim, J.-P., Baek, S.-Y., Ryu, J.: A force bounding approach for multi-degree-of-freedom haptic interaction. Mech. IEEE/ASME Trans. **20**(3), 1193–1203 (2015)
10. Franken, M., Stramigioli, S., Misra, S., Secchi, C., Macchelli, A.: Bilateral telemanipulation with time delays: a two-layer approach combining passivity and transparency. IEEE Trans. Robot. **27**(4), 741–756 (2011)
11. Vulliez, M., Zeghloul, S., Khatib, O.: Design strategy and issues of the delthaptic, a new 6-DOF parallel haptic device. Mech. Mach. Theory **128**, 395–411 (2018)
12. Najmaei, N., Asadian, A., Kermani, M.R., Patel, R.V.: Design and performance evaluation of a prototype MRF-based haptic interface for medical applications. IEEE/ASME Trans. Mech. **21**(1), 110–121 (2016)

13. Jafari, A., Ryu, J.H.: Input-to-state stable approach to release the conservatism of passivity-based stable haptic interaction. In: 2015 IEEE International Conference on Robotics and Automation (ICRA), pp. 285–290, May 2015
14. Desai, I., Gupta, A., Chakraborty, D.: Rendering stiff virtual walls using model matching based haptic controller. IEEE Trans. Haptics **12**(2), 166–178 (2019)
15. Mitra, S.K.: Analysis and Synthesis of Linear Active Networks. John Wiley & Sons, Cambridge (1969)
16. Llewellyn, F.: Some fundamental properties of transmission systems. Proc. IRE **40**(3), 271–283 (1952)
17. van Beek, F.E., Heck, D.J., Nijmeijer, H., Bergmann Tiest, W.M., Kappers, A.M.: The effect of damping on the perception of hardness. In: World Haptics Conference (WHC), 2015 IEEE, pp. 82–87. IEEE (2015)
18. Lawrence, D., Pao, L.Y., Dougherty, A.M., Salada, M., Pavlou, Y., et al.: Rate-hardness: a new performance metric for haptic interfaces. Robot. Autom. IEEE Trans. **16**(4), 357–371 (2000)
19. Singh, H., Janetzko, D., Jafari, A., Weber, B., Lee, C.-I., Ryu, J.-H.: Enhancing the rate-hardness of haptic interaction: successive force augmentation approach. IEEE Trans. Ind. Electron. **67**(1), 809–819 (2019)

Machine Learning

Tactile-Based Self-supervised Pose Estimation for Robust Grasping

Padmaja Kulkarni(✉), Jens Kober, and Robert Babuska

Department of Cognitive Robotics, TU Delft, Delft, The Netherlands
{P.V.Kulkarni,J.Kober,R.Babuska}@tudelft.nl

Abstract. We consider the problem of estimating an object's pose in the absence of visual feedback after contact with robotic fingers during grasping has been made. Information about the object's pose facilitates precise placement of the object after a successful grasp. If the grasp fails, then knowing the pose of the object after the grasping attempt is made can also help re-grasp the object. We develop a data-driven approach using tactile data that computes the object pose in a self-supervised manner after the object-finger contact is established. Additionally, we evaluate the effects of various feature representations, machine learning algorithms, and object properties on the pose estimation accuracy. Unlike other existing approaches, our method does not require any prior knowledge about the object and does not make any assumptions about grasp stability. In experiments, we show that our approach can estimate object poses with at least 2 cm translational and 20° rotational accuracy despite changed object properties and unsuccessful grasps.

1 Introduction

Touch is a prime human sense, which enables humans to grasp in dynamic environments with low lighting, occlusions, and shadows. Similarly, for robot hands, the use of tactile information has proven to be essential for grasping [1–4]. In prior works, tactile information was used to estimate object properties, object-classes, and grasp stability [1–4]. This interpretation aids in the determination of, for example, grasping strategy or contact-force computation. However, it adds little insight into the pose of an object after robot-object interaction happens.

Not being able to have reliable object pose detection after the robot-fingers are in contact with the object is one of the major reasons for the still moderate grasping performance of robot hands [5–7]. When fingers make contact with an object, the force applied by them can cause the object's location to change. Furthermore, it may no longer be possible to use a camera to monitor the object's pose in hand due to occlusion. Knowing this pose helps to place the object precisely after picking it up or to re-grasp if the grasp fails. Thus having a method for reliable in-hand object pose estimate is essential.

B. Siciliano et al. (Eds.): ISER 2020, SPAR 19, pp. 277–284, 2021.
https://doi.org/10.1007/978-3-030-71151-1_25

In this paper, we address the problem of estimating object pose, especially when visual feedback is not possible *during or after the grasp.* We develop a data-driven approach based on tactile data to estimate an object's pose (position and orientation) relative to the robotic gripper.

Tactile sensors provide only local information about the object in contact, whereas vision sensors provide extensive environmental information. Hence, to simplify the problem of object pose estimation with tactile sensors, either assumptions like grasp stability or known object properties (e.g., mesh-model, geometry) are made, or the object is fixed to restrict its motion under gripper forces. For example, Bimbo et al. [8] estimated the object's pose in a robotic gripper using tactile and force data assuming the object's model or geometrical information is accurately known. They minimized the difference of angles between the normals on the object's surface obtained from the geometric model and the observed force-normals to estimate the object's pose using evolutionary algorithms. Corcoran et al. [9] used a particle filter to estimate object pose in the robot gripper. The authors' model computed the likelihood of true contact measurement over possible contact positions, assuming a known object model. The approaches in [10, 11] used *haptic exploration* for object pose prediction with tactile data for a fixed object and assumed the object model to be known. The authors of [5] computed the object's pose with data-driven techniques, using an under-actuated robotic gripper and assuming a stable grasp.

In contrast to these approaches, our paper focuses on directly estimating the object pose based on tactile data without making any assumptions about the object model or grasp stability, and without fixing the object. Furthermore, in the training phase, we use a camera to estimate the ground truth object pose, and thereby generate the target object pose in a self-supervised manner. Note that, for the trained model, the camera is only used to estimate the initial object-pose and *it is not used thereafter during or after the grasp attempt.* The initial object pose facilitates the estimation of changes in the object pose, especially for the symmetric objects, as tactile sensing would not discriminate if the object were rotated by, for example, 180°.

This paper's contribution is two-fold: i) a novel approach for computing Tactile-based Self-supervised Pose Estimation (T-SPoE), and ii) evaluation of the effect of various feature representations, machine learning algorithms, and changing object properties on the accuracy of the pose estimation.

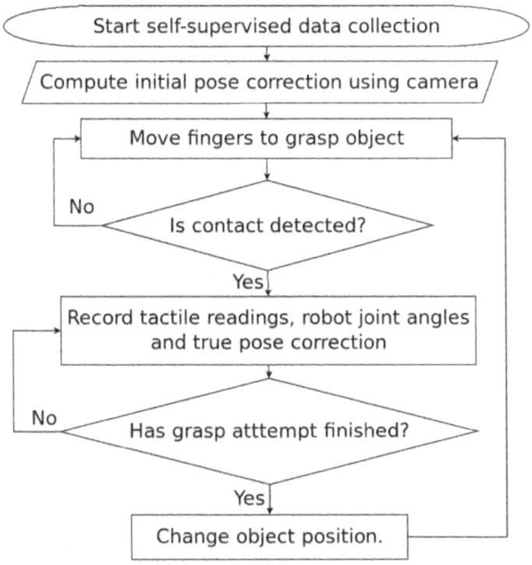

Fig. 1. Data collection with T-SPoE method. Data is recorded continuously during each grasp attempt, and the labels are generated in a self-supervised manner.

2 Tactile-Based Self-supervised Pose Estimation

T-SPoE estimates the object pose based on tactile data. It consists of the following three phases:

1. **Data Collection**
   Figure 1 shows our data collection approach. We use a gripper equipped with tactile sensors and design an experimental setup allowing object grasping. We use a camera to estimate the target object pose in a self-supervised manner. The gripper is in the open position and tries to grasp the object until it is successful (shown in Fig. 2) and then places the object at the same place. During this phase, we record true object pose, tactile data, and finger joint angles at each time step. For recording the true object pose, we mount Fiducial markers on the object and use the ArUco library [12] to compute the marker pose.
2. **Model Training**
   In this phase, based on the collected data, we train models to estimate the pose using machine learning. A different model is trained for every object. The tactile data is preprocessed to obtain data-representations explained in Sect. 3. The input to the model is the preprocessed tactile data, the finger joint angles, and the object's pose before grasping (*initial object pose*). The output of the model is the 6D object pose relative to the gripper.
3. **Pose Estimate**
   In this phase, we process the current tactile data frame to obtain data-representations, as explained in Sect. 3. We then use the processed tactile data frame as input to the trained model to predict the object pose.

Fig. 2. Setup for collecting data. Two fingers of the SDH hand are used to grasp an object. The Asus Xtion PRO camera tracks the marker positions continuously during grasp attempts.

3 Feature Representations

To train the model using machine learning, we use tactile data, finger joint angles, and initial object pose as input features. The initial object pose is a 6D vector representing the initial object pose with respect to the hand, and it is estimated using a camera. The object pose consists of the 3D Cartesian position and three roll, pitch and yaw Euler angles with XYZ version. We use two fingers of the Schunk Dexterous Hand (SDH),[1] which have in total four degrees of freedom. Along with this 10D vector, we use the tactile data from two fingertips of the SDH hand. Each fingertip is equipped with a Weiss Tactile pad with 13×6, i.e., 78 taxels. These total of 156 taxels output *pressure* at the fingertips, which we use as our tactile input.

The tactile feature length of 156 is large and could be potentially redundant. To reduce the dimensionality, along with raw tactile-values, we use the following feature representations and train three different machine learning models per technique:

1. **Tactile Data as a List**
   Here, we use tactile data as a list, where one data point has a dimension of 156×1, and the input feature vector has a total dimension of 166.
2. **Max-Pooled Tactile Data**
   We use this operation to reduce the dimensionality of tactile data, with a rectangular filter of size 1×6 and stride 1. This entails that we take the maximum tactile value from each taxel row. As there are 13 rows, we reduce the dimensions to 13×1 for one finger. The input tactile feature vector, in this case, has a total dimension of $13 \times 2 = 26$.
3. **Principal Component Analysis (PCA)**
   Here, we take co-variance (C) of N tactile data vectors (T) [13]: $C = \sigma(T_{1 \ldots N})$.

[1] https://schunk.com/nl_en/gripping-systems/series/sdh/.

The most significant feature vectors, which explain 80% of data, are computed by computing the largest eigenvalues (λ) as: $v^T C v = \lambda$, with v as the eigenvector of matrix C. In our case, PCA was able to reduce the dimensions to 20×1.

4 Experiments

A setup for data collection is shown in Fig. 2. The Asus Xtion PRO camera is used to track the object's relative position with respect to the SDH hand.

1. **Self-supervised Data Collection**
   We used two objects, a Nescafe box and a ceramic cup for data collection. The object is kept between the hand's open fingers such that object-finger interaction or grasping is possible. The orientation of the object is allowed to change by 360° about the vertical axis. The SDH grasps the object by closing the fingers until either the desired force value is reached or the finger joint-limits are reached. *A grasp is successful when the gripper grasps an object and fails when the object slips out of the gripper.* The objects used for the experimentation are listed in Table 2. For training, the Nescafe box and cup is used. We record the data for 40 successful and 40 unsuccessful grasps for each object. After the training data-collection is complete, the Nescafe box and the cup are enveloped in 1 cm thick foam. Enveloping an object with foam changes the object's dimensions and surface properties, e.g., friction coefficient. The aim of experimenting with the objects covered with foam is to evaluate if the method is robust to such property changes. *The foam-covered objects are only used for evaluation purposes and not during training.*
2. **Model Training**
   For model training, we use four machine learning methods; namely, i) Neural Networks (NN), ii) Support Vector Regression (SVR), iii) K-Nearest Neighbours (K-NN), and iv) Random Forest (RF). The parameters used for these learning algorithms are listed in Table 1. These parameters are decided by trial and error using the highest model score as a performance indicator across 5-fold cross-validation. The parameter details are available in Scikit-learn library[2].

Table 1. Parameters used for the evaluation of various machine learning approaches.

Algorithms	Parameters
NN	Hidden layer sizes: (100, 75, 50, 25, 20, 10)
SVR	C: 1.0, epsilon: 0.2
KNN	Nearest neighbours: 3
RF	Max. depth: 6

[2] https://scikit-learn.org/stable/.

Table 2. Objects and their sizes used for the experimentation with SDH hand.

Objects	Sizes in cm
Box	13 × 13 × 13 (h × w × l)
Cup	8 × 10 (d × h)
Box with foam layer	13 × 15 × 15 (h × w × l)
Cup with foam layer	10 × 10 (d × h)

Table 3. RMSE in pose prediction. Translation (T-error) is in centimeters, and Rotation (R-error) is in degrees.

Features	Type	NN		SVR		KNN		RF	
		T-error	R-error	T-error	R-error	T-error	R-error	T-error	R-error
Tactile data as a list	Box grasped	0.46	1.01	0.82	0.27	0.15	0.8	0.39	1.4
	Cup grasped	1.14	9.4	0.76	0.42	0.8	5.1	0.65	8.7
	Box grasp failed	1.3	6.89	1.09	0.36	0.55	8.9	0.94	4.4
	Cup grasp failed	2.2	9.7	3.03	1.01	2.07	11.06	1.91	8.76
Max pooled tactile data	Box grasped	0.79	1.9	0.8	0.26	0.66	2.6	0.59	1.60
	Cup grasped	1.1	9.6	0.85	0.28	0.46	5.2	0.36	3.5
	Box grasp failed	1.3	6.2	0.97	0.32	0.6	9.5	0.8	4.2
	Cup grasp failed	2.1	9.4	2.7	0.92	1.9	9.6	1.7	8.2
PCA compressed tactile data	Box grasped	2.2	1.8	0.62	0.21	0.34	2.1	0.9	2.7
	Cup grasped	1.1	8.2	0.7 1	2.5	0.79	5.6	0.70	8.5
	Box grasp failed	1.7	3.1	0.74	2.4	0.6	13.34	0.69	1.69
	Cup grasp failed	2.2	8.7	2.3	0.78	1.9	10.2	1.07	8.16

5 Results and Discussion

The pose is predicted using the data collected in Phase 1 and models from Phase 2 of T-SPoE. These results are divided into two grasp modes, depending upon whether the SDH hand grasped the object. According to feature representations and algorithms, the results are listed in Table 3. We use Root Mean Square Error (RMSE) to measure the prediction accuracy. For translational RMSE, RMS (Root Mean Square) distance between the predicted and the target position is computed. For rotational error, RMS error between the predicted and the target Euler angles is calculated. Here, we could use RMS distance between angles in Euler space because we vary the object orientation only about the vertical axis.

The average RMSE of each feature representation does not significantly vary. Hence, we use raw tactile-values in the list form as features. The lowest average RMSE for each feature representation is obtained by using Random Forest Regression. Therefore we use RF for the training phase for T-SPoE. The results obtained by changing the object surface properties and deformability are listed in Table 4. *Note that when the size and deformability of an object are varied, no additional training is performed.*

We calculated the error between the initial (pre-grasp) object pose and the actual object pose after gripper-object interaction to justify our method's usability. This error indicates how much the object has shifted due to the robot-object interaction. In the case of the box in the three scenarios—box grasped, box grasp failed, and box with foam—this average RMSE was found to be 13.64 cm in translation and 21.67° in rotation. For the cup, this error was 12.80 cm and 39.43° in translation and rotation, respectively. Thus this pose-shift is significant, and estimating the object pose after contact with robotic fingers during grasping has been made is necessary for successful manipulation.

In summary, T-SPoE can estimate pose relative to the gripper with RMSE less than 0.5 cm in translation and 3.5° in the rotation when the object is successfully grasped. When the gripper is unable to grasp an object, RMSE increases to 2 cm in translation and 8° in rotation. Moreover, for object size and property changes, T-SPoE was able to maintain the translational RMSE up to 1.9 cm. However, the rotational RMSE increases to 20°. This can be attributed to the fact that the finger-object contact dynamics change significantly for deformable objects. However, this error in pose prediction is less than the error that would occur without using T-SPoE if the object is assumed to be at the initial pre-grasp pose. We conclude that training with varying object-properties is required to make the method robust against the object-property variations and to reduce the error further.

Table 4. RMSE in prediction of the pose with variations in object properties using T-SPoE. Random Forest Algorithm is chosen, as it has the lowest average RMSE with known objects.

Objects	RF	
	T-error (cm)	R-error (deg)
Box with foam	1.90	20.1
Cup with foam	1.74	10.11

6 Conclusion and Future Work

We proposed a self-supervised method T-SPoE for robust grasping tasks using tactile feedback. We evaluated self-supervised object pose prediction using data collected from a fixed SDH hand with two known objects. Our approach does not require any prior object knowledge and does not make any assumptions about the grasp stability. Additionally, we evaluate the effect of various feature representations, machine learning algorithms, and changing object properties on the proposed approach. With the known objects, our method predicted the object pose with RMSE less than 0.5 cm in translation and 3.5° in the rotation when the object is successfully grasped. With changed object properties, this error increases to 1.9 cm and 20° in translation and rotation, respectively. The next step to reduce this error would be to use more training data with changing object properties to make T-SPoE robust against object property variations.

Furthermore, in the future, we plan to evaluate the complete manipulation pick-and-place pipeline with more day-to-day objects of irregular natural shapes, like lettuces, tomatoes, and chicken pieces.

Acknowledgment. This research is funded by the Netherlands Organization for Scientific Research project Cognitive Robots for Flexible Agro-Food Technology, grant P17-01.

References

1. Calandra, R., Owens, A., Upadhyaya, M., Yuan, W., Lin, J., Adelson, E.H., Levine, S.: The feeling of success: does touch sensing help predict grasp outcomes? IEEE Robot. Autom. Lett. (RA-L) **3**(4), 3300–3307 (2017)
2. Liarokapis, M.V., Calli, B., Spiers, A.J., Dollar, A.M.: Unplanned, model-free, single grasp object classification with underactuated hands and force sensors. In: IEEE International Conference on Intelligent Robots and Systems (IROS), vol. 2015, pp. 5073–5080. Institute of Electrical and Electronics Engineers Inc. (2015)
3. Kwiatkowski, J., Lavertu, J.S., Gourrat, C., Duchaine, V.: Determining object properties from tactile events during grasp failure. In: IEEE International Conference on Automation Science and Engineering (CASE), vol. 2019, pp. 1692–1698 (2019)
4. Chebotar, Y., Hausman, K., Su, Z., Sukhatme, G.S., Schaal, S.: Self-supervised regrasping using spatio-temporal tactile features and reinforcement learning. In: IEEE International Conference on Intelligent Robots and Systems (IROS), pp. 1960–1966 (2016)
5. Liarokapis, M., Dollar, A.M.: Learning the post-contact reconfiguration of the hand object system for adaptive grasping mechanisms. In: IEEE International Conference on Intelligent Robots and Systems (IROS), pp. 293–299 (2017)
6. Bütepage, J., Cruciani, S., Kokic, M., Welle, M., Kragic, D.: From visual understanding to complex object manipulation. Ann. Rev. Control Robot. Auton. Syst. **2**(1), 161–179 (2019)
7. Okamura, A., Smaby, N., Cutkosky, M.: An overview of dexterous manipulation. In: IEEE International Conference on Robotics and Automation (ICRA), vol. 1, pp. 255–262. IEEE (2000)
8. Bimbo, J., Kormushev, P., Althoefer, K., Liu, H.: Global estimation of an objects pose using tactile sensing. Adv. Robot. **29**(5), 363–374 (2015)
9. Corcoran, C., Platt, R.: A measurement model for tracking hand-object state during dexterous manipulation. In: Proceedings of the IEEE International Conference on Robotics and Automation (ICRA), pp. 4302–4308 (2010)
10. Klaas, G., Bruyninckx, H.: Markov techniques for object localization with force-controlled robots. In: International Conference on Advanced Robotics (ICAR), pp. 91–96 (2001)
11. Schaeffer, M.A., Okamura, A.M.: Methods for intelligent localization and mapping during haptic exploration. In: Proceedings of the IEEE International Conference on Systems, Man and Cybernetics (SMC), vol. 4, pp. 3438–3445 (2003)
12. Romero-Ramirez, F.J., Muñoz-Salinas, R., Medina-Carnicer, R.: Speeded up detection of squared fiducial markers. Image Vis. Comput. **76**, 38–47 (2018)
13. Laaksonen, J., Kyrki, V., Kragic, D.: Evaluation of feature representation and machine learning methods in grasp stability learning. In: IEEE-RAS International Conference on Humanoid Robots, Humanoids, pp. 112–117 (2010)

Learning Visual Servo Policies via Planner Cloning

Ulrich Viereck[1(✉)], Kate Saenko[2], and Robert Platt[1]

[1] Khoury College of Computer Sciences, Northeastern University, Boston, USA
{uliv,rplatt}@ccs.neu.edu

[2] Department of Computer Science, Boston University, Boston, USA
saenko@bu.edu

Abstract. Learning control policies for visual servoing in novel environments is an important problem. However, standard model-free policy learning methods are slow to learn. This paper explores planner cloning: using behavior cloning to learn policies that mimic the behavior of a full-state motion planner in simulation. We propose Penalized Q Cloning (PQC), a new behavior cloning algorithm. We show that it outperforms several baselines and ablations on some challenging problems involving visual servoing in novel environments while avoiding obstacles. Finally, we demonstrate that these policies can be transferred effectively onto a real robotic platform, achieving approximately an 87% success rate both in simulation and on a real robot.

1 Motivation, Related Work, Problem Statement

Visual servoing in novel environments is an important problem. Whereas classical approaches to the problem [1,3,17] make strong assumptions about the environment, deep learning methods solve these problems in more unstructured settings [5,6,8–10,14,16,18,19]. Standard model-free reinforcement learning methods such as DQN [11], do poorly because they treat this as an unstructured model-free problem. Most current approaches to visual servoing in unstructured domains involve some form of imitation learning or behavior cloning [5,10,14,16,18,19]. However, there are substantial differences between these methods and their relative performance is unclear.

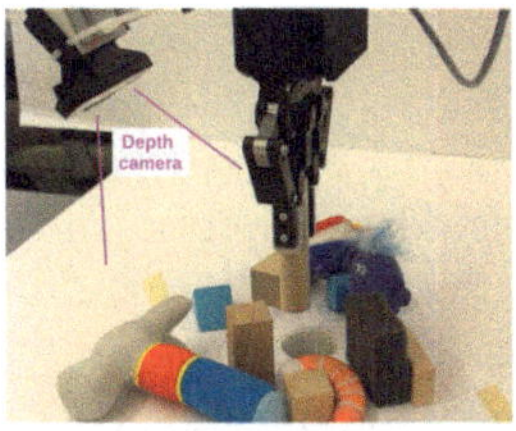

Fig. 1. One of the visual servoing test scenarios: Inserting a peg into a hole with collision avoidance.

This work focuses on a class of imitation learning approaches that we call *planner cloning*, where the expert is an approximately optimal motion planner that generates estimates of the optimal value function and optimal policy given full state feedback. This information is cloned onto a value function or policy over the observation space (i.e. camera images) that can be used at test time where full state feedback is unavailable.

B. Siciliano et al. (Eds.): ISER 2020, SPAR 19, pp. 285–295, 2021.
https://doi.org/10.1007/978-3-030-71151-1_26

The main contribution of this work is a new behavior cloning algorithm that we call *Penalized Q Cloning* (PQC). It has the following characteristics:

- It learns a value function using supervised value targets provided by the planner, similar to AGGREVATE [12].
- It incorporates a penalty into the value targets for suboptimal actions, similar to DQfD [2].
- For every experience, it updates the values for all feasible actions from that state, not just the action experienced.

Algorithm 1 Batch Penalized Q Cloning (Batch PQC)

$\mathbb{D} \leftarrow \varnothing$
while more episodes to execute **do**
  $s_0 \sim \rho_0$
  for $t \in [0, T-1]$ **do** ▷ iterate over time steps
    $a_t = \pi_E(s_t)$
    for $\forall a \in A$ **do**
      if $a = a_t$ **then**
        $q \leftarrow Q_E(s_t, a)$
      else
        $q \leftarrow c$ ▷ apply penalty
      $\mathbb{D} \leftarrow \mathbb{D} \cup \{(s, a, q)\}$
    $s_{t+1} \leftarrow T(s_t, a_t)$
Find Q that minimizes $\mathcal{L}_{PQC}$
Return Q

This algorithm differs from AGGREVATE because it incorporates the value penalties and from DQfD because it uses supervised targets rather than TD targets. We compare PQC with several baselines and algorithm ablations and show that it outperforms all these variations on two challenging visual servoing problems in novel cluttered environments (i.e. Fig. 1). The resulting policies achieve high success rates (approximately 87%) on challenging visual servo tasks.

Finally, we make two additional minor contributions:

1. We show that an approximately optimal motion planning algorithm can be used to produce a near optimal value function that can be used for behavior cloning.
2. We show that, in the visual servo setting at least, it is often advantageous to "finetune" the cloned policy using TD learning as a separate postprocessing step.

Visual Servo Problem: We assume we are given a discrete time system that includes a robotic end-effector, a camera, randomly placed clutter and random object in gripper and target configuration. At the beginning of each episode the agent must move an object grasped into a desired pose relative to the environment. The agent must find a policy $\pi : Z \mapsto A$ over the observation space that minimizes the expected time to reach a goal state for the given system.

Novel Environment Assumption: A key assumption is that each experience with the world will take place with a novel configuration of "clutter objects" (Fig. 1). At the beginning of each episode, several objects uniformly randomly sampled from a database of clutter objects are placed in random positions and orientations in the scene (Fig. 4). The task objective remains the same – to move the grasped object into a desired pose with respect to the environment – but the presence of clutter makes the problems of visual perception and obstacle avoidance more challenging.

Train/Test Information Asymmetry Assumption: We assume that the agent has access to a fully modeled MDP and a simulator during training. During training, the simulator produces state and simulated camera observations. At test time, we assume that the agent does *not* observe the full state. Instead, the agent only has access to camera observations $z \in Z$.

2 Technical Approach – Penalized Q Cloning (PQC)

In view of the train/test information asymmetry assumption, notice that there are really two relevant MDPs: 1) a fully modeled MDP over the underlying state space S; and 2) an unmodeled MDP over the observation space Z. We want to find a policy for the unmodeled MDP, but we can only plan in the modeled MDP. Our approach is therefore to use a full state motion planner to generate an "expert" policy and value function for the fully modeled MDP and then to project those solutions onto the unmodeled MDP using behavior cloning.

Full State Motion Planner as the Expert: First, we need to solve the fully modeled MDP for an expert value function and policy. In our particular case this corresponds to a collision free motion planning problem. We use Djikstra's algorithm over a regular 3D grid of end-effector positions. Edges in the graph correspond to Euclidean distances plus a penalty for approaching obstacles. Shortest paths to goals in the graph correspond to optimal solutions to the modeled MDP. Solutions found by the planner provide to the behavior cloner: an approximately optimal policy π_E and an approximately optimal value function Q_E. Figure 2 shows an example of a trajectory found using this method.

Cloning a Full State Motion Planner: We project these solutions generated by the planner onto the unmodeled MDP using behavior cloning. We do the following:

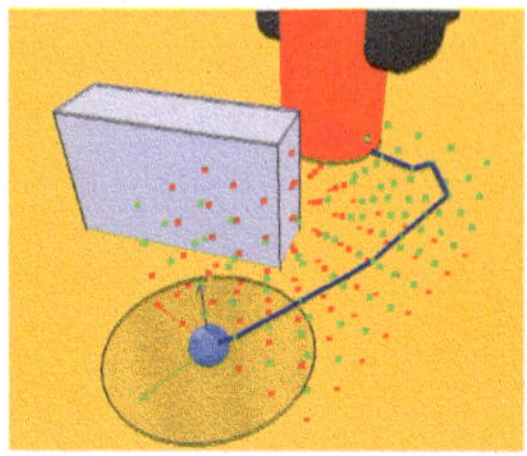

Fig. 2. Illustration of full state motion planning scenario.

(a) Sample Trajectories from Expert: First, we roll out the expert π_E repeatedly. For each action, we simulate the resulting next state and query the expert for an approximate q-value, $Q_E(s, a)$. We store the sequence of experiences $(s_1, a_1, q_1), \ldots, (s_n, a_n, q_n)$ in a dataset $\mathbb{D}$, where $q_i = Q_E(s_i, a_i)$.

(b) Augment Dataset with Supervision for all Actions: Because we have a small number of actions and the expert is implemented by a planner, we can generate supervision for all feasible actions from a given state, not just those executed by the simulator. Specifically, for each experience $(s, a, q) \in \mathbb{D}$, we generate $|A| - 1$ additional experiences $\{(s, a', Q_E(s, a'))|a' \in A \setminus a\}$ and add these to $\mathbb{D}$.

(c) Apply a Penalty to Non-expert Actions: Unlike an approach like DAGGER [13] which clones the policy directly, here we are cloning the value function. This exposes us to a key failure mode: we may learn a good estimate of Q_E while still estimating π_E poorly. Under ideal conditions, our estimate Q is exactly equal to Q_E: $Q(s, a) = Q_E(s, a), \ \forall s, a \in S \times A$. In this case, the greedy policy of the learned value function is equal to expert policy: $\text{argmax}_a Q(s,a) = \text{argmax}_a Q_E(s,a) = \pi_E(s)$. However, since we are using a deep neural network to approximate Q_E, we can expect small errors. This is a problem because even small errors can result in a substantial divergence between $\text{argmax}_a Q(s,a)$ and $\pi_E(s)$. To combat this, we set the action values of non-expert actions to a fixed value c.

We call this *batch penalized Q cloning* (batch PQC), see Algorithm 1. For more details see [15].

Relationship of this Cloning Method to Prior Work: This approach to behavior cloning draws elements from at least two different pieces of prior work. First, since we are cloning the value function rather than the policy, our method can be viewed as a form of AGGREVATE [12]. Second, the fact that we administer a penalty to non-expert actions is similar to what is done in DQfD [2] and ADET [7]. However, since both of those methods use TD learning, they must add an additional term into the loss function in order to achieve this. DQfD adds the relatively complex large margin loss term. ADET adds a cross entropy term between the policy implied by the value function and the expert policy. In contrast, since our method uses a fully supervised target, we can simply reduce the supervised target q value without the additional loss term. We experimentally compare our approach to AGGREVATE and DQfD in Sect. 3.

Algorithm Variations: We also consider a couple of dimensions of variation on the basic algorithm described above.

- Online PQC: Instead of training over a large batch of expert rollouts, *online* PQC uses a DAGGER-like rollout schedule, i.e. it alternates between rollouts from the expert policy and rollouts from the learned policy according to a schedule. Early in training, most rollouts come from the expert. Later, most rollouts come from the learned policy. Our experiments indicate that this method outperforms on the training set but underperforms in novel environments due to overfitting.
- Relative Penalty PQC: In this variant, we replace the constant penalty (line 10 of Algorithm 1) with a relative penalty: $q \leftarrow Q_E(S_t, a) - l$ where l is the margin by which to penalize the non-expert action. This version of the algorithm is more similar to DQfD [2]. Our experiments indicate that it outperforms on the training set but underperforms in novel environments due to overfitting.

Caching Visual Observations: One of the challenges of training in simulation is that it can be computationally expensive to simulate depth images of simulated worlds. In principle, each transition added to the dataset $\mathbb{D}$ requires simulating an image of a novel scene from a novel perspective. In order to speed this up this process, we precompute observations for a large set of clutter configurations. Specifically, we precompute the image that would be observed from each node in the motion planning graph. Then, when rolling out a trajectory during training, we simply recall the observation that corresponds to a given end-effector pose rather than recomputing it. This enables us to roll out as many trajectories as we want from a given scene. As long as this graph is not too large, this approach works well. Using this procedure, since we no longer create a new scene each time a new episode executes, we must select some number of scenes in advance for which to generate data. In our experiments, we simulate observations for at least 500 different scenes where each scene contains randomly selected clutter in a random task setting. In the experiments, we evaluate the effect of increasing the number of scenes used to create this training set (Fig. 6).

3 Experiments

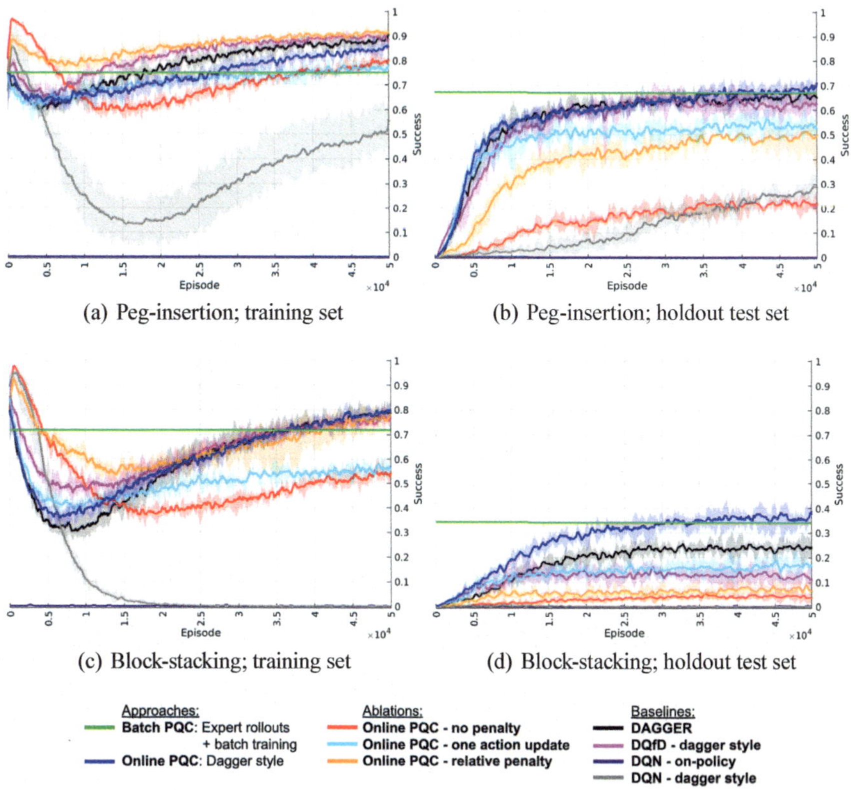

(a) Peg-insertion; training set
(b) Peg-insertion; holdout test set
(c) Block-stacking; training set
(d) Block-stacking; holdout test set

Fig. 3. Success rate as a function of training episode during cloning. (a,b) peg-insertion, (c,d) are block-stacking. (b,d) show success rates on the holdout test set. Results shown for training on 500 scenes. The shaded area indicates the minimum and maximum success rates of 5 experiments with random seeds. Note that PQC (blue) significantly outperforms on the block stacking task on a holdout set (d). Not shown: The test success rate for batch PQC (blue) increases to about 87%(b) and 67%(d) if pretrained on 4k scenes and finetuned using TD learning.

Experimental Setup: We evaluate in simulation PQC against several algorithm variations and baselines on two visual servoing tasks: a peg insertion task and a block stacking task (Figs. 4). In peg insertion, the robot starts execution with the peg in its hand and must move it until it reaches a goal pose just above the hole. The block stacking task is similar except that the robot starts with a block in its hand and must move it to a goal pose just above a second block. This is more challenging because it requires the policy to determine *which* block to stack upon.

In both tasks, the scene is populated by randomly placed clutter that serves both as visual complexity and as obstacles that must be avoided during servoing. In each scene of peg insertion, both peg and hole size are sampled uniformly randomly (peg smaller than hole) and the peg is displaced in the hand with a small random offset. Similarly, in each scene of block stacking, both block sizes are sampled uniformly randomly, the grasped block is offset by a random displacement, and a small uniform random rotation is applied to the block to be stacked upon. Figure 4 illustrates the depth images observed by the agent. Note that they include both the grasped object and the scene. The expert is implemented by a planner that uses Djikstra's algorithm over a $21 \times 21 \times 12$ grid of positions (fixed orientation) with vertices at 1cm intervals (a total of $5k$ vertices, see Fig. 2).

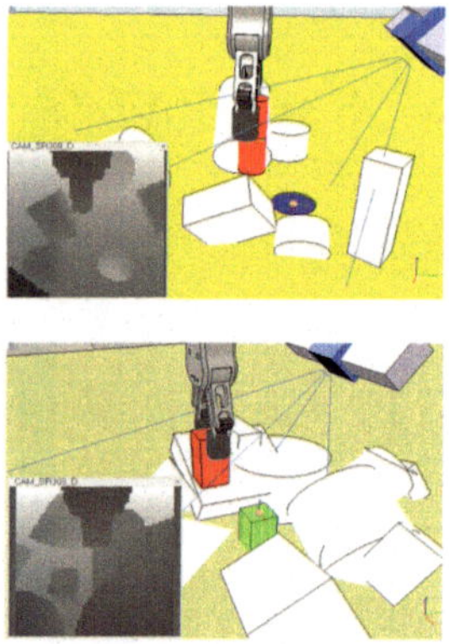

Fig. 4. Top: peg insertion scenario. Bottom: block stacking scenario.

Training Details: For each task, we created a dataset with 50k episodes by generating 500 scenes and rolling out 100 episodes per scene (about 12 steps per episode on average). Each scene was populated by random clutter and random peg/hole/block sizes and positions. For testing, we created a holdout dataset (50k episodes from 100 novel scenes).

Comparisons with Ablations and Baselines: Figure 3 compares the performance of fixed penalty batch PQC with a variety of ablations and baselines.
1. Batch PQC (green): Version of Batch PQC shown in Algorithm 1 (with $c = -0.5$).
2. Online PQC (blue): Same as batch PQC above except that it is trained online using a DAGGER-like rollout schedule.
3. Online PQC with no penalty (red): Ablation. Same as online PQC but without penalty for non-expert actions.
4. Online PQC one action update (cyan): Ablation. Same as online PQC except that we only update the value for the action taken.

5. Online PQC, relative penalty (orange): Same as online PQC except that we use a relative penalty: $q \leftarrow Q_E(s_t, a) - l$, where $l = 0.2$ (the same margin used by DQfD).
6. DAGGER (black): Baseline. Classic DAGGER algorithm implemented using the standard cross entropy loss.
7. DQfD with DAGGER schedule (magenta): Baseline. This version of DQfD uses a single TD loss term plus the large margin loss (margin equal to 0.2 in this experiment) weighted at one tenth the contribution of the TD term (this combination gave us the best performance). It was trained using the DAGGER schedule.
8. DQN on-policy (purple): Baseline. DQN trained on-policy.
9. DQN with DAGGER schedule (grey): Baseline. DQN trained off-policy using the DAGGER schedule.

Finetuning with TD Learning: After cloning, we "finetune" the policy using TD learning (standard DQN) on a new 500 scene dataset. The dataset used for TD learning is distinct from both the training set used for cloning and the holdout test set. Figure 5 show that finetuning results in a small improvement in performance on the holdout test set for both tasks.

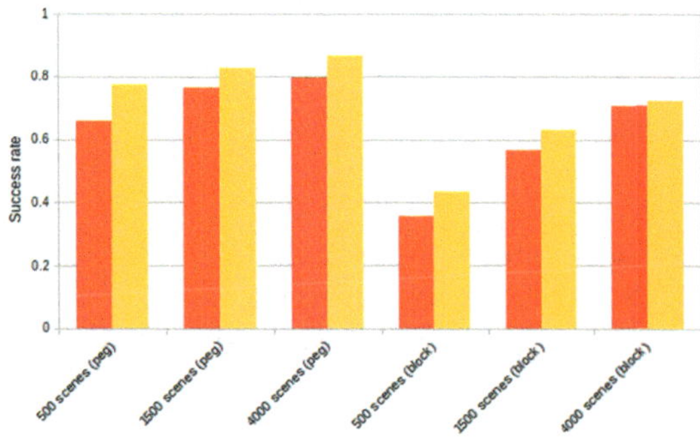

Fig. 5. Effect of finetuning: Success rates fixed penalty batch PQC agents on holdout test set before (red) and after (yellow) finetuning using TD-learning. The left three pairs of bars are for peg insertion. The right three are for block stacking.

Why does TD learning help? First, the expert policy that is cloned during pretraining is only approximately optimal – so there is room for TD learning to improve upon it. Second, recall that the penalty term used in many of the cloning methods causes the agent to learn an incorrect value function, albeit one that supports a near-optimal policy. TD learning probably does a better job finding values that are nearly correct *and* that support a near-optimal policy.

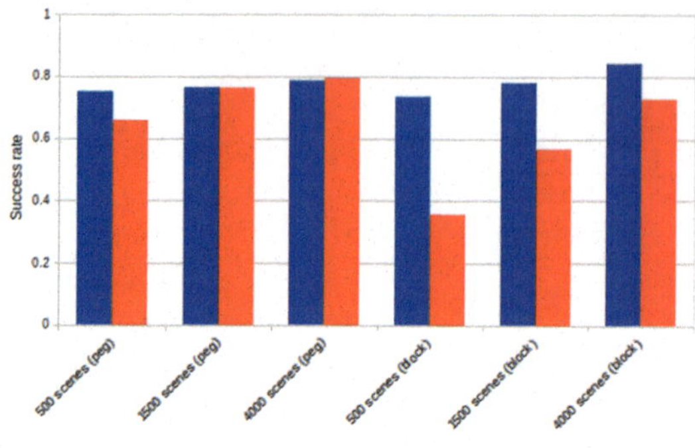

Fig. 6. Effect of training on more scenes: Success rate achieved by fixed penalty batch PQC when training with different numbers of scenes. Blue: performance on training set. Red: performance on holdout test set. The left three pairs of bars are for peg insertion. The right three are for block stacking.

In general, we would expect the success rate to approach 100% after enough episodes using TD learning. However, in our setup this is not the case due to two main reasons. The first issue is the partial observability. In our setup, the agent does not have a memory of previous sensor images and actions. In our experiments we only guarantee that the initial and final states are occlusion and collision free. For example, partial observability could be an issue if the agent can not remember the location of the goal if it becomes occluded during the execution of the task. Occlusion along the trajectories might be unavoidable for some specific scenes and initial poses. The expert is not affected by this restriction because it has full state access. The second issue is the generalization to new scenes. In our case finetuning uses a limited number of scenes and therefore might not generalize enough to novel scenes as discussed in the next section.

Effect of Adding more Training Scenes: Recall from Sect. 2 that we performed online rollouts using cached visual observations. This accelerated online learning because it reduced the number of (expensive) visual rendering operations performed by the simulator during training. However, comparing the results of Fig. 3a and c with those in b and d suggests that there is a significant performance drop when evaluating on holdout test data rather than training data, especially on the blocks task. This suggests that 500 different scenes are not enough to generalize to scenes with new clutter configurations. To test this, we repeated the fixed penalty batch PQC trial with training sets generated with 1500 and 4000 novel scenes instead of just 500. Figure 6 shows the result. Performance on the training set improved marginally but performance on the holdout test set improved dramatically, especially for the block stacking task. Why does cloning seem to require so many novel scenes? If the only effect of clutter was to create scenes that were more complex visually, then 500 should be sufficient. However, our tasks require learning a policy that can avoid colliding with the clutter objects. The blocks task is particularly challenging because the agent must match the image of the grasped block with a block in the scene. These tasks appear to require training data generated from a large number of unique scenes.

Validation on a Physical Robot:

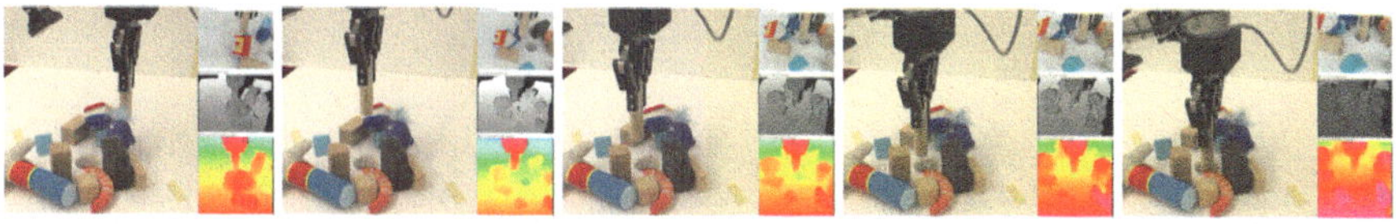

Fig. 7. Example of a single trial on our robot. Small images show sensor images (from top to bottom): RGB (not used), raw depth, and transformed depth (pix2pix GAN).

We performed 100 proof-of-concept trials on the robotic system for the peg-insertion task on five novel scenes containing novel objects placed arbitrarily (example scene in Fig. 7). We used a Robotiq 2F-85 gripper mounted on a UR5 arm in a tabletop setting. An Intel SR300 depth sensor was mounted near the robotic hand as illustrated in Fig. 7. We apply a transform (pix2pix GAN [4]) to sensor images to appear simulated, for details see [15].

Table 1 shows the results of 86% success rate over all trials. Each trial began with the manipulator in a randomly selected pose (not in collision) continued until either reaching the hole, colliding with an object or timing out. We used the same policy for all trials – one that was trained in simulation using the full batch PQC cloning followed by TD learning (finetuning).

4 Main Experimental Insights

Based on these results, a few things are immediately clear. First, Fig. 3 shows that on-policy DQN (purple) underperforms significantly. This justifies our fundamental choice to clone an expert rather than use model-free learning. Second, our proposed methods, batch PQC (green) and online PQC (blue) both perform similarly or slightly worse than two baselines, DQfD (magenta) and DAGGER (black), on the training set. However, they outperform these two baselines on the holdout test set. This is particularly true for the block stacking task which is harder than peg insertion because it requires observing the size of the grasped block in order to determine where to stack, see Fig. 4-b. This suggests that PQC generalizes better to new scenes with different clutter configurations. Third, the

(a) Scene 1 (b) Scene 2 (c) Scene 3 (d) Scene 4 (e) Scene 5

Fig. 8. Different scenes for robot trials. Diameters of peg/hole: (a,b) 2.2 cm/4.2 cm, (c) 2.2 cm/3.9 cm, (d) 1.9 cm/4.2 cm, Scene 5 (no clutter): 2.54 cm/3.9 cm

Table 1. Results from physical robot trials. Success rates and failure modes for each of five different scenes.

		Failure mode		
Scene	Success rate	Collision	Not recognize hole	Missed hole
1	14/20	2	3	1
2	19/20	–	1	–
3	17/20	1	2	–
4	18/20	–	1	1
5	18/20	–	–	2

results indicate that the two ablations, online PQC with no penalty (red) and online PQC with one action update (the two ablations of online PQC), underperform, suggesting that both these elements of the algorithm are important. Finally, if we compare online (fixed penalty) PQC versus relative penalty online PQC, we see that the relative penalty version outperforms on the training set but significantly underperforms on the test set. This suggests that the fixed penalty version generalizes better to novel scenes (Fig. 8).

5 Conclusion

We explore *planner cloning*, an approach that leverages the asymmetry in information that is available to the agent at train and test time. We generate an "expert" policy and q function in simulation with full state information and project these plans onto a policy that the agent can execute via behavior cloning. We propose Penalized Q Cloning (PQC) that outperforms several algorithm ablations and baselines in simulation. Finally, we demonstrate that the resulting policies have similarly good performance on a real robotic system.

References

1. Espiau, B., Chaumette, F., Rives, P.: A new approach to visual servoing in robotics. IEEE Trans. Robot. Autom. **8**(3), 313–326 (1992)
2. Hester, T., Vecerik, M., Pietquin, O., Lanctot, M., Schaul, T., Piot, B., Horgan, D., Quan, J., Sendonaris, A., Osband, I., et al.: Deep q-learning from demonstrations. In: Thirty-Second AAAI Conference on Artificial Intelligence (2018)
3. Hutchinson, S., Hager, G.D., Corke, P.I.: A tutorial on visual servo control. IEEE Trans. Robot. Autom. **12**(5), 651–670 (1996)
4. Isola, P., Zhu, J.-Y., Zhou, T., Efros, A.A.: Image-to-image translation with conditional adversarial networks. In: Proceedings of the IEEE conference on computer vision and pattern recognition, pp. 1125–1134 (2017)
5. James, S., Davison, A.J., Johns, E.: Transferring end-to-end visuomotor control from simulation to real world for a multi-stage task. CoRR, abs/1707.02267 (2017)

6. Kalashnikov, D., Irpan, A., Pastor, P., Ibarz, J., Herzog, A., Jang, E., Quillen, D., Holly, E., Kalakrishnan, M., Vanhoucke, V., et al.: QT-Opt: scalable deep reinforcement learning for vision-based robotic manipulation. arXiv preprint arXiv:1806.10293 (2018)
7. Lakshminarayanan, A.S., Ozair, S., Bengio, Y.: Reinforcement learning with few expert demonstrations. In: NIPS Workshop on Deep Learning for Action and Interaction, vol. 2016 (2016)
8. Levine, S., Finn, C., Darrell, T., Abbeel, P.: End-to-end training of deep visuomotor policies. J. Mach. Learn. Res. **17**(1), 1334–1373 (2016)
9. Levine, S., Pastor, P., Krizhevsky, A., Quillen, D.: Learning hand-eye coordination for robotic grasping with deep learning and large-scale data collection. In: Proceedings International Symposium on Experimental Robotics (ISER), Tokyo, Japan (2016)
10. Mahler, J., Goldberg, K.: Learning deep policies for robot bin picking by simulating robust grasping sequences. In: Conference on Robot Learning, pp. 515–524 (2017)
11. Mnih, V., Kavukcuoglu, K., Silver, D., Rusu, A.A., Veness, J., Bellemare, M.G., Graves, A., Riedmiller, M., Fidjeland, A.K., Ostrovski, G., et al.: Human-level control through deep reinforcement learning. Nature 518(7540), 529 (2015)
12. Ross, S., Bagnell, J.A.: Reinforcement and imitation learning via interactive no-regret learning. CoRR, abs/1406.5979 (2014)
13. Ross, S., Gordon, G., Bagnell, D.: A reduction of imitation learning and structured prediction to no-regret online learning. In: Proceedings of the Fourteenth International Conference on Artificial Intelligence and Statistics, pp. 627–635 (2011)
14. Sadeghi, F., Toshev, A., Jang, E., Levine, S.: Sim2real view invariant visual servoing by recurrent control. CoRR, abs/1712.07642 (2017)
15. Viereck, U., Saenko, K., Platt, R.: Learning visual servo policies via planner cloning. arXiv preprint arXiv:2005.11810 (2020)
16. Yan, M., Frosio, I., Tyree, S., Kautz, J.: Sim-to-real transfer of accurate grasping with eye-in-hand observations and continuous control. arXiv preprint arXiv:1712.03303 (2017)
17. Yoshimi, B.H., Allen, P.K.: Active, uncalibrated visual servoing. In: Proceedings of the 1994 IEEE International Conference on Robotics and Automation, pp. 156–161. IEEE (1994)
18. Zhang, F., Leitner, J., Milford, M., Corke, P.: Sim-to-real transfer of visuo-motor policies for reaching in clutter: domain randomization and adaptation with modular networks. CoRR, abs/1709.05746 (2017)
19. Zhu, Y., Wang, Z., Merel, J., Rusu, A., Erez, T., Cabi, S., Tunyasuvunakool, S., Kramár, J., Hadsell, R., de Freitas, N., et al.: Reinforcement and imitation learning for diverse visuomotor skills. arXiv preprint arXiv:1802.09564 (2018)

Sampling Training Data for Continual Learning Between Robots and the Cloud

Sandeep Chinchali[1(✉)], Evgenya Pergament[2], Manabu Nakanoya[3], Eyal Cidon[2], Edward Zhang[4P], Dinesh Bharadia[5], Marco Pavone[6], and Sachin Katti[1]

[1] Departments of Computer Science, Stanford University, Stanford, CA 94305, USA
{csandeep,skatti}@stanford.edu

[2] Departments of Electrical Engineering, Stanford University, Stanford, CA 94305, USA
{evgenyap,ecidon}@stanford.edu

[3] NEC Corporation, Tokyo, Japan
nakanoya@nec.com

[4] Beijing, China

[5] Department of ECE, UC San Diego, San Diego, CA, USA
dineshb@eng.ucsd.edu

[6] Departments of Aeronautics and Astronautics, Stanford University, Stanford, CA 94305, USA
pavone@stanford.edu

Abstract. Today's robotic fleets are increasingly measuring high-volume video and LIDAR sensory streams, which can be mined for valuable training data to continually improve robotic perception models. However, re-training perception models on growing volumes of rich sensory data in central servers (or the "cloud") incurs significant systems costs for network transfer, cloud storage, human annotation, and cloud computing resources. Hence, we introduce HARVESTNET, an intelligent sampling algorithm that resides on-board a robot and reduces system bottlenecks by only storing rare, useful events to continually improve and re-train perception models. HARVESTNET significantly improves the accuracy of perception models on our novel dataset of road construction sites, field testing of self-driving cars, and face recognition, while reducing cloud storage, dataset annotation time, and cloud compute time by between 65.7–81.3%. Further, it is between 1.05–2.58× more accurate than baseline algorithms and runs on low-power deep learning hardware.

1 Introduction

Learning from rare events, such as construction roadblocks or hazardous weather conditions, can significantly enhance the safety of robotic perception and decision-making models. We envision that fleets of robots, ranging from delivery drones to self-driving cars, will *distributedly* collect and filter valuable training data from their high-volume video and LIDAR sensory streams in order to continually improve the robotic autonomy stack. Despite the benefits of continually learning from field data, they come with severe systems costs that are largely under-addressed in today's robotics literature. Specifically, these bottlenecks (shown in Fig. 1a) stem from the cost of (1) network data transfer to compute servers, (2) limited robot and cloud storage, (3) dataset annotation with

S. Chinchali and E. Pergament—Contributed equally to this work.

B. Siciliano et al. (Eds.): ISER 2020, SPAR 19, pp. 296–308, 2021.
https://doi.org/10.1007/978-3-030-71151-1_27

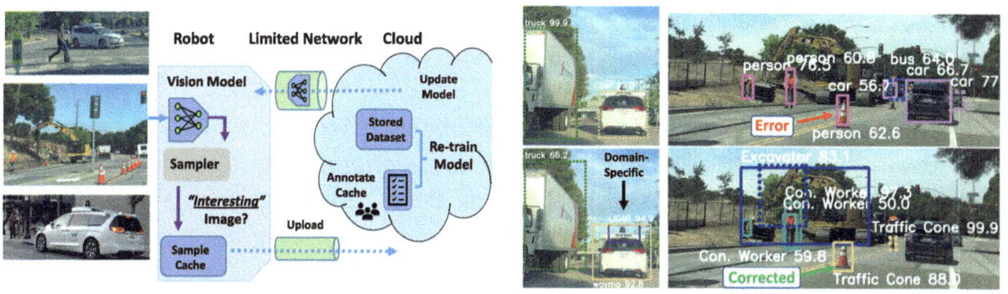

Fig. 1. The Robot Sensory Sampling Problem. (Left, a) What are a minimal set of "interesting" training samples robots can upload to periodically improve perception model accuracy while minimizing network transfer, cloud storage, human annotation, and cloud-computing costs? (Right, b) From only a minimal set of training examples, HARVESTNET adapts vision models trained on the standard COCO dataset [10] (top images) to better, domain-specific models (bottom).

human supervision, and (4) cloud computing for training models on massive datasets. Indeed, it is estimated that a *single* self-driving car will produce upwards of 4 Terabytes of data per day [1].

Motivated by the significant benefits and costs of continual robotic fleet learning, this paper introduces a novel *Robot Sensory Sampling Problem* (Fig. 1a), which asks how a robot can maximally improve model accuracy via cloud re-training, but with minimal end-to-end systems costs. To address this problem, we introduce HARVESTNET, a compute-efficient sampler that reduces systems bottlenecks by "harvesting" only task-relevant training examples for continual learning.

In general, robotic field data contains both *known weakpoints* of a perception model as well as *unknown weakpoints* that have yet to be discovered. A known weakpoint is shown in Fig. 1b, where a fleet operator identifies that construction sites are common errors of a perception model and wishes to collect more of such training examples. In contrast, unknown weakpoints could be rare visual concepts that are not even anticipated by a roboticist. To provide a focused contribution, this paper addresses how robotic fleets can economically collect more training data for *known weakpoints* to maximize model accuracy. Our key technical insight is to design a compute-efficient robot sampler that selectively leverages powerful feedback from the cloud to continually re-train models and calibrate their uncertainty on task-relevant images, thereby adaptively improving the sampling process. By emphasizing a resource-efficient design, we ensure that HARVESTNET applies to a variety of robotic systems with diverse compute constraints and does not interfere with computation for real-time control. Moreover, we discuss how our algorithm can be extended to collecting training data for unknown model weakpoints in Sect. 5.

Literature Review: HARVESTNET is closely aligned with cloud robotics [9], where robots offload computation tasks to the cloud in *real-time* to improve robotic perception, but with minimal network latency [7, 16]. In contrast, we address overall system cost (from network transfer to human annotation time) and instead focus on the open problem of continual cloud-assisted learning by intelligently sampling field data. Our work is also inspired by active learning [8, 14], novelty detection [6, 11], and few-shot learning [12, 17, 18], but differs significantly since we use novel algorithms, distributed between the robot and cloud, that are scalable to run on even resource-constrained robots. Many active learning algorithms use compute-intensive *acquisition functions* to decide which limited samples a learner should label to maximize model accuracy, which often involve Monte Carlo Sampling to estimate uncertainty over a new image. Indeed, the most similar work to ours [8] uses Bayesian Deep Neural Networks (DNNs), which have a distribution over model parameters, to take *several* stochastic forward passes of a DNN to decide whether an uncertain image needs labeling. Such methods are infeasible for resource-constrained robots, which often do not have the memory to store, or time to run, several distributions over model parameters for a *single* image. Indeed, by eschewing standard active learning algorithms, we decide whether to sample an image in only ≈ 0.3 ms with a low-power DNN accelerator that can only store a *single set* of pre-compiled model parameters.

Statement of Contributions: In light of prior work, our contributions are three-fold. First, we collect months of novel video footage to quantify the systems bottlenecks of processing sensory data. Second, we formalize a compute-efficient sampling algorithm that leverages feedback from the cloud to adaptively improve its data-driven filter for task-relevant images, which yields significant accuracy gains on real-world datasets. Third, we demonstrate a prototype sampler and perception models that run efficiently on the low-power Google Edge Tensor Processing Unit (TPU) [4], which are made available to the research community.

Organization: This paper is organized as follows. Section 2 motivates the *Robot Sensory Sampling Problem* by quantifying our system costs for collecting three novel perception datasets. Then, Sect. 3 formulates a general sampling problem and introduces the HARVESTNET system architecture (Fig. 2a). Next, Sects. 4 and 5 quantify HARVESTNET's accuracy improvements for perception models, key reductions in system bottlenecks, and performance on the Edge TPU. Finally, we conclude with future applications in Sect. 6.

2 Motivation

We now describe the costs and benefits of continual learning in the cloud by collecting and processing three novel perception datasets.

Benefits: A prime benefit of continual learning is to correct weak points of existing models or specialize them to new domains. As exemplified in Fig. 1b (right, top), standard vision DNNs pre-trained on the benchmark MS-COCO [10] dataset can misclassify traffic cones as humans but also miss the large *excavator* vehicle, which neither looks like a standard truck nor a car. However, as shown in Fig. 1b (right, bottom), a specialized model re-trained in the cloud using automatically sampled, *task-relevant* field data can correct such errors. More generally, fleets can "crowd-source" salient road scenes to automatically update high-definition (HD) maps.

Costs: The above benefits of cloud re-training, however, often come with understated costs. Figure 2b estimates monthly systems costs for a small fleet of 10 self-driving cars, each of which is assumed to capture 4 h of data daily. In the first scenario,

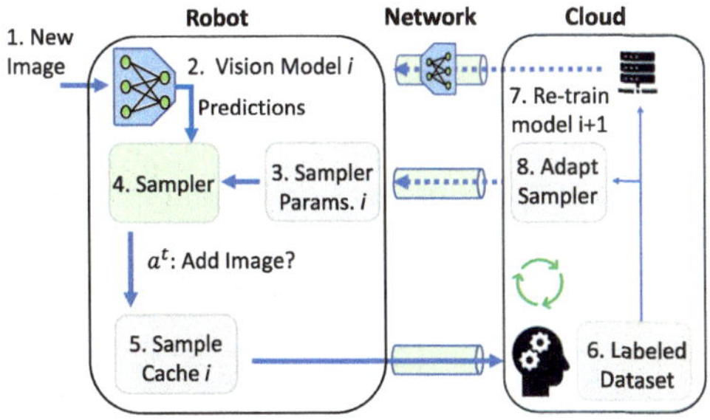

Bottleneck	SDV	Dashcam
Storage	$31,200	$172.50
Upload (hr)	8.88	0.05
Label Cost	$95,256	$15,876
Label Time	22 days	18 days

Fig. 2. (Left, a) **System Architecture:** Our core contribution is a robot sensory sampler in green, which uses predictions from an on-board vision DNN and cloud-selected image targets. (Right, b) **Systems Costs:** Monthly storage/annotation cost and daily network upload time are significant for self-driving car data even if we annotate *1% of video frames*.

called "Dashcam", we assume a single car has only one dashcam capturing H.264 compressed video, as in our collected dataset. In the second scenario, called "SDV", we *extrapolate* from an Intel report [1] that estimates each SDV will generate 4 TB of sensory data in 1.5 h.

We conservatively assumed all field data is uploaded over fast 10 Gbps ethernet, which might be realistic for SDVs, but is much slower for network-limited robots. Importantly, we assume *only 1% of image data* is interesting and needs to be annotated, specifically by the state-of-the-art Google Data Labeling service [2], which has standard rates for object detection bounding boxes. Further, to roughly estimate annotation time, we hand-labeled 1000 bounding boxes in 1.4 h. Though conservative, our estimates show significant costs, especially for annotation. To quantify systems costs first-hand, we collected over three months of video clips from a dashboard iPhone 6 camera ("dashcam"), comprising over 346,788 frames of video. The following datasets, as well as re-trained vision models, are available at https://sites.google.com/view/harvestnet:

1. Identifying Construction Sites: Our footage shows safety-critical scenarios and model drift on roads that were first excavated (Fig. 1b), filled with tar, and rolled back to normal, often by human workers.

2. Identifying Self-driving Vehicles (SDVs): We captured rare footage of SDVs being tested on public roads, predominantly of the Waymo van (Fig. 1b), which can be easily generalized to other rare image classes.

3. Streaming Face Recognition (FaceNet): To test HARVESTNET's robustness to domains outside of road scenes, we use a video dataset of streaming faces of consenting volunteers from a prior robotics paper [7].

3 Problem Statement

We now formulate a general *Robot Sensory Sampling Problem* that is widely applicable, from self-driving cars to mining and space robots. As a common theme, these robots must *prioritize* sensory samples due to constraints in storage, network connectivity, human annotation, and/or cloud-computing time that preclude naïvely using all field data. We now define general system abstractions shown in Fig. 2a:

1. *Sensory Input:* The robot's sensory stream on a day or learning "round" i is denoted by $\{x^{t,i}\}_{t=0}^{T}$, where each $x^{t,i}$ is an image or LIDAR point cloud. After a round's duration of T steps, samples are uploaded for annotation and model learning.

2. *Robot Perception Models:* Vision DNN f_{DNN} maps an image x^t to a prediction $\hat{y}^t$ (e.g. image class or object detections) and associated confidence $\texttt{conf}^t$, typically from softmax layers. The DNN also provides embeddings $\texttt{emb}^t$, which are used to compare the similarity between images as shown in Fig. 3 (right). Our goal is to make the perception DNN learned at the end of round i, denoted by f^i_{DNN}:

$$\mathbf{y}^t = [\hat{y}^t, \texttt{conf}^t, \texttt{emb}^t] = f^i_{\text{DNN}}(x^t), \tag{1}$$

as accurate as possible for increasing i, with the fewest samples uploaded to the cloud. f^i_{DNN} is fixed within a round i of data collection.

3. *Target Image Set:* The scope of this paper is to collect more examples of *known weakpoints* of a model, which we refer to as task-relevant image classes, such as the mis-labeled traffic cones and excavators in Fig. 1b. The set of task-relevant image classes, denoted by `TargetClassSet`, can be identified by a fleet operator who periodically reviews model performance on historical data. Then, the robotic fleet operator can collect a few initial, historical examples of images in `TargetClassSet`, denoted as `TargetImageSet`, to seed the initial cloud dataset and train nominal model f^0_{DNN}, which is used to sample more field examples of model weakpoints. Further, the cloud dataset can already have sufficient training data for well-understood scenes from previous field tests or public datasets.

4. *Sampling Policy:* Our goal is to develop a lightweight sampling policy π_{sample} (green unit in Fig. 2a) that decides whether or not to store a new image x^t based on predictions from the on-board perception DNN $\mathbf{y}^t = f^i_{\text{DNN}}(x^t)$ and the desired target-set `TargetClassSet` of images to look for:

$$a^t = \pi_{\text{sample}}(x^t, \mathbf{y}^t, \texttt{TargetClassSet}, \theta^i_{\text{sampler}}), \tag{2}$$

where $\theta^i_{\text{sampler}}$ are sampler hyper-parameters that, in general, can be adapted per round i. Here, $a^t = 1$ indicates storing the image to later send to the cloud and $a^t = 0$ indicates dropping it, such as for well-understood scenes.

5. *Limited Robot Sample Cache:* In round i, the robot saves sampled images in a small cache denoted by $\texttt{Cache}^i = \{x^\tau\}_{\tau=0}^{N_{\text{cache}}}$, where $0 \leq \tau \leq T$. These limited images of size $N_{\text{cache}} \ll T$ are small enough to store on-board a robot.

6. *Annotated Cloud Dataset:* After $\texttt{Cache}^i$ is uploaded to the cloud, a human, or trusted machine, provides ground-truth annotations $y^\tau_{\text{GroundTruth}}$ for the selected images x^τ, which could be different from the robot's *predictions* $\hat{y}^\tau$. We store an annotated cloud dataset $\mathscr{D}^i = \{x^\tau, y^\tau_{\text{GroundTruth}}\}$ that grows with rounds i:

$$\mathscr{D}^i = \mathscr{D}^{i-1} \cup \{(x^\tau, y^\tau_{\text{GroundTruth}}), \forall x^\tau \in \texttt{Cache}^i\}. \tag{3}$$

Initial dataset $\mathcal{D}^0$ has very few examples of the target images, such as self-driving cars or construction sites, but is seeded with a large set of "common" training examples, such as from MS-COCO [10]. We split the full cloud dataset into train, validation, and test components denoted by $\mathcal{D}^i_{\text{train}}$, $\mathcal{D}^i_{\text{val}}$, and $\mathcal{D}^i_{\text{test}}$.

7. Independent Test Dataset $\mathcal{D}_{\text{test, final}}$: Model accuracy is always reported on an *independent, held-out* test set denoted $\mathcal{D}_{\text{test, final}}$, which can be collected by designated robots in a fleet tasked with collecting purely test data. To assess the sampler's efficacy in accruing *task-relevant* data that yields steady performance gains on the target classes, $\mathcal{D}_{\text{test, final}}$ should have sufficient representation of, or even be biased towards, target classes, alongside common images from ImageNet.

8. Model Re-training and Evaluation: Once the new annotated cache $\texttt{Cache}^i$ is added to the dataset $\mathcal{D}^i$, model re-training function g_{retrain} takes current robot model f^i_{DNN} and the updated annotated dataset $\mathcal{D}^{i+1}$ and outputs an updated model f^{i+1}_{DNN}. The training objective is to minimize loss function $\mathcal{L}(\hat{y}^t, y^t_{\text{GroundTruth}})$ penalizing prediction errors, such as the standard cross-entropy loss for image classification or mean average precision (mAP) for object detection. Evaluation function g_{evaluate} quantifies the accuracy improvements of new model f^{i+1}_{DNN}, denoted by:

$$f^{i+1}_{\text{DNN}} \leftarrow g_{\text{retrain}}(f^i_{\text{DNN}}, \mathcal{D}^{i+1}_{\text{train}}), \quad \text{and} \quad \mathcal{L}^i = g_{\text{evaluate}}(f^i_{\text{DNN}}, \mathcal{D}_{\text{test, final}}). \tag{4}$$

Then, the robot downloads new model f^{i+1}_{DNN} over a network. Having defined our problem formulation, we can now formally define our general sampling problem.

Problem 1 (Robot Sensory Sampling Problem). Given a target set of image classes $\texttt{TargetClassSet}$, initial cloud dataset $\mathcal{D}^0$, and initial robot model f^0_{DNN}, find the best sampling policy π^*_{sample} that minimizes full held-out test set loss $\mathcal{D}_{\text{test, final}}$ for the final robot DNN $f^{N_{\text{round}}}_{\text{DNN}}$ trained over N_{round} days:

$$\pi^*_{\text{sample}} \in \underset{\pi_{\text{sample}}}{\operatorname{argmin}} \mathcal{L}(f^{N_{\text{round}}}_{\text{DNN}}, \mathcal{D}_{\text{test, final}}), \tag{5}$$

subject to the constraint that a robot can only sample N_{cache} images daily.

Problem Complexity: Deriving an optimal solution to Problem 1 is challenging, since it is hard to model the high-dimensional stream of T samples a robot can possibly measure, as well as which limited subset will maximize final model performance. Indeed, even a drone that must select 2000 images from *only 15* min of video has a combinatorial set of $\binom{T}{N_{\text{cache}}} = 4.21 \times 10^{1340}$ images to select from.

Wide Practicality of the Sampling Problem: Though complex, our problem is built on simple abstractions of limited on-board compute, storage, network connectivity, and cloud computing resources. These abstractions apply to diverse robots that measure more data than they can sustainably process for continual learning.

4 The HarvestNet Intelligent Sampler

While obtaining a provably optimal solution to Problem 1 is infeasible, we can significantly constrain the sampler design space by leveraging three empirical insights that

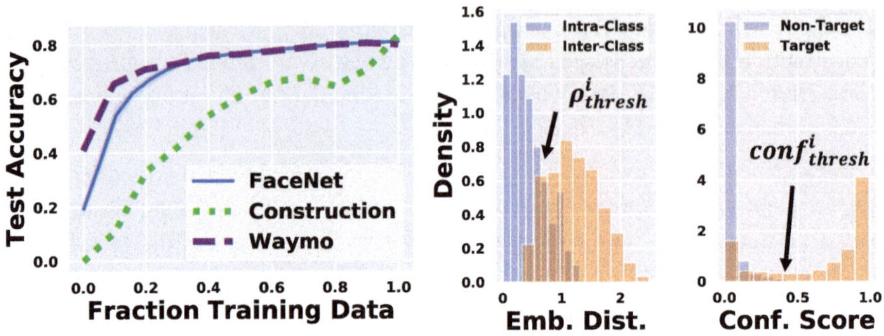

Fig. 3. Empirical insights guiding HarvestNet: (Left) For relatively static targets, test accuracy saturates with more images used to fine-tune a vision DNN. (Right) We use either embedding distances or softmax confidence scores to signal if a new image is likely a target of interest.

directly lead to a compute-efficient sampler. These insights, which are shown in Fig. 3 and detailed below, allow our sampler to run on robots with a low-power Raspberry Pi and Google Edge TPU, as shown in Fig. 6.

Insight 1: *Use transfer learning for high model accuracy with few sampled images.* Many pre-trained CNNs' initial network layers serve as general-purpose feature extractors that yield compact embeddings [15] which can be used to fine-tune final linear classification layers to recognize new classes of interest. HarvestNet exploits transfer learning by using pre-trained feature extraction layers in initial robot model f^0_{DNN} and only re-training final layers to rapidly adapt to novel visual concepts with minimal data. Indeed, Fig. 3 (left) shows that final model test accuracy often saturates quickly after acquiring only a small set of training images. Since the *dynamic* construction scenes rapidly change from day to day, we observe steady accuracy gains even after months of data are acquired.

Insight 2: *Filter task-relevant images using DNN embeddings and softmax scores.*

Embedding distances: Many CNNs have the property that the L_2 distance between embeddings of two images is a proxy for image similarity. In particular, as shown for the FaceNet DNN in Fig. 3 (right), two pictures of the *same person* x^{t_1} and x^{t_2} typically have a lower embedding-distance $\rho(x^{t_1}, x^{t_2}) = \|\texttt{emb}^{t_1} - \texttt{emb}^{t_2}\|_2^2$ than unrelated faces [13]. Using this insight, we compare the embedding distance between a robot's new sensory input x^t to each of the target images in `TargetImageSet`, and cache the sample if the median distance is below an empirical threshold ρ^i_{thresh}, which serves as the hyperparameter $\theta^i_{\text{sampler}} = \{\rho^i_{\text{thresh}}\}$ in Eq. 2:

$$\begin{aligned} a^t &= \pi_{\text{sample}}(x^t, \mathbf{y}^t, \texttt{TargetImageSet}, \theta^i_{\text{sampler}}) \\ &= \mathbb{1}\left(\textsc{Median}\left\{\rho(x^t, x^{\text{targ.}}),\ \forall x^{\text{targ.}} \in \text{TargImSet}\right\} < \rho^i_{\text{thresh}}\right). \end{aligned} \tag{6}$$

In this variant of the sampling policy, the cloud sends a limited `TargetImageSet` of images and their embeddings to the robot to compare with the embedding of a new image x^t. We now describe another sampling variant.

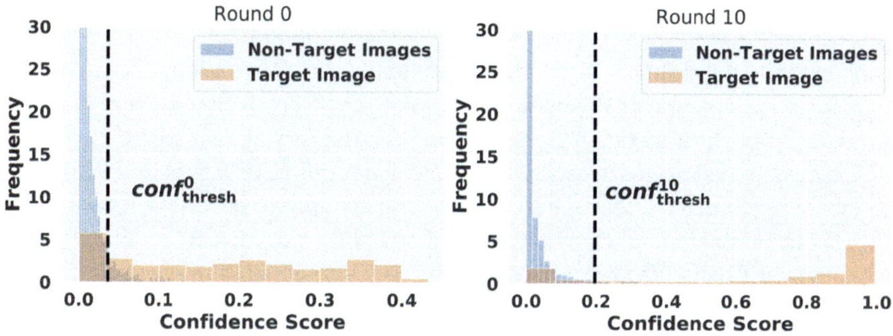

Fig. 4. Role of Cloud Feedback: After a human annotates more samples on each round i, the softmax confidence distributions between target (red) and irrelevant (blue) images become more distinct. Thus, HARVESTNET can improve sampling by adapting the threshold $\texttt{conf}^i_{\text{thresh}}$.

Softmax confidence thresholds: While softmax confidence scores do not yield a *formal* probability distribution over CNN class predictions, we empirically note that images that are indeed truly a class ("true positives") often have a higher softmax confidence element for that class of interest compared to "true negatives". As exemplified in Fig. 3 (right), images that are indeed of Waymo cars (red) have a higher confidence score for the Waymo class output compared to negatives (in blue), showing that confidence scores above a separating threshold are a useful signal to store a sample of interest. Specifically, given a new image x^t, we see if the confidence element *for a target class*, denoted by $\texttt{conf}[\texttt{TargetClassSet}]$, is above an empirically determined threshold $\texttt{conf}^i_{\text{thresh}}$ (Fig. 4), which serves as the hyper-parameter $\theta^i_{\text{sampler}} = \{\texttt{conf}^i_{\text{thresh}}\}$ in Eq. 2. Thus, the sampling policy becomes:

$$a^t = \pi_{\text{sample}}(x^t, \mathbf{y}^t, \texttt{TargetClassSet}, \theta^i_{\text{sampler}}) = \mathbb{1}\left(\texttt{conf}^t[\texttt{TargetClass}] > \texttt{conf}^i_{\text{thresh}}\right). \quad (7)$$

Importantly, when a robot's perception model is already being run for a high-level task, such as navigation, the sampler can easily leverage the already-computed embeddings and confidence scores to efficiently discern if a new image is worth sampling. For example, Eq. 7 runs in $\approx .3$ ms on a Raspberry Pi CPU.

Insight 3: *Use cloud feedback to refine the target image filter (Fig. 4).*
As shown in Fig. 4, our final insight is that DNN softmax confidence distributions between target images (red) and irrelevant images (blue) become more distinct as more human-annotated images are added to the cloud dataset at every round i. Specifically, the `AdaptThresh` function uses the new cloud dataset and improved model at the *end of round* i to best adapt the sampler hyper-parameters $\theta^{i+1}_{\text{sampler}} = \{\texttt{conf}^{i+1}_{\text{thresh}}, \rho^{i+1}_{\text{thresh}}\}$ to improve the sampler's task-relevant image filter:

$$\theta^{i+1}_{\text{sampler}} \leftarrow \texttt{AdaptThresh}(\mathscr{D}^{i+1}, f^{i+1}_{\text{DNN}}). \quad (8)$$

In practice, `AdaptThresh` can be any learning algorithm, such as a support vector machine, that creates a simple decision boundary between confidence scores or embeddings that differentiate task-relevant and non-target images.

4.1 The Lightweight HarvestNet Sampling Algorithm

Having explained insights 1–3, we now present HarvestNet's sampler in Algorithm 1, which serves as a practical heuristic solution to Problem 1. In each round i, a robot has a fixed perception model f^i_{DNN} and observes a sensory stream $\{x^{t,i}\}_{t=0}^{T}$. Sampling policy π_{sample} stores images in cache $\texttt{Cache}^i$, which is uploaded, annotated with ground-truth labels, appended to cloud dataset $\mathscr{D}^i$, and finally used to re-train perception model f^{i+1}_{DNN}. This new model and an updated set of sampling parameters, obtained by Eq. 8, are then downloaded to the robot. Learning repeats for a fixed set of rounds N_{round} until DNN validation accuracy saturates. Key algorithm steps are colored.

Algorithm 1: HarvestNet Sensory Stream Sampler

Assemble Targets $\texttt{TargetClassSet}$, Cloud Dataset $\mathscr{D}^0$
Pre-train Initial Robot Perception DNN f^0_{DNN} on $\mathscr{D}^0$
Init. $\{\texttt{conf}^0_{\text{thresh}}, \rho^0_{\text{thresh}}\} \leftarrow \texttt{AdaptThresh}(\mathscr{D}^0, f^0_{\text{DNN}})$ `// Init. Sampler Params.`
for *Learning Round* $i \leftarrow 0$ **to** N_{round} **do**
  $\texttt{Cache}^i \leftarrow \{\emptyset\}$ `// Init. Empty Cache`
  for $t \leftarrow 0$ **to** T **do**
    $\mathbf{y}^t \leftarrow [\hat{y}^t, \texttt{conf}^t, \texttt{emb}^t] = f^i_{\text{DNN}}(x^t)$ `// Run Vision Model`
    $a^t \leftarrow \pi_{\text{sample}}(x^t, \mathbf{y}^t, \texttt{TargSet}, \{\texttt{conf}^i_{\text{thr.}}, \rho^i_{\text{thr.}}\})$ `// Run Sampler`
    $\texttt{Cache}^i.\text{APPEND}(x^t)$ **IFF** $a^t = 1$ `// Can be priority queue`
  end
  $\{(x^\tau, y^\tau_{\text{GroundTruth}})\} \leftarrow \text{ANNOTATE}(\texttt{Cache}^i)$ `// Label Data`
  $\mathscr{D}^{i+1} \leftarrow \mathscr{D}^i \cup \{(x^\tau, y^\tau_{\text{GroundTruth}})\},\ \forall x^\tau \in \texttt{Cache}^i$ `// Add to Cloud Dataset`
  $\mathscr{D}^{i+1}_{\text{train}}, \mathscr{D}^{i+1}_{\text{val}} \leftarrow \text{TRAINVALSPLIT}(\mathscr{D}^{i+1})$
  $f^{i+1}_{\text{DNN}} \leftarrow g_{\text{retrain}}(f^i_{\text{DNN}}, \mathscr{D}^{i+1}_{\text{train}})$ `// Re-train`
  $\{\texttt{conf}^{i+1}_{\text{thresh}}, \rho^{i+1}_{\text{thresh}}\} \leftarrow \texttt{AdaptThresh}(\mathscr{D}^{i+1}, f^{i+1}_{\text{DNN}})$ `// Improve Sampler`
end
Result: Final DNN $f^{N_{\text{round}}}_{\text{DNN}}$, Cloud Dataset $\mathscr{D}^{N_{\text{round}}}$

4.2 Benchmark Sampling Algorithms

We compare HarvestNet with the following benchmarks:

1. Random Sampling: This scheme randomly fills the limited cache from T possible samples in round i.

2. Non-adaptive cloud feedback: Instead of adapting thresholds using Eq. 8, this scheme simply caches an image based on a probability that is proportional to the softmax confidence. Since it does *not* calibrate itself using a threshold, it often misses true positives that exhibit low confidence scores at early rounds (Fig. 4, left).

3. HarvestNet-PriorityQueue: This scheme caches an image into a priority queue ranked by either softmax confidence scores or embedding distances and drops low-ranked items when the queue is full.

4. Oracle Sampling: This un-realizable upper bound *perfectly* identifies task-relevant images at the robot. The oracle takes the place of an "all-cloud" approach, since uploading all data is infeasible for network-limited robots.

5 HarvestNet's Experimental Performance

We now show HarvestNet (1) achieves high model accuracy for a minimal number of sampled images (Figs. 5a, 5b), and (2) produces compute-efficient vision models. Every benchmark that we evaluate has a fixed cache limit of N_{cache} and thus obeys the constraints of Prob. 1. Further, our results metric of model accuracy is equivalent to minimizing model loss, which is the objective of Prob. 1.

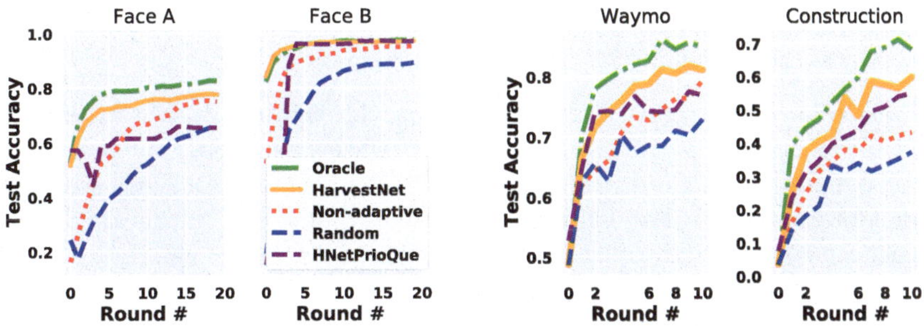

Fig. 5. FaceNet Results: (Left) We achieve steady accuracy gains with minimal face examples, even matching the oracle performance for Face B, shown by the curve overlap. (Right) **Road-Scene Results**: HarvestNet outperforms baselines that sample randomly or without cloud feedback.

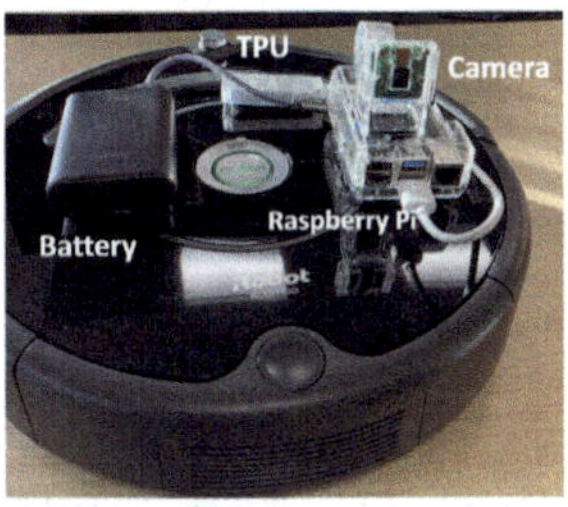

Device	Model	Accuracy (mAP)	Inference Time (ms)	Size (MB)
TPU	MN2	0.17	6.9	5.5
CPU	MN2	0.167	285.6	4.7
Cloud GPU	MN2	0.242	56.8	20
Cloud GPU	Faster R-CNN	0.284	447.5	183

Fig. 6. Prototype: (Left) The lightweight sampler runs on an embedded CPU, such as on the Raspberry Pi, which uses an Edge TPU USB to accelerate DNN inference. **Model Performance**: (Right) HarvestNet re-trained object-detectors flexibly run on the Edge TPU or cloud GPUs.

<u>**FaceNet Results:**</u> Our key result for the FaceNet scenario is shown in Fig. 5a (left), where HarvestNet, in orange, achieves upto 1.68× the final accuracy of benchmarks and no lower than 92% the accuracy of an upper-bound oracle solution in green. Figure

5a also shows HarvestNet can flexibly adapt to different target faces of interest, since it shows steady model improvements for two randomly selected targets, referred to as Face A and B in the plots. Importantly, the robot is only provided 18 initial examples of the target face and only caches $N_{\text{cache}} = 10$ images per round, much less than the thousands of video frames it observes. We used the widely-adopted OpenFace `nn4.small2.v1` DNN for face embeddings [5], which were passed to an SVM for face-classification.

Generalization to Dynamic Road Scenes: HarvestNet can also flexibly use softmax confidence scores in the sampling policy (Eq. 7), alongside the embedding distance variant in the FaceNet case (Eq. 6), to identify challenging scenes of self-driving cars and construction sites. Figure 5b shows that, for the Waymo domain, HarvestNet achieves upto $1.11\times$ the final accuracy of benchmarks and 93.5% the accuracy of an upper-bound oracle solution in green. Despite construction sites exhibiting significant model drift with daily changes, HarvestNet outperforms non-adaptive sampling by over $1.38\times$ and is within 87% of the oracle solution.

Priority Queue Variant: The priority queue benchmark, dubbed HNetPriority Queue in Fig. 5a, does worse than HarvestNet when the proportion of target images per day is highly non-uniform, leading it to cache irrelevant images below the empirical threshold. Depending on how uniform the distribution of sensory inputs are per day, a designer can use the priority queue variant, which is also compute-efficient.

Benefits of re-training: To quantify the benefits of using field data, we consider the final model accuracy if we trained *solely* on high-quality public images of Waymo cars and construction sites and tested on our road dataset. We found less than 200 Google Images for each scenario, which yielded an extremely poor accuracy of only 41.6% for Waymo and 0% for construction. Indeed, public images were often zoomed in, clear pictures that were not representative of real road scenes with partially occluded targets, illustrating the benefit of continual learning.

Systems Bottleneck Savings: The principal benefits of sampling derive from limiting the number of images cached in each learning round (e.g. $N_{\text{cache}} = 60$). Thus, by re-training on between only 18–34% of possible images, we decrease storage, annotation cost and time, and cloud-computing time by 65.7–81.3%. These estimates will vary considerably based on systems infrastructure.

Edge Tensor Processing Unit (TPU) Prototype: Figure 6 (right) shows that samples collected by HarvestNet can be used to flexibly re-train mobile-optimized models, such as MobileNet 2, or more accurate, but slower "cloud-grade" models such as Faster R-CNN (Fig. 1b). Thus, one can trade-off speed, size, and accuracy, quantified by the standard COCO mAP metric [3], based on robot compute resources. For example, for the robot in Fig. 6, our sampler running on a Raspberry Pi CPU decides whether to cache a new image with a mean speed of just $\approx .31$ ms, which is $\sim 22\times$ faster than accelerated TPU inference.

Limitations: HarvestNet might fail in scenarios with extremely rare visual data, such as cases where we cannot obtain enough samples to train the initial robot model on round 0. In future work, we can consider whether *unknown weakpoints* of a model can be cached by storing examples with high confidence score entropy.

6 Discussion and Conclusions

The key contribution of this paper is to introduce a new, widely-applicable sensory sampling problem to the robotics community and experimentally show the gains of our compute-efficient sampler. Since we anticipate the volume of rich robotic sensory data to grow, thereby making systems bottlenecks more acute, we encourage the robotics community to significantly build upon our work. As such, we have open-sourced our dataset, Edge TPU vision DNNs, and sampling algorithm code at https://sites.google.com/view/harvestnet.

In the future, we hope to enhance HarvestNet by fusing data from several cameras and LIDAR sensors on-board a robot. Further, the sampling algorithm can be modified to select adversarial training examples for vision DNNs. Lastly, HarvestNet's sampler can be enhanced to explicitly not store private training data by rejecting such samples. We believe HarvestNet is a vital first step in keeping pace with surging robotic sensory data by insightfully filtering video streams to select actionable training examples for continual cloud-assisted learning.

References

1. Data is the new oil in the future of automated driving (2016). https://newsroom.intel.com/editorials/krzanich-the-future-of-automated-driving/#gs.LoDUaZ4b. Accessed 30 Jan 2019
2. Ai platform data labeling service (2019). https://cloud.google.com/data-labeling/docs/
3. Coco: Common objects in context: Detection evaluation (2019). http://cocodataset.org/#detection-eval
4. Edge tpu (2019). https://cloud.google.com/edge-tpu/. Accessed 01 Sep 2019
5. Amos, B., Ludwiczuk, B., Satyanarayanan, M.: Openface: a general-purpose face recognition library with mobile applications. Technical report, CMU-CS-16-118, CMU School of Computer Science (2016)
6. Bodesheim, P., Freytag, A., Rodner, E., Kemmler, M., Denzler, J.: Kernel null space methods for novelty detection. In: Proceedings of the IEEE Conference on Computer Vision and Pattern Recognition, pp. 3374–3381 (2013)
7. Chinchali, S., Sharma, A., Harrison, J., Elhafsi, A., Kang, D., Pergament, E., Cidon, E., Katti, S., Pavone, M.: Network offloading policies for cloud robotics: a learning-based approach. arXiv preprint arXiv:1902.05703 (2019)
8. Gal, Y., Islam, R., Ghahramani, Z.: Deep bayesian active learning with image data. In: International Conference on Machine Learning (2017)
9. Kehoe, B., Patil, S., Abbeel, P., Goldberg, K.: A survey of research on cloud robotics and automation. IEEE Trans. Automation Sci. Eng. **12**(2), 398–409 (2015)
10. Lin, T.-Y., Maire, M., Belongie, S., Hays, J., Perona, P., Ramanan, D., Dollár, P., Zitnick, C.L.: Microsoft coco: Common objects in context. In: European Conference on Computer Vision, pp. 740–755. Springer (2014)
11. Perera, P., Patel, V.M.: Deep transfer learning for multiple class novelty detection. In: Proceedings of the IEEE Conference on Computer Vision and Pattern Recognition, pp. 11544–11552 (2019)

12. Sanakoyeu, A., Tschernezki, V., Buchler, U., Ommer, B.: Divide and conquer the embedding space for metric learning. In: Proceedings of the IEEE Conference on Computer Vision and Pattern Recognition, pp. 471–480 (2019)
13. Schroff, F., Kalenichenko, D., Philbin, J.: Facenet: a unified embedding for face recognition and clustering. In: Proceedings of the IEEE Conference on Computer Vision and Pattern Recognition, pp. 815–823 (2015)
14. Settles, B.: Active learning literature survey. Technical report, University of Wisconsin-Madison Department of Computer Sciences (2009)
15. Sharif Razavian, A., Azizpour, H., Sullivan, J., Carlsson, S.: CNN features off-the-shelf: an astounding baseline for recognition. In: Proceedings of the IEEE Conference on Computer Vision and Pattern Recognition Workshops, pp. 806–813 (2014)
16. Tanwani, A.K., Mor, N., Kubiatowicz, J., Gonzalez, J.E., Goldberg, K.: A fog robotics approach to deep robot learning: Application to object recognition and grasp planning in surface decluttering. arXiv preprint arXiv:1903.09589 (2019)
17. Wang, X., Yu, F., Wang, R., Darrell, T., Gonzalez, J.E.: Tafe-net: task-aware feature embeddings for low shot learning. In: Proceedings of the IEEE Conference on Computer Vision and Pattern Recognition, pp. 1831–1840 (2019)
18. Zhu, P., Wang, H., Saligrama, V.: Generalized zero-shot recognition based on visually semantic embedding. In: Proceedings of the IEEE Conference on Computer Vision and Pattern Recognition, pp. 2995–3003 (2019)

Playing with Food: Learning Food Item Representations Through Interactive Exploration

Amrita Sawhney(✉), Steven Lee, Kevin Zhang, Manuela Veloso, and Oliver Kroemer

Carnegie Mellon University, Pittsburgh, PA 15232, USA
{amritasa,stevenl3,klz1,mmv,okroemer}@cs.cmu.edu

Abstract. A key challenge in robotic food manipulation is modeling the material properties of diverse and deformable food items. We propose using a multimodal sensory approach to interact and play with food that facilitates the ability to distinguish these properties across food items. First, we use a robotic arm and an array of sensors, which are synchronized using ROS, to collect a diverse dataset consisting of 21 unique food items with varying slices and properties. Afterwards, we learn visual embedding networks that utilize a combination of proprioceptive, audio, and visual data to encode similarities among food items using a triplet loss formulation. Our evaluations show that embeddings learned through interactions can successfully increase performance in a wide range of material and shape classification tasks. We envision that these learned embeddings can be utilized as a basis for planning and selecting optimal parameters for more material-aware robotic food manipulation skills. Furthermore, we hope to stimulate further innovations in the field of food robotics by sharing this food playing dataset with the research community.

1 Introduction

Knowledge of an object's material properties is important for robots learning to perform tasks that involve physical interactions. However, obtaining material properties for deformable objects can be difficult and time consuming. Food items, in particular, vary widely between and within food types depending on factors such as how they were grown, how they were stored, and whether they have been cooked [10]. As a result of our prior knowledge, humans typically can ascertain the basic properties of many different food items using only vision. For properties that are not always clearly conveyed through vision, humans will touch or interact with objects in order to disambiguate its internal material properties, such as knocking on a watermelon to determine its ripeness. Analogously, we

A. Sawhney, S. Lee, K. Zhang—Equal Contribution.

This work was supported by Sony AI.

B. Siciliano et al. (Eds.): ISER 2020, SPAR 19, pp. 309–322, 2021.
https://doi.org/10.1007/978-3-030-71151-1_28

believe that the ability to distinguish properties between food items can be learned using multimodal sensor data from interactive robot exploration [15].

In this paper, we present a unique multimodal food interaction dataset consisting of vision, audio, proprioceptive, and force data acquired autonomously through robot interactions with a variety of food items. Additionally, we use this dataset to explore a self-supervised method for learning embeddings that encode various food material properties and use visual data as input. The network used to learn these embeddings is trained using a triplet loss formulation, which groups similar and dissimilar samples based on the different types of interactive sensor data. In this manner, the robot can learn complex representations of these items autonomously without the need for subjective and time-consuming human labels.

To demonstrate the utility of the learned representations, we subsequently use the learned embeddings as input features to train regressors and classifiers for different tasks and compare their performances with baseline vision-only and audio-only approaches. Our experiments show that regressors and classifiers that use our learned embeddings outperform similar baseline networks on a variety of tasks, indicating that the visual embedding network encodes additional material property information from the different modalities, without requiring the robot to interact with the object at test time. Our project website is located here[1] along with a link to our dataset here[2].

2 Related Work

Many recent works have utilized simulations in order to learn dynamics models and material properties of deformable objects [20,29]. For example, Matl et al. [21] collect data on granular materials and compare the visual depth information with simulation results in order to infer their properties. Yan et al. [32] learn latent dynamics models and visual representations of deformable objects by manipulating them in simulation and using contrastive estimation, which is similar to our approach. In comparison, we worked with real-world robots and objects to collect data since representations of food that are learned through a simulation environment may not accurately transfer to real world due to variable behaviors during complex tasks, such as large plastic deformations during cutting. There are simulators that can simulate large elasto-plastic deformation [4], but they are computationally expensive, unavailable to the public, and have not yet shown their efficacy in this particular domain.

Other works also use a variety of multimodal sensors to inform a robot of deformable object properties in order to better manipulate them [3,13,17]. Erickson et al. [8] use a highly specialized near-infrared spectrometer and texture imaging to classify the materials of objects. Meanwhile, Feng et al. [9] use a visual network and forces to determine the best location at which to skewer a variety of food items for bite acquisition. Finally, Zhang et al. [35] use forces and

[1] https://sites.google.com/view/playing-with-food.
[2] https://tinyurl.com/playing-with-food-dataset.

contact microphones to classify the hardness of ingredients in order to adjust the cutting parameters for slicing actions. In contrast to these works, we only use overhead images as input to our embedding network during inference time, while still incorporating multi-modal data during training.

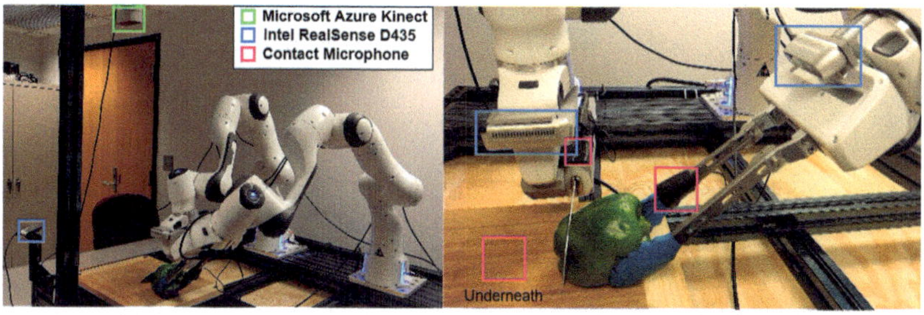

Fig. 1. Our cutting experimental setup.

Fig. 2. Images from the 4 cameras spread around our cutting experimental setup.

Numerous works have focused on using interactive perception to learn about objects in the environment [2]. Katz and Brock [14] used interactive perception to determine the location and type of articulation on a variety of random objects. Sinapov et al. [26–28] had a robot interact with objects using vision, proprioception, and audio in order to categorize them. Chu et al. [5] utilized 2 Biotac sensors on a PR2 along with 5 exploratory actions in order to learn human labeled adjectives from haptic feedback. In our work, we focus our exploratory data collection on deformable food objects instead of rigid objects. Unlike the Epic Kitchens dataset [6], which captures a first person view of humans cooking, we capture proprioceptive information using a robot and repeat consistent interactive actions across food items.

Finally, researchers have been interested in learning deep embeddings for objects in order to create simpler representations for a variety of tasks [1,30]; however, performing the same task for food items has not been studied extensively. Sharma et al. [25] learn a semantic embedding network to represent the size of food slices in order to plan a sequence of cuts. On the other hand, Isola et al. [12] used human labeled adjectives to generalize the transformation of object states, such as the ripeness in fruit, over time. Although the above works learn embeddings for food objects, both of them focus on using solely visual inputs during both train and test time, while we incorporate additional synchronized multi-modal sensory information during the training of our vision-based embedding networks.

3 Robotic Food Manipulation Dataset

3.1 Experimental Setup

Our data collection pipeline involves two different experimental setups: one for robotic food cutting and another for food playing, wherein the robot interacts with the food slices that have already been cut. We collected multi-modal sensor data during the cutting and playing data collection processes using our Franka robot control framework [34]. The sections below detail the setups for both cutting and playing in more detail.

Robot Cutting: Our experimental setup for cutting data collection consists of two Franka Emika Panda Arms mounted on a Vention frame with a cutting board in the center, as shown in Fig. 1. One arm is grasping a custom knife attachment while the other has a set of 8" kitchen tongs mounted to its fingers. There are four cameras attached to the setup: an overhead Microsoft Azure Kinect Camera, a side-view Intel Realsense D435, another D435 mounted above the wrist of the robot holding the knife, and a third D435 mounted on the wrist of the robot holding the tongs. Sample images from each camera are shown in Fig. 2. In addition, there are 3 contact microphones: one mounted underneath the center of the cutting board, another mounted on the knife attachment, and the last mounted on the tongs.

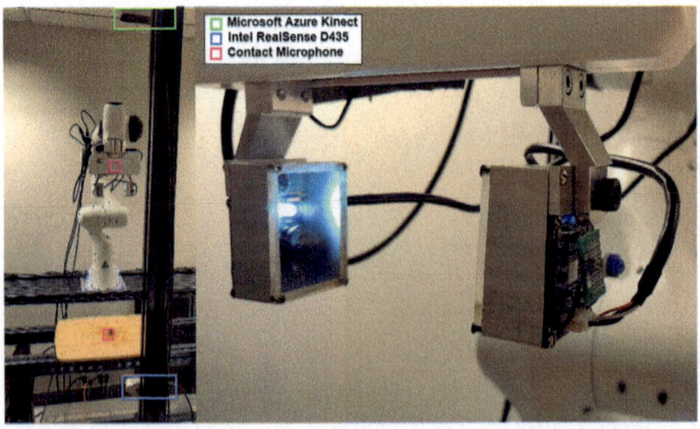

Fig. 3. Our playing experimental setup.

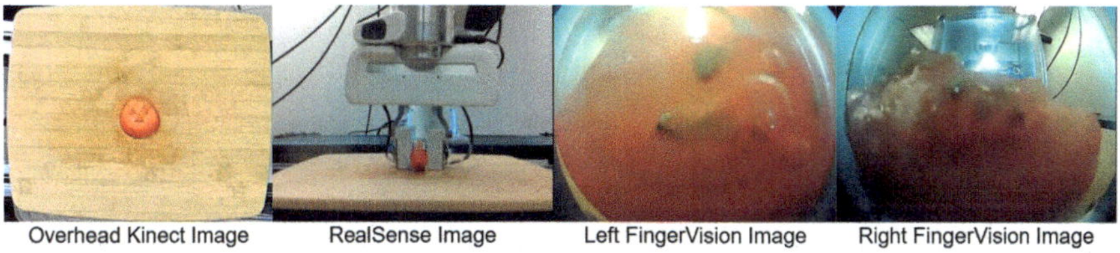

Fig. 4. Examples of Kinect, Realsense, and FingerVision images of a cut tomato.

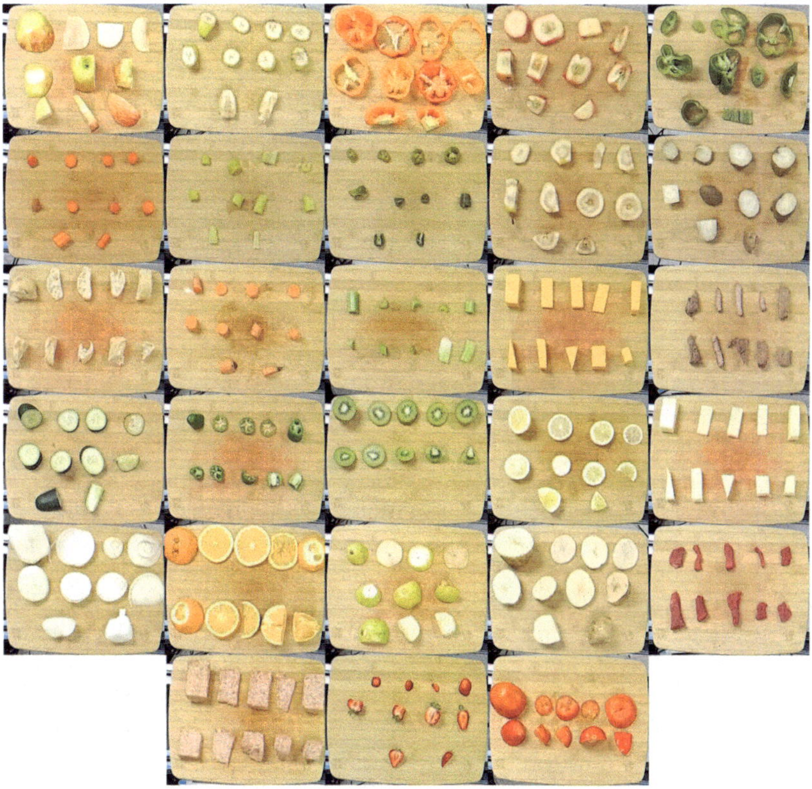

Fig. 5. Food item slices. From left to right, top to bottom: apple, banana, bell pepper, boiled apple, boiled bell pepper, boiled carrot, boiled celery, boiled jalapeno, boiled pear, boiled potato, bread, carrot, celery, cheddar, cooked steak, cucumber, jalapeno, kiwi, lemon, mozzarella, onion, orange, pear, potato, raw steak, spam, strawberry, and tomato.

Robot Playing: For our playing setup, we have a single Franka Emika Panda Arm mounted on a Vention frame with a cutting board in the center, as shown in Fig. 3. We mounted an overhead Microsoft Azure Kinect Camera and a frontal Intel Realsense D435 that faces the robot. We attach a fisheye 1080P USB Camera[3] to each fingertip as in FingerVision [31], except we use a laser-cut clear acrylic plate cover instead of a soft gel-based cover over the camera. This acrylic plate allows us to observe the compression of the object being grasped relative to fingertips. We also added a white LED to better illuminate the object while it is being grasped. Images from all of the cameras are shown in Fig. 4. We have 2 contact microphones on the setup: one mounted underneath the center of the cutting board and the other mounted on the back of the Franka Panda hand. The Piezo contact microphones[4] from both setups capture vibro-tactile feedback through the cutting board, fingers, and tools. The audio from the contact microphones of both the cutting and playing setup are captured using a Behringer UMC404HD Audio Interface[5] and synchronized with ROS using sounddevice_ros [33].

3.2 Data Collection

Using our robot cutting setup mentioned in Sect. 3.1, we taught the robot simple cutting skills using Dynamic Movement Primitives (DMPs) [16,23]. Through ridge regression, we fitted DMP parameters to trajectories collected using kinesthetic human demonstrations as in [35]. Afterwards, we chained DMPs into multiple slicing motions until the specified food item was cut completely through.

In total, we cut 10 slices each from 21 different food types which include: apples, bananas, bell peppers, bread, carrots, celery, cheddar, cooked steak, cucumbers, jalapenos, kiwis, lemons, mozzarella, onions, oranges, pears, potatoes, raw steak, spam, strawberries, and tomatoes. The slices are enumerated from 1 to 14 and were created using similar skill parameters across the food types. The skill parameters vary in slice thickness from 3mm to 50mm, angles from ±30 degrees, and slice orientation where we had normal vs. in-hand cuts when the knife robot cut between the tongs. There are more than 10 slice types because food items are shaped differently and not all cuts can be executed on every food item, especially the angled cuts. While the robot is cutting the food items, we collect audio, image, force, and proprioceptive data. The resulting slices from various food items are shown in Fig. 5.

After the 10 different types of slices have been created, we transfer them to the playing robot setup and begin the data collection process. We run 5 trials on each of the 10 slices in order to capture variations in the objects' behaviors due to differing initial orientations, positions, and changes over time. In each trial, we

---

[3] https://www.amazon.com/180degree-Fisheye-Camera-usb-Android-Windows/dp/B00LQ854AG/.

[4] https://www.amazon.com/Agile-Shop-Contact-Microphone-Pickup-Guitar/dp/B07HVFTGTH/.

[5] https://www.amazon.com/BEHRINGER-Audio-Interface-4-Channel-UMC404HD/dp/B00QHURLHM.

capture a RGBD image from the overhead Kinect and specify the center of the object. Afterwards, the robot closes its fingers and pushes down on the object until 10N of force is measured. We record the robot's position and forces during this action. Next, the robot resets to a known position, grasps the object, and releases it from a height of 15cm. During the push, grasp, and release actions, we record videos from the Realsense and audio from the two contact microphones as mentioned in Sect. 3.1. Videos are only recorded from the FingerVision cameras during the grasp and release actions. Additionally, we record the gripper width when the grasp action has finished and save RGBD images from the overhead Kinect before and after each trial.

Our full dataset is available for download here[6]. The data is located in the appropriately named folders and are sorted first by food type, then slice type, and finally trial number. Additionally, we provide food segmentation masks in the silhouette data folder where we used Deep Extreme Cut (DEXTR) [19] to obtain hand labeled masks of the objects in the overhead and side view images. Then we fine-tuned a PSPNet [36] pre-trained on the Ade20k [37] dataset with our manually labeled masks to generate additional neural network labeled segmentation masks. Finally, we have additional playing data in the old playing data folder, but those slices were all hand cut, and the data was collected in a different environment.

3.3 Data Processing

To train the embedding networks, we first extract features from the data. We transform the raw audio data from both the cutting data and playing data (during the release action, push-down action, and grasp action) into Mel-frequency cepstrum coefficient (MFCC) features [18] using Librosa [22]. These features have been shown to effectively differentiate between materials and contact events [35]. Subsequently, we use PCA to extract a lower-dimensional representation of the cutting ($\mathcal{A}_{cut}$) and playing ($\mathcal{A}_{play}$) audio features.

To form proprioceptive features ($\mathcal{P}$), we use the robot poses and forces from the push down and grasp actions. More specifically, for the push down action we extract the final z position (z_f) of the robot's end-effector once 10N of force has been reached. Using this value, we also find the change in z position between the point of first contact and z_f as Δz, which is an indication of the object's stiffness. We retrieve the final gripper width (w_g) during the grasping action once 60N of force has been reached. These three values are combined to form $\mathcal{P}$. Labels for each data sample, such as food class label ($\mathcal{F}$) and slice type label ($\mathcal{S}$), are also created according to the food type and slice type performed during cutting.

4 Learning Food Embeddings

We train convolutional neural networks, in an unsupervised manner, to output embeddings from overhead images. Our architecture is comprised of

[6] https://tinyurl.com/playing-with-food-dataset.

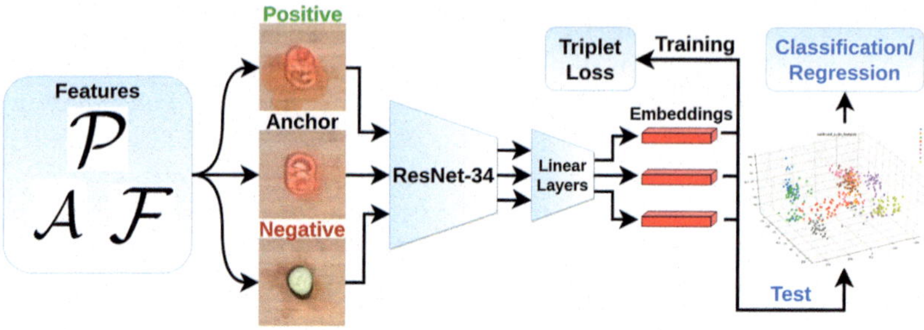

Fig. 6. An overview of our approach. The different features (modalities) defined in Sect. 3.3 are used to form triplets to learn embeddings in an unsupervised manner, which are used for supervised classification and regression tasks (blue text).

ResNet34 [11], which is pretrained on ImageNet [7] and has the last fully connected layer removed. We add an additional three hidden layers, with ReLU activation for the first two, to reduce the dimensionality of the embeddings. We use a triplet loss [24] to train the network, so similarities across food types are captured in the embeddings.

The different modalities of data mentioned in Sect. 3.3 are used as metrics to form the triplets. More specifically, the food class labels ($\mathcal{F}$), slice type labels ($\mathcal{S}$), playing audio features ($\mathcal{A}_{play}$), cutting audio features ($\mathcal{A}_{cut}$), proprioceptive features ($\mathcal{P}$), and combined audio and proprioceptive features ($\mathcal{A}_{play}$+$\mathcal{P}$) are used as metrics. For each sample in the training set, we define the n nearest samples in the PCA feature space (using the L2 norm) as the possible positive samples in a triplet and all other samples as possible negative samples, where n is a hyperparameter (n = 10 was used here). At training time, triplets are randomly formed using these positive/negative identifiers. Figure 6 shows an overview of our approach.

To evaluate the usefulness of these learned embeddings, we train multiple 3-layer multilayer perceptron classifiers and regressors for a variety of tasks, using the learned embeddings as inputs. These results are presented in Sect. 5. When combining multiple modalities, we concatenate the embeddings output from each separate network.

5 Experiments

As mentioned in the previous section, embedding networks were trained using the food class label ($\mathcal{F}$), slice type label ($\mathcal{S}$), playing audio features ($\mathcal{A}_{play}$), cutting audio features ($\mathcal{A}_{cut}$), proprioceptive features ($\mathcal{P}$), and combined audio and proprioceptive features ($\mathcal{A}_{play}$+$\mathcal{P}$) for creating triplets. These embeddings were then used to train the multi-layer perceptrons, mentioned in Sect. 4, to predict the labels or values for five different tasks: classifying food type (21 classes), predicting slice width (the width of the gripper after grasping), classifying the

Table 1. Baseline and multi-layer perceptron results on 5 evaluation tasks using different learned embedding networks that were trained on our full dataset.

Embeddings	Food Type Accuracy - 21 classes (%)	Hardness Accuracy - 3 classes (%)	Juiciness Accuracy - 3 classes (%)	Slice Type Accuracy - 14 classes (%)	Slice Width RMSE (mm)
$\mathcal{F}$	92.0	40.7	36.6	12.9	10.9
$\mathcal{S}$	17.1	37.0	34.9	**40.5**	11.8
$\mathcal{A}_{play}$	85.7	35.0	**46.0**	17.1	9.9
$\mathcal{A}_{cut}$	93.5	33.5	45.6	16.8	11.3
$\mathcal{P}$	49.5	**47.1**	37.0	20.0	**7.9**
$\mathcal{A}_{play}+\mathcal{P}$	83.8	36.4	40.2	21.4	9.5
ResNet	**98.9**	34.9	36.5	30.0	13.9
Classifier w/ $\mathcal{A}_{play}$ as input	84.4	40.8	34.0	30.1	34.4

hardness (3 human-labeled classes - hard, medium, and soft), classifying juiciness (3 human-labeled classes - juicy, medium, dry), and classifying slice type (14 different classes based on the type of cuts the cutting robot performed to generate the slice). For our two baselines, we trained convolutional neural networks, with a ResNet34 architecture, that use only visual data and another set of 3-layer mutli-layer perceptrons that use only $\mathcal{A}_{play}$data as input to generate predictions for each of the tasks.

To assess the generalizability of our approach, we evaluated the hardness and juiciness classification tasks based on leave-one-out classification, where we left an entire food class out of the training set and evaluated the trained classifiers on this class at test time. We then averaged the results across the 21 leave-one-out classification trials. The performance of all the trained networks on each of the tasks are shown in Table 1.

As illustrated in Table 1, the purely visual baseline outperforms our embeddings in the food type classification task due to the vast number of labeled images in the ImageNet dataset that ResNet was pre-trained on. However, ResNet performs worse on the other 4 tasks as ImageNet did not contain prior relevant information on the physical properties that are important for these tasks. Additionally, ResNet was trained to differentiate object classes instead of finding similarities between classes, so when an entire food category was left out of the training dataset, it most likely had no way of extrapolating the correct answer from previous data. Meanwhile, our embeddings contained auxiliary information that encoded the similarity of slices through various multimodal features, without ever being given explicit human labels. This indicates that our interactive multimodal embeddings provided the neural networks with a greater ability to generalize to unseen data as compared to the supervised, non-interactive baselines.

Additionally, the results show that the audio embeddings provide some implicit information that can help the robot distinguish vegetable types. On the other hand, it makes sense that the proprioceptive embeddings are more useful at predicting hardness and slice width as their triplets were generated using similar information. However, absolute labels were never provided when training the embedding networks, so the learned embeddings encoded this relative information themselves. It should also be noted that in the hardness and juiciness leave-one-out classification tasks, some food types, such as tomatoes, were more difficult to classify when left out of the training dataset than others, such as carrots. This may be due to the small size of our diverse dataset, which has few items with similar properties.

Finally, with respect to the slice type prediction task, there were poor performances across the board due to the inherent difficulty of the task. Due to the variability of shapes between food items, the resulting slices generated by the cutting robot, while executing the same actions, differed greatly at times. Thus, it was to be expected that only the embeddings trained using slice type labels performed relatively well on this classification task. Overall, the results of the evaluations above show that certain embeddings performed better on relevant tasks, which supports the hypothesis that they are encoding information on different material properties and can be applied in different use cases.

Table 2. Results on 3 evaluation tasks for different learned embedding networks that were trained using the auxiliary cooked vs. uncooked dataset.

Embeddings	Hardness Accuracy (%)	Juiciness Accuracy (%)	Cooked Accuracy (%)
$\mathcal{A}_{play}$	98.0	62.9	98.9
$\mathcal{P}$	63.0	68.4	60.6
$\mathcal{A}_{play}$+$\mathcal{P}$	**99.7**	**70.6**	**99.1**
ResNet	90.5	66.1	90.4
Classifier w/ $\mathcal{A}_{play}$ as input	82.1	67.4	88.8

Auxiliary Study with Additional Cooked vs. Uncooked Food Data: As an addendum to the 21-food class dataset described in Sect. 3.2, we collected an additional dataset of boiled food classes to further explore and evaluate our method's ability to detect whether a food item is cooked or not through interactive learning. The additional boiled food classes collected were: apples, bell peppers, carrots, celery, jalapenos, pears, and potatoes. Each item was boiled for 10 min. Note that for these additional cooked food classes, we did not have the robot cut the food slices due to difficulties grasping the objects. We combined these boiled classes with their uncooked counterparts from the full dataset to form a 14-class dataset and conducted a subset of evaluations on the embeddings learned from this dataset, shown in Table 2.

The auxiliary study with the boiled food dataset shows that the playing audio data is effective at autonomously distinguishing between cooked and uncooked food items without being provided any human-provided labels. This is likely because cooking food significantly changes its material properties. The high performance of the ($\mathcal{A}_{play}$+ $\mathcal{P}$) embeddings on the hardness, juiciness, and cooked leave-one-out classification tasks on this smaller dataset demonstrates the ability of our approach to generalize to new data when a food class with similar properties was present during training (apples and pears, potatoes and carrots, bell peppers and jalapenos).

Figure 7a shows the playing audio features ($\mathcal{A}_{play}$) of the different cooked and uncooked food classes in this dataset using the top 3 principal components. Figure 7b visualizes the learned embeddings based on $\mathcal{A}_{play}$, also in PCA space. As shown in the plots, there is a distinct separation between the boiled and raw foods in the audio feature space and also in the learned embedding space. Within the cooked and uncooked groupings, there are certain food types that tend to cluster together. For example, the uncooked pear and apple cluster close together, which makes sense given the similarity of these two fruits. Interestingly, they remain clustered close to one another even after they are cooked, even though there is a shift in the feature space between cooked and uncooked.

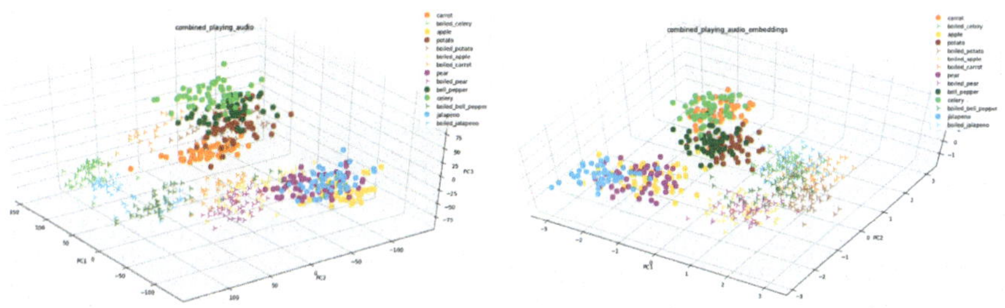

(a) Playing audio PCA features. (b) Playing audio learned embeddings.

Fig. 7. a visualizes playing audio features using PCA. b visualizes the embeddings learned using the playing audio features also in PCA space.

6 Conclusions and Future Work

In this work, we have presented a novel dataset consisting of autonomously collected audio, proprioceptive, force, and visual data that was recorded while a robot played with a variety of slices from 21 unique food types that have different shapes and properties. In addition, we learned visual embedding networks that utilized our multimodal dataset to encode properties of food items using a triplet loss formulation. These learned embeddings were shown to encode similarities between food types without explicit human labeling and outperformed normal visual-only and audio-only baselines on a variety of tasks. We hope others can utilize our publicly available dataset to explore novel ways of deciphering and encoding the material properties of food items in order to further propel research on food robotics forward.

For example, an interesting extension to this work would be to apply state of the art computer vision tracking algorithms in order to observe the deformations and movements of the slices using the videos captured from the RealSense and FingerVision cameras. This could serve as an additional metric for creating triplets or possibly be used as a basis for creating dynamics models for different types of food. In addition, since we recorded videos when cutting the slices, it may be possible to predict the shape of resulting slices given an action or vice versa.

For us, the next challenges we hope to tackle using this dataset include: monitoring the progression of food as it is being cooked in order to inform the robot when to intervene, and using the learned embeddings to better execute manipulation tasks such as cutting, flipping, mixing, picking, and placing food items. If we collect data that is complementary to this dataset then we will append the additional data to the original dataset. We also welcome others to contribute as well. Overall, we believe that this work is an exciting step towards autonomously learning about deformable food items through robotic interactions and play.

References

1. Bengio, Y., Courville, A., Vincent, P.: Representation learning: a review and new perspectives. Trans. Pattern Anal. Mach. Intelli. **35**(8), 1798–1828 (2013)
2. Bohg, J., Hausman, K., Sankaran, B., Brock, O., Kragic, D., Schaal, S., Sukhatme, G.S.: Interactive perception: leveraging action in perception and perception in action. IEEE Trans. Rob. **33**(6), 1273–1291 (2017)
3. Boonvisut, P., Çavuşoğlu, M.C.: Estimation of soft tissue mechanical parameters from robotic manipulation data. IEEE/ASME Trans. Mechatron. **18**(5), 1602–1611 (2012)
4. Chentanez, N., Müller, M., Macklin, M.: Real-time simulation of large elasto-plastic deformation with shape matching. In: Proceedings of the ACM SIGGRAPH/Eurographics Symposium on Computer Animation, pp. 159–167 (2016)
5. Chu, V., McMahon, I., Riano, L., McDonald, C.G., He, Q., Perez-Tejada, J.M., Arrigo, M., Fitter, N., Nappo, J.C., Darrell, T., et al.: Using robotic exploratory procedures to learn the meaning of haptic adjectives. In: 2013 IEEE International Conference on Robotics and Automation, pp. 3048–3055. IEEE (2013)
6. Damen, D., Doughty, H., Farinella, G., Fidler, S., Furnari, A., Kazakos, E., Moltisanti, D., Munro, J., Perrett, T., Price, W., et al.: The epic-kitchens dataset: collection, challenges and baselines. IEEE Comput. Archit. Lett. **01**, 1 (2020)

7. Deng, J., Dong, W., Socher, R., Li, L.-J., Li, K., Fei-Fei, L.: Imagenet: a large-scale hierarchical image database. In: CVPR, pp. 248–255. IEEE (2009)
8. Erickson, Z., Xing, E., Srirangam, B., Chernova, S., Kemp. C.C.: Multimodal material classification for robots using spectroscopy and high resolution texture imaging. arXiv:2004.01160 (2020)
9. Feng, R., Kim, Y., Lee, G., Gordon, E.K., Schmittle, M., Kumar, S., Bhattacharjee, T., Srinivasa, S.S.: Robot-assisted feeding: Generalizing skewering strategies across food items on a realistic plate. *arXiv preprint*arXiv:1906.02350 (2019)
10. Figura, L., Teixeira, A.A.: Food physics: physical properties-measurement and applications. Springer Science & Business Media (2007)
11. He, K., Zhang, X., Ren, S., Sun, J.: Deep residual learning for image recognition. In: CVPR, pp. 770–778 (2016)
12. Isola, P., Lim, J.J., Adelson, E.H.: Discovering states and transformations in image collections. In: CVPR (2015)
13. Jia, B., Hu, Z., Pan, J., Manocha, D.: Manipulating highly deformable materials using a visual feedback dictionary. In: 2018 IEEE International Conference on Robotics and Automation (ICRA), pp. 239–246. IEEE (2018)
14. Katz, D., Brock, O.: Manipulating articulated objects with interactive perception. In: 2008 IEEE International Conference on Robotics and Automation, pp. 272–277. IEEE (2008)
15. Kroemer, O., Niekum, S., Konidaris, G.: A review of robot learning for manipulation: Challenges, representations, and algorithms. *arXiv preprint*arXiv:1907.03146 (2019)
16. Kroemer, O., Sukhatme, G.: Meta-level priors for learning manipulation skills with sparse features. In: International Symposium on Experimental Robotics, pp. 211–222. Springer (2016)
17. Li, Q., Kroemer, O., Su, Z., Veiga, F.F., Kaboli, M., Ritter, H.J.: A review of tactile information: Perception and action through touch. IEEE Trans. Robot. **36**(6), 1619–1634 (2020)
18. Logan, B., et al.: Mel frequency cepstral coefficients for music modeling. Ismir **270**, 1–11 (2000)
19. Maninis, K., Caelles, S., Pont-Tuset, J., Van Gool, L.: Deep extreme cut: from extreme points to object segmentation. In: CVPR (2018)
20. Matas, J., James, S., Davison, A.J.: Sim-to-real reinforcement learning for deformable object manipulation (2018)
21. Matl, C., Narang, Y., Bajcsy, R., Ramos, F., Fox, D.: Inferring the material properties of granular media for robotic tasks. arXiv:2003.08032 (2020)
22. McFee, B., Raffel, C., Liang, D., Ellis, D.P., McVicar, M., Battenberg, E., Nieto, O.: librosa: Audio and music signal analysis in python (2015)
23. Schaal, S., Peters, J., Nakanishi, J., Ijspeert, A.: Learning movement primitives. In: Robotics Research the Eleventh International Symposium, pp. 561–572. Springer (2005)
24. Schroff, F., Kalenichenko, D., Philbin, J.: Facenet: a unified embedding for face recognition and clustering. In: CVPR, pp. 815–823 (2015)
25. Sharma, M., Zhang, K., Kroemer, O.: Learning semantic embedding spaces for slicing vegetables. arXiv:1904.00303 (2019)
26. Sinapov, J., Bergquist, T., Schenck, C., Ohiri, U., Griffith, S., Stoytchev, A.: Interactive object recognition using proprioceptive and auditory feedback. Int. J. Robot. Res. **30**(10), 1250–1262 (2011)

27. Sinapov, J., Schenck, C., Staley, K., Sukhoy, V., Stoytchev, A.: Grounding semantic categories in behavioral interactions: Experiments with 100 objects. Robot. Autonomous Syst. **62**(5), 632–645 (2014)
28. Tatiya, G., Sinapov, J.: Deep multi-sensory object category recognition using interactive behavioral exploration. In: ICRA, pp. 7872–7878. IEEE (2019)
29. Wu, Y., Yan, W., Kurutach, T., Pinto, L., Abbeel, P.: Learning to manipulate deformable objects without demonstrations (2020)
30. Xie, J., Girshick, R., Farhadi, A.: Unsupervised deep embedding for clustering analysis. In: ICML, pp. 478–487 (2016)
31. Yamaguchi, A., Atkeson, C.G.: Combining finger vision and optical tactile sensing: Reducing and handling errors while cutting vegetables. In: Humanoids, pp. 1045–1051. IEEE (2016)
32. Yan, W., Vangipuram, A., Abbeel, P., Pinto, L.: Learning predictive representations for deformable objects using contrastive estimation (2020)
33. Zhang, K.: https://github.com/firephinx/sounddevice_ros
34. Zhang, K., Sharma, M., Liang, J., Kroemer, O.: A modular robotic arm control stack for research: Franka-interface and frankapy. arXiv preprintarXiv:2011.02398 (2020)
35. Zhang, K., Sharma, M., Veloso, M., Kroemer, O.: Leveraging multimodal haptic sensory data for robust cutting. In: Humanoids, pp. 409–416 (2019)
36. Zhao, H., Shi, J., Qi, X., Wang, X., Jia, J.: Pyramid scene parsing network. In: CVPR, pp. 2881–2890 (2017)
37. Zhou, B., Zhao, H., Puig, X., Fidler, S., Barriuso, A., Torralba, A.: Scene parsing through ade20k dataset. In *CVPR*, pages 633–641, 2017

Data-Driven Design of Energy-Shaping Controllers for Swing-Up Control of Underactuated Robots

Wankun Sirichotiyakul(✉) and Aykut C. Satici

Mechanical and Biomedical Engineering Department, College of Engineering, Boise State University, Boise, ID 83725, USA
wankunsirichotiyakul@u.boisestate.edu

Abstract. We propose a novel data-driven procedure to train a neural network for the swing-up control of underactuated robotic systems. Our approach is inspired by several recent developments ranging from nonlinear control theory to machine learning. We embed a neural network indirectly into the equations of motion of the robotic manipulator as its control input. Using familiar results from passivity-based and energy-shaping control literature, this control function is determined by the appropriate gradients of a neural network, acting as an energy-like (Lyapunov) function. We encode the task of swinging-up robotic systems through the use of transverse coordinates and goal sets; which drastically accelerates the rate of learning by providing a concise target for the neural network. We demonstrate the efficacy of the algorithm with both numerical simulations and experiments.

1 Introduction

Motivation. The swing-up control design of pendulum systems has been the subject of innumerable research papers until today. Design approaches ranging from bang-bang control, energy-based approaches, and many others have been extensively studied [1,8,9,11]. These approaches require precise knowledge of the mathematical model describing the system. For systems with high complexity, development of controllers using these methods can be intractable.

Data-driven approaches are more flexible in the sense that the same framework can be applied to many different systems. In this category, the main idea is to use data to find a control policy such that the closed-loop behavior of the system optimizes a certain objective given by some notion of accumulative reward/cost. One of the most commonly used techniques in this area of research is reinforcement learning [18], which seeks a direct mapping from the system states to the control inputs.

Problem Statement. The majority of data-driven approaches treat the closed-loop system as a black-box, i.e. containing no predetermined structure from the governing laws of physics. While this is a flexible solution to many control problems, the vast amount of training data required for training is often prohibitive.

B. Siciliano et al. (Eds.): ISER 2020, SPAR 19, pp. 323–333, 2021.
https://doi.org/10.1007/978-3-030-71151-1_29

Recent advances in machine learning research have shown that it is advantageous to combine available knowledge of the system with the learning framework. In this paper, we discuss the data-driven control design which directly incorporates the dynamics of the system, drastically increasing the rate of learning. The quintessential control problem that we tackle is the swing-up control of pendulum systems.

Related Work. In [5], the dynamics of the system is first learned and then later incorporated into a policy search. This addresses the poor sample complexity of model-free reinforcement learning, but does not allow physical structures of the system to be directly incorporated. Neural ordinary differential equation (ODE) [2], is a recently developed framework that connects deep neural networks to continuous-time dynamical systems. This approach has provided researchers with a modeling basis for incorporating physical structures into their machine learning problems, e.g. using neural ODEs to learn the Hamiltonian dynamics of physical systems [19].

The very recent extension to neural ODE [14] enables the direct incorporation of a neural network into the differential equation governing the evolution of system. We combine this neural-network-embedded ODE with passivity-based control techniques to tackle the swing-up control design problem. The main difference in our work and [19] is that we do not learn the Hamiltonian dynamics using a neural ODE. Rather, we leverage the known dynamics and train a neural network to learn an energy-like function, from which the controller is derived, thereby imposing desirable characteristics on the closed-loop system.

2 Technical Approach

The form of the controllers we design in this paper is based on passivity-based/energy-shaping control. We will borrow from [10,15] to describe the general outline of this method. Let $x \in \mathbb{R}^{2n}$ denote the state of the robot. The state x may be represented in terms of the generalized positions and velocities $x = (q, \dot{q})$ or positions and momenta $x = (q, p)$. The system's equations of motion is

$$\begin{bmatrix} \dot{q} \\ \dot{p} \end{bmatrix} = \begin{bmatrix} 0 & I_n \\ -I_n & 0 \end{bmatrix} \begin{bmatrix} \nabla_q H \\ \nabla_p H \end{bmatrix} + \begin{bmatrix} 0 \\ G(q) \end{bmatrix} u.$$

where $H(q,p) = \frac{1}{2} p^\top M^{-1}(q) p + V(q)$ is the Hamiltonian of the robot, M is the mass matrix, and $V(q)$ represents the potential energy. Energy-shaping control, which is closely tied to passivity-based control [15], uses the control input $u \in \mathscr{U} \subseteq \mathbb{R}^m$ to impose a desired (closed-loop) energy-like function: $H_d : \mathbb{R}^{2n} \to \mathbb{R}$. In [10], this function was chosen to be a quadratic function of the system momenta:

$$H_d(q,p) = \frac{1}{2} p^\top M_d^{-1}(q) p + V_d(q),$$

however, in this work we do not constrain ourselves to desired Hamiltonians of this particular form. Rather, we use a neural network to represent H_d and devise a learning framework to obtain the desirable characteristics automatically. From H_d, we then derive the controller, comprised of an energy-shaping term and a damping injection term as follows

$$u(x) = u_{es}(q,p) + u_{di}(q,p) = -(G^\top G)^{-1}G^\top \frac{\partial H_d}{\partial q} - K_d G^\top \frac{\partial H_d}{\partial p}. \tag{1}$$

2.1 Main Learning Problem

To train the neural net, we formulate the control design for swinging-up robotic mechanisms as an optimization problem of $H_d(x;\theta)$. In particular, we have

$$\begin{aligned} \underset{\theta \in \mathbb{R}^p}{\text{minimize}} \quad & J(\theta) = \int_{t_0}^{t_f} r(\phi, u(\phi;\theta))\, dt, \\ \text{subject to} \quad & \dot{x} = f(x) + g(x)u(x;\theta), \quad x(t_0) = x_0. \end{aligned} \tag{2}$$

where $\theta \in \mathbb{R}^p$ denotes the neural network parameters, $\phi = \phi(t; x_0, \theta)$ is the flow of the equations of motion, the time horizon $t_f \in (0, \infty)$ is a hyperparameter and $r : \mathbb{R}^{2n} \times \mathbb{R}^m \to \mathbb{R}$ is a running cost function.

The running cost function $r(\phi, u)$ that yields the loss function $J(\theta)$ consists of 1) distance, $r_\perp$, to some preferred orbit, 2) set distance, r_{set}, between the current prediction and the goal set, 3) Double-hinge loss, r_{hinge}, to penalize multiple swings of the first link, and 4) the regularization term, r_{reg}, to avoid overtraining the neural network parameters. The running cost is found by the sum of these components:

$$r(\theta) := r_\perp(\phi, u) + r_{\text{set}}(\phi, u) + r_{\text{hinge}}(\phi, u) + r_{\text{reg}}(\theta),$$

where each term depends on θ through the dependence of the trajectory ϕ on θ. We now elaborate on the details of each component of r as follows.

1) Distance to preferred orbit: The first term that makes up the running loss is the distance to a preferred orbit, $\phi^\star$. In Sect. 3, where we apply this framework to the simple pendulum and inertial wheel pendulum, this preferred orbit was chosen to be the homoclinic orbit of the pendulum. In general, such a homoclinic orbit may not be available in closed-form. In this case, the preferred orbit may be computed using one of several methods that are available in the literature, such as trajectory optimization [4], virtual holonomic constraints [16], etc.

Once the preferred orbit has been chosen, the next step is to compute the *transverse coordinates* along this orbit, using the ideas outlined in [7,17]. This method reparametrizes the states of the dynamical system by assigning one state, τ, to be a parameter that determines where along the specific trajectory the state of the system is. The remaining states, $x_\perp \in \mathbb{R}^{2n-1}$, are chosen transversal to the trajectory, with the condition $\phi \to \phi^\star$ if and only if $x_\perp \to 0$ as $t \to \infty$. We penalize the the prediction trajectory by incurring a quadratic loss on $x_\perp$ and on the control effort u:

$$r_\perp(\phi, u) = \frac{1}{2} x_\perp^\top Q x_\perp + \frac{1}{2} u^\top R u \tag{3}$$

2) Set distance: The ultimate goal of the framework is to swing-up the robotic system to an upward (unstable) equilibrium point. We encourage the controller to move the system states to a set around the upward equilibrium point by penalizing whenever the prediction trajectory spends time away from this set.

The penalty, r_{set} is constructed by defining a convex open neighborhood $\mathcal{S}$ of the upward equilibrium point and computing the set distance of the predicted closed-loop trajectory $\phi(t)$ to $\mathcal{S}$, i.e.

$$r_{\text{set}}(\phi) = \text{dist}(\gamma, \mathcal{S}) = \inf\left\{\|x-y\| : x \in \phi(t),\ t \in (t_0, t_f),\ y \in S\right\}. \tag{4}$$

For instance, the set $\mathcal{S}$ can be selected as a ball around the upward equilibrium point of radius $d > 0$ in the standard norm topology. In this case, d is to be selected by the user before the training. If any point along the prediction $\phi(t)$ gets closer than d to the upward equilibrium point, no additional loss is incurred.

3) Double-hinge loss: We want penalize those prediction trajectories ϕ which overshoot the upward equilibrium point and require the system to swing multiple times before reaching the set S using a double-hinge loss term, r_{hinge}. With q_i corresponding to the angle of the ith link, we have

$$r_{\text{hinge}}(\phi, u) = \begin{cases} 0, & |q_i| \leq 2\pi, \\ |q_i| - 2\pi, & \text{otherwise}. \end{cases} \tag{5}$$

2.2 Imposing Symmetry Constraints

For dynamical systems that are naturally symmetric about the origin, i.e. $f(x) = -f(-x)$, it often is desirable to maintain symmetry in the closed-loop system by designing a symmetric control policy. To obtain this in the learned control function of the form in Eq. (1), where H_d is a neural network, we encode this information directly into the neural net by applying a symmetric function h to the input. To demonstrate, a two-layer neural net and its gradient with respect to the input x is are

$$\begin{aligned} H_d(x) &= W_2\,\sigma(W_1 h(x) + b_1) + b_2, \\ \nabla_x H_d(x) &= W_2\,\sigma'(W_1 h(x) + b_1) \odot W_1 \odot h'(x). \end{aligned}$$

Selecting $h(x) = h(-x)$ and hence $h'(x) = -h'(-x)$, we obtain the desired symmetry, i.e. $\nabla_x H_d(x) = -\nabla_x H_d(-x)$, which is directly reflected in the control $u(x)$ in Eq. 2.

Imposing this symmetry constraint in the control function reduces the sample space over which the neural network has to train. This speeds up the training process and increases consistency in the behavior of the closed-loop system.

2.3 Gradients Computation Through ODE Solutions

Solving the optimization problem (2), requires the gradient computation of the $J(\theta)$ through the solution $\phi(t; x_0, \theta)$ of an ODE. The main challenge ultimately lies in the computation of $\frac{\partial \phi}{\partial \theta}$. In the recent literature [2], adjoint sensitivity methods have been used to compute this derivative by first solving a backwards ODE, known as the adjoint problem:

$$\frac{\mathrm{d}\lambda}{\mathrm{d}t} = -\lambda \frac{\partial \left(f(x) + g(x)u \right)}{\partial x}.$$

Then, the desired gradient of the loss function is computed through the expression

$$\nabla_\theta J(\theta) = \lambda(t_0) \frac{\partial \left(f(x) + g(x)u \right)}{\partial x}.$$

This approach is based on Pontryagin's maximum principle [12] and is quite mathematically elegant. However, it requires multiple forward solutions of the ODE to implement; hence, it quickly becomes quite costly to execute.

Recent advances in automatic differentiation (AD) and differentiable programming has allowed more efficient computations of the desired gradients. In [14], a number of existing automatic differentiation schemes and adjoint methods are generalized to enable efficient gradient computation of solutions to ODEs with neural-net architectures embedded in them. We use this very recent framework, which is implemented in the programming language Julia within the package `DiffEqFlux.jl`, to compute the gradients of the loss function $J(\theta)$ and use it in conjunction with ADAM [6] to solve our optimization problem (2).

3 Results

We apply our approach to robots commonly used as benchmarks: the pendulum and the cart-pole. The data-driven control designs for both systems are performed and evaluated in simulations. The controllers are also tested on real systems.

We use a simple fully-connected neural network with two hidden layers and ELU activations [3]. The controller derived from H_d is used for swing up, and a standard LQR is used to stabilize the upward equilibrium. Simulation studies are conducted from a range of initial conditions near the downward equilibrium. If the linear controller is able to catch the swing and stabilize the upward equilibrium, then the simulation is considered successful. This requires the swing-up controller to bring the state into the region of attraction of the linear controller. Performance of the swing-up controller is evaluated by the success rate of the simulations.

3.1 Pendulum

The dynamics of the pendulum is given as

$$I\ddot{\theta} = -mgl\sin(\theta) - b\dot{\theta} + u \tag{6}$$

where $m = 0.377$ kg is the combined mass, $l = 23.2$ cm is the length from the pivot point to the center of gravity of the mechanism, $I = 0.02583$ kg-m^2 is the

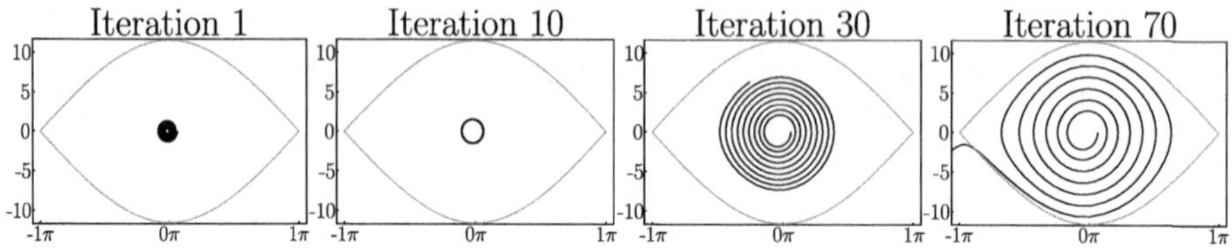

Fig. 1. Convergence to the homoclinic orbit (dashed line) of the learned controller for the physical pendulum. The horizontal and vertical axes are the pendulum angle and angular velocity, respectively.

moment of inertia of the whole mechanism, g is the gravitational constant, u is the torque generated by the actuator and is limited by $|u| \leq 0.57$ Nm. The position $\theta = 0$ corresponds to the downward equilibrium.

The architecture of the neural net used to represent H_d for this system is (2, 48, 36, 1). The first number indicates the dimension of the input layer. The last number indicates the dimension of output layer. The dimensions of each hidden layer are shown in between. The preferred orbit for calculation of $r_\perp$ in Eq. (3) is the homoclinic orbit characterized by

$$\dot{\theta}^\star(\theta) = \sqrt{\frac{2}{I}\left(E_0 - mgl(1 - \cos\theta)\right)}, \tag{7}$$

where $E_0 = 2mgl$ is the potential energy of the pendulum at the unstable equilibrium. With this orbit, the transverse coordinate is $x_\perp = \min(|\dot{\theta} \pm \dot{\theta}^\star|^2)$. The goal set $\mathcal{S}$ used to compute the set distance loss r_{set} described is defined as

$$\mathcal{S} = \left\{(\theta, \dot{\theta}) : \left\| [\theta\ \dot{\theta}]^\top - [\pm\pi\ 0]^\top \right\| \leq 0.1 \right\}.$$

A demonstration of the training progress is shown in Fig. 1. The controller learns to bring the trajectories toward the homoclinic orbit and swing up the pendulum. The learned controller is evaluated by simulating the pendulum for 15 s from a range of initial states. We perform 2,500 simulations with initial angles in the interval $[-0.9\pi, 0.9\pi]$ rad, and initial velocities in the interval $[-10, 10]$ rad/s. The data-driven controller swings up the system with a 100% success rate.

We further bolster the efficacy of our approach by implementing the learned controller on an experimental system. We start the pendulum at a random initial condition close the origin and record the angle and the commanded torque. Figure 2 shows the results of the experiment. These results demonstrate the robustness of the swing-up controller, as the system parameters in Eq. 6 are estimated. Further, the viscous friction model is only an approximation of the true friction in the bearings of the real system. Velocity measurements are also not directly available. They are estimated through the use of a digital low-pass filter.

3.2 Cart-Pole

The cart-pole experiment is conducted using the Linear Servo Base Unit with Inverted Pendulum, manufactured by Quanser [13]. It consists of an unactuated pendulum mounted (via revolute joint) to a linear cart that is actuated. The position of the pendulum is denoted by the angle θ, with the upright position as $\theta = 0$. The position of the cart, denoted by x, is constrained to $|x| < 0.4$ m. The equations of motion are

$$\begin{aligned} I_2\ddot{x} &= I_1\left(F_c - B_{eq}\dot{x} - ml\dot{\theta}^2 s_\theta\right) + ml\left(mglc_\theta s_\theta - B_p\dot{\theta}c_\theta\right), \\ I_2\ddot{\theta} &= (M+m)\left(mgls_\theta - B_p\dot{\theta}\right) + ml\left(F_c c_\theta - B_{eq}\dot{x}c_\theta - ml\dot{\theta}c_\theta s_\theta,\right), \end{aligned} \tag{8}$$

where s_θ and c_θ denote sine and cosine of θ, respectively; $I_1 = I + ml^2$; $I_2 = (M+m)I + ml^2\left(M + m\sin^2\theta\right)$; F_c is the actuator force acting on the cart; and g is gravity. The model parameters can be found on the manufacturer's website.

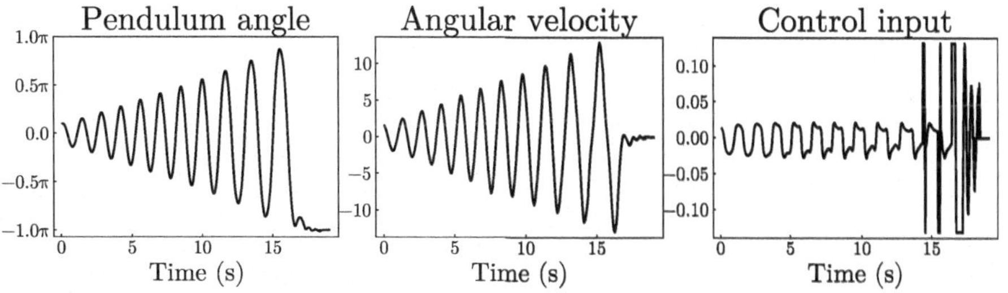

Fig. 2. Experimental results for the physical pendulum. The angle measurement contains a small amount of noise, and the velocity measurement is approximated. The learned controller is able to swing up despite these inaccuracies.

We use a neural network with the following dimensions: $(4, 64, 64, 1)$. Since the cart-pole system is symmetric, we apply the procedure described in Sect. 2.2 to impose symmetry. In particular, we process the input with $g(x) = x^2$ before feeding it to the neural network. The goal set S in the computation of the set distance loss is chosen as

$$S = \left\{\left(x, \dot{x}, \theta, \dot{\theta}\right) : \sqrt{(1-\cos\theta)^2 + \dot{\theta}^2 + \dot{x}^2} \leq 0.125\right\}.$$

We evaluate a total of 2,500 simulations starting from a range of initial states with $\theta \in \pm[30, 150]$ degrees, $\dot{\theta} \in [-8, 8]$ rad/s, and zero in other dimensions. The swing-up controller achieves a 100% success rate for the cart-pole system. The contours of $H_d(x)$ and the controller $u(x)$ derived from it are shown in Fig. 3. To demonstrate the advantages of our approach over with traditional energy-shaping control, we compare the accumulated quadratic cost to perform the swing-up on the cart-pole in Fig. 4. The same cost matrices are used for both the learned controller and the energy-shaping controller. Our approach resulted in lower accumulated cost and successful swing-up from a wide variety of initial conditions.

The results from an experimental trial are shown in Fig. 5. A video accompanying this submission also shows a recording of this specific trial. The learned controller performs the swing up without going over cart track limits. A standard LQR controller successfully catches the swing and stabilizes the upright

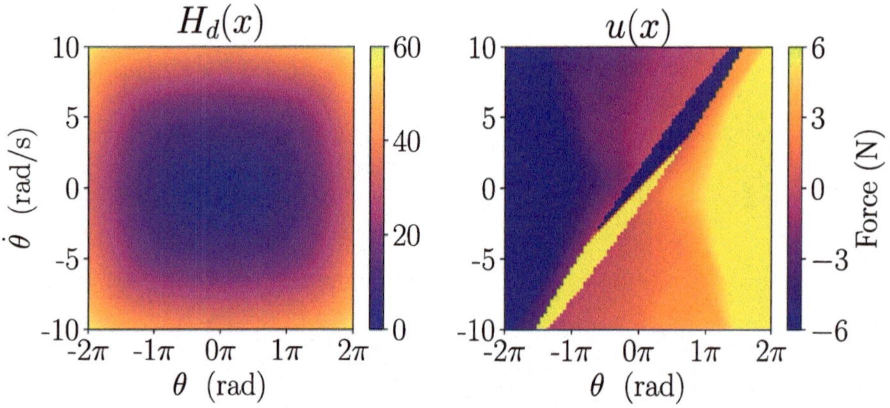

Fig. 3. (Left) The learned desired Hamiltonian $H_d(x)$, and (right) the resulting control policy $u(x)$ for the cart-pole system, projected onto the θ-$\dot{\theta}$ plane with $x = \dot{x} = 0$. The neural network used to learn H_d consists of approximately 4,500 parameters to be trained, which is relatively small in the world of machine learning.

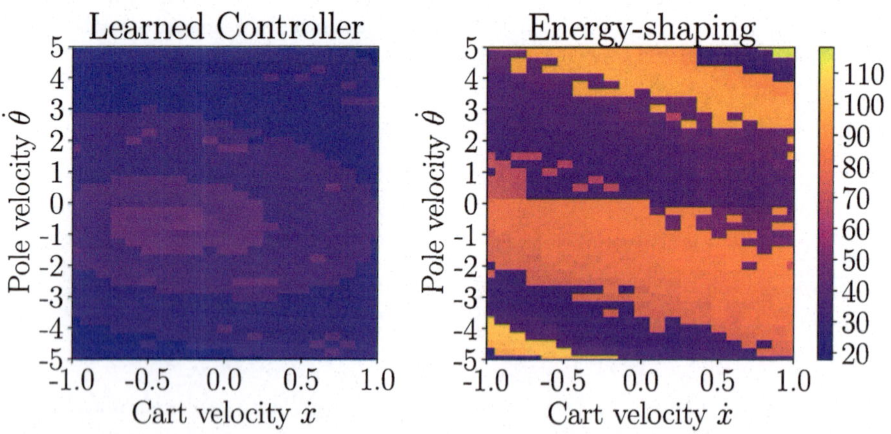

Fig. 4. Each grid represents the accumulated quadratic cost over an 8-s simulation from the corresponding initial cart and pole velocities. The initial cart position is $x = 0$ and the pole starts at the downward position. The learned controller performed the swing-up with lower cost than the traditional energy-shaping controller.

position. These results show that the learned controller is robust against the inaccuracies in velocity approximation, system parameters, and friction model.

4 Main Experimental Insights

The successful experiments bolster the efficacy and robustness of our approach. In both the pendulum and the cart-pole experiments, the system parameters and the viscous friction model used during training are approximations of the true system models. The use of a digital low-pass filter to approximate the velocities further deviates the the system from the conditions used during training. The experimental results show that our approach is robust against these uncertainties.

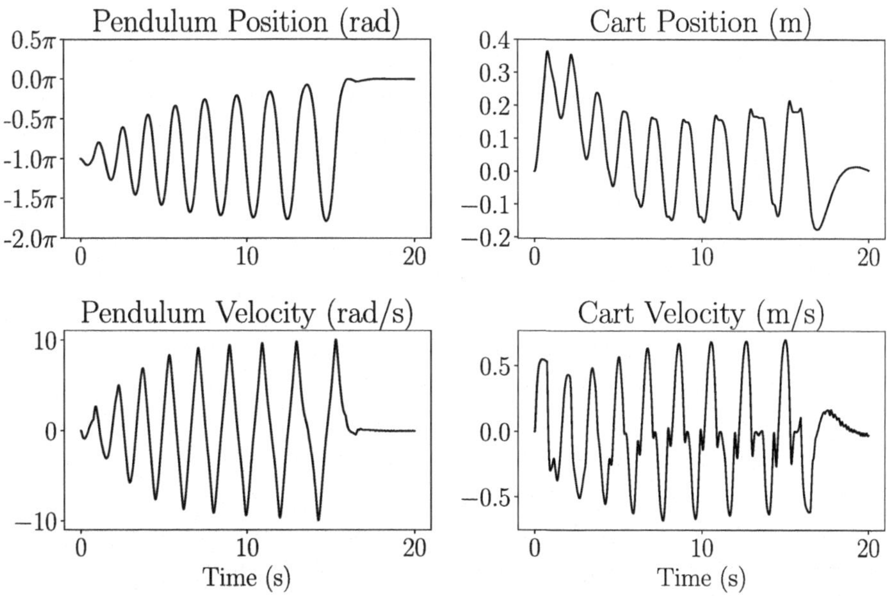

Fig. 5. Experimental results showing state evolution of the cart-pole. The learned controller is able to swing up the pendulum without violating the cart track limits.

Implementation of the traditional energy-shaping control requires careful tuning of the swing-up gain. On the cart-pole system, a gain that is too large resulted in the cart going over track limits, while one that is too small did not swing up the pole. Similar behavior was observed in the simple pendulum experiment. Our data-driven designs do not require gain tuning, simplifying the implementation of the controllers.

One challenge revealed during experiments is the computation of the gradient of H_d. This computation can be more costly than that of traditional swing-up controllers. In real-time applications where update frequency of a kilohertz or higher, the computation power requirement is a valid concern. This issue can be minimized by avoiding large neural network structures or by the use of parallel computing.

We show, through both simulations and experiments, that our data-driven approach can achieve performance levels similar to or exceeding that of traditional swing-up control designs. The incorporation of physical knowledge about the system combined with techniques from dynamics and control theory, such as transverse dynamics and energy-shaping control, lead to relatively quick training sessions. Our approach also takes into account the symmetry of the system's dynamics, resulting in a more consistent closed-loop behavior across the state space. We plan to extend this framework to accommodate systems with contacts in future work.

References

1. Åström, K.J., Furuta, K.: Swinging up a pendulum by energy control. Automatica **36**(2), 287–295 (2000)
2. Chen, R.T.Q., Rubanova, Y., Bettencourt, J., Duvenaud, D.: Neural ordinary differential equations (2018)
3. Clevert, D.A., Unterthiner, T., Hochreiter, S.: Fast and accurate deep network learning by exponential linear units (elus). arXiv preprint arXiv:1511.07289 (2015)
4. Conway, B.: Practical methods for optimal control using nonlinear programming (2002)
5. Deisenroth, M., Rasmussen, C.E.: Pilco: a model-based and data-efficient approach to policy search. In: Proceedings of the 28th International Conference on Machine Learning (ICML 2011), pp. 465–472 (2011)
6. Kingma, D.P., Ba, J.: Adam: a method for stochastic optimization. arXiv preprint arXiv:1412.6980 (2014)
7. Manchester, I.R.: Transverse dynamics and regions of stability for nonlinear hybrid limit cycles. arXiv preprint arXiv:1010.2241 (2010)
8. Mason, P., Broucke, M., Piccoli, B.: Time optimal swing-up of the planar pendulum. IEEE Trans. Autom. Control **53**(8), 1876–1886 (2008)
9. Mori, S., Nishihara, H., Furuta, K.: Control of unstable mechanical system control of pendulum. Int. J. Control **23**(5), 673–692 (1976)
10. Ortega, R., Spong, M.W., Gómez-Estern, F., Blankenstein, G.: Stabilization of a class of underactuated mechanical systems via interconnection and damping assignment. IEEE Trans. Autom. Control **47**(8), 1218–1233 (2002)
11. Park, M.S., Chwa, D.: Swing-up and stabilization control of inverted-pendulum systems via coupled sliding-mode control method. IEEE Trans. Ind. Electron. **56**(9), 3541–3555 (2009)
12. Pontryagin, L.S., Mishchenko, E., Boltyanskii, V., Gamkrelidze, R.: The mathematical theory of optimal processes (1962)
13. Quanser: Linear Servo Base Unit with Inverted Pendulum (2013). Rev. 1.0
14. Rackauckas, C., Ma, Y., Martensen, J., Warner, C., Zubov, K., Supekar, R., Skinner, D., Ramadhan, A.: Universal differential equations for scientific machine learning (2020)

15. van der Schaft, A.J., Van Der Schaft, A.: L2-gain and Passivity Techniques in Nonlinear Control, vol. 2. Springer, Heidelberg (2000)
16. Shiriaev, A., Perram, J.W., Canudas-de Wit, C.: Constructive tool for orbital stabilization of underactuated nonlinear systems: virtual constraints approach. IEEE Trans. Autom. Control **50**(8), 1164–1176 (2005)
17. Shiriaev, A.S., Freidovich, L.B.: Transverse linearization for impulsive mechanical systems with one passive link. IEEE Trans. Autom. Control **54**(12), 2882–2888 (2009)
18. Sutton, R.S., Barto, A.G.: Reinforcement Learning: An Introduction. MIT press, Cambridge (2018)
19. Zhong, Y.D., Dey, B., Chakraborty, A.: Symplectic ode-net: Learning hamiltonian dynamics with control. arXiv preprint arXiv:1909.12077 (2019)

A Recurrent Neural Network Approach to Roll Estimation for Needle Steering

Maxwell Emerson[1(✉)], James M. Ferguson[1], Tayfun Efe Ertop[1], Margaret Rox[1], Josephine Granna[1], Michael Lester[2], Fabien Maldonado[2], Erin A. Gillaspie[2], Ron Alterovitz[3], Robert J. Webster III[1], and Alan Kuntz[4]

[1] Department of Mechanical Engineering, Vanderbilt University, Nashville, USA
Maxwell.emerson@vanderbilt.edu
[2] Department of Medicine and Thoracic Surgery, Vanderbilt University Medical Center, Nashville, USA
[3] Department of Computer Science, University of North Carolina at Chapel Hill, Chapel Hill, USA
[4] Robotics Center and School of Computing, University of Utah, Salt Lake City, USA

1 Motivation

Steerable needles are a promising technology for delivering targeted therapies in the body in a minimally invasive fashion via controlled, actively steered insertions. These robotically actuated needles usually leverage an asymmetric tip [1] to take curved paths through the body, avoiding anatomical obstacles and honing in on a target (see Fig. 1, left). Methods such as duty cycling and sliding mode control enable safe, accurate, and automatic controlled steering of these needles to anatomical targets or along predetermined trajectories in the body [2,3]. These controllers require knowledge of the full 6 degrees of freedom (DOF) pose of the steerable needle's tip as it is steered through the body. To acquire this information for feedback during control, electromagnetic sensors can be embedded in the tip of the needle. However, these sensors typically fill the internal working channel of the needle, precluding the use of the needle for therapy delivery. External sensors, such as ultrasound and bi-plane fluoroscopy (a type of continuous X-ray) can sense the needle tip's position and heading (5 DOF), but are not able to sense its axial orientation (roll angle) [4]. Without full 6-DOF state measurement, an alternative method is needed to estimate the full state of the needle tip for effective steering.

Model-based observer methods have been developed to estimate the orientation of the needle during steering [5–8]. These methods work well when the system behaves similarly to the modeled system they rely on. However, when unmodeled effects dominate the system dynamics these methods can perform poorly. For flexible needles, effects such as unpredictable friction forces due to tissue interactions and long needle lengths create nonlinear torsional dynamics that are difficult to accurately model [4–6].

By contrast, model-free, data-driven approaches have recently been of great research interest in state estimation during control in other domains [9]. However the investigation of such methods has so-far been relatively limited in

B. Siciliano et al. (Eds.): ISER 2020, SPAR 19, pp. 334–342, 2021.
https://doi.org/10.1007/978-3-030-71151-1_30

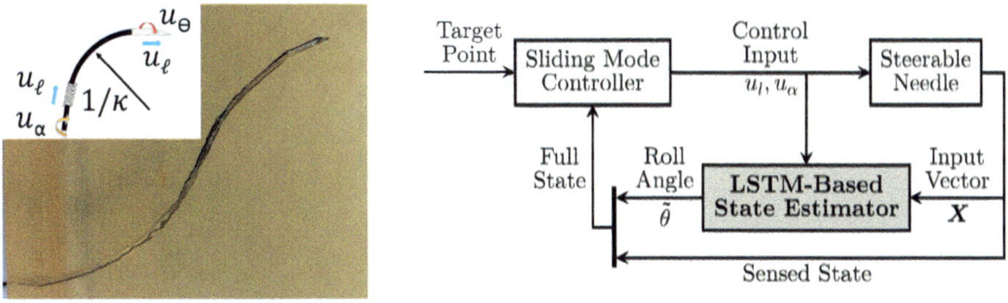

Fig. 1. *Left:* Steerable needle inserted into a gelatin tissue simulant. *Left Inset:* Kinematic diagram of the steerable needle model with un-modeled torsional compliance. This results in rotational lag between the base of the needle and tip of the needle. *Right:* Block diagram showing our LSTM-based estimator in the control loop.

needle steering, e.g., to predicting needle deflection behavior for set insertion depths [10].

In this work, we overcome the limitations present in model-based observers for needle steering by presenting an estimator that *learns* the behavior of the needle from a set of training insertions in a tissue phantom. Our method is capable of accurately estimating the needle's roll angle during new insertions and in tissues that are different from those that it trained on.

Using a model-free representation that learns the nonlinear effects from the training data, we achieve accurate state estimation that abstracts to multiple types of tissue for a system that is subject to significant modeling errors.

Here, we propose a Long-Short-Term Memory (LSTM)-based recurrent neural network (RNN) [11] to estimate the needle's roll angle during steering. Our LSTM-based network takes as input the sensed, partial state of the needle tip and recurrently estimates the needle's unsensed roll angle at each time step. We use the network to estimate the needle state in a sliding mode controller and demonstrate highly accurate steering in multiple mediums—including *ex vivo* ovine brain and *ex vivo* porcine inflated lung—significantly outperforming a traditional Extended Kalman Filter (EKF) estimation method reliant on a kinematic model that does not account for torsional effects in the system.

The key contributions of this paper are (i) an accurate, model-free, learning-based method for steerable needle roll angle estimation and (ii) the integration of the method into sliding mode control for highly accurate steering in multiple mediums, including *ex vivo* tissue that the model was not trained on. As such, our learned method can be trained in advance in gelatin using an internal sensor which can then be removed from the needle prior to the needle's use during clinical deployment in a patient. The needle can then be controlled using external partial sensing and our learned method. In this way, our learning-based method overcomes a key limitation in needle steering, namely the requirement for bulky 6-DOF sensors embedded in the needle itself during clinical deployment. This opens the door for external needle state sensing, enabling accurate tip orientation estimation (and subsequently safe and accurate needle steering) in a way that does not interfere with the needle's ability to deliver therapy to the patient.

2 Technical Approach

We use a kinematic, non-holonomic needle model [1] to define the state of a bevel-tip steerable needle as it moves through tissue. In this model, the needle's behavior is parameterized by its forward motion, transmitted from actuation at its base; the plane in which its bevel lies and in which it curves, changed by rotating the needle at its base; and the curvature of the arc (κ) the needle takes as it is inserted. The control inputs of this model are u_ℓ and u_θ, needle tip insertion velocity and angular velocity, respectively, as shown in Fig. 1, left inset. Most models assume the needle is infinitely rigid in torsion such that u_α, the angular control velocity applied at the needle's base, is perfectly and immediately applied to the needle's tip. In reality, there is a lag in transmission of the rotational velocity applied at the actuator to the needle tip, i.e. $u_\alpha \neq u_\theta$, an effect that is particularly pronounced for long and/or highly flexible needles. Our method overcomes this limitation by learning to estimate the tip angle θ so that the controller can accurately steer the needle.

To do so, we propose a learning-based recurrent neural network with the following layers: (i) input sequence layer (5 units), (ii) LSTM layer (30 units), (iii) fully-connected layer (30 units), and (iv) output regression layer (2 units).

Our network takes as input the vector $\boldsymbol{X} = [\hat{\boldsymbol{p}} \;\; \hat{\boldsymbol{\eta}} \;\; \sin\alpha \;\; \cos\alpha]^T$, where $\hat{\boldsymbol{p}} = (\hat{x} \;\; \hat{y} \;\; \hat{z})$ is the 3-DOF sensor position, isotropically scaled by a predefined maximum workspace component (e.g., the needle's maximum insertion length in tissue) using min-max feature scaling. The 2-DOF sensor axis measurement of the needle tip is given by $\hat{\boldsymbol{\eta}} = (\eta_x \;\; \eta_y \;\; \eta_z)$, representing the needle's heading (i.e., its orientation but without the roll angle). The rotational actuator position at the base of the needle is given by $\alpha = \int u_\alpha dt$. The network then outputs the vector $\boldsymbol{Y} = [\sin\tilde{\theta} \;\; \cos\tilde{\theta}]^T$ where $\tilde{\theta}$ is the estimated roll angle. We parameterize the input and output roll angles via sin and cos, as these continuous representations nicely bound the variables from $[-1, 1]$.

To utilize the network's estimated rotational angle, $\tilde{\theta}$, we integrate the network into a sliding mode controller [2], as shown in Fig. 1, right. $\tilde{\theta}$ is then applied to the needle's sensed heading $\hat{\boldsymbol{\eta}}$ at each time step of the control loop which, in combination with the sensed position $\hat{\boldsymbol{p}}$, enables the controller to accurately steer the needle via the estimated knowledge of the needle tip's full orientation.

3 Experiments and Results

Data Collection and Network Training: We implemented and evaluated our learning-based method on a robotic needle steering system previously presented in [12], designed to perform lung tumor biopsy through a bronchoscope. The steerable needle used was manufactured out of superelastic Nitinol measuring

1.24 mm OD, 1.0 mm ID, 1.3 m long (EuroFlex GmbH), fabricated using the method described in [13], and deployed through a clinical bronchoscope (Ambu USA). To collect the training data for our method, in the form of time series sequences of paired input and output vectors $\boldsymbol{X}$ and $\boldsymbol{Y}$, we performed targeting insertions using a sliding mode controller [2] with a 6-DOF electromagnetic (EM) tracker embedded in the needle tip (Aurora NDI, Inc.). We collected a dataset of 270 insertions in a tissue simulant of 10% gelatin (a tissue phantom frequently used in the needle steering literature) [14]. Each insertion targeted a point sampled uniformly at random within the needle's reachable workspace defined by a cone with bounding curvature of $200\,\text{mm}^{-1}$, and along an insertion interval of 40–75 mm (Fig. 2, left). Every insertion used the controller parameters $\lambda_1 = 5\,\text{mm/sec}$ and $\lambda_2 = 2\pi\,\text{rad/s}$, a 40 Hz controller rate, and achieved less than 1 mm targeting error (Fig. 2, right).

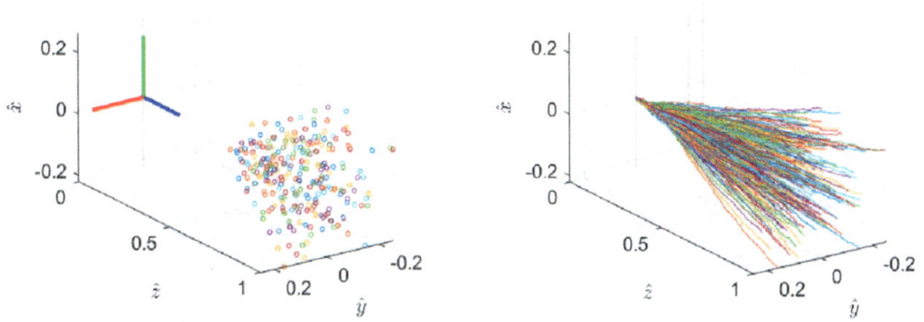

Fig. 2. Experimental dataset of 270 insertions in gelatin for network training. *Left:* The target points. *Right:* The trajectories taken to the target points. Data is isotropically scaled based on z_{max}, which we choose to be an insertion length of 75 mm.

We then normalized and partitioned the data into two subsets: training (240 insertions) and validation (30 insertions). The network was trained using a root-mean-squared error loss function on an error defined over all timesteps in all trajectories in the training dataset. It was trained on an Intel i9-7900X 3.3 GHz 10-core CPU with an NVIDIA Quadro P4000 GPU using ADAM optimization [15] and dropout regularization [16] to prevent over-fitting. The trained network achieved a final RMSE of 0.0457 on the validation set.

Online Estimation and Control in Gelatin and Ovine Brain: After the network was trained, we integrated it into the sliding mode controller to evaluate the system's ability to leverage our method to steer accurately to targets. We performed 30 new insertions in 10% gelatin to evaluate its performance in the tissue phantom in which it was trained and 10 new insertions in ovine brain preserved in 4% Formalin (Carolina Biological Supply, Inc.) to evaluate its performance in biological tissue, a different medium and one in which it was *not* trained. The target points in both mediums were sampled uniformly at random from a similar volume as described in the training sets. At each time step our

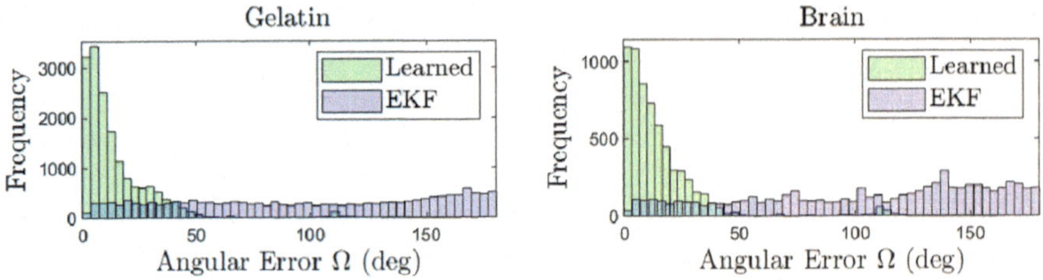

Fig. 3. Histograms of angular error over all insertions in gelatin and brain.

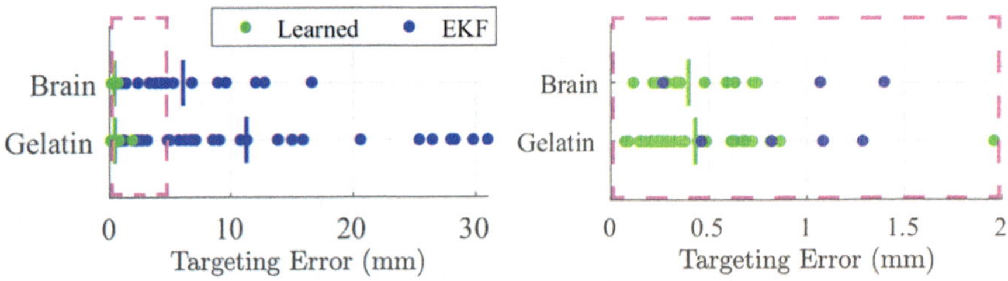

Fig. 4. Targeting errors in gelatin and ovine brain. *Left:* Targeting errors from 10 insertions in gelatin and 5 insertions in brain for our method (green) and the EKF (blue), with the straight lines depicting the average error for each. *Right:* An expanded view of the 0 to 2 mm range of the left figure.

method estimated the orientation angle θ, which was combined with the sensed needle axis and position, forming the full state vector used for sliding mode control.

For comparison to a state-of-the-art method, we implemented an Extended Kalman Filter [17], a model-based observer relying on the non-holonomic kinematic needle model with no model of the torsional dynamics. The EKF had knowledge of the same measured 5-DOF position and axis of the sensorized needle tip and the control input velocities applied at the actuators. The EKF used the sensed information and kinematic needle model to estimate the state of the needle throughout the insertion, its estimated state being incorporated into the same sliding mode controller as our method.

In Fig. 3, we show histograms of the angular error compared with the ground truth measured by the sensor in the needle's tip for each time step over all insertions. The angular error is defined as $\Omega = \arccos\left((tr(\boldsymbol{R}) - 1)/2\right)$, where $\boldsymbol{R}$ is the difference rotation relating the estimated orientation to the ground truth orientation. Our method has error values distributed much closer to zero than the EKF method, indicating superior estimation through each of the insertions.

To demonstrate our method's ability to control the needle to its intended target, in Fig. 4 we show targeting errors (the Euclidean distance between the final needle tip position and the intended target) for each method in each medium. Our method achieved mean targeting errors of 0.43 mm in gelatin and 0.40 mm in brain tissue, while the EKF method achieved mean targeting errors of 11.34 mm in gelatin and 6.12 mm in brain tissue.

Online Steering in *Ex Vivo* Porcine Lung. In addition to evaluating the estimator performance in gelatin and ovine brain, we performed a sequence of online steers in statically-inflated *ex vivo* porcine lung (Animal Technologies, Inc.). The *ex vivo* lung was inflated and accessed using an 8.0 mm endotracheal tube (Smiths Medical ASD, Inc.). We placed custom 3D printed (Formlabs, Inc.) pre-calibrated fiducials on the lung surface with cyanoacrylate glue and used them for point-based registration of the EM tracker frame to the CT frame [18]. A preoperative CT scan was taken using a mobile ENT cone-beam CT scanner (xCAT Xoran Technologies). We loaded the scan into 3D Slicer [19] and manually segmented the fiducial points (sphere centers) in the CT frame. We then manually thresholded the reconstructed CT data (0.4 mm isotropic voxel size) to yield a segmentation of the lung anatomy, see Fig. 5. We then registered the segmented anatomy in the CT frame to the EM tracker frame using the EM-tracked fiducials mounted on the lung. We used this registered segmentation to inform the piercing sites and the intended target points.

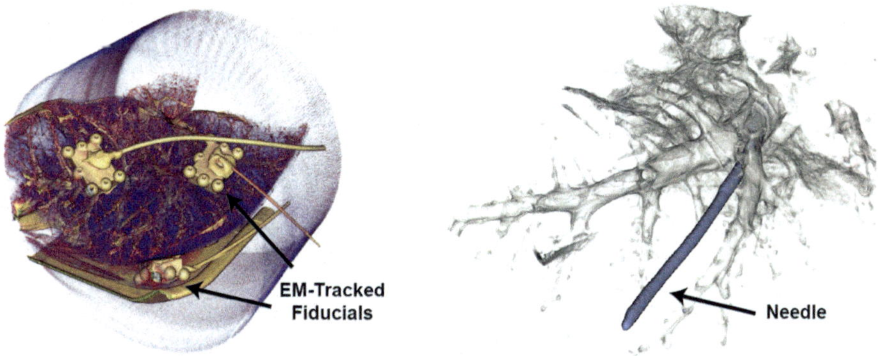

Fig. 5. Rendered CT Scan Volume of the post-steered needle system using the learned estimation method. *Left:* The fiducials are shown in the scan; each contains a 6DOF EM sensor and is glued to the surface of the lung, and is used for point-based registration of the CT frame to the EM tracker frame. *Right:* The same thresholded scan showing the segmented needle deployed post-steer.

Using a clinical bronchoscope, we navigated down to several airways in the left lower lobe and pierced through the airway wall in each trial using a piercing stylet—a 0.9 mm OD superelastic Nitinol tube sharpened to a needle point. We inserted a nitinol tube, 1.5 mm OD, 1.3 mm ID, over the stylet to hold the opening in the airway wall. The piercing stylet was removed and exchanged for the steerable needle.

After loading the needle into the piercing site, we visualized the needle's trumpet-shaped reachable workspace with respect to the segmentation to identify feasible target points for each steer that were collision-free with respect to blood vessels and other surrounding airways. This target point, specified in the CT scanner RAS (right, anterior, superior) coordinates, was transformed into the EM tracker frame using the registration transform acquired from the fiducials. We fed the target point to the sliding mode controller, and the needle was steered to the target with the estimated roll angle as input to the controller. A total of 8 trials were executed, with collision-free steers in 2/8, 1 with each method. We determined the steer was collision-free by inspection of a CT scan of the post-steered needle, prior to retracting the needle back to its starting pose. It is important to note that while we attempted to pick target points in the CT scan that were not visibly in-collision with other parts of the airway and large vessels, we do not consider obstacle avoidance in this work and the steers were performed without consideration for obstacles en-route to the target point. As such, we limit this evaluation to the collision-free trials. In the collision-free trial of the learned estimator, the steer was accurate, achieving a targeting error of 0.46 mm and 19.0° average angular error of the estimate, as measured in the EM tracker frame, not accounting for registration error (see Fig. 6). Conversely, the EKF performed poorly, consistent with our prior experimental results in the other tissues, with a target error of 17.6 mm and 156.7° average angular error. A rendered CT scan of the post-steered needle is shown for the learned estimator method trial in Fig. 5, right.

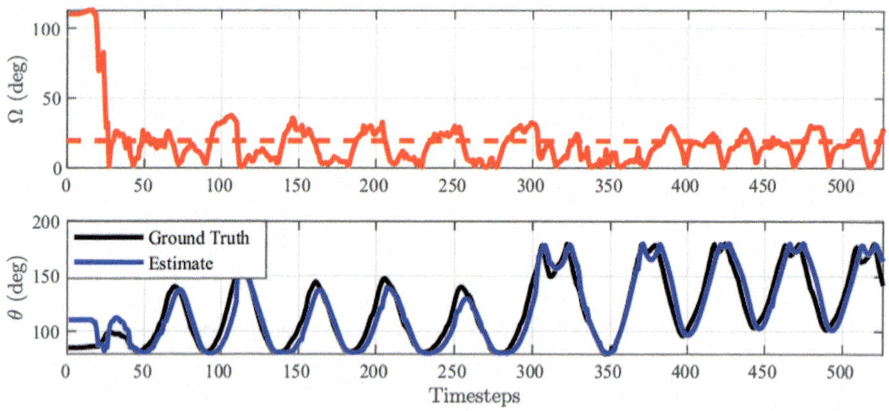

Fig. 6. Time series of the estimate tracking the needle roll angle θ with angular error Ω from a steering trial in ex-vivo porcine lung using the learned estimator. The dashed line shows the mean angular error over the steer.

4 Experimental Insights and Future Work

In this work, we leveraged and validated the performance of a learned estimator for needle steering, paving the way for other learned methods in this application.

Our results show that a learning method can accurately estimate the roll angle of a long, flexible steerable needle in the presence of torsional compliance and unmodeled frictional effects.

Additionally, we show that a network trained on tissue simulant extrapolates to actual tissue in the form of *ex vivo* ovine brain and porcine lung. It will be a subject of future research to investigate the application to other tissues such as liver and kidney, relevant tissues with applications for steerable needles.

In the *ex vivo* lung experiments, the learned estimator performed well, though there were steers which collided with anatomy. This further elucidates the need for better registration methods, improved segmentation algorithms, and planned trajectories that consider these obstacles, instead of point targets. Future research will leverage our method when executing trajectories generated via motion planning in these environments.

Overall, we show that our learning-based method outperforms a model-based observer whose model does not capture the significant torsional dynamics. We demonstrate that neural network-based estimation can result in effective tracking of unsensed state variables, enabling accurate steering to point targets in gelatin, ovine brain, and porcine lung.

Acknowledgments. This research was supported in part by the National Institutes of Health under award R01EB024864.

References

1. Webster III, R.J., Kim, J.S., Cowan, N.J., Chirikjian, G.S., Okamura, A.M.: Nonholonomic modeling of needle steering. IJRR **25**(5–6), 509–525 (2006)
2. Rucker, D.C., Das, J., Gilbert, H.B., Swaney, P.J., Miga, M.I., Sarkar, N., Webster III, R.J.: Sliding mode control of steerable needles. IEEE TRO **29**(5), 1289–1299 (2013)
3. Minhas, D.S., Engh, J.A., Fenske, M.M., Riviere, C.N.: Modeling of needle steering via duty-cycled spinning. In: EMBS Proceedings, pp. 2756–2759 (2007)
4. Swensen, J.P., Lin, M., Okamura, A.M., Cowan, N.J.: Torsional dynamics of steerable needles: modeling and fluoroscopic guidance. IEEE TBME **61**(11), 2707–2717 (2014)
5. Reed, K.B., Okamura, A.M., Cowan, N.J.: Modeling and control of needles with torsional friction. IEEE TBME **56**(12), 2905–2916 (2009)
6. Kallem, V., Cowan, N.J.: Image guidance of flexible tip-steerable needles. IEEE TRO **25**(1), 191–196 (2009)
7. Fallahi, B., Rossa, C., Sloboda, R., Usmani, N., Tavakoli, M.: Partial estimation of needle tip orientation in generalized coordinates in ultrasound image-guided needle insertion. In: IEEE AIM, pp. 1604–1609 (2016)
8. Wood, N.A., Shahrour, K., Ost, M.C., Riviere, C.N.: Needle steering system using duty-cycled rotation for percutaneous kidney access. In: EMBS Proceedings, pp. 5432–5435 (2010)

9. Kuutti, S., Bowden, R., Jin, Y., Barber, P., Fallah, S.: A survey of deep learning applications to autonomous vehicle control. arXiv, 1–23 (2019)
10. Avila-Carrasco, C., Ruppel, M., Persad, R., Bahl, A., Dogramadzi, S.: Analytical vs data-driven approach of modelling brachytherapy needle deflection. IEEE TMRB **2**(4), 519–528 (2020)
11. Greff, K., Srivastava, R.K., Koutnik, J., Steunebrink, B.R., Schmidhuber, J.: LSTM: a search space odyssey. IEEE Trans. Neural Netw. Learn. Syst. **28**(10), 2222–2232 (2017)
12. Amack, S.R., Margaret, M., Jason, E., Tayfun, E.E., Maxwell, K., Alan, M., Fabien, A., Jason and Gafford, J. Alterovitz, R.: Design and control of a compact modular robot for transbronchial lung biopsy. In: SPIE Medical Imaging: Image-Guided Procedures, Robotic Interventions, and Modeling (2019)
13. Rox, M., Emerson, M., Ertop, T.S., Fried, I., Fu, M., Hoelscher, J., Kuntz, A., Granna, J., Mitchell, J., Lester, M., Maldonado, F., Gillaspie, E.A., Akulian, J.A., Alterovitz, R., Webster, R.J.: Decoupling steerability from diameter: helical dovetail laser patterning for steerable needles. IEEE Access **8**, 181411–181419 (2020)
14. Swaney, P.J., Burgner, J., Gilbert, H.B., Webster III, R.J.: A flexure-based steerable needle: high curvature with reduced tissue damage. IEEE TBME **60**(4), 906–909 (2013)
15. Kingma, D.P., Ba, J.L.: Adam: a method for stochastic optimization. In: 3rd International Conference on Learning Representations Proceedings, pp. 1–15 (2015)
16. Gal, Z., Ghahramani, Y.: A theoretically grounded application of dropout in recurrent neural networks. In: NIPS Proceedings (2016)
17. Choset, H.M., Hutchinson, S., Lynch, K.M., Kantor, G., Burgard, W., Kavraki, L.E., Thrun, S.: Principles of Robot Motion: Theory, Algorithms, and Implementation. MIT press, Cambridge (2005)
18. Fitzpatrick, J.M., West, J.B., Maurer, C.R.: Predicting error in rigid-body point-based registration. IEEE Trans. Med. Imaging **17**(5), 604–702 (1998)
19. Kikinis, V.K., Pieper, S.D.: 3D slicer: a platform for subject-specific image analysis, visualization, and clinical support. In: Intraoperative Imaging and Image-Guided Therapy, pp. 277–289 (2014)

Mapping and Localization

Monte-Carlo Localization on Metal Plates Based on Ultrasonic Guided Waves

Othmane-Latif Ouabi[1(✉)], Pascal Pomarede[1], Matthieu Geist[2], Nico F. Declercq[1,3], and Cédric Pradalier[1]

[1] UMI 2958 Georgia Tech-CNRS, Metz, France
[2] Google Research, Paris, France
[3] Georgia Institute of Technology, Atlanta, GA 30332-0250, USA

Abstract. In this work (This work is part of the BugWright2 project. This project is supported by the European Commission under grant agreement 871260 - BugWright2.), we propose a method based on a particle filter for the localization of an industrial robot on a large metal structure that leverages first-order reflections of acoustic waves on a metal plate edges. In our approach, the acoustic measurements are acquired in a (pseudo) pulse-echo mode using a co-localized emitter/receiver pair of piezoelectric transducers, and we assume a known size of the metal plate. To validate the method, the acquisition of acoustic data is made manually, but it is aimed to be performed by a robotic platform soon. The results demonstrate that with our approach, it is possible to recover the robot localization to a precision of a few millimeters.

Keywords: Localization · Ultrasonic Guided Waves · Particle filtering

1 Introduction

Maintaining the integrity of marine vessels structure is a significant stake in ensuring safe travels on the sea. When not removed, bio-fouling that forms on the hull can increase fuel consumption, while concealed damages like corrosion patches can lead to catastrophic failures. Hence, in the marine industry, inspection and cleaning of ship hulls are to be carried out regularly.

Current inspection methods of ship hulls are mainly based on visual inspection, and human operators perform cleaning while the ship is in dry docks. These methods are sub-optimal and time-consuming, as the ship can be immobilized for up to 8 d, with significant financial impacts for shipowners [1]. Autonomous robotic inspection and cleaning of large metal structures have not yet been established. In contrast, they have the potential to be faster, safer, more accurate, and less expensive than conventional methods. The mentioned thought is yet to be demonstrated, but it holds tremendous potential and major benefits to shipowners [1].

As localization is one of the critical features necessary for a proper autonomous application, we address this problem for a robot (namely a magnetic crawler) moving on a rectangular metal panel, which represents the essential element of ship hulls. In combination with conventional localization systems

B. Siciliano et al. (Eds.): ISER 2020, SPAR 19, pp. 345–353, 2021.
https://doi.org/10.1007/978-3-030-71151-1_31

(laser theodolite, Ultra-Wide Band (UWB) beacons...), accurate localization on a metal structure may be achieved by using Lamb waves [8]. Lamb waves are Ultrasonic Guided Waves (UGWs) able to propagate in metal plate structures in a direction parallel to the surface. They can be generated in the material and received by using piezoelectric transducers. These waves can potentially travel over large distances, and when encountering the plate boundaries, they are reflected perpendicularly.

In the literature, prior work was addressing the problem of acoustic-based localization for a mobile agent in indoor environments using the acoustic reflections of sound waves in air [3,6]. Furthermore, studies of Lamb waves mainly focus on the detection and localization of damages by a static network of sensors, consider only the direct transmission of the incident waves (e.g. [2,5,10]) and alleviate the effect of the reflections due to the boundaries that can also be received [7]. However, the localization of a mobile unit on a metal structure using Lamb waves is yet to be thoroughly investigated.

In this work, we show that by leveraging the waves' reflections on the boundaries of a rectangular panel and a basic wave propagation model, a particle filter [4,9] can estimate the robot position despite the complex nature of guided waves [8], the sensing symmetries intrinsic to the plate geometry and the highly reverberating propagation media. The performance of our approach is assessed with real acoustic measurements that are acquired by hand, and results suggest that the position can be recovered with sufficient precision.

2 Technical Approach

Given a rectangular metallic panel of known size $l \times h$, we aim at estimating, for any time step t, the robot state $\mathbf{x}_t = [x_t, y_t, \theta_t]^T$ (2D cartesian coordinates and heading) from acoustic measurements and odometry data $\mathbf{u}_t = [\Delta r_t, \Delta\theta_t]^T$ (translation and rotational movements). The estimation is done in the metal panel frame, *i.e.* the origin is assumed to be the bottom left corner of the plate. The x and y axis are respectively aligned with the edges of size l and h. An illustration of the adopted notations is provided on Fig. 1. In our setup, the acoustic waves are generated in the structure by a piezoelectric transducer, and the reflections are received by a second transducer co-localized with the emitter. This aims at approximating sonar-like pulse-echo sensing, though constrained to an acoustic wave guide.

2.1 Echo Detection

Detecting the acoustic reflections from the measurements is essential to enable the localization of the mobile agent on the metal plate. To this aim, we rely on a basic wave propagation model [5] to model the signal $\hat{\mathbf{s}}(r) \in \mathbb{R}^N$ sensed after a single orthogonal reflection of the incident wave excited by the emitter, on the border, at a distance r from the transducer pair. Under the assumptions

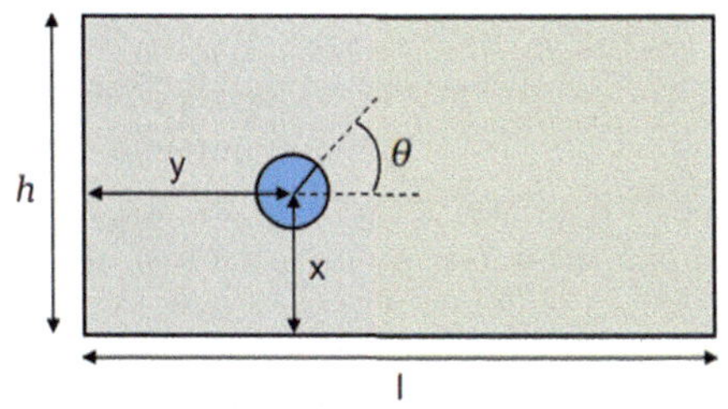

Fig. 1. Illustration representing an agent at the position $\mathbf{x} = [x, y, \theta]^T$ on a metal plate of dimensions $l \times h$.

that the plate is made of a homogeneous and isotropic material, and that the propagation is linear, $\hat{\mathbf{s}}(r)$ can be expressed by:

$$\hat{\mathbf{s}}(r) = \hat{\mathbf{g}}(r) * \mathbf{e}$$

where $\mathbf{e} \in \mathbb{R}^N$ is the discretized excitation signal, $*$ denotes the convolution product and $\hat{\mathbf{g}}(r)$ is the propagation model. From acoustic measurements $\mathbf{s}_t \in \mathbb{R}^N$, the echoes are extracted by computing the correlation signal:

$$\bar{\mathbf{z}}_t[k] = \frac{\hat{\mathbf{s}}(r_k) \cdot \mathbf{s}_t}{||\hat{\mathbf{s}}(r_k)||_2||\mathbf{s}_t||_2} \quad (1)$$

for pre-defined ranges $r_1, r_2, ..., r_K$. '$\cdot$' is the simple dot product. We next determine the envelope $\mathbf{z}_t \in \mathbb{R}^K$ of the resulting correlation signal $\bar{\mathbf{z}}_t \in \mathbb{R}^K$ from the Hilbert transform of the latter. The envelope is assumed to yield the likelihood to observe, within the measurements $\mathbf{s}_t$, an orthogonal reflection of the incident wave at the distances r_k from the transducer pair with the probabilistic model:

$$p(r_k|\mathbf{s}_t) = \eta \cdot e^{\beta \mathbf{z}_t[k]} \quad (2)$$

where β is a tunable parameter and η a normalization constant. In the next, we leverage the probabilistic model built to extract the echoes from the acoustic measurements in a particle filter to estimate the robot position. This approach is also known under the name of Monte-Carlo Localization (MCL) [9].

2.2 Particle Filter

Bayesian filtering is a recursive estimation technique used to infer a probability distribution over unobserved variables from measurements using Bayes rule. Particle filters are approximations of the Bayes filter. They approach the target-probabilistic-distribution with a set of M particles, each of them carrying, at each measurement step, hypothetical values of the hidden variables. The particles are also provided with a weight indicating how their hypothesis on the values of the variables accounts for the measurements. After each time step, the particles are resampled with a probability proportional to their respective weight. Consequently, particles that carry a hypothesis that is consistent with the measurements will tend to survive, while those who do not, will tend to disappear. For the localization problem, each particle represents a belief on the robot state at any measurement step t: $\mathbf{x}_t^{[m]} = [x_t^{[m]}, y_t^{[m]}, \theta_t^{[m]}]^T, m = 1, 2, ..., M$. Our implementation of the filter using guided waves measurements is then the following:

***Initialization*:** The particle filter is initialized by sampling M particles on the metal plate with a uniform distribution.

***Prediction*:** The motion model is applied to every particle:

$$\mathbf{x}_{t+1}^{[m]} \sim p(\mathbf{x}_{t+1} | \mathbf{x}_t^{[m]}, \mathbf{u}_t).$$

If one limits the prediction to this single step, the particle filter will be likely to fail when there are no particles in the vicinity of the actual position. A simple remedy to this problem consists of disturbing the proposal distribution [9]. Here, we will randomly disturb a particle state by drawing, after the prediction and with a low probability α, a new sample:

$$\mathbf{x}_{t+1}^{[m]} \longleftarrow \mathcal{N}(\mathbf{x}_{t+1}^{[m]}, \boldsymbol{\Sigma}),$$

where $\mathcal{N}$ represents the normal distribution and $\boldsymbol{\Sigma}$ is a parameter of the filter.

***Update and Resampling*:** We compute, for every particle, the importance weight by determining the likelihood of presence of the four first-order echoes that should be contained within the measurement $\mathbf{s}_{t+1}$ given the particles belief of the position, and using the model in (2):

$$w_{t+1}^{[m]} = p(x_{t+1}^{[m]} | \mathbf{s}_{t+1}) p(y_{t+1}^{[m]} | \mathbf{s}_{t+1}) p(l - x_{t+1}^{[m]} | \mathbf{s}_{t+1}) p(h - y_{t+1}^{[m]} | \mathbf{s}_{t+1}).$$

Finally, we create a new set of particles by drawing the m-th particle with a probability proportional to $w_{t+1}^{[m]}$. This procedure is then repeated for the next measurement step.

3 Results

We acquired acoustic data by placing an emitter/receiver pair of transducers at 108 different positions on a rectangular grid. Each measurement contains $N = 500$ samples collected at a sampling rate of 1.25 MHz. The metal plate used for the experiment has a size of $l \times h = 60 \times 45$ cm and is made of aluminium with a thickness of 6 mm. The experimental setup can be seen on Fig. 2.

To generate Lamb waves in the plate, we use an excitation signal that corresponds to two tone-bursts of a sinusoidal waveform at 100 kHz as shown on Fig. 3.a. The signal is amplified to up to 100 V peak-to-peak before being sent to the emitter. We use coupling gel to ensure good contact between the transducers

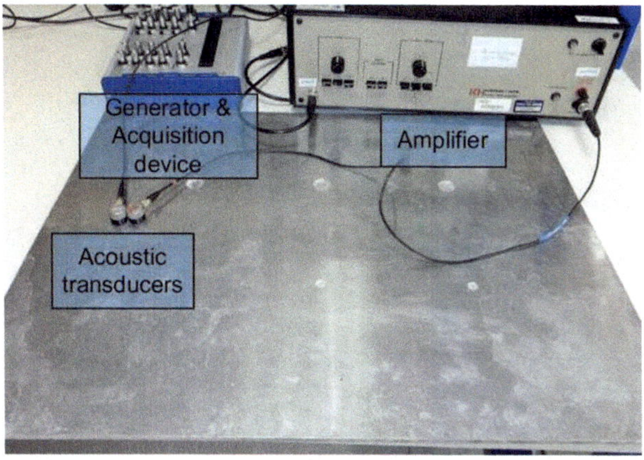

Fig. 2. Picture of the experimental setup. Two acoustic transducers are co-localized to approximate pulse-echo sensing.

and the plate surface. We also remove the very first received wave packet from the measurements, as it is due to the excitation of the emitter near the receiver. Figure 3.b depicts a typical acoustic measurement obtained in this configuration and Fig. 3.c shows the envelope of the resulting signal $\mathbf{z}_t$ obtained by correlating the measurement with the propagation model as in (1). It can be observed that the peaks of the correlation signal match with the expected ranges for the first-order reflections.

Next, to test our framework, we simulate a robot following a lawnmower path along the y-axis by presenting the corresponding sequence of measurements to the particle filter along with the theoretic displacement of the transducers used as the odometry data. At the end of the path, the "simulated" robot is taking the same path but backwards. We use, however, the same measurements. This provides us with a total of $T = 2 \times 108 = 216$ measurement steps. The particle filter is then run in this scenario using $M = 500$ particles. We first illustrate the performance of our approach in Fig. 4, which depicts the position of the particles during the measurement steps 2, 12, 108 and 216. During Step 2, the particles are concentrated in the bottom-left part of the plate as, for establishing the uniqueness of the position, the filter shall be initialized with particles only in one quarter of the surface due to the plate symmetry. At this stage, both the position and orientation of the fictional robot is still very uncertain. After only a few measurement steps, the position is well recovered, as shown during Step 12. The state estimation remains correct during the entire trajectory as the particles remain globally in the vicinity of the true position until the very last step.

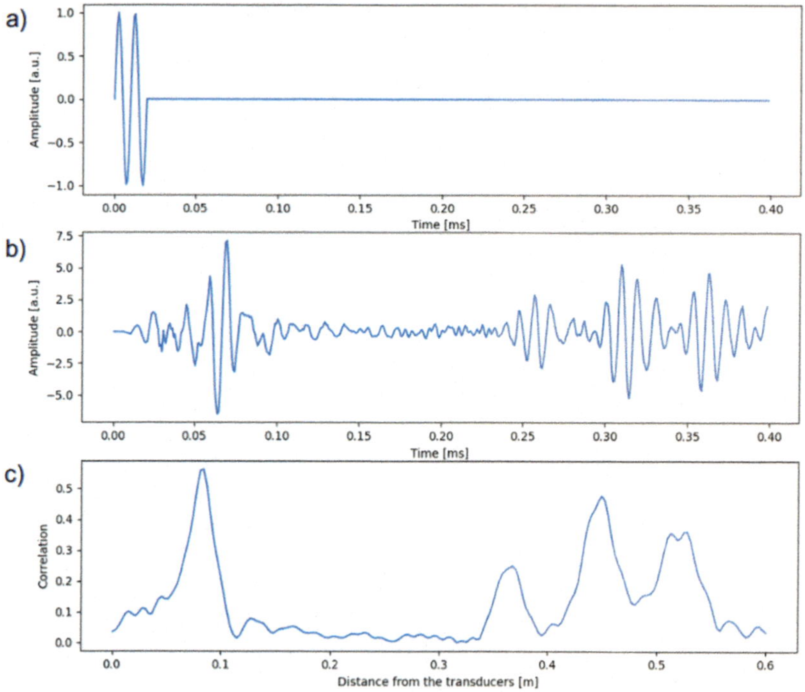

Fig. 3. a) Excitation signal used to generate Lamb waves. b) Acoustic measurement obtained for the transducer position $[x, y] = [0.08, 0.08]$. c) Signal $\mathbf{z}_t$ obtained from the measurement shown in b). The peaks related to the 4 first orders echoes and which should be observable for the ranges $[x, y, l-x, h-y] = [0.08, 0.08, 0.52, 0.37]$, are clearly visible. The peak present at 45 cm is due to a higher-order reflection.

To quantitatively assess the accuracy of the estimation, we show in Fig. 5 the median of the belief on the position and the ground truth. The results show that the estimation of the y coordinate is accurate as only a few measurement steps were necessary to reach an estimation error lower than 1 cm, and this level of accuracy remains stable for all the rest of the trajectory. It took a few more measurement steps for the filter to converge towards a correct estimation of x. Nevertheless, still, it eventually achieves a comparable level of precision on the estimation. Overall, the results suggest that our approach has the potential for the estimation of 2D coordinates of an industrial robot in the plate frame, and it should be further investigated.

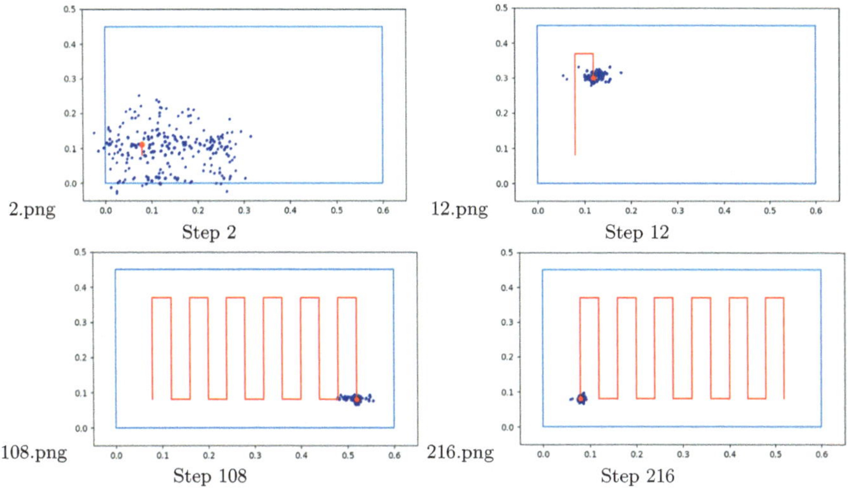

Fig. 4. Position estimated by all the particles (blue dots) during Steps 2, 12, 108 and 216 for a lawn-mower path on the plate. The true outline of the plate and true sensor positions correspond to the blue rectangle and red dot respectively.

4 Experiments Completed or Scheduled

The experiment that was carried out aimed at demonstrating the applicability of guided waves for the localization of an inspection robot on a metal structure. Although the results showed that acoustic measurements might contain sufficient information for that objective, the presented approach raises some questions:

The first limitation lies in the use of two transducers which are not rigorously at the same position. In contrast, the algorithm assumes co-localization of the emitter/receiver pair—the error due to such an approximation increases when the spatial periodicity of the waves decreases. The reason is that the measured signal will be more sensitive to small displacements. Hence, in our future experiments, we will use specific equipment enabling the use of a single transducer in a pulse-echo mode, as this could also reduce the estimation error.

Besides, further investigations are required to assess the stability of the filter using different scenarios. Indeed, a common issue of particle filters is that, when the number of particles is not sufficient to cover the whole state space, there might not be any particle that will approach the right position and the estimation could fail. Moreover, our approach only makes use of first-order reflections, whereas high order echoes can also be sensed, as seen in Fig. 3.c. This may also hinder the filter's convergence.

Finally, using a robot to acquire the acoustic measurements will be implemented as well, with the corresponding odometry readings. Taking into account the speed of the acoustic waves, which can typically be higher than 2000 m/s, the performances of our algorithm could also be assessed with a higher data acquisition rate.

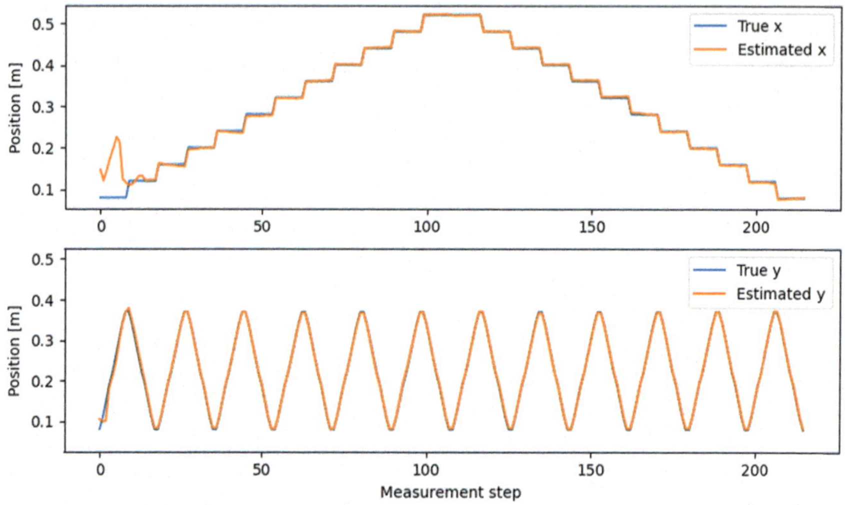

Fig. 5. Results of the robot localization using the particle filter.

5 Main Experimental Insights

We aim to achieve practical implementation of the filter on the field, but this may bring its challenges. Here, the experiment was done in a laboratory environment, which resulted in the acquisition of clean acoustic measurements with a high signal-to-noise ratio. However, noisy signals are to be expected on a real ship hull in outdoor environments, due, for example, to challenging environmental conditions and more inferior surface quality of the metal plates. Besides, metal plates on ship hulls may be larger than the one use in the experiment, which would also participate in the attenuation of the signals, and not perfectly rectangular. Rough metal plates may cause complex scattering of the waves, which is likely to prevent a proper localization of the robot as well.

Future work shall also focus on the problem of Simultaneous Localization and Mapping. Indeed, accurate knowledge of the shape of the plates may not always be available. Hence, estimating both the plate geometry and the robot localization from acoustic data is a desirable next step.

Finally, the fact that the acquisition has to be done on an embedded system will be a source of additional difficulties. For instance, the robotic system will have to automatically ensure a good coupling between the transducers and the surface while the device itself might be a source of noise due to the mechanical displacement of the robot and/or the electronics.

References

1. Bugwright2, autonomous robotic inspection and maintenance on ship hulls and storage tanks, description of the innovative action (2019)
2. Alleyne, D.N., Cawley, P.: The interaction of lamb waves with defects. IEEE Trans. Ultrason. Ferroelectr. Freq. ControlControl **39**(3), 381–397 (1992)
3. Bordoy, J., Schindelhauer, C., Höflinger, F., Reindl, L.M.: Exploiting acoustic echoes for smartphone localization and microphone self-calibration. IEEE Trans. Instrum. Meas. **69**(4), 1484–1492 (2020)
4. Chen, H., Liu, Z.: Application and Challenges of Signal Processing Techniques for Lamb Waves Structural Integrity Evaluation: Part A-Lamb Waves Signals Emitting and Optimization Techniques (09 2018)
5. Jeong, H., Cho, S., Wei, W.: A baseline-free defect imaging technique in plates using time reversal of lamb waves. Chinese Phys. Lett. **28**(6), 064301 (2011). https://doi.org/10.1088%2F0256-307x%2F28%2F6%2F064301
6. Kreković, M., Dokmanić, I., Vetterli, M.: Echoslam: Simultaneous localization and mapping with acoustic echoes. In: 2016 IEEE International Conference on Acoustics, Speech and Signal Processing (ICASSP), pp. 11–15 (2016)
7. Sedlak, P., Hirose, Y., Enoki, M.: Acoustic emission localization in thin multi-layer plates using first-arrival determination. Mech. Syst. Signal Process. **36**(2), 636–649 (2013). http://www.sciencedirect.com/science/article/pii/S0888327012004669
8. Su, Z., Ye, L.: Identification of Damage Using Lamb Waves: From Fundamentals to Applications, Springer, vol. 48 (01 2009)
9. Thrun, S., Burgard, W., Fox, D.: Probabilistic Robotics (Intelligent Robotics and Autonomous Agents). The MIT Press, Cambridge (2005)
10. Zhang, Y., SIDIBE, Y., Maze, G., LEON, F., Druaux, F., Lefebvre, D.: Detection of damages in underwater metal plate using acoustic inverse scattering and image processing methods. Appl. Acoust. **103**, 110–121 (2016). https://hal-normandie-univ.archives-ouvertes.fr/hal-02503535

Crowd-Driven Mapping, Localization and Planning

Tingxiang Fan, Dawei Wang, Wenxi Liu, and Jia Pan(✉)

The University of Hong Kong, Pokfulam, Hong Kong, SAR
jpan@cs.hku.hk

Abstract. Navigation in dense crowds is a well-known open problem in robotics with many challenges in mapping, localization, and planning. Traditional solutions consider dense pedestrians as passive/active moving obstacles that are the cause of all troubles: they negatively affect the sensing of static scene landmarks and must be actively avoided for safety. In this paper, we provide a new perspective: the crowd flow locally observed can be treated as a sensory measurement about the surrounding scenario, encoding not only the scene's traversability but also its social navigation preference. We demonstrate that even using the crowd-flow measurement alone without any sensing about static obstacles, our method still accomplishes good results for mapping, localization, and social-aware planning in dense crowds.

1 Introduction

Navigating a mobile robot in complex, cluttered, and dynamic human scenarios has a wide variety of applications, including last mile delivery, service robot in malls, and surveillance. However, most existing navigation algorithms only work well in scenarios with only a few or no person due to two main challenges in dense crowds.

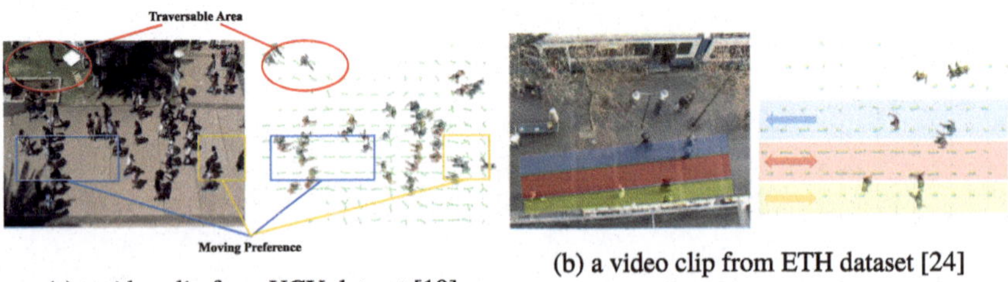

(a) a video clip from UCY dataset [19]

(b) a video clip from ETH dataset [24]

Fig. 1. Two real-world examples illustrate two types of latent social information in crowd flow: (a) crowd flow indicates the region traversability and scene's moving preference; (b) crowd flow encodes the scene's social navigation rules.

First, in a dense crowd, a robot is difficult to figure out the layout of surrounding environment and its accurate location using the classical simultaneous

B. Siciliano et al. (Eds.): ISER 2020, SPAR 19, pp. 354–368, 2021.
https://doi.org/10.1007/978-3-030-71151-1_32

localization and mapping (SLAM) framework, because the existence of moving obstacles like pedestrians can occlude the robot's observation and the landmark features may be confused with features on moving objects, resulting in mistakes in correspondence calculation. Thus, prior works often treat moving objects as interference and intend to filter them out from the sensory observation, which unfortunately is difficult to be implemented robustly even with the recent progress in computer vision. Such filtering also discards the rich human-robot interaction information that is helpful for achieving social-friendly robot behaviors.

Second, in a dense crowd, a robot is difficult to accomplish safe and effective navigation without getting stuck in crowds. Some recent works consider pedestrians as moving obstacles and use either reactive policies [9,10,20] or short-horizon planning [3] to trade-off collision avoidance and goal approaching. However, these methods do not model social-awareness and thus the resulting robot motion may be safe but intimidating, e.g., the robot may abruptly accelerate or stop towards its neighbors' tiny movements. Such overly responsive behavior will disturb the normal crowd behavior and significantly reduce the robot's social-friendliness and navigation effectiveness.

In this paper, we solve above challenges by repurposing the role of pedestrians in autonomous robotic navigation. Rather than treating pedestrians as noises to static landmarks or obstacles to be actively avoided, we use the observation of nearby pedestrian crowds as another type of sensory measurement about the surrounding environment with crowds. In particular, we consider the following navigation problem in the extreme case: suppose the robot has no sensor mechanism for measuring geometry of static obstacles and can only perceive the movement of its nearby pedestrians, can the robot accomplish mapping, localization, and planning? Motivated by human's behaviors in Fig. 1, we find the answer is YES, because the observed vector field of local crowd flow can be used to infer the *traversability* of different locations in the unstructured scene and the crowd's social *moving preferences*, including rules of left/right hand traffic, crossing and overtaking [4]. These two types of crowd latent information are of great importance for achieving mapping, localization, and social-friendly navigation in dense crowd scenes.

Main Contributions: We propose a novel framework, called *CrowdMapping*, for robotic navigation in dense crowds. It includes three modules:

- *CrowdMapper* uses local crowd observation as sensory measurement and applies clustering algorithm to reconstruct a crowd-flow map capturing the movement pattern of pedestrians in the scenario;
- *CrowdLocalizer* uses a flow-matching metric to leverage the crowd-flow map for localization;
- *CrowdPlanner* allows a robot to follow the movement pattern in the crowd-flow map and accomplish social-aware planning.

2 Related Work

2.1 Crowd-Flow Estimation

To capture the crowd-flow pattern of the environment, many estimation methods have been developed. According to the observation perspective, we categorize existing crowd-flow estimation methods into two types, global and local methods. The global estimation methods use global cameras or drones to obtain observation of the crowd flow in the entire scene [11,27]. However, the global observation is often unattainable for robots due to the limited viewports of on-board sensors. Thus, local estimation methods are proposed to estimate the crowd flow around the robot. Many exiting work have achieved real-time pedestrian flow tracking by on-board sensors [19,21,26]. Compared to the global estimation approaches, however, these local methods with limited perceptive are more likely to encounter the occlusion issue. To overcome it, some studies propose the more robust occlusion-aware estimation approaches for tracking systems [8,15].

2.2 Crowd-Flow Navigation

Navigation in crowd-flow involves the complex human-robot interaction. Some previous works on crowd navigation usually consider pedestrians as dumb moving obstacles that need to be avoided actively [5,10,20]. Thus, the resulting robot behavior can be too sensitive to human's subtle movement, which may disturb the pedestrian flow and reduce the throughput of the entire crowd. To eliminate this issue, some methods formulate the social flow factors inside the human-robot interaction model. [22] presents an iterative A* algorithm to follow a pedestrian along the crowd flow, but it works only when the pedestrian to be followed and the robot share the same goal. In [14], the robot uses inverse reinforcement learning to imitate the pedestrian's behavior in a crowd for reducing the disturbance between the robot and the crowd. It uses Gaussian processes to online estimate the pedestrian flow vector field in the scenario, but is sensitive to the time-varying crowd mode and thus not appropriate for mapping and localization. More importantly, the mimic policy is not optimal in terms of minimizing the robot-crowd disturbance. [1] introduces a crowd-aware planner to plan a trajectory that conforms to the social-flow direction and also avoids congestion, but it assumes the global crowd flow field to be static and known beforehand, which unfortunately is often impractical.

3 CrowdMapping Framework

Our fundamental assumption is that the pedestrian crowd flow can provide sufficient time-invariant features for mapping the navigation scenario. In particular, the moving patterns in the crowd flow reflect the resistance of static obstacles to the crowd movement along different directions: the high resistance along a direction indicates the existence of obstacles along that direction and vice versa.

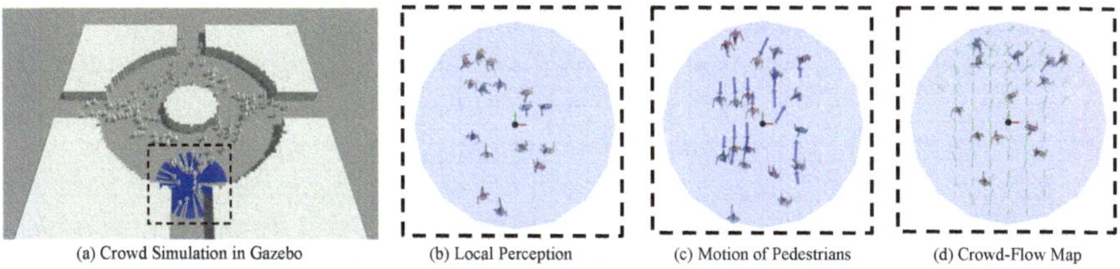

Fig. 2. The *CrowdMapper* pipeline. (a): the simulated crowded scenario, with the local region around the robot highlighted; (b): the robot's observation, where the blue circle indicates the perception region; (c): the captured motion vector field of the robot's nearby pedestrians; (d): the recovered the time-invariant crowd-flow feature that will be fused into the global crowd-flow map.

Such resistance effect depends on map layout and pedestrians' moving preference, and thus is mostly time-invariant, making it suitable for map reconstruction. For instance, in the sidewalk scenario shown on the right of Fig. 1, there are two types of crowd flow patterns. In the yellow box region, there is no resistance in all directions, thus humans can move freely and their paths may intersect. In the blue box region, people prefer moving along the east-west direction, implying the existence of street boundary that resists human motion in the south-north direction. Such directional resistance can also be used to encode social navigation rules, e.g., the right-traffic rule in Fig. 1b. Such crowd flow patterns are time-invariant and can be used to reconstruct a map reflecting the navigation constraints posed by the scenario obstacles and social rules.

We now describe our novel mapping framework and its applications in three steps. First, Sect. 3.1 presents the building of global crowd-flow map from the robot's local perception of the pedestrian crowd. The obtained crowd-flow map is then used for localization in Sect. 3.2, and for crowd-aware planning in Sect. 3.3. We call them *CrowdMapper*, *CrowdLocalizer* and *CrowdPlanner*, respectively.

3.1 CrowdMapper

In *CrowdMapper*, the robot leverages the local crowd distribution around itself as the sensory measurement for mapping, which can be reliably collected nowadays using pedestrian detection and tracking [2] technique, thanks to the progress of computer vision. In particular, at time step t, the robot's local observation about the crowd is $\{\mathbf{m}^i_{t-\Delta t:t}\}_{i=1}^N$, where N is the number of pedestrians within the robot's perception range and i denotes the pedestrian ID in the tracking system. $\mathbf{m}^i_{t-\Delta t:t} = [\bar{p}_x, \bar{p}_y, l, \theta] \in \mathbb{R}^4$ is a vector encoding the i-th pedestrian's state in terms of position and velocity at time step t, where l is the moving distance in Δt duration, θ is the moving direction, and $\bar{p}_x, \bar{p}_y$ are the starting coordinates for the movement. In this paper, we choose $\Delta t = 1\,\mathrm{s}$ to make a good trade-off between tracking fast movements and capturing stable movement patterns in the crowd.

Given local observations $\{\mathbf{m}_{t-\Delta t:t}^i\}$, the robot next uses K-means clustering [13] to recover time-invariant crowd-flow patterns locally around. In particular, we first discretize the 2D plane into grid cells $\mathcal{C}_r$ at a certain resolution. Then, we assign $\mathbf{m}_{t-\Delta t:t}^i$ to different cells using the location $[\bar{p}_x, \bar{p}_y]$ as index. After that, we apply K-means to all velocities $[l, \theta]$ assigned to the same cell to identify the possible navigation directions in each cell. Since the number of clusters is unknown beforehand, inspired by the elbow method [16], we attempt the cluster number from 1 to the maximum cluster number N_c, and choose the final cluster number as the one that does not reduce the total within-cluster sum of square (WSS) value dramatically or whose WSS is below a certain threshold WSS_m. Finally the robot can recover the local crowd-flow map.

To construct a global crowd-flow map, the robot needs to fuse the local maps obtained in an online manner when moving around in the scenario. The entire *CrowdMapper* pipeline is shown in Fig. 2. Our constructed crowd-flow map can be applied in many navigation tasks, including but not limited to the localization and social-aware planning tasks that will be described below.

3.2 CrowdLocalizer

We now demonstrate the robot can localize itself using the online constructed crowd-flow map instead of the traditional geometry map. Our approach is general and can be integrated in any traditional localization framework, but here we use the particle-filter based localization framework [7] as an example to describe our method.

In concrete, the particle-filter localization includes three repetitive steps: 1) generating n_p hypothesis samples for robot locations, 2) computing the weight for each sample by evaluating the hypothesis quality in terms of the matching quality between the observation and map locally around the hypothesis, and

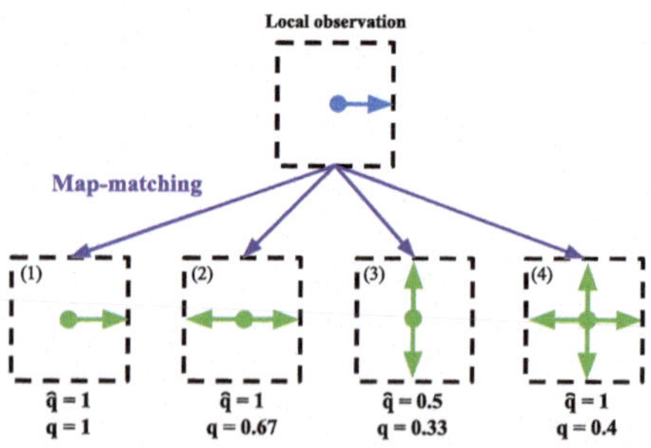

Fig. 3. Example of the flow-matching metric. Suppose that the observed crowd flow in the current grid cell is moving toward right, and there are four hypothesis cells in the map as shown in the bottom. The computed matching quality q when the hyper-parameter $\gamma = 0.5$ is listed for each candidate, where a larger q indicates a higher matching quality.

3) important sampling to update the hypothesis samples. When using the crowd-flow map for localization, we leave step 1) and 3) unchanged, because they are irrelevant to crowd-flow details. We modify step 2) by designing an appropriate metric to evaluate the matching quality between the local crowd observation and the crowd-flow map locally around a given location hypothesis. In particular, we first compute the minimum difference θ_m between an observed direction θ and all the possible directions in the grid cell where the location hypothesis locates. Next, we normalize θ_m to $[0, 1]$ and get the score $\hat{q} = \frac{\pi - \theta_m}{\pi}$. In addition, among different hypothesis cells with the same $\hat{q}$ score, we prefer the cell with the minimum number of possible directions, since the corresponding matching is most robust. Thus, we discount $\hat{q}$ with the factor $\gamma \in [0, 1]$ and get the final matching score $q = \frac{1}{1+(n-1)\gamma}\hat{q}$, where n is the number of all directions in a hypothesis cell. Note that tuning γ can trade off the preference between the diversity and error of the directions. We illustrate the flow-matching metric using a toy example in Fig. 3. Moreover, it is worth noting that our flow-matching metric is complementary with the traditional geometry-matching metric, and both metrics can be employed simultaneously to fuse the crowd and geometry information. In particular, the crowd-matching metric is desirable in highly dynamic scenarios like human crowds while the geometry-matching metric is more helpful in static scenarios.

3.3 CrowdPlanner

We now describe how to utilize rich social navigation information encoded in the constructed crowd-flow map to accomplish socially-compliant robotic navigation in crowds, which is non-trivial when using traditional geometry map. Although there exist some approaches that can take into account crowd flow into the navigation policy [1,14], most of them assume an ideal localization, which is difficult to hold in real-world dense pedestrian scenarios. Our solution separates planning into two parts: in the long time horizon, the robot follows the global patterns in the existing crowd-flow map and plans a minimum cost trajectory connecting the robot's current location and its goal; in the short time horizon, the robot uses efficient re-planning to update the long horizon plan with online observed local crowd flow data, which is also beneficial for imperfect localization. Our planning algorithm is formulated as a search over the crowd-flow grid map, whose topology is defined by eight-neighborhood, i.e., the robot can move in eight different directions. Since imperfect localization needs frequent replanning, we choose D* [18] as the default planner, though other solvers such as A* can also be used.

For the search-based planner, we need to specify cost and cost-to-go functions over the grid map for designing search heuristics. The cost function defines the transition cost when the robot moves between cells after executing a selected action, and the cost-to-go function estimates the shortest distance from the robot's current position to the goal. We use the same cost-to-go function as the geometry grid map, but we design a new cost function to encourage the robot following crowd flow patterns. In particular, besides the traditional cost

Algorithm 1. Navigation cost computation for a robot's possible movement

Input: The crowd-flow map $\mathcal{M}$, the robot's current cell location $\mathbf{p}$, the robot's one possible movement $\mathbf{v}^{robot}$, weights w_{rc} and w_{lc} for resistance and lubricating costs respectively.
$Cost = \|\mathbf{v}^{robot}\|$
$[RC_{\mathbf{p}}, LC_{\mathbf{p}}] = [0, 0]$
// Compute the crowd-flow cost at the current location
for $\mathbf{v}^{flow}$ **in** $\mathcal{M}(\mathbf{p})$ **do**
  $RC_{\mathbf{p}} \mathrel{+}= \mathbf{v}^{robot} \cdot (-\mathbf{v}^{flow})$, $LC_{\mathbf{p}} \mathrel{+}= \|\mathbf{v}^{robot} \times \mathbf{v}^{flow}\|$
end for
$RC = RC_{\mathbf{p}}$, $LC = LC_{\mathbf{p}}$
// Sum up all cost with weights
$Cost \mathrel{+}= w_{rc} \cdot RC + w_{lc} \cdot LC$
Output: $Cost$

penalizing the long-distance movement, we consider another two costs, the *resistance cost* (RC) and *lubricating cost* (LC). Intuitively, RC penalizes the motions that are not consistent with the crowd flow, while LC formulates the fact that moving at directions perpendicular to the possible crowd-flow directions in a cell is likely to meet obstacles and thus should be avoided. To compute the RC cost for the robot at the position $\mathbf{p}$ taking the motion $\mathbf{v}^{robot}$, we project $\mathbf{v}^{robot}$ in the negative direction of every crowd-flow $-\mathbf{v}_{\mathbf{p}}^{flow_i}$ in the cell locating at $\mathbf{p}$. In this way, we get a scalar cost for each direction, i.e., $RC_{\mathbf{p}}^{i} = \mathbf{v}^{robot} \cdot (-\mathbf{v}_{\mathbf{p}}^{flow_i})$. We define the total RC cost at $\mathbf{p}$ as $RC_{\mathbf{p}} = \sum_i RC_{\mathbf{p}}^{i}$. For LC cost, we compute a vector $LC_{\mathbf{p}}^{i}$ as the cross product of $\mathbf{v}^{robot}$ and $\mathbf{v}_{\mathbf{p}}^{flow_i}$, i.e. $LC_{\mathbf{p}}^{i} = \|\mathbf{v}^{robot} \times \mathbf{v}_{\mathbf{p}}^{flow_i}\|$. The total LC cost is defined as $LC_{\mathbf{p}} = \sum_i LC_{\mathbf{p}}^{i}$.

Intuitively, the negative value of RC implies that the robot movement is consistent with the crowd flow and thus the resulting navigation behavior is social-friendly, and vice-versa. LC is always positive, and a higher LC value means that the movement in this cell has lower lubrication, i.e. has a higher chance to collide with static obstacles. We finally combine these two costs with the traditional distance-based cost with suitable blending weights (i.e. w_{rc} and w_{lc}) and the entire procedure for computing the crowd-aware navigation cost is shown in Algorithm 1.

4 Experiments

In this section, we first describe the implementation details of our approach. Next, we analyze our mapping algorithm in different scenarios and then validate the applicability of the crow-flow map for localization and planning in a crowded scenario. Finally, we compare our proposed approach with two baseline methods.

4.1 Implementation Details

Our simulation experiments are implemented on a PC with an Intel i7-9700 CPU and an NVIDIA RTX 2060 GPU. As shown in Fig. 2, Gazebo [17] is used for simulating the mobile Turtlebot platform with a 5 m perception range; Menge [6] is used for simulating realistic crowd behaviors. And we summarize the hyper-parameters used in our experiments as follows. For *CrowdMapper*, we set the elbow clustering threshold WSS_m as 0.5. For *CrowdLocalizer*, we set the particle number n_p to 500 and the crowd-flow matching metric parameter γ as 0.5. For *CrowdPlanner*, we set w_{rc} to 1.0 and w_{lc} to 0.5 because *RC* should be more important than *LC* for planning in crowds.

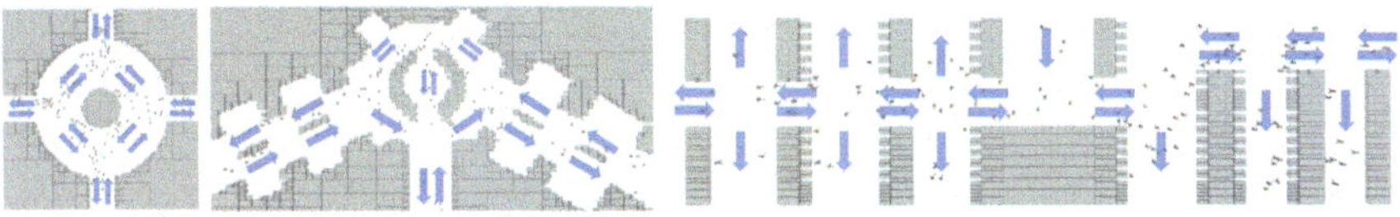

(a) Fountain (b) Conference Venue (c) Canteen

Fig. 4. Different benchmark scenarios. The blue arrows indicate crowd motion patterns. (a) is a 50×50 m^2 fountain scene whose crowd density is 0.130 m^2; (b) is a 100×50 m^2 conference venue scene whose crowd density is 0.045 m^2; (c) is a 100×25 m^2 canteen scene whose crowd density is 0.081 m^2.

4.2 Results for Crowd-Driven Mapping

To evaluate the performance of our proposed algorithms, we set up several indoor and outdoor scenarios with different crowd densities, as shown in Fig. 4. To quantitatively measure *CrowdMapper*'s quality, we use the crowd information captured by an imaginary bird-view camera that globally and perfectly monitors the entire scenario to compute the ground-truth crowd flow field as the baseline. The errors between the baseline and our reconstructed crowd-flow map are used to investigate whether the time-invariant movement patterns of pedestrians are well captured. For the error metric, since we are measuring the difference in flow, we borrow the flow-matching metric in Sect. 3.2. However, since the direction variance in a cell is not important, we use $\hat{q}$ rather than q as the error metric for a single grid cell. We finally calculate the average error over all cells as the error over the entire grid-based map.

We analyze the mapping result in two separate stages, the mapping stage and testing stage, as shown in Fig. 5b. In the mapping stage, the mapping quality increases when the robot explores the scenarios but eventually it converges to a fixed value, marked by the blue dash line in Fig. 5b. Apparently the mapping quality is not perfect (i.e., <1), which can be explained by two main challenges. First, we only discretize the crowd flow in a cell into eight directions and thus the robot cannot accurately model regions with complex crowd flow. Second, the robot only has partial observation to the scenario and may miss areas with a few pedestrians or crowds appearing in a period when it is absent.

After the mapping stage completes, we compare the obtained global crowd-flow map with the pedestrian motion field real-time captured by the imaginary bird-view camera for a series of frames. As shown in Fig. 5b, the mapping quality is unstable at the beginning, and after a while gradually decreases until it converges to another fixed value marked by the red dashed line in Fig. 5b. The slow declined tail in the procedure can also be explained by robot's partial observation, since the global camera can observe crowd flow appearing in regions with rare human activity while the robot cannot. We provide more detailed analysis about this phenomenon in the next paragraph. Note that there is a small gap between the red and blue dashed lines. This is because the robot will affect the movement of surrounding pedestrians during the mapping stage but not during the testing stage.

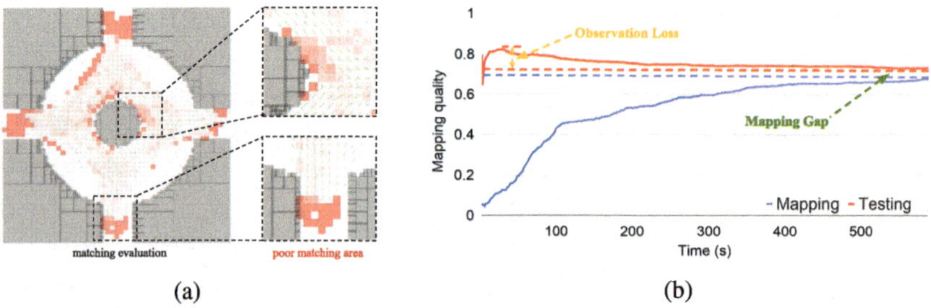

Fig. 5. (a) Heatmap for fountain scene's map quality. The darkness of the red color indicates the loss magnitude. (b) The map quality of the mapping and testing stages over different time steps.

Now we provide a detailed analysis about the error distribution using the fountain scenario as an example. We plot in Fig. 5a the heatmap of the converged quality loss (i.e. 1 - matching quality) in the testing stage. We can observe two main types of poor matching regions. First, in some areas, the robot can

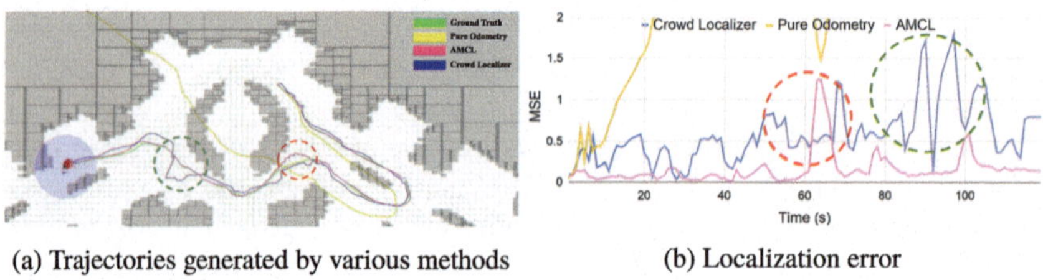

Fig. 6. A localization example in the *Conference Venue* scene.

hardly observe any crowd, which leads to a significant inconsistency with the observation from the bird-view camera. These areas are often near the boundary of crowd-flow map, as shown by the zoomed-in region in the bottom-right of Fig. 5a. Second, the robot indeed observes the crowd motion in some areas, but the observed data is insufficient for computing accurate time-invariant features, e.g., the region shown in the zoomed top-right in Fig. 5a. However, as shown in the heatmap, our mapping approach can obtain a high quality crowd-flow map correctly reflecting the layout of the entire scenario.

4.3 Results for Crowd-Driven Localization and Planning

After constructing a crowd-flow map, we next investigate whether *CrowdLocalizer* and *CrowdPlanner* can leverage the map to achieve localization and planning. Here we evaluate these two modules separately, i.e., evaluating first the localization accuracy, and then the planning performance assuming perfect localization. We will discuss planning under imperfect localization in Sect. 4.4.

We compute the localization accuracy as the Mean Square Error (MSE) between the ground truth and estimated paths. To better understand the effectiveness of *CrowdLocalizer*, we compare it with two baseline methods: the pure odometry method and a geometry-based particle filter method AMCL [23], both are implemented in ROS.

One qualitative comparison between these methods is shown in Fig. 6 on the conference venue benchmark. We can observe that both *CrowdLocalizer* and AMCL can well amend odometry sensor's cumulative errors. However, since they use different sensory observations (geometry for AMCL and crowd for *CrowdLocalizer*), their localization error distributions are different. For AMCL, the localization drift may occur around the intersection of crowds where people occlude most static landmarks, as shown by the red dashed circle in Fig. 6. In contrast, the localization drift of *CrowdLocalizer* usually appears in areas with sparse pedestrians, as shown by the green dashed circle in Fig. 6. When the observed pedestrian number increases, the drift can be effectively corrected, as shown in Fig. 7.

Similar results are also observed in quantitative experiments, where we let the robot move for 300 s in the benchmark scenarios and compute the average MSE of each trajectory. The results in Table 1 indicate that, as the crowd density increases, the localization error of AMCL gradually increases while the error of *CrowdLocalizer* is relatively stable and even slightly decreases. In general, traditional localization approaches is more advantageous in regions with a few people, while our method outperforms in dense crowds. By combining traditional approaches, our method can solve localization tasks in general scenarios with varying crowd density.

We then evaluate the performance of the planning module in term of social-friendly crowd navigation under perfect localization. Given a planning result, the robot will follow the plan and meanwhile actively avoid collisions with nearby pedestrians using a state-of-the-art reactive collision avoidance policy [10]. A high-quality plan is expected to minimize the inter-disturbance with the crowd.

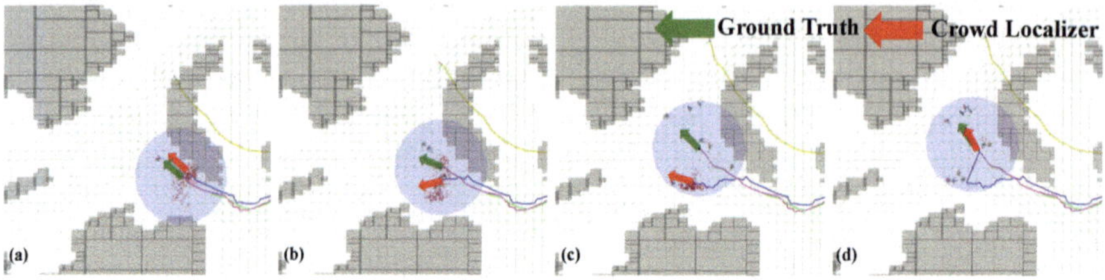

Fig. 7. Our localization quickly adjusts the pose estimation when crowds are observed.

Table 1. The MSE of trajectories in the testing scenarios: Fountain, Canteen, and Conference Venue.

Method	Fountain	Canteen	Conf. venue
Pure odometry	3.470	18.915	6.188
AMCL	9.029	0.610	**0.105**
CrowdLocalizer	**0.586**	**0.599**	0.613

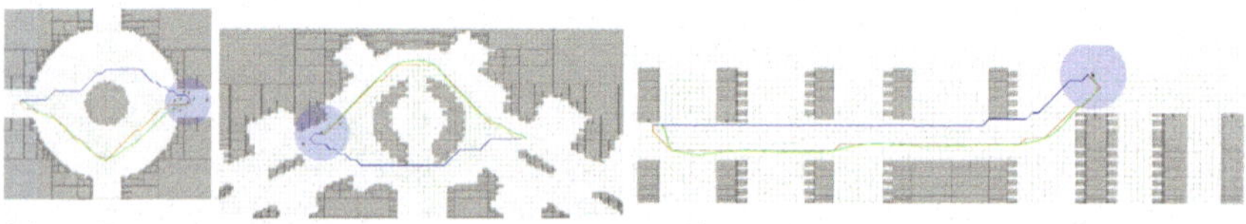

(a) Fountain scene (b) Conference Venue scene (c) Canteen scene

Fig. 8. Trajectories planned by different approaches. The blue curve is the result of *A* (shortest)*; the orange curve is result of *A* (social)*, and is also the plan for *CrowdPlanner* using D* at time 0; the green curve is the final trajectory of *CrowdPlanner* after D* replanning.

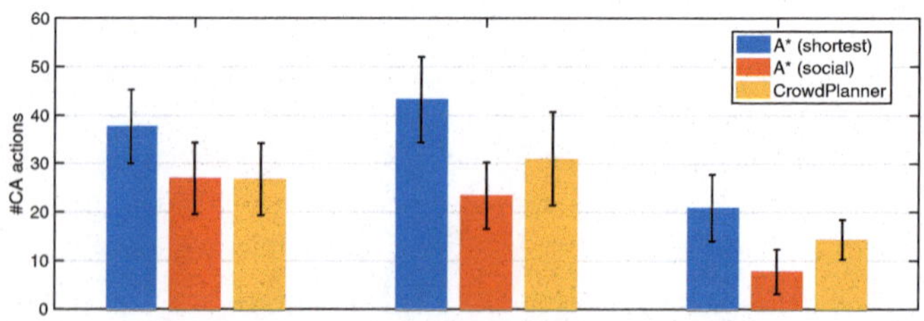

Fig. 9. The numbers of collision avoidance actions in trajectories planned using different approaches in benchmark scenarios (from right to right: fountain, conference, and canteen.)

Thus we use the number of the robot's active collision avoidance actions (i.e. #CA actions) during the navigation to quantify whether the planned path is socially compliant. Motivated by the fact that a plan passes through open space, which has the minimum inter-disturbance with the crowd, should be a smooth path with limited acceleration, we use the acceleration in the path over a short period of t_{traj} as a signal to detect CA actions. If the acceleration exceeds a predefined threshold $\dot{v}_{thres}$, the robot is considered as taking a CA action at that time. Through our experiments, we find that CA actions can be well detected with $t_{traj} = 0.5\,\text{s}$ and $\dot{v}_{thres} = 0.15\,\text{m/s}^2$. Furthermore, we compare the shortest trajectory planned by A* [12] (i.e. *A* (shortest)*), with trajectories that take into account the crowd-flow map but using different solvers A* (i.e. *A* (social)*) and D* (i.e. *CrowdPlanner*).

Table 2. Comparison experiments measured by the success rate and the number of executed collision avoidance actions (mean/std).

Method		Fountain		Canteen		Conf. venue	
Localizer	Planner	Success rate	# CA actions	Success rate	# CA actions	Success rate	# CA actions
AMCL	A* (shortest)	92%	39.26 (11.34)	**90%**	36.20 (7.83)	**100%**	22.00 (8.34)
	A* (social)	34%	**34.38 (8.11)**	86%	31.91 (8.73)	**100%**	**11.40 (6.24)**
	CrowdPlanner	52%	43.87 (12.92)	80%	40.36 (7.69)	**100%**	19.16 (8.78)
CrowdLocalizer	A* (shortest)	90%	54.62 (14.31)	82%	40.96 (12.92)	90%	25.06 (7.44)
	A* (social)	86%	44.71 (15.36)	84%	30.71 (10.56)	92%	11.85 (10.14)
	CrowdPlanner	**96%**	40.38 (12.08)	**90%**	**29.58 (9.64)**	84%	25.44 (7.68)

Since the crowd motion is random in simulation, we run each planning algorithm for 50 times in each testing scenario and then compare the different trajectory plans, as demonstrated in Fig. 8. We can observe that the planners (i.e. *CrowdPlanner* and *A* (social)*) considering the social flow is prone to generate a crowd-following trajectory, though their paths can be longer than that of *A* (shortest)* planner that is not social aware. In addition, we count #CA actions a robot will execute during the navigation and the results are summarized in Fig. 9. We can observe that the social-aware planners outperform the *A* (shortest)* planner.

4.4 Comparison of Complete Navigation Systems

Finally, we compare our crowd-based navigation framework with two baseline navigators to evaluate the overall performance. Similar to our previous experiments, we employ the #CA actions to measure the social compliance of a navigation behavior. In addition, we also compute the navigation success rate in terms of arriving goal without collisions to evaluate the navigator's robustness.

From the results in Table 2, we find that the AMCL localizer is sensitive to crowd density and planner type, whose navigation success rate is significantly downgraded in dense crowd scenarios such as Fountain and Canteen. The only exception is when AMCL is working with *A* (shortest)* whose navigation behavior (as shown by the blue trajectory in Fig. 8a) appears to be similar to a coastal planner [24], i.e., whenever the robot is about to lose its localization, it can walk towards nearby static landmarks to correct the error. In contrast, the success rate of *CrowdLocalizer* is stable in all three scenes, though in Conference scene with sparse pedestrians, it performs less perfect than AMCL. Besides, *CrowdPlanner* can dynamically follow the movement of nearby pedestrians and thus can improve the success rate compared to the *A* (social)* planner. In general, the combination of *CrowdLocalizer* and *CrowdPlanner* is superior in scenes with dense crowds, while in less crowded scenes, AMCL with social-aware planner achieves better results.

4.5 Real-Robot Experiments in a Rush Hour Crowd

We have deployed our crowd-driven navigation system on a wheeled robot to validate its feasibility and performance in the real world. The experiments are conducted on a university campus at rush hour to make a challenging and real scenario with dense pedestrians. We first teleoperate the robot to navigate within the crowd to construct the crowd-flow map. Next, we test the performance of our proposed localization and planning methods. To make the navigation more robust, we also complement our crowd-driven methods with the traditional SLAM method [25], as we discussed in Sect. 4.3. For more details, please refer to our supplementary multimedia materials.

5 Conclusion

We propose to utilize the movement information of pedestrians, which is commonly treated as harmful noises in traditional SLAM, for mapping, localization, and social-aware planning. Compared with the geometry map in traditional SLAM, our crowd-flow map provides a new perspective for robots better understanding its surrounding dynamic scenarios. Our approach has demonstrated preliminary but promising navigation performance on both simulated and real-world scenarios.

Acknowledgement. This work is supported by Innovation and Technology Fund (ITF) ITS/457/17FP, and General Research Fund (GRF)11207818, 11202119, NSFC-RGC joint grant N_HKU103/16.

References

1. Aroor, A.: Online learning and planning for crowd-aware service robot navigation. Ph.D. thesis, The City University of New York (2019)

2. Bertoni, L., Kreiss, S., Alahi, A.: Monoloco: Monocular 3d pedestrian localization and uncertainty estimation. In: ICCV, pp. 6861–6871 (2019)
3. Cai, P., Luo, Y., Saxena, A., Hsu, D., Lee, W.S.: Driving in a crowd by learning from tree search. In: RSS, Lets-drive (2019)
4. Chen, Y.F., Everett, M., Liu, M., How, J.P.: Socially aware motion planning with deep reinforcement learning. In: IROS, pp. 1343–1350 (2017)
5. Chen, Y.F., Liu, M., Everett, M., How, J.P.: Decentralized non-communicating multiagent collision avoidance with deep reinforcement learning. In: ICRA, pp. 285–292 (2017)
6. Curtis, S., Best, A., Manocha, D.: Menge: a modular framework for simulating crowd movement. Collective Dynamics **1**, 1–40 (2016)
7. Dellaert, F., Fox, D., Burgard, W., Thrun, S.: Monte carlo localization for mobile robots. ICRA **2**, 1322–1328 (1999)
8. Dong, X., Shen, J., Dajiang, Yu., Wang, W., Liu, J., Huang, H.: Occlusion-aware real-time object tracking. IEEE Trans. Multimed. **19**(4), 763–771 (2016)
9. Everett, M., Chen, Y.F., How, J.P.: Motion planning among dynamic, decision-making agents with deep reinforcement learning. In: IROS, pp. 3052–3059 (2018)
10. Fan, T., Cheng, X., Pan, J., Long, P., Liu, W., Yang, R., Manocha, D.: Getting robots unfrozen and unlost in dense pedestrian crowds. IEEE RAL **4**(2), 1178–1185 (2019)
11. Gajjar, V., Gurnani, A., Khandhediya, Y.: Human detection and tracking for video surveillance: a cognitive science approach. In: ICCV, pp. 2805–2809 (2017)
12. Hart, P.E., Nilsson, N.J., Raphael, B.: A formal basis for the heuristic determination of minimum cost paths. IEEE Trans. Cybernetics **4**(2), 100–107 (1968)
13. Hartigan, J.A., Wong, M.A.: Algorithm as 136: a k-means clustering algorithm. J. R. Soc. Series C **28**(1), 100–108 (1979)
14. Henry, P., Vollmer, C., Ferris, B., Fox, D.: Learning to navigate through crowded environments. In: ICRA, pp. 981–986 (2010)
15. Judd, K.M., Gammell, J.D.: Occlusion-robust mvo: Multimotion estimation through occlusion via motion closure. arXiv preprint arXiv:1905.05121 (2019)
16. Ketchen, D.J., Shook, C.L.: The application of cluster analysis in strategic management research: an analysis and critique. Strateg. Manag. J. **17**(6), 441–458 (1996)
17. Koenig, N., Howard, A.: Design and use paradigms for gazebo, an open-source multi-robot simulator. IROS **3**, 2149–2154 (2004)
18. Koenig, S., Likhachev, M.: D* lite. In: AAAI, pp. 476–483 (2002)
19. Linder, T., Breuers, S., Leibe, B., Arras, K.O.: On multi-modal people tracking from mobile platforms in very crowded and dynamic environments. In: ICRA, pp. 5512–5519 (2016)
20. Long, P., Fan, T., Liao, X., Liu, W., Zhang, H., Pan, J.: Towards optimally decentralized multi-robot collision avoidance via deep reinforcement learning. In: ICRA, pp. 6252–6259 (2018)
21. Miller, J., Hasfura, A., Liu, S.-Y., How, J.P.: Dynamic arrival rate estimation for campus mobility on demand network graphs. In: IROS, pp. 2285–2292 (2016)
22. Müller, J., Stachniss, C., Arras, K.O., Burgard, W.: Socially inspired motion planning for mobile robots in populated environments. In: COGSYS (2008)
23. Pfaff, P., Burgard, W., Fox, D.: Robust Monte-carlo localization using adaptive likelihood models. In: European Robotics Symposium, pp. 181–194 (2006)
24. Roy, N., Burgard, W., Fox, D., Thrun, S.: Coastal navigation-mobile robot navigation with uncertainty in dynamic environments. In: ICRA, pp. 35–40 (1999)

25. Shan, T., Englot, B.: Lego-loam: lightweight and ground-optimized lidar odometry and mapping on variable terrain. In: IROS, pp. 4758–4765 (2018)
26. Yan, Z., Duckett, T., Bellotto, N.: Online learning for human classification in 3d lidar-based tracking. In: IROS, pp. 864–871 (2017)
27. Zhang, Q., Chan, A.B.: Wide-area crowd counting via ground-plane density maps and multi-view fusion CNNs. In: CVPR, pp. 8297–8306 (2019)

Visual Semantic Mapping and Localization Using Parameterized Road Lanes

Sheng Yang(✉), Yiming Chen, and Mingyang Li

Alibaba Group, Hangzhou, China
{shengyang,yimingchen,mingyangli}@alibaba-inc.com

Abstract. We present a novel method for outdoor monocular visual mapping and localization, by detecting and probabilistically parameterizing road lanes. To present road lane information, we use cascaded deep models for detection on keyframes, and rely on piecewise cubic Catmull-Rom splines for parameterization in complicated environments. Additionally, we propose a maximum-a-posteriori estimation framework to conduct offline mapping, by jointly optimizing the lanes' geometric appearance and vehicle poses. The computed maps can be used to perform semantic localization, without relying on traditional visual point features. The underlying key design motivation is that, compared to visual point features, road semantic information is naturally highly compact and of long-term consistency (in terms of existence, appearance, and so on). Results from both simulated and real-world experiments show that our method is able to enhance estimation accuracy with cost-effective maps by wide margins.

Keywords: SLAM · Lane detection · Catmull-Rom spline

1 Introduction

Upgrading handcrafted point features [7,8,14] to high-level semantic entities [12] yields interpretable and stable maps for localizing vehicles at the city scale. As one of the most commonly seen components by vehicle-mounted cameras, road lane represents strong and long-term persistent information for localization and mapping. Additionally, compared to point features whose amount can easily reach millions in a town-size map, road lane is under naturally compressed representation and thus requires significantly less memory and disk resources. Those properties make the road lane an appealing entity to explore in robotic research.

To the best of authors' knowledge, the methodology of road lane based mapping and localization via a monocular camera remains *un-explored*. Straightforwardly representing road lanes as straight lines [3,4] will lose the generality, i.e., unable to fully characterize bending or undulating roads. On the other hand, several existing methods choose to parameterize lanes as spline curves, and demonstrate appealing performance. However, those algorithms rely on alternative sensor configurations with direct capability of capturing depth information: For laser scanners, Pedraza et al. use B-spline to fit noisy data of obstacles [10].

B. Siciliano et al. (Eds.): ISER 2020, SPAR 19, pp. 369–379, 2021.
https://doi.org/10.1007/978-3-030-71151-1_33

For stereo cameras, Rao et al. propose using the epipolar geometry of stereo cameras to reconstruct planar Bézier curves, and Wang et al. parameterize object contours as polybéziers [15]. These solutions, however, cannot be directly applied to monocular based use cases.

To allow monocular mapping and localization using road lanes, we design a novel and complete algorithmic pipeline, with the following main contributions:

- We characterize road lanes by piecewise cubic Catmull-Rom spline curves, which are jointly estimated with local ground coefficients and camera poses via a maximum-a-posteriori estimation framework.
- We explicitly formulate curve association parameters during the initialization and optimization, to ensure that none-probabilistic approximations are not involved when constructing visual cost functions.
- We design a customized learning based lane detection module, which, to the best of our knowledge, is the first attempt of using a learned model to replace traditional image-based contour extraction in mapping and localization.
- We propose the first *complete* monocular pipeline for mapping and localization using road lanes in urban environments.

2 Mapping and Localization with Splines

Figure 1 illustrates our algorithmic pipeline, which is to build an offline road lane map and subsequently use that for map based localization, similar to [13]. We note that, compared to online simultaneous localization and mapping (SLAM), our pipeline suits better for a variety of applications. e.g., autonomous driving and community robots. Compared to existing point based systems [8,18], our work contains three major innovative modules: Lane detection and tracking (Sect. 2.1), parameterization and initialization (Sect. 2.2), and recursive estimation (Sect. 2.3). The lane detection module uses a cascaded neural network to find out lane contours represented as sample and corner pixels (Fig. 1-a). During offline mapping, we perform a full detection on all involved keyframes, to gather sufficient multi-view observations and reconstruct lanes into a 3D lane map (Fig. 1-b). On the other hand, in online localization, we reduce the frequency of detection to allow real-time operations on resource-constrained processors, and use these sparse detected keyframes to construct projective association (Fig. 1-c) for vehicle pose estimation and cumulative drifts recovery.

2.1 Lane Detection and Tracking

To reliably detect and utilize road lanes, we develop two cascaded deep models for *lanes* and *dashed lane corners* hierarchically. Our perception module takes a monocular image as input, to firstly compute sample pixels for the left and right contour of each lane instance, and secondly corners of these dashed lanes. To allow efficient multi-task detection process, we build our network using the

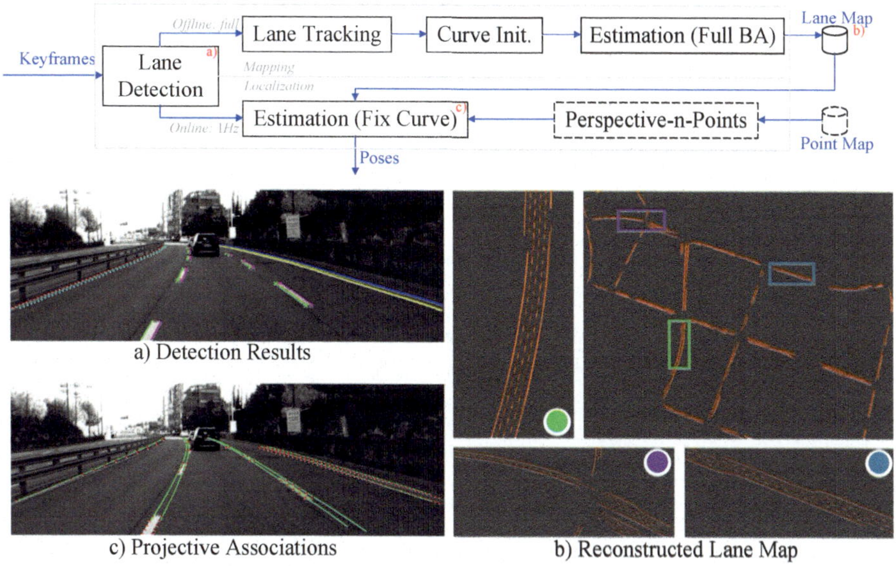

Fig. 1. Major modules and representative results of our proposed algorithm. a) Lanes are detected and represented as sample pixels on its contour, visualized by colored dots. Additionally, the detected corners of dashed lanes are shown by cross-marks. b) Constructed semantic lane map with zoomed-in views. c) Association of detected lanes (red) and map lanes (green) during localization.

architecture similar to that of CenterNet [19]. This anchor free detection method separates the backbone feature extraction process from customizable top-level heads, and thus adapt to multiple tasks based on an efficient network structure. We choose the DLA-34 [16] and the DCN module as our backbone, and use the following heads computing their losses as listed in Table 1 via the following steps:

1. **Lane instance as sample pixels.** The shared heatmap head H_{hm} for both samples and corners here is to output the category and a rough foreground. The boundary regression head H_{e_bnd} outputs the left-right and top-bottom boundary points relative to each foreground point. The output type head H_{e_att} is a three-category classification branch, which determines whether the left-right or top-bottom (or neither) boundaries of a lane point should be kept. We integrate the embedding head H_{e_emb} introduced by LaneNet [9], to reduce the dimension of feature representation of each location. These feature representations are later clustered via DBSCAN, and thereby different lane types can be distinguished.
2. **Dashed lane corners in pairs.** On these detected dashed lanes, we scan along with each lane dash, to extract 64×64 image patches at its two ends, for performing another round of keypoint extraction and robustly detecting corners in pairs. Based on the original implementation of the CenterNet keypoint detection model [19], we additionally generate a line segment based on

Table 1. Corresponding loss terms L and weights for different top-level heads H.

Note	Definition	Weight	Samples	Corners
L_{hm}	Focal loss of the center position	1.0	√	√
L_{e_bnd}	L1 distance to four sub-boundaries	0.2	√	
L_{e_att}	Cross entropy loss for sub-boundary type	1.0	√	
L_{e_emb}	Instance embedding loss [9]	1.0	√	
L_{c_iof}	L1 offset by the output stride (instance)	1.0		√
L_{c_hm}	Focal loss of the keypoint	1.0		√
L_{c_eof}	L1 offset by the output stride (each point)	1.0		√
L_{c_ang}	L1 angular diff. of the corner line segment	1.0		√
L_{c_len}	L1 length diff. of the corner line segment	1.0		√

such a pair, and compare it to the annotated corners, so as to get two additional heads H_{c_ang} and H_{c_len} for assessing angular and length differences, respectively.

During offline mapping, we track consecutively detected frames to establish multi-view geometry optimization problems, by assuming that the local road surface is relatively flat, and the corresponding ground surface parameters $\mathbf{G}$ (later introduced in Sect. 2.3) are known. The tracking module first accumulates a relative transformation between these two frames with IMU measurements [2,17], and then uses the Hungarian matching strategy to associate road lanes cascadingly in the pixel space. Specifically, we first reproject these sample pixels to the other frame through their ray-ground intersected position, and then compute the intersection over union (IoU) of formed polylines, assuming 5.0 pixel line width for holding an instance-wise reciprocal matching. After such instance-wise associations are established, we continue to reproject dashed lane corners for another round of reciprocal matching through single pixels. During these two stages, matches with IoU percentage less than 50%, or pixel distance larger than 5.0 pixels, are ignored.

2.2 Lane Parameterization and Initialization

In this work, we choose piecewise cubic Catmull-Rom spline curves [1] to represent the 3D lane contours through a series of control points $\mathbf{C}_k \in \mathbb{R}_3$. The shape of each piece $\mathbf{C}(t') \in \mathbb{R}_3$ is defined as:

$$\mathbf{C}(t') = \begin{bmatrix} t'^3 \\ t'^2 \\ t' \\ 1 \end{bmatrix}^\top \begin{bmatrix} -\tau & 2-\tau & \tau-2 & \tau \\ 2\tau & \tau-3 & 3-2\tau & -\tau \\ -\tau & 0 & \tau & 0 \\ 0 & 1 & 0 & 0 \end{bmatrix} \begin{bmatrix} \mathbf{C}_{k-1} \\ \mathbf{C}_k \\ \mathbf{C}_{k+1} \\ \mathbf{C}_{k+2} \end{bmatrix}^\top, t' \in [0,1]. \quad (1)$$

where $\tau = 0.5$ is a curvature parameter called tension that describes how sharply such a curve bends at its control points. Traversing t' through its domain allows us to interpolate between $\mathbf{C}_k$ and $\mathbf{C}_{k+1}$ smoothly. In practice, a successfully computed lane can easily reach hundreds of meters long, consisting of tens of control points. For notation simplicity of a piecewise spline, we denote $t \triangleq k + t' \in [2, M+1]$ for a whole piecewise curve $\mathbf{C}(t)$, which consists of M segments controlled by $M+2$ points $\mathbf{C}_k, k \in [1, M+2]$.

To iteratively refine control point estimates, initial 3D guesses are required. Compared with cubic B-spline [10] and cubic Bézier [11,15] curves, which do not cross all intermediate control points, piecewise cubic Catmull-Rom passes through all intermediate control points which ensures C1 continuity. To initialize 3D control points, we first convert detected lane sample pixels into 3D locations $\hat{\mathbf{C}}_{ci}$ as our tracking module did through ground surface coefficients $\mathbf{G}$. Subsequently, we employ a RANSAC framework, which randomly picks M samples ($\mathbf{C}_2$ to $\mathbf{C}_{M+1}$) from these intersected positions as control points, and use the loss function for refinement:

$$\underset{\mathbf{C}_k}{argmin} \sum_c \sum_i \left\| \hat{\mathbf{C}}_{ci} - \mathbf{C}(t_{ci}) \right\|^2_{\boldsymbol{\Omega}^{dis}_{ci}} + \sum_{k=3}^{M+1} \left\| |\mathbf{C}_k - \mathbf{C}_{k-1}| - \lambda_1 \right\|^2_{\boldsymbol{\Omega}^{ext}}, with:$$
$$\mathbf{C}_1 \leftarrow \begin{cases} |\mathbf{C}_2 - \mathbf{C}_1| = |\mathbf{C}_3 - \mathbf{C}_2| \\ \alpha_2 \cdot \mathbf{N}_2 = \tau(\mathbf{C}_3 - \mathbf{C}_1) \end{cases}, \quad \mathbf{C}_{M+2} \leftarrow \begin{cases} |\mathbf{C}_{M+2} - \mathbf{C}_{M+1}| = |\mathbf{C}_{M+1} - \mathbf{C}_M| \\ \alpha_{M+1} \cdot \mathbf{N}_{M+1} = \tau(\mathbf{C}_{M+2} - \mathbf{C}_M) \end{cases}. \tag{2}$$

In Eq. 2, the first part defines the 3D residual of curve fitting, and the second part is a regularization term for ensuring these control points are sampled evenly. $\boldsymbol{\Omega}^{dis}_{ci} = (0.1m)^2 \mathbf{I}_3$ and $\boldsymbol{\Omega}^{ext} = (0.25 \cdot \lambda_1)^2 \mathbf{I}_1$ are covariance matrices, with $\lambda_1 = 40$ meters as the expected sampling distance between control points.

For determining two external control points $\mathbf{C}_1$ and $\mathbf{C}_{M+2}$, we first estimate the local 3D direction $\mathbf{N}$ of the chosen control point $\mathbf{C}_2$ and $\mathbf{C}_{M+1}$ from its neighbor sample pixels, and then use the terms in Eq. 2 to preserve the curve direction at these two control points. Such equations mean to solve possible intersections of a sphere and a line. If we find more than one solutions, we accept the candidate located at the opposite side of $\mathbf{C}_3$ and $\mathbf{C}_{M-1}$, respectively.

To efficiently determine the 3D association parameter t_{ci} for each sampled point during the calculation of Eq. 2, we solve the equation $\mathbf{C}'(t_{ci})(\mathbf{C}(t_{ci}) - \hat{\mathbf{C}}_{ci}) = 0$ of t_{ci} to determine possible stationary points on each piece. If we get real roots t_{ci} in each corresponding domain, we also judge them together with all integers between $[2, M+1]$ to pick up the best association with the minimum point-to-curve distance. The above polynomial equation is quintic for each piecewise spline, and such 3D associations are only performed in the curve initialization phase for efficient sample consensus attempts.

For repeatedly picking M sequential samples as middle control points during such a sample-consensus initialization, we first randomly pick one sample, and recursively expand on its two sides by randomly picking another sample with their distance between $[0.5\lambda_1, 1.5\lambda_1]$. We pick at most 500 random sequences for each spline, and evaluate the loss in Eq. 2 to pick the best one.

2.3 Estimator Design

Lane-aware Bundle Adjustment (LBA). Considering these initialized curves as optimizable landmarks, our BA method concurrently refines:

1. Standard BA poses $\mathbf{T}_c \in \mathbb{SE}_3$, proprioceptive sensor constraints, and traditional point features as used in other visual BA framework [18].
2. Ground coefficients θ, ϕ, and d. Any ground 3D point $\mathbf{x}$ should be constrained by $\mathbf{G}(\theta, \phi, d) \cdot \mathbf{x} \triangleq [\cos\theta\cos\phi, \cos\theta\sin\phi, \sin\theta]^\top \mathbf{x} + d = 0$, with $\theta \in [0, \pi], \phi \in [0, 2\pi], d \in \mathbb{R}_3$.
3. The position of control points $\mathbf{C}_{ak} \in \mathbb{R}_3$ for each curve a.
4. The association parameters $t_{aci} \in [2, M_a + 1]$ for all associated sample and corner pixels, for scrolling along the curve and determine the actual associated position.

To optimize lanes, we build a camera observation cost function for each curve $\mathbf{C}_a$ as:

$$\mathbf{f}^{cam}_{a,c,i} = \left\| p_{aci} - \pi(\mathbf{T}_c^{-1} \cdot \mathbf{C}_a(t_{aci})) \right\|^2_{\mathbf{\Omega}^{cam}_{a,c,i}}, \tag{3}$$

where $p_{aci} \in \mathbb{R}_2$ is the location of a detected pixel, $\pi(\cdot)$ the camera projection operator to convert 3D point in camera's local coordinate frame into 2D pixel space, and $\mathbf{\Omega}^{cam}_{a,c,i} = {\sigma_1}^2 \mathbf{I}_2$ the noise covariance assigned w.r.t. the Median pixel error (Mp.) of sample and corner detection as reported in Fig. 2.

In Eq. 3, we dedicately consider these association parameters t_{aci}, because the analytical form of residual between observed pixels and projected curves cannot be precisely known in advance due to the projection operation $\pi(\cdot)$ of high-order curves, i.e., explicitly representing a cubic 3D shape projected on the image is difficult than points and straight lines because it is not an affine transformation. Hence, to initialize such parameters and jointly optimize it during LBA, we first assign t_{aci} with the solved 3D association parameter t_{ci} in Eq. 2, and then perform a standalone non-linear optimization of t_{aci} through $argmin\,\mathbf{f}^{cam}_{a,c,i}$. We filter those samples having multiple association hypotheses, i.e., with sub-optimal association whose residual is less than 1.5 times of the best association. Compared to the existing methods, explicitly optimizing these association parameters t_{aci} eliminates the approximation during both perspective projection [11] and on-curve sampling [15], thus our method constructs a fully probabilistic maximum-a-posteriori estimation.

Practically, to avoid domain constraints $t_{aci} \in [2, M_a + 1]$ during optimization, we extend every curves at their two endpoints $\mathbf{C}_1$ and $\mathbf{C}_{M+1}$ using a ray represented as Eq. 4 keeping the C1 continuity:

$$\mathbf{C}(t) = \begin{cases} (t-2) * \tau(\mathbf{C}_3 - \mathbf{C}_1) + \mathbf{C}_2, & t \in (-\infty, 2) \\ (t-(M+1)) * \tau(\mathbf{C}_{M+2} - \mathbf{C}_M) + \mathbf{C}_{M+1}, & t \in (M+1, \infty) \end{cases}. \tag{4}$$

In addition to sampled lanes points, our detection head also computes corners of dash lanes, which are indexed and point-wise tracked, providing reliable information for estimation. We associate tracked dash corners to exactly one association parameter, and initialize it with the non-linear optimization through multiple observations formulated as $argmin \sum_c \mathbf{f}^{cam}_{a,c,i}$. We note that as a replacement from treating dashed lane corners as regular point landmarks, processing corners within the curve optimization reduces the corresponding DOF (1 v.s. 3), and utilizes these measurements to solve the shape of reconstructed curves.

On the other hand, we assumed that observed lane samples within each keyframe are locally and approximately planar, to provide prior information and solve the involved ground surface coefficients as:

$$\mathbf{f}^{pl}_{a,c,k} = \left\| \mathbf{G} \cdot (\mathbf{T}_c^{-1} \cdot \mathbf{C}_{ak}) \right\|^2_{\mathbf{\Omega}^{pl}_{a,c,k}}, \tag{5}$$

where $\mathbf{\Omega}^{pl}_{a,c,k} = (0.2m)^2$ is assigned according to the thickness or noise of the corresponding planar objects. We use classical feature points within the convex-hull of detected lanes on each frame, to perform a RANSAC 3D plane fitting for initializing these surface parameters.

Offline Mapping and Map Structure. After these variables are initialized, we construct a factor graph optimization problem, involving basic variables and constraints in our previous work [18], and above introduced factors (Eq. 3 and 5) with Cauchy loss functions for numerical stability. After optimization, we store the representative control points $\mathbf{C}_{ak}$ of each optimized curve a, and association parameters t_{ack} of dashed lane corners as a lane map (Fig. 1-b), for online localization usage.

Online Localization with a Lane Map. During online localization, the shape of these curves, i.e. their control points $\mathbf{C}_{ak}$, is fixed, and we use Eq. 3 to construct localization hints, by replacing $\mathbf{T}_c$ with the corresponding camera poses in the map coordinate, after perspective-n-points are successfully solved. To obtain reliable association parameters t_{aci}, we again perform the curve projective association algorithm for detected sample pixels. For dashed lane corners, their association parameters are directly loaded from the map, and fixed for converting Eq. 3 to the camera observation cost of regular points.

3 Experiments and Evaluations

To demonstrate the advantages of the proposed method in practical applications, we conducted experiments using our road lane based pipeline in the combination of point features. We note that, as shown in Fig. 1, traditional point features can be used as optional additive components in our estimation framework, to allow more stable computation and handling possible degenerated case, i.e., a vehicle moves along a straight road without dash lanes. In such a case, the estimation uncertainty along the motion direction would grow without bounds if we only use curve landmarks. Our objective is to show that, due to road lanes' properties of being naturally compressed and distinguishable, the point-lane map will be of significantly improved quality: *higher* localization precision with *less* map size.

Datasets. We use several simulated sequences and a publicly available real-world dataset KAIST [5] to evaluate our algorithm. For generating simulated data, we collected timestamped real vehicle trajectories and ground coefficients, to randomly generate road lanes on the ground and project them on observable frames. After that, we added Gaussian noise to these lane sample pixels, whose standard derivation set to σ_1 (Fig. 2) as our reached position. Classical point landmarks were also simulated through such a generation and projection manner, while dashed lane corners are not simulated. The track length distribution of these points and lanes reported in real sequences was followed to avoid an idealistic everlasting tracking of landmarks, i.e., to simulate occlusion, failure detection, and sequential matching cases. For KAIST sequences, we use its four urban road sequences (26, 28, 38, 39) to examine our proposed method.

Perception Model Training. To train the proposed lane detection models, we manually annotated 3207 images extracted from these four KAIST sequences (4.4% of the whole sequence), including lane contours and corners, to four lane categories as listed in Fig. 2. Based on the DLA-34 pre-trained model provided by CenterNet [19], we randomly divide labeled images into a training set and test set with the ratio of 85% and 15%, and augment them through scaling and color enhancement to generate 512×512 model inputs. We use the Adam optimizer [6] and set the initial learning rate to 1.25×10^{-4}, which drops by 0.1 at 100 and 200 epochs, to train the model separately in two tasks.

Perception Model Performance. Figure 2 shows the performance of our trained deep model on the KAIST dataset. We evaluate these tasks separately on the test set, to summarize their classification precision, detection recall, and additionally the pixel error of extracted keypoints w.r.t. manual annotations on these labeled images. The median pixel error (Mp.) is then used as σ_1 for both our state estimator and simulated sequences generation. Detailed distributions of detected targets are also reported as the listed histograms.

Vehicle Localization Performance. To evaluate online localization performance, we use GNSS (not always available through data sequences), IMU measurements, and monocular images, to perform mapping and localization respectively. A representative reconstructed lane map can be seen in Fig. 1-b. During online localization, we disabled GNSS measurements to concentrate on evaluating the contribution of visual associations.

We choose to evaluate the proposed method against a point-only baseline [18]. The only difference between our proposed method and this baseline is the integration of curve perception and estimation. On simulated data, we acquire these listed errors in Table 2, by registering and comparing the localization trajectory

	Pre. (%)	Rec. (%)	Ap. (px)	Mp. (px)
Samples	93.3	89.0	2.97	0.52
Corners	98.1	94.2	0.86	0.72

Lane Categories	
solid yellow	solid white
dash yellow	dash white

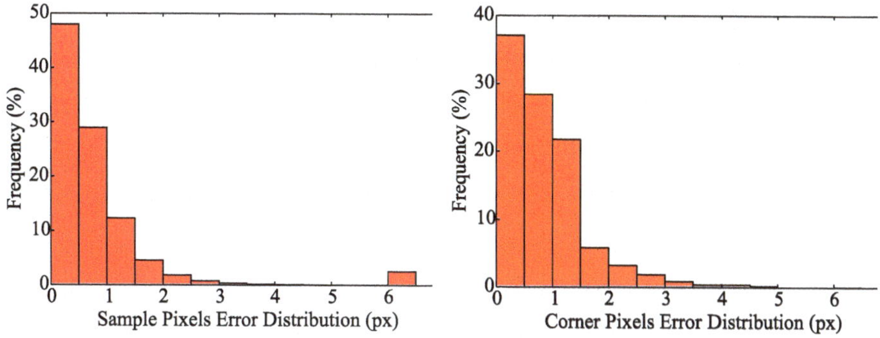

Fig. 2. Statistics of keyframe detection performance. Top-Left: Precision (Pre.), Recall (Rec.), Average/Median pixel error (Ap./Mp.) of different elements. Top-right: Categories of lane instances. Bottom: Error distribution of sample and corner pixels, with exceeded error (>6px) trimmed into the last bin.

Table 2. Localization errors on simulated data. Lane map size is **0.027** MB/km. P/C. Size: Point/Lane map size, in MB.

	Baseline [18]		Ours	
P. Size	T.-E	R.-E	T.-E	R.-E
1.05	17.80	**0.613**	**11.83**	0.686
1.57	15.55	1.611	**9.56**	**0.551**
2.10	6.47	0.515	**3.76**	**0.238**
2.62	5.55	0.240	**3.70**	**0.223**
5.24	3.58	**0.199**	**2.68**	**0.199**

Table 3. Localization errors on KAIST sequences. T/R.-E: Translational/Rotational Error, in meters and degrees.

	Baseline [18]		Ours	
Seq.	T.-E	P. Size	T.-E	C. Size
26	2.223	2.55	**1.319**	+0.07
28	**0.588**	8.97	0.641	+0.26
38	0.817	9.52	**0.637**	+0.29
39	1.076	8.86	**0.578**	+0.27
Mean	1.043	29.90	**0.711**	+0.89

with the ground truth using different sparsity of classical point landmarks in the visual map. Similarly, on KAIST sequences, Table 3 reflects the localization errors by registering and comparing the localization trajectory with the corresponding mapping trajectory. Results on both datasets have shown that, compared to point representation, memory cost for lanes is one order of magnitude smaller, while it still leads to significantly improved accuracy.

Efficiency. During localization, although the most time-consuming perception module takes 17.6 ms/693 MB GPU memory and 7.5 ms/422 MB for detecting samples and corners, respectively, such a module does not become a bottleneck of our algorithm, since the localization query attempts merely run at

1.0 Hz for supporting 5.0 Hz sliding window optimization with sparsified queried keyframes [18]. Experiments were performed on a desktop PC with i7-8700 CPU at 3.20 Hz and an NVIDIA GTX 1080 Ti.

References

1. Catmull, E., Rom, R.: A class of local interpolating splines. In: Computer Aided Geometric Design, pp. 317–326. Elsevier (1974)
2. Chen, Y., Zhang, M., Hong, D., Deng, C., Li, M.: Perception system design for low-cost commercial ground robots: sensor configurations, calibration, localization and mapping. In: IEEE/RSJ International Conference on Intelligent Robots and Systems, pp. 6663–6670 (2019)
3. Gomez-Ojeda, R., Moreno, F., Zuñiga-Noël, D., Scaramuzza, D., Gonzalez-Jimenez, J.: PL-SLAM: a stereo SLAM system through the combination of points and line segments. IEEE Trans. Rob. **35**(3), 734–746 (2019)
4. Huang, S., Ma, Z., Mu, T., Fu, H., Hu, S.: Lidar-monocular visual odometry using point and line features. In: IEEE International Conference on Robotics and Automation, pp. 1091–1097 (2020)
5. Jeong, J., Cho, Y., Shin, Y.S., Roh, H., Kim, A.: Complex urban dataset with multi-level sensors from highly diverse urban environments. Int. J. Rob. Res. **38**(6), 642–657 (2019)
6. Kingma, D.P., Ba, J.: Adam: a method for stochastic optimization. arXiv preprint arXiv:1412.6980 (2014)
7. Lowe, D.G.: Distinctive image features from scale-invariant keypoints. Int. J. Comput. Vis. **60**(2), 91–110 (2004)
8. Mur-Artal, R., Tardós, J.D.: ORB-SLAM2: an open-source SLAM system for monocular, stereo, and RGB-D cameras. IEEE Trans. Rob. **33**(5), 1255–1262 (2017)
9. Neven, D., De Brabandere, B., Georgoulis, S., Proesmans, M., Van Gool, L.: Towards End-to-End lane detection: an instance segmentation approach. In: IEEE Intelligent Vehicles Symposium, pp. 286–291 (2018)
10. Pedraza, L., Rodriguez-Losada, D., Matia, F., Dissanayake, G., Valls Miro, J.: Extending the limits of feature-based SLAM with B-Splines. IEEE Trans. Rob. **25**(2), 353–366 (2009)
11. Rao, D., Chung, S., Hutchinson, S.: CurveSLAM: an approach for vision-based navigation without point features. In: IEEE/RSJ International Conference on Intelligent Robots and Systems, pp. 4198–4204 (2012)
12. Sadat, A., Ren, M., Pokrovsky, A., Lin, Y., Yumer, E., Urtasun, R.: Jointly learnable behavior and trajectory planning for self-driving vehicles. In: IEEE/RSJ International Conference on Intelligent Robots and Systems, pp. 3949–3956 (2019)
13. Schneider, T., Dymczyk, M., Fehr, M., Egger, K., Lynen, S., Gilitschenski, I., Siegwart, R.: Maplab: an open framework for research in visual-inertial mapping and localization. IEEE Rob. Autom. Lett. **3**(3), 1418–1425 (2018)
14. Song, Y., Cai, L., Li, J., Tian, Y., Li, M.: Sekd: self-evolving keypoint detection and description. arXiv preprint arXiv:2006.05077 (2020)
15. Wang, Z., Kneip, L.: Fully automatic structure from motion with a spline-based environment representation. arXiv preprint arXiv:1810.12532 (2018)
16. Yu, F., Wang, D., Shelhamer, E., Darrell, T.: Deep layer aggregation. In: IEEE/CVF Conference on Computer Vision and Pattern Recognition, pp. 2403–2412 (2018)

17. Zhang, M., Chen, Y., Li, M.: Vision-aided localization for ground robots. In: IEEE/RSJ International Conference on Intelligent Robots and Systems, pp. 2455–2461 (2019)
18. Zhang, M., Zuo, X., Chen, Y., Liu, Y., Li, M.: Pose estimation for ground robots: On manifold representation, integration, re-parameterization, and optimization. IEEE Trans. Rob., 1–19 (2020)
19. Zhou, X., Wang, D., Krähenbühl, P.: Objects as points. arXiv preprint arXiv:1904.07850 (2019)

LION: Lidar-Inertial Observability-Aware Navigator for Vision-Denied Environments

Andrea Tagliabue[1(✉)], Jesus Tordesillas[1], Xiaoyi Cai[1], Angel Santamaria-Navarro[2], Jonathan P. How[1], Luca Carlone[1], and Ali-akbar Agha-mohammadi[2]

[1] Massachusetts Institute of Technology, Cambridge, MA, USA
{atagliab,jtorde,xyc,jhow,lcarlone}@mit.edu

[2] Jet Propulsion Laboratory, California Institute of Technology, Pasadena, CA, USA
{angel.santamaria.navarro,aliagha}@jpl.nasa.gov

Abstract. State estimation for robots navigating in GPS-denied and perceptually-degraded environments, such as underground tunnels, mines and planetary sub-surface voids [1], remains challenging in robotics. Towards this goal, we present LION (Lidar-Inertial Observability-Aware Navigator), which is part of the state estimation framework developed by the team CoSTAR [2] for the DARPA Subterranean Challenge [3], where the team achieved second and first places in the Tunnel and Urban circuits in August 2019 and February 2020, respectively. LION provides high-rate odometry estimates by fusing high-frequency inertial data from an IMU and low-rate relative pose estimates from a lidar via a fixed-lag sliding window smoother. LION does not require knowledge of relative positioning between lidar and IMU, as the extrinsic calibration is estimated online. In addition, LION is able to self-assess its performance using an observability metric that evaluates whether the pose estimate is geometrically ill-constrained. Odometry and confidence estimates are used by HeRO [4], a supervisory algorithm that provides robust estimates by switching between different odometry sources. In this paper we benchmark the performance of LION in perceptually-degraded subterranean environments, demonstrating its high technology readiness level for deployment in the field.

Video: https://youtu.be/Jd-sqBioarI

1 Introduction

Accurate and efficient odometry estimation for motion planning and control of robots in GPS-denied, perceptually-degraded, unknown underground environments still remains challenging in robotics. Common visual and visual-inertial

This research work was partially carried out at the Jet Propulsion Laboratory, California Institute of Technology, under a contract with the National Aeronautics and Space Administration. Government sponsorship acknowledged.

A. Tagliabue, J. Tordesillas and X. Cai—Contributed equally.

B. Siciliano et al. (Eds.): ISER 2020, SPAR 19, pp. 380–390, 2021.
https://doi.org/10.1007/978-3-030-71151-1_34

Fig. 1. Ground robots using LION to explore the Arch Coal Mine (West Virginia, USA), at approximately 275 m (900 ft) underground.

solutions [5–8] are not suitable for this task due to the potential absence of light or the strong illumination changes caused by mounting a light source on a moving robot. State estimation strategies for these environments [9,10] usually rely on proprioceptive sensing (e.g., IMU, wheel-encoders) and/or active exteroceptive methods (e.g., lidar, SoNaR, Time-of-Flight cameras or ultrasonic sensors). To compute odometry estimates in subterranean environments, the robots of the team CoSTAR relied on the complementarity of a lidar and an IMU as preferential sources of information (Fig. 1).

Lidar odometry (LO) and Lidar-Inertial odometry (LIO) present multiple challenges, especially related with the trade-offs between estimation accuracy and computational complexity, and scene observability. Most of the current approaches compute lidar odometry via the Iterative-Closest Point (ICP) algorithm [11–15] or by extracting and matching features [16–18]. While the first approach is generally less computationally expensive, the second is more practical for the purpose of creating a sparse map used for efficient re-localization. In both cases, the odometry estimates can be refined by aligning lidar-scans or lidar-features with a local map [16,18]. Fusing the information from an IMU generally improves the estimation accuracy [17,19–21], since it further constrains the estimate and guarantees a high-output rate, thanks to techniques such as IMU pre-integration [22]. Regardless of the scan-alignment technique used, lidar-based methods produce poor estimates in those scenarios that are not sufficiently geometrically-rich to constrain the motion estimation [23–25]. This is especially true in tunnel-like environments [9,10], where it is not easy to constrain the relative motion along the main shaft of the tunnel. Techniques employed to mitigate these issues consist of observability-aware point-cloud sampling [23], degeneracy detection and mitigation [24,26,27] and the fusion of other sources of information, such as inertial data from an IMU.

This work presents LION, a Lidar-Inertial Observability-Aware algorithm for Navigation in perceptually-degraded environments, which was adopted by the team CoSTAR for the first year of the DARPA Subterranean Challenge. Our solution relies on the fusion of pre-integrated IMU measurements [22] with pose estimates from scan-to-scan ICP alignments, taking advantage of a fixed-lag

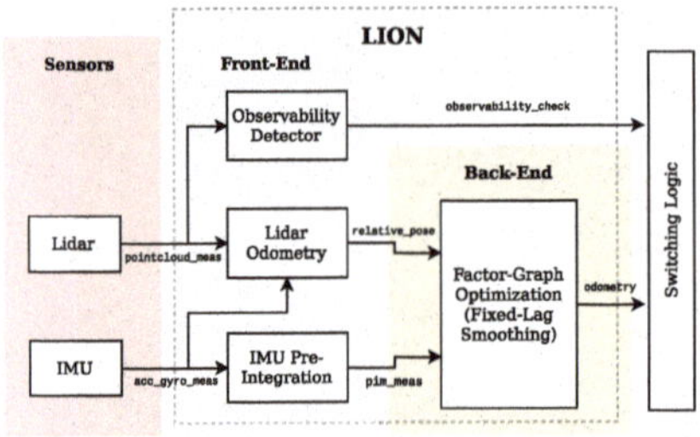

Fig. 2. Front-end and back-end of LION. The front-end runs ICP and IMU pre-integration, whose outputs are used as factors in the back-end pose-graph optimization

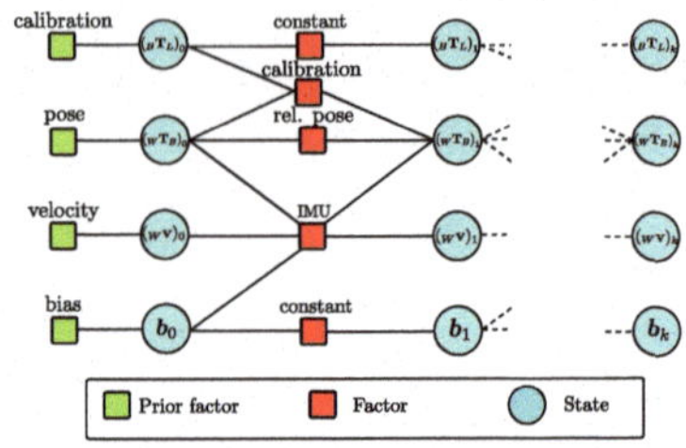

Fig. 3. Factor graph representing state and measurements used by LION. In this figure we use the shortcut $\boldsymbol{b}_j := \{\mathbf{b}^a, {}_B\mathbf{b}^g\}_j$ for the biases.

smoother architecture [28,29]. We simultaneously calibrate online the extrinsics (translation and rotation) between the lidar and IMU sensors. To address potential observability issues, we use a geometric observability score [23] that allows LION to predict potential degradation in its output given the geometric structure of the observed scene. By reporting the proposed score, we can switch to a different state estimation algorithm (e.g., from wheel encoders, visual-inertial or thermal-inertial) via a supervisory algorithm, such as HeRO [4]. This approach guarantees a continuous, reliable and gravity-aligned state estimate to the cascaded planning and control algorithms [30].

LION has been extensively tested in two coal mines of Beckley (West Virginia), two NIOSH experimental mines in Pittsburgh (Pennsylvania), a gold mine in Idaho Springs (Colorado), and many office-like environments at the Jet Propulsion Laboratory (California), showing its high readiness level for deployment in the field.

2 Technical Approach

Lidar-Inertial Odometry: LION is a sliding window estimator divided into two parts: a front-end consisting of the lidar-odometry and IMU pre-integration, and a factor-graph back-end (see Fig. 2). In the following, we use an odometry reference frame W, a body reference frame B, attached to the center of the IMU, and the lidar frame L, attached to the center of the lidar. The rigid transformation from a point expressed in a frame A to a point expressed in frame B is represented as the 4×4 matrix ${}_B\mathbf{T}_A$, which contains the rotation matrix ${}_B\mathbf{R}_A$ and translation vector ${}_B\mathbf{t}_A$.

The *front-end* of LION consists of three modules: lidar odometry, IMU pre-integration and the observability detector (see Sect. 2). The lidar odometry module obtains relative pose measurements ${}_{L_{k-1}}\mathbf{T}_{L_k}$ by using the Generalized Iterative Closest Point algorithm [31] on two lidar point clouds captured from consecutive lidar frames L_k and L_{k-1} at time stamps k and $k-1$ respectively.

To ease the convergence of the ICP algorithm, every incoming point cloud is pre-rotated in a gravity aligned-frame (i.e., setting the rotation estimated by the IMU as initial guess for ICP). The IMU pre-integration module leverages the state-of-the-art on-manifold pre-integration theory to summarize the high-rate IMU measurements into a single motion constraint [22,32]. Alternatively, the scan-to-scan based front-end can be replaced by LOCUS [33], which additionally aligns the incoming scan to an incrementally built local map and performs a refinement step.

In the *back-end*, the relative pose measurements produced by the front-end are stored in a factor graph in conjunction with the IMU measurements. A representation of the states and factors used in the factor graph can be found in Fig. 3. Let us define the state vector $\mathbf{x}_j$ at the j-th time step as

$$\mathbf{x}_j := \{\, {}_W\mathbf{T}_B,\ {}_W\mathbf{v},\ {}_B\mathbf{b}^a,\ {}_B\mathbf{b}^g,\ {}_B\mathbf{T}_L\}_j$$

where ${}_W\mathbf{T}_B$ is the IMU-to-world transformation, ${}_W\mathbf{v}$ is the linear velocity, ${}_B\mathbf{b}^a$ and ${}_B\mathbf{b}^g$ are the accelerometer and gyroscope biases of the IMU, and ${}_B\mathbf{T}_L$ is the lidar-to-IMU transformation. Following the notation from [32], let $\mathcal{K}_k := \{k-m+1, ..., k\}$ denote the m time steps inside the sliding window, and let $\mathcal{X}_k := \{\mathbf{x}_j\}_{j\in\mathcal{K}_k}$ and $\mathcal{Z}_k$ denote respectively the states and measurements in this sliding window. The factor graph optimization aims to solve therefore the following program [32]: $\mathcal{X}_k^\star := \arg\min_{\mathcal{X}_k} (-\log_e p(\mathcal{X}_k|\mathcal{Z}_k))$, where $p(\mathcal{X}_k|\mathcal{Z}_k)$ is the posterior distribution. We model the optimization problem using GTSAM [29] and solve this optimization problem using iSAM2 [28].

Observability Metric: In subterranean environments, it is crucial for a state estimation algorithm to determine how well the geometry of the scene is able to constrain the estimation in all the translational directions. Following [23,34], and assuming that the rotation is small, the Hessian of the Point-To-Plane ICP cost is given by $2\boldsymbol{A}$, where $\boldsymbol{A} := \sum_{i=1}^{M} \boldsymbol{H}_i^\top \boldsymbol{H}_i := \begin{bmatrix} \boldsymbol{A}_{rr} & \boldsymbol{A}_{rt} \\ \boldsymbol{A}_{rt}^\top & \boldsymbol{A}_{tt} \end{bmatrix}$, $\boldsymbol{H}_i := [-(\boldsymbol{p}_i \times \boldsymbol{n}_i)^\top,\ -\boldsymbol{n}_i^\top]$, and $\boldsymbol{n}_i$ is the surface unit normal vector based at a point $\boldsymbol{p}_i$. The eigenvector associated with the smallest eigenvalue of $\boldsymbol{A}$ is the *least observable* direction for pose estimation. The translational part is usually the most challenging part of the pose to estimate, mainly because of the presence of long shafts and corridors. We propose therefore to use the condition number $\kappa(\boldsymbol{A}_{tt}) := \frac{|\lambda_{max}(\boldsymbol{A}_{tt})|}{|\lambda_{min}(\boldsymbol{A}_{tt})|}$ as the observability metric. The larger the condition number $\kappa(\boldsymbol{A}_{tt})$ is, the more poorly constrained the optimization problem is in the translational part. When this condition number is above a user-defined threshold, LION issues a warning to the switching logic HeRO [4], so that other more reliable odometry sources can be used.

3 Experimental Results

3.1 LION Estimation Performance in the 2019 Tunnel Competition

We first evaluate the performance of LION (translation and rotation errors, and repeatability of the results) in the two runs of the two different tracks of

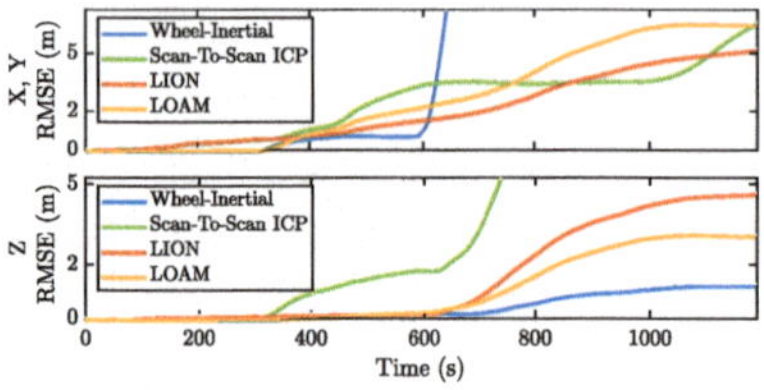

Fig. 4. Position estimation RMSE. (Track A, Run 2)

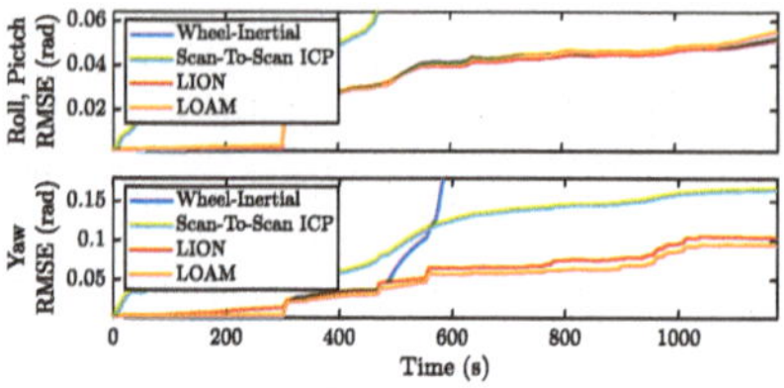

Fig. 5. Attitude estimation RMSE. (Track A, Run 2)

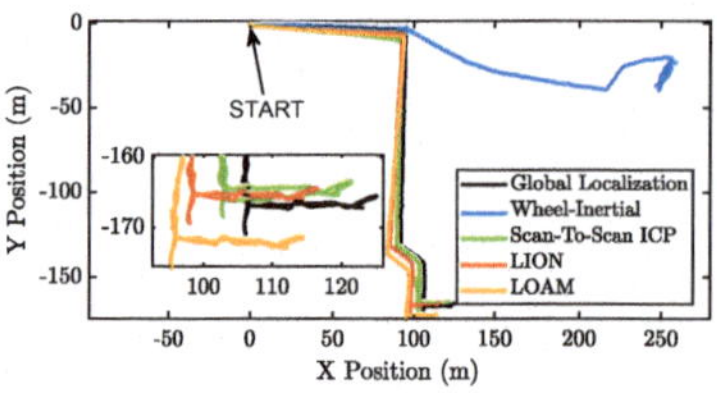

Fig. 6. x, y position estimates. (Track A, Run 2, first 1000 s)

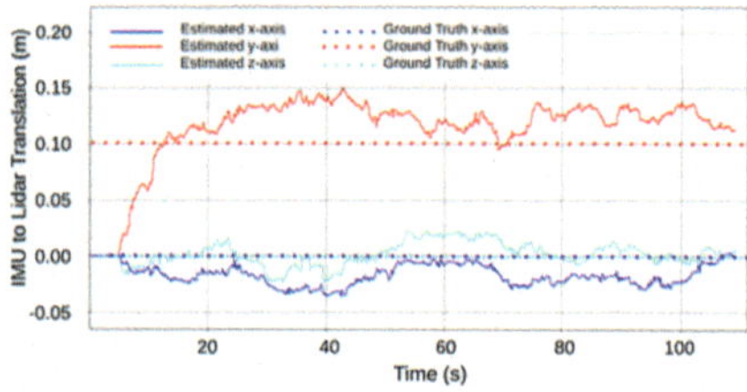

Fig. 7. Estimation of the translation of the extrinsics.

the Tunnel Competition, held in the NIOSH experimental mines in Pittsburgh, USA. Lidar odometry was computed 10 Hz, while IMU and LION output could be provided at up 200 Hz. The sliding window of LION used is 3 s and LION back-end was tuned to use approximately 30% of one CPU core of an i7 Intel NUC. For reference, we compare its performance with Wheel-Inertial odometry (wheel odometry fused with an IMU via an Extended Kalman Filter) and Scan-To-Scan odometry (the relative pose input of LION), with LAMP [35] as ground truth.

The results are summarized in Table 1. We can see that fusing inertial data with the odometry from the front-end (LION) significantly reduces the drift of the pure lidar-odometry method (Scan-To-Scan). This is especially evident from the estimation error along the z-axis shown in Fig. 4. Additionally, LION reliably estimates the attitude of the robot (Fig. 5), achieving small roll and pitch errors (since they are observable in the IMU via the gravity vector) and yaw. Figure 6 shows that LION outperforms the baseline approach (Wheel-Inertial) especially in terms of drift in yaw and horizontal translation. To showcase LION's auto-calibration capabilities, we generate in simulation a dataset where the lidar is translated of 0.1 m with respect to the IMU, along the IMU's y-axis. In Fig. 7 we observe that after approximately 20 s, LION (continuous lines) estimates the correct extrinsics (dotted lines). Last, we compare LION with the state-of-the-art method LOAM [18]. The comparison, shown in Figs. 4, 5 and 6 and Table 1, highlights that LION performs comparably to LOAM for short trajectories ($t < 600$ s in Fig. 4, and Table 1, Track B, Run 2), while LOAM (thanks to the presence of a map which compensates for drifts) achieves lower position and yaw drift for longer trajectories. LION obtains comparable or slightly lower roll and pitch errors thanks to the fusion of IMU data (Fig. 5), which guarantees a gravity-aligned output provided at IMU rate. The output rate of LOAM is instead limited to the rate of the lidar.

3.2 Observability Module

We first show how the condition number can detect geometrically-unconstrained scenes. To test this and to build intuition, we first use the dataset `JPL-Corridor`, recorded in an office at JPL, whose main challenge is the lack of geometric features along the direction of the corridor. We also use the dataset `Arch-Coal-Mine`, recorded in the Arch Coal Mine (see Fig. 8), which consists of a straight tunnel followed by an intersection. The plots of the condition number and the eigenvectors scaled with the eigenvalues for the `JPL-Corridor` dataset are shown in Fig. 8. At the beginning and the end of the corridor, there are enough features in all the directions, and the condition number is $\kappa(\boldsymbol{A}_{tt}) \approx 2$.

Table 1. Estimation error of Wheel-Inertial Odometry, Scan-To-Scan Matching, LION, and LOAM for two runs of the two tracks of the Tunnel competition, computed for one of the robots deployed. The second row shows the total distance and time traveled in each run. Note $\boldsymbol{t}$(m) and $\boldsymbol{R}$(rad) indicate the RMSE for position and attitude estimation, and $\boldsymbol{t}$(%) indicates the percentage drift in position.

	Track A						Track B					
	Run 1 (685 m, 1520 s)			**Run 2** (456 m, 1190 s)			**Run 1** (467 m, 1452 s)			**Run 2** (71 m, 246 s)		
Algorithm	$\boldsymbol{t}$(m)	$\boldsymbol{t}$(%)	$\boldsymbol{R}$(rad)	$\boldsymbol{t}$(m)	$\boldsymbol{t}$(%)	$\boldsymbol{R}$(rad)	$\boldsymbol{t}$(m)	$\boldsymbol{t}$(%)	$\boldsymbol{R}$(rad)	$\boldsymbol{t}$(m)	$\boldsymbol{t}$(%)	$\boldsymbol{R}$(rad)
Wheel-Inertial	130.50	19.05	1.60	114.00	25.00	1.28	78.21	16.75	0.99	6.91	9.79	0.12
Scan-To-Scan	105.47	15.40	0.90	18.72	4.11	0.18	56.6	12.14	0.79	4.55	6.45	0.27
LION	56.92	8.31	0.36	7.00	1.53	0.10	17.59	3.77	0.27	3.78	5.36	0.05
LOAM	10.99	1.60	0.14	7.22	1.58	0.08	13.21	2.83	0.21	5.55	7.87	0.03

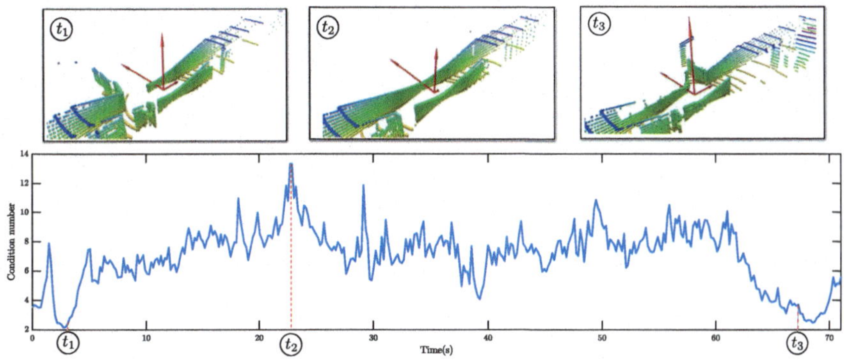

Fig. 8. Plot of the condition number $\kappa(\boldsymbol{A}_{tt})$ in the `JPL-Corridor` dataset (bottom), together with three snapshots of the eigenvectors of $\boldsymbol{A}_{tt}$ scaled with their corresponding eigenvalues (top).

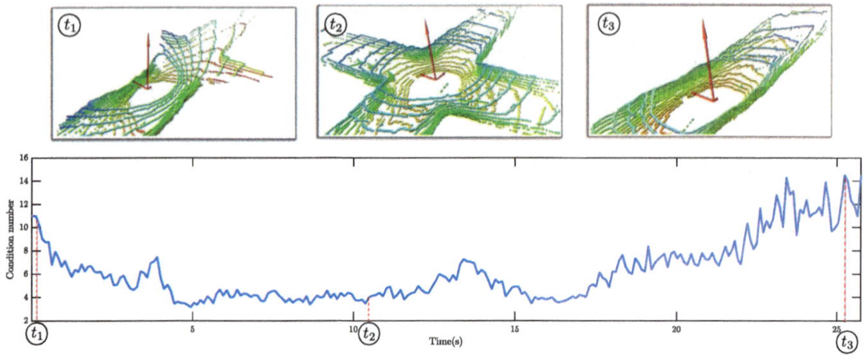

Fig. 9. Plot of the condition number $\kappa(\boldsymbol{A}_{tt})$ in the `Arch-Coal-Mine` dataset (bottom), together with three snapshots of the eigenvectors of $\boldsymbol{A}_{tt}$ scaled with their corresponding eigenvalues (top)

However, in the middle of the corridor, the condition number reaches values $\kappa(\boldsymbol{A}_{tt}) > 13$, which happens when the eigenvalue associated with the eigenvector along the direction of the tunnel becomes really small. This big condition number makes it hard to determine if the robot is moving based on changes in the geometry of the incoming point clouds. A similar situation happens in the `Arch-Coal-Mine` dataset (see Fig. 9). Before and after the intersection, $\kappa(\boldsymbol{A}_{tt}) > 10$ indicates the low observability along the main shaft. The observability is improved when the robot reaches the intersection, in which the condition number is $\kappa(\boldsymbol{A}_{tt}) \approx 4$. Note also the symmetry in the intersection of the two scaled eigenvectors that point to the two different corridors.

Using this condition number as the observability metric, HeRO can decide to switch to other odometry sources different from LION (like WIO) when there are not enough lidar features. This behavior is shown in Fig. 10 for a real experiment carried out in an office-like environment where a section of the first corridor does not have enough lidar features. If the observability module is not used, this lack of features in the corridor creates a *lidar slip*, where the movement of the robot is not well observed by LION. This produces a total error of ≈ 9 m when the robot goes back to the original position. However, when the observability module is used, the lack of observability in the direction of the corridor is detected, and WIO is used for that section of the corridor, instead of LION. This produces a total error of ≈ 1 m, which is small enough to trigger a successful loop-closure in the employed global mapping solution [35].

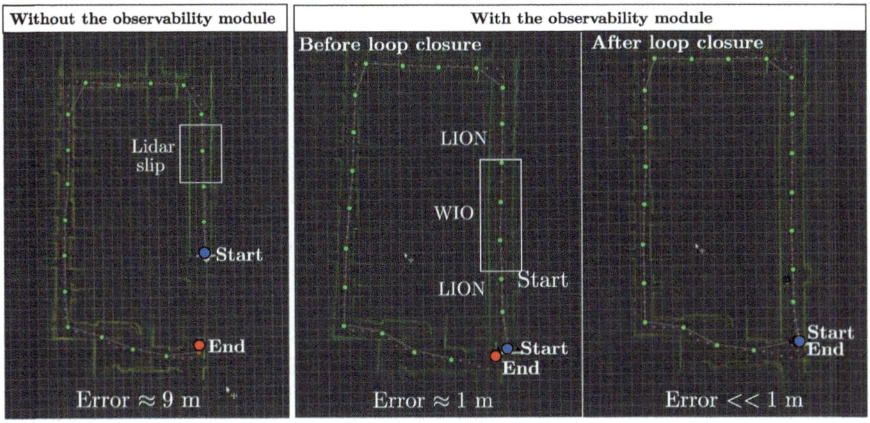

Fig. 10. Comparison of the translation error with and without the observability module in an office-like environment. On the left, the observability module is not used, which creates a *lidar slip* in the first part of the corridor, producing an error of ≈ 9 m. On the right, the observability module is used, and the switching-logic HERO switches to WIO instead of LION for the section of the corridor without lidar features. The total error is ≈ 1 m ("Before loop closure"). Improved state estimation (reduced drift in the output of HeRO) benefits the global mapping solution [35], which can now correctly detect a loop closure (Fig. 10 "After loop closure"), further reducing the drift.

4 Experimental Insights and Conclusions

This work presented LION, a Lidar-Inertial Observability-Aware algorithm for Navigation of robots in vision-denied environments. LION has been extensively tested in different subterranean environments, including the Tunnel Competition of the DARPA Subterrranean Challenge in August 2019. In the following we review the insights and reasons behind some of the choices made, as well as the main outstanding technical challenges.

- **Local state estimator**: The main goal of LION is to provide a high-rate, continuous and smooth output to the downstream algorithms. As such, LION does not build any map and does not perform any loop closures, and thus its inertial reference frame will slowly drift over time. The slow drifts in the output of LION are compensated by the mapping strategy LAMP [35], which is running within the localization framework.
- **Loosely-coupled architecture**: In contrast to other state-of-the-art works [16–18], the lidar front-end and back-end of LION are loosely-coupled (i.e., we do not explicitly add features point or scans in the state of the estimator). The main reason behind this was the desire of sharing the computational resources required by the lidar front-end with the mapping solution [35]. In addition, such architecture was dictated by the need of modularity in choosing front-end/back-end algorithms, and distributing the risk between several estimation engines to remove the one-point failure in case of a single estimation engine.

- **Not feature-based**: In LION, and contrary to [18], no features are extracted to match two point clouds because of two reasons. On one hand, feature extraction is usually computationally expensive, which would lead to a reduction of the performance of other modules of LION due to the limited computational resources onboard. Moreover, LION is meant for exploration of completely unknown environments, where there could be human-made structures (full of corners, planes and lines) or completely unstructured terrains. The use of feature extraction for such uncertain environments poses a prior on the environment the robot is going to navigate through, leading therefore to its own risks or failures.
- **Automatic extrinsic calibration**: lidar to IMU calibration is critical, especially in the relative orientation of the two sensors, since small errors can quickly integrate and cause a large drift. Offline calibration methods require a calibration target [36] and/or specific motion sequences [37], and can be impractical in a field setup where the sensor can be re-positioned at the last minute. This is the reason why we choose to estimate the extrinsics directly with LION.
- **Supervisory Algorithm**: LION has been designed to be one of the many odometry sources used by HeRO [4], a switching logic that multiplexes different odometry sources such as wheel encoders, visual-inertial or thermal-inertial odometry. The selection of one of these odometry sources is done according to the reliability of their output as state estimate. The confidence metrics for LION include its output rate, the detection of the potential failures in its inputs (lidar and IMU), and the observability score detailed in Sect. 2.

Acknowledgments. The authors would like to thank Kasra Khosoussi (ACL-MIT), Benjamin Morrell (JPL) and Kamak Ebadi (JPL) for helpful insights and discussions. Part of this research was carried out at the Jet Propulsion Laboratory, California Institute of Technology, under a contract with the National Aeronautics and Space Administration.

References

1. Agha, A., Mitchell, K.L., Boston, P.: Robotic exploration of planetary subsurface voids in search for life. AGU Fall Meet. Abs. **2019**, P41C–3463 (2019)
2. Team costar. https://costar.jpl.nasa.gov/. Accessed 29 June 2020
3. Darpa subterranean challenge. https://subtchallenge.com/. Accessed 29 June 2020
4. Santamaria-Navarro, A., Thakker, R., Fan, D.D., Morrell, B., Agha-mohammadi, A.-A.: Towards resilient autonomous navigation of drones. In: The International Symposium on Robotics Research (2019)
5. Bloesch, M., Omari, S., Hutter, M., Siegwart, R.: Robust visual inertial odometry using a direct EKF-based approach. In: IEEE/RSJ International Conference on Intelligent Robots and Systems (IROS), vol. 2015, pp. 298–304. IEEE (2015)
6. Qin, T., Li, P., Shen, S.: Vins-mono: a robust and versatile monocular visual-inertial state estimator. IEEE Trans. Rob. **34**(4), 1004–1020 (2018)

7. Mur-Artal, R., Montiel, J.M.M., Tardos, J.D.: ORB-SLAM: a versatile and accurate monocular SLAM system. IEEE Trans. Rob. **31**(5), 1147–1163 (2015)
8. Forster, C., Zhang, Z., Gassner, M., Werlberger, M., Scaramuzza, D.: svo: semidirect visual odometry for monocular and multicamera systems. IEEE Trans. Rob. **33**(2), 249–265 (2016)
9. Neumann, T., Ferrein, A., Kallweit, S., Scholl, I.: Towards a mobile mapping robot for underground mines. In: Proceedings of the 2014 PRASA, RobMech and AfLaT International Joint Symposium, Cape Town, South Africa, pp. 27–28 (2014)
10. Papachristos, C., Khattak, S., Mascarich, F., Alexis, K.: Autonomous navigation and mapping in underground mines using aerial robots. In: IEEE Aerospace Conference, vol. 2019, pp. 1–8. IEEE (2019)
11. Besl, P.J., McKay, N.D.: Method for registration of 3D shapes. In: Sensor Fusion IV: Control Paradigms and Data Structures, vol. 1611, pp. 586–606. International Society for Optics and Photonics (1992)
12. Chen, Y., Medioni, G.: Object modelling by registration of multiple range images. Image Vis. Comput. **10**(3), 145–155 (1992)
13. Zhang, Z.: Iterative point matching for registration of free-form curves and surfaces. Int. J. Comput. Vis. **13**(2), 119–152 (1994)
14. Magnusson, M., Lilienthal, A., Duckett, T.: Scan registration for autonomous mining vehicles using 3D-ndt. J. Field Rob. **24**(10), 803–827 (2007)
15. Pomerleau, F., Colas, F., Siegwart, R., Magnenat, S.: Comparing icp variants on real-world data sets. Auton. Rob. **34**(3), 133–148 (2013)
16. Shan, T., Englot, B.: LeGO-LOAM: lightweight and ground-optimized lidar odometry and mapping on variable terrain. In: IROS 2018 International Conference on Intelligent Robots and Systems (IROS), pp. 4758–4765. IEEE (2018)
17. Ye, H., Chen, Y., Liu, M.: Tightly coupled 3D lidar inertial odometry and mapping. arXiv preprint arXiv:1904.06993 (2019)
18. Zhang, J., Singh, S.: LOAM: lidar odometry and mapping in real-time. In: Robotics: Science and Systems, vol. 2, p. 9 (2014)
19. Qin, C., Ye, H., Pranata, C.E., Han, J., Zhang, S., Liu, M.: LINS: a lidar-inertial state estimator for robust and efficient navigation. arXiv preprint arXiv:1907.02233 (2019)
20. Hemann, G., Singh, S., Kaess, M.: Long-range GPS-denied aerial inertial navigation with lidar localization. In: IROS 2016 International Conference on Intelligent Robots and Systems (IROS), pp. 1659–1666. IEEE (2016)
21. Lin, J., Zhang, F.: Loam_livox: a fast, robust, high-precision lidar odometry and mapping package for lidars of small fov. arXiv preprint arXiv:1909.06700 (2019)
22. Forster, C., Carlone, L., Dellaert, F., Scaramuzza, D.: IMU preintegration on manifold for efficient visual-inertial maximum-a-posteriori estimation. Georgia Institute of Technology (2015)
23. Gelfand, N., Ikemoto, L., Rusinkiewicz, S., Levoy, M.: Geometrically stable sampling for the ICP algorithm. In: Fourth International Conference on 3-D Digital Imaging and Modeling, 2003. 3DIM 2003. Proceedings, pp. 260–267. IEEE (2003)
24. Zhang, J., Kaess, M., Singh, S.: On degeneracy of optimization-based state estimation problems. In: ICRA on Robotics and Automation (ICRA), vol. 2016, pp. 809–816. IEEE (2016)
25. Censi, A.: An accurate closed-form estimate of ICP's covariance. In: Proceedings of the IEEE International Conference on Robotics and Automation, vol. 2007, pp. 3167–3172. IEEE (2007)

26. Zhang, J., Singh, S.: Enabling aggressive motion estimation at low-drift and accurate mapping in real-time. In: ICRA 2017 Conference on Robotics and Automation (ICRA), pp. 5051–5058. IEEE (2017)
27. Hinduja, A., Ho, B.-J., Kaess, M.: Degeneracy-aware factors with applications to underwater slam. In: IROS, pp. 1293–1299 (2019)
28. Kaess, M., Johannsson, H., Roberts, R., Ila, V., Leonard, J.J., Dellaert, F.: iSAM2: incremental smoothing and mapping using the bayes tree. Int. J. Rob. Res. **31**(2), 216–235 (2012)
29. Dellaert, F.: Factor graphs and gtsam: A hands-on introduction. Technical report. Georgia Institute of Technology (2012)
30. Thakker, R., Alatur, N., Fan, D.D., Tordesillas, J., Paton, M., Otsu, K., Agha-mohammadi, A.-A.: Autonomous traverse of off-road extreme terrains in dark and dust: an experimental perspective on physical mobile robots. In: ISER (2020)
31. Segal, A., Haehnel, D., Thrun, S.: Generalized-icp. In: Robotics: Science and Systems, vol. 2, no. 4, p. 435. Seattle, WA (2009)
32. Forster, C., Carlone, L., Dellaert, F., Scaramuzza, D.: On-manifold preintegration theory for fast and accurate visual-inertial navigation. IEEE Transactions on Robotics, pp. 1–18 (2015)
33. Palieri, M., Morrell, B., Thakur, A., Ebadi, K., Nash, J., Carlone, L., Guaragnella, C., Agha-mohammadi, A.: Locus-a multi-sensor lidar-centric solution for high-precision odometry and 3D mapping in real-time. Under Review (2020)
34. Bonnabel, S., Barczyk, M., Goulette, F.: On the covariance of ICP-based scan-matching techniques. In: American Control Conference (ACC), vol. 2016, pp. 5498–5503. IEEE (2016)
35. Ebadi, K., Chang, Y., Palieri, M., Stephens, A., Hatteland, A., Heiden, E., Thakur, A., Morrell, B., Carlone, L., Aghamohammadi, A.: LAMP: large-scale autonomous mapping and positioning for exploration of perceptually-degraded subterranean environments. In: ICRA 2020 International Conference on Robotics and Automation (ICRA) (2020)
36. Le Gentil, C., Vidal-Calleja, T., Huang, S.: 3D lidar-imu calibration based on upsampled preintegrated measurements for motion distortion correction. In: ICRA Automation (ICRA), vol. 2018, pp. 2149–2155. IEEE (2018)
37. Della Corte, B., Andreasson, H., Stoyanov, T., Grisetti, G.: Unified motion-based calibration of mobile multi-sensor platforms with time delay estimation. In: IEEE Robotics and Automation Letters, vol. 4, no. 2, pp. 902–909 (2019)

The DARPA SubT Urban Circuit Mapping Dataset and Evaluation Metric

John G. Rogers III[1(✉)], Arthur Schang[2], Carlos Nieto-Granda[1], John Ware[1], John Carter[1], Jonathan Fink[1], and Ethan Stump[1]

[1] CCDC Army Research Laboratory (ARL), Adelphi, MD 20783, USA
{john.g.rogers59,jonathan.r.fink3,ethan.a.stump2}.civ@mail.mil
[2] Parsons Corporation, Lorton, VA 22079, USA
arthur.schang@parsons.com

1 Introduction

Software solutions to the Simultaneous Localization and Mapping (SLAM) problem have successfully transitioned from the laboratory to the commercial sector in recent years; however, these systems often operate under ideal conditions and in structured environments. In this paper, current state-of-the-art approaches to SLAM will be evaluated in more realistic conditions mimicking a disaster recovery scenario.

The setting for existing datasets is typically more benign: on-road driving [3], or in indoor spaces such as a hospital or an office building[1]. These types of settings are characterized by: favorable lighting, regular man-made structures, nearly flat topography, free from standing water, and may have access to GPS. The urban underground presents a starkly different setting: poor to no lighting, significant debris or clutter, complex 3D topography including stairwells, standing water, and no access to GPS. By providing a dataset which contains some of these characteristics, the effectiveness of current technology can be evaluated.

To address this lack of publicly available datasets for subterranean mapping, we present a new dataset taken from the DARPA Subterranean (SubT) Challenge Urban Circuit. This dataset consists of sensor data captured from both a ground and an air robot teleoperated through two separate courses inside a partially completed nuclear reactor building, which has been abandoned for decades. The ground robot's sensor payload was selected to support a diverse set of approaches to SLAM. In addition, the robot's start locations were instrumented with alignment markers, the site was professionally surveyed, and the ground truth locations of key artifacts placed around the courses were recorded. Finally, a set of support code is also provided which enables users to evaluate their own mapping approaches against the provided objective metrics.

In designing this dataset, we have followed the structure of other well known and publicly available datasets, such as the KITTI Vision Benchmarking

---

[1] These datasets can be found at the RADDISH SLAM benchmark at http://radish.sourceforge.net/ and http://ais.informatik.uni-freiburg.de/slamevaluation/datasets.php.

B. Siciliano et al. (Eds.): ISER 2020, SPAR 19, pp. 391–401, 2021.
https://doi.org/10.1007/978-3-030-71151-1_35

Suite [3], that contain a redundant set of sensory inputs in order to support multiple algorithmic techniques. The ground datasets contain stereo as well as two LiDAR sensors. This style of benchmark dataset enables ablation studies, where algorithmic components and sensor inputs are switched off or degraded to objectively determine marginal effects on performance.

This paper extends our previous work [7] by introducing the SubT Urban Circuit dataset, incorporating a new relative error metric, and extending our evaluated SLAM systems to include visual approaches on an additional aerial platform. These extensions improve this contribution's utility as an evaluation metric to compare a diverse set of mapping algorithms in difficult environments.

2 Technical Approach

The ground portion of the SubT Urban dataset[2] was collected using a modified iRobot Packbot Explorer, as seen in Fig. 1. This robot was equipped with a variety of sensors which are commonly used by researchers. High-resolution point clouds were collected with an Ouster 64 LiDAR. In addition, a front-facing point cloud was collected using the spinning Hokuyo on a Carnegie Robotics Multisense SL, which also provided stereo images and illumination for these cameras. The data collected by the two LiDAR sensors is complementary; the Ouster data is more dense, while the Multisense data is a complete frontal field-of-view, as can be seen in Fig. 2. For this challenge, an new artifact type representing a CO_2 gas leak was introduced. To detect the leaking CO_2, we included an SCD30 CO2 sensor (not pictured) which measures the CO_2 level in addition to temperature and humidity. The signal strength to nearby WiFi access points is also logged (this is useful to detect the cell phone artifact). The robot platform encoders are fused with the orientation estimate from a Microstrain GX5-25 IMU to generate an odometric estimate of motion. This dataset is used to test a SLAM algorithm by re-processing the robot data offline. The optimized trajectory solution of the SLAM system under test is composed with a relative artifact measurement. This global estimate of the artifact's position is then compared to the ground-truth position to determine accuracy. The entire procedure can be seen in Fig. 3. The ground truth position of a Vent artifact can be seen relative to a measurement estimated using a SLAM algorithm under test which scores a point in Fig. 4.

Each challenge course includes an entry portal with fiducial markings and their surveyed ground truth locations, as can be seen in Fig. 5. These markers and their locations are used by the frame alignment node to establish a global frame of reference. In each course, a set of 20 landmarks consisting of backpacks, cell phones, vents, survivors (rescue dummies), and small room with elevated CO2 gas are professionally surveyed in this global reference frame. We have noted the relative pose to the robot for each artifact as it is seen in the course data file.

[2] Instructions for downloading the SubT challenge datasets and supporting code repository can be found at https://bitbucket.org/jgrogers/subt_reference_datasets.

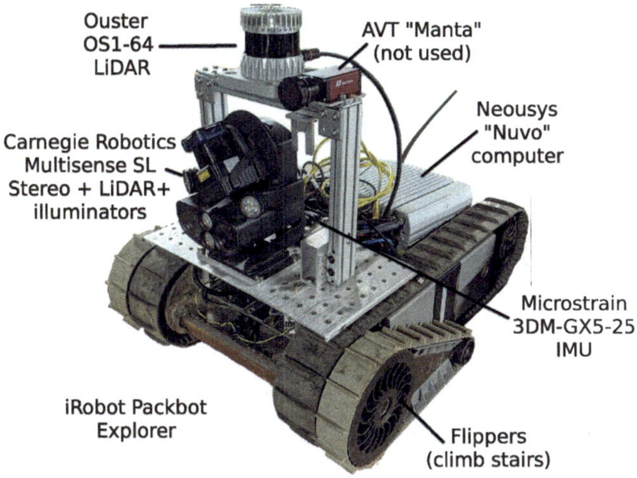

Fig. 1. The robot used to collect the SubT Urban challenge reference dataset.

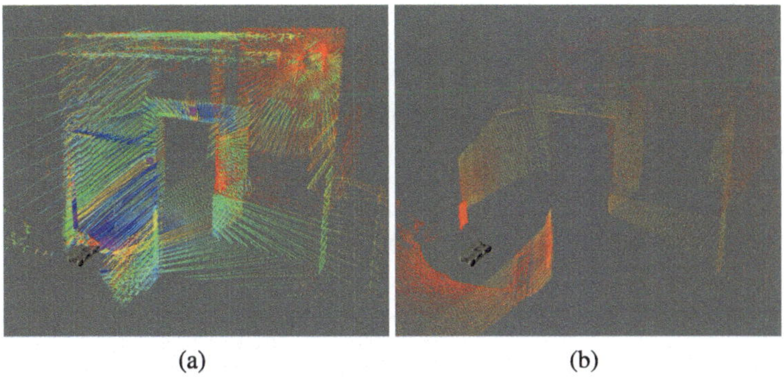

Fig. 2. Comparison of MultiSense SL (a) and Ouster OS1-64 (b) LiDAR

Additionally, for this circuit challenge we have introduced several runs which using a custom quadrotor frame equipped with an Intel RealSense D435i camera. These data sets feature faster frame rates on stereo and structured light vision from the D435i camera which makes it more suitable for evaluating visual SLAM techniques.

This dataset includes an analysis of several techniques available, including LiDAR-based techniques OmniMapper [12], Cartographer [4], and LeGO-LOAM [10], as well as visual SLAM techniques ORB_SLAM2 [6], MSCKF [11], and Kimera [8].

Many prior SLAM evaluation metrics follow the approach of [1], which introduced an incremental relative error metric designed to equally penalize errors near the start of a trajectory with those near the end. The SubT challenge has instead selected an absolute metric for evaluation of teams since absolute posi-

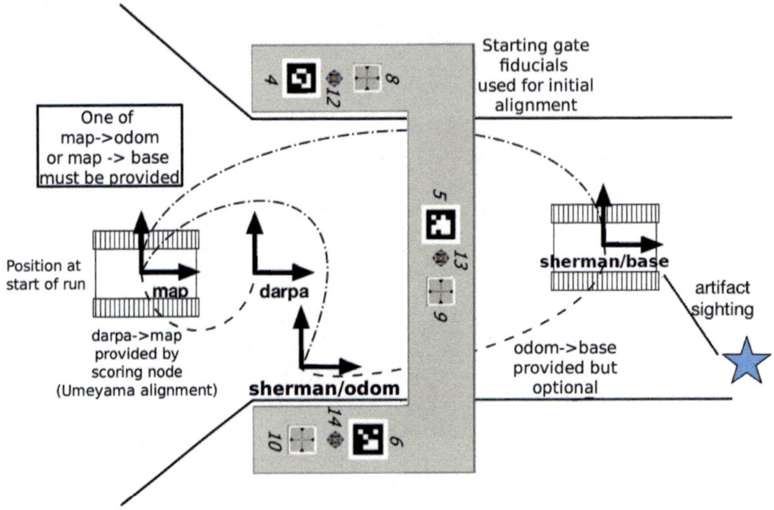

Fig. 3. The frames of reference used by the scoring node to evaluate a SLAM algorithm. The scoring node determines the alignment between the 'darpa' frame and the robot's starting position, designated as the 'map' frame by solving for the relative pose from known fiducial landmark positions in the starting gate and distal fiducial using the Umeyama algorithm. The integrated robot odometry is provided in the optional 'odom' → 'base' transform. A SLAM algorithm can provide corrections through a 'map' → 'odom' transform, or it can provide the complete transform from 'map' → 'base' by disabling the output of the integrated odometry. When an artifact is observed, its relative pose is composed with the current estimate of the robots position in the 'darpa' frame for comparison with the known ground-truth position of that artifact.

tion information would be needed to drill access passages from the surface for rescue operations in cases of structural collapse. Here we replicate the SubT scoring metric; a point is given for each artifact localized within 5 m of its absolute position. In addition to the raw point score, we evaluate SLAM algorithms in the root-mean squared error (RMSE) to the ground truth artifact locations. For this paper, we also introduce an additional relative error metric which can be used when global localization is not achieved. This was added to make use of additional test runs which do not observe the global frame fiducials due to omission or hardware failures.

Several factors have limited the number of usable runs that can be evaluated by the absolute metric. Due to time constraints and scheduling during this complex event, some segments were captured without observing the entry portal alignment fiducials. In addition, some algorithms were unable to proceed past the stair descent portions. To increase the number of usable runs and evaluated algorithms in the SubT Urban dataset, we have included a new *relative metric* in addition to the absolute metric. The relative metric is computed via a branch-and-bound data association solver which finds correspondence between

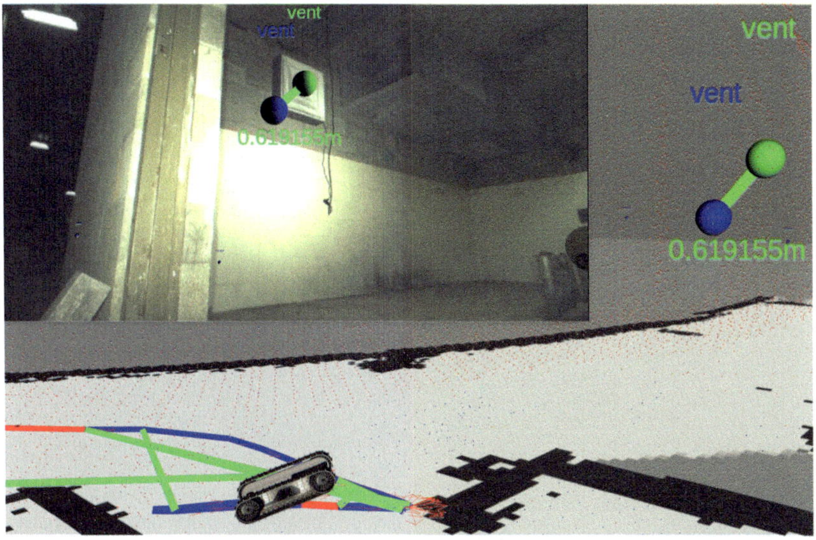

Fig. 4. The robot is using its flippers to pitch up to observe a heating vent artifact in the Alpha course. The inset in the upper left is the live camera view including the projection of relevant markers including the ground truth location of this artifact, the current measurement, and the error. The ground truth location is represented by a blue sphere, and the current measurement being evaluated by the scoring system is shown in green for measurements which are within 5 m (and score a point), or would be shown in red if the measurement is more than 5 m off (and would not score a point). The error on this measurement is shown in meters.

artifact ground-truth positions and map frame artifact reports. For a candidate data association, a rigid-body transform is computed using the Umeyama multi-point correspondence transform estimation algorithm. The RMSE error is then computed from these transformed artifact reports.

3 Experiments

The collection of data from all environments was performed by direct teleoperation of the robot through a game controller joystick attached to a wearable computer which sends control commands and receives camera video via WiFi. The reinforced concrete construction in the Urban circuit course limits the range of WiFi radio signals; therefore, the operators must remain in close proximity to the robot, and may briefly appear in sensor data. The majority of operator interference occurs in the laser scan data which can be mitigated by utilizing the entire scan as the interference is insignificant over the entire field of view of the sensor. The SLAM approaches selected for evaluation include a variety of modern open-source techniques alongside our own technique. The variety in sensor modalities utilized by these techniques is intended to highlight the challenges

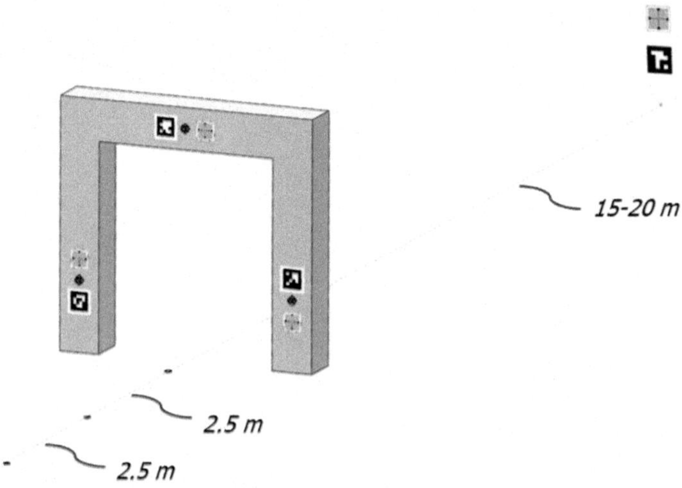

Fig. 5. The alignment 'portal' present at the entrance to each DARPA SubT challenge course. Each portal includes multiple fiducial landmark types in three positions, in addition to a distal landmark at a known location further within the course. The scoring node provided in the software distribution accompanying this paper utilizes the April tag markers to measure the global frame alignment from the robot's starting position. Image taken from DARPA Competition Rules: Urban Circuit document.

of this dataset and to differentiate between the performance of LiDAR-based SLAM in comparison to visual SLAM. To understand the importance of SLAM techniques to maintain global positional accuracy, raw proprioceptive odometry trajectory estimation is included in the results section alongside the aforementioned SLAM techniques.

OmniMapper
The first SLAM algorithm evaluated with this dataset is an approach from our previous work that was developed based upon the open-source *OmniMapper* [12]. OmniMapper is a modular framework capable of integrating sensory measurements from a variety of sensing modalities. For this series of experiments, OmniMapper is setup to consume LiDAR data from the Ouster OS1-64, odometry from the GVRBot's motors, and orientation data from the gyroscope. A *pose graph* is constructed along the robot's trajectory where the transform between adjacent poses is computed using Iterative Closest Point (ICP) [9] if ICP's residual error is below an acceptable threshold, otherwise filtered odometric data is substituted. ICP is bootstrapped with the filtered odometric transform between adjacent poses to improve convergence speed. *Loop closures* are attempted when a robot revisits an area previously traversed using ICP. If ICP returns low residual error when comparing non-adjacent pose graph measurements, additional constraints are added to the pose graph. This graph of relative frame transformations is optimized via iSAM2 [5] in the GTSAM solver [2].

Cartographer

Cartographer [4] is an open-source mapping suite which can incorporate 2D and 3D LiDAR, landmark measurements, odometry, and IMU measurements to generate a 2D or 3D representation of an environment. Cartographer uses a fast correlation-based scan matcher instead of ICP for loop closure and frame-to-frame matching. The nonlinear optimization framework *Ceres* is used to compute the robot's trajectory.

The ROS launch files and configuration files used to generate the reported Cartographer results are included in the associated repository. As Cartographer is capable of operating without platform odometry, it is evaluated here in two configuration modes. The first configuration is similar to the way OmniMapper is evaluated, incorporating odometry in addition to LiDAR and IMU measurements. The second configuration omits odometry, and is designated *CartographerNO* for "No-Odometry".

LeGO-LOAM

LeGO-LOAM [10] is another LiDAR-based SLAM approach. Whereas both of the above techniques compare laser scan point clouds directly, LeGO-LOAM operates instead on extracted features from laser scans. Two types of features are extracted which enable measurements of components of motion. Planar features (from an assumed ground plane) are used to estimate motion in the z, *roll* and *pitch* directions. Edge features are used to estimate motion in the x, y, and *yaw* directions. As this approach requires the presence of a ground plane in all scans, it does not work on datasets during stair descent sections (**alpha 2** and **beta 2**).

ORB_SLAM2

ORB-SLAM2 [6] is a lightweight, accurate mapping system run on monocular, stereo, or depth camera (RGB-D) data. This technique utilizes *visual odometry* to construct local keyframe models which are used to triangulate landmark locations. ORB-SLAM2 is also capable of performing loop closure using bag-of-words descriptors and for relocalization when tracking fails. This technique's promising results in aboveground datasets, such as state-of-the-art accuracy on the KITTI [3] dataset [6], warrants an assessment against a subterranean dataset.

S-MSCKF

S-MSCKF [11] is a lightweight, accurate state estimator run on IMU and stereo camera data. This technique utilizes filter-based stereo *visual inertial odometry* to achieve a high degree of accuracy at consistently low computational cost. The filter-based approach is more computationally efficient than competing optimization based methods. This allows for S-MSCKF to employ a stereo camera configuration that rivals both the accuracy and computational footprint of state-of-the-art monocular VIO solutions.

Kimera

Kimera [8] is an open-source library for real-time metric-semantic visual-inertial SLAM. This library has four key components: visual-inertial odometry (VIO) module for fast and accurate state estimation, a robust pose graph optimizer for global trajectory estimation, a lightweight 3D mesher module for fast mesh reconstruction, and a dense 3D metric-semantic reconstruction module. Kimera runs in real-time on a CPU and is equipped with a suite of continuous integration and benchmarking tools.

4 Results

The SubT Urban Circuit presents many challenges to robotic systems which are unique. Current state-of-the-art SLAM approaches have met many of these challenges but have also fallen short in others. Results from running the dataset segments on each of the algorithms described in Sect. 3 can be seen in Table 1. Overall, the ground segments **alpha 1** and **beta 1** yield high scores among many of the tested algorithms; however, scores are much lower where the robot descends a stairway in segments **alpha 2** and **beta 2**. Aerial segments **aerial alpha 1** and **aerial beta 1** are similar in complexity and extent however the latter has significantly worse lighting and the flight path is more constrained.

The most challenging features of the ground dataset are the segments where the robot is driven down a 10 m staircase. By adjusting Cartographer's loop closure mechanism, that algorithm was able to maintain accurate trajectory estimation during these sections. As the ground plane is not seen for these stairwell sections, LeGO-LOAM is unable to continue tracking (and in fact locks onto the wall instead, introducing a 90° roll in the map which is catastrophic. For **alpha 2**, LeGO-LOAM was restarted after the stair descent and was able to continue to produce results (useful only in the relative metric since the entry portal was not observed due to skipping ahead) until another catastrophic failure ended the run.

The second most challenging aspect of this dataset is poor lighting. As this course represents a disaster scenario, much of it is unlit. The ground robot is equipped with onboard illuminators; however, the resulting illumination was often insufficient to maintain feature tracks for visual SLAM. In the case of the ground datasets, the low frame rate of the stereo camera (10 Hz) coupled with poor illumination yielded no useful results for visual SLAM approaches. On the aerial robot, operation was limited to the lit areas of the courses. Upon traversing into rooms with poor lighting, visual SLAM algorithms often failed to track even with a higher frame rate on the RealSense camera (30 Hz).

Another significant challenge present in the **Beta** course is the lengthy entry corridor. In this case, there are a number of features throughout the corridor that provide potent landmarks for LiDAR-based SLAM techniques. This type of environment challenged Cartographer without the aid of an external odometry estimate as repeated local loop closures occurred to shorten the hallway when LiDAR odometry confidence measuring the distance traversed down the hallway fell. Reducing the maximum distance over which to perform loop closures to a reasonable threshold for the scale of the environment prevented Cartographer from performing erroneous loop closures in repeated feature-sparse segments of the hallway. The stairway descent in **alpha 2** consisting of four stacked flights of stairs also prompted erroneous loop closures from Cartographer due to the limited field-of-view and perception aliasing on the staircase. This case also required

Table 1. Mapping score, RMSE, relative RMSE, minimum, and maximum error for each algorithm on the Urban circuit dataset. In this table, a dash indicates that the algorithm was unable to be run on the dataset. The † superscript indicates a run had to be terminated early due to a sensor failure present in this run (Ouster LiDAR stopped working during both alpha 2 and beta 2 runs, and only Omnimapper was able to continue with only the Multisense SL LiDAR). The $*$ superscript indicates that a modified version of a run was made which starts after the stairs are descended and therefore has no absolute frame of reference and can only report relative RMSE. The ‡ superscript indicates results are reported on a partial run due to a mapping system failure or catastrophic divergence. On both of the runs where the robot descended the stairwell, reported results are based upon Multisense SL LiDAR. Ouster LiDAR data was not available due to vibration damage. The "aerial" prefix on a run name designates that this run involved flying an UAV as opposed to driving a UGV. UAV runs suffer from insufficient observations of portal fiducials so only relative RMSE error statistics are useful.

Run			Algorithm	Score	RMSE [m]	Min err [m]	Max err [m]	$RMSE_{rel}$ [m]
Run	Length [m]	Levels						
Alpha 1	605	1	OmniMapper	11/11	1.78	0.60	3.45	1.12
			Cartographer‡	10/10	1.44	0.31	2.34	0.77
			Cartographer NO	11/11	1.56	0.20	3.55	1.22
			LeGO-LOAM	8/11	5.06	0.37	12.08	2.65
			Odometry	3/11	6.87	1.91	10.63	3.42
Alpha 2	404	2	OmniMapper	6/6	2.37	0.21	3.68	1.24
			Cartographer†	3/4	4.19	2.33	5.25	2.35
			Cartographer NO†	0/4	48.40	13.04	66.93	1.55
			LeGO-LOAM†$*$	2/2	–	–	–	0.71
			Odometry	4/6	4.08	1.15	7.03	2.35
Beta 1	473	1	OmniMapper	11/11	1.91	0.37	3.36	0.81
			Cartographer‡	9/9	2.32	0.86	4.66	1.28
			Cartographer NO	11/11	2.03	0.67	4.61	1.49
			LeGO-LOAM‡	3/4	3.89	0.37	7.42	1.44
			Odometry	2/11	7.05	1.38	10.16	3.63
Beta 2	676	2	OmniMapper	6/10	5.73	0.59	10.42	2.87
			Cartographer†	–	–	–	–	–
			Cartographer NO†	–	–	–	–	–
			LeGO-LOAM†	–	–	–	–	–
			Odometry	1/10	7.35	3.73	9.08	3.25
Aerial alpha 1	404	1	ORB-SLAM2	4/5	–	–	–	7.27
			MSCKF	4/5	–	–	–	8.40
			Kimera	3/5	–	–	–	26.69
Aerial beta 1	356	1	ORB-SLAM2	3/4	–	–	–	7.75
			MSCKF	4/4	–	–	–	2.27
			Kimera‡	2/3	–	–	–	15.86

a similar reduction in the maximum distance over which to perform loop closures in order to localize accurately in the global reference frame.

The most challenging aspects of the aerial dataset are periods where the rate of motion increases dramatically and areas where illumination deteriorates. These scenarios appear in both datasets however illumination is significantly worse in **aerial beta 1**. There is a scenario during **aerial alpha 1** where the

UAV performs an aggressive maneuver in a hallway with moderate lighting that causes a significant jump in tracking for ORB-SLAM2 and Kimera. This jump in tracking builds up enough error in Kimera to prevent scoring additional points on **aerial beta 1** while ORB-SLAM2 is able to stay within the maximum error threshold by performing a loop closure to continue scoring.

Data collected from **Beta** course yields a tracking error for ORB-SLAM2 approximately 135 s into the run due to the UAV venturing too close to a low-contrast wall where ORB-SLAM2 cannot identify enough distinct features within the severely reduced field-of-view. Due to ORB-SLAM2's inability to handle this scenario and Kimera having a similar tracking error earlier in the run, results from the **aerial beta 1** run begin after 140 s into the data collect. Kimera fails to maintain accurate localization in poorly illuminated areas. ORB-SLAM2's relative RMSE in the **aerial beta 1** run remains within the scoring threshold until traversing through an area with low-illumination. ORB-SLAM2 accrues more error per meter of tracking than S-MSCKF for this case but utilizes loop closures to offset this disadvantage. S-MSCKF is able to score all artifacts in **aerial beta 1**, including the fourth artifact, preceded by an area of exceptionally poor illumination, highlighting the robustness of its filter-based approach over other techniques' weighting of visual odometry for areas with inadequate illumination.

References

1. Burgard, W., Stachniss, C., Grisetti, G., Steder, B., Kümmerle, R., Dornhege, C., Ruhnke, M., Kleiner, A., Tardös, J.D.: A comparison of SLAM algorithms based on a graph of relations. In: 2009 IEEE/RSJ International Conference on Intelligent Robots and Systems, pp. 2089–2095. IEEE (2009)
2. Dellaert, F.: Factor graphs and GTSAM: A hands-on introduction. Technical report, Georgia Institute of Technology (2012)
3. Geiger, A., Lenz, P., Stiller, C., Urtasun, R.: Vision meets robotics: the KITTI dataset. Int. J. Rob. Res. **32**, 1231–1237 (2013)
4. Hess, W., Kohler, D., Rapp, H., Andor, D.: Real-time loop closure in 2D LIDAR SLAM. In: 2016 IEEE International Conference on Robotics and Automation (ICRA), pp. 1271–1278 (2016)
5. Kaess, M., Johannsson, H., Roberts, R., Ila, V., Leonard, J., Dellaert, F.: iSAM2: incremental smoothing and mapping with fluid relinearization and incremental variable reordering. In: 2011 IEEE International Conference on Robotics and Automation, pp. 3281–3288. IEEE (2011)
6. Mur-Artal, R., Tardós, J.D.: ORB-SLAM2: an open-source SLAM system for monocular, stereo, and RGB-D cameras. IEEE Trans. Rob. **33**(5), 1255–1262 (2017)
7. Rogers, J.G., Fink, J.R., Stump, E.A.: Test your SLAM! the subt-tunnel dataset and metric for mapping. In: IEEE International Conference on Robotics and Automation (2020)
8. Rosinol, A., Abate, M., Chang, Y., Carlone, L.: Kimera: an open-source library for real-time metric-semantic localization and mapping. In: IEEE International Conference on Robotics and Automation (ICRA) (2020). https://github.com/MIT-SPARK/Kimera

9. Segal, A., Haehnel, D., Thrun, S.: Generalized-ICP. In: Robotics: Science and Systems, Seattle, WA,vol. 2, p. 435 (2009)
10. Shan, T., Englot, B.: Lego-loam: lightweight and ground-optimized LiDAR odometry and mapping on variable terrain. In: IEEE/RSJ International Conference on Intelligent Robots and Systems (IROS), pp. 4758–4765. IEEE (2018)
11. Sun, K., Mohta, K., Pfrommer, B., Watterson, M., Liu, S., Mulgaonkar, Y., Taylor, C.J., Kumar, V.: Robust stereo visual inertial odometry for fast autonomous flight. IEEE Rob. Autom. Lett **3**(2), 965–972 (2018). https://doi.org/10.1109/LRA.2018.2793349
12. Trevor, A.J., Rogers, J.G., Christensen, H.I.: Omnimapper: a modular multimodal mapping framework. In: 2014 IEEE International Conference on Robotics and Automation (ICRA), pp. 1983–1990. IEEE (2014)

Experimental Validation of Energy-Optimal Turning Radii for Skid-Steer Rovers

Meysam Effati(✉), Jean-Sebastien Fiset, and Krzysztof Skonieczny

Electrical and Computer Engineering, Concordia University, Quebec, Canada
{m_effati,j_fise,kskoniec}@encs.concordia.ca

Abstract. For skid-steer rovers, a point turn consumes a lot of power compared to a straight-line motion. Hence, turning radius, in path planning for this kind of rover, is a factor that should be considered explicitly. In this article, the energy-optimal paths for skid-steer rovers are presented, based on a method using Pontryagin's minimum principle. The finite number of extremals are compared and an optimal map is generated that shows which path type requires minimum energy depending on start/end pose for a particular example. Moreover, the turns in optimal paths (aside from a small number of special cases where start and end points are extremely close) are to be circular arcs of a particular turning radius, R', equal to half of a skid-steer rover's slip track. R' is the turning radius at which the inner wheels of a skid-steer rover are not commanded to turn. Experiments with a Husky UGV rover validate the energy-optimality of using R' turns on hard ground. Moreover, the experimental results confirm that circular arc - line - circular arc (CLC) paths save energy compared to point turn - line - point turn (PLP) paths, on both hard ground and sand. It should be noted that PLP is the most commonly used maneuver to go from a starting to an end pose for skid-steer rovers.

1 Introduction

Motivation. Skid-steer rovers, such as the Clearpath Husky, are very commonly used in robotic experimentation. Due to their mechanical simplicity, maneuverability and robustness, they are also widely used in various field robotics applications. Energy-efficient driving is an important consideration, especially when autonomously planning paths in power-starved environments. For example, path planning for skid-steer lunar rovers (including the Soviet Union's 8-wheel Lunokhod rovers [1] or series of rovers recently developed for the Canadian Space Agency (CSA) [2]) require efficient power/energy consumption during their maneuvers because of the constrained amount of solar power they can collect while mobile.

Skid-steer rovers consume a lot more power in point turns compared to straight line motion. As energy is the integral of power over time, lower instantaneous power consumption for wider arcs must be traded off against shorter

B. Siciliano et al. (Eds.): ISER 2020, SPAR 19, pp. 402–410, 2021.
https://doi.org/10.1007/978-3-030-71151-1_36

Fig. 1. A CLC path performed by a Husky rover.

traversal distance for tighter arcs by evaluating the total energy consumed when following different paths.

Problem Statement. The objective of this work is to find a trajectory from an arbitrary $[x_0, y_0, \theta_0]$ to an arbitrary $[x_f, y_f, \theta_f]$ that minimizes the energy consumption for a skid-steer rover. The control inputs are V_r and V_l, the right and left skid-steer wheel velocities, respectively. A flat 2-dimensional plane without obstacles is assumed; note that this constitutes a 'Local Planning Method' (LPM) [3].

Related Work. Analytical solutions to the optimization problem above constitute an LPM for energy-optimal skid-steer rover trajectory planning. LaValle argues that LPMs provide useful motion primitives for global planning algorithms (which may include obstacles and are likely numerical) because (a) they can lead to more optimal (according to the objective function) overall paths, and (b) they can be more computationally efficient compared to executing global algorithms with ad hoc motion primitives [3]. Prior work by the authors has made preliminary progress on LPMs for energy-efficient skid-steer rovers [4], which this work expands upon.

LPMs have been successfully derived for minimum-time and minimum-length path planning problems for various types of robots [5–8], but none to date have been found for minimum-energy path planning. Energy-efficient path planning has been studied for car-like rovers [9]; a full LPM solution was not derived.

Sampling Based Model Predictive Optimization (SBMPO) [10] is popular for path planning with skid-steer rovers. Because SBMPO samples from the feasible space of control inputs, it automatically satisfies dynamic constraints. Attempts to determine energy-efficient paths for skid-steer rovers all use SBMPO [11, 12]. Challenges in practice related to suboptimality and computational cost [4] align exactly with the above mentioned points raised by LaValle [3], motivating optimal local paths.

The key contributions of this work are the derivation of LPM for energy-optimal skid-steer rover paths, and an experimentation campaign designed to validate and test the limitations of this approach.

2 Technical Approach

In prior work, we had focused on optimizing the parameters of paths selected from among a pre-defined class of paths: consisting a turn followed by a straight line and finally a second turn [4]. One commonly used example of this type of path is a PLP. It was shown numerically, using a popular skid-steer power model for hard ground [13], that the optimal radii for the turns was not in fact point turns (i.e. $R = 0$) but instead circular arcs (i.e. CLC paths) of radius $R = R'$ where R' is the radius at which the inner wheels are not commanded to turn. In Sect. 4, we summarize experiments validating this result on hard ground.

In the present work, Pontryagin's Minimum Principle is used to find minimum-energy paths from among all dynamically admissible paths (i.e. without restricting to a pre-defined class of CLC paths, for example). The technical approach is built upon the procedures outlined in prior literature [7,8]. The details of the procedure are too numerous to expand here, but a general outline of the technical approach to find the minimum-energy path connecting any arbitrary given start and end pose follows:

1. Define the Hamiltonian and find all control inputs that minimize it.
2. Generate level sets of the Hamiltonian for each minimizing control, in a parameterized space, and identify intersections as control switching points.
3. Analyze the level sets to enumerate all feasible types of extremal paths (i.e. sets of minimizing controls, ordered according to the switching points encountered in the parameterized space) and their subsets.
4. Evaluate the energy cost of each of the finite number of extremal paths to find that which gives lowest cost.

3 Results

As was discussed in the Related Work section, prior work in LPMs has focused on time-optimal problems. If it were possible to formulate the optimal-energy problem (Step 1 in the technical approach) as an equivalent optimal-time problem, then prior results and procedures could be applied. The formulation of such an equivalency is the first result presented here.

First, we define a **general path** as any sequential set of connected point turns, straight lines, and/or curves parameterized by $R(s)$, where R is the turning radius as a function of distance along the path, s. Next, we define a **proper constraint** on the control inputs as one that does not place a limit on a skid-steer rover's turning radii; i.e. one that allows point turns, straight lines, and $0 < R < \infty$.

In practical cases, including when assuming a popular skid-steer power model for hard ground [13], it turns out that the energy consumed along any path segment (point turn, straight line, or circular arc of arbitrary radius) can be expressed purely as a function of path geometry parameters (i.e. $|\Delta\theta|$ for a point turn, $|\Delta s|$ for a straight line, $R(s)$ otherwise), with no explicit dependence on the control inputs otherwise. A key result is the realization that this means an optimal path found for energy optimization with a proper velocity constraint is also an optimal path for energy optimization with any other proper velocity constraint.

Any general path can be achieved by a skid-steer rover with a proper velocity constraint. Moreover, by definition there are no general paths that lead to lower energy consumption than an optimal path. A proper velocity constraint ensures that all general paths *have been considered.* An energy-optimal path is thus energy-optimal regardless of input velocity constraints, as long as the velocity constraints are proper. Only the particular optimal controls used to follow such an optimal path vary between problems with different proper velocity constraints.

Finally, one can find a set of proper constraints on the input velocities that enforces constant instantaneous power consumption. For the popular skid-steer power model for hard ground, these constrains are shown in Fig. 2. Under a constant power condition, optimal-energy is of course equivalent to optimal-time. Steps 2 - 4 of the technical approach can then be applied according to all existing theory for optimal-time problems. This is not a trivial problem given the unique set of input velocity constraints that must be considered (i.e. Fig. 2), but can be completed systematically. The results of such systematic application of theory to the equivalent time-optimal problem are summarized in Fig. 3. The figure shows the optimal paths to take from the starting pose $[x, y, \pi/4]$ to the end pose $[0, 0, 0]$, where $< x, y >$ is any point in the x-y plane.

As can be seen in the optimal map in Fig. 3, CLC paths cover the largest portion of the x-y plane and are thus crucial path types to consider for energy-efficient skid-steer rover operation. Other important path types that cover significant portions of the x-y plane include CCLC (circular arc - circular arc - line - circular arc) and CLCC. In these cases the 2 "C" segments are distinct due to a change in driving direction (e.g. clockwise and forward switching to clockwise and backward). It is important to note that for all "C" segments shown in the optimal map, the turning radius is R'. For starting points near the origin, sometimes more complex parallel-parking-like maneuvers (e.g. CLCCLC) or whirls (sequences of only turns [8]) are optimal.

As the procedure described above results in an LPM, potential difficulties with utilizing it in a global planning method are discussed in the following. For the global path, the nodes should be determined, and then, the optimal local paths should be implemented between the nodes. However, the optimal initial and final angles at intermediate nodes are not pre-defined. A preliminary approach is to design a global path using sampling-based methods, and setting the angles based on the data obtained for the global path (e.g. at the average between the line segments pointing from the current node to the previous node and to the next node). However, such heuristics do not guarantee global optimality. Further investigation of efficient incorporation of LPMs into global methods could be a fruitful direction of future work.

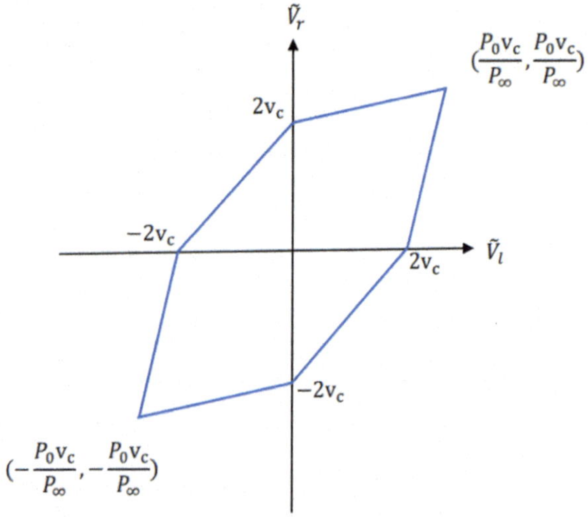

Fig. 2. The control space of the *equivalent time-optimal* problem. P_0 and P_∞ are power consumption for point turns and straight lines, respectively. $2v_c = |V_r| + |V_l|$.

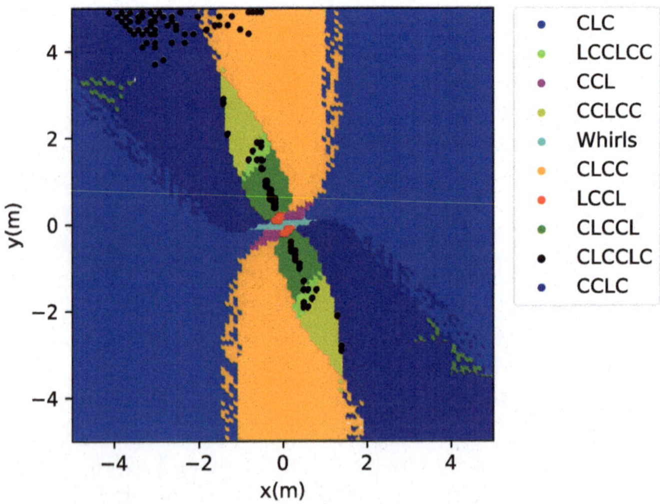

Fig. 3. Map indicating the optimal paths to the origin when starting at different $(x, y, \frac{\pi}{4})$ on the $x - y$ plane.

4 Experiments

Experiments studying optimal CLC paths have been completed on hard ground (concrete) and on granular terrain (sand). These are highly pertinent because CLC paths make up the largest portion of the energy-optimal map in Fig. 3. Further experiments are planned on both surfaces in summer/fall of 2020 to study CCLC and CLCC paths, which also make up an important portion of the optimal map.

A PLP path from $[-10m, 0, \pi/4]$ to $[0, 0, \pi/4]$ is tested against of a set CLC paths with various turning radii using a Husky rover. For a fair comparison, the start-to-end displacement of the PLP path is ensured to be within $\pm 2\%$ of that of the CLC. Rover power consumption (and thus energy consumption over each path) is measured with two TI INA226 bidirectional current/power sensors, measuring the power to each Husky motor. After computing E_{CLC} and E_{PLP}, the energy consumption of the CLC and PLP path respectively, the energy difference, $\Delta E\%$ (see Fig. 4 and Fig. 5), is computed by the following equation:

$$\Delta E\% = \frac{E_{CLC} - E_{PLP}}{E_{PLP}} \times 100 \tag{1}$$

A negative $\Delta E\%$ indicates that the CLC path consumed less total energy than its corresponding PLP path. The $\Delta E\%$ results are plotted against R_{ACT}/R'_{ACT} where R_{ACT} is the measured turning radius (by circular fit of position measurements) and R'_{ACT} is the measured turning radius when the commanded turning radius is R'.

Figure 4 shows that, on hard concrete ground, R' is the turning radius where the biggest energy reduction is observed. This validates all the theoretical and numerical results reported earlier (above), using the hard ground skid-steer power model.

As it can be seen in Fig. 5, the biggest energy reduction on sand is observed at around $1.5 < R_{ACT}/R'_{ACT} < 2.5$. A two-tailed t-test comparing the reduction in energy consumption for CLC paths with this range of turning radii to their corresponding PLP paths confirms statistical significance ($p < 0.001$).

5 Main Experimental Insights

The optimal map in Fig. 3 confirms the importance of using CLC paths in optimizing energy consumption for skid-steer rovers. This has important implications for experimental robots, such as the Clearpath Husky, that are routinely commanded to drive sub-optimal PLP paths. It was further observed in experimentation that in addition to being energy-optimal, CLC paths also saved time because the rover made some forward progress during the turns.

The experimental results confirm that CLC paths save energy compared to PLP paths, on both hard ground and sand. On hard ground, the biggest energy savings (approx. 10%) occur at turning radii of R', as predicted by theory. On sand, there are also significant energy savings (approx. 10–20%), but they occur closer to $2R'$. Note that as the distance between start and end points increases, energy saving percentages will diminish as the turns make up smaller portions of the overall path. Prior work by the authors has also shown that there can be an *increase* in energy consumption of CLC paths relative to PLP for $R < R'$ [14], which is important to keep in mind.

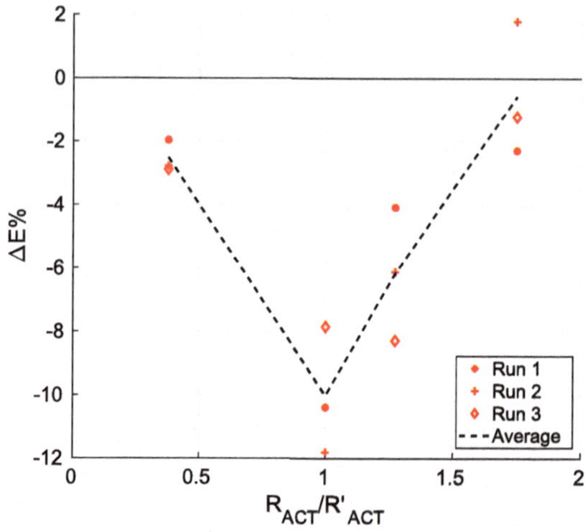

Fig. 4. Energy difference percentage ($\Delta E\%$) as a function of $R_{ACT}\ /R'_{ACT}$ for a 45°-10m-45° path on hard concrete ground.

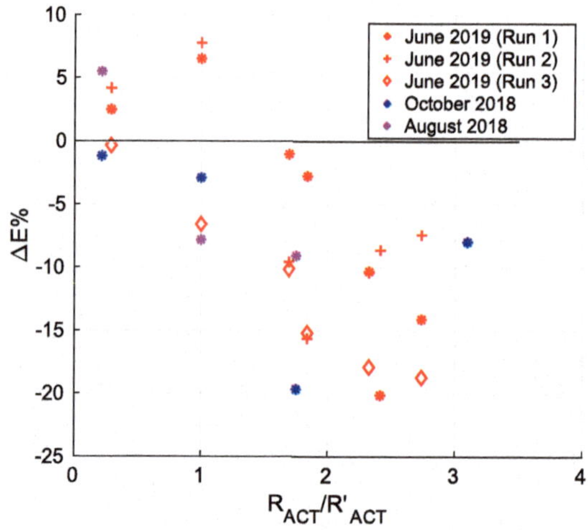

Fig. 5. Energy difference percentage ($\Delta E\%$) as a function of $R_{ACT}\ /R'_{ACT}$ for a 45°-10m-45° path on sand.

6 Conclusions and Future Work

Skid-steer rovers consume a lot more power in point turns compared to straight-line motions. As energy is the integral of power over time, the turning radius should be considered explicitly for this type of rover. Hence, the focus of analysis in this paper is to address the energy-efficient path planning problem while the rover moves on obstacle-free hard ground.

Using the Pontryagin's minimum principle for skid-steer rovers, the answer to the energy-efficient path planning problem is found. Moreover, a scenario, going from a starting point in 2D x-y space with the orientation of $\pi/4$ to the origin with orientation of 0, is provided to demonstrate the method where the extremal paths are utilized to obtain the optimal path. The results indicate that there are 3 types of path that covers the most area of the optimal map: CLC, CCLC, and CLCC. We would expect similar results for other scenarios besides the mentioned example. Furthermore, the experiments on both hard ground and sand for CLC paths indicate at least 10% energy saving compared to PLP paths. It should be mentioned that the "C"s for hard-ground experiments occur at turning radii of R', which confirm the theoretical analysis. However, the turning radii of the "C"s for sand experiments happens at around $2R'$.

The discrepancy between results on sand and on hard ground highlight an important area for future work: the development of new power models for skid-steer rovers turning on granular material. Phenomena including wheels digging into the soil and pushing against accumulating piles of sand introduce differences relative to friction contact that are apparent in the experimental data.

Acknowledgments. The authors acknowledge funding and technical collaboration from Mission Control Space Services Inc. as well as financial support from the Natural Sciences and Engineering Research Council of Canada (NSERC).

References

1. Kassel, S.: Lunokhod-1 soviet lunar surface vehicle. Tech. Rep, RAND CORP SANTA MONICA CA (1971)
2. Farhat, M., Lamarche, T., Dupuis, E., Salerno, A., Lake, S.: Developing pud–lunar rover test platform. ASTRA (2011)
3. LaValle, S.M.: Planning Algorithms. Cambridge university press (2006)
4. Effati, M., Fiset, J.S., Skonieczny, K.: Considering slip-track for energy-efficient paths of skid-steer rovers. J. Intell. Robot. Syst. **100**(1), 1–14 (2020)
5. Dubins, L.E.: On curves of minimal length with a constraint on average curvature, and with prescribed initial and terminal positions and tangents. Am. J. Math. **79**(3), 497–516 (1957)
6. Reeds, J., Shepp, L.: Optimal paths for a car that goes both forwards and backwards. Pac. J. Math. **145**(2), 367–393 (1990)
7. Balkcom, D.J., Mason, M.T.: Time optimal trajectories for bounded velocity differential drive vehicles. Int. J. Robot. Res. **21**(3), 199–217 (2002)
8. Furtuna, A.A.: Minimum time kinematic trajectories for self-propelled rigid bodies in the unobstructed plane. Ph.D. dissertation, Dartmouth College, United States (2011). https://rlab.cs.dartmouth.edu/publications/FurtunaPhD.pdf
9. Tokekar, P., Karnad, N., Isler, V.: Energy-optimal trajectory planning for car-like robots. Auton. Robot. **37**(3), 279–300 (2014)
10. Dunlap, D.D., Caldwell, C.V., Collins, E.G., Chuy, O.: Motion planning for mobile robots via sampling-based model predictive optimization. Recent Adv. Mob. Robot. **1** (2011)
11. Gupta, N., Ordonez, C., Collins, E.G.: Dynamically feasible, energy efficient motion planning for skid-steered vehicles. Auton. Robot. **41**(2), 453–471 (2017)

12. Pentzer, J., Reichard, K., Brennan, S.: Energy-based path planning for skid-steer vehicles operating in areas with mixed surface types. In: 2016 American Control Conference (ACC), pp. 2110–2115. IEEE (2016)
13. Morales, J., Martinez, J.L., Mandow, A., García-Cerezo, A.J., Pedraza, S.: Power consumption modeling of skid-steer tracked mobile robots on rigid terrain. IEEE Trans. Robot. **25**(5), 1098–1108 (2009)
14. Fiset, J.S., Effati, M., Skonieczny, K.: Effects of turning radius on skid-steered wheeled robot power consumption on loose soil. In: 12th Conference on Field and Service Robotics (FSR), pp. 115–129 (2019)

Multi-robots

Swarm of Inexpensive Heterogeneous Micro Aerial Vehicles

Dinesh Thakur[1(✉)], Yuezhan Tao[1], Rebecca Li[1], Alex Zhou[1], Alex Kushleyev[2], and Vijay Kumar[1]

[1] GRASP Lab, University of Pennsylvania, 3330 Walnut Street, Philadelphia, PA 19104, USA
{tdinesh,yztao,robot,alexzhou,kumar}@seas.upenn.edu
[2] ModalAI, 10855 Sorrento Valley Rd, San Diego, CA 92121, USA
akushley@modalai.com

1 Introduction

A team of Micro Aerial Vehicles (MAVs), or a Swarm, is theoretically able to accomplish more complex tasks than a single robot by covering more area, gathering more data, and ensuring resilience to single-robot failure. In practice, however, deploying a large swarm of MAVs is often extremely difficult due to challenges in scaling communication, estimation, and control systems. A survey of aerial swarm robotics and the corresponding technological challenges can be found in [1].

One of the greatest challenges to aerial swarm robotics is robust state estimation. In outdoor settings, GPS has seen great success in [2] and [3], where 30 and 52 drones, respectively, were flown using GPS. However, GPS still suffers from a lack of precision and is prone to heavy interference in many settings, limiting these swarms to operate outdoors and while staying far apart.

Real-Time Kinematic (RTK) GPS is an alternative that provides position measurements 10x as precise as normal GPS. Hence, it has been extensively used by the entertainment industry for outdoor aerial shows, as shown by Intel [4] and Verge Aero [5]. However, in addition to needing a specialized receiver/data-link on-board the MAV, it requires the user to set up a bulky base-station with known coordinates and line of sight to the MAVs.

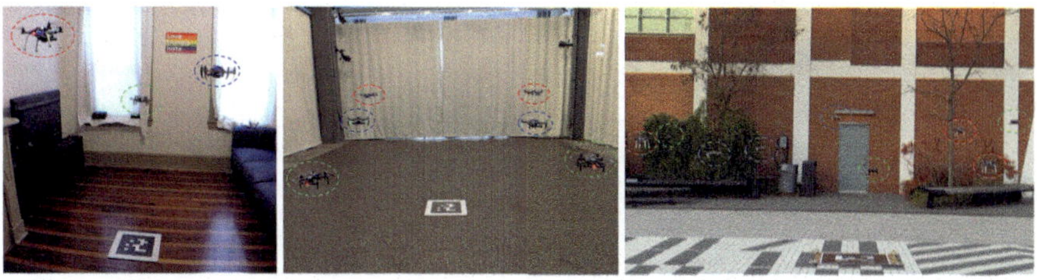

Fig. 1. Heterogeneous VIO swarm of MAVs in formation flight. Left: Three MAVs in a living room. Middle: Six MAVs in the laboratory. Right: Nine MAVs outdoors.

B. Siciliano et al. (Eds.): ISER 2020, SPAR 19, pp. 413–423, 2021.
https://doi.org/10.1007/978-3-030-71151-1_37

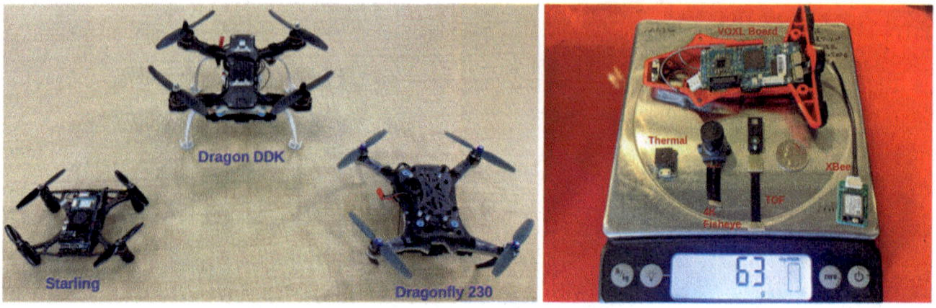

Fig. 2. Three different MAVs designs (left), VOXL board, sensing and communication payloads (only 63 g) (right)

For indoor robotic systems, where GPS is not available, motion capture systems provide high-precision robot tracking. Motion capture was instrumental in improving the early control and navigation systems of aerial vehicle teams [6–8]. While useful, motion capture systems are expensive, time consuming to set up, and limiting as they force the swarm to operate within fixed tracking boundaries. Ultra-wideband (UWB) based solutions provide an inexpensive robot-tracking alternative to high-accuracy motion capture systems [9–11]. While there have been many successes with UWB, as outlined in [12], they have a significant drawback in that they require the installation and calibration of fixed nodes or anchors in predefined positions. This leads to long setup times for UWB systems, as well as limiting the swarm's operation to the airspace around the anchors.

On-board Visual Odometry offers a promising indoor and outdoor solution to robot localization, without the setup drawbacks of RTK or UWB. With on-board visual odometry, robots perceive their environment with cameras to estimate their own state. Previous vision-based swarm systems utilized extensive patterns of fiducial markers to serve as a map of the environment in order to navigate and avoid obstacles [13–15]. In these experiments, robots are restricted an area where a sufficient number of tags are always visible in order to determine their locations. Other works utilize fiducial markers placed on robots to accomplish relative formation control [16,17], but are sensitive to the field of view of the cameras. This also requires the robots to carry additional tag payloads. A vision-based strategy to perform a formation flight based on visual inertial odometry (VIO) is provided in [18]. However, this system requires a known initial position of each drone, resulting in drift of the tracking estimate over time.

More recently, combining on-board sensor measurements with on-board UWB relative localization has become a popular approach to onboard localization [12]. The authors in [19] show an accurate relative localization between multiple flying robots that combine sensory information (e.g., velocity, yaw rate, height) with range measurements from UWB antennas to give relative pose estimates. The overall system requires an autonomous initialization maneuver of up to 60 s before the filter converges, which introduces a large setup time. [20] proposed a fully decentralized visual-inertial-UWB fusion framework for relative

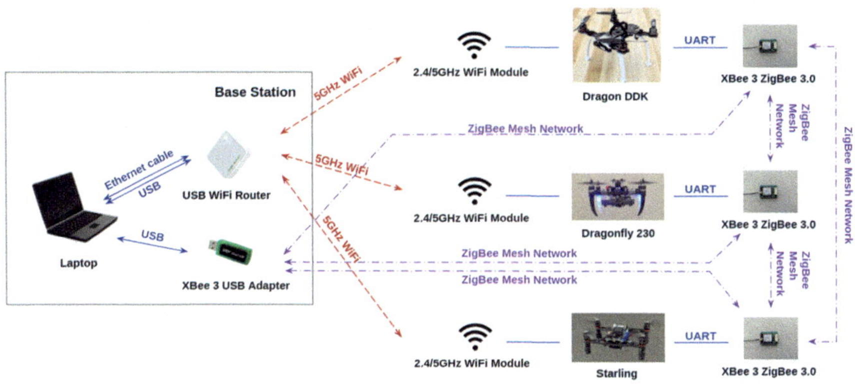

Fig. 3. Heterogeneous MAV swarm system architecture.

state estimation and was tested on five drones. This framework makes an inconvenient assumption for filter initialization, which is that the initial difference in yaw between the observed drones is minimal.

All of the multi-robot systems mentioned previously consist of *homogeneous* teams of robots. Heterogeneous teams have a higher potential to complete a wider variety of tasks more quickly. While there is active research in controlling heterogeneous teams of robots, until now, heterogeneous multi-robot algorithms have only been tested in simulation. This is because we lack inexpensive heterogeneous robotic test beds. These are important to the research community as they allow for generating interesting datasets that can be used for benchmarking various algorithms.

Towards this end, we propose a framework that will not only bridge the gap between simulation and experimentation, but also will address some of the limitations in previously discussed methods of deploying multi-robot systems. Most previous approaches do not consider setup or initialization time, which is critical for rapid deployments. In this work we present multiple contributions around a heterogeneous multi-robot test bed. First, we combine perception, planning, and control for autonomous navigation of multiple aerial vehicles. The vehicles are accessible in that they are inexpensive and rely on smartphone-grade hardware. Our method does not require any external knowledge, such as the relative initial positions between drones, or a prior setup. Our method solely relies on on-board sensors for localization. Second, we show heterogeneity among MAVs in terms of size, agility, sensing, and communication links. We interface multiple different sensors on the same smartphone-grade computer. We develop an auxiliary communication network to compliment the WiFi communication for the MAVs using the Digi XBee® 3 to form ZigBee mesh networks that enable the running of distributed and decentralized algorithms on MAV Swarms. Finally, we release as open-source[1] the following: the code for control on individual MAVs;

[1] https://github.com/tdinesh/hetro_swarm.

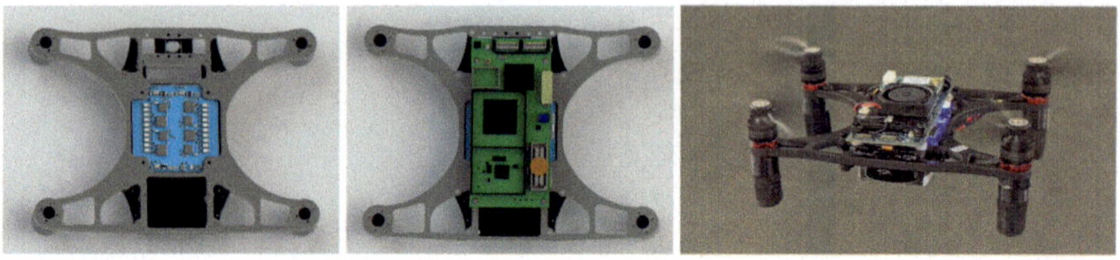

Fig. 4. Starling assembly CAD render showing inline ESC (left). Render showing VOXL board integration (middle). Starling MAV in flight (right)

the code for controlling a team of MAVs, including user interface GUI; open-hardware design of different MAVs and design files for custom-designed boards for communications. We also demonstrate the application of our system to sensor coverage of an area in a completely decentralized manner. To our knowledge, this is the first fully autonomous, heterogeneous aerial system capable of formation control without any external infrastructure.

2 Technical Approach

2.1 Heterogeneous Platform Description

To realize heterogeneity in physical size among agents, we employ three different platforms. The first is the Starling, a 0.12 m diameter, 185 g quadrotor with 3 in. propellers utilizing a 2-cell (2S), 650 mAh Lithium Polymer (LiPo) battery. The second is the Dragonfly 230, a 0.15 m diameter, 250 g quadrotor with 4 in. propellers and a 2S 910 mAh LiPo battery. The third is the Dragon DDK, a 0.22 m diameter, 410 g quadrotor with 5 in. propellers and a 3S 1350 mAh battery. Figure 2 displays the three MAV platforms. In terms of agility, the Starling is the most agile with a 4:1 thrust-weight ratio, while the DDK is the least agile with a thrust-weight ratio of 2:1. All three of these MAVs utilize a highly-integrated VOXL [21] flight board, a successor to the now End of Life (EOL) Qualcomm®, Snapdragon™Flight™. The VOXL board supports UART, I2C, SPI, and GPIO functions for other peripheral connections.

Given the Dragonfly 230 is a proprietary design and the Dragon DDK is a discontinued platform [22], we developed a new platform "Starling" as shown in Fig. 4 that is designed with several improvements over the predecessors. In order to reduce complexity and increase reliability, the vibration damping system is simplified. The reduction in overall vibrations is achieved by making the carbon frame very stiff, with 4mm thick carbon fiber and other mechanical design improvements. This reduces the twist of the arms in flight due to the propeller imbalance. A new Electronic Speed Controller (ESC) from ModalAI is integrated and recessed into the carbon frame in order to reduce overall thickness of the component stack. A power module from ModalAI (APM) has been modified to fit in line with ESC to help achieve the low profile of the assembly. Additionally, the motor wires are routed in special notches in carbon fiber in order to protect them from damage. This design is completely open-source, and will soon be available to purchase as a kit.

Utilizing the compact VOXL board as the sole computing platform allows us to have heterogeneous sensing capabilities while keeping platform weight to a minimum. The VOXL board allows for connection of up to three physical MIPI ports. A downward facing VGA camera with a fish-eye lens uses one dedicated channel for state estimation with VIO. The other two MIPI channels can be configured as desired. The board comes pre-configured with a 4K camera and a stereo pair of cameras. We utilized a compact Lepton 3.5 infrared camera, a Time of Flight (ToF) sensor, and a 4K camera with an M12 mount that allows different lenses to be used, as shown in Fig. 2. We also designed custom mounts for each of these sensors that can be easily swapped depending on the application.

To realize heterogeneity in communication, we design a custom carrier board for Digi XBee® 3 that connects to the VOXL over UART. This allows for another complementary communication mode in addition to the on-board 2.4/5 GHz WiFi Fig. 3. A ZigBee mesh network allows for intra-swarm communication, and thus enables truly distributed algorithms to be run on-board.

All of these components can be fit onto the smallest platform, the Starling, while keeping the overall weight under 185 g. To our knowledge, this would be the smallest sub-200 g MAV capable of showing heterogeneity in sensing and communication.

2.2 System Architecture

The individual MAV control and estimation is based on our previous work [23, 24]. The system combines the Inertial Measurement Unit (IMU) data and the VGA downward camera image data in an Unscented Kalman Filter (UKF) framework in order to localize the vehicle in the environment. Previously, we demonstrated the ability to avoid obstacles in completely dark environments with the Dragonfly 230 platform using a smart-phone grade processor and an on-board lighting payload (all under 250 g).

The multiple MAV framework builds upon [18], which provides a scalable and extensible architecture for controlling multiple MAVs from a single user. However, the framework requires a precise, known initial offset between each MAV. To achieve this, the starting location of the robots needed to be initialized from takeoff plates that had been carefully installed. This led to a significant increase in setup time, limiting the framework's ability to be used with large swarms.

We remove the limitation of having to install a takeoff plate by estimating the relative starting offset via the detection of a common fiducial tag. Upon takeoff, each robot using the downward angled camera can detect a single April-Tag [25] placed at a global reference. An Extended Kalman Filter (EKF) is implemented to estimate the transformation between the odometry reference frame of an individual MAV and the shared global frame. Each MAV estimates on-board T_w^o, which is the transformation of the odometry frame to the global

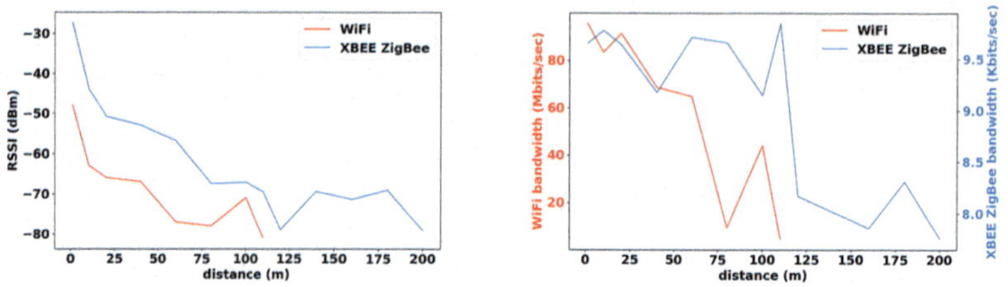

Fig. 5. Outdoor tests for WiFi and ZigBee network measurement. Left: RSSI. Right: Bandwidth.

frame, and relays it the ground-station. The ground station then uses Centralized Concurrent Assignment and Planning of Trajectories Algorithm (C-CAPT) [8] for goal assignment and trajectory generation for each robot. More details on exact implementation can be found in [18].

3 Experimental Results

3.1 Experiment on WiFi and XBee® ZigBee® Network Performance

We evaluated the performance of WiFi and Zigbee mesh networks via their bandwidth and packet losses. The router was set up in IEEE 802.11ac mode at 5 GHz. Our computer was wired to the router and served as the ground station. A Digi XBee® 3 RF module connected to the ground station with universal serial bus (USB), served as ZigBee coordinator and formed the ZigBee mesh network. The VOXL board was equipped with two wireless modules working independently. The first module was the on-board WiFi module, which connects to the router. The other was the XBee® 3 RF module, which connected with the VOXL board through UART. The module served as the ZigBee router and joined the ZigBee mesh network formed by the XBee module connected to the ground station. As an illustration of real world communication performance, detailed statistics on RSSI and bandwidth of WiFi network and XBee ZigBee network can be seen in Fig. 5.

In order to demonstrate the WiFi network performance in various use cases, we designed four test scenarios and monitored the round trip time (RTT) and loss of packets for each of them. As a base case, all MAVs sent odometry information to the ground station at 30 Hz, which was required for centralized formation control. As shown in Fig. 6, there was no obvious change in RTT and packet loss when one of the MAVs in the team adds compressed image (80% compressed JPEG) at 30 Hz to ground station. While in the case that one of the MAVs in the team transmit additional raw image data of VGA resolution (640×480) besides odometry, packet loss appears and the network congests as the scale of the swarm team increased. In the context of mapping with a swarm of MAVs, where all MAVs transmit odometry and compressed image data to ground station, the RTT and packet loss increases significantly.

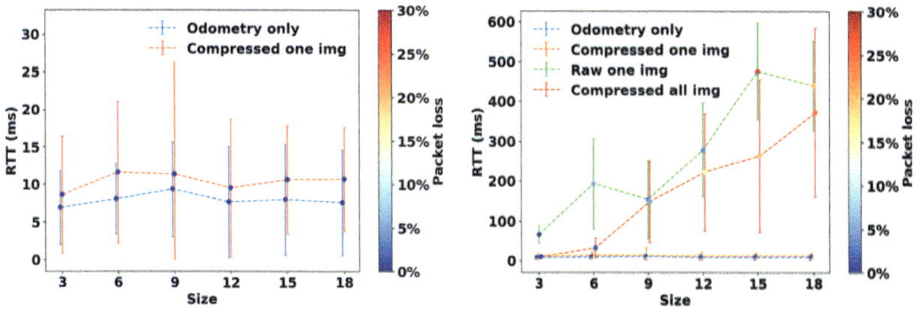

Fig. 6. Round Trip Time and packet loss for packets in WiFi network under four scenarios. The RTT (mean and SD) and packet loss are collected for 100 packets with a 100 bytes payload. Blue Line: MAVs transmit odometry to ground station only. Orange Line: MAVs transmit odometry and one of them transmits compressed images to ground station. Green Line: MAVs transmit odometry and one of them transmits raw images to ground station Red Line: MAVs transmit odometry and compressed images to ground station.

Based on our experimental results, the WiFi network and the ZigBee network in our system provided complementary support on bandwidth and coverage, with the WiFi providing higher bandwidth and ZigBee providing higher range. Moreover, the increased communication loss over WiFi as the scale of swarm team increased led to centralized formation control failure due to the loss of control messages. This reinforces the importance of having ZigBee mesh network serve as the auxiliary network in our system.

3.2 Experiment with Sensor Coverage Using Decentralized GNN Controllers

We demonstrated that a team of robots can perform decentralized sensor coverage without GPS or other global information. Sensor coverage, or the task of distributing robots to monitor areas of interests, is a primary application of swarms of robots. Often it may be done in places where GPS is not available, and thus each robot must rely on local information to determine its next control action. While there has been much algorithmic progress in sensor coverage over a variety of scenarios in simulation [26–28], few of these have actually been tested on physical hardware.

For this experiment, we utilize GNNs in order to construct decentralized controllers for the robots. GNNs not only generate decentralized controllers, but because they are learned, they can learn to overcome problems such as noise in VIO or neighbor connectivity. Decentralized controllers with GNNs have been previously demonstrated in [29]. Our aggregation GNN controllers, derived from [30] also combine information from the immediate neighbors of each robot to calculate its control action, which takes advantage of the fact that information may need to be communicated across the swarm but can only be done in a local manner.

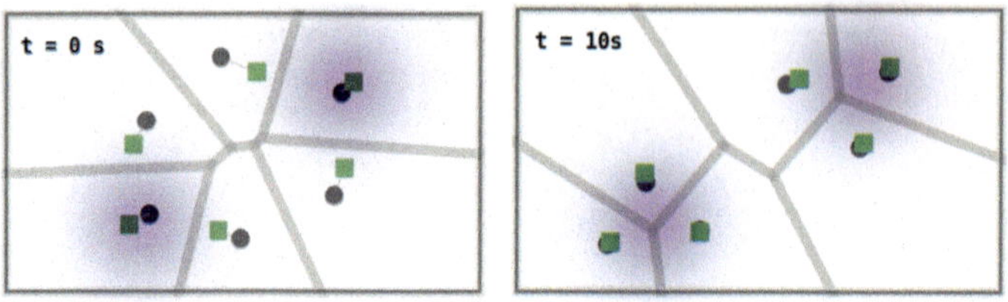

Fig. 7. Visualization of the sensor coverage configurations at the start (left) and end (right) of the sensor coverage problem. The black dots are the robots, the green dots are the centers of the Voronoi cell of each robot, outlined in grey, and the areas of importance in purple. Note as the time goes on in the experiments, the robots converge upon their Voronoi cell centers, which is locally optimal sensor coverage.

We demonstrated our GNN controlling six robots in a 4m by 6m testing arena. In Fig. 7, we can see how the robots group move from a start configuration to a final configuration that is closer to the areas of interest. We can see that the GNNs controllers cause the robots to converge to the center of their Voronoi cells, similar to Lloyd's algorithm which is commonly used as an analytical method for locally optimal sensor coverage.

3.3 Experiments with System Deployment

With very limited access to motion capture systems due to the COVID-19 pandemic flight experiments with three heterogeneous MAVs were conducted in a living room (3 sq.m). Later, these experiments were extended to six MAVs in laboratory settings and nine MAVs in outdoor environment as shown in Fig. 1. Setup or initialization time is critical for rapid deployments of MAVs. Figure 8 shows the setup or initialization time required for our system by a single operator.

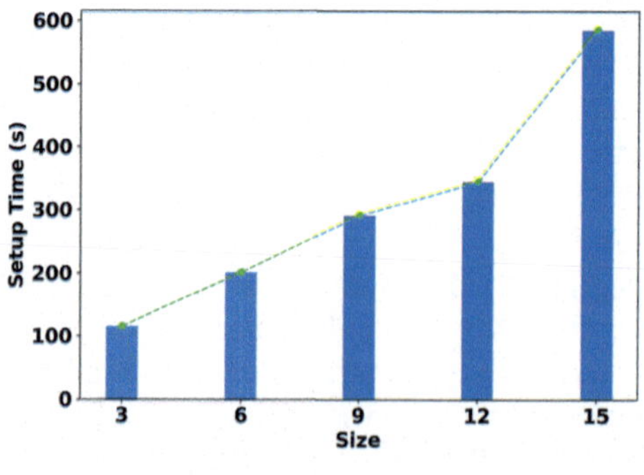

Fig. 8. System deployment time. The deployment includes battery installation, placement of MAVs, system boot and MAV software launch for all MAVs in the experiment.

4 Main Experimental Insights

The COVID-19 crisis has highlighted the need to develop inexpensive robotic test-beds outside of the laboratory environments. Currently, MAV swarms cannot easily be tested outside of lab settings without access to external sensing infrastructure. Our early experiments demonstrate the effectiveness of our work in meeting that goal with a heterogeneous swarm system that is quick and easy to deploy.

The experiments on WiFi and Zigbee highlight the complementary nature of both systems. As the number of nodes in a WiFi network increases, so too does contention for medium access, leading to bandwidth drops, and interference from neighboring transmissions (also known as the hidden node problem). Cumulatively, these factors result in a poor network performance. Having an additional mesh network that does not degrade drastically as the number of nodes increases opens up the possibility of carrying out purely decentralized types of algorithms on-board swarms of MAVs with heterogeneous sensing.

In order to provide our platform as a test bed to the research community, we release our platform as open-source and make all the tools widely available to be reused and adapted to new research problems. We anticipate our tools can be used to generate interesting datasets that can be used in benchmarking multi-robot algorithms. We demonstrated the effectiveness of our system as a test bed on a sensor coverage problem with six robots. As shown in the supplementary video, these experiments qualitatively demonstrate the validity of our method where the swarm has been deployed in a living room, lab environment, as well as outdoors. This work takes the first step towards a quick, easy-to-deploy heterogeneous swarm system by a single human operator without high overhead costs[2].

In future experiments, we plan to conduct studies on scalability by further increasing the number of nodes in the system shown in Fig. 9

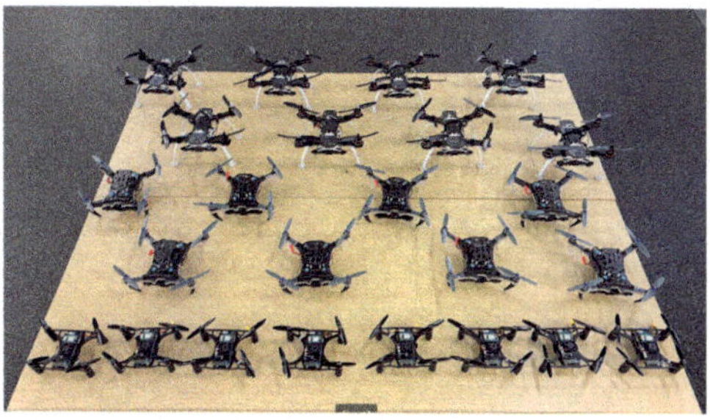

Fig. 9. Heterogeneous swarm of 24 MAVs.

[2] https://youtu.be/25ZvSxXsGMw

Acknowledgments. We gratefully acknowledge the support from ARL Grant DCIST CRA W911NF-17-2-0181, NSF Grant CNS-1521617, ARO Grant W911NF-13-1-0350, ONR Grant N00014-20-1-2822, ONR Grant N00014-20-S-B001 and Qualcomm Research.

References

1. Chung, S., Paranjape, A.A., Dames, P., Shen, S., Kumar, V.: A survey on aerial swarm robotics. IEEE Trans. Rob. **34**(4), 837–855 (2018)
2. Vásárhelyi, G., Virágh, C., Somorjai, G., Nepusz, T., Eiben, A.E., Vicsek, T.: Optimized flocking of autonomous drones in confined environments. Sci. Rob. **3**(20) (2018)
3. Balázs, B., Vásárhelyi, G., Vicsek, T.: Adaptive leadership overcomes persistence-responsivity trade-off in flocking. J. Roy. Soc. Interface **17**(167), 20190853 (2020)
4. Intel. Intel Drone Light Show. https://tinyurl.com/y8zsn7te
5. VergeAero. Verge Aero X1 Light Show. https://verge.aero/
6. Kushleyev, A., Mellinger, D., Powers, C., Kumar, V.: Towards a swarm of agile micro quadrotors. Auton. Rob. **35**(4), 287–300 (2013)
7. Preiss, J.A., Honig, W., Sukhatme, G.S., Ayanian, N.: Crazyswarm: A large nano-quadcopter swarm. In: 2017 IEEE International Conference on Robotics and Automation (ICRA), pp. 3299–3304. IEEE (2017)
8. Turpin, M., Michael, N., Kumar, V.: Capt: concurrent assignment and planning of trajectories for multiple robots. IJRR **33**(1), 98–112 (2014)
9. Alarifi, A., Al-Salman, A., Alsaleh, M., Alnafessah, A., Al-Hadhrami, S., Al-Ammar, M.A., Al-Khalifa, H.S.: Ultra wideband indoor positioning technologies: analysis and recent advances. Sensors **16**(5), 707 (2016)
10. Ledergerber, A., Hamer, M., D'Andrea, R.: A robot self-localization system using one-way ultra-wideband communication. In: 2015 IEEE/RSJ International Conference on Intelligent Robots and Systems (IROS), pp. 3131–3137. IEEE (2015)
11. Mueller, M.W., Hamer, M., D'Andrea, R.: Fusing ultra-wideband range measurements with accelerometers and rate gyroscopes for quadrocopter state estimation. In. IEEE ICRA, vol. 2015, pp. 1730–1736 (2015)
12. Shule, W., Almansa, C.M., Queralta, J.P., Zou, Z., Westerlund, T.: Uwb-based localization for multi-uav systems and collaborative heterogeneous multi-robot systems: a survey (2020)
13. Afanasov, M., Mottola, L., Whitehouse, K.: Poster: testbed for aerial drone applications. In: 2017 EWSN, pp. 192–193. Junction Publishing, USA (2017)
14. Sanchez-Lopez, J.L., Pestana, J., de la Puente, P., Carrio, A., Campoy, P.: Visual quadrotor swarm for the imav 2013 indoor competition. In: Armada, M.A., Sanfeliu, A., Ferre, M. (eds.) ROBOT2013: First Iberian Robotics Conference, pp. 55–63. Springer, Cham (2014)
15. Sanchez-Lopez, J.L., Pestana, J., de la Puente, P., Campoy, P.: A reliable open-source system architecture for the fast designing and prototyping of autonomous multi-UAV systems: Simulation and experimentation. JIRS **84**(1), 779–797 (2016)
16. Faigl, J., Krajník, T., Chudoba, J., Přeučil, L., Saska, M.: Low-cost embedded system for relative localization in robotic swarms. In: 2013 IEEE ICRA, pp. 993–998 (2013)
17. Lightbody, P., Krajník, T., Hanheide, M.: A versatile high-performance visual fiducial marker detection system with scalable identity encoding. In: Proceedings of the Symposium on Applied Computing, ser. SAC 2017, pp. 276–282 (2017)

18. Weinstein, A., Cho, A., Loianno, G., Kumar, V.: Visual inertial odometry swarm: an autonomous swarm of vision-based quadrotors. IEEE RAL **3**(3), 1801–1807 (2018)
19. Li, S., Coppola, M., Wagter, C., de Croon, G.C.H.E.: An autonomous swarm of micro flying robots with range-based relative localization (2020)
20. Xu, H., Wang, L., Zhang, Y., Qiu, K., Shen, S.: Decentralized visual-inertial-uwb fusion for relative state estimation of aerial swarm (2020)
21. ModalAI. VOXL Flight board. https://www.modalai.com/
22. Dragon Drone Development Kit. DDK. https://worldsway.com/product/dragon-drone-development-kit/
23. Thakur, D., Loianno, G., Liu, W., Kumar, V.: Nuclear Environments Inspection with Micro Aerial Vehicles: Algorithms and Experiments. In: Xiao, J., Kröger, T., Khatib, O. (eds.) Proceedings of the 2018 ISER, pp. 191-200. Springer, Cham (2020)
24. Thakur, D., Loianno, G., Jarin-Lipschitz, L., Zhou, A., Kumar, V.: Autonomous inspection of a containment vessel using a micro aerial vehicle. In: IEEE SSRR, vol. 2019, pp. 1–7 (2019)
25. Wang, J., Olson, E.: Apriltag 2: efficient and robust fiducial detection. In. IEEE/RSJ International Conference on Intelligent Robots and Systems (IROS), vol. 2016, pp. 4193–4198 (2016)
26. Santos, M., Mayya, S., Notomista, G., Egerstedt, M.: Decentralized minimum-energy coverage control for time-varying density functions. In: International Symposium on Multi-Robot and Multi-Agent Systems (MRS), vol. 2019, pp. 155–161. IEEE (2019)
27. Pimenta, L.C., Kumar, V., Mesquita, R.C., Pereira, G.A.: Sensing and coverage for a network of heterogeneous robots. In: 47th IEEE Conference on Decision and Control, vol. 2008, pp. 3947–3952. IEEE (2008)
28. Xu, X., Diaz-Mercado, Y.: Multi-robot control using coverage over time-varying non-convex domains. In: ICRA (2020)
29. Tolstaya, E., Gama, F., Paulos, J., Pappas, G., Kumar, V., Ribeiro, A.: Learning decentralized controllers for robot swarms with gnn. CoRR, vol. abs/1903.10527 (2019)
30. Gama, F., Marques, A.G., Leus, G., Ribeiro, A.: Convolutional neural network architectures for signals supported on graphs. IEEE Trans. Signal Process. **67**(4), 1034–1049 (2018). https://doi.org/10.1109/TSP.2018.2887403

Sailing a Boat Through a Macroscopic Smart-Fluid Composed of a Robot Swarm

Paul Motter(✉), Zachary Hoover, and Michael Otte

Department of Aerospace Engineering, University of Maryland, College Park, MD 20742, USA
otte@umd.edu

Abstract. We investigate how a physically inert vehicle, which we call a *"boat,"* can locomote through a smart fluid by communicating with the particles of a smart fluid, and thereby inducing the fluid to exert forces and moments on the boat. Our smart fluid is composed of a swarm of Kilobot robots that move on a flat surface. The boat communicates with the fluid particles through wireless messages sent from its bow and stern. By sending different pairs of messages from its bow and stern the boat directs the subsets of the smart fluid in different locations to move in different ways. This enables the boat to maneuver forward, turn left, and turn right. In experiments on a physical testbed, we evaluate how the presence and size of the boat's rudder affects motion efficacy. We find that using a small rudder provides better overall motion than using no rudder or using a large rudder.

1 Introduction

Smart fluids—fluids with properties which can be controlled by an artificial process—have the potential to enable amazing breakthroughs in manufacturing, medicine, and other tasks that require micro-manipulation. In this paper we demonstrate proof-of-concept that a object with no moving parts can sail through a smart fluid by altering the physical forces exerted by fluid particles on that object. The object—which we refer to stylistically as the *boat*—broadcasts infrared messages into the fluid from both its bow and stern. The fluid particles behave in different ways based on the messages they receive. Thus, the boat is able to maneuver by broadcasting different pairs of messages from its bow and stern simultaneously. The boat moves forward by asking the smart fluid to move away from the bow and toward the stern. A boat equipped with a (static) rudder turns by asking the fluid to exert a rotational moment via the boat's rudder. Meanwhile, a rudderless boat could turn by asking the smart fluid to create a clockwise or counterclockwise vortex.

In a series of hardware experiments we: (1) investigate how different rudder lengths of the boat affect the smart fluid's ability to propel and turn the boat, (2) demonstrate repeatability of both forward and turning maneuvers, and (3) demonstrate the repeatability of performing non-trivial maneuver sequences—in particular, we study the boat's ability to sail a 'Z' shaped path.

B. Siciliano et al. (Eds.): ISER 2020, SPAR 19, pp. 424–435, 2021.
https://doi.org/10.1007/978-3-030-71151-1_38

Microrobots that use magnetic fields have been studied for their potential to revolutionize medicine [1,5] as well as their potential for micro-manipulation [11]. Some robots use a global magnetic field [4,14] or other global field [2,6] to achieve motion. Other work leverages *local* interactions created by a magnetic local force [3,13]. We are inspired by this second class of methods and investigate a solution where the transported object (our boat) moves in a variety of ways by creating a local force-field gradient within the smart fluid.

Previous work has demonstrated that a robotic swarm can transport an object and/or use a robot swarm as a macroscopic fluid. For example, [6] and [2] use a homogeneous swarm control to transport an object using global commands. [15] and [7] use a swarm to move an object while passively communicating motion information from the object to the swarm. Aggregate motion is used to move an object in a direction indicated by a light source in [8]. Granular convection is used in [12] to create a pressure gradient for the purposes of object transportation; achieved by having the fluid globally repulsed from the transported object's destination. Our work differs from previous work in that the transported object (boat) creates ego-centric localized force field gradients in the smart fluid. Also, unlike [10], this work focuses mainly in the transport of an object, not in the transport of an amalgamated group of robots.

We style our work as a boat sailing through a smart fluid. The boat has no moving parts and one or more transmitters; we use two throughout this paper. The fluid used is made of particles that have the ability to run, store a program, and sense their distance from each transmitter. They are able to exert a local force on demand.

We define the *smart fluid sailing problem* that we study as follows: Given a smart fluid and a boat with a means of affecting the properties of that fluid, achieve transportation of the boat from a start location to a goal location by having the boat direct a local force-field gradient in the smart fluid.

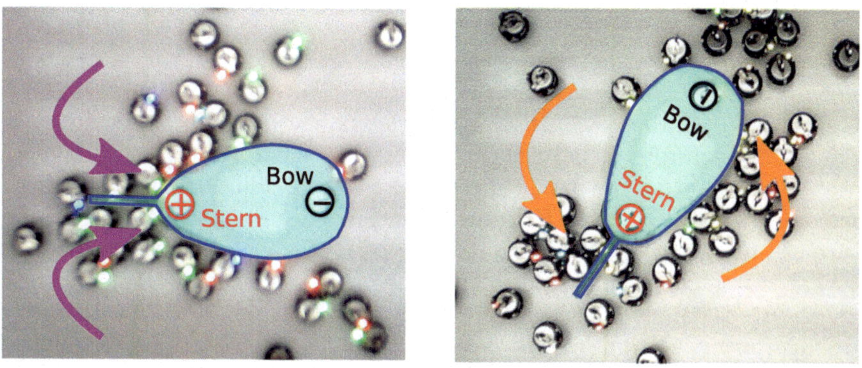

Fig. 1. A boat (highlighted blue) with no moving parts maneuvers by sending signals from two transponders (highlighted red/black and labeled ±, - in the bow and + in the stern) to a smart fluid composed of Kilobot robot fluid particles. Forward (towards the - pole) and turning motion (left and right sub-figures) are achieved by causing a local asymmetry in physical fluid forces (purple/orange arrows).

2 Technical Approach

The boat has two transmitters, one each in the front and rear of the boat. It is convenient to think of the two transmitters as creating oppositely charged "poles." We refer to the front (the "bow") and rear ("stern") transmitters (and their poles) as negative b_- and positive b_+, respectfully. The boat maneuvers with 3 *actions*: forward a_F, turns right a_R, and turns left a_L by sending messages to the smart fluid from its transmitters.

Each smart fluid particle r_i has seven *states*. The default state is *random motion* (s_R). States *move towards* (s_T) and *move away* (s_A) involve movement to or from the transmitters, respectively. Four turning-related states are named for each combination of turning direction (left, right) and boat end (front/negative, rear/positive). For example, a bot that is near b_- (the *negative* or forward most or bow transmitter) and has received the a_L (turn left action) command would be in state *left turn negative* (s_{L-}). Depending on a particle's state, its interaction with the two poles may be that of attraction, repulsion, or orbital motion around the pole (clockwise and/or counterclockwise).

Each smart fluid particle uses Algorithm 1 to determine its (non-random) action each time a message is received from the boat. A particle changes state based on a combination of which transmitter b_i it last received a message from (lines 2 and 11), the distance to that transmitter (lines 1, 5, 14, 20), and the particle's previous state (lines 2 and 11). If a particle has not received a message after a set period of time, its state is reset to s_R, the $r.turnDirection$ is reassigned randomly, and $r.turnFlag$ is reset (not shown in an algorithm here). Thus, most smart-fluid particles in the environment move randomly, but those near to the boat have their state temporarily altered by the boat's transmitted messages.

Forces exerted on different sides of the boat will cause it to move in a direction or rotate in a manner directly related to the force vectors applied to it. If the boat desires to move forward (in the direction of b_-), then an accumulation of particles (and forces) at b_+ is sufficient. For a turn, opposed forces on the sides of the boat at front/rear are sufficient (see Figs. 1 and 2).

Algorithm 2 is used by particles interacting with boats that have a static rudder. From the boat's point of view, particles at both the positive and negative poles of the boat orbit towards the same side of the boat. Particles hitting the rudder create a torque that turns the boat, while particles in front of the boat move to the outside of the intended rotation to create additional torque near the front of the boat (Fig. 2-C). If a particle receives a message from b_i requesting a turning state, then the particle moves towards the boat until its distance d is less than $r.orbitDist$ from b_i (lines 1 and 5). A flag $r.turnFlag$ indicates that a turning state is happening (lines 1 and 3), and the particle orbits the transmitter at radius $r.orbitDist$ (lines 1–14).

Boats lacking a rudder could turn by inducing the smart-fluid to create a vortex. All robots rotate either clockwise or counterclockwise around the boat (Fig. 2-B). Pseudo-code for the corresponding particle algorithm is very similar to that in Algorithm 2 for a boat with a rudder, except lines 6 and 11 read **if** $r.s \in \{s_{R+}, s_{R-}\}$, switching the direction of rotation around b_- during a

Algorithm 1: `set_action`(r_i, a_i, b_i)

```
d ← estimate_distance(r_i, b_i)
if b_i = b_+ and r.s ∉ {s_A, s_R−, s_L−}
    if a_i = a_F
        r.s ← s_T
        move_towards(d)
        return
    else if a_i = a_L
        r.s ← s_L+
    else if a_i = a_R
        r.s ← s_R+
else if b_i = b_− and r.s ∉ {s_T, s_R+, s_L+}
    if a_i = a_F
        r.s ← s_A
        move_away(d)
        return
    else if a_i = a_L
        r.s ← s_L−
    else if a_i = a_R
        r.s ← s_R−
turn(d)
```

turning maneuver (ompare Fig. 2-B and C). The authors chose the algorithms for turning each style of boat (rudder and rudderless) by selecting which method seemed most effective for each during pre-trial development.

Particle routines for moving toward or away from the boat's receivers are presented in Algorithms 3 and 4, respectively. Particles are assumed to have distance sensors but not directional sensors. Each particle moves in a circle in a direction randomly defined at startup (*r.turnDirection*), until the transmitter distance sensed is above or below a threshold $r.d_{last}$ (Algorithm 3 or 4, respectively).

3 Results

Our work makes three main contributions. First, using a robotic testbed we demonstrate that it is possible and repeatable to sail through a smart fluid by creating a local force gradient. Second, we show that rudder size affects mean forward and rotational motion duration, as seen in Table 1, and that a rudder allows for repeatable turning motion. Third, we demonstrate that it is possible to chain multiple maneuvers to meet a non-trivial navigation requirement. A summary of these results appears in Table 2. 'N/A' indicates that no data was collected due to inability of a design to complete the scheduled trials. The experiments themselves are described in the next section.

Algorithm 2: `turn`(d)

```
if r.turnFlag = false and d > r.orbitDist
    move_towards(d)
    r.turnFlag ← true
else
    if d < r.orbitDist
        if r.s ∈ {s_R+, s_L-}
            turn left
        else
            turn right
    else
        if r.s ∈ {s_R+, s_L-}
            turn right
        else
            turn left
```

Algorithm 3: `move_toward`(d)

```
if d ≤ r.d_last
    r.d_last ← d
    move forwards
else
    r.d_last ← weighted_ave(d, r.d_last)
    turn r.turnDirection
```

Algorithm 4: `move_away`(d)

```
if distance > r.d_last
    r.d_last ← d
    move forwards
else
    r.d_last ←
    weighted_ave(d, r.d_last)
    turn r.turnDirection
```

4 Hardware Experiments

We test our methods using a smart fluid constituted of Kilobot robots [9]. Kilobot robots measure approximately 3 cm high (disregarding the charging hook) and 3 cm in diameter and resemble a thick coin resting horizontally upon three 2 cm legs evenly spaced apart (see Fig. 6). They have the ability to move forwards, turn right and left, and sense local visible light intensity. Nine-byte messages, sent by bouncing an infrared signal off of the surface a bot stands on, can be broadcast from each Kilobot to all bots within about a 20 cm radius. Kilobots are able to measure the relative distance to each other based on the signal strength of received messages.

The boat we use is an approximately 19.5 × 12.5 cm teardrop-shaped piece of styrofoam. The first boat design was a simple rectangle. Through informal iteration, we discovered that with the simplistic forward motion model applied, it was very difficult for the swarm to move a rectangular boat forwards. A teardrop design allows bots moving towards b_+ from anywhere in its neighborhood to exhibit forward motion on the boat, as a concentration of particles attracted to b_+ "squeeze" the boat forwards. Any left/right asymmetry in the design was unintentional (Fig. 3).

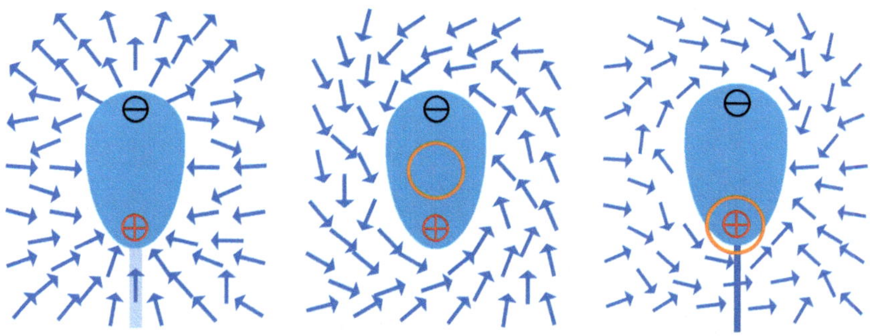

Fig. 2. The boat induces particles in its local area to move as indicated by the arrows, exerting local forces where a particle's motion path intersects with the boat. Forward motion (A) is the same with and without a rudder. Turing requires different smart fluid interaction without (B) and with (C) a rudder. The center of rotation is outlined with an orange circle. In this example, the intended result of the actions in (B) and (C) would be counterclockwise rotation of the boat.

Table 1. Maneuver duration, as well as Y axis (perpendicular to the intended vector of travel) error and angle error from the intended vector of travel for the forward action tests. These values were calculated by averaging the magnitude of the measurements for the respective error for each successful test.

Rudder	Action	Mean (s)	Std. (s)	Attempts	Success	Y err (cm)	Angle err
None	Forward	609.57	285.56	9	7	12.55	41.23°
Short	Forward	454.40	118.79	9	5	16.39	52.10°
Long	Forward	610.00	147.60	6	5	19.33	26.46°
None	Turn	N/A	N/A	6	0	N/A	N/A
Short	Turn	378.40	153.92	6	5	N/A	N/A
Long	Turn	423.20	220.64	6	5	N/A	N/A

Table 2. "Z-Pattern" statistics Over 5 trials (in minutes:seconds)

Mean	Std.	Shortest	Longest
60:59	18:33	42:59	91:56

Table 3. Boat weights, with and without embedded Kilobots

Rudder	Empty(g)	With bots(g)
None	2.71	36.67
Short	4.44	38.43
Long	5.35	39.32

Two inert (not physically actuating) Kilobots are embedded in the boat and used as the transmitters b_+ and b_-. The transmitter bots are wedged into the boat, suspending the foam about 2 cm above the ground, with only the six legs

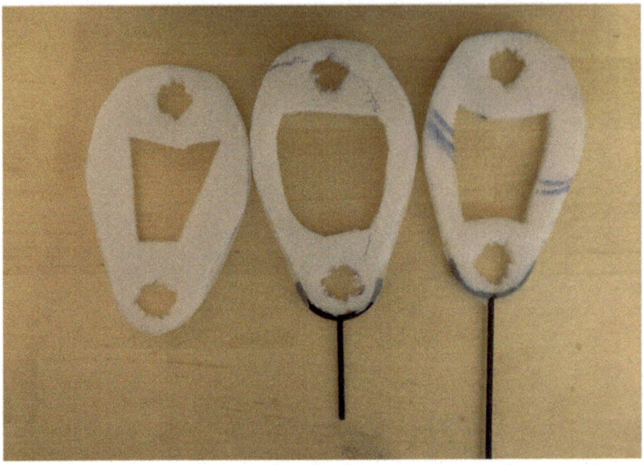

Fig. 3. Rudders: none, short (7.5 cm), long (15 cm).

from the two bots contacting the ground. This is far enough from the ground and the bots are close enough to the edge of the boat to not adversely effect message transmission from the boat bots to swarm bots nearest the respective poles. By altering the visual light intensity in the environment with a series of switches, a user remotely controls which action messages the boat's transmitters broadcast into the smart fluid. For example, bright, medium, and dark light indicate that the transmitters should collectively ask the smart fluid to create a forward, left turn, or right turn maneuver. The swarm itself does not exhibit behavior based on light level, only the instructions it receives from the transmitters.

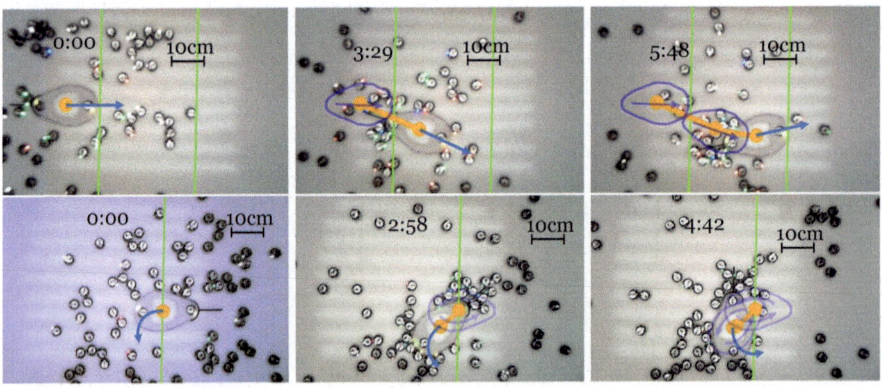

Fig. 4. Top row: Forward motion test. The boat starts perpendicular to and just behind the starting line; we measure the time until the bow breaks the finish line (lines are marked with dry erase during the tests, and highlighted green here). **Bottom Row:** Turning Test. The boat starts perpendicular to a straight line (green); we measure the duration of time until it rotates parallel to that line.

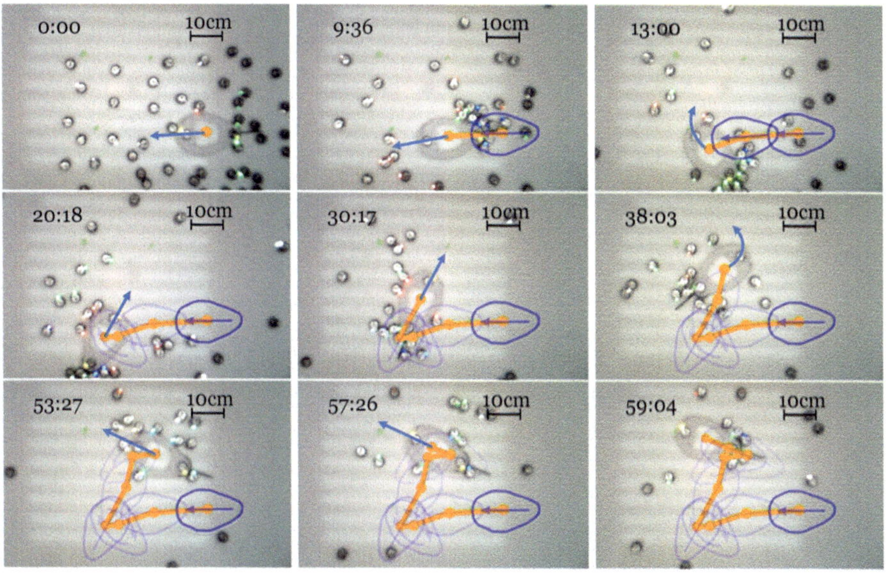

Fig. 5. In the "Z-test" trial the boat progresses through the sequence: forward, turn right, forward, turn left, forward. The boat's progress over the course of a "Z-test" trial is highlighted in orange. Note the time intervals between steps as swarm density decreases and time increases.

Experiments are performed on a 90 cm by 120 cm whiteboard placed on top of a professionally-leveled billiards table. LED light strips suspended one meter above the whiteboard are used to control visual light intensity. A camera mounted one meter above the workspace records each experiment.

We performed repeated trials to evaluate the forward and turning performance of three boat designs. Due to the significance of the teardrop shape, reverse was not considered, though pre-trial development testing with a teardrop-shaped boat proved reverse to be difficult. Forward motion is measured by the time duration required for the boat to cross a 30.5 cm (or one foot in standard units, which the authors chose for familiarity) region (see the top row of Fig. 4). Only the 'forward' command is transmitted to the smart fluid during this experiment, so trials are disregarded if the boat is unable to cross the span, e.g., due to turning sideways. Presumably, we could have used the left and right commands to keep the boat on course. However, in order to measure only effectiveness in forward motion and not contaminate these results with the a design's turning ability, we dropped these trials. Trials alternate between moving opposite directions across the workspace to help control for surface irregularities. At least five successes were recorded for each rudder type. Turning ability is assessed by measuring the time duration required for the boat to rotate 90° in a desired direction (see the bottom row of Fig. 4). Trials are disregarded if the boat has not turned 90 degrees after fifteen minutes, or if the boat turns 90° in the wrong direction.

Fig. 6. A test in progress. Note that the boat's body is suspended by the Kilobots embedded in it.

Trials are performed in different workspace locations and in different turning directions to help control for discrepancies in boat shape and workspace surface. Five successful trials are recorded.

The trials described above demonstrate that the short ruder configuration has the best performance of the three models we evaluate. We tested the ability of the short rudder boat to perform a sequence of maneuvers by having it sail in a 'Z' pattern to visit four locations, requiring forward motion and both left and right turns (Fig. 5). Five trials are recorded. A human operator controlled the light levels, doing his best to guide the boat through the Z pattern as quickly as possible. Unlike the straight line tests, turning commands were allowed to correct for inconsistencies in forward motion, demonstrating the controllability of our solution.

5 Experimental Insights

An important contribution of our work is proof of concept that 'sailing' a boat through a smart fluid, by inducing the fluid to create a local force gradient around the boat, is both possible and repeatable. The 'Z-path' trials show that single maneuvers can be chained together to enable more complex motion over a non-trivial path. They also showed that motion is possible (albeit at a slower pace) with only a few bots. Our work also shows that a smart fluid can be reconfigured on-the-fly to suit the local needs of an agent.

Our experiments demonstrate that rudder size affects both forward and turning motion ability. A short rudder has both improved straight line speed and rotation speed over no rudder and long rudder configurations (see Table 1). Both short and long rudders were found to improve turning motion, as we did not observe a rudderless boat turn at all during formal trials. While we developed a process for turning a rudderless boat design as described above (see Fig. 2B) in informal iteration before the formal trials, the results were never consistent. This prompted us to explore other options and led to experimentation on boats with rudders, which seemed more consistent in rotational motion during the pre-trial stages of this work. Therefore, we were unsurprised when the trials for rotational motion of a rudderless boat failed, either by turning 90° the wrong way or failing to achieve 90 degrees of rotation in the correct direction in 15 min.

We note that none of the boat designs went particularly straight during the forward action tests, as seen in Table 1. However, as directional correction by the testers was disallowed during these tests, the path of the boat was highly susceptible to imperfections in the testing surface as well as the random placement of the swarm bots. While disallowing correctional commands during the forward motion tests were necessary for accurate measurements, our ability to control the boats during the Z test, where correctional commands *were* allowed for the straight portions of the pattern, assuaged our fears over the boat's inability to move in a perfectly straight line.

We believe that a rudder provides the smart fluid particles a place to physically engage with the boat during forward motion, as well as a lever that can assist rotation. We also believe that a rudderless boat was unable to achieve turning motion as, without the lever the rudder provides, our simplistic turning method was not precise enough to push on the boat in exactly the right places to rotate repeatedly or quickly.

6 Conclusions

We present a method for a vehicle or *boat* able to affect a surrounding smart fluid to demonstrate two-dimensional motion by inducing that fluid to exert local forces in a controlled manner. We then demonstrate this method using a micro-robotic swarming platform, the Kilobot, showing that repeatable motion and control of the boat over a complex path is possible. We also demonstrate that with our boat design, a rudder is necessary for consistent rotational motion of the boat, and that a boat with a shorter rudder results in more rapid motion compared to a boat with a longer rudder.

Acknowledgements. We would like to thank Nicholas Casertano for helping with the Kilobot code, Abigail Motter for assisting with the testing process, Brian Motter for the use of his billiards table, Christoph Neisess for printing the rudders, and Aaron Becker for his conversations with Michael Otte in which he helped develop the concept for this paper.

References

1. Alapan, Y., Yasa, O., Yigit, B., Yasa, I.C., Erkoc, P., Sitti, M.: Microrobotics and microorganisms: biohybrid autonomous cellular robots. Ann. Rev. Control Robot. Auton. Syst. **2**, 205–230 (2019)
2. Becker, A., Habibi, G., Werfel, J., Rubenstein, M., McLurkin, J.: Massive uniform manipulation: controlling large populations of simple robots with a common input signal. In: 2013 IEEE/RSJ International Conference on Intelligent Robots and Systems, pp. 520–527. IEEE (2013). ISSN 2153-0866
3. Carlson, J.D., Jolly, M.R.: MR fluid, foam and elastomer devices. Mechatronics **10**(4–5), 555–569 (2000). http://www.sciencedirect.com/science/article/pii/S0957415899000641
4. Ceylan, H., Giltinan, J., Kozielski, K., Sitti, M.: Mobile microrobots for bioengineering applications. Lab Chip **17**(10), 1705–1724 (2017). https://pubs.rsc.org/en/content/articlelanding/2017/lc/c7lc00064b
5. Erkoc, P., Yasa, I.C., Ceylan, H., Yasa, O., Alapan, Y., Sitti, M.: Mobile microrobots for active therapeutic delivery. Adv. Ther. **2**(1), 1800064 (2019). https://onlinelibrary.wiley.com/doi/abs/10.1002/adtp.201800064
6. Groß, R., Dorigo, M.: Towards group transport by swarms of robots. Int. J. Bio-Inspired Comput. **1**(1–2), 1–13 (2009). https://www.inderscienceonline.com/doi/abs/10.1504/IJBIC.2009.02277
7. Kube, C.R., Bonabeau, E.: Cooperative transport by ants and robots. Robot. Auton. Syst. **30**(1), 85–101 (2000). http://www.sciencedirect.com/science/article/pii/S0921889099000664
8. Li, S., Batra, R., Brown, D., Chang, H.D., Ranganathan, N., Hoberman, C., Rus, D., Lipson, H.: Particle robotics based on statistical mechanics of loosely coupled components. Nature **567**(7748), 361–365 (2019). https://www.nature.com/articles/s41586-019-1022-9
9. Rubenstein, M., Ahler, C., Nagpal, R.: Kilobot: a low cost scalable robot system for collective behaviors. In: 2012 IEEE International Conference on Robotics and Automation, pp. 3293–3298. IEEE (2012). ISSN 1050-4729
10. Savoie, W., Berrueta, T.A., Jackson, Z., Pervan, A., Warkentin, R., Li, S., Murphey, T.D., Wiesenfeld, K., Goldman, D.I.: A robot made of robots: emergent transport and control of a smarticle ensemble **4**(34), eaax4316 (2019). https://robotics.sciencemag.org/content/4/34/eaax4316
11. Soo Kim, P.S., Becker, A., Ou, Y., Julius, A.A., Kim, M.J.: Swarm control of cell-based microrobots using a single global magnetic field. In: 2013 10th International Conference on Ubiquitous Robots and Ambient Intelligence (URAI), pp. 21–26. IEEE (2013)
12. Sugawara, K., Correll, N., Reishus, D.: Object transportation by granular convection using swarm robots. In: Ani Hsieh, M., Chirikjian, G. (eds.) Distributed Autonomous Robotic Systems, vol. 104, pp. 135–147. Springer, Heidelberg (2014). http://link.springer.com/10.1007/978-3-642-55146-8_10
13. Wang, G., Ma, Y., Tong, Y., Dong, X.: Development of manganese ferrite/graphene oxide nanocomposites for magnetorheological fluid with enhanced sedimentation stability. J. Ind. Eng. Chem. **48**, 142–150 (2017). http://www.sciencedirect.com/science/article/pii/S1226086X16305317

14. Wang, X., Chen, X.Z., Alcâtara, C.C.J., Sevim, S., Hoop, M., Terzopoulou, A., Marco, C.d., Hu, C., Mello, A.J.d., Falcaro, P., Furukawa, S., Nelson, B.J., Puigmartí Luis, J., Pané, S.: MOFBOTS: metal–organic-framework-based biomedical microrobots. Adv. Mater. **31**(27), 1901592 (2019). https://onlinelibrary.wiley.com/doi/abs/10.1002/adma.201901592
15. Wang, Z., Schwager, M.: Force-amplifying N-robot transport system (force-ANTS) for cooperative planar manipulation without communication **35**(13), 1564–1586 (2016). https://doi.org/10.1177/0278364916667473

Rapid and High-Fidelity Subsurface Exploration with Multiple Aerial Robots

Kshitij Goel(✉), Wennie Tabib, and Nathan Michael

The Robotics Institute, Carnegie Mellon University, Pittsburgh, PA 15213, USA
{kgoel1,wtabib,nmichael}@andrew.cmu.edu

Abstract. This paper develops a communication-efficient distributed mapping approach for rapid exploration of a cave by a multi-robot team. Subsurface planetary exploration is an unsolved problem challenged by communication, power, and compute constraints. Prior works have addressed the problems of rapid exploration and leveraging multiple systems to increase exploration rate; however, communication considerations have been left largely unaddressed. This paper bridges this gap in the state of the art by developing distributed perceptual modeling that enables high-fidelity mapping while remaining amenable to low-bandwidth communication channels. The approach yields significant gains in exploration rate for multi-robot teams as compared to state-of-the-art approaches. The work is evaluated through simulation studies and hardware experiments in a wild cave in West Virginia.

1 Introduction

Planetary exploration has benefited from advancements in robotics through automation of data collection for planetary science and robotic precursor missions for human space exploration [1]. To date, robotic precursor missions have engaged in surface exploration of Mars [2] but have not explored subsurface environments despite the potential geological and astrobiological significance of these domains [3,4]. As a result, robotic subsurface exploration has been identified as a key technology for future missions to these planets [5]. Autonomous navigation and high-resolution perceptual modeling are critical needs in the context of subsurface planetary exploration [6]. A challenge of operating in subsurface environments is communicating to a surface station. Communication may be limited or impossible due to the inability of radio waves to penetrate rock, impeding data relay to Earth, so compact data transmission is critical. Operating on planets far from Earth introduces additional restrictions on power and compute that may be mitigated by leveraging multiple robots to increase coverage in spatially expansive environments [7]. This work addresses a key challenge for planetary exploration: enabling rapid multi-robot exploration in subsurface environments by leveraging a perceptual modeling framework amenable to low-bandwidth communication while remaining high-fidelity.

Exploration frameworks cannot assume *a priori* knowledge about the structure of the environment so the exploration system must operate with unknown

B. Siciliano et al. (Eds.): ISER 2020, SPAR 19, pp. 436–448, 2021.
https://doi.org/10.1007/978-3-030-71151-1_39

Fig. 1. Cave exploration with two aerial robots in West Virginia, USA.

locomotion constraints. Aerial robots have recently been leveraged to mitigate these constraints in the subterranean domain [8] and considered for subsurface mapping on Mars [3]. In this work, we consider aerial robots operating in a cave on Earth (Fig. 1) as an analog scenario for subsurface exploration on Mars. These robots are often limited by size, weight, and power (SWaP) constraints [8]. Energy constraints on these platforms impose limits on flight endurance necessitating rapid exploration, since the existence of a replenishment infrastructure in the planetary exploration context cannot be guaranteed at these sites [6]. Several frameworks for rapid exploration have been proposed that either use a single fast-moving aerial robot [9–12] or multiple slow-moving aerial robots [13,14]; however, a real-world deployable framework that combines the elements from both is desirable. Such a deployment can potentially be realized by sending multiple aerial robots ("daughtercraft") from a lander ("mothership") to perform rapid, effective, and affordable high-resolution mapping of the target environment, similar to the concept surface mission studied by Matthies [15] for Titan. To this end, a distributed perceptual modeling framework that provides communication-efficient map sharing can enable the daughtercraft team to improve the rate of exploration while transmitting scientific data to the mothership and Earth. The experimental evaluation in this work (Sect. 3) is motivated by this concept of operations.

Related Work: With the ongoing DARPA Subterranean Challenge [16], there is an increased interest in deploying a team of robots in cave networks. Although existing systems have not shown autonomous operations in a cave, multi-robot exploration systems have been proposed for mine and tunnel environments. These environments differ from cave environments because in most cases the terrain is flat. Notably, teams competing in the challenge have identified that accounting for limited availability of communication resources within the exploration framework is a key milestone for future work [17–19]. Ebadi et al. [17] state that the communication bottlenecks faced during mapping with a multi-robot system due to the use of downsampled point clouds can be addressed by map compression techniques or compact representations for motion planning. Dang et al. [18] use the state-of-the-art, memory-efficient OctoMap [20] approach for map representation

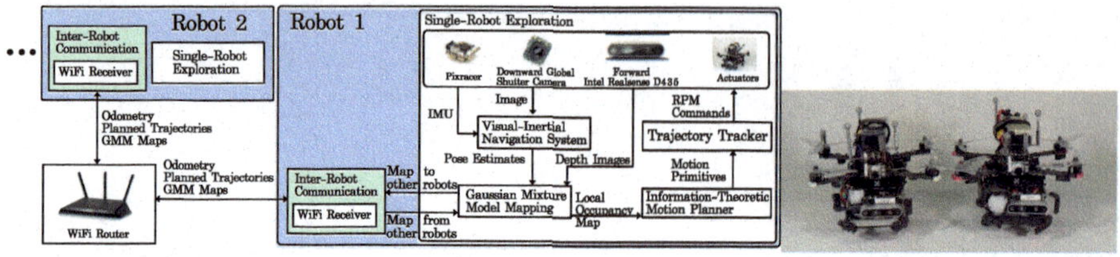

Fig. 2. (Left) Overview of the rapid multi-robot exploration framework and (Right) aerial systems used in experiments in this work.

but mention efficient map sharing as one of the future challenges. Rouček et al. [19] use elevation maps for mapping but only on wheeled and ground robots because the transmission of these maps requires a physically large communication module. Furthermore, the speeds of robots are constrained for these systems. These shared challenges indicate a gap in the state-of-art for communication-efficient distributed mapping methods in rapid aerial multi-robot exploration systems for subterranean domains. Corah et al. [14] highlight the benefits of a distributed mapping strategy that exploits the compactness of Gaussian Mixture Models (GMMs) relative to the occupancy grid approach [21]. However, the approach is computationally prohibitive for real-world deployment, limits robot speeds, and the effects of communication constraints on the exploration performance of the robot team are not discussed.

Contributions: We build upon prior single-robot works [8,10] to develop multi-robot exploration with the following contributions: (1) a GMM-based distributed mapping approach and occupancy reconstruction for information-theoretic motion planning; (2) an evaluation of the fidelity and memory consumption of the approach against OctoMap [20] and occupancy grid mapping [21]; and (3) a simulation study on the effects of constrained communication on the exploration rate of a two-robot team. All contributions are presented in the context of caves.

2 Technical Approach

An overview of the system is shown in Fig. 2. Each robot is equipped with single-robot exploration and inter-robot communication modules. The exploration module consists of four major subsystems: GMM mapping, information-theoretic motion planning, visual-inertial state estimation, and trajectory tracking. The inter-robot communication module enables sharing information between robots or other computers on the network. The GMM mapping and planning subsystems together with the communication module constitute distributed mapping (Sect. 2.1) and multi-robot planning (Sect. 2.2), respectively. In this section, the following mathematical notation is used: lower-case letters represent scalar values, lower-case bold letters represent vectors, upper-case bold letters represent matrices, and script letters represent sets.

2.1 GMM-Based Distributed Mapping

This section details the distributed mapping approach to share environment models between robots. Consider a team of N robots. At timestep t robot $i \in N$ receives the depth sensor observation, $\mathcal{Z}_t^i$, which represents a set of points. A Gaussian mixture model (GMM) is learned from these points following the approach from [8]. The GMM is parameterized by $\Theta = \{\pi_m, \boldsymbol{\mu}_m, \boldsymbol{\Sigma}_m\}_{m=1}^M$ where $\boldsymbol{\mu}_m \in \mathbb{R}^3$ is a mean, $\boldsymbol{\Sigma}_m \in \mathbb{R}^{3\times3}$ is a covariance, and $\pi_m \in \mathbb{R}$ is a weight such that $\sum_{m=1}^M \pi_m = 1$. A GMM representing point set $\mathcal{Z}_t^i$ is denoted as $\Theta_{\mathcal{Z}_t^i}$.

Keyframe GMMs: To reduce redundant observations keyframe GMMs are identified for transmission to other robots. A keyframe GMM, $\hat{\Theta}_{\mathcal{Z}_t^i}$, is determined by approximating the field of view for the current sensor observation as a rectangular pyramid and calculating the overlapping volume with other keyframe fields of view. If the volume is smaller than a user-defined threshold, λ, the sensor observation is considered to be a keyframe. $\hat{\Theta}_{\mathcal{Z}_t^i}$ and the sensor pose, $\mathbf{S}_t^i \in$ SE(3), are transmitted to the other robots or computers on the network.

Each robot maintains its own environment representation and relative initial transforms between robots are assumed to be known. When robot j receives $\hat{\Theta}_{\mathcal{Z}_t^i}$, it is received in the frame of robot i. To transform it into the frame of robot j, the relative initial rotation $\mathbf{R}_0^{ji} \in \mathbb{R}^{3\times3}$ and translation $\mathbf{x}_0^{ji} \in \mathbb{R}^3$ parameters are applied to the means and covariances of the distribution using the following equations.

$$\boldsymbol{\mu}^j = \mathbf{R}_0^{ji}\boldsymbol{\mu}^i + \mathbf{x}_0^{ji} \qquad \boldsymbol{\Sigma}^j = \mathbf{R}_0^{ji}\boldsymbol{\Sigma}^i(\mathbf{R}_0^{ji})^T, \tag{1}$$

The transformed GMM is incorporated into robot j's existing GMM map following the approach from [8,22].

Occupancy Reconstruction: A local occupancy grid map $\mathbf{m}_t^i$ is maintained and centered around the robot's current position $\mathbf{x}_t^i$ for use in information-theoretic motion planning. To generate $\mathbf{m}_t^i$, a number of points $\mathbf{p} \in \mathbb{R}^3$ equal to the support size, or number of points used to learn the distribution, is sampled and raytraced to the sensor pose $\mathbf{x}_t^i$. The probability of occupancy along the ray is updated.

Multi-robot Map Updates: Care must be taken to update $\mathbf{m}_t^j$ when receiving $\hat{\Theta}_{\mathcal{Z}_t^i}$. In addition to applying the transformation parameters so that $\hat{\Theta}_{\mathcal{Z}_t^i}$ is transformed into the frame of robot j, $\mathbf{m}_t^j$ must also be updated by sampling points from the transformed $\hat{\Theta}_{\mathcal{Z}_t^i}$ and raytracing through $\mathbf{m}_t^j$ to the sensor pose, $\mathbf{S}_t^i$, which must also be transformed into the frame of robot j. This ensures the occupancy is updated with observations from both robots. A visualization of this is shown in Fig. 3. Robot i takes a sensor observation (left) and learns $\hat{\Theta}_{\mathcal{Z}_t^i}$ (center). This keyframe GMM is transmitted to robot j, transformed into the frame of robot j, and then used to update $\mathbf{m}_t^j$ (right).

Fig. 3. Overview of the distributed mapping approach.

2.2 Planning for Rapid Multi-robot Exploration

Robot i uses $\mathbf{m}_t^i$ for information-theoretic receding-horizon planning via the strategy presented in [10] that accounts for perception latencies and kinodynamic constraints of the robot. The approach uses Monte Carlo tree search (MCTS) [23] to evaluate the Cauchy-Schwarz Quadratic Mutual Information (CSQMI) [24] for a set of motion primitives over a user-specified time horizon. An informative primitive sequence is selected that maximizes the CSQMI over the MCTS tree. Safety is ensured by checking for collisions with the environment.

The informative trajectories are shared with other robots and inter-robot collision avoidance is enabled through a standard priority-based collision checker assuming a cylindrical robot model [25]. The priorities are assigned manually before the exploration run and remain constant throughout. To reduce the computational complexity for lower priority robots, three optimizations are applied. First, the collision checking is only active when a pair of robots are within a pre-specified radius. To enable this on each robot without assuming a centralized oracle, the robots share odometry information at a sufficiently high rate (10 Hz) compared to the planning frequency (1 Hz). Second, the number of cylinders sampled over the planned trajectory is limited to a pre-specified maximum to cap the number of cylinder-cylinder collision checks. This maximum value and the associated cylinder collision radius are selected conservatively based on the length of the motion primitive assuming the robot starts at hover and achieves a top speed at the endpoint. Third, for each robot the collision checks are performed only with the candidate motion primitive and the associated stopping motion primitive at the first depth of the MCTS tree because each depth of the tree is of a sufficiently long duration (2 s) as compared to the planning time (1 s). The inter-robot collision checker is used in the constrained-bandwidth simulation study (Sect. 3.3).

3 Experimental Design and Results

The experimental evaluation is motivated through a concept of operations for a multi-robot exploration mission in a Martian cave. Two robotic systems explore a Martian cave, transmit their maps to a surface station, which serves as a relay to an orbiter, and the orbiter transmits the data to operators on Earth. Three evaluations are conducted to quantify the system performance through this concept of operations: first, the perceptual fidelity and memory usage of the map is compared to state-of-the-art approaches in a representative cave environment (Sect. 3.1); second, a hardware experiment is demonstrated with two rapidly exploring aerial systems and the communication requirement for each mapping approach is compared (Sect. 3.2); and third, a simulation study is conducted to study the effects of the bandwidth constraints on exploration performance (Sect. 3.3).

To correctly analyze the performance of the simulation study, the bottleneck in data transmission rate is identified and bounds on the rates are determined. In this scenario, data is transferred between the subterranean robot and surface station[1], surface station to orbiter[2], and orbiter to Earth[3]. The bottleneck in communication is between the subterranean robot and surface station when the robot is transmitting at depths between 20–25 m below ground, so the results in Sect. 3.3 are presented for the rates 0.1–0.25 Mbit/s, which are in line with data transmission rates at these depths. Throughout this section the shorthand OG is used to refer to the occupancy grid mapping approach [21] while OM refers to OctoMap [20].

3.1 Perceptual Detail Evaluation

The first evaluation compares the perceptual fidelity of different environment representations in the context of memory usage. An RGB image and point cloud of a crevice in the cave are shown in Figs. 4a and 4b. It is not clear from the image and depth information if the passage continues or there is a lack of data due to insufficient accuracy in the sensor observation. In either case, additional views are required to determine the exact nature of the passage. Figure 4c demonstrates that as the resolution of the OG and OM approaches increases, the memory demands also substantially increase. By comparison, the GMM approach requires substantially less memory. When using the GMM approach, the resulting resampled point cloud is shown in Fig. 4d, where a hole in the data is visible. This approach is compared to OM with varying leaf sizes in Figs. 4e to 4g.

To obtain these results, a GMM was learned consisting of 100 components. Each component requires 10 floating point numbers which includes 6 floating

[1] Whittaker et al. [6] suggest the use of either very low frequency (VLF) radios or magneto-inductive (MI) links to achieve limited data rate through thick layers of rock. The MI links in particular can provide approximately 20–25 m dry soil penetration at channel capacity ranging from 0.1–0.25 Mbit/s when using small antennas (coils) [26]. In the results presented in Sect. 3.3, it is assumed that the robots could be equipped with these MI links.

[2] Orbiters can communicate at approximately 0.208–0.521 Mbit/s with a surface station for 8 min per sol, or Martian day [27].

[3] To transmit from the orbiter to Earth, the communication rate depends on which orbiter is above the lander to relay the data to Earth. The simulation study in Sect. 3.3 assumes the lowest data rate from the Mars Odyssey orbiter, which ranges from 0.128–0.256 Mbit/s [27].

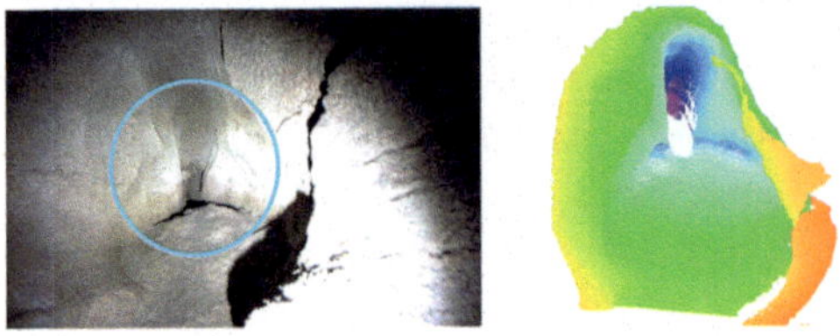

(a) RGB Image (b) Point Cloud

	0.025 m	0.05 m	0.1 m
(bytes)	(bytes)	(bytes)	(bytes)
GMM	**Occupancy Grid (OG)**		
4028	1.3×10^6	1.8×10^5	2.7×10^4
GMM	**OctoMap (OM)**		
4028	2.2×10^5	5.8×10^4	1.4×10^4

(c) Memory

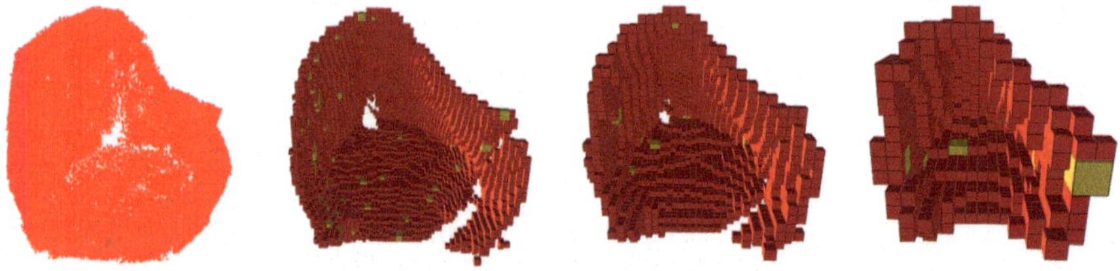

(d) Resampled GMM (e) OM (0.025 m) (f) OM (0.05 m) (g) OM (0.1 m)

Fig. 4. Fidelity and memory usage evaluation of several mapping approaches. (a) and (b) illustrate data from a representative environment the robot may encounter in the cave. A potential passage is circled in cyan. (c) highlights significant reduction in memory usage required by the GMM approach as compared to the OG and OM approaches. (d) Resampled points from the GMM are shown in red. (e)–(g) illustrate the OctoMap representation with leaf sizes varying from 0.025 m to 0.1 m. Leaf voxels are shown in red and larger voxels in yellow.

point numbers to represent the symmetric covariance, 3 floating point numbers for the mean, and 1 floating point number to represent the mixing weight. Additional memory was used to represent the pose via six floating point numbers (three each for translation and rotation) where each floating point number is assumed to be four bytes. A 32-bit unsigned integer (four bytes) is also used to represent the support size of the GMM. In the OG case, 1 floating point number is used to store the logodds value and one unsigned integer (4 bytes) is used to represent the index for each voxel in the change set. The total change set of N voxels is transmitted along with meta-data to reconstruct the grid. The meta-data consists of three unsigned integers to represent the dimensions of the grid in width, height, and length as well as three floating point numbers to represent the origin for a total of 24 bytes. The total data required to represent the sensor observation with an OG is $8N + 24$ bytes. For OM, the full probabilistic model is serialized and stored to disk. The size of the file is reported in the table. The motivation for retaining the logodds values in the OG and OM representations is to enable information-theoretic planning. The advantage of the GMM approach is that the probability of occupancy can be reconstructed at an arbitrary voxel resolution [22,28], which significantly reduces the memory requirements as compared to the OG and OM approaches. The OG and OM approaches must retain the probability of occupancy to enable information-theoretic exploration [24,29].

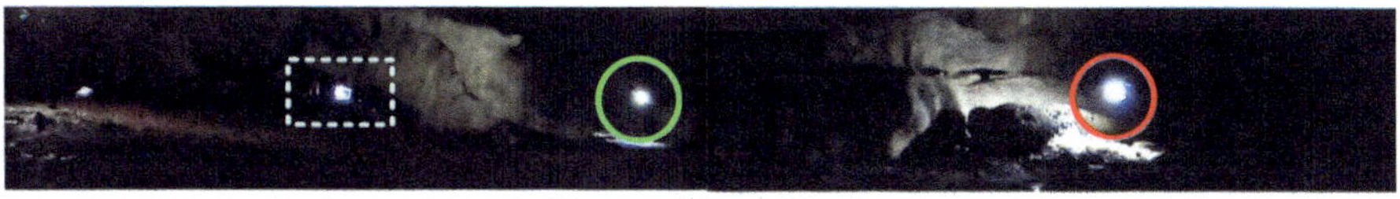

(a) Robots (circled) deployed in a cave. Communication router shown via dotted line.

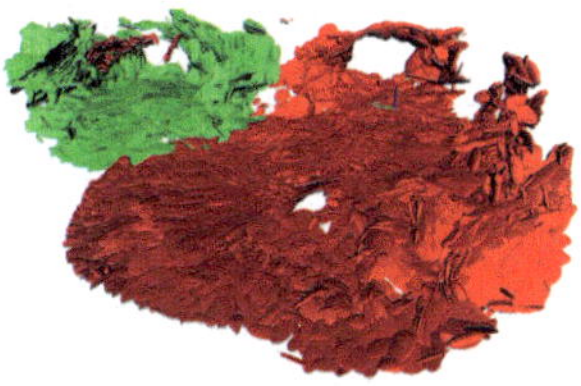

(b) Combined GMM map

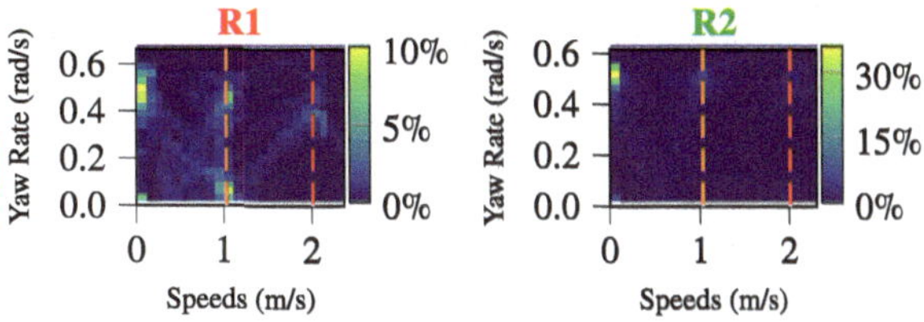

(c) Speed bounds shown by dashed lines.

Completion %:	45%		65%		85%	
Case	Map Size (Mbit)	Time (hours)	Map Size (Mbit)	Time (hours)	Map Size (Mbit)	Time (hours)
GMM	$\mathbf{0.7 \times 10^1}$	$\mathbf{8.0 \times 10^{-3}}$	$\mathbf{1.4 \times 10^1}$	$\mathbf{1.6 \times 10^{-2}}$	$\mathbf{1.9 \times 10^1}$	$\mathbf{2.2 \times 10^{-2}}$
OG (0.1 m)	9.2×10^1	1.0×10^{-1}	1.57×10^2	1.7×10^{-1}	2.0×10^2	2.3×10^{-1}
OG (0.05 m)	5.4×10^2	6.0×10^{-1}	9.1×10^2	0.1×10^1	1.2×10^3	0.1×10^1
OG (0.025 m)	3.9×10^3	0.4×10^1	6.7×10^3	0.7×10^1	8.9×10^3	0.9×10^1
OM (0.1 m)	2.4×10^2	2.6×10^{-1}	3.9×10^2	4.4×10^{-1}	5.2×10^2	5.8×10^{-1}
OM (0.05 m)	1.6×10^3	0.2×10^1	2.6×10^3	0.3×10^1	3.4×10^3	0.4×10^1
OM (0.025 m)	9.8×10^3	1.1×10^1	1.6×10^4	1.8×10^1	2.1×10^4	2.4×10^1

(d) Communication-Efficiency Comparison

Fig. 5. Rapid and communication efficient exploration a cave with a team of two aerial robots. (a) illustrates the environment with the two robots (R1 and R2) and the WiFi router used for communication. (b) illustrates the final GMM maps generated on the base-station. (c) shows the percentage density plots for linear speeds and yaw rates as measured by the visual-inertial navigation system during flight. (d) highlights that the GMM approach requires significantly less memory to represent the combined map as compared to the state of the art approaches. In the context of transmitting this data using a channel with capacity 0.25 Mbit/s, it would take significantly less time for the GMM approach as compared to the other approaches. A video of the flight can be accessed here: https://youtu.be/osko8EKKZUM.

3.2 Hardware Experiments

The second evaluation consists of hardware experiments for two aerial systems exploring the cave. The experiment demonstrates (1) each robot generates informative plans with linear speeds up to 2.37 m/s and yaw rates up to 0.6 rad/s while maintaining safety and (2) the communication required to transmit the map from robots to a base station is substantially less as compared to the OG and OM approaches. For the purposes of this experiment, the robots are deployed in disjoint bounding boxes and the coordination between robots is not studied.

What follows is a description of the experimental setup (including the implementation details) and results.

Each robot in the multi-robot system employs the navigation and control technique outlined in prior work [8]. The robots communicate with other computers on the network via WiFi and use the User Datagram Protocol (UDP) to transfer packets over the network. Before the start of each experiment, the SE(3) transform between the takeoff positions of the robots is measured manually using the navigation approach. The relative initial transform is used by the distributed mapping subsystem to align the GMM map fragments in the frames of other robots to the current robot's local frame.

The maximum speed[4] of the robots in the xy-plane is 2.0 m/s, the maximum speed towards unknown space is 1.0 m/s, the maximum z-direction speed is 0.25 m/s, and the maximum yaw rate is constrained to 0.5 rad/s. One of the metrics used to assess the planning performance is quantifying the maximum speed and yaw rate achieved by the robot while ensuring collision free operation. Both linear and yawing motions are exploratory actions for an aerial robot equipped with a limited field of view depth sensor [8,10]. The data transmitted from the robots to the base station is used to quantify the success of the mapping approach. The GMM results of Fig. 5 are generated in flight during an actual trial in the cave. To enable a fair comparison, the depth images collected from the GMM exploration trial in the cave are post-processed using the OG and OM approaches. This ensures that variation in the other subsystems does not unduly affect the results. An analysis to quantify the memory required for each approach similar to Sect. 3.1 is presented. The OG and OM results are generated by updating the map using the depth information for the current image and publishing the change set. For the OM approach, the change set is serialized to file as the full probabilistic model to enable the base station and other robot to exactly recreate the map for information-theoretic exploration.

The two deployed robots are denoted by **R1** and **R2** in Fig. 5. The robots achieve high exploration rates by selecting actions that enable safe operation at linear speeds up to 2.37 m/s and yaw rates up to 0.6 rad/s, which are of the same order as state-of-the-art fast exploration works[5] [10–12]. Moreover, note that since **R1** operates in a relatively open space compared to **R2**, a larger percentage of high speed actions are selected (Fig. 5c). In contrast, the planner selects the yawing motion and slow linear actions towards frontiers more often for **R2** to allow for safe operation in a constrained space (Fig. 5c). Both of these behaviors in the multi-robot system arise automatically due to the choice of the action representation for single-robot planning in [10]. These behaviors show that the same action representation can be used on every robot in the team without

[4] The speed limits and the operational volumes were chosen based on the cave passage dimensions. The authors worked with cave management to select a test site that contained neither actively growing speleothems or bats. Possible effects of imperfect trajectory tracking and state estimation were also taken into account.

[5] The attained speeds exceed the limits slightly due to imperfect trajectory tracking and state estimation.

any change in parameters and still allow for intelligent speed adaptation for rapid and safe exploration.

The combined map from R1 and R2 requires significantly less time to transmit under the bandwidth constraint when measuring at various points during exploration (Fig. 5d). An implication of this in the context of the concept of operations is that at 100% exploration completion it will take about 104.40 s to transmit the GMM map, 12.30 h to transmit the 0.025 m resolution OG map, and 1.25 days to transmit the 0.025 m resolution OM map to Earth. It is important to note why the OM approach requires more memory than the OG approach for this result while it required less memory than the OG approach in Fig. 5c. The change set must be encoded as an OctoMap before serializing to file. The approach presented by Hornung et al. [20] requires that the spatial relationships between nodes be implicitly stored in the encoding. This means that the serialized stream does not contain any 3D coordinates and additional data must be stored to preserve the structure of the octree. This is in contrast to the OG approach that stores a logodds value and index from which 3D coordinates can be recovered. Therefore, for small change sets, the OM approach has much higher overhead than the OG approach.

3.3 Effects of Constrained Communications

For this study the assumption on the robots operating in disjoint spaces is relaxed and a priority-based inter-robot collision checker is implemented for shared space operation. The simulation consists of a two-robot team that explores the cave environment. Two approaches are tested: GMM and OG. The OM approach is not compared for this experiment because to the best of our knowledge there is no existing open-source implementation of the Shannon mutual information used for planning by Zhang et al. [29]. Further, this enables us to retain the same planning subsystem for a fair comparison of the GMM and OG approaches. The communication rate is varied among 0.1 Mbit/s, 0.25 Mbit/s, and unconstrained. Each configuration is tested in 40 experiments with a 700 s duration. The duration of the exploration is chosen based on the top speed of the robots and the spatial dimensions of the environment. The exploration software is run on separate computers in a distributed fashion over a wired connection. The simulations are run on two desktop computers running Ubuntu 18.04 with Intel i7-6700K CPUs. One computer has 32 GB RAM and the other has 16 GB of RAM. For the wired connection, the data rate is limited via the network traffic control tool in Linux that uses the Token Bucket Filter (TBF) to maintain the specified rate value [30]. Figure 6 illustrates the results from the simulation study. As the communication bandwidth is reduced from no limit in Fig. 6a to 0.25 Mbit/s the OG approach begins to drop packets and the exploration performance of the multi-robot approach decreases as compared to the GMM approach (see Fig. 6d). At this rate, the GMM approach achieves 85% environment coverage in less than 80% of the time that it takes the OG approach. However, as the communication rate decreases further to 0.1 Mbit/s the GMM approach also suffers though it is able to outperform the OG approach.

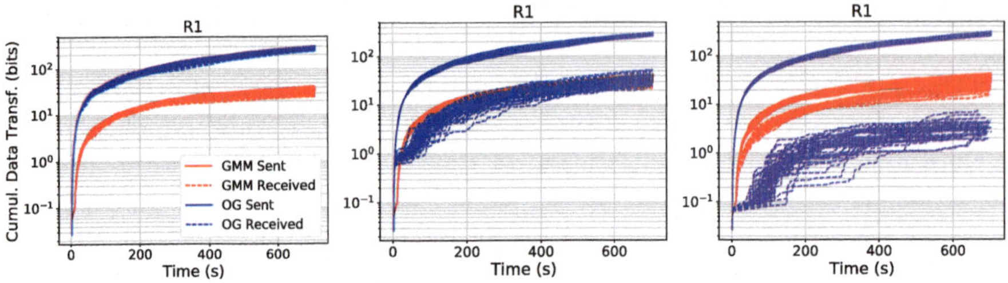

(a) No limit (b) 0.25 Mbit/s (c) 0.1 Mbit/s

Completion %:	45%			65%			85%		
Comm. Limit	GMM (s)	OG (s)	Δ (%)	GMM (s)	OG (s)	Δ (%)	GMM (s)	OG (s)	Δ (%)
No limit	80.19	81.51	1.62	130.73	131.98	0.95	225.8	237.51	4.93
0.25 Mbit/s	79.91	92.38	**13.5**	129.1	160.15	**19.39**	214.86	282.11	**23.84**
0.1 Mbit/s	86.51	93.43	7.41	142.95	165.88	13.82	247.83	270.15	8.27

(d) Exploration completion times

Fig. 6. Variation of exploration performance with inter-robot communication limits. (a), (b), and (c) plot the cumulative map data sent and received for the GMM and OG approaches under different data rate constraints (the plots are shown for R1 only for brevity). The received data is impacted significantly for the OG approach at 0.25 Mbit/s while both approaches are affected at 0.1 Mbit/s. Note that in all experiments the planning and coordination methodology is kept the same for a fair comparison. (d) compares the time to achieve a certain percentage of environment coverage. We observe that at the 0.25 Mbit/s constraint, the GMM approach improves the performance of the team by up to 23.84%.

4 Conclusion and Future Work

This work leveraged the compactness of Gaussian mixture models for high-fidelity perceptual modeling to increases the rate of multi-robot exploration in reduced bandwidth scenarios such as autonomy in caves. The mapping approach enables retention of environment details while remaining amenable to low-bandwidth transmission. The advantage of this mapping strategy is that it enables a substantial increase in exploration rate of the multi-robot team as compared to state-of-the-art mapping techniques even as the communication bandwidth of the connection between robots decreases. Future work will improve perceptual detail in the environment and develop hierarchical strategies that adapt the fidelity of the model based on the sensor data. Multi-modal mapping (for example, thermal, RGB, etc.) may also be beneficial in these scenarios. Finally, coordination strategies can be developed to enable robots to share communication-efficient policies and improve the rate of exploration.

References

1. Board, S.S., Council, N.R., et al.: Vision and Voyages for Planetary Science in the Decade 2013–2022. National Academies Press, Washington (2012)
2. Maimone, M., Cheng, Y., Matthies, L.: Two years of visual odometry on the mars exploration rovers. J. Field Robot. **24**(3), 169–186 (2007)
3. Phillips-Lander, C., et al.: Macie: Mars astrobiological caves and internal habitability explorer (a new frontiers mission concept). LPI **2326**, 1142 (2020)
4. Stamenkovic, V., et al.: The next frontier for planetary and human exploration. Nature Astron. **3**(2), 116–120 (2019)
5. Miranda, D.: 2020 nasa technology taxonomy (2020)
6. Whittaker, R., et al.: Exploration of planetary skylights and tunnels (2014)
7. Cortés, J., Egerstedt, M.: Coordinated control of multi-robot systems: a survey. SICE J. Control, Meas. Syst. Integr. **10**(6), 495–503 (2017)
8. Tabib, W., Goel, K., Yao, J., Boirum, C., Michael, N.: Autonomous cave surveying with an aerial robot. arXiv preprint arXiv:200313883 (2020)
9. Cieslewski, T., Kaufmann, E., Scaramuzza, D.: Rapid exploration with multirotors: a frontier selection method for high speed flight. In: Proc. of the IEEE/RSJ International Conference on Intelligent Robots and System Vancouver, Canada (2017)
10. Goel, K., Corah, M., Boirum, C., Michael, N.: Fast exploration using multirotors: analysis, planning, and experimentation. In: Conference on Field and Service Robot., Tokyo, Japan (2019)
11. Dai, A., Papatheodorou, S., Funk, N., Tzoumanikas, D., Leutenegger, S.: Fast frontier-based information-driven autonomous exploration with an mav. arXiv preprint arXiv:200204440 (2020)
12. Dharmadhikari, M., et al.: Motion primitives-based path planning for fast and agile exploration using aerial robots. In: 2020 IEEE International Conference on Robotics and Automation (ICRA), pp. 179–185. IEEE (2020)
13. Cesare, K., Skeele R., Yoo, S.H., Zhang, Y., Hollinger, G.: Multi-UAV exploration with limited communication and battery. In: 2015 IEEE International Conference on Robotics and Automation (ICRA), pp. 2230–2235. IEEE (2015)
14. Corah, M., O'Meadhra, C., Goel, K., Michael, N.: Communication-efficient planning and mapping for multi-robot exploration in large environments. IEEE Robot. Autom. Lett. **4**(2), 1715–1721 (2019). https://doi.org/10.1109/LRA.2019.2897368
15. Matthies, L.: Niac phase 1 final study report on titan aerial daughtercraft (2017)
16. DARPA Darpa subterranean challenge. https://www.subtchallenge.com/ (2020)
17. Ebadi, K., et al.: Lamp: Large-scale autonomous mapping and positioning for exploration of perceptually-degraded subterranean environments. arXiv preprint arXiv:200301744 (2020)
18. Dang, T., Tranzatto, M., Khattak, S., Mascarich, F., Alexis, K., Hutter, M.: Graph-based subterranean exploration path planning using aerial and legged robots. J. Field Robot. **37**(8), 1363–1388 (2020)
19. Rouček, T., Pecka, M., Čížek, P., Petříček, T., Bayer, J., Šalanskỳ, V., Heřt, D., Petrlík, M., Báča, T., Spurnỳ, V., et al.: Darpa subterranean challenge: Multi-robotic exploration of underground environments. In: International Conference on Modelling and Simulation for Autonomous Systems, Springer, Cham, pp. 274–290 (2019)
20. Hornung, A., Wurm, K.M., Bennewitz, M., Stachniss, C., Burgard, W.: Octomap: an efficient probabilistic 3d mapping framework based on octrees. Auton. Robots **34**(3), 189–206 (2013)

21. Elfes, A.: Using occupancy grids for mobile robot perception and navigation. IEEE Comput. Soc. **22**(6), 46–57 (1989)
22. Tabib, W., Goel, K., Yao, J., Dabhi, M., Boirum, C., Michael, N.: Real-time information-theoretic exploration with gaussian mixture model maps. In: Proc. of Robotics: Science and Systems FreiburgimBreisgau, Germany (2019) https://doi.org/10.15607/RSS.2019.XV.061
23. Chaslot, G.: Monte-Carlo tree search. PhD thesis, Universiteit Maastricht (2010)
24. Charrow, B., Liu, S., Kumar, V., Michael, N.: Information-theoretic mapping using Cauchy-Schwarz quadratic mutual information. In: Proceedings of the IEEE International Conference on Robotics and Automation, Seattle, WA (2015)
25. Cai, C., Yang, C., Zhu, Q., Liang, Y.: Collision avoidance in multi-robot systems. In: 2007 International Conference on Mechatronics and Automation, pp. 2795–2800. IEEE (2007)
26. Kisseleff, S., Akyildiz, I.F., Gerstacker, W.H.: Survey on advances in magnetic induction-based wireless underground sensor networks. IEEE Int. Things J. **5**(6), 4843–4856 (2018)
27. NASA Mars science laboratory data rates/returns (2019). https://marsmobile.jpl.nasa.gov/msl/mission/communicationwithearth/data/
28. O'Meadhra, C., Tabib, W., Michael, N.: Variable resolution occupancy mapping using gaussian mixture models. IEEE Robot. Autom. Lett. **4**(2), 2015–2022 (2019). https://doi.org/10.1109/LRA.2018.2889348
29. Zhang, Z., Henderson, T., Karaman, S., Sze, V.: Fsmi: fast computation of shannon mutual information for information-theoretic mapping. Int. J. Robot. Res. **39**(9), 1155–1177 (2020)
30. Hubert, B.: Linux advanced routing & traffic control. In: Ottawa Linux Symposium, sn, vol. 213 (2002)

Nonlinear Model Predictive Control for Formations of Multi-Rotor Micro Aerial Vehicles: An Experimental Approach

I. K. Erunsal[1,2(✉)], R. Ventura[2], and A. Martinoli[1]

[1] Distributed Intelligent Systems and Algorithms Laboratory, School of Architecture, Civil and Environmental Engineering, École Polytechnique Fédérale de Lausanne, Lausanne, Switzerland
{kagan.erunsal,alcherio.martinoli}@epfl.ch

[2] Institute for Systems and Robotics, Instituto Superior Técnico, Lisbon, Portugal
rodrigo.ventura@isr.tecnico.ulisboa.pt

Abstract. In a world where the complexity and performance requirements of the tasks requested from micro aerial vehicles are continuously increasing, smooth design and deployment of multi-robot systems are gaining more significance. This paper tackles such challenging requirements by firmly adopting an architecture based on Nonlinear Model Predictive Control (NMPC). In order to efficiently design such architecture, we propose an approach emphasizing a closure of the reality gap between algorithmic design and physical experiments. More specifically, we use canonical system identification methods combined with additional calibration effort to enhance the faithfulness of our model in a high-fidelity simulation environment. By employing the accurate model obtained, we prototype our decentralized NMPC algorithm in a real-time iteration scheme. To improve further the performance, multi-modal, multi-rate, decentralized extended Kalman filters are integrated to the architecture. While experiments involving up to three quadrotors in high-fidelity simulation and reality outlined the approach's validity, they also pointed out its limitations when subtle effects generated by aerodynamic interactions among quadrotors are not taken into account in the control design.

Keywords: Formation control · Multi-rotor micro aerial vehicles · Nonlinear model predictive control · Real-time iteration

1 Introduction

The last decade has been characterized by an increasing research effort on the control and coordination strategies for Micro Aerial Vehicles (MAVs) performing challenging missions such as scientific exploration, remote sensing, construction and search and rescue [1,2]. Among the main design blocks of multi-MAV coordination strategies, formation control has a crucial role and it has been widely studied [3]. One of the promising methods to perform this task effectively

B. Siciliano et al. (Eds.): ISER 2020, SPAR 19, pp. 449–461, 2021.
https://doi.org/10.1007/978-3-030-71151-1_40

leverages Nonlinear Model Predictive Control (NMPC), due to its architectural flexibility and ability to simultaneously satisfy the performance requirements and constraints of complex nonlinear systems [4]. Furthermore, if the number of robots in the system is large, the decentralized or distributed versions of NMPC (D-NMPC) are more suitable to subdivide the problem into decoupled pieces and reduce the interaction requirements [12]. In addition, in order to promote further autonomy, several researchers have investigated the formation control relying exclusively on relative localization and attempted to solve its limitations [15]. However, not only because of the implementation complexity but also the safety concerns related to multi-MAV systems, it is currently very time-consuming and challenging to iteratively prototype and validate any control and coordination strategy in reality. One of the techniques that makes such iterative prototyping framework more efficient is to bridge physical experiments and algorithmic design, typically carried out with mathematical formalism and highly abstracted simulations (e.g., bodiless agents implemented in Matlab) with an embodied, high-fidelity simulation environment. However, choosing an appropriate high-fidelity simulator is not enough: additional calibration efforts are typically needed to increase the faithfulness of the simulation. This paper studies the effectiveness of such framework in the design and validation of decentralized NMPC-based control architectures for a multi-MAV system. As a concrete case study for our approach, we have chosen a leader-follower formation control architecture due to its implementation simplicity and considered three MAVs in total. While the leader is responsible for steering the formation to a desired point, the followers' duty is to maintain the formation as depicted in Fig. 1. There are various successful implementations of MPC-based strategies for a single MAV performing trajectory tracking [5], robust trajectory following [6] and perception-aware motion generation [7]. Furthermore, there exist different successful validations of multi-robot formation control concerned with the formations in motion [8], encirclement [9], rendezvous [10], outdoor flocking [18] and formation maintenance [19]. However, although some of them introduce basic system identification techniques, none of them present a complete framework for closing the reality gap, including system identification, calibration of simulation and validation of control and estimation algorithms in reality. Additionally, none of them focuses on the intersection between D-NMPC and formation control of MAVs leveraging exclusively relative localization systems.

2 Technical Approach

Any MPC-based strategy requires a realistic model of the system dynamics to be controlled. To this purpose, we adopted the approach proposed in [5] given the fact that the hardware and software setup is very similar to ours. Note that, for our vehicle model, the same notation as in [11] is used for the vectors, coordinate frames and rotation matrices. However, there are a few main differences as well. First, the state $\boldsymbol{s}$ is defined as in Eq. (1):

$$\boldsymbol{s} = [\boldsymbol{x}_{b/n}^{n}{}^{T}\ \boldsymbol{v}_{b/n}^{n}{}^{T}\ \boldsymbol{t}_{b/n}^{n}{}^{T}]^{T} \tag{1}$$

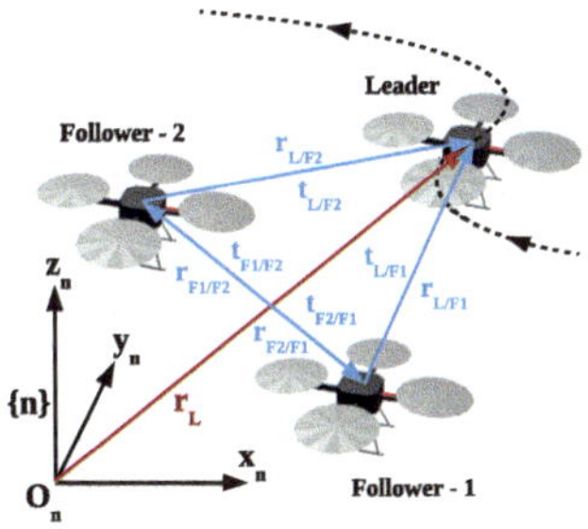

Fig. 1. Formation control problem with one leader and two followers with the inter-vehicle positions and trajectory followed

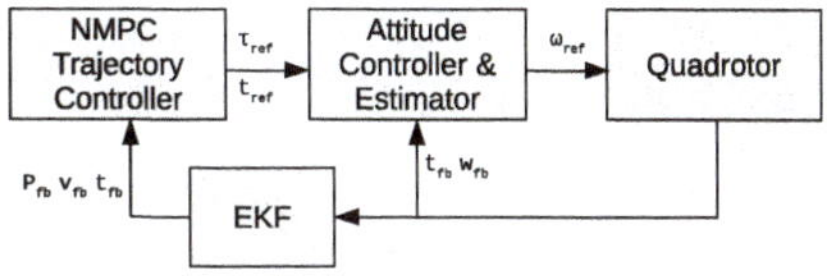

Fig. 2. Cascaded control/estimation architecture for the individual quadrotor

where $\boldsymbol{x}^n_{b/n}$ is the absolute position, $\boldsymbol{v}^n_{b/n}$ is the velocity and $\boldsymbol{t}^n_{b/n}$ is the attitude of the quadrotor in the body-fixed frame $\{b\}$ with respect to the inertial frame $\{n\}$ (lower index) expressed in the same frame $\{n\}$ (upper index). Second, the model now includes attitude dynamics as three first-order closed-loop system equations. The complete model, which is valid for the leader, is given in Eqs. (2)–(4).

$$\dot{\boldsymbol{x}}^n_{b/n} = \boldsymbol{v}^n_{b/n} \tag{2}$$

$$m_b \dot{\boldsymbol{v}}^n_{b/n} = m_b \boldsymbol{g} + \boldsymbol{R}^n_b \boldsymbol{F}^b_{b/n} \tag{3}$$

$$\dot{\boldsymbol{t}}^n_{b/n,i} = \frac{1}{\tau_i}(k_i \boldsymbol{t}_{ref,i} - \boldsymbol{t}^n_{b/n,i}) \quad \text{for} \quad i = 1, 2, 3 \tag{4}$$

where m_b is the mass, $\boldsymbol{g}$ is the gravitational acceleration, $\boldsymbol{R}^n_b$ is the rotation matrix, $\boldsymbol{F}^b_{b/n}$ is the applied body forces, τ_i's are the time constants and k_i's are the gains of the first order dynamical model for roll, pitch and yaw angles. In order to predict the system evolution in NMPC, the parameters to be identified are the mass m_b, gains k_i, time constants τ_i and the conversion parameters from required thrust to attitude commands (throttle) as a function of battery voltage. Furthermore, in order to obtain an accurate simulation of the closed-loop system, the following characteristics should also be acquired: the torque constant T_q and thrust constant T_k of the propellers, the inertia I_b of the quadrotor, the lumped Position-Derivative (PD) attitude controller's parameters K_p and K_d for each Euler angle to achieve the identified attitude response. The corresponding experiments to obtain these parameters will be explained in Sect. 3.

Once the vehicle model has been selected and calibrated, the optimization problem to be tackled with NMPC can be generated for the trajectory tracking quadrotors. The problem is formulated in Eq. (5), for additional details, refer to [12]. The controller structure is designed in a cascaded way as shown in Fig. 2, inspired by [5]. This allows us to separate a high(er) frequency task (attitude control) from a low(er) frequency one (trajectory control).

For this problem, the outputs of NMPC, $\boldsymbol{u}$, are the reference thrust, roll and pitch angles. In addition, the cost functions $\boldsymbol{c_k}$ and $\boldsymbol{c_N}$ minimize the trajectory errors, non-gravitational desired thrust, i.e. the thrust spent on maneuvers, roll and pitch angles and their switching rates based on the selected vehicle model f_k. The stage inequality constraints h_k include the input and state bounds corresponding to the actuation limits and comfort constraints. In addition, the terminal constraint g_N is selected as a state set by considering the stability analysis.

$$
\begin{aligned}
\underset{\substack{\boldsymbol{x}(1),\ldots,\boldsymbol{x}(N) \\ \boldsymbol{u}(0),\ldots\boldsymbol{u}(N-1)}}{\text{minimize}} \quad & \sum_{k=0}^{N-1} c_k(\boldsymbol{x}(k), \boldsymbol{u}(k), \boldsymbol{r}(k)) + c_N(\boldsymbol{x}(N), \boldsymbol{u}(N), \boldsymbol{r}(N)) \\
\text{subject to} \quad & \boldsymbol{x}(k+1) = f_k(\boldsymbol{x}(k), \boldsymbol{u}(k)), \quad k = 0, .., N-1 \\
& h_k(\boldsymbol{x}(k), \boldsymbol{u}(k)) \leq 0, \quad k = 0, .., N-1 \\
& g_N(\boldsymbol{x}(N)) \leq 0
\end{aligned}
\tag{5}
$$

To conduct the experiments accurately, a multi-modal, multi-rate Extended Kalman Filter (EKF) has been designed to filter out the noise on the perception. This filter is multi-modal because different dynamics are valid while flying and landed; it is multi-rate because the sampling rates of the various sensors used for navigation are different. The parameters of this filter are tuned based on the relative magnitudes of the measured process and sensor noise levels.

Considering the designed setup, various trajectory tracking experiments are carried out by generating trajectories between successive way points. These experiments will be presented in detail in Sect. 3.

Once the vehicle model and simulation are calibrated, the prototyping framework is ready for the multi-robot experiments. As introduced in Sect. 1, the leader-follower formation of multiple quadrotors is controlled by fully decentralized NMPC with the help of multi-modal, multi-rate local EKF. Since the architecture does not involve any inter-vehicle communication, the role of EKF is to filter out the noise in the relative sensing measurements and to estimate the linear and yaw velocity of the neighbor vehicles accurately. Note that the controller of the followers has also a cascaded structure as that of the leader. However, the followers include not only their own system dynamics but also the constant-velocity dynamics of all their neighbors in the formation. This assumption is valid if the acceleration of the vehicles is limited to low values, which is the case, for instance, during indoor flights. Another important point is that all states are now written with respect to a stationary, inertial frame, which we call MPC inertial frame $\{m\}$, where we assume that the initial pose of each vehicle corresponds to the zero vector at the beginning of the motion. This frame also facilitates the transfer of variables from estimator to controller since the EKF needs an inertial frame. In addition, two more frames needed to be included in order to write the model and controlled variables completely. The first one is the MPC body frame $\{d\}$, which is anchored to the body throughout the MPC horizon, and the second is the MPC control frame $\{c\}$, whose orientation is roll- and pitch-free while still being anchored to the body frame $\{d\}$. The latter frame

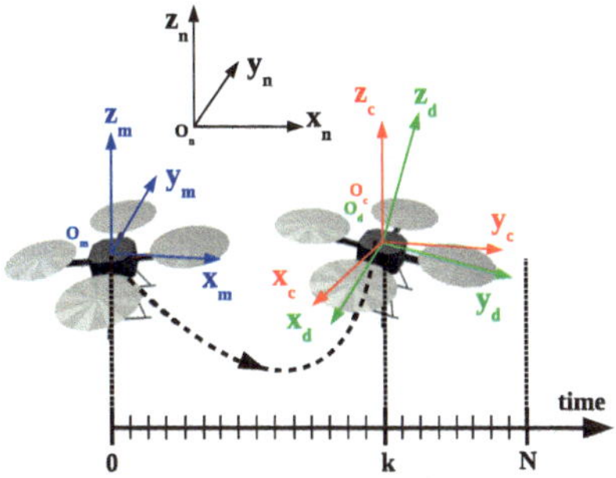

Fig. 3. MPC coordinate frames in a horizon of length N: $\{m\}$ (blue), $\{d\}$ (green), $\{c\}$ (red).

Fig. 4. Helipal Storm Drone-4 v3

is considered due to the fact that roll and pitch angles of a quadrotor cannot be controlled directly when performing 3D formation control, consequently, the control variables (relative positions) are defined here. All frames are represented in Fig. 3. The described model for the followers is given in Eqs. (6)–(11).

$$\dot{\boldsymbol{v}}^m_{d/m} = \frac{\boldsymbol{R}^m_d}{m_b}\boldsymbol{F}^d_{d/m} + \boldsymbol{g} \tag{6}$$

$$\dot{\boldsymbol{t}}^m_{d/m,i} = \frac{1}{\tau_i}(k_i \boldsymbol{t}_{ref,i} - \boldsymbol{t}^m_{d/m,i}) \quad \text{for} \quad i = 1, 2, 3 \tag{7}$$

$$\dot{\Delta\boldsymbol{x}}^m_{d/m,j} = \boldsymbol{v}^m_{d/m,j} - \boldsymbol{v}^m_{d/m} \quad \text{for} \quad j = 1, ..., N_{nh} \tag{8}$$

$$\dot{\boldsymbol{v}}^m_{d/m,j} = \boldsymbol{0} \quad \text{for} \quad j = 1, ..., N_{nh} \tag{9}$$

$$\dot{\psi}_j = w_{z,j} \quad \text{for} \quad j = 1, ..., N_{nh} \tag{10}$$

$$\dot{w}_{z,j} = 0 \quad \text{for} \quad j = 1, ..., N_{nh} \tag{11}$$

where Δx, v_j, ψ_j, w_j and N_{nh} are the relative positions, absolute velocity, yaw angle and yaw rate of the neighbor vehicles expressed in $\{m\}$ and the number of neighbors in the formation, respectively. For the definitions of variables in Eqs. (6) and (7), please refer to Eqs. (3) and (4). Note that, for formation control, the cost function, has a similar structure as Eq. (5), as it minimizes the relative position and velocity errors, non-gravitational desired thrust, roll and pitch angles. Additionally, the inequality constraints include the state and input bounds according to the actuation and safety limits. Note that, the relative positions are included in the state vector and can therefore be bounded in order to obtain safe formations.

Here, it is worth mentioning the solution strategy used for NMPC. We chose a strategy based on Real-Time Iteration (RTI), as it is arguably the most successful and widely exploited approach to efficiently solve NMPC problems in real-time [20]. To apply this method, initially, a discretization is performed on the system model, constraints and cost function in order to obtain a structured Non-Linear

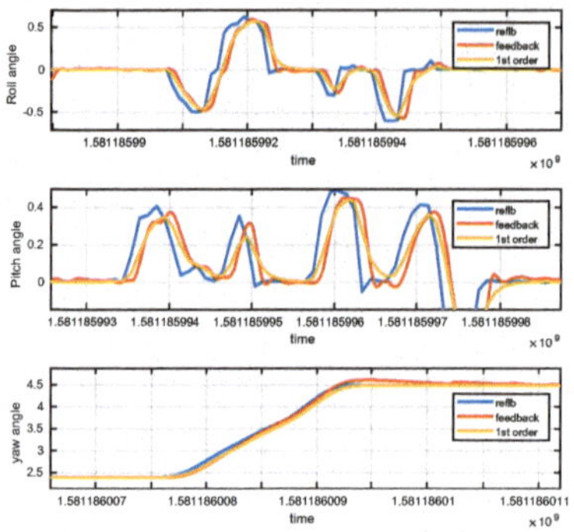

Fig. 5. Attitude subsystem identification results: reference, feedback and first order fitted model

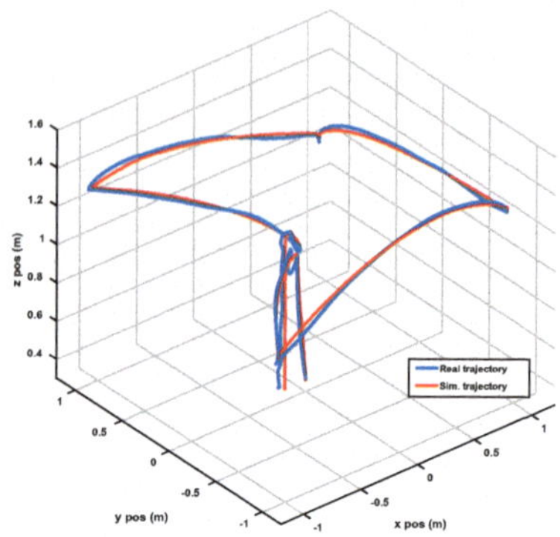

Fig. 6. Trajectory comparison between reality and simulation

Algorithm 1. RTI scheme

$X^g \leftarrow x^g$ {Initial guess of evolution of states}
$U^g \leftarrow u^g$ {Initial guess of evolution of inputs}
$X_0 \leftarrow x_0$ {Feedback obtained, initial condition}
$P_s \leftarrow p_s$ {Parameters and online data}
while $(t == 0 \; || \; t_{CPU} \leq t_s)$ && $KKT \geq threshold$ **do**
    {Initially iterate until obtain a low KKT result, for the next iterations check CPU time, control sample time and KKT to finish}
    $(X^*, U^*) \leftarrow SQP\_step(X^g, U^g, X_0, P_s)$ {Obtain optimal states and inputs by solving QP}
    if (X^*, U^*) is infeasible || solver_error **then**
        $sflag \leftarrow true$ {Solver error flag}
        break
    end if
    $(X^g, U^g) \leftarrow (X^*, U^*)$ {Warm start for next sub-iteration}
end while
if $sflag == true$ **then**
    Apply $U^g(1)$ {Apply the solution found in previous iteration for this time step}
    $(X^g, U^g) \leftarrow X^g_s, U^g_s$ {Warm start for next iteration: time-shifted version of the previous guess}
else
    Apply $U^*(0)$ {Apply the first input of optimal solution}
    $(X^g, U^g) \leftarrow X^*_s, U^*_s$ {Warm start for next iteration: time-shifted version of the optimal solution}
end if

Program (NLP). Next, the NLP is sequentially approximated by Quadratic Programs (QPs) using linearization techniques. After this step, a condensing method is applied to the QP to take advantage of fast condense QP solvers. Generally, a warm initialization is leveraged considering the fact that the optimal solution to the current iteration is very similar to the previous one. At this stage, a linear algebra solver exploits the QP to obtain a Newton direction for the solution. The resulting output is selected as an initial guess to the next QP and this procedure continues until the convergence is achieved. The optimality of the solution is measured by the Karush-Kuhn-Tucker (KKT) value. The details of the RTI scheme which is repeated for each control iteration and tailored for this problem are explained in Algorithm 1.

3 Experiments and Results

The quadrotors employed in the experiments are Helipal Storm Drone-4 v3 endowed with a Pixhawk Cube-2 autopilot, a Raspberry PI 3B onboard computer, an IMU, an optical flow sensor and a external digital compass as shown in Fig. 4. The 3D position information is generated by a Motion Capture System (MCS) and the optic-flow velocity is emulated by the same system due to its reliability. The employed MCS allows for millimeter accuracy in the measurement of the vehicles pose with an update rate 60 Hz. All computations are carried out onboard by leveraging the Robot Operating System (ROS). As solver, the nonlinear OCP solver ACADO [21] is used with the active set QP solver qpOASES [22]. A control sample time of 0.05 s and a horizon length of 15 are chosen, resulting in a 0.75 s of total prediction horizon.

3.1 System Identification

The propulsion subsystem of the drone (T_q, T_k and Thrust-Throttle-Voltage characteristics) is identified by using a RC Benchmark 1585 propeller setup [16]. Next, the mass is simply measured by a digital scale and the inertia is estimated by adopting a double-arm pendulum setup designed according to [13]. Furthermore, the attitude subsystem parameters are found based on the closed-loop position tests by fitting first-order responses to attitude references. Figure 5 shows sample results of such system identification campaign. By leveraging this information, the lumped PD parameters of the autopilot are characterized and integrated into the simulation. Finally, all noise levels of sensors are measured in steady state and introduced into the simulation. The results of the parameter identification procedure are summarized in Table 1. Note that the full parameter identification is carried out only for one quadrotor since the other ones employ exactly the same equipment (e.g., frame, motors, payload etc.). Only the mass is re-measured for the other drones. Although we are aware of the fact that there might be slight manufacturing variations among the vehicles, we observed that our NMPC architecture is not sensitive to them.

Table 1. Results of parameter identification

Parameter	Value	Unit
Mass, m	1.37	kg
Inertia, I_{xx}, I_{yy}, I_{zz}	0.0123, 0.0136, 0.0107	kgm^2
Thrust constant, T_k	$9.803 * 10^{-8}$	N/rpm^2
Torque constant, T_q	$1.507 * 10^{-9}$	Nm/rpm^2
Thrust-Throttle-Voltage characteristics	3rd order polynomial	-
First order time constants, τ_i, i=1,2,3	0.1203, 0.1303, 0.0651	-
First order gains, k_i, i=1,2,3	1.01, 1.02, 0.99	-
Lumped proportional gains, $K_{p,i}$, i=1,2,3	3.0, 3.0, 0.5	-
Lumped derivative gains, $K_{p,i}$, i=1,2,3	0.4, 0.4, 0.4	-

3.2 Calibration of Simulation

With the fully identified vehicle model at hand, the trajectory tracking of the drone is achieved by generating six way points in a 5 m^3 volume and operating the decentralized NMPC-EKF architecture implemented in ROS. Data are collected and compared with the simulation performed in Webots [14], an open-source high-fidelity robotics simulator. By substituting the Euler angle biases with data measured on a real quadrotor and manually fine-tuning the thrust bias, good levels of trajectory matching between reality and simulation can be obtained, as illustrated in Fig. 6. The average error between the simulated and real closed-loop trajectories is about 0.06 m. In addition, the control inputs, i.e. thrust, roll and pitch references, for the real and simulated quadrotors are depicted in Fig. 7 for the sake of completeness. As can be observed, they are quite synchronous. The accompanying video further illustrates the faithfulness of the simulation.

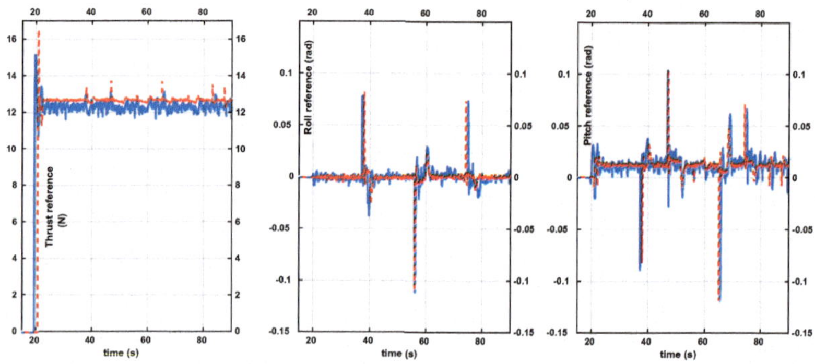

Fig. 7. Thrust, roll and pitch references sent to autopilot, generated by NMPC. Blue-solid: real data, red-dashed: simulated data

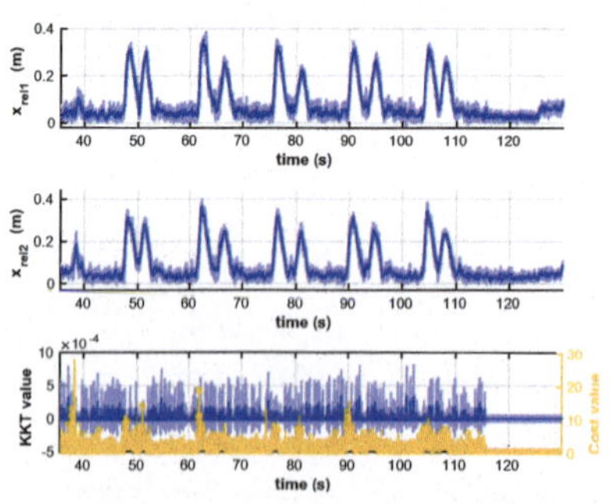

Fig. 8. Formation control results for multiple runs

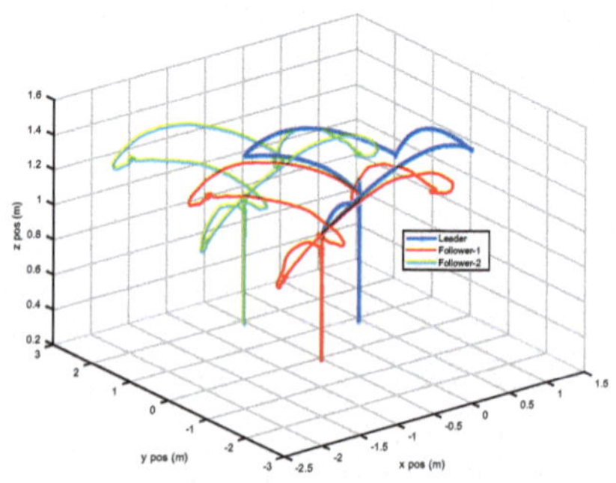

Fig. 9. Leader and follower trajectories in simulation

3.3 Multi-robot Formation Control in Simulation

We have carried out various formation control experiments in Webots, all involving one leader and two followers, as illustrated in Fig. 11. Note that in these simulations, 3D relative position and orientation information is generated by a 3D relative range and bearing sensor emulating the real system proposed in [17]. The trajectories in one experiment are shown in Fig. 9: the leader follows a prescribed trajectory and followers try to maintain a triangular formation by only relative sensing and estimation. The formation control errors, i.e. follower-leader and follower-follower, together with the solver's quality outputs, i.e. the cost and KKT values, over ten runs are shown in Fig. 8 for one of the followers. As can be seen, the average steady-state norm-error is around 0.05 m while the solver always converges. The quality of the formation control is also visualized in the enclosed video leveraging the calibrated high-fidelity simulation tool.

3.4 Multi-robot Formation Control in Reality

Two drones, i.e. one leader and one follower, are employed in order to validate the concept. This setup is illustrated in Fig. 10. Two main experiments are designed and executed in order to clearly distinguish the effects of aerodynamic formation disturbance: formation control with a manually translated leader and with an autonomously flying leader. They will be elaborated in the following sections.

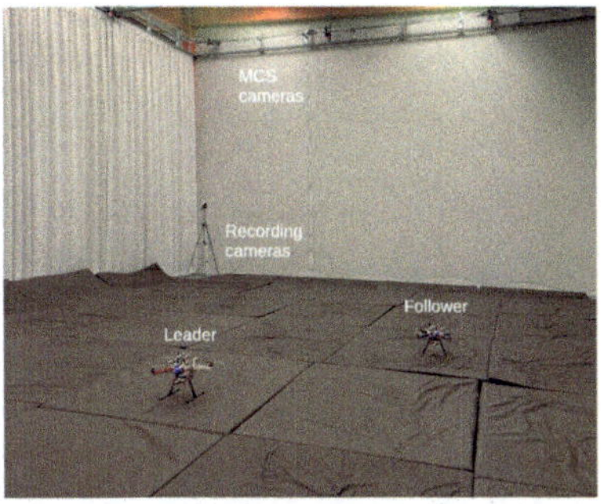

Fig. 10. Flight arena with one leader and one follower vehicles

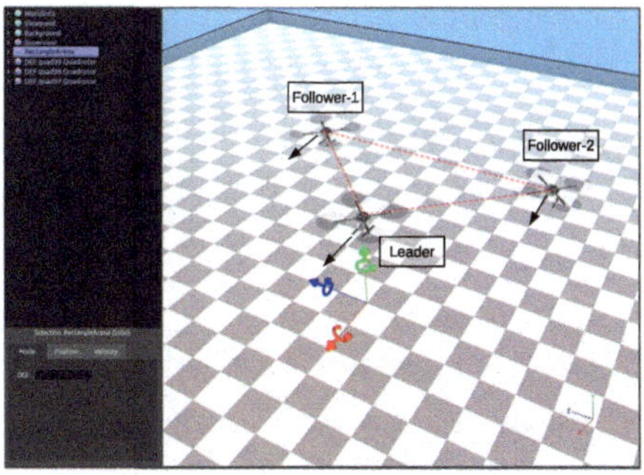

Fig. 11. Webots simulator with one leader and two follower vehicles

3.4.1 Formation Control with Manually Translated Leader

For this test, the leader is initially placed at a take-off height of 1 m. Since the follower's formation control algorithm is agnostic to absolute position sensing, the take-off is performed by a separate PID controller. Then, a switching algorithm accomplishes a smooth transition to the NMPC and the follower vehicle engages to the leader. The engagement relative position is selected as [2.5 –2.5 0] m. Next, the leader is manually and randomly moved in a confined space of 1.5 m^3. The tests

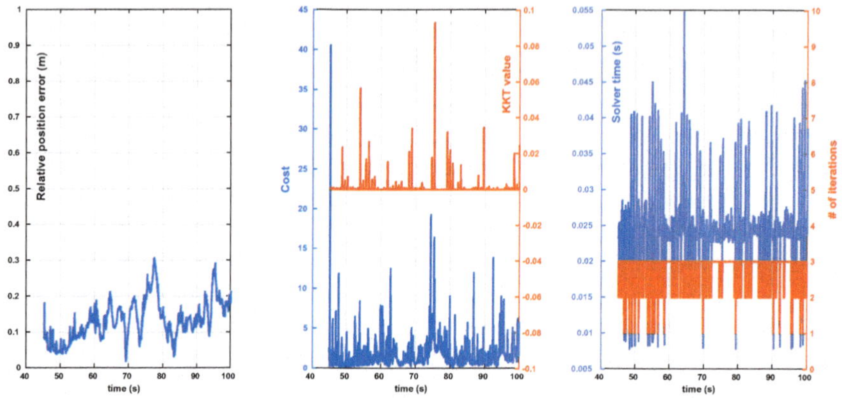

Fig. 12. Performance of formation with manually translated leader

are conducted three times and no significant differences among them are observed. The result of one of the trails is shown in Fig. 12. Note that the enclosed video also demonstrates this experiment. As can be observed from the left-most plot, the average error in 3D is about 0.14 m. In transition, this error can go up to 0.3 m; however, convergence is achieved when the leader stops and the average stationary error of 0.11 m is maintained. Concerning the solution quality, the KKT value is always lower than 0.1, and except occasional peaks, it stays smaller than 0.01. For real experiments with high dynamics such as that of a multi-MAV system, these values are acceptable. It is also clear that the cost function value converges towards zero after each perturbation. Lastly, thanks to the RTI algorithm, the solver time almost always remains lower than the control sampling period, which is 0.05 s. The overall results show that a satisfactory following performance is achieved in the absence of mutual aerodynamic disturbance.

3.4.2 Formation Control with Autonomously Flying Leader

The leader vehicle is controlled by the NMPC controller and follows a predefined 3D trajectory consisting of six way points in a limited space of 3.5 m^3. This trajectory is initialized after the follower is engaged to the leader with a relative position of [3.6 −2.8 0] m. Since the lateral directions of a quadrotor are highly sensitive to disturbances, the first observation is that the air flow generated by the neighbor drone creates a stochastic oscillatory effect in these directions. The results of one representative experiment are depicted in Fig. 13. This experiment is also included in the video enclosure. The main difference compared to the previous experiment is that the relative position errors significantly rise, reaching an average error of 0.44 m (in spite of the fact that convergence is maintained). There are two main causes of this increase: first, the follower is vastly affected by the wind disturbance generated by the leader. These perturbations appear to be very turbulent and are not included in the model of the follower and leader. Second, due to the high gain control of the leader, i.e. in order to reject

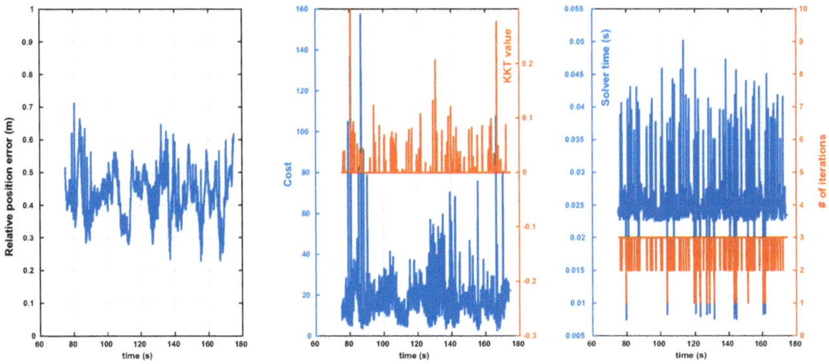

Fig. 13. Performance of formation with autonomously flying leader

disturbances as much as possible, the leader does no longer operate in a low-acceleration regime, which leads to a further invalidation of the neighbor vehicle model in the NMPC computations. Finally, we can observe that both the average cost function and KKT values grow due to the highly nonlinear motion. Despite this, it is clear that the RTI scheme is able to keep the solver time under the control sample time, which is crucial for real-time control.

4 Conclusion and Outlook

Carrying out this work has allowed us to gain the following experimental insights. *Firstly*, during the real experimental validation on trajectory tracking of a single drone, it was clear that even an uncalibrated simulation performed in the high-fidelity simulator Webots coupled with ROS was already very beneficial from a software prototyping perspective (the very same code can be run in simulation and reality). Nonetheless, due to the model-based approach chosen, the differences between the trajectories generated in simulation and those of the real system were still significant. We managed to reduce the reality gap by following a two-stage calibration strategy: first, by identifying the physical parameters of the vehicle model underlying our NMPC strategy with canonical system identification algorithms for both real and simulated vehicles; second, by manually tuning simulation parameters in order to match the produced closed-loop trajectories. At the end of such two-stage process, it was straightforward to design, iteratively optimize, and validate control and estimation algorithms both in high-fidelity simulation and physical reality. *Secondly*, projecting these findings to multi-robot systems, the transition was found to be very smooth. In simulation, the followers are successful in tracking the leader, especially in smooth trajectory regions. However, due to the constant velocity assumption of the predicted trajectories in NMPC, one can see transient divergence on the sharp turns. This is an expected outcome and a drawback of a fully decentralized architecture which by design does not leverage communication. The impact of such choice is, however, insignificant for low acceleration flights, a valid assumption for some scenarios such as patrolling. On the positive side, since the computation and communication overhead of decentralized NMPC is considerably lower than that of distributed architectures, such choice shows good scalability towards systems involving larger number of vehicles. *Thirdly*, the experiments in reality carried out with both quadrotors flying outlined that the mutual aerodynamic disturbances significantly affect the reality gap, and in turn the formation performance, since they are not modeled in the controllers and they invalidate the constant velocity assumption. *Finally*, we believe that these results can be improved by integrating a robust framework to the system, for instance, a disturbance mapping/observer or a robust version of NMPC. This measure should reduce the model uncertainties and the impact of disturbances (causing a decrease in overshoots and steady-state errors) and in turn results in a diminution of formation errors. An additional performance improvement can be achieved through a distributed architecture by leveraging communication between vehicles. Not only the predicted trajectories but also the estimated disturbance information would be shared to enhance the model synchronization.

Acknowledgement. This work has been partially sponsored by the FCT grant [PD/BD/135151/2017], the FCT doctoral program RBCog and the FCT project [UID/EEA/50009/2013] (Additional information about this project can be found here: https://www.epfl.ch/labs/disal/research/quadrotorformationmpc/).

References

1. Bemporad, A., Rocchi, C.: Decentralized linear time-varying model predictive control of a formation of unmanned aerial vehicles. In: 50th IEEE Conference on Decision and Control, pp. 7488–7493 (2011)
2. Monteriù, A., Freddi, A., Longhi, S.: Nonlinear decentralized model predictive control for unmanned vehicles moving in formation. Inf. Technol. Control **44**(1), 89–97 (2015)
3. Chao, Z., Zhou, S.L., Ming, L., Zhang, W.G.: UAV formation flight based on NMPC. Math. Probl. Eng. **2012**, 1–15 (2012)
4. Eren, U., Prach, A., Koçer, B., Raković, S., Kayacan, E., Açıkmeşe, B.: MPC in aerospace systems: current state and opportunities. J. Guidance Control Dyn. **40**(7), 1541–1566 (2017)
5. Kamel, M., Burri, M., Siegwart, R.: Linear vs nonlinear MPC for trajectory tracking applied to rotary wing MAVs. IFAC-PapersOnLine **50**(1), 3463–3469 (2017)
6. Kamel, M., Stastny, T., Alexis, K., Siegwart, R.: Model predictive control for trajectory tracking of UAV using ROS. In: Koubaa A. (ed.) Robot Operating System (ROS): The Complete Reference, vol. 2, pp. 3–39. Springer, Cham (2017)
7. Falanga, D., Foehn, P., Lu, P., Scaramuzza, D.: PAMPC: perception-aware model predictive control for quadrotors. In: IEEE/RSJ International Conference on Intelligent Robots and Systems, pp. 1–8 (2018)
8. Van Parys, R., Pipeleers, G.: Distributed MPC for multi-vehicle systems moving in formation. Robot. Auton. Syst. **97**, 144–152 (2017)

9. Hafez, A.T., Marasco, A.J., Givigi, S.N., Iskandarani, M., Yousefi, S., Rabbath, C.A.: Solving multi-UAV dynamic encirclement via MPC. IEEE Trans. Control Syst. Technol. **23**(6), 2251–2265 (2015)
10. Gowal, S., Martinoli, A.: Real-time optimization of trajectories that guarantee the rendezvous of mobile robots. In: IEEE/RSJ International Conference on Intelligent Robots and Systems, pp. 3518–3525 (2012)
11. Erunsal, I.K., Ventura, R., Martinoli, A.: NMPC for 3D Formation of Multirotor MAVs with Relative Sensing in Local Coordinates. arXiv.org (2019) e-print 1904.03742
12. Erunsal, I.K., Martinoli, A., Ventura, R.: Decentralized NMPC for 3D formation of MAVs with relative sensing and estimation. In: IEEE International Symposium on Multi-Robot and Multi-Agent Systems, pp. 176–178 (2019)
13. Quan Q.: Introduction to Multicopter Design and Control, Ch. 6, pp. 134–138. Springer, Beijing (2017)
14. Michel, O.: Webots: professional mobile robot simulation. Int. J. Adv. Robot. Syst. **1**(1), 39–42 (2004)
15. Schiano, F., Franchi, A., Zelazo, D., Giordano, P.R.: A rigidity-based decentralized bearing form. controller for groups of UAVs. In: IEEE/RSJ International Conference on Intelligent Robots and Systems, pp. 5099–5106 (2016)
16. RC Benchmark Series 1585 Thrust Stand. https://www.rcbenchmark.com/ (2020)
17. Dias, D., Ventura, R., Lima, P., Martinoli, A.: On-board vision-based 3D relative localization system for multiple quadrotors. In: IEEE International Conference on Robotics and Automation, pp. 1181–1187 (2016)
18. Yuan, Q., Zhan, J., Li, X.: Outdoor flocking of quadcopter drones with decentralized model predictive control. ISA Trans. **71**, 84–92 (2017)
19. Huck, S., Rueppel, M., Summers, T., Lygeros, J.: Rcopterx-experimental validation of a distributed leader-follower MPC approach on a miniature helicopter test bed. In: European Control Conference, pp. 802–807 (2014)
20. Gros, S., Zanon, M., Quirynen, R., Bemporad, A., Diehl, M.: From linear to nonlinear MPC: bridging the gap via the real-time iteration. Int. J. Control **93**(1), 62–80 (2020)
21. Quirynen, R., Vukov, M., Zanon, M., Diehl, M.: Autogenerating microsecond solvers for nonlinear MPC: a tutorial using ACADO integrators. Optimal Control Appl. Methods **36**(5), 685–704 (2015)
22. Ferreau, H.J., Kirches, C., Potschka, A., Bock, H.G., Diehl, M.: qpOASES: a parametric active-set algorithm for quadratic programming. Math. Program. Comput. **6**(4), 327–363 (2014)

Distributed Heterogeneous Multi-robot Source Seeking Using Information Based Sampling with Visual Recognition

Carlos Nieto-Granda[1(✉)], Shenghye Wang[2], Vikas Dhiman[2], John Rogers III[1], and Henrik I. Christensen[2]

[1] U.S. Army Combat Capabilities Development Command Army Research Laboratory (ARL), Adelphi, MD, USA
[2] University of California San Diego, San Diego, CA, USA

1 Introduction

Search and rescue robots often need to find the source of a signal like radio-active material, heat signature, or gas leak. This problem is known as stochastic source seeking. To plan an optimal exploration trajectory, many researchers [1] have suggested using information gain metrics like Mutual Information. However, in partially observable environments, computing information gain of a trajectory becomes challenging if parts of the map are unknown. Stochastic source seeking has been typically addressed in fully observable environments, with a single target [1]. On the other hand, exploration in a partially observable environment [2] has dealt with exploration without a source seeking objective. However, the existing approaches that depend upon fully observable state do not generalize to all kinds of maps and obstacles. The work closest to our approach is Shreshta et al. [8]. However, they attempt to predict large regions in the map, which are unlikely to generalize to new maps. Moreover, Shreshta et al. [8] uses a flood-fill over the predicted map to estimate the information gain for each action. Flood filling can be brittle and lead to misleading information gains with even small holes in the map walls.

Contribution. We present our state of the art architecture for multi-robot source seeking visual recognition. Our approach has been tested in simulation and real experiments at the University of California San Diego. Besides, we provide a mathematical model of the problem and an algorithm solution to the source seeking problem.

C. Nieto-Granda—This work was done during the author's Ph.D. Thesis at the University of California San Diego.

B. Siciliano et al. (Eds.): ISER 2020, SPAR 19, pp. 462–471, 2021.
https://doi.org/10.1007/978-3-030-71151-1_41

2 Technical Approach

We deployed a heterogeneous multi-robot team with UGVs and UAVs which mission is to locate, identify, and neutralize (defuse) an unknown number of IEDs hidden in the environment. The robots can measure radioactivity emitted by the IEDs. Besides, the robots navigate and perform the source seeking task autonomously. The robots are capable of requesting or providing service to use all the available resources to minimize the time to defuse all the IEDs. Besides, each robot can negotiate with the other members of the team and re-compute the exploration regions.

Problem 1 ***(Model-based source seeking).*** Consider N targets, whose locations $\{y_i\}_{i=1}^N$ are unknown. We are given a robot with pose x_t that moves according to a given dynamics model $x_{t+1} = f(x_t, u_t, m) + \eta_t$. The problem is to find the policy π that deactivates all the targets in minimum time,

$$\begin{aligned} \pi^* &= \arg\min_{\pi} T \\ \text{s.t} \quad a_{i,T} &= 0 \quad \forall i \in \{1, \dots, N\} \\ a_{i,t+1} &= g(a_{i,t}, x_t, m, u_t) \\ x_{t+1} &= f(x_t, u_t, m) + \eta_t \\ z_t &= \sum_{i=1}^{N} \phi(x_t, y_i, m) + \epsilon_{i,t} \\ u_t &= \pi(z_{1:t}) \end{aligned} \tag{1}$$

Problem 2 ***(Information Gathering).*** The Information Gathering problem is the problem of seeking to maximize the information about the unknown map m using the observations $l_{1:t}$,

$$\begin{aligned} \pi^* &= \arg\min_{\pi} \mathbb{H}(m \mid l_{1:t}, x_{1:t}) \\ \text{s.t} \quad x_{t+1} &= f(x_t, u_t, m) + \eta_t \\ l_t &= \psi(x_t, m) + \omega_t \\ u_t &= \pi(l_{1:t}), \end{aligned} \tag{2}$$

where $\mathbb{H}(m \mid l_{1:t}, x_{1:t})$ is an information measure over the map m. Typically it is the Shanon entropy of the map random variable.

In this work, we combine the two problems, Problem 1 and Problem 2, to address them jointly. We also extend this problem to a heterogeneous multi-robot setting.

Distributed Heterogeneous Multi-robot Source Seeking (DHMRSS) We consider pairs of Heterogeneous robots with each pair containing a ground robot and an aerial vehicle. The ground robots have a large battery back-up and can cover large areas on the ground while aerial vehicles that ride on top of ground robots have short battery back-up, but they can reach places were ground robots cannot.

Problem 3 (**DHMRSS**). Under the transition model and observation model, we want to design a policy $\pi_k : \mathcal{Z}^t \times \mathcal{A}^{Kt} \to \mathcal{U} \times \mathcal{U}^{(a)} \times \{0,1\} \times \mathcal{B}$ for each robot $k \in \{1,\ldots,K\}$. The kth robot takes an action $u_{k,t}, u_{k,t}^{(a)}, u_{k,t}^{\chi}, \beta_{k,t} = \pi_k(\zeta_{k,1:t}, \boldsymbol{\alpha}_{k,1:t})$ at each time step in order to deactivate all the signal sources in least time possible,

$$
\begin{aligned}
\{\pi_k^*\}_{k=1}^K &= \arg \min_{\{\pi_k\}_{k=1}^K} T \\
\text{s.t} \quad a_{i,T} &= 0 \quad \forall i = \{1,\ldots,N\} && \text{// Terminal condition} \\
B_k &\geq \sum_{t=1}^{T-1} \chi_{k,t} && \text{// Battery constraint} \\
s_{k,t+1} &= \mathbf{F}(s_{k,t}, m, u_{k,t}, u_{k,t}^{(a)}, u_{k,t}^{(\chi)}) && \text{// State transition model} \\
\zeta_{k,t} &= \mathbf{O}(s_{k,t}, m) && \text{// Observation model} \\
u_{k,t}, u_{k,t}^{(a)}, u_{k,t}^{(\chi)}, \beta_{k,t} &= \pi_k(\zeta_{k,1:t}, \boldsymbol{\alpha}_{k,1:t}) && \text{// policy}
\end{aligned}
\tag{3}
$$

where B_k is the battery constraint for each of the aerial vehicle.

Note that Problem 3 is a combination of Problem 1 and Problem 2 but the objective of information gathering is implicit in the problem. Since we want to defuse *all* the signal-sources, we need to find all of them which requires exploration of the environment. We consider two specific instantiations of $bel_0(s_0)$ in Problem 3, one with uniform prior over the map space and another with given map of the building but unknown locations of the furniture and obstacles.

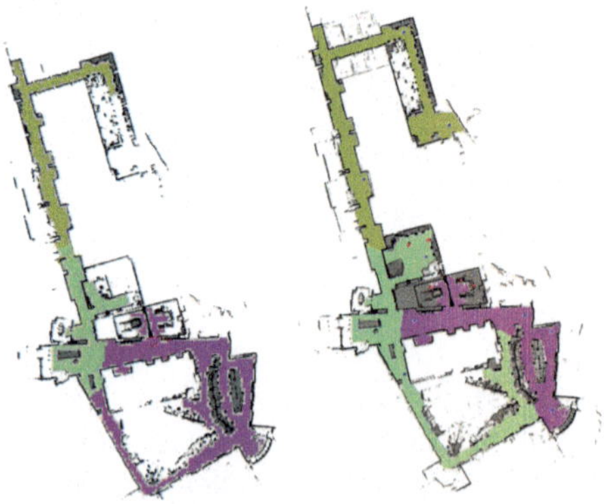

Fig. 1. Comparison between the exact segmentation with the new dynamic reconfiguration after the robots agreed to explore regions from the other ones.

New Coordination Strategy

In [6], our robot team is capable of performing their tasks and accomplish their mission. However, we did not take care of the management of resources.

For example, when a robot finishes a task, it just continues mapping and navigating even if the assigned region for exploration was explored entirely. We extend our framework to provide the robots with the capability to broadcast two new messages: *a*) *Task accomplished - Offering assistance.* *b*) *Requesting assistance to accomplish a task.*

In our preliminary experiments, the time to explore a floor plan decreased comparing to our previous exploration strategy. However, the robots need to be able to communicate with the other members of the team. To enable the capability to communicate with the other robots even if one of them is far from the team, we developed a new module that allows the robot that finished the task to find other members. As we explained in [3], the environment's segmentation is based on graph cuts. Therefore, we recompute the robots' initial knowledge to navigate the trajectory or exploration regions that the other robots follow.

One of the requirements to use this module is that the robot team has to know the map. The comparison between our previous and the current system is shown in Fig. 2. In this scenario, the robot A, which exploration area is in color green, (Fig. 1) sent a broadcast message *Offering Assistance* to the robot B (exploration area color purple). Robot A and Robot B negotiated and recomputed their exploration areas, and robot A continues exploring and searching for targets while robot B is provides paths to the drone inside the rooms.

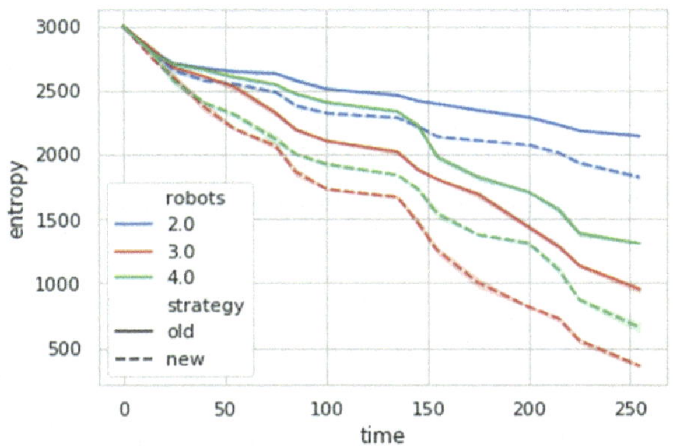

Fig. 2. An example of running our exploration methods with and without the new coordination strategy module. The entropy decreases faster, because the robots are constantly exploring the environment.

Shape Adaptation of Gaussian Regions

Our exploration module, called "*Infosploration*" [4], allows the robots to explore an unknown environment using frontiers and identify rooms and hallways. We improve our Gaussian regions module to adapt to the shape of the rooms and hallways. This improvement allows the robots to label the rooms and enable a voxel exploration module called "*Voxelsploration.*" The details of this application are found in [5]. These regions are labeled by our place categorization system previously presented in [7].

Exploration Paths and Areas for Target Localization
Infosploration computes all the trajectories that the robot can travel considering the robot dynamics, power consumption, distance to the frontiers or regions, and the probability of the target's location. However, all the trajectories are computed in a 2D environment.

Input: VoxelMap *region*, Robot Full Trajectory: X_k, Voxels Uncertaintiy: *tau*
if $UGV \longleftarrow \emptyset$ *No reachable area* **then**
  Compute a plane with high probability to find the target given by radio signal z_t;
  Create an exploration region enclosed of the computed planes A_m;
  if *Robot Full Trajectory* X_k *provides high values of radio signal* z_t **then**
    Map the trajectory X_k to 3D and substract it as a polygon;
    from the $A_m \longrightarrow S_{area}$;
  end
  Run our planner to provides a trajectory that satisfy the full coverage of S_{area};
end
Output: Searching Plan

Algorithm 1: Segmentation of Searching Areas

Voxelsploration computes the paths in 3D, which allows us to have a heterogeneous team with different dynamics. Our system has been tested in 3 different platforms: packbots (capable of going upstairs), jackals (only navigate in one floor), and quadrotors (capable of navigating on different heights).

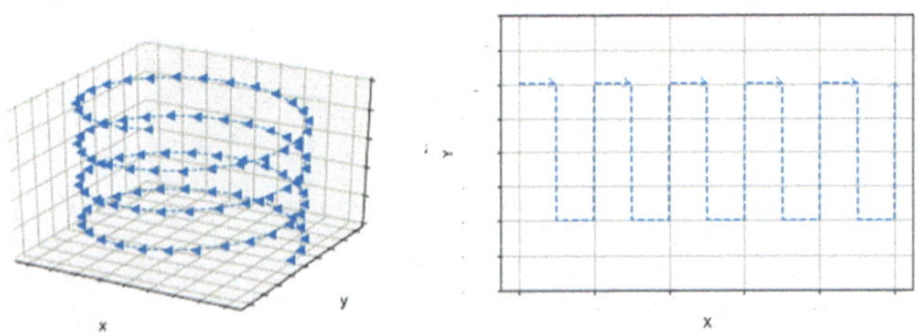

Fig. 3. Visualization of two searching trajectories provided to the drone to localize a target.

Our new exploration module provides the capability to reach areas where the mobile robot cannot reach. Algorithm 1 explains the steps that our method follows to segment and decrease the search region on the environment. Figure 3 shows two different trajectories in which the UGV after navigating on the environment computes the probability of the target location and provides the search path that the drone has to follow to find the target.

In each experiment, the ZigXbees are tuned randomly, so the robot might listen to some from a far distance while others the robot has to be in the room to detect them. Besides, some of the rooms have obstacles that deny access to the UGV so it needs to request the assistance of the UAV. One of the environments that we explored is the Atkinson Hall First Floor. This scenario is shown in Fig. 6 has different rooms with long hallways. Inside, most of the rooms have glass, and the semantic label provided is hallway, office, or lab. In this series of experiments, we only run the adaptable Gaussian Regions to determine if the robot was in another area different than a hallway. The size of this floor and the long hallways made a perfect exploration scenario. The communication among the robots is very constrained so that it most of the cases there is no communication (Fig. 4).

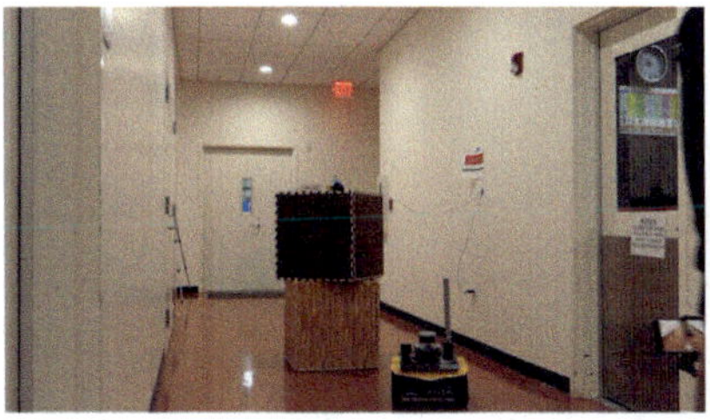

(a) UGV looking for the fake IED which is impossible for it to reach it.

(b) After the UGV failed. The UAV assistance is required and it fly with the trajectory provided by the UGV.

Fig. 4. Source seeking scene where the UGV is not capable to reach the fake IED and requests the assistance of the UAV.

3 Experiments

We deploy a team of ground robots (Fig. 5) paired with an aerial vehicle to explore the environment. Each ground robot runs its own mapping system (Omnimapper [9]) and broadcasts messages among the team members. Each ground robot can listen to the radio signals. These signals are simulating radioactivity or gas emission in the area. The robot behavior on this mission is the following:

When the task of the UGV is to find the target's location, it will navigate to the highest probability location of the signal and make visual identification of the target. However, there are many scenarios where the UGV will not be able to identify the target base on its dynamics, hardware, or reachability of the sensors. Then, the UGV will compute a trajectory with specific viewpoints that the UAV is requested to navigate to identify the target.

Input: Robot pose: X_k, Radio signal: z_t
Output: Task
while $z_t = 0$ **do**
| Explore and look for targets
end
Find the location of the target

Algorithm 2: Robot tasks

Live Robot Experiments. A series of experiments were performed with a DJI drone which was augmented with a pozyx device and a Zigxbee, and 3 Clearpath Jackals (UGV) which were augmented with a 360 Velodyne sensor, an RGB-d camera, an IMU LORD Microstrain, a Ubiquiti 5 GHz bullet antenna, a ZigXbee and a Pozyx device (to test different sensors for seeking RSSI), a GoPro Camera and an additional 2TB SSD for data storage. The UGVs are autonomous in these experiments. However, in the situation when the UGV requires assistance from the UAV. The UGV will stop for setting up the UAV (Fig. 4).

Fig. 5. Our hetereogeneous UAV/UGV robot team used in these experiments.

4 Results

The main mission of this work is to locate, identify, neutralize threats on the environment. In this case, we simulate explosives installed and hidden on different floors of our university buildings. We are interested in optimizing the time that the robot team will take to accomplish the task. We ran our exploration and source seeking system in each UGV on board. Each UGV is autonomous and has a UAV that is teleoperated following the path planning computed and visualized to the operator. However, our system can send navigation commands to a UAV with a computer onboard.

Atkinson Hall - 1st Floor. This scenario consists of big hallways, two rooms, and a big atrium. Unfortunately, in this scenario, our place categorization only provides two labels. Therefore, it was not relevant to use it to label the exploration regions. Figure 6b shows one UGV and UAV exploring Atkinson Hall 1st Floor. We can observe that the UGV finds the IED in the hallway, and the drone fly over the table and the cube to localize the other IED.

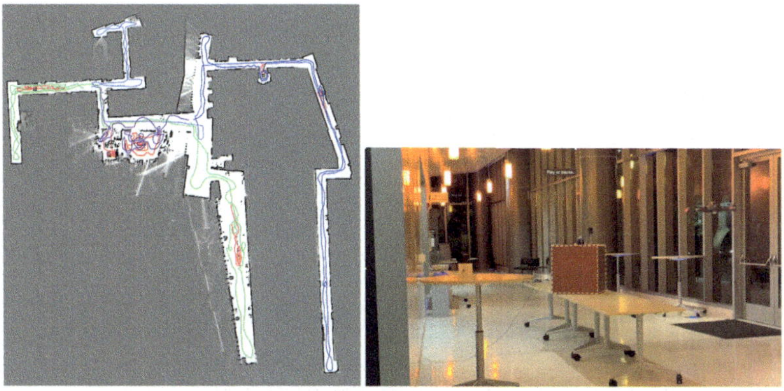

(a) Full robot trajectories from the 3 teams UAV/UGV. The UAV trajectory is shown in red and the rest of the robots in green, blue and purple.

(b) Our robot team exploring and defusing IEDs at Atkinson Hall 1st Floor.

Fig. 6. Atkinson Hall 1st Floor explored with drone trajectories and targets.

The visualization of the trajectory of each UGV: blue, green, and purple, and the UAV trajectory in red is shown in Fig. 6. The IEDs are classified into two groups: only reached by UGV (blue star) and only reached by UAV (red star).

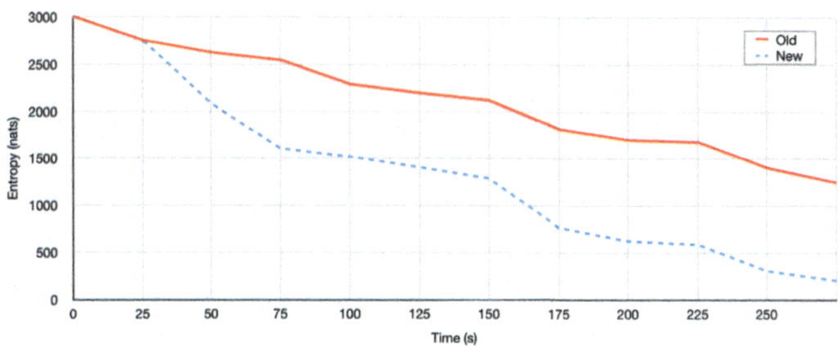

Fig. 7. A comparison of the entropy overtime of the robot team at Atkinson Hall 1st floor.

In this experiment, it is clear that the robot with a blue trajectory explored more than the other two. The reason is that the exploration with the UAV was faster than the other two robot teams. We compute the entropy as a metric to guarantee that the robot team explores all the floor map while at the same time is defusing the IEDs. The comparison of our framework with and without the new coordination strategy is shown in Fig. 7. As expected, the entropy decreases faster when we use the new coordination strategy.

Atkinson Hall - 6th Floor. This experiment was performed on the 6th Floor of Atkinson Hall. The floor map consists of hallways, offices with different dimensions. The targets were set up on the top of tables where the robot has to avoid various obstacles during the exploration.

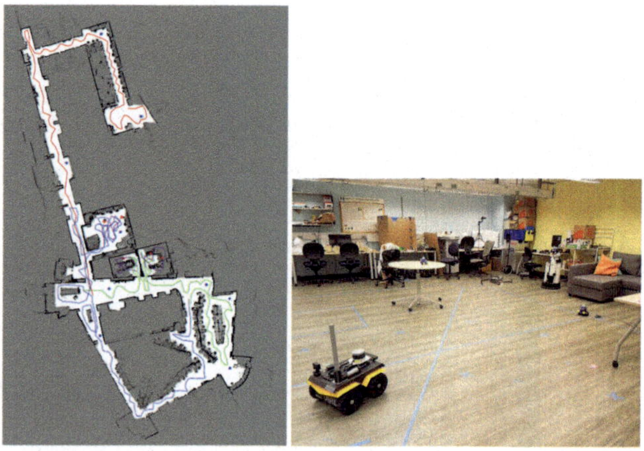

(a) Full robot trajectories from the 3 teams UAV/UGV. The UAV trajectory is shown in red and the rest of the robots in green, blue and purple.

(b) Our UGV/UAV exploring and defusing IEDs at Atkinson Hall 6th Floor.

Fig. 8. Atkinson Hall 6th Floor explored with drone trajectories and targets.

This scenario is entirely semantically labeled using DEDUCE [7], as is shown in Figure 3.18. Figure 8 shows the trajectory of each UGV: red, blue, and green, and the UAV trajectory in purple. The IEDs are classified into two groups: only reached by UGV (blue star) and only reached by UAV (red star). Besides, the trajectories in blue and green shows that the new coordination strategy recomputed to explore and find the IEDs without waiting for the other robot. The initial and the new segmentation are shown in Fig. 1.

We compute the entropy as a metric to guarantee that the robot team explores all the floor map while at the same time is defusing the IEDs. The comparison of our framework with and without the new coordination strategy is shown in Fig. 9. As expected, the entropy decreases faster when we use the new coordination strategy.

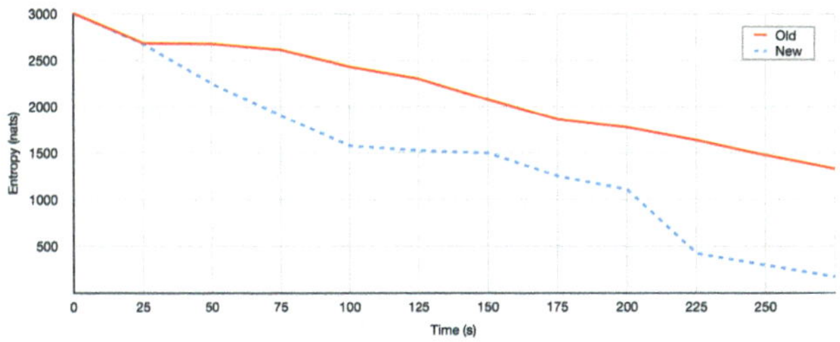

Fig. 9. A comparison of the entropy overtime of the robot team at Atkinson Hall 6th floor

5 Discussion

In this work, we incorporated all the techniques presented in the previous chapters of this thesis. We aim that our framework is robust to find policies that satisfy the exploration and source seeking missions. This capability enables the possibility of performing real scenarios missions. For example, the robot is performing a surveillance task when it detects a leak of gas. Then, the robot switches to a source seeking task to prevent a possible explosion. Besides, when the robot has prior knowledge of the environment and its history, the robot can make decisions faster and effectively.

References

1. Atanasov, N.A., Le Ny, J., Pappas, G.J.: Distributed algorithms for stochastic source seeking with mobile robot networks. J. Dyn. Syst. Meas. Control **137**(3), 031004 (2015)
2. Choudhury, S., Kapoor, A., Ranade, G., Scherer, S., Dey, D.: Adaptive information gathering via imitation learning. In: RSS, Massachusetts (2017)
3. Fung, N., Rogers III, J.G., Nieto-Granda, C., Christensen, H., Kemna, S., Sukhatme, G.: Coordinating multi-robot systems through environment partitioning for adaptive informative sampling. In: IEEE ICRA, Montreal (2019)
4. Fung, N.C., Nieto-Granda, C., Gregory, J.M., Rogers, J.G.: Autonomous exploration using an information gain metric. US Army Research Laboratory Technical report. ARL (2016)
5. Nieto-Granda, C.: Autonomous exploration of unknown environments using voxel maps. ARL Technical report. ARL (2019)
6. Nieto-Granda, C., Rogers III, J.G., Fung, N., Kenma, S., Christensen, H.I., Sukhatme, G.: On-line coordination tasks for multi-robot systems using adaptive informative sampling. In: ISER, Argentina (2018)
7. Pal, A., Nieto-Granda, C., Christensen, H.I.: DEDUCE: diverse scene detection methods in unseen challenging environments. In: IEEE IROS, Macau (2019)
8. Shrestha, R., Tian, F., Feng, W., Tan, P., Vaughan, R.: Learned map prediction for enhanced mobile robot exploration. In: IEEE ICRA, Montreal (2019)
9. Trevor, A.J., Rogers, J.G., Christensen, H.I.: Omnimapper: A modular multimodal mapping framework. In: 2014 IEEE International Conference on Robotics and Automation (ICRA), pp. 1983–1990. IEEE (2014)

To Boldly Dive Where No One Has Gone Before: Experiments in Coordinated Robotic Ocean Exploration

José Pinto[1(✉)], Maria Costa[1], Keila Lima[1], Paulo Dias[1], João Pereira[1], Manuel Ribeiro[1], Renato Campos[1], Zara Mirmalek[2], Renato Mendes[1,3,4], Francisco López Castejón[5], Javier Gilabert[5], Maria Paola Tomasino[4], Catarina Magalhães[4], José C. B. da Silva[4], Paulo Relvas[6], Trent Lukaczyk[7], Kay Arne Skarpnes[8], Martin Ludvigsen[8], Alexander Chekalyuk[9], Bruno Loureiro[10], Ian G. Brosnan[11], Jing Li[11], Ami Hannon[11], João Borges de Sousa[1], and Kanna Rajan[1,12]

[1] Laboratório de Sistemas e Tecnologia Subaquática (LSTS), Faculdade de Engenharia da Universidade do Porto, Porto, Portugal
`zepinto@lsts.pt`
[2] Harvard University, Cambridge, MA, USA
[3] CESAM, Department of Physics, Univ. of Aveiro, Aveiro, Portugal
[4] CIIMAR, Univ. of Porto, Porto, Portugal
[5] Technical University of Cartagena (UPCT), Cartagena, Spain
[6] CCMAR and Univ. of Algarve, Faro, Portugal
[7] FlightWave Aerospace Systems, Santa Monica, CA, USA
[8] Department of Marine Technology, NTNU, Trondheim, Norway
[9] Columbia University, New York, USA
[10] CIBIO - Centro de Investigação em Biodiversidade e Recursos Genéticos, Vairão, Portugal
[11] NASA Ames Research Center, Moffett Field, CA, USA
[12] SIFT LLC, Minneapolis, MN, USA
`https://www.lsts.pt`

Abstract. Study of ocean processes is important to understanding climatic variability especially on the productive upper water-column. Ocean currents regulate the climate, it captures CO_2 from the atmosphere and oxygen is generated by its plankton communities, all of which are part of the global environmental cycle which are being impacted by anthropogenic change. Much of the ocean, however, remains unexplored especially the bio-geochemical processes in the water-column which need to be examined at scale. Satellite remote sensing captures only surface effects while expensive research vessels can only make discrete observations in finite periods of time. Our work with *networked* marine robotics in the aerial, surface and underwater domains is at the vanguard of a new approach to scientific observation, which brings together technology to enable vessels and robots to work in tandem for capturing synoptic views of open ocean phenomena. We describe a cruise in the Spring of 2018 in the open waters of the Pacific where we employed a fleet of autonomous robots for simultaneous observations of mesoscale and

B. Siciliano et al. (Eds.): ISER 2020, SPAR 19, pp. 472–487, 2021.
https://doi.org/10.1007/978-3-030-71151-1_42

sub-mesoscale features of an unexplored frontal zone. We articulate our approach to multi-vehicle coordination and challenges that lie ahead for research in this harsh domain.

Keywords: Ocean exploration · Networked maritime robotics · Multi-vehicle operations

1 Introduction

Due to its vastness and inaccessibility, the oceans remain mostly unexplored. Even though its processes are highly dynamic and govern everyday life, forecasts rely mostly on satellite imagery and moored/drifting sensors which are sparse in number. Typically sensors are outfitted on research vessels, buoys or Lagrangian drifters and are constrained by location, subsample at discrete locations or for optical remote sensing, rely on the absence of clouds. "Tracking" a dynamic phenomenon to observe its evolution in space and time is a significant challenge in the sciences, although critical for bio-geochemical process studies. Fronts, internal waves, plumes, slicks, lagrangian coherent structures, anoxia and hypoxia are critical to our understanding the bio-geochemistry; yet these ocean phenomena

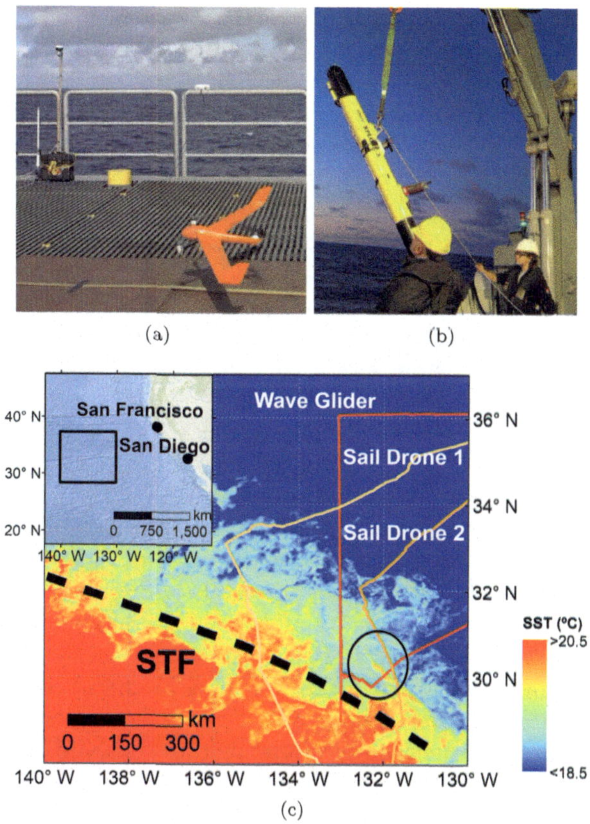

Fig. 1. (a) An Unmanned aerial robotic vehicle (UAV) used in the cruise. (b) Launch of an autonomous underwater vehicle, a powered robotic platform with a range of sensors, being deployed from the research vessel. (c) The eastern Pacific region of exploration (inset). Sea Surface Temperature (SST) imagery product from June 8th with Wave Glider and Saildrone ASV total trajectories from May 1st to June 10th. The operational AUV area is marked with a circle. The STF (dashed black line) is associated with a rapid change in surface temperature.

and their dynamics have proven to be hard to quantify with traditional scientific methods.

The challenges of open ocean exploration are even more daunting. Sampling is driven in large part by scales – phenomena which are vast (1000's of km^2) yet driven by ocean physics and biology in fine scale (in the micro-meters). Connecting the dots between the two extremes has become critical as our understanding of the changing oceans has improved. Traditional ship-based scientific observation is slowly being augmented by the use of autonomous robotic platforms such as that being described in this work, but this increases operations complexity by requiring very different assets to be coordinated in space and time. This and other challenges are addressed by our approach, which is at the forefront of ocean science and marine robotics.

In this work, we describe a study to explore a frontal zone with a range of sensors using heterogeneous platforms. Our target was the Eastern spur of the Northern Pacific Subtropical Front (STF), $\sim$ 1500 Km from the California coastline (Fig. 1(c)). This mid-Pacific region was last explored in some detail in the 1980s [1] when its dominant salinity gradient was characterized. The region often has extensive cloud cover with consequent sparsity in remote sensing data to seed the exploration, making *in-situ* sampling the only effective alternative to study the STF. Fronts are regions of horizontal gradients of physical, chemical, and biological processes, with distinct water masses. They can occur at different spatial scales, from several hundred meters up to many thousand kilometers [2] and not usually defined as a single occurrence but more as a phenomenon in a cascade of scales. STF is commonly observed as a large-scale ($\sim$ 1000 Km) feature which can catalyze the generation of mesoscale ($\sim$ 100 Km) meanders, 'eddies' and rings, and sub-mesoscale ($\sim$ 10 Km) 'filaments' and other smaller structures. Exploration therefore requires a judicious mix of assets with information derived from multiple sources with different levels of synopticity.

Although ocean fronts affect all aspects of marine ecosystems, such as biological productivity [3], their real impact and influence remains unknown due to the practical difficulty of simultaneously obtaining physical and biological data in high resolution in space and time. Fronts in coastal regions, especially those driven by temperature gradients can often be observed by the naked eye under certain conditions; however finding those offshore, like the STF which while more persistent, amounts to finding a 'needle in a haystack'. Open water exploration with an expensive ship-based survey is infeasible; satellite remote sensing, when available, might not be timely enough to geolocate even such slow moving water masses. As a result, open water exploration is challenging even in the best of circumstances.

We describe an exploration methodology using networked multi-vehicle operations of aerial, surface and underwater vehicles for scientific exploration onboard the R/V *Falkor*. Sparse remote sensing data provided an approximate area in the $\sim$ 1000 Km^2 scale. The challenge posed in locating the front was solved by the cascading use of autonomous surface vehicles (ASVs) followed by the use of the research vessel with straight line surveys, in turn followed by exploration in high-resolution by slow moving autonomous underwater vehicles (AUVs) and unmanned aerial vehicles (UAVs). The novelty of this work is multifold; first, it demonstrates the applicability of coordinated autonomous robotic observation across aerial, surface and underwater domains with low cost platforms. Second, it shows how such heterogeneous platforms can be controlled with increased situational awareness on ship and shore. Third, it demonstrates the use of embedded machine intelligence for operational effectiveness, along with responsiveness to human-in-the-loop augmentation of robotic goals. All with intermittent connectivity between multiple robots which can adapt to dynamic environmental conditions over often unreliable and low-bandwidth communication links. The ensemble of assets besides the R/V *Falkor*, included multiple UAVs (Fig. 1(a)), two Saildrones [4], a Liquid Robotics WaveGlider [5] ASV, and multiple Light AUVs [6] (Fig. 1(b))[1].

This paper is structured as follows. A brief Sect. 2 places this effort in the context of other fielded and relevant marine robotic deployments with multiple vehicles. Section 3 describes the core of the technical effort involving embedded methods related to waypoint dispatching for AUV control. Section 4 describes the experimental setup and results, while Sect. 5 wraps up with the challenges and lessons learned.

2 Related Work

Inter-disciplinary science driven robotic ocean exploration in the open ocean while becoming more prevalent, is still sparse; those where multiple and heterogenous robotic vehicles have been deployed for scientific exploration, sparser still. Consequently, our efforts have been at the forefront of exploration and include a range of experiments [7–9] with multi-vehicle deployments. Scientific exploration with AUVs especially in the coastal ocean in such inter-disciplinary settings is more common especially in the context of specific phenomenon of scientific interest [10–15]. The coordinated use of multiple vehicles in the harsh open ocean environment is still novel, however. Typically when multiple vehicles are fielded, individual entities such as universities and research institutions, bring one or more vehicles that are operated separately while being manually coordinated [16,17]. What is unusual in this work is the breadth and depth of control of vehicles from a single laboratory, with the exception of Saildrone and WaveGlider ASVs which were already in the area making measurements. Such networked systems allow the backbone infrastructure to be designed and applied systematically, impacting situational awareness and ultimately scientific data gathering. The refinement of such past efforts is at the core of this paper.

[1] https://schmidtocean.org/cruise/exploring_fronts_with_multiple_aerial-surface-underwater-vehicles/.

3 Technical Approach

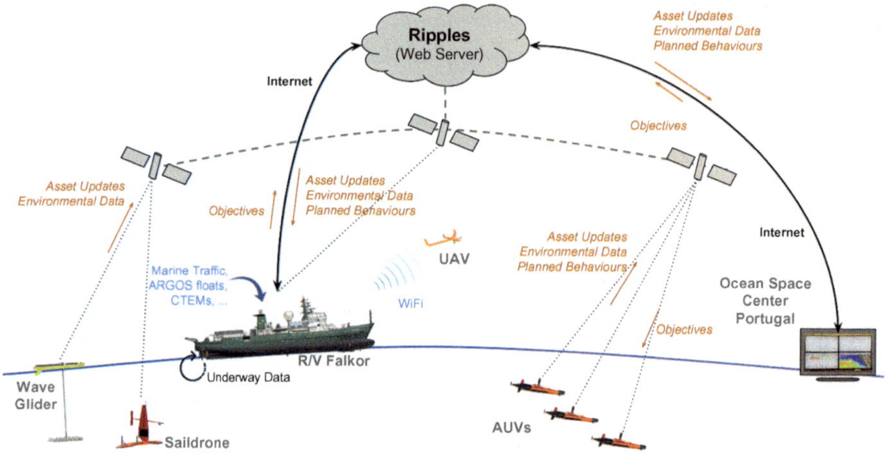

Fig. 2. Overall system architecture and infrastructure for 2018 open ocean cruise on the R/V *Falkor*. Robotic assets deployed operated up to dozens of kilometers from the vessel.

In open water exploration, finding a target phenomenon to sample is in and of itself a challenge. To mitigate that, our *modus operandi* was to extend the 'human senses' as we set out to use an ensemble of assets to narrow down the spatial extent of the search. ASVs, including the WaveGlider and Saildrones, provided the synoptic view that was needed. In our case, the STF was mainly driven by salinity gradients, and therefore rapid assimilation of data from the ASVs proved to be critical.[2] Once a salinity gradient in the near proximity of our hypothesized front was established, it was critical to sample with in-situ vehicles like the AUVs, to obtain high resolution data. By using this approach of systematically narrowing down the spatial scales of observation, we were then able to establish the presence of the front and to map it underwater. Consequently, the AUVs became the primary focus of data gathering in high resolution. The focus in this section, therefore, is structured towards the onboard temporal dispatch mechanism on AUVs designed to work with rapid human-in-the-loop environmental assessment.

For the operator, however, the lack of observability of AUVs underwater, results in a unique challenge related to situational awareness; AUVs can primarily communicate when at the surface over satellite. It is important for the operator to estimate the vehicles' behavior and when/where they will resurface as these are the opportunities to assess the outcome of execution, obtain more science and engineering data and alter high-level objectives, all in an iterative cycle. By tightly coupling onboard adaptation on the vehicles with user expectations, it is possible to maintain user awareness of system state even with the

[2] Temperature gradients can be observed by optical remote sensing from space.

sparsity in communication links. This was the principal reason an improved situation awareness and an early warning collision system coupled with encoding communication gaps in planning and execution were incorporated into a temporal dispatching scheme we discuss below (see Sect. 4).

Table 1. LSTS toolchain components and their description with citations to past work.

Component	Description	
DUNE	Low-level onboard navigation and control software embedded on all mobile platforms	[18]
NEPTUS	Shore/ship command and control system used to visualize multiple autonomous heterogeneous vehicles. It can integrate data from external assets and other sources of information such AIS receivers, meteorological forecasting services and satellite imagery	[19]
IMC	This binary protocol is used by all of the above components both for inter-module and intra-module communications	[20]
Ripples	Cloud infrastructure used to disseminate data from ongoing field deployments, as well as providing simplified web interfaces for following the operations using a browser; it is also a stripped down alternative to NEPTUS	[21]

Figure 2 depicts the system design of our approach. It builds upon existing open source software using the LSTS toolchain [18], with a brief summary in Table 1. Ripples is used as a communications hub/router accessed using a web API or directly over satellite when an internet connection is not available. It provides a web front-end that integrates data from vehicles and other services such as Marine Traffic[3], ARGOS floats[4], CMEMS.[5] Multiple users can access Ripples simultaneously from ship/shore for command and control. Doing so allows scientists on ship/shore to follow and interact with the operations team while providing scientific context from available information sources, critical in open water exploration.

Planning and Execution

Recent work on embedded planners on marine robots [22,23] have been used to define the behavior of vehicles in response to the trifecta of user-defined objectives, environmental conditions and platform constraints. In our previous efforts, multi-vehicle plans have been created in a distributed manner, with an off-board planner dividing workload among vehicles and having the vehicles in turn, adapt the synthesized plans onboard [7,8]. In doing so, the central and onboard planners shared execution timelines in part, with synchronization of timelines occurring opportunistically whenever the vehicle(s) is/are within communication range.

In [23] a plan defined by a central planner explicitly included communication objectives for synchronizing state between the onboard and off-board planners

[3] https://www.marinetraffic.com/.
[4] https://argo.ucsd.edu/.
[5] https://marine.copernicus.eu/.

but temporal execution of the plan was relaxed, allowing the onboard planner to alter allocated tasks without reporting these changes to the user. Such planning flexibility is useful in scenarios with a substantial behavior variability such as when mapping terrains with unknown morphology, but impacts predictability and overall user situational awareness. Further, our previous operations using onboard deliberation were focused along operationally different modes namely mapping an unknown and dense feature in the water-column with the need for adaptation with no human intervention necessary [14,15]. However in open water exploration, the need to cover a large spatial extent and in the process to "discover" a hard to find feature with sensors, requires human-in-the-loop engagement especially when operating substantially away from any infrastructure and continuously for multiple days, such as in our deployment.

In our approach, high-level objectives in the form of waypoint locations and *optional* time of arrival for some of these locations are provided by users or automated planners. On the vehicles, an onboard plan dispatcher creates and validates a schedule of arrival at all user-defined waypoints and decides how to move between waypoints to arrive on time while acquiring data. At every waypoint, the onboard planner checks to see if there are any modifications in the high-level objectives and alters its plan if needed. While submerged, the onboard planner may alter the plan in the face of unexpected situations such as currents, sensor failures and other unplanned events; and on emerging, synchronizes these changes with a central planner. This results in the use of a simple *opportunistic* temporal dispatch for waypoint execution, consistent with the need to deal with such environmental uncertainty. In other words, no complex trade-offs or deliberations were necessary for our mode of exploration.

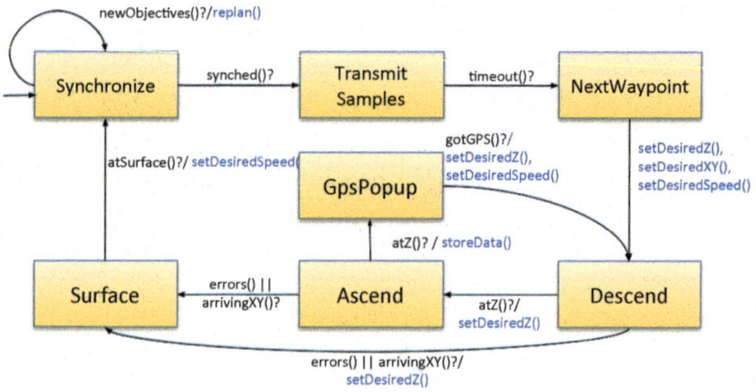

Fig. 3. A simplified state diagram of onboard execution dispatch using temporal propagation of waypoints alongside a model of the AUVs' operational constraints.

Onboard, the high-level objectives or partial plans, are received and then grounded into a complete temporal plan, as shown in Algorithm 1. The partial plan provided to the onboard planner consists of a sequence of scheduled and unscheduled waypoints. The planner uses the current location of the vehicle and current time as the initial state. The algorithm then finds sequences of unscheduled waypoints (delimited by those scheduled) and assigns a speed-of-travel according to the total distance to be traveled and time available, while assuming no adverse environmental conditions. For any final waypoint(s) of the plan that are not scheduled, the algorithm assigns a nominal speed. After scheduling all user-defined waypoints, if any two waypoints of the plan are too distant, additional waypoints are inserted in order to provide opportunities for the vehicle to communicate periodically with the central planner. Doing so greatly improves awareness and human decision-making on ship/shore.

During plan execution, the user can modify goals by adding and removing waypoints via `Ripples`. On the next communication opportunity, the plan changes are sent to the vehicle ensuring that the onboard planner will go through the new list of waypoints, update the schedule of execution and validate it by balancing feasibility according to capabilities and estimated battery charge. If the plan is invalid given these constraints, it is replaced by an empty plan tantamount to idling at the surface. This process therefore, results in a fine balance between what is traditionally noted as 'fail operational' (where an onboard executive is able to 'patch' a plan) and 'fail safe' (where the executive relies on more expensive human intervention) in spacecraft operations [24]. Additionally, whenever a new plan is grounded by the onboard executive, it is sent back over a satellite link to be visualized by `Ripples`, as a means of confirmation and tracking by the operator(s) on ship/shore.

For oceanographic applications, the vehicle travels in a saw-tooth 'yo-yo' pattern surveying the water column. Periodically, the planner forces the vehicle to surface to acquire a new geo-located GPS fix. With every GPS update, the

Algorithm 1: The onboard waypoint scheduling and temporal dispatch algorithm central to open water exploration.

```
Input: plan[] : a partially scheduled list of waypoints
plan.insertAt(0, createWaypoint(getCurrentPosition(), getCurrentTime()));
lastScheduledWaypoint ← 0;
dist ← 0;
for i ← lastScheduledWaypoint + 1 to plan.size() − 1 do
    dist ← dist + distance(plan[i − 1], plan[i])
    if isScheduled(plan[i]) then
        deltaTime ← plan[i].eta − plan[lastScheduledWaypoint].eta
        speed ← dist/deltaTime
        for j ← lastScheduledWaypoint + 1 to i do
            plan[j].eta ← plan[j − 1].eta + speed ∗ distance(plan[j − 1], plan[j])
        lastScheduledWaypoint ← i
for i ← lastScheduledWaypoint + 1 to plan.size() − 1 do
    plan[i].eta ← plan[i − 1].eta + nominalSpeed ∗ distance(plan[j − 1], plan[j]
i ← 0
while i < plan.size() do
    if plan[i + 1].eta − plan[i].eta > commTimeout then
        n ← (plan[i + 1].eta − plan[i].eta)/commTimeout
        insertIntermediateWaypoints(plan[i], plan[i + 1], n)
        i ← i + n
    i ← i + 1
```

executive recalculates the speed of travel to reach the subsequent targeted waypoint at the planned time of arrival. Figure 3 depicts a state diagram of the vehicle's behavior between waypoints. All future planned waypoints, which are opportunity windows for the operator to change the behavior of the vehicle, are known a priori on ship/shore; planned behaviors between waypoints cannot be preempted. Doing so also allows partitioning responsibility of plan execution with multiple operators spanning different operational shifts on ship/shore, resulting in a decrease in operator cognitive burden while increasing the possibility of opportunistic intervention.

4 Field Experiment and Results

Our cruise, between May 28th and June 18th targeted the study of the Eastern spur of the North Pacific Sub-Tropical Front (STF) which evolves in space and time ranging from days to minutes and from hundreds of Kms to meters. A brief overview of the cruise is shown in Table 2. From the outset, the fundamental aim of the cruise was to **find**, **tag** and **follow** the STF so as to characterize the water mass and its environment. Finding the front in of itself was challenging and was initiated by looking at remote sensing data. Sea Surface Temperature (SST) is typically used as a *prior* so as to create a synoptic view of the study area. However, continuous cloud cover including a month prior to the cruise precluded that possibility. Infrared and color remote sensing products can provide strong validation and indirect measurements of open ocean processes, but need cloud free access to the upper ocean while large-scale synthetic ocean models cannot resolve sub meso-scale and small-scale dynamics of ocean phenomenon, including fronts. We therefore resorted to a mix of the two approaches. A qualitative analysis based on monthly averages of previous years indicated that the STF should be located roughly 1500 Km to the west of San Diego (30 °N–35 °N, 130 °W–135 °W). Predictions from CMEMS, a large-scale ocean model, augmented this search to fill in the gaps left by remote sensing (Table 3) .

Onboard the *Falkor*, data collected by Wi-Fi connected robotic platforms' were available in near real-time, as were the observations from underway hull-mounted thermosalinograph, which sampled salinity and temperature from the water surface. Since AUV connectivity is typically sparse, scientists on ship/shore used `Ripples` to access collected data, supervise execution and alter AUV plans when able to communicate. Their observations were augmented by scientists on shore, also connected via `Ripples` in Portugal who were able to weigh in, on critical junctures of decision-making.

Operational Results

Based on climatological analysis of the remote sensing data and daily predictions from large-scale ocean models, ASVs were deployed ahead of the ship to try locate the front and narrow down the survey area; typically in oceanographic cruises such search is performed with the research vessel itself. Using robotic platforms spread over hundreds of Km^2 allowed a large spatial coverage within a

Table 2. A brief overview of the 2018 cruise to sample the Sub-Tropical Arctic Front in the Pacific Ocean.

Date	Event description
April 28th	Analysis of remote sensing SST imagery and numerical model data to confirm STF location
May 1st	Deployment of WaveGlider towards the estimated study region
May 27th	Two Saildrone ASVs tasked to transit to the study area from a previous mission
May 28th	The *Falkor* departs from San Diego; the ASVs spot the apparent frontal zone
May 30th	Arrival of the *Falkor* in the study area and deployment of AUVs
June 8th	First viable SST remote sensing images confirm existence of a frontal jet
June 16th	Start of *Falkor's* return transect to shore

shorter period of time. In this case, data from autonomous platforms indicated frontal crossings at different locations, tens of Kms apart. Using such prior information from ASVs, allowed the scientists to circumscribe the very large area of initial interest to a smaller 100 $\times$ 100 Km^2.

Table 3. Cruise assets that complimented our approach to exploration. A diverse array of platforms and sensors for observation of a number of key ocean variables, including temperature, salinity, depth, planktonic density, chlorophyll concentration, turbidity and fluorescent dissolved organic matter, dissolved oxygen, pH, and water current velocities.

Asset	Description
R/V *Falkor*	Underway water sampling, High-Performance computing cluster (HPC), towed CTD (conductivity, temperature, density), CTD Rosette, ADCP (Acoustic Doppler Current Profiler).
Multiple AUVs	3 equipped with CTDs and long-endurance (>50 h) and 3 other vehicles equipped with CTDs and physical and bio-chemical sensors, with an endurance >24 h (Fig. 1(b))
UAVs	3 FlightWave Edge-130 UAVs with Dimethyl sulfide (DMS) sensors, infra-red or multiSpectral cameras and 1 UAV carrying visible light cameras for outreach purposes (Fig. 1(a))
ASVs	Two SailDrones and a WaveGlider
Other sensors	An underwater plankton imager [25] and an Advanced Laser Fluorometer (ALF) for spectrum analysis of the water-column in near real-time

Upon arrival at the study area (Fig. 1(c)), 3 AUVs were deployed to begin a high-resolution (∼ 2 Km) mapping of the expected frontal zone (Fig. 4(a) – box #1), while the *Falkor* explored the area using an underway thermosalinograph. This first assessment, complemented by the data previously collected by ASVs, and a qualitative analysis of remote sensing data available and predictions from ocean models, worked as a "prior" to inform decision making for the next set of observations.

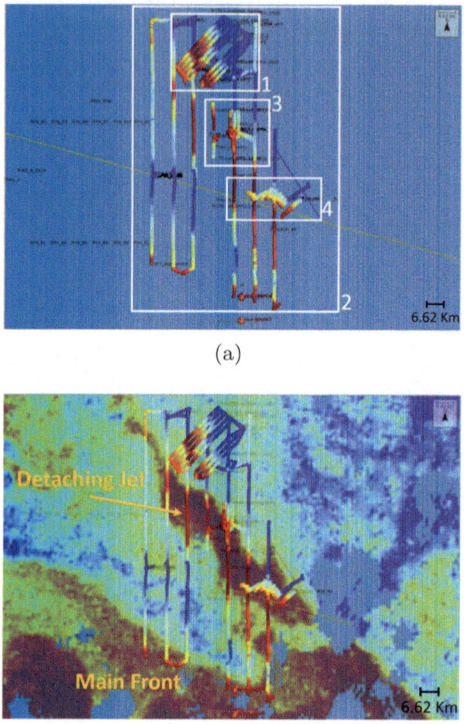

(a)

(b)

Fig. 4. (a) Mapping the front with AUVs. Color gradient for surface salinity measurements (0–5 m depth) acquired by CTDs mounted on AUVs during the yo-yo paths. (b) Surface salinity collected by AUVs and, in background, a SST satellite imagery from June 8th, which confirmed the presence of a frontal jet, partially detached from the main front.

The data collected thus far showed that the initial location scouted by the ASVs may not have been on/near the large-scale front, as the salinity thresholds were crossed twice (in and out of the front). At this point, the numerical models running on the *Falkor*'s HPC proved useful to shed some light on the data collected especially given lack of remote sensing data. Then, several AUVs on parallel tracks in a radiator pattern centered at the front and covering an area of 75 Km × 90 Km enabled an unprecedented sub mesoscale horizontal resolution of 5 Km between cross-frontal sections (Fig. 4(a) – box #2) and a resolution of 800 m along each cross-section. 3D maps of the feature under study were produced using the data sampled by the three AUVs synchronously traveling through the feature of interest (Fig. 5(b)).

During this more traditional straight-line survey, the *Falkor* made observations using shipboard instrumentation over selected areas pinpointed by scientists on ship and shore, with ADCP current profiles, CTD casts, as well as water samples. The data collected along with a no-cloud SST remote-sensing image on June 8th enabled the formulation of a new picture of local ocean dynamics. The data showed a direct correlation between temperature and salinity collected by the AUVs and confirmed the assessments made regarding the existence of a small structure partially detached from the main front, shown in Fig. 4(b) showing surface velocities in the order of 2 knots.

The rapid decision-making during the cruise supported by the software infrastructure and the interaction between onboard and onshore scientists showed that the initially theorized frontal location actually corresponded to a frontal jet, a sub-mesoscale spiral-like feature (Fig. 4(b) and Fig. 5(b)). Such a feature has rarely been observed and never with this level of detail at such large-scale. Responding to this new information, the assets were then refocused on its study. High-resolution surveys were then performed with other AUVs, equipped with physical and biological sensors, augmented with ship-based measurements and water sampling.

Coordinated ship-robotic surveys were later conducted at the boundary of the jet (Fig. 5(a)). UAVs were used in this scenario to collect thermal infrared imagery and test a novel dimethyl sulfide (DMS) sensor over this frontal zone with the sharpest contrast. The sensor was integrated and tested over the front and also areas where high biological activity was identified by the science team. The purpose of these surveys was to evaluate the capability of this early-stage prototype to detect DMS from the air, since it serves as a proxy for biological activity as a trace gas produced by microscopic marine organisms. The results of these surveys are still being analyzed.

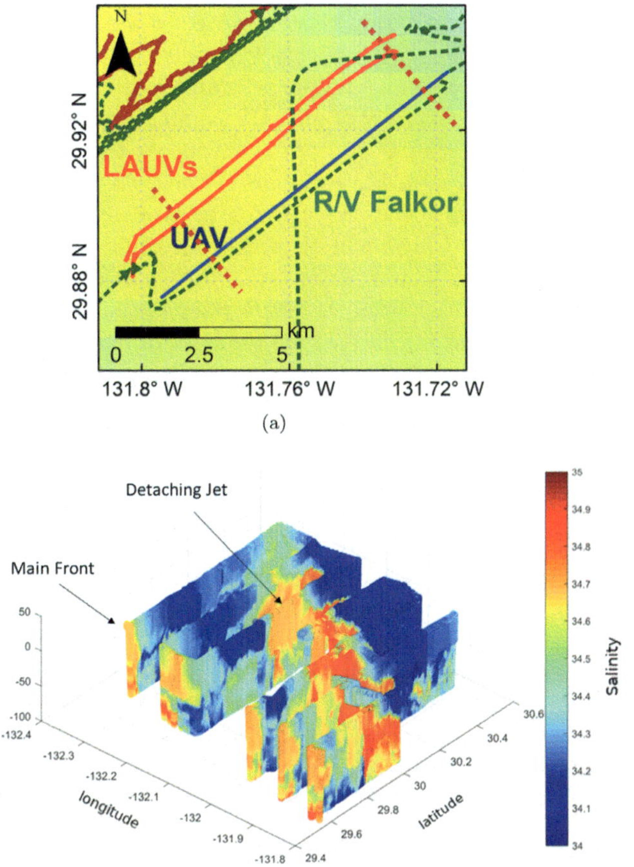

(a)

(b)

Fig. 5. (a) Location of the second coordinated ship-robot survey conducted at the northwestern boundary of the frontal jet (4(a), – box #4). The dashed red lines marked the approximate edges of the front boundary. (b) 3D view of the STF using AUV salinity data (4(b), – box #2). Both the jet detaching from the front and its boundary are visible. Z axis corresponds to depth in meters.

After an exhaustive small scale survey with multiple assets, an automated front tracking algorithm (box #4) [26] was designed and tested initially with the WaveGlider and the *Falkor* and subsequently embedded on an AUV. Survey transects were generated onboard the AUVs according to locally perceived salinity variations and every change of plan was synchronized with `Ripples`. Figure 5(b) shows detailed of the STF generated by the ensemble of AUVs and other assets.

On Situational Awareness. Using the waypoint plan execution methods articulated above, a situational awareness engine was designed which collates temporal projections of all deployed vehicles in our multi-vehicle scenario. Along with Automatic Identification System (AIS), a standard international vessel tracking system, this engine projects waypoints with ship-traffic overlays to warn the AUV operator of impending cross traffic and the possibility of collisions in part to deal with the danger of an AUV coming near the surface on a 'yo-yo'. While the plan projection is an approximation of the actual behavior of the vehicle, it provides a reasonable estimate of the vicinity of vehicle's position at every waypoint. These checks typically occur over a span of 10 h; whenever the engine detects that a vehicle will be within proximity of a ship, it triggers a warning to the operator. This feature, while less critical for open ocean conditions, was developed in anticipation of future work in coastal zones. However, its utility in providing operator awareness in the light of multiple-assets in the water-column was highly effective.

Sub-sampled Data Collection. Oceanographic data collected by the AUVs are transmitted opportunistically over satellite communications. At each waypoint, after synchronization with the central `Ripples` server, the vehicles stay at the surface to send *sub-sampled* data for rapid ship/shore environmental assessment by scientists. It does so by encoding collected data and binning it into a fixed-size array of measurements stored together with a timestamp, approximate GPS position and depth. The resulting encoded vertical profile has size $4 \times n + 4$ bytes (where n is the number of averages encoded) – all the averaged values, the maximum depth, latitude and longitude are encoded as 32-bit floating point numbers, while the timestamp is encoded as a 32-bit integer. After encoding the profile, LZ4 compression is applied and then stored for sending on the next communications opportunity.

5 Discussions and Lessons Learned

The outcomes of this open water cruise were not only the advancement of scientific knowledge or technology advancement, but also how future oceanographic expeditions could leverage the use of autonomous robotic platforms. Not only do such vehicles provide high resolution data across a vastly expanded spatial and temporal scale, but they have significantly altered the *way we explore* by extending the 'human senses'.

The dichotomy of 'exploration versus exploitation' for instance, was an ongoing discussion during daily decision making, driven by admittedly a simple question: 'where and when to sample the water column'. Yet it hides a complex set of technical, scientific and even intuitive biases designed into such an undertaking. For instance, initial exploration efforts leaned towards being risk averse, while the adopted workflow was being tested. As events progressed, operationally riskier outcomes were adopted organically; this is common in exploration of any kind [27]. In addition, scientific interests and the push and pull of disciplinary interests in a culturally diverse group also shaped the cruise's outcomes and decision making.

Exposure to failure is a constant in such endeavors, yet surprisingly our hardware proved to be robust. Six AUVs were deployed for multiple missions while completing a total of more than $\sim$ 1800 km of in-water transit and UAVs performed 21 successful flights. Persistent operations were achieved with 4 daily shifts with 2 operators per shift. The only other failure, of the ship's winch, was compensated by data collection from our ensemble of AUVs and ASVs.

The technology fielded for this cruise was critical for how we worked. Temporal coordination of the assets from the ship, coupled with the dispatching scheme and coupling that to collision avoidance brought increasing predictability. Events planned for the upcoming hours or even days ahead aided the determination of rendezvous points for recovering AUVs. This helped too, in designing, implementing and subsequently fielding a new front tracking algorithm [26] with an AUV without losing track of its execution.

The collision avoidance feature provided cognitive downscaling with added situational awareness to know well in advance of any effective trajectory intersections with the *Falkor* itself, aiding the ships crew as also the team. Not only were the operators able to follow the state of every deployed asset but the full complement of the *Falkor*'s crew were able to be aware of findings throughout the cruise. Such a methodology is even more important in circumstances where a team is physically separated, resulting in a shared understanding of the evolution of activities past, present and future. The use of `Ripples`, which was aggregating data from multiple fielded assets, also contributed significantly. During satellite internet blackouts, `Ripples` was configured to use satellite messaging (for both commands and asset updates). Although this alternate mode of communication was effective allowing execution monitoring and modification of plans, the latency of communication was substantially higher even if it demonstrated the efficacy of operations in the absence of connectivity.

Finally, another outcome was the use of an onshore command/control Ocean Space Center (OSC) playing a key role on this coordination and persistent monitoring, not only aiding in the decisions of where (and when) to go and what to do, but also in the orchestration of the operations with *Falkor*'s bridge. Near real-time situational awareness was not only available to the vehicle operators with `NEPTUS`, but also to all crew members and science team distributed throughout the ship, allowing them to follow the movement of assets that were not within sight for most of the time. This work was a step forward in ocean exploration, demonstrating scientific value in combining traditional and new methods and the ability to explore and sample environmental phenomena using a coordinated multi-vehicle approach.

Acknowledgments. We are grateful to the Schmidt Ocean Institute (SOI) for supporting the 3 week cruise on the R/V *Falkor* and to its crew and captain. We are also grateful to SOI for providing access to the Wave Glider and to Prof. Barbara Block, Stanford University, for access to Saildrones in the mid-Pacific. RM was supported by a Post-Doctoral grant from FCT (Portuguese Foundation for Science and Technology) # SFRH/BPD/115093/2016. LSTS acknowledges support from the ENDURANCE program funded under the NORTE 2020 through the European Regional Development Fund (ERDF).

References

1. Lynn, R.J.: The subarctic and northern subtropical fronts in the eastern north pacific ocean in spring. J. Phys. Oceanogr. **16**(2), 209–222 (1986)
2. Belkin, I.M., Cornillon, P.C.: Fronts in the world ocean's large marine ecosystems. ICES CM **500**(130), 21 (2007)
3. Chapman, C.C., Lea, M.A., Meyer, A., Sallée, J.B., Hindell, M.: Defining southern ocean fronts and their influence on biological and physical processes in a changing climate. Nat. Clim. Change **10**, 209–219 (2020)
4. Meinig, C., Lawrence-Slavas, N., Jenkins, R., Tabisola, H.M.: The use of Saildrones to examine spring conditions in the Bering sea: vehicle specification and mission performance. In: OCEANS 2015-MTS/IEEE Washington, pp. 1–6. IEEE (2015)
5. Hine, R., Willcox, S., Hine, G., Richardson, T.: The wave glider: a wave-powered autonomous marine vehicle. In: OCEANS 2009, MTS/IEEE, Biloxi, Miss, pp. 1–6. IEEE (2009)
6. Sousa, A., Madureira, L., Coelho, J., Pinto, J., Pereira, J., Sousa, J.B., Dias, P.: LAUV: the man-portable autonomous underwater vehicle. IFAC Proc. Volumes **45**(5), 268–274 (2012)
7. de Sousa, J.B., Pereira, J., Pinto, J., Lourenfo, P.C., Galocha, J.M., Fontes, J., Silva, M., Rajan, K., Johansson, T.A., Alves, J., Munafò, A., Pelekanakis, K., Petroccia, R., Silva, M.C., Incze, M.: Rapid environmental picture Atlantic exercise 2015: a field report. In: OCEANS 2016 MTS/IEEE Monterey, pp. 1–6, September 2016
8. Ferreira, A.S., Costa, M., Py, F., Pinto, J., Silva, M.A., Nimmo-Smith, A., Johansen, T.A., de Sousa, J.B., Rajan, K.: Advancing multi-vehicle deployments in oceanographic field experiments. Auton. Rob. 43, 1555–1574 (2018). https://doi.org/10.1007/s10514-018-9810-x
9. Costa, M.J., Pinto, J., Dias, P.S., Pereira, J., Lima, K., Ribeiro, M., Sousa, J.B., Lukaczyk, T., Mendes, R., Tomasino, M.P., Magahlaes, C., Belkin, I., Lopez-Castejon, F., Gilabert, J., Skarpnes, K.A., Ludvigsen, M., Rajan, K., Mirmalek, Z., Chekalyuk, A.: Field report: exploring fronts with multiple robots. In: IEEE AUV, Porto (2018)
10. Schmidt, H., Bellingham, J.G., Johnson, M., Herold, D., Farmer, D.M., Pawlowicz, R.: Real-time frontal mapping with AUVs in a coastal environment. In: OCEANS 96 MTS/IEEE Conference Proceedings. The Coastal Ocean-Prospects for the 21st Century, vol. 3. IEEE, pp. 1094–1098 (1996)
11. Gottlieb, J., Graham, R., Maughan, T., Py, F., Elkaim, G., Rajan, K.: An experimental momentum-based front detection for autonomous underwater vehicles. In: IEEE International Conference on Robotics and Automation (ICRA), St. Paul, MN (2012)

12. Smith, R.N., Schwager, M., Smith, S.L., Jones, B.H., Rus, D., Sukhatme, G.S.: Persistent ocean monitoring with underwater gliders: adapting sampling resolution. J. Field Rob. **28**(5), 714–741 (2011)
13. Smith, R., Py, F., Cooksey, P., Sukhatme, G., Rajan, K.: Adaptive path planning for tracking ocean fronts with an autonomous underwater vehicle. In: ISER, Morocco, June 2014
14. Das, J., Py, F., Harvey, J.B., Ryan, J.P., Gellene, A., Graham, R., Caron, D.A., Rajan, K., Sukhatme, G.S.: Data-driven robotic sampling for marine ecosystem monitoring. Int. J. Rob. Res. **34**(12), 1435–1452 (2015)
15. Fossum, T.O., Fragoso, G.M., Davies, E.J., Ullgren, J., Mendes, R., Johnsen, G., Ellingsen, I., Eidsvik, J., Ludvigsen, M., Rajan, K.: Towards adaptive robotic sampling of phytoplankton in the Coastal ocean. Sci. Rob. **4**(27) (2018). https://doi.org/10.1126/scirobotics.aav3041
16. CANON: Controlled, Agile and Novel Observing Network. http://www.mbari.org/canon/
17. Ramp, S.R., Davis, R.E., Leonard, N.E., Shulman, I., Chao, Y., Robinson, A., Marsden, J., Lermusiaux, P., Fratantoni, D., Paduan, J.D., et al.: Preparing to predict: the second autonomous ocean sampling network (AOSN-II) experiment in the Monterey Bay. Deep Sea Res. Part II **56**(3), 68–86 (2009)
18. Pinto, J., Martins, P.S.D.R., Fortuna, J., Marques, E., Sousa, J.: The LSTS toolchain for networked vehicle systems. In: MTS/IEEE Oceans, pp. 1–9. IEEE (2013)
19. Dias, P.S., Fraga, S.L., Gomes, R.M., Goncalves, G.M., Pereira, F.L., Pinto, J., Sousa, J.B.: NEPTUS- a framework to support multiple vehicle operation. In: MTS/IEEE Oceans, Brest, France, pp. 963–968 (2005)
20. Martins, R., Dias, P., Marques, E., Pinto, J., Sousa, J., Pereira, F.: IMC: a communication protocol for networked vehicles and sensors. In: OCEANS 2009 - EUROPE, pp. 1–6 (2009)
21. Pinto, J., Mendes, R., da Silva, J.C.B., Dias, J.M., de Sousa, J.B.: Multiple autonomous vehicles applied to plume detection and tracking. In: MTS/IEEE OCEANS–Techno-Oceans, Kobe, pp. 1–6 (2018)
22. Rajan, K., Py, F., Berreiro, J.: Towards deliberative control in marine robotics. In: Seto, M., (ed.) Marine Robot Autonomy. Springer Verlag (2012). https://doi.org/10.1007/978-1-4614-5659-9_3
23. Chrpa, L., Pinto, J., Ribeiro, M., Py, F., Sousa, J., Rajan, K.: On mixed-initiative planning and control for autonomous underwater vehicles. In: 2011 IEEE/RSJ International Conference on Intelligent Robots and Systems (IROS), Hamburg, Germany (2015)
24. Muscettola, N., Nayak, P., Pell, B., Williams, B.: Remote agent: to boldly go where no AI system has gone before. AI J. **103**, 5–48 (1998)
25. Davies, E.J., Brandvik, P.J., Leirvik, F., Nepstad, R.: The use of wide-band transmittance imaging to size and classify suspended particulate matter in seawater. Mar. Pollut. Bull. **115**(1–2), 105–114 (2017)
26. Belkin, I., de Sousa, J.B., Pinto, J., Mendes, R., López-Castejón, F.: A new front-tracking algorithm for marine robots. In: IEEE Autonomous Underwater Vehicle (AUV), vol. 2018, pp. 1–3. IEEE (2018)
27. Bellingham, J., Rajan, K.: Robotics in remote and hostile environments. Science **318**(5853), 1098–1102 (2007)

Perception

Composing Pick-and-Place Tasks by Grounding Language

Oier Mees[1(✉)] and Wolfram Burgard[1,2]

[1] University of Freiburg, Freiburg, Germany
meeso@informatik.uni-freiburg.de
[2] Toyota Research Institute, Toyota, USA

Abstract. Controlling robots to perform tasks via natural language is one of the most challenging topics in human-robot interaction. In this work, we present a robot system that follows unconstrained language instructions to pick and place arbitrary objects and effectively resolves ambiguities through dialogues. Our approach infers objects and their relationships from input images and language expressions and can place objects in accordance with the spatial relations expressed by the user. Unlike previous approaches, we consider grounding not only for the picking but also for the placement of everyday objects from language. Specifically, by grounding objects and their spatial relations, we allow specification of complex placement instructions, e.g. "place it behind the middle red bowl". Our results obtained using a real-world PR2 robot demonstrate the effectiveness of our method in understanding pick-and-place language instructions and sequentially composing them to solve tabletop manipulation tasks. Videos are available at http://speechrobot.cs.uni-freiburg.de.

1 Introduction

As robots become ubiquitous across human-centered environments the need for natural and effective human-robot communication grows. Natural language provides a rich and intuitive way for humans and robots to interact due to the possibility of referring to abstract concepts. Moreover, many real-world tasks can be effectively described by a series of language instructions. In this work, we aim to develop an approach that enables a robot to solve complex manipulation tasks by understanding a series of unconstrained language expressions characterizing pick-and-place commands. To do so, the robot has to locate unconstrained object categories based on arbitrary natural language expressions, known as referring expression comprehension, and understand spatial relations to generate object placing locations. In other words, the robot needs to "ground" the referred objects and their spatial relations from language in its world model.

However, understanding unconstrained language instructions is challenging due to the complexity and wide variety of abstract concepts expressed via human language, e.g. "fetch the yellow thing" and "place it left of the bottom object".

B. Siciliano et al. (Eds.): ISER 2020, SPAR 19, pp. 491–501, 2021.
https://doi.org/10.1007/978-3-030-71151-1_43

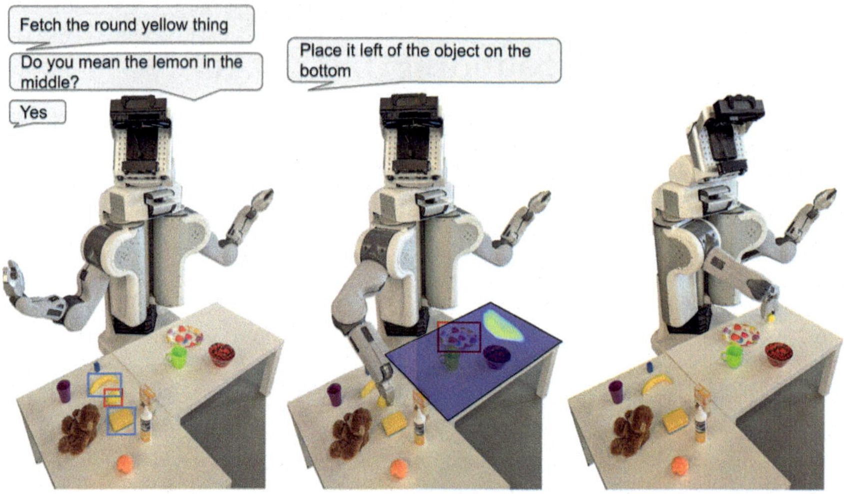

Fig. 1. The goal of our work is to control a robot to perform tabletop manipulation tasks via natural language instructions. Our approach is able to segment objects in the scene, locate the objects referred to in language expressions, solve ambiguities through dialog and place objects in accordance with the spatial relations expressed by the user.

Moreover, the expression might contain ambiguities because there are several "yellow things" in which case the robot should be able to resolve the ambiguity through dialogue, as shown in Fig. 1. Finally, the robot needs to reason about where to place the "yellow thing" relative to the "leftmost container" in order to reproduce the spatial relation "right", which is inherently ambiguous as natural language placement instructions do not uniquely identify a location in a scene.

In this paper, we propose the first comprehensive system for controlling robots to perform tabletop manipulation tasks by sequentially composing unconstrained pick-and-place language instructions. Our approach consists of two neural networks. The first network learns to segment objects in a scene and to comprehend and generate referring expressions. The second network estimates pixelwise object placement probabilities for a set of spatial relations given an input image and a reference object. The interplay between both networks allows for an effective grounding of object semantics and their spatial relationships, without assuming a predefined set of object categories. We demonstrate the effectiveness of our approach by enabling non-expert users to instruct tabletop manipulation tasks to a robot, based on sequences of pick-and-place speech commands.

2 Related Work

Our work is primarily concerned with the task of grounding natural language instructions and spatial relations in the context of the robot's world model [1].

Locating entities in images based on language is closely related to object recognition. Previous works in robotics [2,3] have addressed semantic object retrieval by training classifiers to recognize predefined object categories. These approaches are limited in real-world scenarios as they are not capable of handling variation in the users natural language descriptions and are restricted to a small number of objects.

Spatial relations also play a crucial role in understanding natural language instructions [4,5], as objects are often described in relation to others in tasks such as object placing [6–8] or human robot interaction [2,9,10]. Concretely, spatial relations help the robot disambiguate multiple instances of the same object and to define target areas for placing the picked objects. In our previous work, we introduced a novel method to predict pixelwise object placement probability distributions for a set of commonly used prepositions in natural language [7]. In contrast, we relax the assumption of having a single reference object on the tabletop and add a grounding model to effectively place arbitrary objects in a scene that contains multiple objects.

Recently, there has been significant progress made towards systems that can demonstrate their visual understanding by generating or responding to natural language in the context of images [11–15]. To learn joint visual-linguistic representations, state-of-the-art approaches use convolutional neural networks to encode visual features and recurrent neural networks to process language, replacing traditional handcrafted visual features and language parsers. We leverage advances in modular networks [16–18] for referential expression comprehension. This allows decomposing language into modular components related to subject appearance, location, and relationship to other objects, flexibly adapting to expressions containing different types of information in an end-to-end fashion.

Most related to our approach are the works by Shridhar *et al.* [9] and Hatori *et al.* [4], as both use an interactive fetching system to localize objects mentioned in referring expressions with bounding boxes. We tackle temporally more extended tasks, using our model which enables complex object placement commands such as "place the cup on top of the leftmost box". Notably, sequentially composing pick-and-place language instructions can lead to desirable high-level behaviours, such as tidying up a tabletop or table setting for example. Finally, in contrast to the template-based picking approaches of prior interactive fetching systems [4, 9] we leverage state-of-the-art methods for grasping novel objects with 6-DOF grasps [19].

3 Method Description

In this section we describe the technical details of our method to control a robot to perform tabletop manipulation tasks via natural language instructions. Our approach relies on two models: a grounding model that identifies the most likely object referred by a language instruction and a neural network that predicts object placing locations conditioned on a set spatial relation. An overview of the system is given in Fig. 2.

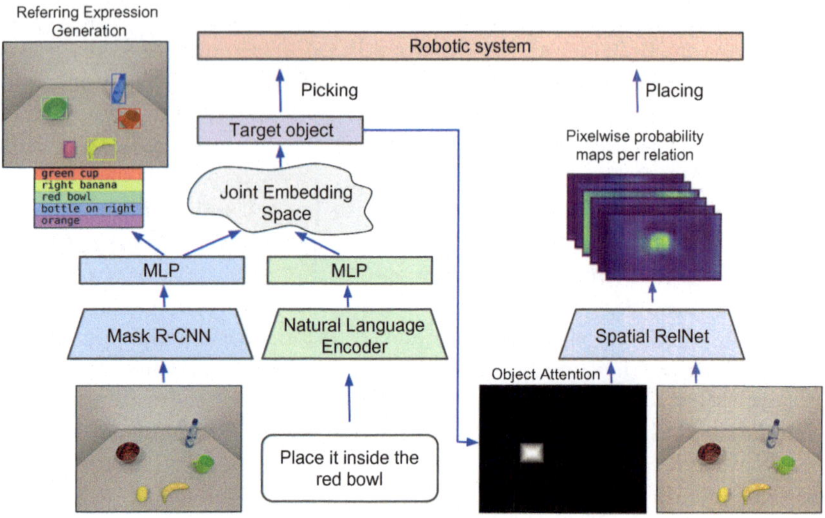

Fig. 2. Overview of the system architecture. Our grounding network processes the input sentence and visual object candidates detected with Mask-RCNN [20] and performs referential expression comprehension. Additionally, it generates referential expressions for each object candidate to disambiguate unclear instructions. Once the reference object of a relative placement instruction has been identified, a second network predicts object placing locations for a set of spatial relations.

3.1 Target Object Selection

We start off by detecting and segmenting all objects in the scene. We train a semantic segmentation network based on Mask-RCNN [20] with a Resnet-101 backbone, which extracts a set of region proposals or object candidates o_i from an image. After all objects on the scene are recognized, we need to identify which object the user is referring to in its language instruction. Given an input image I and expression r, the target object selection is formulated as a task to find the best bounding box from the set of predicted candidate boxes $O = \{o_i\}_{i=1}^{N}$. Our grounding model is based on MAttNet [16], a modular referring expression comprehension network. To enable human-robot communication in cases of ambiguous instructions, we have extended it to support the generation of self-referential expressions, described in Sect. 3.2.

The candidate regions are encoded by a neural network consisting of three modular grounding components related to subject appearance, location and relationship to other objects. These modules combine image features encoded by a Resnet-101 network with relational and geometric features pertaining to the neighborhood or context of each candidate region. The language expression r is encoded in a word embedding layer, which encodes each word in the input sentence to a vector representation, followed by a bi-directional Long Short-Term Memory (LSTM) and a fully-connected (FC) layer. Additionally, the language network learns two types of attention: attention weights that are computed on each word for each module and are summarized as phrase embedding $q^m \mid m \in \{\text{subj, loc, rel}\}$, and module weights $[w_{subj}, w_{loc}, w_{rel}]$ that estimate how much each module contributes to the overall expression score. Each visual module computes scores for each object candidate by calculating the cosine similarity between the vector representation of the instruction, and that of the candidate image region. Finally, the output module takes a weighted average of these scores to get an overall matching score $S(o_i \mid r) = w_{subj}S(o_i \mid q^{subj}) + w_{loc}S(o_i \mid q^{loc}) + w_{rel}S(o_i \mid q^{rel})$. During training, we sample triplets consisting of a positive match (o_i, r_i) and two random negative samples (o_i, r_j) and (o_k, r_i), where o_k is some other object and r_j is an expression describing some other object in the same image to apply a hinge loss:

$$\begin{aligned}\mathcal{L}_1 = \sum_i [&\lambda_1 \max(0, m_1 + S(o_i \mid r_j) - S(o_i \mid r_i)) \\ &+ \lambda_2 \max(0, m_1 + S(o_k \mid r_i) - S(o_i \mid r_i))].\end{aligned} \tag{1}$$

3.2 Resolving Ambiguities

If the referred object cannot be uniquely identified by the grounding model, the system needs to ask for clarification from the human operator. Inspired by recent advances in image caption generation and understanding [13,21,22], we incorporate a LSTM based captioning module to our grounding network that allows the robot to describe each detected object with a natural language description. Our referring expression generation module is jointly trained with our grounding network and shares the features used in the three modules related to subject appearance, location and relationship to other objects. Specifically, the visual target object representation v_i^{vis} is modeled by a concatenation of the ResNet-101 C3 and C4 features, followed by one FC layer which is shared with the comprehension network and one exclusive FC layer. To facilitate the generation of referential expressions that contain location information, such as "the cup in the middle", we leverage the representation learned by the location module v_i^{loc}. This module combines a 5-d vector representing the top-left position, bottom-right position and relative area to the image for the candidate object, together with a relative location encoding of up to five surrounding objects of the same category. Finally, we integrate the output of the relationship module v_i^{rel}, which encodes the appearance and localization offsets of up to five category-agnostic objects in the targets surroundings to enable modeling sentences such as "the teddy bear on top of the box". The final visual representation for the target object is then a concatenation of the above features $v_i = [v_i^{vis}, v_i^{loc}, v_i^{rel}]$. The model is trained to generate sentences r_i by minimizing the negative log-likelihood:

$$\mathcal{L}_2 = -\sum_i \log P(r_i \mid v_i). \tag{2}$$

To generate discriminative sentences, we use a Maximal Mutual Information constraint proposed by Mao *et al.* [21] that encourages the generated expression to describe the target object better than the other objects within the image. Concretely, given a positive match (o_i, r_i) we sample a negative (o_k, r_i), where o_k is some other object, and optimize the following max-margin loss:

$$\mathcal{L}_3 = \sum_i [\lambda_3 \max(0, m_2 + \log P(r_i \mid v_k) - \log P(r_i \mid v_i)). \tag{3}$$

In order to detect if an instruction is ambiguous, we leverage the max-margin loss the comprehension model is trained with. Concretely, during training the max-margin loss aims to guarantee that every correct pair of a sentence and an object has scores by a margin m_1 than any other pair with a wrong object or sentence. Therefore, if at test time there are more than one objects within that threshold, we consider them potential targets. For each candidate we generate multiple self-referential expressions via beam search and use the comprehension module to rerank these expressions and select the least ambiguous expression, similar to Yu *et al.* [22]. We then let the system ask the human "Do you mean ...?". After asking the question, the user can respond "yes" to choose the referred object or "no" to continue iterating through other possible objects. Alternatively, the user can provide a specific correcting response to the question, e.g., "no, the banana on the right", in which case we re-run our grounding module.

3.3 Relational Object Placement

Once an object has been picked, our system needs to be able to place it in accordance with the instructions from the human operator. We combine referring expression comprehension with the grounding of spatial relations to enable complex object placement commands such as "place the ball inside the left box". Given an input image I of the scene and the location of the reference item, identified with our aforementioned grounding module, we generate pixelwise object placement probabilities for a set of spatial relations by leveraging the Spatial-RelNet architecture we introduced in our previous work [7]. We consider pairwise relations and express the subject item as being *in relation to* the reference item. We model relations for a set of commonly used natural language spatial prepositions $C = \{\texttt{inside}, \texttt{left}, \texttt{right}, \texttt{in front}, \texttt{behind}, \texttt{on top}\}$. As natural language placement instructions do not uniquely identify a location in a scene, Spatial-RelNet predicts non-parametric distributions to capture the inherent ambiguity. A key challenge to learning such pixelwise spatial distributions is the lack of ground-truth data. Spatial-RelNet overcomes this problem by leveraging a novel auxiliary learning formulation, as shown in Fig. 3. During training, pixel locations (u, v) are sampled according to the probability maps Γ produced by Spatial-RelNet. To get the learning signal, high level features of objects are implanted into a pretrained auxiliary classifier f_φ to compute a posterior class probability over relations. This way, we can reason over what relation would most likely be formed if we placed an object at the given location.

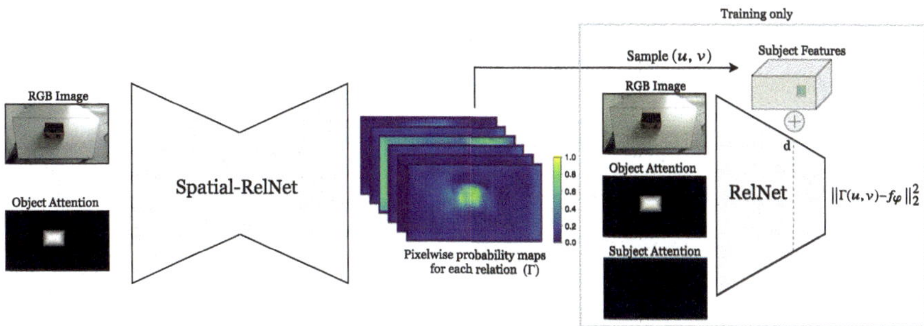

Fig. 3. Our Spatial-RelNet [7] network processes the input RGB image and an object attention mask to produce pixelwise probability maps Γ over a set of spatial relations. During training, we sample locations (u, v) according to Γ, implant inside an auxiliary classifier network at the sampled locations high level features of objects and classify the hallucinated scene representation to get a learning signal for Spatial-RelNet. At test time the auxiliary network is not used.

4 System Implementation

4.1 Machine Learning Setup

During training, we sample the same triplets for both the object comprehension module and the expression generation module. We set the margin $m_1 = 0.1$ for the comprehension ranking and $m_2 = 1.0$ for the generation loss. We additionally use MAttNet's auxiliary visual attribute classification loss. We use the Adam optimizer to train the joint model with an initial learning rate of 0.0004. For the contrastive pairs, we set $\lambda_1 = 1$, $\lambda_2 = 1$ and $\lambda_3 = 0.1$. We make the word embedding of the comprehension and generation modules shared to reduce the number of parameters. For implementation details of Spatial-RelNet, we refer to the original paper [7].

4.2 Robot Setup

To pick an object from language, we first identify the object with our grounding model and extract the corresponding segmentation mask of the selected object. We use an Amazon Echo Dot device to synthesize the voice instructions. We localize the object in 3D space and generate grasp poses with Grasp Pose Detection (GPD) [19], which predicts a series of 6-DOF candidate grasp poses given a 3D point cloud for a 2-finger grasp. The reachability of the proposed candidate grasps are checked using MoveIt!, and the highest quality reachable grasp is executed with the PR2 robot. For placing the object, we first sample a location from the spatial distribution predicted by our Spatial-RelNet model. We rely on keyword spotting to select the corresponding predicted distribution. Next, we localize the pixel coordinate in 3D space and plan a top-down grasp pose to the calculated 3D point. Finally, the end-effector is moved above the desired location and then the gripper is opened to complete the placement.

5 Experiments

We evaluate our approach under two settings. First, we evaluate the capability of our approach to comprehend and generate referring expressions for a wide variety of objects on the RefCOCO dataset [11]. Next, we evaluate the ability of our robotics system to follow pick-and-place language instructions in human-robot experiments.

5.1 RefCOCO Benchmark

The RefCOCO dataset contains images and corresponding referring expressions that uniquely identify a wide variety of objects in the images. We compare our grounding networks ability to comprehend and generate referring expressions against several strong baselines on Table 1. For evaluating the comprehension, we compute the intersection-over-union (IoU) of the selected region with the ground-truth bounding box, considering IoU > 0.5 a correct comprehension. To evaluate the generation module, we leverage standard machine translation metrics commonly used in image captioning, such as METEOR and CIDEr. We observe that by jointly training the comprehension and language generation modules, they regularize each other and improve their respective performances, demonstrating the effectiveness of multitask learning [22–24].

Table 1. Referring expression comprehension and generation on the RefCOCO dataset, with human-annotated ground-truth object regions.

	RefCOCO comprehension			RefCOCO generation			
	Val	TestA	TestB	TestA		TestB	
				Meteor	CIDEr	Meteor	CIDEr
Mao [21]	–	63.15	64.21	–	–	–	–
INGRESS [9]	77	76.7	77.7	–	–	–	–
SLR [4,22]	79.56	78.95	80.22	0.268	0.697	0.329	1.323
MAttNet [16]	85.65	85.26	84.57	–	–	–	–
Ours	**86.15**	**87.18**	**85.36**	**0.29**	**0.753**	**0.33**	**1.33**

5.2 Robot Experiments

We evaluate our approach on two real-world scenarios: picking and placing objects according to user defined object arrangements and a tidy-up task. We will first describe the setup of the object arrangement experiment. Our study involved 4 participants recruited from a university community[1]. The robots workspace

[1] Further quantitative experiments were infeasible at time of submission due to COVID-19.

contained two tables, as shown in Fig. 1. One table contained previously unseen objects in clutter. The second table contained a single reference object. The average number of objects on the cluttered table was 5.6. The participants were asked to instruct a PR2 robot to arrange a desired target scene by picking objects from the cluttered table and using relational expressions to place them on the second table. In addition to the robots RGB-D camera we placed a second camera in front of the cluttered table and performed online registration to compute a global point cloud. The tidy-up task consisted of iteratively picking 4 colored objects from the cluttered table and placing the same colored objects on the left container and the remaining objects on the right container. In this experiment we were interested in evaluating the number of actions the robot has to take to complete the task, given unambiguous instructions.

Table 2 shows the performance of our approach on a PR2 robot for the first experiment. Our approach achieves a 78.3% target object selection accuracy and a 85.7% accuracy on selecting the reference object the placing will be relative to. The higher accuracy of the latter is due to fewer candidate objects being on the placing table and the participants preferring to use ambiguous expressions for the picking instructions. The robot took $\sim$ 20 s to complete an action from the moment the human started to speak. We report a grasping performance of 74.4% with GPD. We find that some objects such as mugs are particularly difficult for GPD as it often fails to find feasible grasps due to either occluded object parts or noisy measurements on thin structures such as rims. Our object placement approach achieves a success rate of 85.7%. We observe some failure cases for large object placements, because of missing 3D priors of the objects to be placed. Thus, when placing a big box left of a small box, it is possible that the chosen placement results in the big box partially ending up on top of the small box. For the tidy up task, we report a mean task length of 14.4 actions, due to several re-grasp attempts. Overall, our results demonstrate the ability of our approach to allow non-expert users to instruct tabletop manipulation tasks based on sequences of pick-and-place speech commands.

Table 2. Performance of our approach on a real robot platform following natural language instructions to pick and place objects in a tabletop scenario.

	Target object selection	Target object grasping	Placing base grounding	Placing success	Avg. number of feedback	Pick and place
Ours	78.3% (47/60)	74.4% (35/47)	85.7% (30/35)	83.3% (25/30)	0.63 (60/95)	63% (60/95)

6 Conclusions and Discussion

In this paper, we proposed the first robotic system that allows non-expert users to instruct tabletop manipulation tasks by sequentially composing unconstrained pick-and-place language instructions and can clarify a human operator's intention through dialogue. We demonstrate the effectiveness of our approach to encode high-level behaviours in a highly challenging, realistic environment. Even though we are far from achieving robots that can learn to relate human language to their world model, we hope our work is a step in this direction.

While the experimental results are promising, our approach has several limitations. First, relative object placement instructions do not allow for fine-grained target specification due to its inherent ambiguity. Addressing this issue would require learning user preferences from feedback [25]. Second, we observe some failure cases for large object placements, because of missing 3D priors of the objects to be placed. Integrating 3D priors is a natural extension to enable optimizing placement poses [26] and to reason over the effects of actions on the scene [27]. Third, we find that GPD often fails to find feasible grasps due to either occluded object parts or noisy measurements. Integrating methods that can complete occluded scene regions [28,29] or generate more diverse grasps [30] might help alleviating these problems. Finally, our approach is limited to tabletop tasks that can be characterized by pick-and-place actions. An exciting area for future work may be one that not only grounds object semantics and spatial relations, but also grounds actions in order to learn complex behaviours with language conditioned continuous control policies [31,32].

Acknowledgments. This work has been supported partly by the Freiburg Graduate School of Robotics and the German Federal Ministry of Education and Research under contract number 01IS18040B-OML. We thank Henrich Kolkhorst for his contributions to the speech-to-text pipeline and to Andreas Eitel for valuable discussions.

References

1. Clark, H.H., Brennan, S.E.: Grounding in communication. Prespectives on Socially Shared Cognition (1991)
2. Guadarrama, S., Riano, L., Golland, D., Go, D., Jia, Y., Klein, D., Abbeel, P., Darrell, T.: Grounding spatial relations for human-robot interaction. In: IROS (2013)
3. Pangercic, D., Pitzer, B., Tenorth, M., Beetz, M.: Semantic object maps for robotic housework-representation, acquisition and use. In: IROS (2012)
4. Hatori, J., Kikuchi, Y., Kobayashi, S., Takahashi, K., Tsuboi, Y., Unno, Y., Ko, W., Tan, J.: Interactively picking real-world objects with unconstrained spoken language instructions. In: ICRA (2018)
5. Paul, R., Arkin, J., Roy, N., M Howard, T.: Efficient grounding of abstract spatial concepts for natural language interaction with robot manipulators. In: RSS (2016)
6. Jiang, Y., Lim, M., Zheng, C., Saxena, S.: Learning to place new objects in a scene. IJRR **31**(9), 1021–1043 (2012)
7. Mees, O., Emek, A., Vertens, J., Burgard, W.: Learning object placements for relational instructions by hallucinating scene representations. In: ICRA (2020)
8. Mees, O., Abdo, N., Mazuran, M., Burgard, W.: Metric learning for generalizing spatial relations to new objects. In: IROS (2017)
9. Shridhar, M., Hsu, D.: Interactive visual grounding of referring expressions for human-robot interaction. In: RSS (2018)
10. Misra, D.K., Sung, J., Lee, K., Saxena, A.: Tell me dave: context-sensitive grounding of natural language to manipulation instructions. IJRR **35**(1-3), 281–300 (2016)

11. Kazemzadeh, S., Ordonez, V., Matten, M., Berg, T.: Referring to objects in photographs of natural scenes. In: EMNLP, Referitgame (2014)
12. Antol, S., Agrawal, A., Jiasen, L., Mitchell, M., Batra, D., Zitnick, C.L., Parikh, D.: Visual question answering. In: ICCV, Vqa (2015)
13. Johnson, J., Karpathy, A., Fei-Fei, L.: Fully convolutional localization networks for dense captioning. In: CVPR, Densecap (2016)
14. Anderson, P., Qi, W., Teney, D., Bruce, J., Johnson, M., Sünderhauf, N., Reid, I., Gould, S., van den Hengel, A.: Interpreting visually-grounded navigation instructions in real environments. In: CVPR, Vision-and-language navigation (2018)
15. Reed, S., Akata, Z., Yan, X., Logeswaran, L., Schiele, B., Lee, H.: Generative adversarial text to image synthesis. In: ICML (2016)
16. Licheng, Y., Lin, Z., Shen, X., Yang, J., Xin, L., Bansal, M., Berg, T.L.: Modular attention network for referring expression comprehension. In: CVPR, Mattnet (2018)
17. Hu, R., Rohrbach, M., Andreas, J., Darrell, T., Saenko, K.: Modeling relationships in referential expressions with compositional modular networks. In: CVPR (2017)
18. Andreas, J., Rohrbach, M., Darrell, T., Klein, D.: Neural module networks. In: CVPR (2016)
19. Gualtieri, M., Ten Pas, A., Saenko, K., Platt, R.: High precision grasp pose detection in dense clutter. In: IROS (2016)
20. He, K., Gkioxari, G., Dollár, P., Girshick, R.: Mask r-cnn. In: CVPR (2017)
21. Mao, J., Huang, J., Toshev, A., Camburu, O., Yuille, A.L., Murphy, K.: Generation and comprehension of unambiguous object descriptions. In: CVPR (2016)
22. Yu, L., Tan, H., Bansal, M., Berg, T.L.: A joint speaker-listener-reinforcer model for referring expressions. In: CVPR (2017)
23. Caruana, R.: Multitask learning. Mach. Learn. **28**(1), 41–75 (1997)
24. Mees, O., Merklinger, M., Kalweit, G., Burgard, W.: Unsupervised robot skill learning from videos. In: ICRA, Adversarial skill networks (2020)
25. Abdo, N., Stachniss, C., Spinello, L., Burgard, W.: Organizing objects by predicting user preferences through collaborative filtering. IJRR **35**(13), 1587–1608 (2016)
26. Haustein, J.A., Hang, K., Stork, J.A., Kragic, D.: Object placement planning and optimization for robot manipulators. In: IROS (2019)
27. Nematollahi, I., Mees, O., Hermann, L., Burgard, W.: Unsupervised structured dynamics models from physical interaction. In: IROS, Hindsight for foresight (2020)
28. Mees, O., Tatarchenko, M., Brox, T., Burgard, W.: Self-supervised 3d shape and viewpoint estimation from single images for robotics. In: IROS (2019)
29. Varley, J., DeChant, C., Richardson, A., Ruales, J., Allen, P.: Shape completion enabled robotic grasping. In: IROS (2017)
30. Mousavian, A., Eppner, C., Fox, D.: 6-dof graspnet: variational grasp generation for object manipulation. In: ICCV (2019)
31. Lynch, C., Sermanet, P.: Grounding language in play. arXiv preprint arXiv:2005.07648 (2020)
32. Shao, L., Migimatsu, T., Zhang, Q., Yang, K., Bohg, J.: Concept2robot: learning manipulation concepts from instructions and human demonstrations. In: RSS (2020)

Multi-sensory Integration in a Quantum-Like Robot Perception Model

Davide Lanza[1(✉)], Paolo Solinas[2], and Fulvio Mastrogiovanni[1]

[1] Department of Informatics, Bioengineering, Robotics, and Systems Engineering, University of Genoa, Via All' Opera Pia 13, 16145 Genoa, Italy
davide.lanza@eleves.ec-nantes.fr, fulvio.mastrogiovanni@unige.it
[2] Department of Physics, National Institute for Nuclear Physics (Genoa section), University of Genoa, Via Dodecaneso 33, 16146 Genoa, Italy
solinas@fisica.unige.it

Abstract. Formalisms inspired by Quantum theory have been used in Cognitive Science for decades. Indeed, Quantum-Like (QL) approaches provide descriptive features that are inherently suitable for perception, cognition, and decision processing. A preliminary study on the feasibility of a QL robot perception model has been carried out for a robot with limited sensing capabilities [1]. In this paper, we generalize such a model for multi-sensory inputs, creating a multidimensional world representation directly based on sensor readings. Given a 3-dimensional case study, we highlight how this model provides a compact and elegant representation, embodying features that are extremely useful for modeling uncertainty and decision. Moreover, the model enables to naturally define query operators to inspect any world state, which answers quantifies the robot's degree of belief in that state.

1 Introduction and Related Work

Quantum computing has been applied in robotics as a tool for speed up classical tools, or in a framework that is still the classical one [2]. Our approach is quite different: starting from the very properties of quantum systems, we studied how to exploit them in a novel, simpler framework. This approach has been shown effective in quantum perception and cognition modeling, and we argue that this could be extremely useful in Robotics, providing also a way to translate the theoretical models of quantum cognition to practical robotics application. In contrast to the aforementioned research in Quantum Robotics, this approach could be useful even as purely simulated, because its merits are due to quantum system properties rather than merely computational advantages.

Since the early intuitions by Amann [3], quantum cognition research studied the links between perception and quantum dynamics [4–7]. A relevant example is the work in Manousakis [6], which proposed a QL model to describe probability distributions of perceptive dominances in subjects experiencing binocular rivalry.

A preliminary model inspired by the work in Manousakis has been proposed to assess the feasibility of a QL perception model for a robot with limited sensing

B. Siciliano et al. (Eds.): ISER 2020, SPAR 19, pp. 502–509, 2021.
https://doi.org/10.1007/978-3-030-71151-1_44

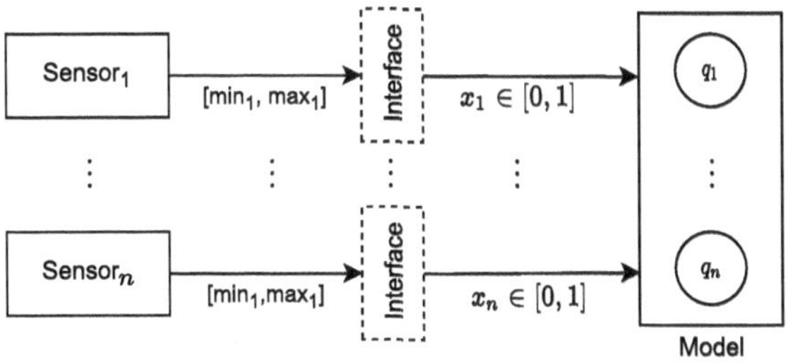

Fig. 1. Sensor and input interfaces scheme for a model made by n qubits q. Each sensor i provides a reading bet $\in [\min_i, \max_i]$ which is then normalized between 0 and 1 by the correspondent interface. The input vector of the model is $\boldsymbol{x} = [x_1, \ldots, x_n]$

capabilities [1]. The reason behind this choice is the great descriptive potential which a quantum formalism inherently provides [8–12].

Indeed, following Caves *et al.* [13], quantum probability theory can be understood within the Bayesian approach, with probabilities quantifying the degree of belief about a certain state. In this case, maximal information for a question does not imply complete knowledge, i.e., it does not allow us to predict which state will be measured (which answer will be given) but provides only each state's probability of being measured (the degree of belief about the possible answers).

This interpretation of the measurement as a *query* given a certain belief (i.e., the quantum system state) can be extremely useful for decision-making [14] and, as we show in the present paper, it allows to extend the model's properties to the whole n-dimensional Hilbert space of the quantum system. Moreover, we argue that this interpretation could be adapted with significant results in robot perception and cognition models as well. A QL model provides a way to deal with uncertain perceptual knowledge and decision-making without an explicit representation. We posit that this is a more elegant, more compact approach because the *state* is not a mere vector state as in current thinking. Representing it as a quantum state, we may be capable of leveraging the properties related to measurement and uncertainty in quantum mechanics. Finally, this approach discloses new perspectives for further investigations, since a wide range of QL cognition models have been discussed in the existing literature and can be applied to Robotics [9,10].

The present work is part of the quantum-robot.org project, which aims to develop a high-level framework for QL cognitive architecture design. The preliminary QL perception model proposed investigated in [1] dealt with one sensory input channel, performing a time integration of the input discriminating between two states. The goal of the present study is to generalize such a first single-qubit model to a multi-qubit approach (see Fig. 1). Moreover, exploiting state superposition and a change of basis in the Hilbert space H, we can significantly extend the considered states range. Indeed, we can virtually deal with any possible

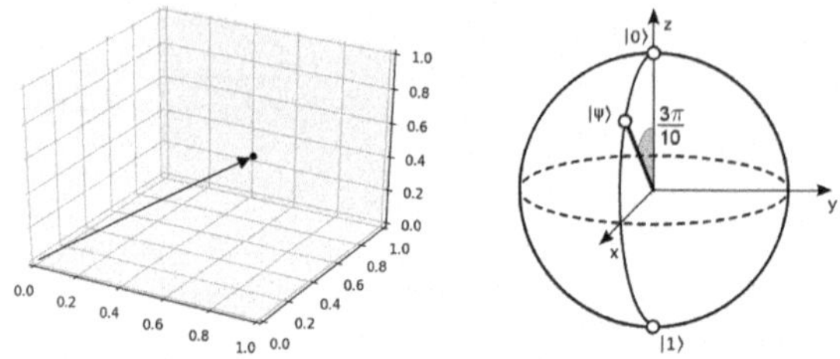

Fig. 2. Three-dimensional representation of a vector input $\boldsymbol{x} = [0.8, 0.3, 0.7]$ for a $n = 3$ model (left) and the corresponding Bloch sphere representation of the second qubit q_2 encoding $x_2 = 0.3$ (right).

state in H[1]. To keep the analysis simpler, we do not consider time integration in this paper, although the model may be easily extended to time windows Δt as illustrated for the single-qubit model in [1].

2 Technical Approach

As illustrated in Fig. 1 we consider n sensors, each one returning a lower- and upper-bounded scalar reading. Each reading is then domain-wise normalized by the correspondent interface. Then, the model receives as input a n dimensional vector composed of real values x_i between 0 and 1, each one corresponding to the i-th sensor. At each reading update, the model has a vector $\boldsymbol{x} = [x_1, \ldots, x_n]$ in input. For example, considering a camera-like sensor able to provide only the three RGB average values of the image, we can decompose it in three sensors, each of them with an interface normalizing the $[0, 255] \in \mathbb{N}$ readings in a $[0, 1] \in \mathbb{R}$ interval. In this case, the input vector $\boldsymbol{x}$ is represented in three dimensions as shown in Fig. 2. Every qubit q_i encodes the sensory information of the corresponding i-th sensor[2]. Following [1] we encode sensory data with a unitary operator U which applies a rotation $R_y(\pi x_i/\tau)$ to $q_i, \forall i \in [1, n]$. The main differences with the previous model rely on the multi-qubit generalization, the lack of temporal integration ($\tau = 1$ for the model proposed here), and the extension to continuous inputs (the previous study assumed x_i being either 0 or 1). Therefore, the information x_i is encoded in the angle of the Bloch sphere representation of q_i, as shown in Fig. 2.

Many indirect methods are available to exploit information encoded in qubits [16]. Here, we consider only state measurements exploiting Caves *et al.* [13] interpretation, as stated in Sect. 1. Measuring the quantum system leads

[1] It is noteworthy that, despite this great freedom in choosing a basis, it is possible to query only one state at a time, since the quantum superposition collapses in one of the basis states after each measurement [15].

[2] This sensor-dimension correspondence holds up for the Hilbert state's dimensions if no change of basis is operated (no query is applied to the model) as seen in the experimental results reported in Table 1.

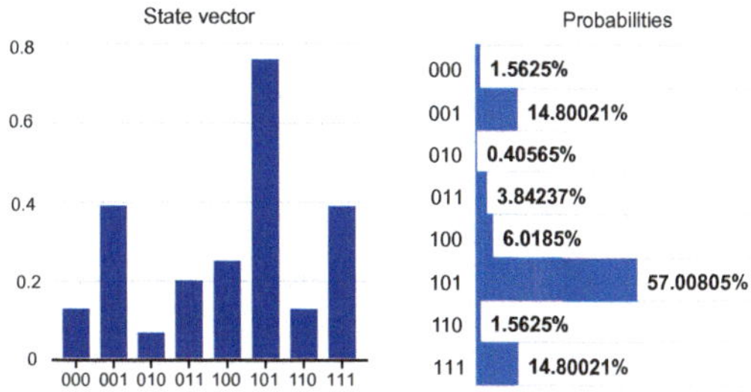

Fig. 3. The state vector $|\psi\rangle$ histogram and the corresponding measurement probabilities of the quantum model encoding the vector input $\boldsymbol{x} = [0.8, 0.3, 0.7]$ of Fig. 2 (IBM Quantum Experience Circuit Composer simulation).

to the collapse of its state in one of its basis states, namely the set of 2^n states composed by all the ordered combinations of the $|0\rangle, |1\rangle$ basis states of each single qubit. The probability for the collapse to produce a certain state as a measurement is given by the current state superposition. For example, considering the input vector $\boldsymbol{x} = [0.8, 0.3, 0.7]$ we saw in Fig. 2, the overall system state $|\psi\rangle$ produced by the application of the rotation operators is $|\psi\rangle = [0.125, 0.385, 0.064, 0.196, 0.245, 0.755, 0.125, 0.385]$.

As illustrated in Fig. 3, the probability of measuring a certain state is the square of the vector coefficient corresponding to that state. For example, for the 6-th state $|101\rangle$[3] we have a coefficient $c_6 = 0.755$ and then the probability of measuring it is $|c_6|^2 = 0.57$ [15]. These coefficients are related to the input vector due to the applied rotation operators [1]. Since every superposition state is itself a state in the Hilbert space and has a physical meaning on its own, we can operate a basis change. This allows us to define a "query operator" $Q(\bar{\boldsymbol{x}})$ which addresses a specific target perceived state $\bar{\boldsymbol{x}}$, applying an inverse rotation $R_y(-\bar{x}_i\pi)$ to $q_i, \forall i \in [1, n]$. This enables us to directly associate the state $|000\rangle$ with this target state. However, by changing the basis we lose the sensor-dimension correspondence and hence the other basis states lose the analogical meaning they retain in the original basis. Nevertheless, if we measure the state after applying Q, we know that obtaining a state containing "two zeros" means measuring a state closer to the target rather then one with just "one zero", while $|111\rangle$ is the opposite state.

3 Experiments and Results

We defined a case study considering an ideal camera-like sensor providing the average RGB values of each recorded image. This RGB color sensor has been

[3] We use the IBMQ Qiskit [17] notation rather than the one usually used in quantum mechanics. This means that, in the Dirac notation, the states which are normally ordered as $|q_1, \ldots, q_n\rangle$ are instead ordered as $|q_n, \ldots, q_1\rangle$, following the usual bits notation $|\text{MSB}, \ldots, \text{LSB}\rangle$.

Table 1. The Model behavior in the original basis and applying $Q(\bar{\boldsymbol{x}})$, $N = 10^6$. For the examples in which a query has been applied, the target and the Euclidean distance d(input, target) are reported.

Input (R,G,B)	Target (R,G,B)	$\|000\rangle$	$\|001\rangle$	$\|010\rangle$	$\|011\rangle$	$\|100\rangle$	$\|101\rangle$	$\|110\rangle$	$\|111\rangle$	d
(0,25,0)	–	97.67%	0%	2.33%	0%	0%	0%	0%	0%	–
(55,0,210)	–	6.67%	8.31%	0%	0%	82.27%	10.23%	0%	0%	–
(10,75,125)	–	41.10%	0.15%	10.22%	0.04%	38.75%	0.14%	9.57%	0.04%	–
(0,200,200)	–	1.22%	0%	9.83%	0%	9.82%	0%	79.13%	0%	–
(230,15,230)	–	0.05%	2.28%	0%	0.02%	2.28%	94.55%	0.02%	0.81%	–
(215,225,220)	–	0.01%	0.14%	0.26%	4.15%	0.19%	3.02%	5.47%	86.77 %	–
(102,18,124)	(132,35,107)	94.5%	3.3%	1.0%	0%	1.0%	0%	0%	0%	38.44
(84,48,38)	(132,35,107)	75.5%	7%	0.5%	0%	15.5%	1.4%	0.1%	0%	85.05
(36,101,84)	(132,35,107)	57.0%	25.7%	10.6%	4.8%	1.2%	0.5%	0.2%	0%	118.75
(239, 239,110)	(132,35,107)	6.0%	3.6%	56.6%	33.8%	0%	0%	0%	0%	230.38

conceptually decomposed in three scalar sensors (Fig. 1). It is noteworthy that RGB scalar values can be thought of as scalar readings coming from different sensors, even based on different physical transduction mechanisms. However, for representation' sake, we opted for the RGB average values because they allow for a compact, color-related vector representation.

We implemented a 3-qubit model (2^3 basis states) relying on the IBMQ Qiskit framework [17]. To collect data about the probable outcomes, for each tested input $\bar{\boldsymbol{x}}$, we simulated $N = 10^6$ measurements. In Table 1 are reported some detailed operative examples, either in the canonical base or after a specific query, i.e., applying $Q(\bar{\boldsymbol{x}})$. In the first case, the basis states maintain a precise meaning, hence the corresponding colors are reported. In the second case, the Euclidean distance d between the input and the target is added[4]. To exhaustively explore the model's behavior through all the possible inputs, we sampled the RGB space with a sampling step of 5. Hence, we tested the model for $(255/5)^3 = 132651$ different inputs[5]. We have not applied any Q operator for these tests since the behavior would not change. Indeed, applying Q changes only the basis states, not the behavior of the system in their regards.

Graphical visualizations of the behavior for three basis states and the $|000\rangle$ state relative to its distance to the input are reported in Fig. 4 and Fig. 5, respectively.

4 Experimental Insights

The model behaves as expected. The confidence curve is shown in Fig. 5 has a sinusoidal shape as observed in [1], which gives a nonlinear, yet definite, correspondence between stored information and measurements. As illustrated in

[4] Implemented using NumPy's norm function on the difference between the input and the target RGB vectors (not their normalized counterpart $\boldsymbol{x}$ and $\bar{\boldsymbol{x}}$).

[5] The simulation took several days, but the average computational time of a single measurement simulation on an Intel Core i5 10210U is $0.5 \cdot 10^{-2}$ s (tests available in [18] via notebooks).

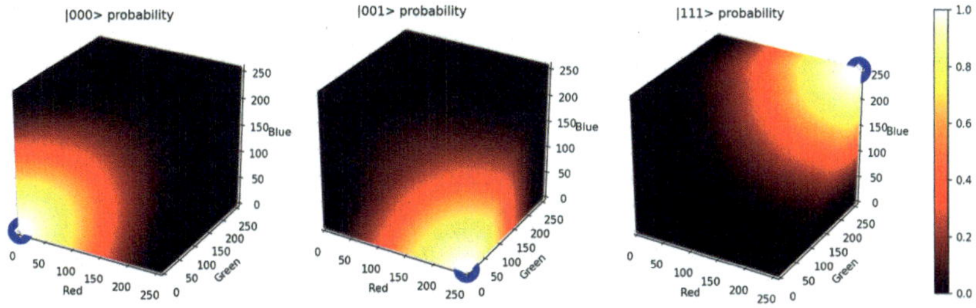

Fig. 4. Probabilities spanned through the whole RGB space. Input $\bar{\boldsymbol{x}}$ are points in the RGB cube sampled with a step of 5. No Q applied. The basis state considered is circled in blue.

Fig. 5, asserting similarity relatively to the euclidean distance between the measured state and $|000\rangle$ (the target state for queries) we notice how the more "zeros" the measured state has, the more is similar to $|000\rangle$. This result is important because it allows us to use the n qubit system as a multidimensional data integrator which inherently provides a coherent representation for the n scalar inputs, a similarity metric'relying on the geometric properties of the Hilbert space H, and a quantum decision process based upon this metric in the knowledge representation space. Moreover, these properties can be extended to every state, allowing a compact and elegant representation. Indeed, even if Fig. 5 refers to the original basis, this behavior can be easily generalized using every state as a target applying $Q(\bar{\boldsymbol{x}})$ accordingly.

The limits of this approach are shown as well in Fig. 5, where the $|111\rangle$ outcome is obtained only for extremely different input-target combinations, e.g., $\bar{\boldsymbol{x}}$ near $|001\rangle$ and $\boldsymbol{x}$ near $|110\rangle$. Targeting states with queries that span the whole RGB cube will likely give readings which have, at most, one "zero", as seen for the results reported in Table 1. Further work should investigate either if it is possible

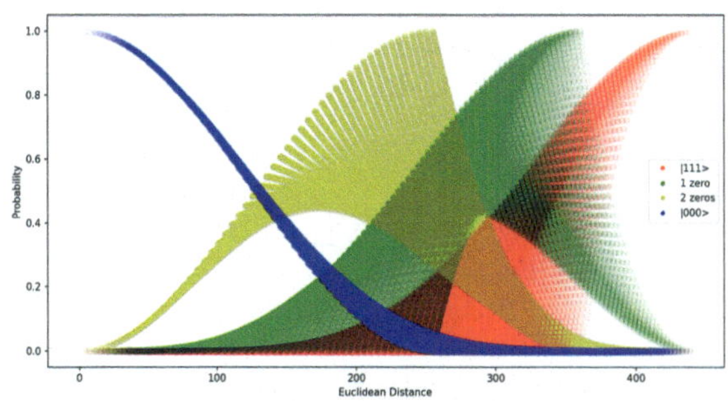

Fig. 5. The probability of measuring a $|000\rangle$ outcome considering $\boldsymbol{x}$ from the full RGB space. Results are plotted along the Euclidean distance between RGB (0, 0, 0) and the input RGB vector. $|001\rangle, |010\rangle, |100\rangle$ grouped as "2 zeros". $|011\rangle, |101\rangle, |110\rangle$ grouped as "1 zero". No Q applied.

to narrow the probability curves for a target state or show how this would not represent an obstacle in practical applications requiring such narrowing. In this regard, multi-stage, cascade models could help in addressing the issue.

It is noteworthy that in our case study colors are just used as a graphical tool rather than expressing an actual concept (e.g., the "redness" of a certain input color). The results provided in Table 1 are not to consider as precise statements about the color "perceived" by the system, rather a decision/classification process based on incomplete data. The measurement probabilities indicate the degree of confidence the model has in answering the query in a certain way, based upon the previously collected knowledge.

Finally, we have to recall that for this study, the knowledge implied in such a quantum decision-making process relates only to a single instant, but this is not a constraint. As already studied in [1], it is possible to generalize the proposed model for extended time windows, taking into account the robot perceptual history on a certain temporal window.

References

1. Lanza, D., Solinas, P., Mastrogiovanni, F.: A preliminary study for a quantum-like robot perception model. *[quant-ph]*. arXiv:2006.02771, June 2020
2. Petschnigg, C., Brandstotter, M., Pichler, H., Hofbaur, M., Dieber, B.: Quantum computation in robotic science and applications. In: 2019 International Conference on Robotics and Automation (ICRA) (2019)
3. Amann, A.: The Gestalt problem in quantum theory: generation of molecular shape by the environment. Synthese **97**(1), 125–156 (1993). ISSN 0039-7857, 1573-0964. https://doi.org/10.1007/BF01255834
4. Conte, E.: Testing Quantum Consciousness. NeuroQuantology **6**(2) (2008). ISSN 1303-5150. https://doi.org/10.14704/nq.2008.6.2.167
5. Conte, E., Khrennikov, A.Y., Todarello, O., Federici, A., Mendolicchio, L., Zbilut, J.P.: Mental states followquantum mechanics during perception and cognition of ambiguous figures. Open Syst. Inf. Dyn. **16**(01), 85–100 (2009). ISSN 1230-1612, 1793-7191. https://doi.org/10.1142/S1230161209000074
6. Manousakis, E.: Quantum formalism to describe binocular rivalry. Biosystems **98**(2), 57–66 (2009). ISSN 0303-2647. https://doi.org/10.1016/j.biosystems.2009.05.012
7. Paraan, M.R., Bakouie, F., Gharibzadeh, S.: A more realistic quantum mechanical model of conscious perception during binocular rivalry. Front. Comput. Neurosci. **8**, 15 (2014). ISSN 1662-5188. https://doi.org/10.3389/fncom.2014.00015
8. Khrennikov, A.Y.: Ubiquitous Quantum Structure. Springer, Heidelberg (2010). ISBN 978-3-642-05100-5, 978-3-642-05101-2. https://doi.org/10.1007/978-3-642-05101-2
9. Busemeyer, J.R., Bruza, P.D.: Quantum Models of Cognition and Decision. Cambridge University Press (2012). ISBN 978-0-511-99771-6
10. Asano, M., Khrennikov, A., Ohya, M., Tanaka, Y., Yamato, I.: Quantum Adaptivity in Biology: From Genetics to Cognition. Springer, Dordrecht (2015). ISBN 978-94-017-9818-1. https://doi.org/10.1007/978-94-017-9819-8

11. Haven, E., Khrennikov, A. (eds.): The Palgrave Handbook of Quantum Models in Social Science: Applications and Grand Challenges. Palgrave Macmillan, London (2017). ISBN 978-1-137-49275-3, 978-1-137-49276-0. https://doi.org/10.1057/978-1-137-49276-0
12. Conte, E.: Algebraic Quantum Theory of Consciousness. Aracne, Rome (2018). ISBN 978-88-255-1707-1
13. Caves, C.M., Fuchs, C.A., Schack, R.: Quantum probabilities as Bayesian probabilities. Physical Review A 65(2), 022305 (2002). ISSN 1050-2947, 1094-1622. https://doi.org/10.1103/PhysRevA.65.022305
14. Busemeyer, J.R., Wang, Z.: What is quantum cognition, and how is it applied to psychology? Curr. Dir. Psychol. Sci. **24**(3), 163–169 (2015). SAGE Publications Inc. ISSN 0963-7214. https://doi.org/10.1177/0963721414568663
15. Nielsen, M.A., Chuang, I.L.: Quantum Computation and Quantum Information. Cambridge University Press, December 2010. ISBN 978-0-511-97666-7
16. Hangos, K.M., Ruppert, L.: State estimation methods using indirect measurements. In: Quantum Probability and Related Topics, Santiago, Chile. World Scientific, pp. 163–180, January 2011. ISBN 978-981-4338-73-8. https://doi.org/10.1142/9789814338745_0009
17. IBM Quantum Experience, *Qiskit*. https://github.com/Qiskit/qiskit
18. Lanza, D.: Quantum-robot Python Package. https://github.com/Davidelanz/quantum-robot

Probabilistic Representation of Objects and Their Support Relations

Fabian Paus(✉) and Tamim Asfour

Institute for Anthropomatics and Robotics, Karlsruhe Institute of Technology (KIT), Adenauerring 2, 76131 Karlsruhe, Germany
{paus,asfour}@kit.edu

Abstract. Understanding uncertainty about objects and their relations in a scene is essential for action selection in robotics. We propose a novel approach for a probabilistic representation of objects and their support relations taking into account pose and shape uncertainty. Starting with a segmented point cloud a probability distribution over the object geometry is estimated, from which samples are drawn to calculate a probability distribution over support relations. To evaluate the approach, we created a new RGB-D dataset, the KIT Support Relation dataset (KIT-SR), consisting of 60 scenes annotated with pixel-wise object labels and ground-truth support relations. Furthermore, we augmented the Object Segmentation Database (OSD) with support relation annotations. We evaluated our proposed probabilistic approach against two state-of-the-art deterministic approaches and show significantly improved precision, recall, F1, and Brier scores.

Keywords: Uncertainty representation · Object relations · RGB-D

1 Introduction

In cluttered scenes, humans are able to intuitively understand which objects are supported by each other and which objects can be safely removed without causing the scene to collapse [2]. To equip a robot with such an intuitive physics understanding, support relations between objects need to be extracted from its sensors. Pure computer vision approaches have been developed to extract support relations between segmented image parts using MAP (Maximum a posteriori) inference [8], energy function minimization [3,10], or stability reasoning [13]. The goal of these approaches is to exploit support relations to improve image segmentation. In robotics, support relations are required to determine manipulation order in clutter [6]. Furthermore, the extracted relations should be physically plausible, enabling the robot to select actions [5] and predict action outcomes [4]. These approaches rely on the robot's knowledge about the camera pose, and therefore the gravity vector. By determining which objects exerts a force on another, support relations are derived. Recently, different learning algorithms

B. Siciliano et al. (Eds.): ISER 2020, SPAR 19, pp. 510–519, 2021.
https://doi.org/10.1007/978-3-030-71151-1_45

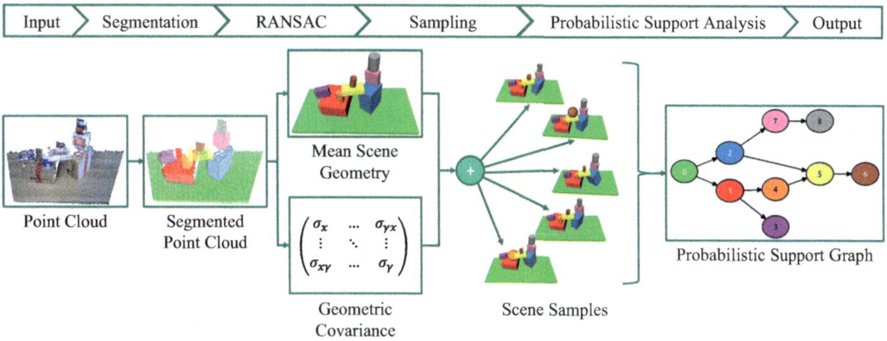

Fig. 1. Overview of the probabilistic analysis of support relations based on point clouds.

using CRFs [11] and CNNs [12] have been proposed to extract manipulation order directly from image data.

While the approaches from the computer vision community have leveraged probabilistic representations for support relations (e.g. [8,12]), the approaches from the robotics community are still limited to deterministic representations (e.g. [4,5]). In this work, we address this research gap by investigating the benefits of a probabilistic representation for support relations to determine manipulation order in cluttered scenes.

In our previous work, we developed a method for extracting physically plausible support relations between unknown objects based on RGB-D images [4]. We used geometric primitive fitting based on RANSAC (Random Sample Consensus) to estimate the pose and shape of unknown objects in the scene. However, this approach did not consider the uncertainty associated with the pose and geometry estimation. Therefore, we will present a novel probabilistic scene representation that models both pose and shape uncertainty. By sampling from the resulting geometric scene distribution, we determine a probability distribution over support relations in the scene. We show that this approach improves the detection rate of support relations significantly and enables the humanoid robot ARMAR-6 [1] to manipulate cluttered table-top scenes safely.

2 Technical Approach

First, we present a novel probabilistic scene and support relation representation, which explicitly models uncertainty in object geometry and support relation existence. Then, we show how to determine the parameters of this representation based on a point cloud of the perceived scene. Figure 1 shows an overview of our approach.

2.1 Probabilistic Scene and Support Relation Representation

We assume that the shape of an object in a scene can be approximated using geometric primitives. In this work, a geometric primitive can be a box, a cylinder, or a sphere, but the set of primitives can be trivially extended. Each primitive $p_i = (t_i, \mathbf{x}_i)$ consists of a primitive type $t_i \in \{\text{Box}, \text{Cylinder}, \text{Sphere}\}$ and a state vector $\mathbf{x}_i \in \mathbb{R}^{N(t_i)}$. The state vector parametrizes the geometry of a primitive, e. g. a sphere is parameterized by four variables: a three-dimensional center point and a radius. The size $N(t_i)$ of the state vector depends on the primitive type:

- $N(\text{Box}) = 10$ (Position, Orientation, Extents)
- $N(\text{Cylinder}) = 8$ (Position, Direction, Radius, Height)
- $N(\text{Sphere}) = 4$ (Position, Radius)

A scene consists of a set of n objects $\mathcal{O} = \{o_1, o_2, \ldots, o_n\}$. The geometry of each object o_i is represented as a joint probability distribution $P(t_i, \mathbf{x}_i \mid o_i)$ over the primitive type t_i and state vector $\mathbf{x}_i$.

$$P(t_i, \mathbf{x}_i \mid o_i) = P(t_i \mid o_i) \cdot P(\mathbf{x}_i \mid t_i, o_i)$$

Since the primitive type and state vector are dependent variables, we can represent the joint probability distribution as the product of a discrete distribution over primitive types $P(t_i \mid o_i)$ and a dependent distribution over state vectors $P(\mathbf{x}_i \mid t_i, o_i)$. We choose a multi-variate Gaussian distribution to represent $P(\mathbf{x}_i \mid t_i, o_i)$:

$$\begin{aligned} P(\mathbf{x}_i \mid t_i = \text{Box}, o_i) &\sim \mathcal{N}(\mu_{i,\text{Box}}, \Sigma_{i,\text{Box}}) \\ P(\mathbf{x}_i \mid t_i = \text{Cylinder}, o_i) &\sim \mathcal{N}(\mu_{i,\text{Cylinder}}, \Sigma_{i,\text{Cylinder}}) \\ P(\mathbf{x}_i \mid t_i = \text{Sphere}, o_i) &\sim \mathcal{N}(\mu_{i,\text{Sphere}}, \Sigma_{i,\text{Sphere}}) \end{aligned}$$

Note that the mean $\mu_{i,t_i} \in \mathbb{R}^{N(t_i)}$ and the covariance $\Sigma_{i,t_i} \in \mathbb{R}^{N(t_i) \times N(t_i)}$ have different dimensions corresponding to the primitive type and its state vector dimension.

Drawing a sample from the joint distribution $P(t_i, \mathbf{x}_i \mid o_i)$ produces a concrete geometric primitive $p_i^s = (t_i^s, \mathbf{x}_i^s)$. To implement this sampling, first, a primitive type t_i^s is sampled from the discrete probability distribution $P(t_i \mid o_i)$. Then, a state vector $\mathbf{x}_i^s$ is generated by sampling from the multivariate Gaussian distribution $\mathcal{N}(\mu_{i,t_i^s}, \Sigma_{i,t_i^s})$. Note that the primitive type t_i^s is a concrete value and not a random variable.

Given a set of objects $\mathcal{O}$, we represent the probability distribution over the complete scene geometry as a joint distribution over independent object geometries. Therefore, we can express the geometric scene distribution as the product of distributions for the set of primitives $\mathcal{P} = \{p_1, p_2, \ldots, p_n\}$:

$$P(\mathcal{P} \mid \mathcal{O}) = \prod_{i=1}^{n} P(p_i | o_i) = \prod_{i=1}^{n} P(t_i, \mathbf{x}_i | o_i)$$

In this work, we analyze which objects are physically supported by other objects in order to determine safe manipulation sequences. A support relation $\texttt{SUPP} \subseteq \mathcal{O} \times \mathcal{O}$ is a binary relation between objects. We define that a support

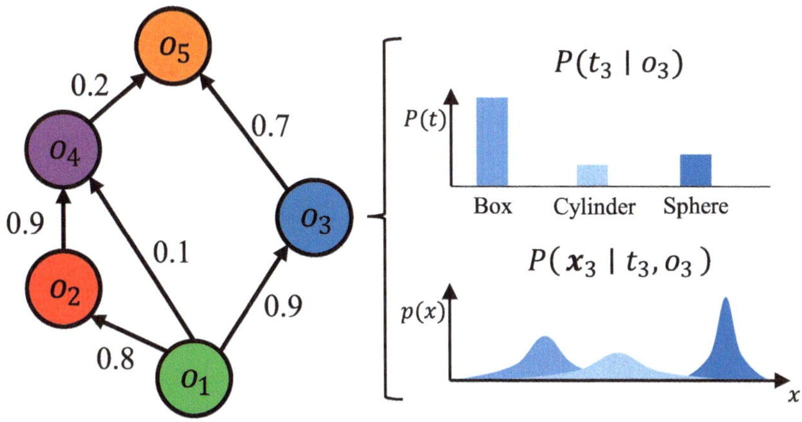

Fig. 2. A probabilistic support graph for the object set $\mathcal{O} = \{o_1, o_2, o_3, o_4, o_5\}$. Each vertex contains the parameters for distributions over primitive type and state vector. Each edge is annotated with the existence probability of a support relation between the corresponding objects.

relation $\mathtt{SUPP}(\mathtt{A}, \mathtt{B})$[1] between objects $\mathtt{A} \in \mathcal{O}$ and $\mathtt{B} \in \mathcal{O}$ exists if and only if removing A causes B to lose its motionless state [5]. The existence probability of a support relation $\mathtt{SUPP}(\mathtt{A}, \mathtt{B})$ depends not only on the objects A and B, but also on other objects in $\mathcal{O}$:

$$P(\mathtt{SUPP}(\mathtt{A}, \mathtt{B}) \mid \mathcal{O}) = P((\mathtt{A}, \mathtt{B}) \in \mathtt{SUPP} \mid \mathcal{O}), \quad \mathtt{A}, \mathtt{B} \in \mathcal{O}$$

We can now represent a scene and the corresponding support relations as a directed graph $\mathcal{G} = (V, E)$, in which the vertices $V = \mathcal{O}$ represent objects and the edges $E = \mathtt{SUPP}$ represent support relations. By considering the probability distributions over scene geometry and support relations, we can create a probabilistic support graph $\mathcal{G}_{\text{prob}} = (V_{\text{prob}}, E_{\text{prob}})$ where each vertex $v_i \in V_{\text{prob}}$ contains the parameters of the joint probability distribution $P(t_i, \mathbf{x}_i \mid o_i)$ and each edge $e_i \in E_{\text{prob}}$ contains the existence probability of a support relation $P(e_i \in \mathtt{SUPP} \mid \mathcal{O})$. Figure 2 shows an example of a probabilistic support graph.

2.2 Extraction from Point Clouds

This subsection describes how the probabilistic scene and support relation representation can be extracted from an input point cloud of a scene. The extraction algorithm consists of four steps: segmentation, geometry estimation, sampling, and support relations analysis (see Fig. 1).

First, we segment the input point cloud using state-of-the-art algorithms [9]. For each segment in the point cloud, we estimate the joint probability distribution $P(t_i, \mathbf{x}_i \mid o_i)$ via a modified RANSAC algorithm. During the geometric

[1] We use the notation $\mathtt{SUPP}(\mathtt{A}, \mathtt{B})$ for $(\mathtt{A}, \mathtt{B}) \in \mathtt{SUPP}$.

primitive fitting using RANSAC, points are randomly drawn from the segment. For each primitive type, the state vector is estimated based on the drawn points. Then, the number of inlier points and the sum of squared errors to all points in the segment is computed. Instead of selecting the primitive parameters which produced the maximum number of inliers, we collect all fitted primitive types $t_{i,j}$, state vectors $\mathbf{x}_{i,j}$, and the corresponding sum of squared errors $\varepsilon_{i,j}$ where j indicates the RANSAC iteration count. Given the total number of RANSAC iterations k, we can now estimate the discrete probability distribution over primitive types $P(t_i \mid o_i)$ by counting the occurrences of each concrete primitive type T:

$$P(t_i = T \mid o_i) = \frac{\sum_{j=1}^{k} (t_{i,j} = T)}{k}$$

The error can be used to calculate the weighted mean and covariance for the state vectors per primitive type. For each object o_i and primitive type t_i in each RANSAC iteration j, we assign a normalized weight $w_{i,j}$ using a Boltzmann distribution over the error $\varepsilon_{i,j}$. The constant β is a parameter, which controls the impact of errors on the weight.

$$w_{i,j} = \frac{e^{-\varepsilon_{i,j}/\beta}}{\eta}, \quad \eta = \sum_{j=1}^{k} e^{-\varepsilon_{i,j}/\beta}$$

To sample a set of concrete primitives from this geometric distribution, we sample a primitive per object o_i. First, the primitive type t_i^s is sampled from the discrete distribution $t_i^s \leftarrow P(t_i \mid o_i)$. Then, the state vector $\mathbf{x}_i^s$ can be drawn from the dependent multi-variate Gaussian distribution $\mathbf{x}_i^s \leftarrow P(\mathbf{x}_i \mid t_i = t_i^s, o_i)$. This results in a set of sampled primitives $\mathcal{P}_s$:

$$\mathcal{P}_s = \{p_1^s, p_2^s, \ldots, p_n^s\}, \quad p_i^s = (t_i^s, \mathbf{x}_i^s)$$

In the last step, we want to compute the existence probabilities of support relations given a geometric scene distribution $P(\mathcal{P} \mid \mathcal{O})$. We approximate the existence probabilities via Monte Carlo simulation. First, we generate m sets of primitives $\{\mathcal{P}_1, \mathcal{P}_2, \ldots, \mathcal{P}_m\}$ by sampling from the geometric scene distributions. Then, we extract pairwise support relations between the sampled primitives using static equilibrium and support polygon analysis presented in [4]. This way, we can estimate the existence probability for a support relation between two objects o_i and o_j by counting the occurrences of concrete support:

$$P(\mathsf{SUPP}(o_i, o_j) \mid \mathcal{O}) \approx \frac{1}{m} \cdot \sum_{k=1}^{m} \mathsf{SUPP}(p_i^k, p_j^k)$$

where $p_i^k, p_j^k \in \mathcal{P}_k$ are primitives sampled from their objects' geometric probability distributions. As $m \to \infty$, this estimate converges to the true value.

$$\lim_{m \to \infty} \left(\frac{1}{m} \cdot \sum_{k=1}^{m} \mathsf{SUPP}(p_i^k, p_j^k) \right) = P(\mathsf{SUPP}(o_i, o_j) \mid \mathcal{O})$$

Fig. 3. A single entry in the KIT Support Relation (KIT-SR) dataset consists of color and depth images of the scene, a colored and labeled point cloud as well as manually annotated support relations.

3 Experiments

We present experimental results for the extraction of support relations from point cloud data. First, we introduce the two datasets and the evaluation procedure. Then, we present and discuss the results. The code and a reproduction guide for the results presented in this section are available online[2].

3.1 Evaluation Datasets

We present the KIT Support Relation (KIT-SR) dataset, which contains table-top scenes with a varying number of objects and support relations. We provide full pixel-wise segmentation and support relation annotations for each of the 60 scenes. The dataset is available online[3]. Figure 3 shows a scene from the dataset and the corresponding available data.

Additionally, we use the Object Segmentation Dataset (OSD, used in [7]) to compare our probabilistic support relation extraction with the deterministic approaches presented in [4,5]. The OSD contains table-top scenarios with varying object geometry and scene complexity, ranging from simple scenes with two objects to cluttered scenes with up to 16 objects. Since the dataset focuses on object segmentation, we added ground truth annotations for support relations to all 111 scenes by hand. These annotations are also available online.

3.2 Evaluation Procedure

The evaluation compares our novel probabilistic support extraction (Prob. Supp. Extr.) with two deterministic approaches. The first deterministic approach is based on static equilibrium analysis (St. Eq., [5]) and serves as a baseline for the comparison. The second deterministic approach extends the static equilibrium

[2] https://gitlab.com/h2t/interactive-scene-exploration/-/wikis/ISER-2020.
[3] https://gitlab.com/h2t/interactive-scene-exploration/-/wikis/KIT-SR.

analysis with support polygon analysis (St. Eq. + Supp. Poly., [4]) to detect top-down support relations, i. e. an object is supported by an object that is geometrically above it.

We evaluate the three approaches on the datasets by computing the metrics precision, recall, F1 score, and Brier score. Given the ground truth support relation $\texttt{SUPP}_{\text{GT}}$ and a hypothesis $\texttt{SUPP}_{\text{Hyp}}$ generated by one of the extraction methods, we can calculate precision and recall as follows:

$$\text{Precision} = \frac{|\texttt{SUPP}_{\text{GT}} \cap \texttt{SUPP}_{\text{Hyp}}|}{|\texttt{SUPP}_{\text{Hyp}}|}, \quad \text{Recall} = \frac{|\texttt{SUPP}_{\text{GT}} \cap \texttt{SUPP}_{\text{Hyp}}|}{|\texttt{SUPP}_{\text{GT}}|}.$$

The F1 score is the harmonic mean of precision and recall:

$$\text{F1 Score} = 2 \cdot \frac{\text{Precision} \cdot \text{Recall}}{\text{Precision} + \text{Recall}}$$

When working with binary classification metrics and a probabilistic support representation, we first need to binarize the probabilities, i. e. decide whether support exists or not. We define support between $\texttt{A}$ and $\texttt{B}$ to exist if the existence probability is above the threshold 0.5:

$$P(\texttt{SUPP}(\texttt{A}, \texttt{B})) > 0.5$$

As a proper score function, the Brier score can be directly used with probabilities. It measures the accuracy of our probabilistic support relation extraction. The Brier score gets smaller when the hypothesis gets more accurate compared to the ground truth. The probabilities $P_{\text{Det}}(\texttt{SUPP}(\texttt{A}, \texttt{B}))$ for the deterministic approaches are set to 1 if support exists between objects $\texttt{A}$ and $\texttt{B}$, 0 otherwise.

$$\text{Brier Score} = \frac{1}{|\mathcal{O} \times \mathcal{O}|} \sum_{\texttt{A},\texttt{B} \in \mathcal{O} \times \mathcal{O}} (P_{\text{Hyp}}(\texttt{SUPP}(\texttt{A}, \texttt{B})) - P_{\text{GT}}(\texttt{SUPP}(\texttt{A}, \texttt{B})))^2$$

3.3 Results

Figure 4 and Table 1 show the evaluation results on the KIT-SR dataset and the OSD. We can see a significant improvement in precision and a moderate increase in recall when comparing the probabilistic with the deterministic approaches. The Brier score is also significantly improved. Furthermore, the addition of support polygon analysis [4] yields better results than the static equilibrium analysis [5] but remains behind the probabilistic approach. The false positives and false negatives of the deterministic approaches are mostly caused by uncertainties about object geometry due to the dataset only containing single viewpoint clouds. The probabilistic approach captures these uncertainties and avoids making hard deterministic decisions early.

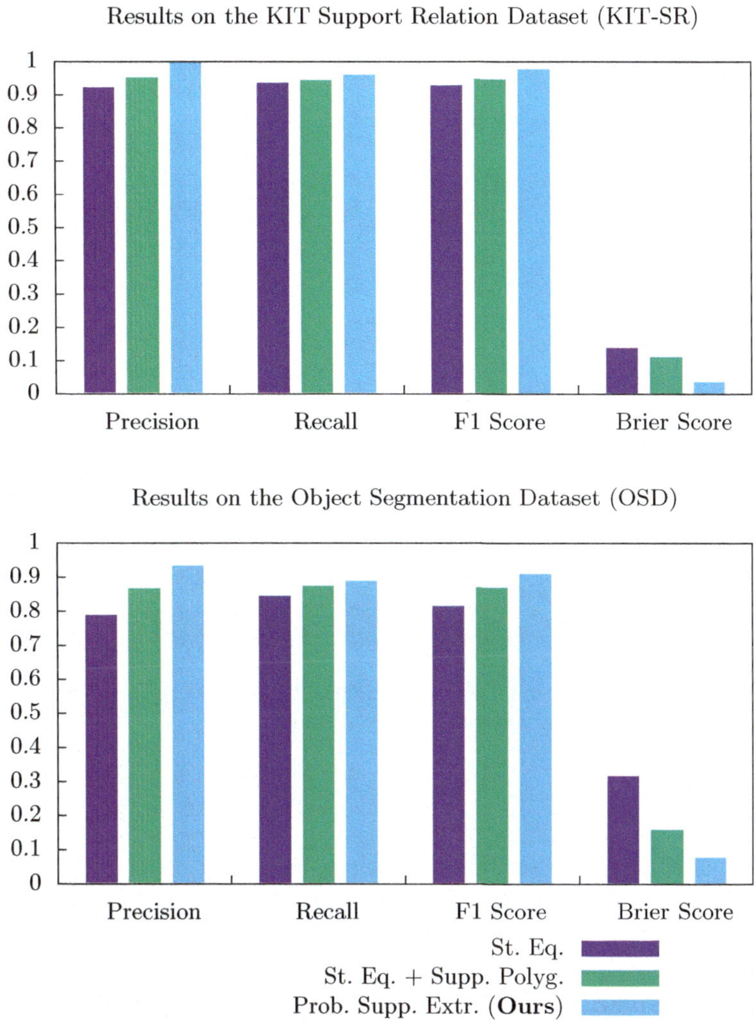

Fig. 4. Evaluation of the three approaches using the evaluation metrics precision, recall, F1 score, and Brier score on the two datasets KIT-SR and OSD. Higher values for precision, recall, and F1 score indicate better results, while lower values for the Brier score indicate better results.

Table 1. Evaluation results on the two datasets.

Dataset	Method	Precision	Recall	F1	Brier score
		(higher is better)			(lower is better)
KIT-SR	St. Eq. [5]	0.919	0.934	0.926	0.138
	St. Eq. + Supp. Poly. [4]	0.949	0.941	0.945	0.109
	Prob. Supp. Extr. (**Ours**)	0.994	0.957	0.976	0.034
OSD	St. Eq. [5]	0.786	0.842	0.813	0.314
	St. Eq. + Supp. Poly. [4]	0.864	0.872	0.868	0.158
	Prob. Supp. Extr. (**Ours**)	0.931	0.886	0.908	0.076

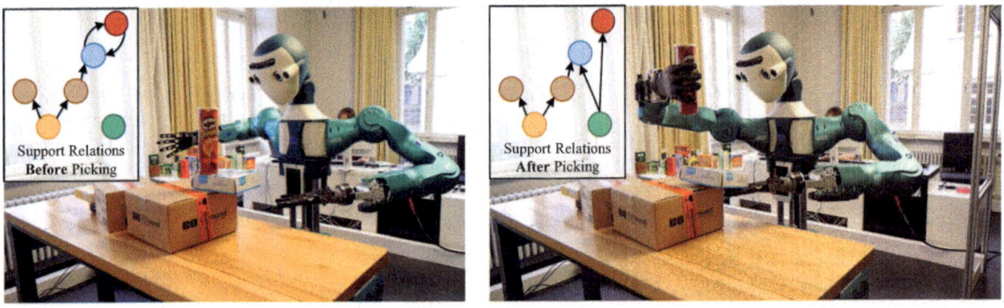

Fig. 5. ARMAR-6 picks a chips' can with the right hand and supports another object with the left hand to prevent the scene from collapsing.

3.4 Robot Experiments

Figure 5 shows such a scenario, where the humanoid robot ARMAR-6 tries to pick the chips can from a stack of objects. Using the probabilistic support relations, the robot is able to infer that the box below the chips can is likely to fall, and prevents this by supporting the box with the left hand. A video of the experiment is available online[4].

4 Conclusion

In this paper, we have shown that a probabilistic representation of perceived object geometry and their support relations improves the support detection rate significantly compared to existing deterministic approaches. Occlusions and partial views on objects cause problems for the deterministic support analysis, which can be overcome by the proposed probabilistic approach. This novel representation allows a robot to reason about the safety and effects of its actions. In future work, we plan to combine this probabilistic scene representation with action effect prediction to support action sequence planning.

[4] https://youtu.be/VBOvr5w7KhA.

Acknowledgment. The research leading to these results has received funding from the Deutsche Forschungsgemeinschaft (DFG, German Research Foundation) – Project Number 146371743 – TRR 89 Invasive Computing and from the European Union's Horizon 2020 Research and Innovation programme under grant agreement No. 731761 (IMAGINE).

References

1. Asfour, T., Wächter, M., Kaul, L., Rader, S., Weiner, P., Ottenhaus, S., Grimm, R., Zhou, Y., Grotz, M., Paus, F.: Armar-6: a high-performance humanoid for human-robot collaboration in real world scenarios. IEEE Robot. Autom. Mag. **26**(4), 108–121 (2019). https://doi.org/10.1109/MRA.2019.2941246
2. Battaglia, P.W., Hamrick, J.B., Tenenbaum, J.B.: Simulation as an engine of physical scene understanding. Proc. Natl. Acad. Sci. **110**(45), 18327–18332 (2013). https://doi.org/10.1073/pnas.1306572110
3. Jia, Z., Gallagher, A.C., Saxena, A., Chen, T.: 3d reasoning from blocks to stability. IEEE Trans. Pattern Anal. Mach. Intell. **37**(5), 905–918 (2014)
4. Kartmann, R., Paus, F., Grotz, M., Asfour, T.: Extraction of physically plausible support relations to predict and validate manipulation action effects. IEEE Robot. Autom. Lett. (RA-L) **3**(4), 3991–3998 (2018)
5. Mojtahedzadeh, R., Bouguerra, A., Schaffernicht, E., Lilienthal, A.J.: Support relation analysis and decision making for safe robotic manipulation tasks. Robotics and Autonomus Systems **71**, 99–117 (2015). https://doi.org/10.1016/j.robot.2014.12.014
6. Panda, S., Hafez, A.A., Jawahar, C.: Single and multiple view support order prediction in clutter for manipulation. J. Intell. Robot. Syst. **83**(2), 179–203 (2016)
7. Richtsfeld, A., Mörwald, T., Prankl, J., Zillich, M., Vincze, M.: Segmentation of unknown objects in indoor environments. In: 2012 IEEE/RSJ International Conference on Intelligent Robots and Systems, pp. 4791–4796. IEEE (2012)
8. Silberman, N., Hoiem, D., Kohli, P., Fergus, R.: Indoor segmentation and support inference from rgbd images. In: European Conference on Computer Vision, pp. 746–760. Springer (2012)
9. Stein, S.C., Schoeler, M., Papon, J., Worgotter, F.: Object partitioning using local convexity. In: Proceedings of the IEEE Conference on Computer Vision and Pattern Recognition, pp. 304–311 (2014)
10. Xue, F., Xu, S., He, C., Wang, M., Hong, R.: Towards efficient support relation extraction from RGBD images. Inf. Sci. **320**, 320–332 (2015)
11. Yang, C., Lan, X., Zhang, H., Zhou, X., Zheng, N.: Visual manipulation relationship detection with fully connected CRFs for autonomous robotic grasp. In: 2018 IEEE International Conference on Robotics and Biomimetics (ROBIO), pp. 393–400. IEEE (2018)
12. Zhang, H., Lan, X., Zhou, X., Tian, Z., Zhang, Y., Zheng, N.: Visual manipulation relationship network for autonomous robotics. In: 2018 IEEE-RAS 18th International Conference on Humanoid Robots (Humanoids), pp. 118–125. IEEE (2018)
13. Zheng, B., Zhao, Y., Yu, J., Ikeuchi, K., Zhu, S.C.: Scene understanding by reasoning stability and safety. Int. J. Comput. Vision **112**(2), 221–238 (2015)

MSL-RAPTOR: A 6DoF Relative Pose Tracker for Onboard Robotic Perception

Benjamin Ramtoula[1], Adam Caccavale[2(✉)], Giovanni Beltrame[3], and Mac Schwager[2]

[1] University of Oxford, Oxford, UK
benjamin@robots.ox.ac.uk
[2] Stanford University, Stanford, USA
{awc11,schwager}@stanford.edu
[3] Polytechnique Montréal, Montréal, Canada
giovanni.beltrame@polymtl.ca

Abstract. Determining the relative position and orientation of objects in an environment is a fundamental building block for a wide range of robotics applications. To accomplish this task efficiently in practical settings, a method must be fast, use common sensors, and generalize easily to new objects and environments. We present MSL-RAPTOR, a two-stage algorithm for tracking a rigid body with a monocular camera. The image is first processed by an efficient neural network-based front-end to detect new objects and track 2D bounding boxes between frames. The class label and bounding box is passed to the back-end that updates the object's pose using an unscented Kalman filter (UKF). The measurement posterior is fed back to the 2D tracker to improve robustness. The object's class is identified so a class-specific UKF can be used if custom dynamics and constraints are known. Adapting to track the pose of new classes only requires providing a trained 2D object detector or labeled 2D bounding box data, as well as the approximate size of the objects. The performance of MSL-RAPTOR is first verified on the NOCS-REAL275 dataset, achieving results comparable to RGB-D approaches despite not using depth measurements. When tracking a flying drone from onboard another drone, it outperforms the fastest comparable method in speed by a factor of 3, while giving lower translation and rotation median errors by 66% and 23% respectively.

Keywords: Perception · Aerial robots · Machine learning

1 Introduction

For a robot to intelligently interact with its environment, it must quickly and accurately determine the relative position and orientation of neighboring robots, people, vehicles, and a variety of other static and dynamic obstacles in the scene. From drone racing [1] to pedestrian avoidance [2], tasks relying on such estimates are numerous and are

B. Ramtoula and A. Caccavale—Equal contribution.

B. Siciliano et al. (Eds.): ISER 2020, SPAR 19, pp. 520–532, 2021.
https://doi.org/10.1007/978-3-030-71151-1_46

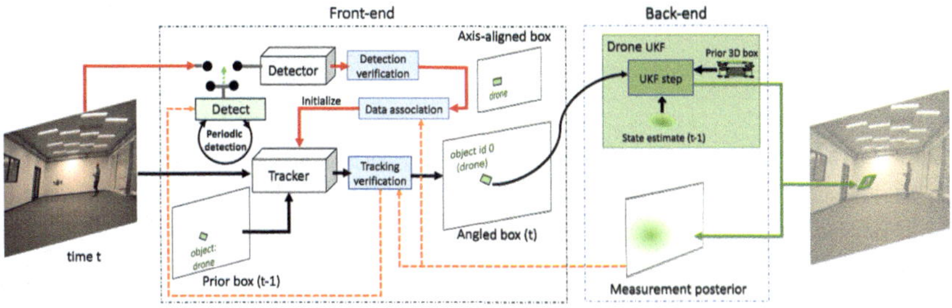

Fig. 1. Overview of MSL-RAPTOR. The front-end generates angled-bounding boxes & class labels from monocular images. The back-end tracks the object's pose with a class-specific UKF, and passes its uncertainty metrics to the front-end to detect when the tracker has lost the object.

essential to creating truly autonomous robots. Regardless, the ability to track the six Degree of Freedom (6DoF) relative pose of a dynamic object in real-time, with onboard sensing and computation, is largely beyond the reach of existing robot perception frameworks. To fill this need, we propose **M**onocular **S**equential **L**ightweight **R**otation **A**nd **P**osition **T**racking **O**n-**R**obots (MSL-RAPTOR), a lightweight, general-purpose robot perception module for tracking the 6DoF relative pose of a dynamic object using only an onboard monocular camera. MSL-RAPTOR performs similarly to state-of-the-art approaches relying on more sensors (e.g. depth) and greater computation capability.

Existing methods for 6DoF relative pose estimation from vision can be divided into model-based, deep learning-based, and hybrid methods. Existing model-based approaches rely on matching the view of an object to a database of templates or visual features. These methods are limited in the face of scenes with low textures or changing conditions, and cannot be used when prior 3D models of objects are not available [3,4]. Deep learning methods such as PoseCNN [5] and LieNets [6] overcome these challenges by directly estimating the relative pose from images, learning this inverse map from a large corpus of training data. Hybrid approaches combine elements of deep neural networks and model-based methods. For example, BB8 [7], SSD-6D [8], and SingleShotPose [9] (SSP) all use deep neural networks to estimate the 2D projections of the 3D bounding box encompassing the object. These projections are then used to solve the Perspective-n-Point (PnP) problem to obtain a pose estimate [10]. Another hybrid method, NOCS [11], estimates pose using pixel correspondences and depth data with category-level canonical representations. However, the above deep and hybrid methods ignore temporal correlations among multiple frames to aid in tracking objects. In contrast, 6-PACK [12] leverages temporal consistency by learning to represent objects using keypoints that are matched between frames. While these methods present impressive pose estimation performance, they are either too slow, too computationally demanding, or require too much prior model knowledge to be used for practical onboard relative pose perception for resource-limited robots.

These same limitations apply to the many pose tracking approaches developed for self-driving cars, for which tracking methods represent a key component of the system [2]. These rely on the use of rich sensors and/or advanced computational capabilities to achieve online performance and often assume planar constraints since they deal solely with land-based objects (e.g. pedestrians, other vehicles), hence are only SE(2) trackers with 3DoF (e.g. [13, 14]).

Contribution. With the requirements for onboard robotic perception in mind, we propose MSL-RAPTOR, which seeks to combine the best aspects of deep learning-based and model-based perception paradigms. MSL-RAPTOR takes raw image data with a dual deep neural network architecture on the front-end to produce and track an angled 2D bounding box around an object in the scene. A single 2D bounding box is not enough information to determine the object's pose (see Fig. 2(a)), so this semantically labeled bounding box is passed to a model-based unscented Kalman filter (UKF) [17, 18] back-end to fuse measurements across time and produce a Bayesian relative pose estimate for the object. The deep front-end uses a YOLOv3 network [15], retrained with robot-specific object classes, for initial object detection and labeling, together with a SiamMask network [16] for quickly tracking the 2D object bounding box from frame to frame once it has been detected. The 2D bounding box corners and semantic object labels (e.g., "person", "UAV", "car", "tree") are then passed to the UKF back-end. The back-end uses the bounding box corners as pseudo-measurements, which are fused over time to infer the full 6DoF relative pose of the object. The UKF propagates the target's motion with a label-specific motion model (since people, cars, and UAVs all move differently). We then close the loop between back-end and front-end with a re-detection trigger. If the Mahalanobis distance between the measurement and its posterior is above a threshold or if the pose estimate is too close to the edge of the field of view (judged using the z-test based on its covariance), the target is considered lost, and we re-trigger the YOLO target detection on the front-end. With this framework, we can track an object through occlusions and temporary absences from the field of view. MSL-RAPTOR currently assumes a single object in the scene, but the extension to multiple objects (and classes) is straightforward.

We demonstrate MSL-RAPTOR running on a Jetson TX2 onboard a standard quadrotor UAV platform while tracking another UAV at 8.7 frames per second. This is 3 times faster than the fastest comparable state-of-the-art method, SSP [9], and produces pose estimates with 66% and 23% lower translation and rotation median errors, respectively. We also compare performance with several state-of-the-art deep learning methods for 6DoF pose tracking using a depth camera. We show on standard datasets that MSL-RAPTOR performs similarly to these existing methods despite only using a monocular camera with no depth information.

2 Problem Formulation

We consider the scenario where a computationally-limited mobile robotic agent with a calibrated monocular camera needs to identify and track the relative pose of a dynamic object O from a series of RGB images $I^t \in \mathbb{R}^{W \times H}$. The state at time t is $\mathbf{x}^t = [\mathbf{p}^t, \mathbf{v}^t, \mathbf{q}^t, \boldsymbol{\omega}^t] \in \mathbb{R}^{13}$ where $\mathbf{p}^t \in \mathbb{R}^3$ is 3D position, $\mathbf{v}^t \in \mathbb{R}^3$ is linear velocity, $\mathbf{q}^t \in \mathbb{R} \times \mathbb{R}^3$ is the orientation as a quaternion, and $\boldsymbol{\omega}^t \in \mathbb{R}^3$ is angular velocity. The position and orientation component of this state can be represented as the object pose $\mathbf{T}^t \in SE(3)$. Henceforth the time superscripts will be dropped for notational convenience unless they are needed for clarity.

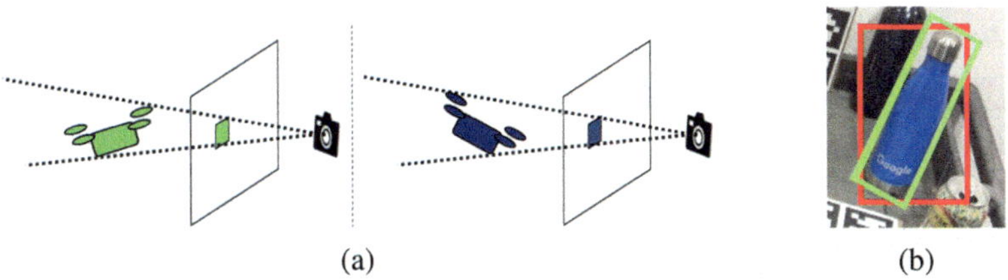

Fig. 2. Illustration of the pose ambiguity from bounding box measurements: (a) same 2D bounding box produced from difference poses; (b) pose ambiguity can be reduced with an angled bounding box, but not eliminated.

The object O has a span in 3D space that can be approximated with a sparse set of n points $\mathbf{B} \in \mathbb{R}^{n\times3}$, most commonly the corners of its 3D bounding box which we use by default. Additionally, each object has a corresponding class label (e.g. `drone`, `bottle`, etc.) denoted c. Each class of object has an associated dynamics model used to propagate the components of the state in the UKF back-end: $\mathbf{x}^{t_2} = \mathbf{F}_c(\mathbf{x}^{t_1})$. By default a simple constant velocity model is used, but more sophisticated class-specific models may be provided. In general we assume no knowledge of the tracked object's control input.

When processing an image, the 2D projections of the object's points $\mathbf{B}$ onto the image I are enclosed in the minimum-area *angled* bounding box: $\mathbf{z} = [x, y, w, h, \alpha]$ where x and y are the column and row of the box center, w and h are the box width and height, and α is the box angle from the horizontal image axis. If instead an *axis-aligned* bounding box is produced (i.e. $\alpha = 0$), it is denoted $\tilde{\mathbf{z}}$ (Fig. 2(b)). The distribution of estimates over states and measurements are both modeled as Gaussians, $\mathcal{N}(\hat{\mathbf{x}}, \hat{\Sigma})$ and $\mathcal{N}(\hat{\mathbf{z}}, \hat{\mathbf{S}})$ respectively. When evaluating the consistency of a measurement $\mathbf{z}^t$ with our prior estimate, we use the Mahalanobis distance defined as:

$$\mathbb{M}(\mathbf{z}^t, \hat{\mathbf{z}}^{t-1}, \hat{\mathbf{S}}^{t-1}) \doteq \sqrt{(\mathbf{z}^t - \hat{\mathbf{z}}^{t-1})^T \left(\hat{\mathbf{S}}^{t-1}\right)^{-1} (\mathbf{z}^t - \hat{\mathbf{z}}^{t-1})}. \quad (1)$$

3 Algorithm

The architecture of our approach is illustrated in Fig. 1. The front-end has two components for processing the images: a detector and a visual tracker. The detector can identify all objects in an image and their class labels, but is slower and only generates axis-aligned bounding boxes. The tracker is faster and produces angled bounding boxes, but requires an accurate 2D bounding box from the previous timestep. The back-end relies on a class-specific UKF, taking in an angled bounding box measurement from the front-end to produce pose estimates. For each image, the decision to use the tracker vs. detection is determined by the consistency of the new bounding box measurement with the measurement posterior generated by the back-end's UKF.

3.1 Object Detection

Upon receiving an image I, the angled bounding boxes must be extracted. If the object has not been previously located, or if re-detection has been triggered, a pre-trained, out-of-the-box object detector outputs the axis-aligned bounding box $\tilde{\mathbf{z}}$ and class label c for each visible object O_i: $\mathbb{D}(I) \rightarrow \{\tilde{\mathbf{z}}_i, c_i\}$.

Only bounding boxes for the class of interest are considered, and if multiple boxes are found, the one closest to our current estimate as measured by Eq. 1 is used. If our estimate's uncertainty is too high, then the box with highest confidence is chosen. This axis-aligned bounding box is then used to initialize the visual tracker (Sect. 3.2) which produces an angled bounding box measurement $\mathbf{z}$. This, along with the class label c from the detector, serves to initialize the UKF (Sect. 3.3). Because the filter requires an initial pose estimate, the initial 6D pose is roughly approximated assuming the width, height, and roll are aligned with the 2D bounding box. This is a poor assumption in general, but the robustness of the filter will compensate for the initialization error. As the class of the object is also identified, class-based initialization rules can also be used (e.g. mugs are usually upright).

3.2 Visual Tracking

Once an object has been detected and the associated tracker and UKF are initialized, MSL-RAPTOR's front-end only relies on the tracker to provide measurements to the UKF. The tracker is an object-agnostic method, which produces an angled bounding box measurement given the previous bounding box and a new image $\mathbb{T}(I^t, \mathbf{z}^{t-1}) \rightarrow \mathbf{z}^t$.

When a new image is available, the tracker generates an angled bounding box that serves as an observation for the UKF predicting the object's pose (Sect. 3.3).

Aside from the speed benefit, we use a visual tracking method that produces angled bounding boxes that are more informative measurements than the axis-aligned boxes as illustrated in Fig. 2. While the angle is still not sufficient to uniquely determine an object's pose, it reduces the set of poses that would result in the given measurement which improves tracking performance.

3.3 Unscented Kalman Filter

To estimate pose from a series of angled bounding box measurements, we use a class-dependent unscented Kalman filter. Probabilistic filters like the UKF consist of a prediction and update step. During prediction, the state is propagated forward using the object dynamics, and the expected measurement from the resulting anticipated state is calculated. In the update step, the difference between the predicted and actual measurement is used to adjust the state estimate and uncertainty. A typical output of a UKF is the mean and covariance of the state, but as part of the update process, the measurement mean and covariance are also calculated. These are passed to the front-end for determining whether to trigger re-detection (Sect. 3.4).

The prediction step of the UKF relies on a dynamics model, which can optionally depend on the object tracked. The MSL-RAPTOR framework gives access to a class label for each object, enabling the use of class-specific dynamics if available, or a simple constant velocity model if not. For systems without significant inertial effects this is a good approximation, otherwise, we rely on the robustness of the filter to correct for the modeling error.

To predict the expected measurement associated to a pose, the knowledge of the 3D points representing the tracked object is used. The vertices $\mathbf{B}$ are projected into the image plane using the known camera intrinsic matrix $\mathbf{K}$ and the estimate of the object's pose in the camera frame $\mathbf{T}$ [19]. The minimum area rectangle that encloses these projected pixels is the predicted measurement $\pi_{\mathbf{K}}(\mathbf{B},\mathbf{T}) \rightarrow \mathbf{z}$.

A UKF is chosen over alternatives such as an EKF or PF due to its efficient handling of the non-linearities inherent in the system. The PF's natural ability to track multiple-hypothesis potentially could allow it to better determine poses that best explain the series of bounding box measurements, but empirical testing showed the computational requirements of this method to be too high to run onboard.

3.4 Updating the Tracked Objects

Occlusions, objects leaving the field of view, and poor measurements (e.g. due to light reflections) can all cause tracking failures that lead to inaccurate bounding boxes. No assumptions are made regarding these scenarios. To address these challenges, detection is triggered periodically to identify and locate when the object has become unoccluded or has entered the image. This has the advantage of compensating for any drift or errors that might occur in the visual tracking. The front-end's verification mechanisms can also trigger detection to correct for uncertain measurements.

Bounding Box Verification

Detected Measurements. Detection returns both the bounding box and the class label of the object. This can be used to filter detections by class and to set class-specific rejection conditions for the measurement that can help prevent false-positive matches. For example, an aspect ratio condition can be set so that a bounding box $\tilde{\mathbf{z}}$ is valid only if it is in a valid class-specific aspect-ratio range:

$$\Gamma_{\min}(c) < \tilde{w}/\tilde{h} < \Gamma_{\max}(c),$$

where the thresholds $\Gamma_{\min}(c)$ and $\Gamma_{\max}(c)$ can be determined from the typical geometry of the class. Additionally, a threshold on the minimum distance $d_{\min}$ to the edge of the image can prevent bad measurements due to truncated bounding boxes:

$$\begin{aligned} d_{\min} &< \tilde{x} - \tilde{w}/2 < \tilde{x} + \tilde{w}/2 < W - d_{\min} \\ d_{\min} &< \tilde{y} - \tilde{h}/2 < \tilde{y} + \tilde{h}/2 < H - d_{\min}. \end{aligned}$$

Tracker Measurements. At each iteration of the UKF, anticipated measurements are predicted for a range of perturbations about the current state estimate. From this set of possible measurements, a mean $\hat{\mathbf{z}}$ and covariance $\hat{\mathbf{S}}$ representing the distribution of observations is calculated. The Mahalanobis distance (Eq. 1) is compared against a threshold to evaluate the likelihood of new measurements. If a mismatch between the state estimate of the back-end and the new measurement is detected, this could indicate drift in the tracker or some other failure, and detection is triggered.

Additionally, to ensure the object is fully in view, z-tests using the diagonal values of $\hat{\mathbf{S}}$ are performed at each edge. The results are compared with a threshold.

The probabilistic thresholds in this section can be chosen from the quantiles of the chi-squared distribution and normal distribution, respectively. When the test fails, the system switches into detection mode to either confirm it should stop tracking the object, or to correct its tracked bounding box.

4 Experimental Results

4.1 Implementation Choices

For our experiments we implement MSL-RAPTOR with YOLOv3 [15] from Ultralytics [20] for our front-end's object detector. We chose YOLOv3 due to its ubiquity, speed, and demonstrated ability to detect a range of objects. To support the objects used in our experiments, we retrain it on the COCO dataset mixed with a custom dataset containing images of a flying quadrotor, which leads to 81 supported object categories. For our visual tracker, we rely on the implementation of SiamMask [16] provided by the authors, which is object-agnostic and therefore does not require retraining. We chose SiamMask for its speed, ability to produce angled bounding boxes, and impressive tracking performance. We use the provided weights trained on the VOT dataset [21].

4.2 Evaluation on the NOCS-REAL275 Dataset

We begin by evaluating MSL-RAPTOR on the NOCS-REAL275 dataset [11] to answer a fundamental question about the method: can the algorithm track pose accurately for a range of objects? In all, the dataset contains seven training videos and six testing videos from a camera moving in different scenes. The scenes contain objects with class labels in {`bowl`, `mug`, `can`, `bottle`, `laptop`, and `camera`}. We tune our class-specific UKF parameters using the training set and present the results evaluated on the testing set. Similarly to the baseline methods, we do not penalize orientation errors about axes of symmetry for the `can`, `bowl`, and `bottle` classes.

Implementation Details. Our re-detection and verification mechanisms presented in Sect. 3.4 serve to make the tracking robust to occlusions and to handle objects exiting and entering the frame. However, sequences of the dataset are relatively short and mostly maintain all objects in the field of view. Moreover, the datasets used to train our detector did not contain instances of all the classes appearing in the scenes, and the training set from NOCS-REAL275 would not be general enough to train the detector to properly detect new instances of objects in the testing set. For these reasons, we disable the probabilistic detection and verification mechanisms for evaluation on this dataset and rely only on visual tracking in the front-end. This is a conservative approach as violations of these assumptions will reduce our performance.

To replace MSL-RAPTOR's filter initialization, which is based on the disabled detection, the same method is used as the baselines [12,22,23]. For a given object O, random uniform translation noise of up to 4cm is added to the ground truth initial pose $\mathbf{T}^0$. From this initialized pose, the first bounding box is calculated using the UKF's measurement prediction function $\pi_{\mathbf{K}}(\mathbf{B}, \mathbf{T}^0)$. We add uniformly sampled noise of up to 5% of the width and height of the box to its row and width, and its column and height respectively. With this first bounding box, the visual tracker will generate future measurements.

For objects in the dataset with shapes not well approximated by a 3D bounding box (bowl, laptop, mug), we add additional vertices to restrict the volume. For example, 12 points are used instead of 8 for the laptop, resulting in an "L" shape when viewed from the side. For mugs and bowls, extra points allow a tighter fit around curved rims. Using these refined volumes reduces the mismatch between predicted and observed measurements in the UKF, which can lead to incorrect adjustments of the pose estimate. This effect is greatest when objects are close to the camera, as in the NOCS dataset.

Table 1. Results on NOCS-REAL275. Translations are in cm and rotations are in degrees.

Method	Modality	Bottle		Bowl		Camera		Can		Laptop		Mug		Overall	
		R_{err}	t_{err}	R_{err}	t_{err}	R_{err}	t_{err}	R_{err}	t_{err}	R_{err}	t_{err}	R_{err}	t_{err}	R_{err}	t_{err}
NOCS [11]	RGB-D	25.6	1.2	4.7	3.1	33.8	14.4	16.9	4.0	8.6	2.4	31.5	4.0	20.2	4.9
ICP [24]	RGB-D	48.0	4.7	19.0	12.2	80.5	15.7	47.1	9.4	37.7	9.2	56.3	9.2	48.1	10.5
Keypoint Net [25]	RGB-D	28.5	8.2	9.8	8.5	45.2	9.5	28.8	13.1	6.5	4.4	61.2	6.7	30.0	8.4
6-PACK [12]	RGB-D	15.6	1.7	5.2	5.6	35.7	4.0	13.9	4.8	4.7	2.5	21.3	2.3	16.0	3.5
Ours	**Mono.**	18.7	9.8	16.2	4.0	23.4	8.5	17.6	9.0	17.4	11.6	35.3	7.7	21.8	8.2

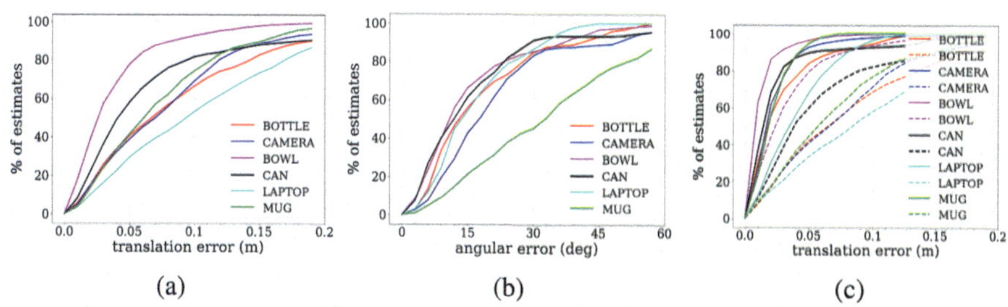

Fig. 3. Empirical cumulative distribution function on errors: (a) translation only; (b) rotation only; (c) translation in-plane error (solid line) and depth error (dashed line)

Finally, as the baseline methods are evaluated using desktop GPUs, for this set of experiments only we use a desktop computer with an Nvidia RTX 2060 GPU.

Baselines. We compare our results with those reported by Wang *et al.* over the same dataset [12] for four methods, **all of which use RGB-D input**:

- NOCS [11], a recent 6D pose estimation solution relying on pixel correspondences to a canonical representation.
- ICP [24], the Open3D implementation of the iterative closest point algorithm.
- KeypointNet [25], a category-based 3D keypoint generator.
- 6-PACK [12], a state-of-the-art method using learned 3D keypoints with an anchoring mechanism incorporating temporal information for improved tracking.

Results. The average rotation and translation errors over each class are presented in Table 1 and the accuracy visualized in Fig. 3. Despite relying only on monocular images, MSL-RAPTOR achieves comparable performance to state-of-the-art methods that use RGB-D. As is expected for a monocular camera, depth estimates are less accurate than translations perpendicular to the agent's camera axis (i.e. parallel to the image plane). This can be seen in Fig. 3(c) where these are broken out into separate errors. This is a problem that comes with the use of monocular images rather than RGB-D data.

4.3 Baseline Evaluation in an Aerial Robotics Scenario

To demonstrate that MSL-RAPTOR provides robust onboard perception, we ran the algorithm onboard a drone with an Nvidia Jetson TX2 to track another drone. The ego quadrotor uses the DJI F330 frame and is equipped with off-the-shelf and custom made components as shown in Fig. 4. The quadrotor uses an RGB camera (640×480 resolution). Experiments were designed so that the ego drone moves about the experimental space causing the image background to change substantially, a challenge which demonstrates the robustness of the system. The flights varied from slow to aggressive, with the tracked object occasionally exiting and reentering the field of view or being occluded by other objects. As would be expected in collision avoidance situations, the distances between the ego and tracked drone routinely reached 8 m (a substantially larger separation distance than what is seen in most existing pose estimation datasets). This provides important validation of the use of this method for planning and decision making onboard agile robots. From our collected data, we use 5500 images over several runs to form a training set for the detector and baseline, and use 5000 different images to form a testing set to evaluate pose predictions.

Fig. 4. Quadrotor used for tracking objects in robotic perception scenarios.

Implementation Details. We do not assume knowledge about the control input of the tracked drones, and a simple constant velocity dynamics model is used. Since we are looking at tracking the other drone at up to 10 m distance from the ego drone, we treat the tracked drone as symmetric about the vertical axis considering the yaw angle does not have an appreciable effect on our measurements, and does not provide useful information about its behavior. For a fair comparison, the same assumption is made for the baseline method. With a higher resolution camera or for tracking at closer ranges this constraint could be removed.

Baseline. While most state-of-the-art methods struggle to achieve real-time speeds on desktop GPUs, SSP [9] reports inference-only speeds of 50 Hz. This, in addition to its impressive accuracy, makes it the only candidate likely to be competitive when running onboard. SSP takes in an image and outputs the estimated locations of the projected 3D bounding box of the object on the image plane. Its network is based on YOLO's architecture and is modified to produce the 3D bounding box projections instead of a 2D bounding box. The relative rigid body transformation is then calculated from the projected vertex locations and the 3D bounding box geometry using a PnP algorithm.

SSP requires the full object pose for each image when training. Outside of laboratory environments with specialized equipment, this is more challenging to acquire than the 2D bounding boxes used by MSL-RAPTOR. Moreover, SSP is presented as an instance-based method, meaning it must be trained on specific objects for which it will infer pose. For this reason, we did not evaluate it on the NOCS dataset that contains unseen objects instances in the testing set.

We train SSP using a desktop GPU, and present results for both SSP and MSL-RAPTOR evaluated on an Nvidia Jetson TX2.

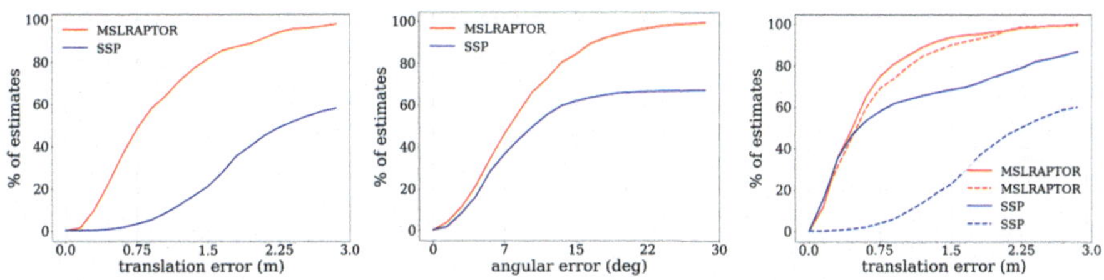

(a) (b) (c)

Fig. 5. Empirical cumulative distribution function on errors: (a) translation only; (b) rotation only; (c) translation in-plane error (solid line) and depth error (dashed line)

Results. Figure 5 shows the better accuracy of MSL-RAPTOR compared to SSP. The median translation and rotation errors are 0.82 m and 8.59° for MSL-RAPTOR compared to 2.45 m and 11.26° for SSP (66% less translation error and 23% less rotation error). When tracking a drone, MSL-RAPTOR runs 3X faster than SSP on a TX2, with an average pose prediction speed of 0.115 s (8.7 Hz) compared to 0.335 s (2.98 Hz) for SSP.

SSP directly outputs the locations of the 3D bounding box projections on the image, so when the drone is far away or moving quickly (causing blur), these points can be warped and scaled (Fig. 6). Regardless, the PnP algorithm will find a rigid body transformation that best matches these projections to the actual 3D bounding box, leading to significant depth errors as seen in Fig. 5(c). Since the in-plane translation is affected by the average of these points, and the rotation by the relative positioning, the difference in performance is less stark for the rotation and in-plane translation errors. As SSP does not incorporate temporal information it cannot identify nor correct for these situations which lead to estimates with large errors, resulting in the apparent asymptotes in Fig. 5.

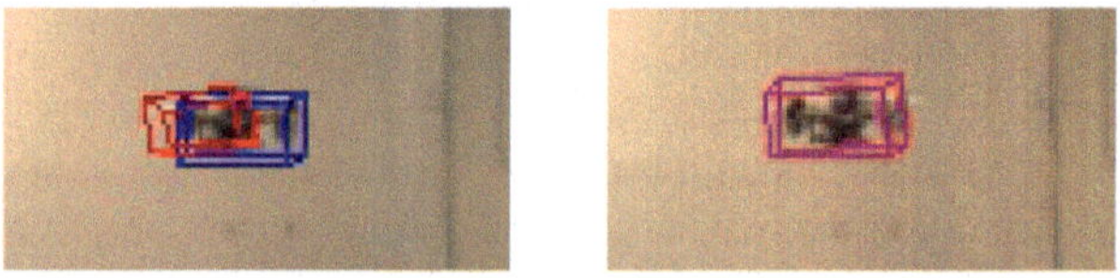

(a) (b)

Fig. 6. Left: SSP's prediction for the projected 3D bounding box (red), and the ground truth box (blue). Note that due to the fast motion of the drone, the blurry image results in a distorted prediction. Right: MSL-RAPTOR's prediction for the same frame (purple).

5 Conclusions

In this paper, we propose MSL-RAPTOR, an efficient monocular relative pose tracking algorithm that combines the image processing capabilities of modern neural networks with the robustness advantages of classical probabilistic filters. The front-end extracts class labels and angled bounding boxes, which are used as observations by the back-end's UKF. The back-end estimates the relative pose and returns measurement uncertainty parameters to the front-end for methodically determining when to trigger the slower, but more accurate, re-detection. When not detecting, the faster visual tracking method is used. The algorithm runs 3 times faster than the quickest state-of-the-art method, while still out-performing it in terms of precision in onboard robotic perception scenarios. Furthermore, despite not using depth measurements, comparable performance to state-of-the-art results is achieved on the NOCS-REAL275 dataset.

Acknowledgements. This research was supported in part by ONR grant number N00014-18-1-2830, NSF NRI grant 1830402, the Stanford Ford Alliance program, the Mitacs Globalink research award IT15240, and the NSERC Discovery Grant 2019-05165. We are grateful for this support.

References

1. Spica, R., Falanga, D., Cristofalo, E., Montijano, E., Scaramuzza, D., Schwager, M.: A real-time game theoretic planner for autonomous two-player drone racing. In: Proceedings of Robotics: Science and Systems, Pittsburgh, Pennsylvania (2018)

2. Badue, C., et al.: Self-driving cars: a survey. Expert Syst. Appl. **165**, 113816 (2021)
3. Schmidt, T., Newcombe, R., Fox, D.: DART: dense articulated real-time tracking. In: Proceedings of Robotics: Science and Systems, Berkeley, USA (2014)
4. Andriluka, M., Roth, S., Schiele, B.: Monocular 3D pose estimation and tracking by detection. In: Conference on Computer Vision and Pattern Recognition, pp. 623–630. IEEE (2010)
5. Xiang, Y., Schmidt, T., Narayanan, V., Fox, D.: PoseCNN: a convolutional neural network for 6D object pose estimation in cluttered scenes. In: Robotics: Science and Systems, Pittsburgh, Pennsylvania (2018)
6. Do, T.T., Pham, T., Cai, M., Reid, I.: Real-time monocular object instance 6d pose estimation. In: British Machine Vision Conference (BMVC), vol. 1, p. 6 (2018)
7. Rad, M., Lepetit, V.: BB8: a scalable, accurate, robust to partial occlusion method for predicting the 3D poses of challenging objects without using depth. In: Proceedings of the IEEE International Conference on Computer Vision, pp. 3828–3836 (2017)
8. Kehl, W., Manhardt, F., Tombari, F., Ilic, S., Navab, N.: SSD-6D: making RGB-based 3D detection and 6D pose estimation great again. In: IEEE International Conference on Computer Vision, pp. 1521–1529 (2017)
9. Tekin, B., Sinha, S.N., Fua, P.: Real-time seamless single shot 6D object pose prediction. In: IEEE Conference on Computer Vision and Pattern Recognition, pp. 292–301 (2018)
10. Lepetit, V., Moreno-Noguer, F., Fua, P.: EPnP: an accurate O(n) solution to the PnP problem. Int. J. Comput. Vis. **81**(2), 155 (2009)
11. Wang, H., Sridhar, S., Huang, J., Valentin, J., Song, S., Guibas, L.J.: Normalized object coordinate space for category-level 6D object pose and size estimation. In: Proceedings of the IEEE Conference on Computer Vision and Pattern Recognition, pp. 2642–2651 (2019)
12. Wang, C., Martín-Martín, R., Xu, D., Lv, J., Lu, C., Fei-Fei, L., Savarese, S., Zhu, Y.: 6-PACK: category-level 6D pose tracker with anchor-based keypoints. In: IEEE Conference on Robotics and Automation (ICRA), Paris, France (2020)
13. Cho, H., Seo, Y.W., Kumar, B.V., Rajkumar, R.R.: A multi-sensor fusion system for moving object detection and tracking in urban driving environments. In: IEEE Conference on Robotics and Automation (ICRA), pp. 1836–1843 (2014)
14. Buyval, A., Gabdullin, A., Mustafin, R., Shimchik, I.: Realtime vehicle and pedestrian tracking for didi udacity self-driving car challenge. In: 2018 IEEE International Conference on Robotics and Automation (ICRA), pp. 2064–2069. IEEE (2018)
15. Redmon, J., Farhadi, A.: YOLOv3: an incremental improvement. arXiv:1804.02767 (2018)
16. Wang, Q., Zhang, L., Bertinetto, L., Hu, W., Torr, P.H.: Fast online object tracking and segmentation: a unifying approach. In: IEEE Conference on Computer Vision and Pattern Recognition, pp. 1328–1338 (2019)
17. Julier, S., Uhlmann, J., Durrant-Whyte, H.F.: A new method for the nonlinear transformation of means and covariances in filters and estimators. IEEE Trans. Autom. Control **45**(3), 477–482 (2000)
18. Wan, E.A., Van Der Merwe, R.: The unscented Kalman filter for nonlinear estimation. In: Proceedings of the IEEE 2000 Adaptive Systems for Signal Processing, Communications, and Control Symposium, pp. 153–158 (2000)
19. Hartley, R., Zisserman, A.: Multiple view geometry in computer vision. Cambridge University Press (2004)
20. Jocher, G., guigarfr, perry0418, Ttayu, Veitch-Michaelis, J., Bianconi, G., Baltacı, F., Suess, D., WannaSea, U., IlyaOvodov: ultralytics/yolov3: Rectangular Inference, Conv2d + Batchnorm2d Layer Fusion, April 2019
21. Kristan, M., Matas, J., Leonardis, A., Vojir, T., Pflugfelder, R., Fernandez, G., Nebehay, G., Porikli, F., Čehovin, L.: A novel performance evaluation methodology for single-target trackers. IEEE Trans. Pattern Anal. Mach. Intell. **38**(11), 2137–2155 (2016)

22. Issac, J., Wuthrich, M., Cifuentes, C.G., Bohg, J., Trimpe, S., Schaal, S.: Depth-based object tracking using a robust gaussian filter. In: 2016 IEEE International Conference on Robotics and Automation (ICRA), Stockholm, Sweden (2016)
23. Wuthrich, M., Pastor, P., Kalakrishnan, M., Bohg, J., Schaal, S.: Probabilistic object tracking using a range camera. In: 2013 IEEE/RSJ International Conference on Intelligent Robots and Systems (2013)
24. Zhou, Q.Y., Park, J., Koltun, V.: Open3D: a modern library for 3D data processing. arXiv preprint arXiv:1801.09847 (2018)
25. Suwajanakorn, S., Snavely, N., Tompson, J.J., Norouzi, M.: Discovery of latent 3D keypoints via end-to-end geometric reasoning. In: Advances in Neural Information Processing Systems, pp. 2059–2070 (2018)

Robust Vision-Based Pose Correction for a Robotic Manipulator Using Active Markers

Lukas Meyer(✉), Klaus H. Strobl, and Rudolph Triebel

German Aerospace Center (DLR), Institute of Robotics and Mechatronics, Wessling, Germany
{lukas.meyer,klaus.strobl,rudolph.triebel}@dlr.de

Abstract. Robots with elastic or lightweight components are becoming common in research, but can suffer from undesired positioning imprecision, which motivates a vision-based pose correction of the manipulator. For robotic manipulators that operate outdoors and under changing illumination conditions, robustness of the vision components is of principal concern. We propose a monocular manipulator pose correction based on active markers which are detected by convergence criteria on the image gradient field. We show the capabilities of the method in several outdoor and indoor experiments, considering the use case of a planetary exploration rover prototype equipped with a lightweight robotic arm. The vision-based manipulator pose correction method proves to be successful despite back light, reflections, and image overexposure and additionally allows continued robot operation in the case of extrinsic camera decalibration.

1 Motivation and Related Work

In 2017, the Lightweight Rover Unit (LRU) of the DLR performed several scientific experiments on the volcano Mt. Etna in Italy during a Moon Analogue campaign [10], as illustrated in Fig. 1.

The successful mission also revealed inaccuracies during manipulation and sensitivity of the system to extrinsic camera decalibration due to the robot's lightweight structure. We develop a vision-based pose correction (VBPC) method for manipulators to overcome these issues with the emphasis on the robustness for robot operations in challenging illumination conditions.

The LRU is scheduled to participate in a second Moon-analog scientific mission on the volcano Mt. Etna in June and July 2021 [15], a sequel to the 2017 campaign. We consider the LRU and the Etna mission scenario as our primary use case for VBPC, addressing three aspects: First, a correction of the manipulator pose is needed, which is applied after having finished a coarse approach towards a target and before starting the dexterous manipulation process. Second, the detected position of the markers within the camera image is used as initialization of a tracking approach with focus on robust visual servoing, which we

B. Siciliano et al. (Eds.): ISER 2020, SPAR 19, pp. 533–542, 2021.
https://doi.org/10.1007/978-3-030-71151-1_47

Fig. 1. Two operation scenarios of the LRU during the 2017 Etna Moon Analogue Campaign [10]: Collection of scientific equipment from a lander (left) and deployment of a geophone in the ground (right). Credit: DLR.

currently develop. Lastly, infrequent comparison of the end effector pose between VBPC and forward kinematics aims at detecting an hand-to-eye decalibration.

In literature, several approaches exist for pose estimation of robotic manipulators using vision. A widely used technique is fitting the three dimensional robot model to available depth information [4,7,14]. Monocular tracking based on rendering CAD models is done by e.g. [3,5,13]. Machine learning based methods are found in [9,11]. The mentioned methods require either dense depth information or extensive computation resources; neither of which is available to us.

We therefore follow an active marker based approach that is more robust to illumination changes than passive markers. Likewise, [1,2] use active markers for pose estimation. However, they either require that the markers are the brightest object in the image or require specific wavelength filters – requirements that are not available to us. We therefore follow a novel approach by detecting the convergence of the image gradients at the marker locations, which turns out to be robust w.r.t. most of the external disturbances.

The LRU (see Fig. 1) has a robotic arm mounted on its rear. It also features a stereo camera pan-tilt head for navigation and two fixed rear monocular cameras for close-up observation of upper and the lower sections of the manipulator's workspace. As end effector, the arm is equipped with the cylindrical *Envicon* docking interface [10] that has a ring of 12 light-emitting diodes (LEDs) which we use as active markers for VBPC. In this work, we consider the principal navigation camera and the upwards facing rear camera. The cameras operate at 14 Hz and 2 Hz respectively. The frequencies of the cameras are determined by other system components, therefore limiting the rear camera to updates of static poses, whereas the navigation camera allows for a continuous pose correction during manipulator movements.

2 Technical Approach

Our approach to correct the manipulator's pose is based on combining a-priori information from the forward kinematics with a marker detection based on the structure of the image gradient field and a subsequent fitting of the marker model.

2.1 Marker Detection

The common optical characteristic of our active markers – which is invariant under a high variability in illumination – is the convergence of the image gradient vectors towards the center of each marker as illustrated in Fig. 2. We calculate the *convergence score* of each pixel in a region of interest (ROI) by applying a modified *IRIS-filter* [8] to the gradient field.

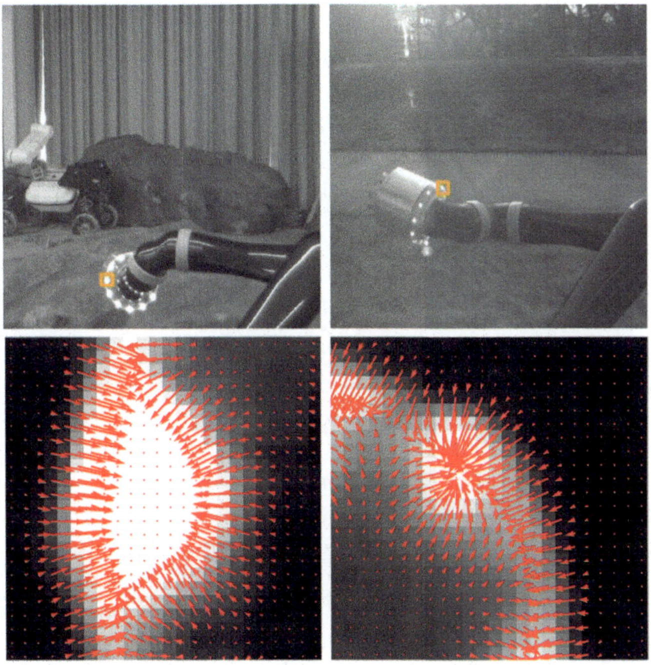

Fig. 2. View of the manipulator from the navigation camera (top row), and detailed view of individual LEDs (orange box, shown in bottom row) with the respective gradient vector field (red). Different illumination conditions are shown: over exposure and reflections, indoors (left); and backlight, outdoors (right).

The modified IRIS-filter calculates a convergence score C for a pixel (u, v) by considering the orientation of the nearby gradient vectors and additionally includes the magnitude of the gradient vectors to focus only on relatively bright regions. We initialize each pose correction step using the forward kinematics to define a ROI in the camera image and to obtain a prediction for the marker positions within the ROI. For each pixel, we define N half-lines over the circular arc with a maximum radius $R_{\max}$. Each half-line has an associated unit-orientation-vector $\boldsymbol{r}_i$ with $i = 1, ..., N$. Along the ith half line, starting with a distance $R_{\min}$ from the center, we sample the gradient vectors $\boldsymbol{g}$, such that

$$C_i(u, v) = \max_n \frac{1}{n} \sum_{m=R_{\min}}^{n} -\cos(\theta_m)\|\boldsymbol{g}_m\| \tag{1}$$

with $R_{\min} < n < R_{\max}$. The angle between $\boldsymbol{g}_m$ and the half-line is denoted by θ_m. The overall score C is the sum of all line-scores C_i. Marker candidates are the pixels with the highest values in the resulting IRIS-score image.

Finally, we compute the variance of the IRIS-score for each pixel as

$$\sigma^2(u, v) = \frac{1}{N} \sum_{i=0}^{N} (C_i(u, v) - \bar{C}(u, v))^2, \tag{2}$$

with $\bar{C}$ denoting the mean IRIS-score over all N half-lines. The variance allows to penalize pixels that have a high IRIS-score along several half-lines, but a low score in others, e.g. pixels that correspond to circular segments or similar. The detected marker candidates are these pixels with the highest score after penalization.

2.2 Robust Pose Correction

The obtained marker candidates are fitted to the respective model of the LED-distribution. In our case, the markers are arranged as a circle, which allows for a random sample consensus (RANSAC) based fitting of an ellipse using the algorithm from [6]. To obtain the ellipse, theoretically, only five out of the 12 LEDs need to be visible. However, our experiments revealed that with five visible markers, the resulting ellipse becomes highly sensitive to noise in the marker position and is therefore rejected by our a-priori informed outlier detection. Therefore, usually six markers need to be detected for a robust fitting, thus allowing for a 50% occlusion rate. Note that the approach can generally be used with any other type of marker pattern requiring only minor adaptation.

We compare the resulting ellipse with the ellipse computed from the a-priori prediction for outlier rejection. The position of the ellipse center is extracted to obtain the lateral location of the end effector w.r.t. the camera frame. The depth of the end effector position is calculated from the observed length of the ellipse major axis.

The initial result of the pose correction allows to obtain an image-to-world correspondence between the IRIS-maxima and the respective LEDs from the known model. This is used for a final refinement step using an iterative implementation of the perspective n-point (PnP) algorithm [12], with the position from the ellipse fitting and the orientation from the forward kinematics for initialization. The result from the PnP iteration provides the final pose of the LEDs w. r. t. the camera frame and is used for robot operations.

For our pose correction, we focus primarily on robustness towards the challenging and varying illumination conditions. We expect that the VBPC will experience increasingly false detections with increasing disturbances. For this purpose, we rely on a strict outlier rejection policy with the tradeoff of a decreased update rate of the pose. As outlier criteria, we mainly consider the fact that the kinematic pose is erroneous, however this error is bounded. We furthermore consider the shape of the ellipse w.r.t. the a-priori kinematic information. To account for a possible low update rate, we map the corrected end effector pose as offset to the end effector frame that is calculated by the forward kinematics. This allows to maintain a locally valid pose correction that only requires infrequent VBPC updates.

2.3 Concept Verification

The results of our algorithm are illustrated in Fig. 3. The circle of the LEDs is detected in all scenarios. The end effector frame from the VBPC (colored) is shown together with the a-priori prediction from the forward kinematics (gray). The first row (a to c) shows the VBPC results considering the principal navigation cameras. LED detection via the upward facing rear camera is shown in (d) to (f), with (f) being generally the most challenging scenario due to bright cluttered background. In all scenarios, VBPC allows for an improved pose estimate.

LED detection and ellipse fitting is successful for most of the frames. With increasing disturbances, false positive LED candidates may appear, which are filtered out by the outlier rejection step.

The current Python-based implementation of the IRIS-filter and VBPC achieves a 2 Hz update rate. With the final C++ version, we expect real-time capabilities with both camera systems.

3 Accuracy

First, we evaluate the accuracy of our VBPC method. For this, we perform experiments with the LRU inside a laboratory environment. We use precise end effector tracking with a Vicon motion capture system to obtain a ground truth of the manipulator pose. The cameras are calibrated intrinsically and extrinsically and the images have been rectified.

During the experiments, the end effector is commanded to move to 15 different poses while being observed by the rover cameras. The target poses are chosen to cover the majority of the field of view of both cameras and to cover most of the manipulator's workspace.

The obtained positional accuracy of VBPC with the principal navigation camera is shown in Fig. 4 together with the accuracy of the forward kinematics, considering continuous movement of the manipulator. First of all, it can be

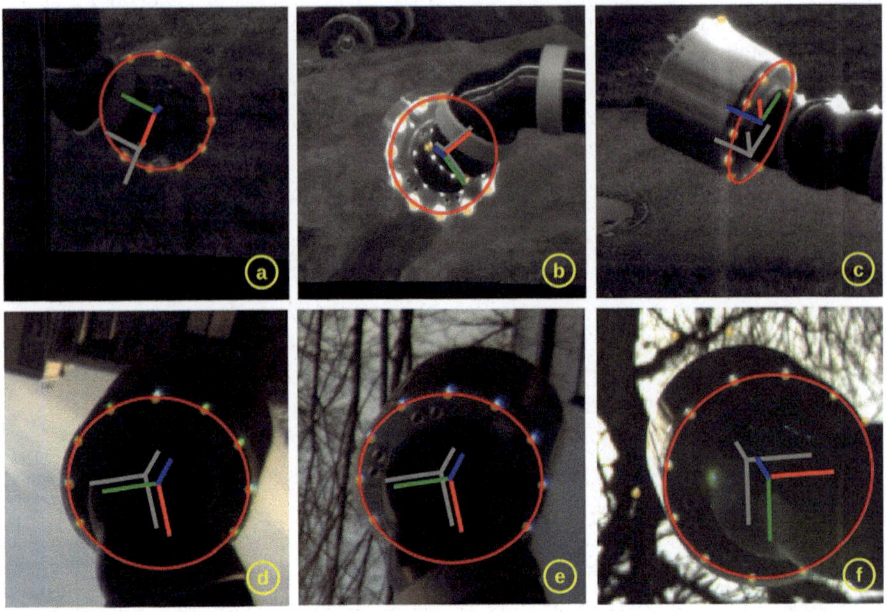

Fig. 3. The end effector under varying illumination conditions with detected LED candidates (orange), the fitted ellipse (red), the center frame from the kinematics (gray), and the center frame by VBPC (RGB) from the navigation camera (top row), and from the rear camera (bottom row).

seen that the positional error of the forward kinematics w.r.t. the ground truth strongly correlates with the horizontal extension of the manipulator, which is additionally shown as reference. The error of the forward kinematics can reach up to 3.5 cm at stretched out manipulator poses, which emphasizes the necessity for a vision based pose correction.

In contrast to the forward kinematics, the VBPC is able to determine the position of the end effector with a significantly higher accuracy, generally around 0.7 cm, infrequently reaching 1 cm. During the experiment, all false positives are rejected by our outlier detection step.

The VBPC fails to obtain valid measurements of the end effector position at four instances, which are marked grey in Fig. 4. The first two periods without valid updates (labeled A and B) are the result of an ill-posed ellipse fitting problem. During these periods, the marker-ring is observed by the navigation camera from the side, resulting in highly eccentric ellipses that also restrict the observability of the individual markers. In the two cases A and B, the eccentricity factors ϵ of the ellipses reach 0.99 or higher. The eccentricity factor is calculated as

$$\epsilon = 1 - \frac{a_{\text{minor}}}{a_{\text{major}}}, \tag{3}$$

where a_{minor} and a_{major} denote the length of the minor and major axis of the ellipse.

The other two periods without valid updates (marked as C and D), are due to partial or complete self-occlusions by the robotic arm (five or less markers are visible). Generally, as long as six or more LEDs are visible during the experiment, the VBPC is able to obtain a correct position measurement.

Regarding the orientation, VBPC is not able to correct the angular error. The average angular error of the forward kinematics lies at 3°, which can be considered as sufficiently accurate for manipulation. Recall that we use the orientation from the forward kinematics as initial guess for the iterative PnP approach. The angular error of the forward kinematics is too small to be further improved from the detected marker positions. Therefore, PnP is not able to improve the pose estimate compared to the initial guess and returns the corresponding initial values.

Finally, we evaluate the accuracy of VBPC considering the 15 static target poses observed by both the rear upwards facing camera and the principal navigation camera. For the principal navigation camera, the positional error is determined as $0.7 \pm 0.4\,\text{cm}$ for the 15 target poses. For the rear camera, the positional error is calculated as 0.5 cm with a standard deviation of 0.2 cm. The difference in accuracy between both cameras can be attributed to the different distances from the cameras to the end effector. Recall, that we evaluate the rear camera only at static poses, due to its low update rate and due to the manipulator frequently leaving the camera image during movement.

4 Experimental Robustness

To verify the robustness of the detection, we consider multiple experiments in indoor and outdoor scenarios. The goal is to cover the range of possible operation scenarios that the rover will face.

The first experiment (Exp-1) is the indoor scenario from Sect. 3 as reference. In general, indoor robot operation requires higher camera exposure time, causing an over-exposure of the LEDs, as for example illustrated in Fig. 3(b). Therefore, Exp-1 can be used to evaluate the robustness w.r.t. over-exposed markers in the image.

With the outdoor scenarios, other aspects challenge the algorithm, which we reflect in the experiments accordingly. We consider different sun positions and the orientation of the rover with respect to the sun, as these cause different types and magnitudes of reflections and influence the appearance of the background. During the experiments, the robotic arm of the LRU moved to several poses, similar to the indoor experiment Exp-1.

The first outdoor experiment (Exp-2) considers a high standing sun. The remaining experiments were performed at low point sun positions (Exp-3 to 5). Examples of the marker appearance in the navigation cameras for high and low sun positions are shown in Fig. 3(a) and (c) respectively and for the rear camera in (d) and (f).

For Exp-3, the rover is oriented with 90° to the sun. During Exp-4, the cameras face away from the sun, but the end effector appears fully illuminated

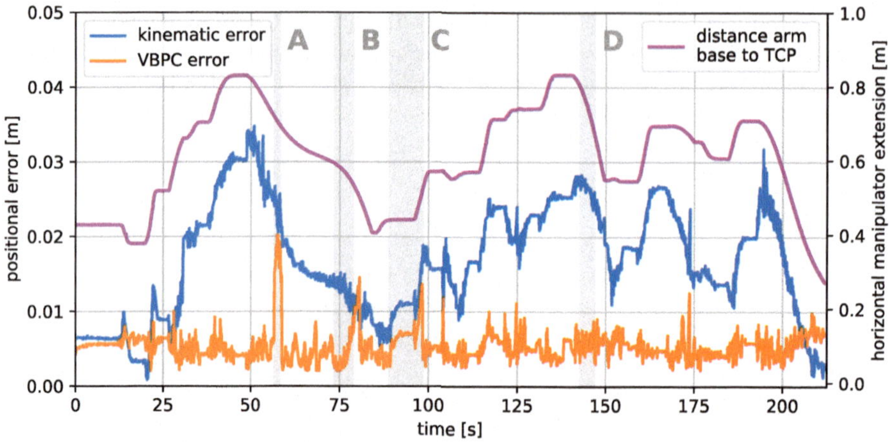

Fig. 4. Results from the principal navigation camera: The kinematic position error and the VBPC position error w.r.t. the ground truth tracking system are shown. The kinematic error correlates with the horizontal extension of the manipulator (shown in magenta). The grey areas show periods of unsuccessful arm pose estimation by the VBPC.

in the camera image. Finally, the rover cameras face directly into the sun during the Exp-5. All selected experiments are listed in Table 1.

Table 1. Robustness experiments for the principal navigation camera. Experiments, where not all false positives are rejected by the outlier detection are marked with *.

	Experiment type	End-effector movement [m]	Static pose success rate	General rate of detection [%]	Rate for $\epsilon < 0.9$ [%]
Exp 1	Indoor reference experiment	4.0	14/15	86.5	92.1
Exp 2	Outdoor, noon sun	4.7	10/10	73.0	88.8
Exp 3	Outdoor, low sun from 90° side	4.65	9/10	90.2	92.3
Exp 4	Outdoor, low sun from 180°	1.4	5/5	47.2*	60.8*
Exp 5	Outdoor, facing towards low sun	4.0	10/10	76.4	88.4

During all experiments, the manipulator is commanded to several target poses, where it remains static for 2–3 s and subsequently moves to the next pose. We evaluate the robustness of our detection regarding two criteria.

The first criterion is the capability to correctly detect the markers while the manipulator remains static at the respective target positions. This is the application that is most relevant for our future operations. Table 1 shows that the detection is successful in all five experiments, with only two exceptions during Exp-1 and Exp-3. In the case of the two unsuccessful detections, only five markers were visible, a correlation that was already described in Sect. 3. Indeed, the unsuccessful detection for Exp-1 is the grey area C in Fig. 4.

The other robustness evaluation criterion considers successful detection of the markers during continuous arm movement along varying trajectories. We consider every frame as a new pose, if the end effector moved more than 1 cm compared to the previously considered pose. The corresponding metric is the rate of successfully detected poses compared to the overall number of poses. In Sect. 3, we established a connection that VBPC is prone to failure for highly eccentric ellipses, therefore we especially investigate the detection rate for well-conditioned ellipses ($\epsilon < 0.9$).

Table 1 shows that VBPC is able to detect the markers during approximately 90% of each experiment considering ellipses with $\epsilon < 0.9$. A notable exception is Exp-4 with only 60% success rate. During this experiment, the protective glass cover of the markers is frequently oriented perpendicular to the sun's direction, causing reflections that make the LEDs indistinguishable. This issue can be overcome by considering the sun orientation during robot operations or modifying the hardware using less reflective, matte materials.

Finally, we consider the upwards facing rear camera with the static poses. During Exp-1 to Exp-3, the end effector is reliably detected using VBPC in 14, 10, and 8 cases respectively. Contrary to this, during Exp-4 and Exp-5, VBPC fails to compute a correct pose estimate for most of the positions (it is only successful in 2/5 and 3/10 cases). This especially holds for end effector positions that appear in the upper parts of the image, where the bright background from the sky is combined with strong reflections. Furthermore, in these two cases, VBPC is susceptible to false positives that are not rejected by the outlier detection step, showing the limitations of our approach.

5 Conclusion and Outlook

We proposed a vision based pose correction to account for kinematic errors in robotic manipulators w.r.t. the observing cameras, which occur due to elasticities, extrinsic camera decalibration, or other external influences. Our VBPC method considers a planetary-analog scenario and therefore focuses primarily on robustness towards outdoor illumination, and was specifically developed for the upcoming Moon-analog mission on the volcano Mt. Etna.

In our work, we showed the accuracy of the VBPC, which is able to significantly improve the pose estimate of the manipulator compared to the forwards kinematics. Our method is robust against external disturbances. This robustness especially holds for static pose corrections, but also allows to determine the manipulator pose during dynamic movements.

VBPC will be used as initialization for a tracking of the manipulator with the intended purpose to be used for a vision based servo control scheme. For future work, we additionally plan to use our detection for continuous camera calibration verification.

Furthermore, a tightly coupled fusion of forward kinematics and vision based observations is envisioned, where the pose estimation from vision and kinematics is coupled with the continuous parameter estimation of an error model.

References

1. Breitenmoser, A., Kneip, L., Siegwart, R.: A monocular vision-based system for 6D relative robot localization. In: IEEE/RSJ International Conference on Intelligent Robots and Systems, pp. 79–85 (2011)
2. Faessler, M., Mueggler, E., Schwabe, K., Scaramuzza, D.: A monocular pose estimation system based on infrared LEDs. In: IEEE International Conference on Robotics and Automation, pp. 907–913 (2014)
3. Fantacci, C., Pattacini, U., Tikhanoff, V., Natale, L.: Visual end-effector tracking using a 3D model-aided particle filter for humanoid robot platforms. In: IEEE/RSJ International Conference on Intelligent Robots and Systems, pp. 1411–1418 (2017)
4. Garcia Cifuentes, C., Issac, J., Wüthrich, M., Schaal, S., Bohg, J.: Probabilistic articulated real-time tracking for robot manipulation. IEEE Rob. Autom. Lett. **2**, 577–584 (2017)
5. Gratal, X., Romero, J., Kragic, D.: Virtual visual servoing for real-time robot pose estimation. IFAC World Congr. **1**, 9017–9022 (2011)
6. Kanatani, K., Sugaya, Y., Kanazawa, Y.: Guide to 3D Vision Computation. Springer, New York (2016)
7. Klingensmith, M., Galluzzo, T., Dellin, C., Kazemi, M., Bagnell, J.A.D., Pollard, N.: Closed-loop servoing using real-time markerless arm tracking. In: IEEE International Conference on Robotics and Automation (Humanoids Workshop) (2013)
8. Kobatake, H., Hashimoto, S.: Convergence index filter for vector fields. IEEE Trans. Image Process. **8**, 1029–1038 (1999)
9. Lambrecht, J.: Robust few-shot pose estimation of articulated robots using monocular cameras and deep-learning-based keypoint detection. In: International Conference on Robot Intelligence Technology and Applications, pp. 136–141 (2019)
10. Lehner, P., Brunner, S.G., Dömel, A., Gmeiner, H., Riedel, S., Vodermayer, B., Wedler, A.: Mobile manipulation for planetary exploration. In: IEEE Aerospace Conference, pp. 1–11 (2018)
11. Liang, C.J., Lundeen, K.M., McGee, W., Menassa, C.C., Lee, S., Kamat, V.R.: A vision-based marker-less pose estimation system for articulated construction robots. Autom. Constr. **104**, 80–94 (2019)
12. OpenCV: Open source computer vision library (2020)
13. Ortenzi, V., Marturi, N., Stolkin, R., Kuo, J.A., Mistry, M.: Vision-guided state estimation and control of robotic manipulators which lack proprioceptive sensors. In: IEEE/RSJ International Conference on Intelligent Robots and Systems, pp. 3567–3574 (2016)
14. Schmidt, T., Hertkorn, K., Newcombe, R., Marton, Z.C., Suppa, M., Fox, D.: Depth-based tracking with physical constraints for robot manipulation. In: IEEE International Conference on Robotics and Automation, pp. 119–126 (2015)
15. Schuster, M.J., Müller, M.G., Brunner, S.G., Lehner, H., Lehner, P., Dömel, A., Vayugundla, M., Steidle, F., Lutz, P., Sakagami, R., Meyer, L., Belder, R., Smisek, M., Stürzl, W., Triebel, R., Wedler, A.: Towards heterogeneous robotic teams for collaborative scientific sampling in lunar and planetary environments. In: IEEE/RSJ International Conference on Intelligent Robots and Systems (Workshop on Informed Scientific Sampling in Large-scale Outdoor Environments) (2019)

Sensing Soft Robot Shape Using IMUs: An Experimental Investigation

Josie Hughes[1(✉)], Francesco Stella[2], Cosimo Della Santina[2,3,4], and Daniela Rus[1]

[1] Distributed Robotics Lab, CSAIL, MIT, Cambridge, USA
josieh@mit.edu

[2] Cognitive Robotics Department, Delft University of Technology, Delft, Netherlands

[3] Informatics and Mathematics Department, Technical University of Munich, Munich, Germany

[4] Institute of Robotics and Mechatronics, German Aerospace Center (DLR), Cologne, Germany

Abstract. Shape estimation of soft robotic systems is challenging due to the range of deformations that can be achieved, and the limited availability of physically compatible sensors. We propose a method of reconstruction using Inertial Measurement Units (IMUs), which are mounted on segments of a deformable manipulator. This approach utilizes the piecewise constant curvature model in combination with the quaternion data from IMUs to allow for accuracy reconstruction and closed-loop control. A key strength of this approach is that it is hardware agnostic, and could be used on any soft structure to provide pose reconstruction and controllability. We explore this approach experimentally on a growing, extendable 3D printed continuum body structure, demonstrating that high accuracy reconstruction that can be achieved.

Keywords: Soft manipulation · Kinematic control · Shape estimation

1 Introduction

Within the domain of soft robotics, continuum body manipulators offer particularly exciting capabilities, with applications ranging from surgery, compliant exploration and human like environmental interactions [4,12]. To achieve such capabilities there is firstly the challenge of the design and fabrication of the system [6]. However, there is then the second key challenge of developing the artificial brains to control the robot [5]. In this regard a central problem is that of perception, since understanding the robot shape is critical for control. However, accurately measuring and reconstructing the shape of a soft robot without relying on exteroceptive sensing techniques is very challenging [9].

There is a significant body of related work focusing on shape and pose reconstruction of continuum body structures. Many proprioceptive techniques for such robots rely on soft sensors which are custom or specific to the particular physical manifestation of the manipulator [8]. The use of learning techniques [14,15] or mixed strategies [16], are often required due to the complex nonlinear characteristics of these sensors. More recently approaches such as model predictive

B. Siciliano et al. (Eds.): ISER 2020, SPAR 19, pp. 543–552, 2021.
https://doi.org/10.1007/978-3-030-71151-1_48

control [2], and computational methods [1] have been shown to perform well for trajectory optimization of soft robots.

In light of this existing work, and the challenges that remain, we focus on the problem statement of providing accurate and reliable proprioception to soft robots. We believe that Inertial Measurement Units (IMUs) are a feasible solution for equipping soft robots with preconception capabilities, without altering the soft nature of the system, and without requiring the implementation of complex machine learning techniques. Existing work has shown how IMUs can be used for accurate reconstruction of rigid robots [13] and UAVs [11], and has also been widely demonstrated for human pose reconstruction [13]. Extending these capabilities to the domain of soft robots, would provide a 'universal' proprioceptive solution which could be used across a wide number of soft arms and continuum structures. To work towards this goal, this work makes the following contributions:

- A theoretical framework for using IMU sensor data to enable the reconstruction of the shape of a soft continuum manipulators, based upon the piecewise constant curvature model.
- Development of a extensible, flexible, multi-segment, continuum body manipulator to demonstrate this approach. We will contextually test and benchmark against the use of encoders for proprioception [7].
- Experimental results that quantify the accuracy that can be achieved when using IMUs for reconstruction, and for closed loop control of a continuum body manipulator.

In the remainder of this paper we first introduce the technical approach and methods for IMU based reconstruction. We introduce the robot platform that has been developed to explore this approach. Following on from this we show the results of applying the approaches to the robot platform demonstrating the effectiveness and accuracy that can be achieved.

2 Technical Approach and Methods

Intertial Measurament Units (IMUs) sense 3-axis linear acceleration and angular velocity. They also include a magnetometer which senses the direction of the magnetic field. Although standard methods to fuse these sources of information are prone to drift over time, recently proposed techniques have proven able to reduce this problematic accumulation of error to a minimum. Such techniques have been shown to be particularly effective in the case of sensing orientation. For example, the Mahony filter in [11] has shown that drifts can be limited to only the local direction of the gravity field. For the case of continuum body robots, the output from these filters is not the only information that we have regarding the shape of the soft robot.

Using the piecewise constant curvature (PCC) approximation we can identify the full shape of each segment from a function which has just two variables. Consider a number of reference frames attached as in Fig. 1, on for the end of each segment plus one at the base. Let R_{i-1}^{i} be the rotation matrix and t_{i-1}^{i} the

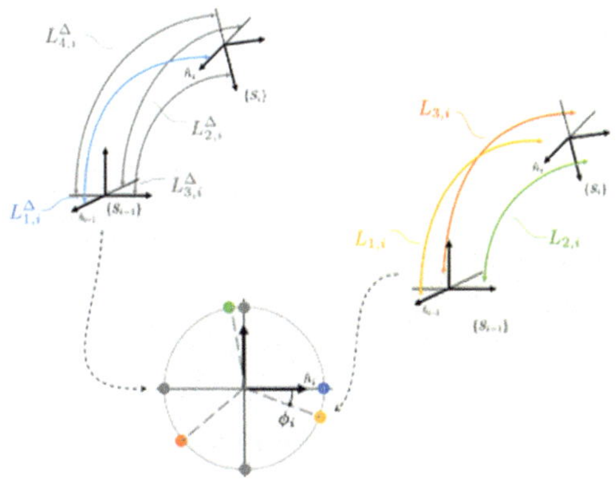

Fig. 1. Model and reference frames for the manipulator.

translation mapping the reference frame at base of the i−th segment to the one attached to its tip.

According to the description proposed in [3], the posture of the i−th segment can be described through the two banding variables $\Delta_{x,i}, \Delta_{y,i} \in \mathbb{R}$ and the total length of the segment L_i. More specifically the orientation is described by the rotation matrix $R^i_{i-1}(\Delta_{x,i}, \Delta_{y,i}, s_i) =$

$$\begin{bmatrix} 1+\frac{\Delta^2_{x,i}}{\Delta^2}\left(1-\cos\left(s_i\Delta_i\right)\right) & \frac{\Delta_{x,i}\Delta_{y,i}}{\Delta^2}\left(1-\cos\left(s_i\Delta_i\right)\right) & -\frac{\Delta_{x,i}}{\Delta}\sin\left(s_i\Delta_i\right) \\ \frac{\Delta_{x,i}\Delta_{y,i}}{\Delta^2}\left(1-\cos\left(s_i\Delta_i\right)\right) & 1+\frac{\Delta^2_{y,i}}{\Delta^2}\left(1-\cos\left(s_i\Delta_i\right)\right) & -\frac{\Delta_{y,i}}{\Delta}\sin\left(s_i\Delta_i\right) \\ \frac{\Delta_{x,i}}{\Delta}\sin\left(s_i\Delta_i\right) & \frac{\Delta_{y,i}}{\Delta}\sin\left(s_i\Delta_i\right) & 1+\left(1-\cos\left(s_i\Delta_i\right)\right) \end{bmatrix}, \tag{1}$$

and the position by the vector

$$t^i_{i-1}(\Delta_{x,i}, \Delta_{y,i}, L_i, s_i) = \frac{L_i}{\Delta^2_i}\begin{bmatrix} \left(1-\cos\left(s_i\Delta_i\right)\right)\Delta_{x,i} \\ \left(1-\cos\left(s_i\Delta_i\right)\right)\Delta_{y,i} \\ \sin\left(s_i\Delta_i\right)\Delta_i \end{bmatrix}, \tag{2}$$

where $\Delta_i = \sqrt{\Delta^2_{x,i} + \Delta^2_{y,i}}$. The local coordinate along the segment is $s_i \in [0, 1]$, with 0 referring to the base, and 1 to the tip (see Fig. 1). Since we have an IMU at the base and tip of each continuum body segment, we can use the posture provided by a stack of Mahony filters [10] to extract the global posture of each reference frame $\hat{R}^i_0$. We then evaluate the relative posture $\hat{R}^i_{i-1} = \left(\hat{R}^{i-1}_0\right)^{-1}\hat{R}^i_0$. Finally, we can extract the parameterization of each segment by properly inverting (1) for $s_i = 1$. We propose to do that by evaluating the configuration of the segment as the solution of the optimization problem

$$\arg\min_{\Delta_{x,i}, \Delta_{y,i} \in \mathbb{R}^2} ||R^i_{i-1}(\Delta_{x,i}, \Delta_{y,i}, 1) - \hat{R}^i_{i-1}||^2_{\mathrm{F}}, \tag{3}$$

where $||\cdot||_{\mathrm{F}}$ is the Frobenius norm, and $R^i_{i-1}(\Delta_{x,i}, \Delta_{y,i}, 1)$ is as in (1). This is evaluated by starting from the initial guess

$$\Delta_{\mathrm{x},i} = \frac{1}{2}(\hat{R}^i_{i-1}[3,1] - \hat{R}^i_{i-1}[1,3]) \arccos(\hat{R}^i_{i-1}[3,3]) / \sin(\arccos(\hat{R}^i_{i-1}[3,3])),$$
$$\Delta_{\mathrm{y},i} = \frac{1}{2}(\hat{R}^i_{i-1}[3,2] - \hat{R}^i_{i-1}[2,3]) \arccos(\hat{R}^i_{i-1}[3,3]) / \sin(\arccos(\hat{R}^i_{i-1}[3,3])), \quad (4)$$

where $\hat{R}^i_{i-1}[j,k]$ is the element of $\hat{R}^i_{i-1}$ positioned at row j and column k. Note that if the PCC hypothesis were perfectly fulfilled, than $R^i_{i-1}(\Delta_{\mathrm{x},i}, \Delta_{\mathrm{y},i}) = \hat{R}^i_{i-1}$ - i.e. this would be the global minimum of (3). Yet the real system will likely have a non perfectly constant bending. Therefore, we use standard gradient descent method to locally refine the guess. In this way the residual drift is tamed by the constraints imposed by the kinematic model. Finally, full postures (i.e. both orientation and position) of each point along the segment can be retrieved by substituting back the parameterization into (1). Note that under PCC hypothesis the total length of the segment L_i can be expressed as average of the length of the three actuated segments, which in turn is know from the motor encoders.

In this way the IMU can be used to evaluate full posture, instead of orientation only. Finally, the extracted configuration can be used to regulate the end effector configuration of the robot through kinematic feedback controllers. We extend here the inverse kinematics approach introduced in [7, Sect. 2C], by using the error between desired tip configuration and the measured one, rather than computing the value off-line.

3 Experimental Platform

To validate this algorithm we have developed a single segment, continuum body manipulator based upon our previous work [7]. The actuation system consists of a compliant tri-axial rack and pinion system formed from flexible 3D printed rack elements (Fig. 2). On this motorised platform we have the three rack and pinion mechanisms, each controlled by a DC motor equipped with a quadrature encoder. A PID controller has been implemented to allow a the length of the legs of the manipulation to be set. By varying the length of flexible track between the tip and the base for each of the flexible backbones, the orientation and location of the tip can be controlled in 3D space. This allows for both growth and change in length of the robot, and also motion of the movement of the tip in 3D space. Using the encoders, the approximate position of the end-effector can be controller using the piecewise constant curvature approximation, as previously shown in [7]. Multiple, independent stages of this manipulator can be combined together to generate larger systems, and the properties of the 3D printed racks can be varied to control the bending properties. We have equipped the manipulator with two 9 Degrees of Freedom (DoF) IMUs (BMO005), mounted on the base and the tip of the manipulator, with the two aligned in the Z axis. The IMUs have been calibrated, and the 4 axis quaternions were sampled 10 Hz. Initial readings were discarded to ensure the magnetometer is calibrated. Patterns of reflective motion tracking markers have been mounted on the two stages of the robot, to allow the ground truth of the position of the robot to be captured using a motion tracking experiments. The location of the sensors and markers is highlighted in Fig. 2.

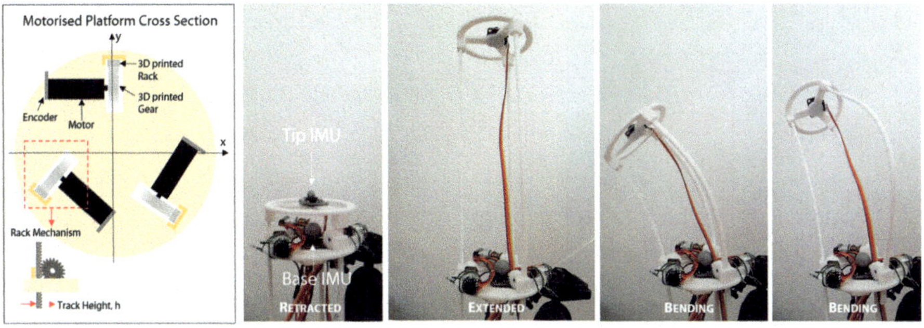

Fig. 2. Design of the actuation stages of the manipulator showing the triaxial motors and 3D printed tracks. Accompanied by pictures showing the range of motions and actions that can be achieved using the manipulator. The placement of the IMUs is highlighted.

It is important to underline here that encoders are a sensing technology that is specific to this design, and are general not feasible for other kind of soft robots. Conversely, implementation of IMUs on a soft robot would be agnostic of the actuation method providing versatility.

4 Results

In this section we first explore the performance of open loop pose reconstruction when using the IMUs on the robot system. Following on from this, we then implemented closed loop control as proposed above to increase the positional accuracy of the robot.

4.1 Pose Reconstruction

In the first experiments we consider the accuracy with which the pose can be reconstructed using the IMU sensor data. The end effector was systematically moved around the working space of the manipulator in the x, y and z axis by varying the demanded encoder positions. During this experiment the true 6 DoF position of the end effector and base of the manipulator was captured using motion capture, and the data from the tip and base IMU was also recorded. Using the methods identified in Sect. 2, the pose of the robot was reconstructed over this period using the IMU data.

In Fig. 3 we show a number of poses of the robot overlaid with a reconstruction of the robot that has been calculated using the IMU data. These results show that there is a close similarity between the shape of the manipulator and the reconstruction. We also see a high accuracy in the estimation of the prediction of the location of the tip of the end effector.

To quantify the performance of the pose reconstruction over the entire workspace of the robot, a comparison between the true pose of the end effector

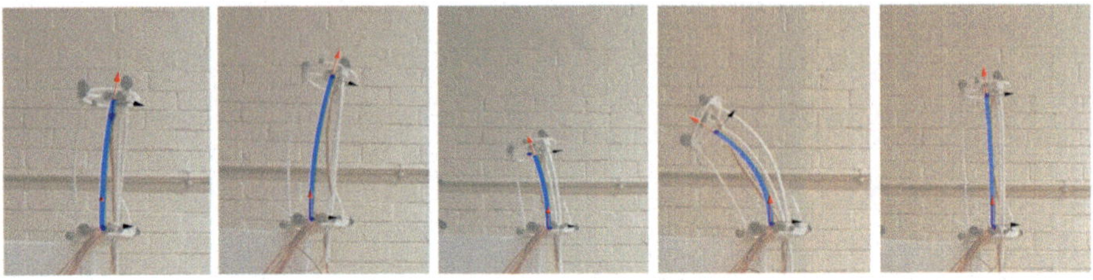

Fig. 3. Comparison between real pose of the robot, and reconstructed pose determined using the IMU data.

(as determined by motion capture), and the pose as estimated using the IMU data has been calculated. The error when reconstructing the location is evaluated by computing the Euclidean distance between the end effector position as reconstructed from the IMU and the true location. Over the entire recording, the reconstruction had a mean accuracy of 2.7 cm, with an accompanying standard deviation of 1.2 cm. For reference, the total length of the robot segment was approximately 20 cm. The error in reconstructing the orientation was evaluated by computing the norm of the relative rotation between the motion capture frame and the IMU based frame, i.e. $|q_{MC} conj(q_{IMU})|$, where q_{MC} and $conj(q_{IMU})$ are the quaternion vectors of the motion capture and the complex conjugate of the IMU quaternion projected on the PCC kinematic model respectively. The recorded accuracy of the orientation has a mean of 0.92 and a standard deviation of 0.12. In this context, a result of 1 represent perfect reconstruction of the orientation. In the context of deformable robots and the variety of different motions that can be achieved with the end effector, this offers an accuracy that would assist in enabling many tasks.

4.2 Closed Loop Control

To experimentally test using the IMU data for closed loop control, the end-effector of the robot has been set to move to a variety of positions along the x, y and z planes within the work-space of the robot. This was performed using encoder only 'open loop control', where the position was determined by setting specific demand encoder values, and also closed loop control using the IMU data in addition. The visual results are shown in Fig. 4 where the motion in each axis is shown for both the open loop and closed loop motion strategies. It can be observed that the IMU based control shows a far closer match to the expected motion path. When only the encoder based approach is used slippage between the gears and rack, and other causes of error lowers the positional accuracy significantly. This is particularly the case for poses which are further from the base of the robot, or poses which require large amounts of bending.

To quantitatively compare the performance of the two different control approaches, the 3 DoF position of the end effector was measured when moved in the three axes (as in Fig. 4), and the euclidean distance error between the

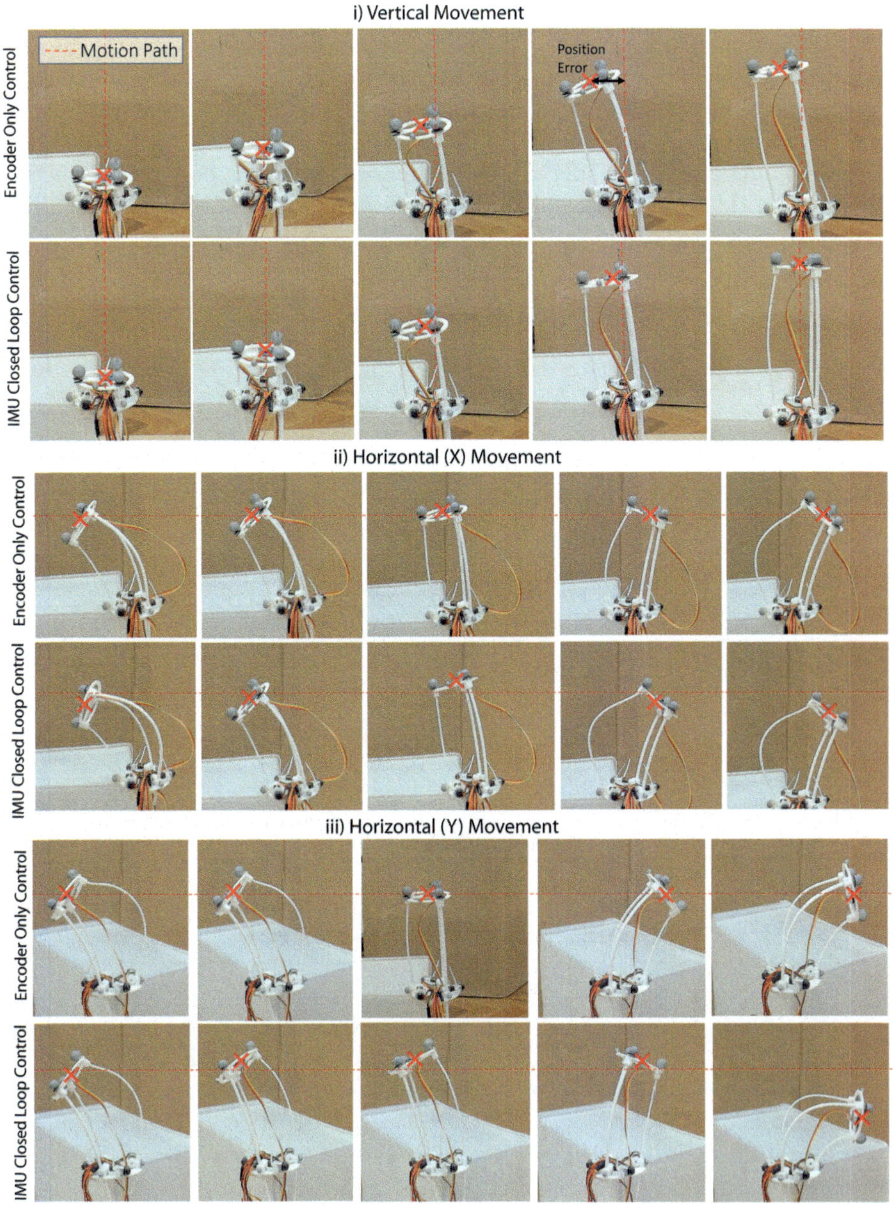

Fig. 4. Graphical comparison of the motion that is achieved when using open loop (encoder only control), and closed loop control using IMU feedback. This is shown for motion in the vertical Z axis, horizontal X axis, and horizontal Y axis.

demand and true position measured. Each pose was repeated 5 times to test and evaluate the reliability, with the results shown in Fig. 5. We see that on average the closed loop controller halves the position error. As seen in the pictorial results this is particularly the case for larger motions where the flexible racks of the robot are most bend. For very low height, or less extreme motions, for example in low heights in the z axis the accuracy for both methods is high, however, for more expansive motions the IMU based closed loop control becomes increasing important. Considering the mean accuracy of the pose reconstruction (2.7 cm), the IMU based closed loop control achieves a precision which is in line with this.

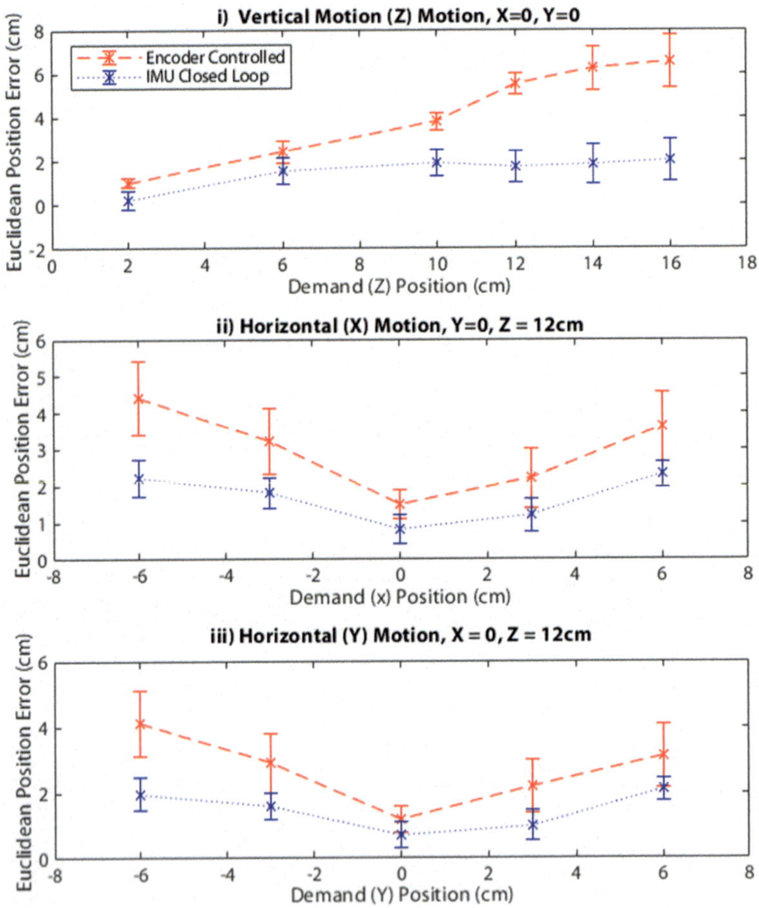

Fig. 5. Quantatative comparison between the performance of the open loop vs. closed loop control for motion in the X, Y, and Z axis.

5 Discussion and Conclusions

In this work we introduced a new method for combining the feedback from IMUs on a deformable manipulator with the PCC model. This constrains the IMU response, minimizing the effects of drift commonly observed when using the response from IMUs. This approach was demonstrated on a growing, deformable manipulator for both pose reconstruction and closed loop control of position. It was shown how the pose could be reconstructed with an mean error of 2.7 cm in position. Using this ability to reconstruct the pose closed loop control using the IMUs was applied to significant reduce the error in positioning.

By introducing this IMU based proprioceptive sensing approach, we provide a universal approach for control and reconstruction of soft, curvature based continuum manipulators. The IMUs can be easily integrated or applied to manipulators of any form factor, including soft of highly performance structures. Using the sensor feedback allows for accurate reconstruction and closed loop control.

Further work could explore the use of IMUs in multi segment robot structure, and also on robots with other actuation methods to validate it across other designs. As this method is highly scaleable it would be possible to create a complex, many-segmented continuum body robot structure. The use of the IMU based pose reconstruction could aid exploration of complex environments when using soft robots. This IMU based method could also be extended to other continuum body structure such as grippers, or soft walkers for proprioceptive sensing.

Acknowledgments. We would like to acknowledge grant NSF EFRI (1830901) which made this work possible.

References

1. Bern, J.M., Banzet, P., Poranne, R., Coros, S.: Trajectory optimization for cable-driven soft robot locomotion. In: Robotics: Science and Systems (2019)
2. Bruder, D., Gillespie, B., Remy, C.D., Vasudevan, R.: Modeling and control of soft robots using the koopman operator and model predictive control. arXiv preprint arXiv:1902.02827 (2019)
3. Della Santina, C., Bicchi, A., Rus, D.: On an improved state parametrization for soft robots with piecewise constant curvature and its use in model based control. IEEE Rob. Autom. Lett. **5**(2), 1001–1008 (2020)
4. Della Santina, C., Catalano, M.G., Bicchi, A.: Soft Robots, pp. 1–14. Springer Berlin Heidelberg, Berlin, Heidelberg (2020). https://doi.org/10.1007/978-3-642-41610-1_146-1
5. Della Santina, C., Katzschmann, R.K., Bicchi, A., Rus, D.: Model-based dynamic feedback control of a planar soft robot: trajectory tracking and interaction with the environment. The International Journal of Robotics Research, p. 0278364919897292 (2019)
6. Hughes, J., Culha, U., Giardina, F., Guenther, F., Rosendo, A., Iida, F.: Soft manipulators and grippers: a review. Front. Rob. AI **3**, 69 (2016)

7. Hughes, J., Della Santina, C., Rus, D.: Extensible high force manipulator for complex exploration. In: 2020 3rd IEEE International Conference on Soft Robotics (RoboSoft), pp. 733–739. IEEE (2020)
8. Hughes, J., Iida, F.: Tactile sensing applied to the universal gripper using conductive thermoplastic elastomer. Soft Rob. **5**(5), 512–526 (2018)
9. Katzschmann, R.K., Della Santina, C., Toshimitsu, Y., Bicchi, A., Rus, D.: Dynamic motion control of multi-segment soft robots using piecewise constant curvature matched with an augmented rigid body model. In: 2019 2nd IEEE International Conference on Soft Robotics (RoboSoft), pp. 454–461. IEEE (2019)
10. Mahony, R., Hamel, T., Pflimlin, J.M.: Nonlinear complementary filters on the special orthogonal group. IEEE Trans. Autom. Control **53**(5), 1203–1218 (2008)
11. Mahony, R., Kumar, V., Corke, P.: Multirotor aerial vehicles: modeling, estimation, and control of quadrotor. IEEE Rob. Autom. Mag. **19**(3), 20–32 (2012)
12. Rus, D., Tolley, M.T.: Design, fabrication and control of soft robots. Nature **521**(7553), 467–475 (2015)
13. Santaera, G., Luberto, E., Serio, A., Gabiccini, M., Bicchi, A.: Low-cost, fast and accurate reconstruction of robotic and human postures via imu measurements. In: 2015 IEEE International Conference on Robotics and Automation (ICRA), pp. 2728–2735. IEEE (2015)
14. Scimeca, L., Hughes, J., Maiolino, P., Iida, F.: Model-free soft-structure reconstruction for proprioception using tactile arrays. IEEE Rob. Autom. Lett. **4**(3), 2479–2484 (2019)
15. Thuruthel, T.G., Shih, B., Laschi, C., Tolley, M.T.: Soft robot perception using embedded soft sensors and recurrent neural networks. Science Robotics, vol. 4, no. 26 (2019)
16. Truby, R.L., Della Santina, C., Rus, D.: Distributed proprioception of 3D configuration in soft, sensorized robots via deep learning. IEEE Rob. Autom. Lett. **5**(2), 3299–3306 (2020)

Planning and Control

Learning to Generate Cost-to-Go Functions for Efficient Motion Planning

Jinwook Huh(✉), Galen Xing, Ziyun Wang, Volkan Isler, and Daniel D. Lee

Samsung AI Center, New York, NY 10111, USA
saic-ny@samsung.com

Abstract. Traditional motion planning can be computationally burdensome for practical robots, due to extensive collision checking and considerable iterative propagation of cost values. We present a novel neural network architecture which can directly generate the cost-to-go (c2g) function for a given configuration space and a goal configuration. The output of the network is a continuous function whose gradient in configuration space can be directly used to generate trajectories in motion planning without the need for protracted iterations or extensive collision checking. This higher order function (i.e. a function generating another function) representation lies at the core of our motion planning architecture, *c2g-HOF*, which can take a workspace as input, and generate the cost-to-go function over the configuration space map (C-map). Simulation results for 2D and 3D environments show that c2g-HOF can be orders of magnitude faster at execution time than methods which explore the configuration space during execution. We also present an implementation of c2g-HOF which generates trajectories for robot manipulators directly from an overhead image of the workspace.

1 Introduction

Motion planning is the task of generating robot commands which take a robot from an initial configuration to a goal configuration while avoiding collisions. It is one of the fundamental problems in robotics and has been extensively studied [4,16,17,19]. Motion planning algorithms reason about the physical *workspace* the robot operates in as well as the robot's *configuration space* which relate the robot pose to the physical space. Workspace and configuration space information can be combined into a free configuration space map (C-map) which partitions the space into regions that either are in collision or are free [17].

For low degree-of-freedom (dof) robots, it is possible to build a C-map explicitly as a grid and use shortest path algorithms such as Dijkstra's or A* for robot motion planning. However, grid-based planning requires intensive collision checks and solving forward kinematics for planning in configuration space of manipulators. Moreover, such methods become infeasible in high dimensional spaces because the grid size grows exponentially with the dimensionality.

J. Huh and G. Xing—The first two authors contributed equally to this work.

B. Siciliano et al. (Eds.): ISER 2020, SPAR 19, pp. 555–565, 2021.
https://doi.org/10.1007/978-3-030-71151-1_49

State of the art approaches to mitigate these challenges include sampling based methods which use configuration space samples to build a graph-based representation of the C-map, and RRTs that explore the C-space by sampling robot inputs and simulating robot motion [12,13,17]. The performance of these methods degrades in complicated environments, since they also require intensive collision checks in high dimensional spaces.

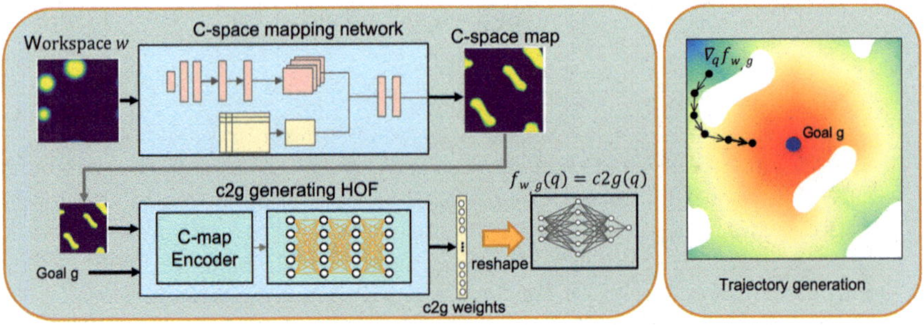

Fig. 1. Left: c2g-HOF architecture. The input is a representation (e.g. an image) of the workspace. The C-space mapping network builds the configuration space map. This C-map along with a goal configuration are the inputs to the higher order function (HOF) network whose outputs are the weights of the cost-to-go function. These weights are reshaped into a neural network which maps robot configurations to their cost-to-go values in a continuous fashion. **Right:** Robot trajectories can be generated simply by following the gradient of the cost-to-go function with respect to the robot configuration.

In this work, we present a learning-based method for generating continuous motion plans. Specifically, we present a neural network which generates a continuous cost-to-go function over the entire configuration space. The key concept underlying our method is the notion of a Higher Order Function (HOF) [20,28] represented as a neural network. In our specific case, we train a HOF to learn to output the cost-to-go function over the C-map. Both the HOF and the cost-to-go function are represented as neural networks.

As shown in Fig. 1, our method starts with a representation of the workspace as input and generates the C-map. The HOF then generates the cost-to-go function for a given goal over this C-map. The HOF network generates the weights of the neural network representing the cost-to-go as a smooth and continuous function. We call our method **c2g-HOF** which stands for *cost-to-go Higher Order Function Network*. The motion plan can be generated by following the gradient of the output cost-to-go function (Fig. 1-right). The main advantage of our method is that, once trained, it can generate a continuous cost-to-go map over the entire C-map much faster than approaches which construct C-map representations during execution.

In summary, the main contributions of the present work are 1) the introduction of a C-space mapping network to generate the probability of collision in the

C-space given a workspace image or voxel; 2) learning the cost-to-go function as represented by a c2g-network using a Higher Order Function (HOF) network; 3) a methodology for fast motion planning with the c2g-network and fast adaptation to new workspaces using the trained HOF network to instantly update c2g-network weights.

We demonstrate the c2g-HOF approach using a novel experimental setup where the workspace is directly represented from an overhead image in which a custom 2-dof or 3-dof manipulator operates. Results from synthetic experiments demonstrate that our method is more than 14 times faster than RRT and 5 times faster than A* for trajectory generation. On average, our generated trajectory lengths are only 5% longer than the optimal A* path.

2 Related Work

There has been much prior research on motion planning to find an optimal trajectory with computational efficiency in high dimensional space [4,17,19]. However, since motion planning approaches for manipulators fundamentally requires intensive collision checks and computation of kinematics, it is very difficult to find an efficient planning approach in high-dimensional space. In low degree-of-freedom (dof) robots, search-based planning approaches use graph search methods over discrete representations [3,18,26]; they build a grid and then compute an optimal shortest path via Dijkstra's or A* algorithm. Potential field methods on the other hand focus on designing local control laws based upon heuristic functions which drive the robot to the goal while steering away from obstacles [14,24]. However, these approaches require heavy computational times for generating the explicit C-map since there is no explicit mapping function between workspace and configuration space.

In order to overcome this difficulty, sampling based methods directly sample nodes in the configuration space and build a graph-based representation of the C-map [1,11–13,17]. However, the performance of sampling based planners strongly depends on sampling methods and local steering algorithms. Although there are many variations of sampling based planners for efficiency, they commonly degrade in complicated environments such as narrow passages.

There are several approaches which can reduce the computation time, such as, storing the previous tree structures and retrieving a path which has the closest vertices to the new start and goal configurations [2,5,8,22]. However, because these approaches need to save the intermediate trajectories, they require extra memory in high-dimensional spaces and are inefficient in finding infeasible or impaired areas from the saved path library when the environment is changed.

Recently, several learning-based approaches for motion planning based on deep neural networks have been proposed [9,21,23]. The approach of [9] is based on a Conditional Variational Autoencoder (CVAE) for sampling policies whereas [23] learns motion planning networks by encoding the workspace into a latent feature space and generating actions.

There are also recent neural network approaches to learn critical points where sampling based planners need to pass through for efficient planning [10,15]. Since these approaches focus on improvement of traditional sampling based planners by a new sampling policy based on neural networks, they require hybrid algorithms by combining the network and traditional planning algorithms. In addition, since these networks are trained based upon many feasible trajectories in C-space for each workspace, they are difficult to converge for various workspaces compared to learning the cost-to-go in C-space.

Another method is [27] which encodes the dynamic programming iterations for motion planning in the layers of a neural network. In contrast, c2g-HOF learns to represent a nearly optimal continuous cost-to-go function as the weights of a multilayer perceptron neural network.

Our method distinguishes itself from these prior methods as it can estimate cost-to-go values over the entire C-space from the workspace without heavy computation. Collision free trajectories can be generated without mapping between the workspace and the C-space using inverse and forward kinematics during execution.

3 Methods

In this section, we will describe the dataset for training, the network architecture, and the training procedures.

3.1 Dataset

In order to train the network, we generated a dataset composed of 30,000 randomly generated workspaces and computed cost-to-go values in C-space using Dijkstra's algorithm. For each workspace, we generate obstacles with random locations and sizes in the 2D plane or 3D workspace. Since Dijkstra's algorithm requires discrete representation, we discretize C-space as a 360×360 grid ($64 \times 64 \times 64$ voxels for 3-dof manipulator). We compute explicit C-map based on robot kinematics and collision checks for each workspace and then use the C-map for Dijkstra's algorithm. We can obtain cost-to-go values in the grid for each workspace.

3.2 Neural Network Architecture

The c2g-HOF architecture, presented in Fig. 1 consists of three components: (1) The C-space mapping network which builds a representation of the C-map from the workspace, (2) HOF network which outputs the weights of the cost-to-go function, and (3) the cost-to-go function represented as a neural network which can map any configuration to a positive real number which represents the distance of that configuration to the goal configuration (i.e., the cost-to-go value).

The C-space mapping network is a functional representation of the C-map as a function $f_c : \mathcal{W} \rightarrow \mathcal{C}$ where $\mathcal{W}$ is the workspace, and the output is a representation of the C-map. In this work, we use a uniform representation with a (soft) collision indicator for each configuration. The C-space mapping network has 7 linear perceptrons (5 linear perceptrons for 2D) followed by a ReLU activation in each layer to estimate shape of manipulator in workspace for a given configuration (a set of joint angles). We train this network by minimizing mean squared errors (MSE) between predicted points and ground truths by traditional forward kinematics. Since the trained C-space mapping network instantly computes all the shapes in the workspace for a whole configuration, it can check collisions for the whole configuration given a workspace image (voxel in 3D) to obtain the C-map.

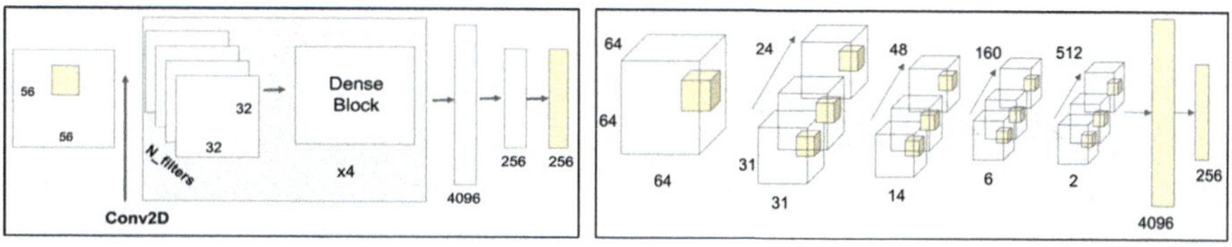

Fig. 2. The specific encoder used as part of the HOF network depends on the dimensionality of the C-space. In this work, 2D image-based (left) and 3D voxel-based (right) encoders are used.

The second component is the HOF network (c2g generating HOF) which is trained to output the cost-to-go function from the input C-map. The specific encoder used in this component depends on the representation of the C-map. In this paper, we apply image and voxel based encoders for 2D and 3D representations as shown in Fig. 2. The 2D encoder has four dense blocks which are composed of four 2D convolution layers followed by batch normalization and ReLU activation. The 3D encoder has four 3D convolution layers followed by batch normalization and ReLU activation. The input of the 3D encoder is $64 \times 64 \times 64$ voxels.

The last component is the c2g-network which represents the cost-to-go function. The c2g-network predicts a cost-to-go value to the goal for a input configuration. In this work, we use a three layer perceptron with 128 neurons at each layer followed by ReLU activations to represent the cost-to-go function.

To summarize, the HOF encoder takes the C-map as input I and generates a cost-to-go function $f_I : \mathcal{Q} \rightarrow \mathcal{C}$, such that for all $q \in \mathcal{Q}$, $c = f_I(q) \in \mathcal{C}$ is the cost-to-go of q. It is worth emphasizing that the output cost-to-go function f_I is a continuous function and is not directly linked to the input representation. This is because, during training, a denser set of samples continuously sampled from the C-map are used to adjust the weights of the neural network representing the cost-to-go function given in the third layer.

3.3 Training

At each iteration, given a set of 2,000 (40,000 for 3D) random samples in C-space, the network is trained by minimizing mean-square error (MSE) between the predicted cost-to-go by network and ground truth cost-to-go by Dijkstra's algorithm in C-space. The c2g generating HOF is trained by backpropagation of loss, since the output of c2g generating HOF is used for the weights of the c2g-network.

3.4 Motion Planning

In order to generate a complete motion planning methodology, we prefix the c2g generating HOF with a C-space mapping network, which takes representations of the workspace and the robot as input and generates the C-map which is the input to the HOF network. The c2g generating HOF outputs the weights of the c2g-network from C-map input. With the c2g-network, motion planning can be performed by following the gradient of cost-to-go values. Since the cost-to-go function depends on the environment, the c2g generating HOF needs to compute the weights of the c2g-network once the environment is changed. When the environment is not changed, it is unnecessary to compute the weights of c2g-network by the c2g generating HOF. Since the inference of cost-to-go values by c2g-HOF is very fast, trajectories are generated almost instantly in the conditions that workspace is changed.

4 Results

In this section, we present results for two representative environments. In order to make various test cases, we have different manipulator configurations (2-link planar and 3-dof) and different workspace (2D plane workspace and 3D workspace).

4.1 Evaluation

We evaluated c2g-HOF in two sets of environments. The first environment consists of a 2-link manipulator operating in a 2D random workspace and the second environment is for a 3-dof manipulator operating in a 3D workspace with randomly generated obstacles on a table. For each test setting, we generated a dataset composed of 10,000 trajectories (200 workspaces $\times$ 50 random trajectories in each workspace) separately.

As shown in Fig. 3, the predicted cost-to-go value is almost the same as the optimal value given by Dijkstra's algorithm. The generated trajectory arrives at the goal point while avoiding obstacles by following the gradient of the cost-to-go values. Both c2g-HOF and RRT have 99% success rates. Figure 4 shows comparison results between c2g-HOF, RRT, and A* in terms of planning time and trajectory length. While RRT and A* require many collision checks, the c2g-network computes cost-to-go without extensive collision checks or space discretization. In the 2D workspace, we see that planning of c2g-HOF is 14 times faster than RRT, and 5 times faster than A* with a smaller variance. The values shown are deviation from the optimal value: the trajectory length is normalized

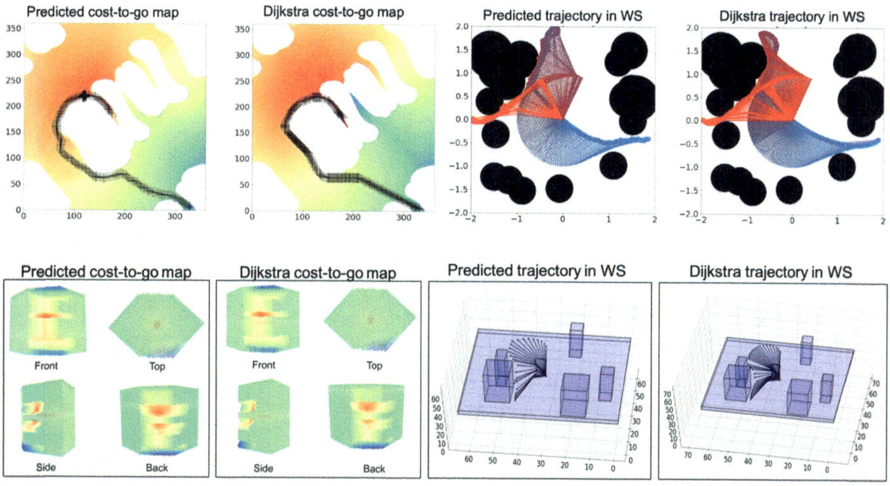

Fig. 3. Comparison of c2g-HOF output against the optimal solution given by Dijsktra's algorithm for 2D (top row) and 3D (bottom row) instances. cost-to-go map and the output trajectory are shown on the left. The same trajectory in the workspace is shown on the right.

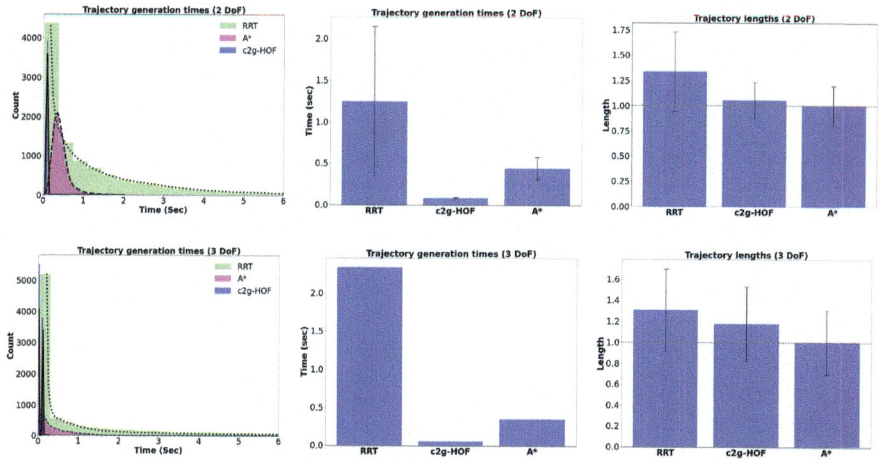

Fig. 4. Results for the 2-dof robot (top row) and 3-dof robot (bottom row). Histogram of planning times are shown on the left. The middle plot shows mean and variances of these distributions whereas the right plot shows the trajectory length mean and variance values. (The trajectory length is normalized by the length of the A*).

by the length of the A* in the same environments. We can see that the trajectory with c2g-HOF is 5% longer than the optimal A* trajectory length but 21% shorter than the RRT trajectory. In the 3D workspace, c2g-HOF is 4 times faster than A* and it generates 17% longer trajectory. It is worth noting that the planning time of c2g-HOF has low variance and hence does not depend heavily on the specific instance.

4.2 Physical Robot Experiments

To verify the performance of the trained c2g-HOF on a real robot, we conducted experiments with real 2-link and 3-dof manipulators. While the 2-link manipulator has no joint limits, the 3-dof manipulator has joint limits at the second and third joints. In our setting, the workspace is represented directly from an overhead Realsense D435 camera. The robot shape and kinematics are learned separately for the C-space mapping network. The HOF network generates the cost-to-go function. The trajectory is generated by following the gradient of cost-to-go values. As shown in Fig. 5, the robot performs well even in the presence of narrow passages: 2-link and 3-dof manipulators insert the link between two obstacles while avoiding collisions successfully.

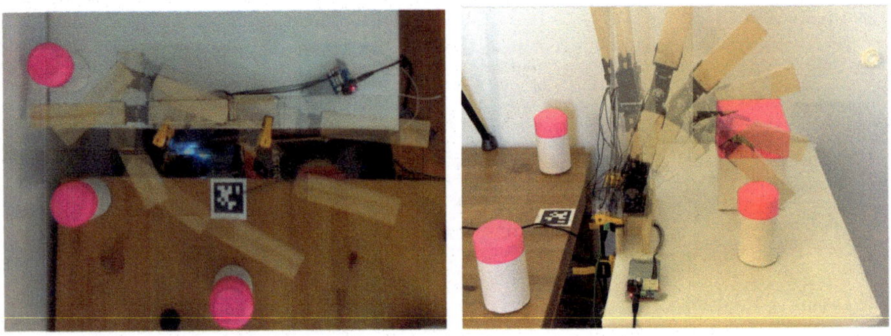

Fig. 5. Our experiment setup and an example robot motion generated by following the gradient of the cost2go network.

5 Discussion

We showed that a neural network can be trained as a higher order function which can output a continuous cost-to-go function over a C-space for a given robot and a workspace. The architecture is fairly general since the same network can be adapted to various settings simply by changing the encoder architecture. In our case, we used the same c2g-network (except for the input dimension) in both two and three dof cases. The performance of the HOF-network indicates that it can learn the cost-to-go function when coupled with an appropriate input representation. More broadly, our results suggest that the exploration efforts of sampling based methods such as RRTs can be shifted to training time.

Successfully training the network to learn cost-to-go functions requires addressing two challenges: The first is to prevent the network bias toward simpler environments which can happen for example when the training set obstacles are placed uniformly at random. To mitigate this issue, we made sure that our dataset has numerous "narrow passage" instances. The second challenge is regarding C-map sampling. The HOF network is conditioned on a discrete map of the configuration space. However, during training, random samples over the entire c-space and their cost-to-go values are used to learn a smooth, continuous cost-to-go function. The sampling strategy during training has direct impact on performance. We leveraged techniques developed for sampling configuration spaces efficiently such as sampling densely around collision boundaries [6,25]. Additional experimentation is needed to assess how these strategies will scale up to more complex environments and robots in the case of c2g-HOF.

In the current c2g-HOF architecture, we used a separately trained network to generate the C-map from workspace and robot descriptions. However, our network is end to end trainable. In theory, it is possible to condition the network on a robot description (or even a picture of the robot) and learn the weights of the c-map network. This capability would allow us to incorporate online calibration/system identification into the architecture and constitutes an interesting avenue for future research.

6 Conclusion and Future Work

We presented a novel learning approach based on a neural network architecture in order to generate the cost-to-go function in a C-space given a workspace. The higher order function network, c2g-HOF, inputs C-map and generates weights of the cost-to-go function represented by the c2g-network. The output of the network is a continuous function for a given goal configuration. The C-map is generated efficiently by C-space mapping network. Since the gradient of the generated cost-to-go function can be directly used to generate trajectories for motion planning. We demonstrated the suggested planning can be orders of magnitude faster at execution time than traditional methods with trajectories near optimal in length.

An important question for future work is whether the cost-to-go function can be constructed directly from the workspace, without going through the C-map. In our ongoing work we show promising results indicating that this is possible [7]. In addition, we are planning to perform additional tests in more complicated environments such as a kitchen shelf. The shelf setting will generate additional challenges such as a higher number of local minima around tight spaces.

References

1. Bekris, K.E., Chen, B.Y., Ladd, A.M., Plaku, E., Kavraki, L.E.: Multiple query probabilistic roadmap planning using single query planning primitives. In: IEEE/RSJ International Conference on Intelligent Robots and Systems, vol. 1, pp. 656–661 (2003)

2. Berenson, D., Abbeel, P., Goldberg, K.: A robot path planning framework that learns from experience. In: Proceedings of the IEEE International Conference on Robotics and Automation (ICRA), pp. 3671–3678 (2012)
3. Bonet, B., Geffner, H.: Planning as heuristic search. Artif. Intell. **129**(1–2), 5–33 (2001)
4. Choset, H.M., Hutchinson, S., Lynch, K.M., Kantor, G., Burgard, W., Kavraki, L.E., Thrun, S.: Principles of Robot Motion: Theory, Algorithms, and Implementation. MIT Press, Cambridge (2005)
5. Coleman, D., Şucan, I.A., Moll, M., Okada, K., Correll, N.: Experience-based planning with sparse roadmap spanners. In: Proceedings IEEE International Conference on Robotics and Automation (ICRA), pp. 900–905 (2015)
6. Hsu, D., Kavraki, L.E., Latombe, J.C., Motwani, R., Sorkin, S.: On finding narrow passages with probabilistic roadmap planners. In: International Workshop on Algorithmic Foundations of Robotics, pp. 141–154 (1998)
7. Huh, J., Isler, V., Lee, D.D.: Cost-to-go function generating networks for high dimensional motion planning. arXiv preprint arXiv:2012.06023 (2020)
8. Hwang, V., Phillips, M., Srinivasa, S., Likhachev, M.: Lazy validation of experience graphs. In: Proceedings of the IEEE International Conference on Robotics and Automation (ICRA), pp. 912–919 (2015)
9. Ichter, B., Harrison, J., Pavone, M.: Learning sampling distributions for robot motion planning. In: Proceedings of the IEEE International Conference on Robotics and Automation, pp. 7087–7094 (2018)
10. Ichter, B., Schmerling, E., Lee, T.W.E., Faust, A.: Learned critical probabilistic roadmaps for robotic motion planning. In: 2020 IEEE International Conference on Robotics and Automation (ICRA), pp. 9535–9541. IEEE (2020)
11. Jaillet, L., Cortés, J., Siméon, T.: Sampling-based path planning on configuration-space costmaps. IEEE Trans. Rob. **26**(4), 635–646 (2010)
12. Karaman, S., Walter, M.R., Perez, A., Frazzoli, E., Teller, S.: Anytime motion planning using the RRT*. In: IEEE International Conference on Robotics and Automation, pp. 1478–1483 (2011)
13. Kavraki, L.E., Švestka, P., Latombe, J.C., Overmars, M.H.: Probabilistic roadmaps for path planning in high-dimensional configuration spaces. IEEE Trans. Robot. Autom. **12**(4), 566–580 (1996)
14. Khatib, O.: Real-time obstacle avoidance for manipulators and mobile robots. In: Autonomous Robot Vehicles, pp. 396–404. Springer (1986)
15. Kumar, R., Mandalika, A., Choudhury, S., Srinivasa, S.S.: LEGO: leveraging experience in roadmap generation for sampling-based planning. arXiv preprint arXiv:1907.09574 (2019)
16. Latombe, J.C.: Robot motion planning (1991)
17. LaValle, S.M.: Planning Algorithms. Cambridge University Press, Cambridge (2006)
18. Likhachev, M., Gordon, G.J., Thrun, S.: ARA*: anytime A* with provable bounds on sub-optimality. In: Advances in Neural Information Processing Systems 16, pp. 767–774. MIT Press (2004)
19. Lynch, K.M., Park, F.C.: Modern Robotics. Cambridge University Press, Cambridge (2017)
20. Mitchell, E., Engin, S., Isler, V., Lee, D.D.: Higher-order function networks for learning composable 3D object representations. In: International Conference on Learning Representations (2019)
21. Molina, D., Kumar, K., Srivastava, S.: Learning critical regions for robot planning using convolutional neural networks (2019)

22. Phillips, M., Cohen, B., Chitta, S., Likhachev, M.: E-graphs: bootstrapping planning with experience graphs. In: Proceedings of the Robotics: Science and Systems, Sydney, Australia, July 2012
23. Qureshi, A.H., Bency, M.J., Yip, M.C.: Motion planning networks. arXiv preprint arXiv:1806.05767 (2018)
24. Rimon, E., Koditschek, D.E.: Exact robot navigation using artificial potential functions. IEEE Trans. Robot. Autom. **8**(5), 501 (1992)
25. Shi, K., Denny, J., Amato, N.M.: Spark PRM: using RRTs within PRMs to efficiently explore narrow passages. In: Proceedings of the IEEE International Conference on Robotics and Automation, pp. 4659–4666. IEEE (2014)
26. Stentz, A., et al.: The focussed D^* algorithm for real-time replanning. IJCAI **95**, 1652–1659 (1995)
27. Tamar, A., Wu, Y., Thomas, G., Levine, S., Abbeel, P.: Value iteration networks. In: Advances in Neural Information Processing Systems, pp. 2154–2162 (2016)
28. Wang, Z., Mitchell, E.A., Isler, V., Lee, D.D.: Geodesic-HOF: 3D reconstruction without cutting corners. arXiv preprint arXiv:2006.07981 (2020)

Using Nonlinear Normal Modes for Execution of Efficient Cyclic Motions in Articulated Soft Robots

Cosimo Della Santina[1,2,3(✉)], Dominic Lakatos[1], Antonio Bicchi[4,5], and Alin Albu-Schaeffer[1,2]

[1] Institute of Robotics and Mechatronics, German Aerospace Center (DLR), Oberpfaffenhofen, Germany
alin.Albu-Schaeffer@dlr.de

[2] Informatics and Mathematics Department, Technical University of Munich (TUM), Garching bei München, Germany

[3] Cognitive Robotics Department, Delft University of Technology, 2628 CD Delft, The Netherlands

[4] "Enrico Piaggio", University of Pisa, Largo Lucio Lazzarino 1, 56126 Pisa, Italy

[5] Department of Advanced Robotics, Istituto Italiano di Tecnologia, via Morego, 30, 16163 Genova, Italy

Abstract. Thanks to their body elasticity, articulated soft robots promise to produce effective and robust oscillations with low energy consumption. This in turn is an important feature which can be exploited in the execution of many tasks, as for example locomotion. Yet, an established theory and general techniques allowing to excite and sustain these nonlinear oscillations are still lacking. A possible solution to this problem comes from nonlinear modal theory, which defines curved extensions of linear Eigenspaces called Eigenmanifolds. Stabilizing these surfaces is equivalent to exciting regular hyper-efficient oscillations in the robotic system. This paper proposes a first experimental validation of the Eigenmanifold stabilization technique. It also proposes a simple yet effective means of injecting energy into the system, so to sustain the oscillations in presence of damping. We consider as experimental setups a single robotic leg, and a full soft quadruped. Preliminary locomotion results are provided with both systems.

Keywords: Articulated soft robotics · Locomotion · Nonlinear oscillations · Feedback control

1 Introduction

Inspired by the vertebrate branch of the animal kingdom, articulated soft robots are robotic systems embedding compliant elements into a classic rigid (skeleton-like) structure [2,9]. Soft robots promise to push their limits far beyond the barriers that affect their rigid counterparts. However, existing control strategies aiming at achieving this

This work is supported by the European Union ERC project 835284 M-Runners.

B. Siciliano et al. (Eds.): ISER 2020, SPAR 19, pp. 566–575, 2021.
https://doi.org/10.1007/978-3-030-71151-1_50

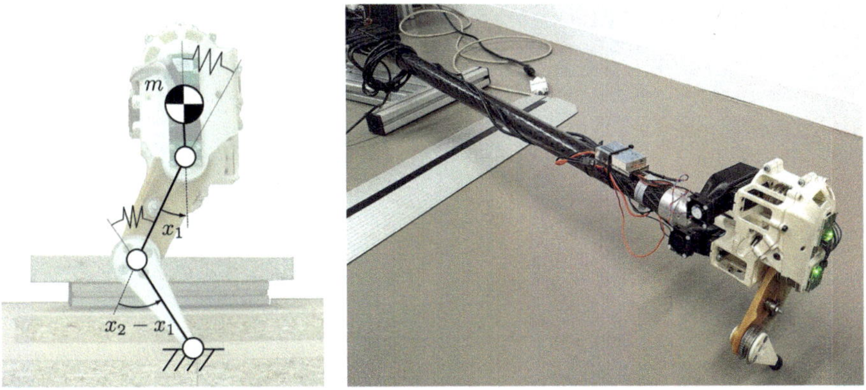

Fig. 1. Experimental setup: a 2-DoF (the upper part is constrained to stay vertical) segmented leg. The left panel shows also a sketch of the robot scheme with main quantities highlighted. The right panel shows the pole system constraining the upper part of the leg.

goal are either tailored on specific examples [11, 14], or rely on model cancellations [5, 10] - thus defeating the purpose of introducing elasticity in the first place [8].

In a series of recent works [1, 3, 6, 7], we have proposed a framework for exciting efficient oscillatory motions in robots immersed in a potential field. The long term aim of this theory is to equip generic soft robots (of both articulated and continuum kind) to execute periodic tasks efficiently and effectively. A main component of this theory are Eigenmanifolds, that we defined in [1] as nonlinear continuations of the classic linear Eigenspaces. When the soft robot is initialized on one of these manifolds, it evolves autonomously while presenting regular - and thus practically useful - evolutions. We call these evolution normal modes. In addition to that, we proposed in [7] a control strategy making modal manifolds attractors for the system, and acting on the total energy of the soft robot to move it from a modal oscillation to the other. In this way, a large class of autonomous behaviors can be excited, which are direct expression of the embodied intelligence of the articulated soft robot. Finally, [1, 7] describe a specialization of the theory to systems with strong (yet implementable) kinetic and potential symmetries.

Despite the core ideas behind our work coming from physical intuition and preliminary experimental validations [12], the formulation that we have provided so far is rather theoretical, and very much in need of an experimental validation. The aim of this paper is to provide such an experimental validation using as testbed the articulated soft leg in Fig. 1. We will introduce a simplified control strategy, and we will test its effectiveness on this system to implement swing-like oscillations. Preliminary application to the articulated soft robotic quadruped Bert [13] is also provided, showing that these techniques can be employed to implement locomotion.

2 Technical Approach

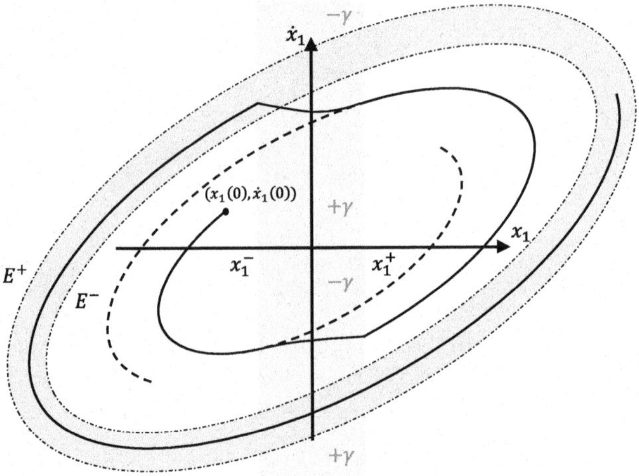

Fig. 2. The state evolves under the control action (6). When the system is in a neighborhood of equilibrium configuration $x_{\mathrm{m}} = 0$ (i.e. when it crosses the gray area), energy is injected by the controller, moving the system to another of its autonomous orbits. Eventually this brings the robot in the region of state space with the desired amount of energy.

2.1 Eigenmanifold: A Very Concise Definition

Building upon a theory laid down by more than one century of research in mathematics, physics, and engineering, in [1] we propose an extension of linear modes to robotics, which is then summarized in a coordinate dependent framework in [7]. We must give those definitions for granted here, due to space limitations, and only provide an intuitive introduction to the concept. Consider a mechanical system in the standard form

$$M(x)\ddot{x} + C(x,\dot{x})\dot{x} + G(x) = \tau, \tag{1}$$

where $x \in \mathbb{R}^n$ are the configuration coordinates of the robot. $M(x), C(x,\dot{x}) \in \mathbb{R}^{n \times n}$ are the usual inertia and Coriolis matrices, and $G(x) \in \mathbb{R}^n$ is the potential field. $\tau \in \mathbb{R}^n$ are the control inputs. The total energy is

$$E(x,\dot{x}) = \frac{1}{2} x^{\mathrm{T}} M(x) x + V(x), \tag{2}$$

where $V(x)$ is the potential associated to $G(x)$.

An Eigenmanifold is a direct extension of an Eigenspace to this kind of nonlinear mechanical systems. It is defined by imposing to a curved surface many of the properties that define an eigenspace in the linear case. It is a two dimensional invariant submanifold of the state space $(x,\dot{x})$. It contains an equilibrium configuration of the robot. Also, it is such that any evolution contained in it is periodic, it has a trajectory which is lineshaped, and it is identified by its energy level. Consider an eigenspace of the linearized system at an equilibrium

$$ES = \text{Span}\{(c,0),(0,c)\}, \tag{3}$$

with $c \in \mathbb{R}^n$. In [1] we show that we can always describe the Eigenmanifold prolonging this linear eigenspace as the set of states such that

$$X(x_\text{m}, \dot{x}_\text{m}) = x, \quad \dot{X}(x_\text{m}, \dot{x}_\text{m}) = \dot{x}, \tag{4}$$

where X and $\dot{X}$ are two functions from ES to $\mathbb{R}^n$ describing the manifold geometry - called coordinate embedding - and $(x_\text{m}, \dot{x}_\text{m}) = (c^\text{T}x, c^\text{T}\dot{x})$. In [1] we discuss how to extract $(X, \dot{X})$ from ES.

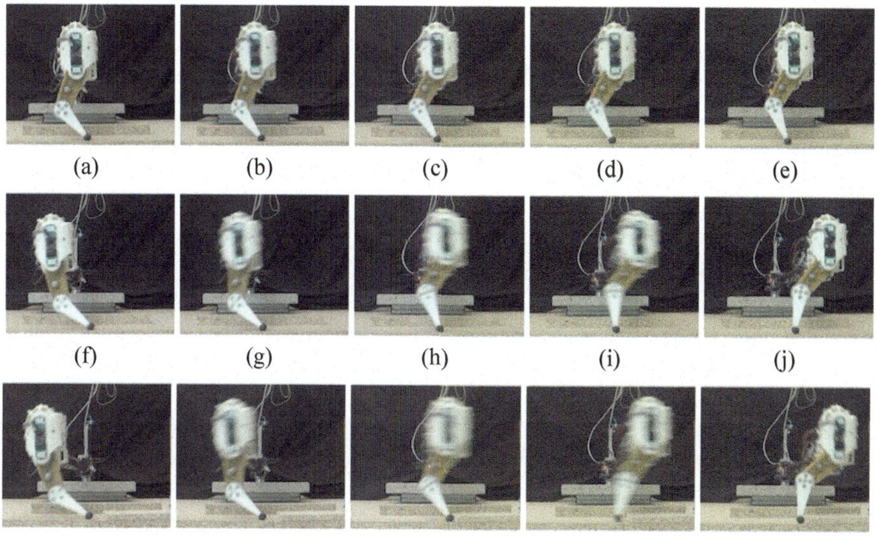

(a) (b) (c) (d) (e)

(f) (g) (h) (i) (j)

(k) (l) (m) (n) (o)

Fig. 3. Nonlinear oscillations induced by the proposed algorithm on a segmented soft leg. Panels (a–e) present one oscillation for $\alpha = 0.3$ Nm, panels (f–j) for $\alpha = 0.7$ Nm, and panels (k–0) for $\alpha = 0.9$ Nm. The oscillations amplitude increase with the increase of the gain α.

2.2 Exciting Nonlinear Oscillations with a Simple Feedback

Let $(X, \dot{X})$ be a coordinate embedding of an eigenmanifold. In [7] we proposed the following feedback control to excite nonlinear oscillations in a robot subject to a potential field

$$\tau(x,\dot{x}) = M(x)\left(\kappa_\text{p}\left(X(x_\text{m},\dot{x}_\text{m}) - x\right) + \kappa_\text{d}\left(\dot{X}(x_\text{m},\dot{x}_\text{m}) - \dot{x}\right) + \alpha\tau_\text{E}(x,\dot{x},\bar{E})\right). \tag{5}$$

The idea is to make the modal manifold an attractor by means of a PD-like action (first two terms), and then pick the right oscillation among all the available ones through energy regulation (implemented by τ_E). The control gains are $\kappa_\text{p}, \kappa_\text{d}, \alpha \in \mathbb{R}$. The PD

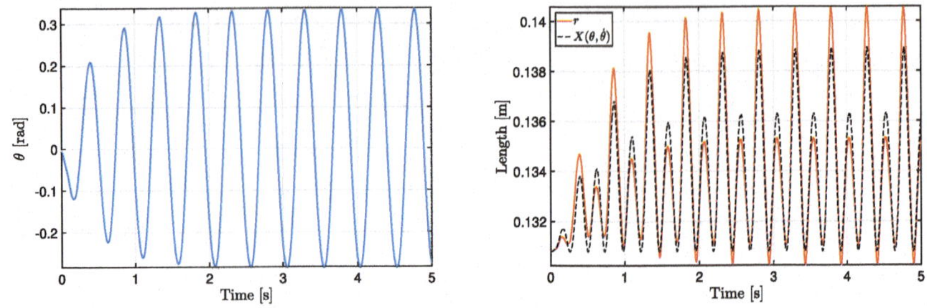

Fig. 4. Experimental evolutions of (θ, r), for $\alpha = 0.5\,\mathrm{Nm}$. The right panel reports also the ideal evolution on the manifold $X(\theta, \dot{\theta})$ for the measured evolution of θ, as a dashed gray line.

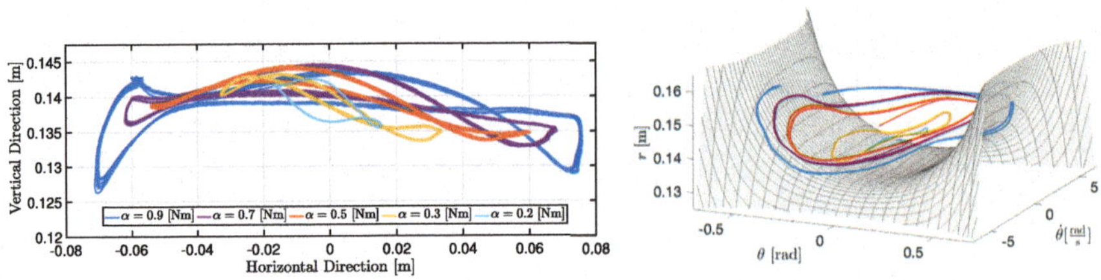

(a) (b)

Fig. 5. Experimental trajectories Panel (a) shows the evolutions in Cartesian coordinates of the leg's center of mass. Panel (b) presents the same evolutions in the space $(\theta, \dot{\theta}, r)$. 15 s of oscillations are considered. The ideal modal manifold $r = X(\theta, \dot{\theta})$ is superimposed.

regulation is quite simple to implement, since X and $\dot{X}$ are available in closed form (e.g. as polynomial). Due to this simplicity, we experienced a high level of robustness when testing the technique in simulation. This has been very consistent across several kinds of uncertainties and many type of systems. With this paper, we want to double-check this insight experimentally.

The possibility of injecting or removing energy from the system allows to select the desired mode within the modal manifold, therefore increasing or decreasing the amplitude of oscillation (and/or its frequency). In [7], $\tau_{\mathrm{E}}(x, \dot{x}, \bar{E})$ realizes energy regulation through the feedback loop $\dot{x}(\bar{E} - E(x, \dot{x}))$. This is a more complex component to implement since the energy is a transcendental function of the state, and fundamentally realizes a form of strongly model dependent feedback. Since we want to bring this abstract theory on an experimental ground, we consider here an even simpler and more robust strategy. This is effectively the energy regulation introduced in [7], reduced to its most essential components

$$\tau_{\mathrm{E}} = \begin{cases} 0 \text{ if } x_{\mathrm{m}} \notin [x_{\mathrm{m}}^{-}, x_{\mathrm{m}}^{+}] \vee E(X, \dot{X}) \in [E^{-}, E^{+}] \\ 1 \text{ if } x_{\mathrm{m}} \in [x_{\mathrm{m}}^{-}, x_{\mathrm{m}}^{+}] \wedge ((E(X, \dot{X}) < E^{-} \wedge \dot{x}_{\mathrm{m}} > 0) \vee (E(X, \dot{X}) > E^{+} \wedge \dot{x}_{\mathrm{m}} < 0)) \\ -1 \text{ otherwise} \end{cases}, \tag{6}$$

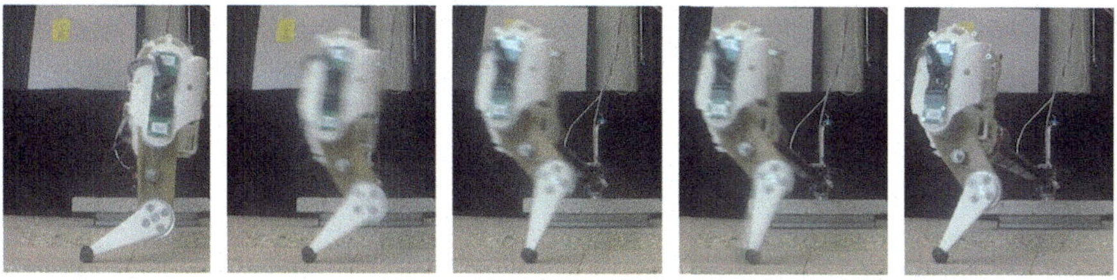

(a) (b) (c) (d) (e)

Fig. 6. Stills of an articulated soft leg jumping forward by exploiting a nonlinear mode excited through a simple controller. Panels (a, b) show the stance phase, Panel (c) the take off, Panel (d) the flight phase, and Panel (e) the landing.

where $E^+ > E^- > 0$, and $x_{\mathrm{m}}^+ > 0 > x_{\mathrm{m}}^-$ are scalar constants. This controller idea is shown in Fig. 2. Similar to a swing which is kept in persistent oscillations by an occasion push in the right direction, the idea is to inject or remove energy in small chunks until $E(x(t))$ reaches $[E^-, E^+]$. Note indeed that the conditions selecting the sign of the torque are such that the change of energy $\dot{E} = \dot{x}^T \tau_{\mathrm{E}}$ is always positive when $E(X, \dot{X}) < E^-$ and negative viceversa. Note that this strategy is akin to the well-known swing up controller proposed in [4].

Although designed with the goal of being intuitive and robust, this control action makes the closed loop system hybrid. Therefore the actual proof of convergence will require some effort which is beyond the scope of the present paper. Our aim here is instead to give experimental substantiation to the whole idea of exciting complex nonlinear oscillations by means of simple feedback control actions stabilizing Eigenmanifolds, and see which kind of lessons we can learn from this validation.

3 Experimental Validation

3.1 Experimental Setup

As a first experimental validation of the proposed strategy, we consider here the soft segmented leg in Fig. 1. It is made of two links with same length b - considered here massless - and a main body - with mass m. Linear revolute springs act on both joints (mechanism not shown in figure). Please refer to [12] for more details on this system. The leg is mechanically constrained to evolve on the Sagittal plane and the main body to remain vertical, by the pole shown in the right part of Fig. 1. We hypothesize infinite friction between the foot and the environment, so that the ground contact behavior is approximated with a revolute joint. Thus the configuration of the robot can be described by the two angles x_1 and x_2 in Fig. 1.

3.2 The Model, the Mode, and the Control Algorithm

We are interested here in generating swing oscillations of the center of mass. We describe the system configuration through polar coordinates of the center of mass expressed w.r.t. the foot frame

$$\theta = x_1/2 + x_2/2, \quad r = b\sqrt{2(1+\cos(x_1 - x_2))}. \tag{7}$$

These two novel coordinates represent respectively the distance between hip and foot, and its angle w.r.t. the vertical. The resulting dynamics has the following form (see [12] for the detailed derivation)

$$\begin{aligned} \ddot{\theta} &= -2\dot{r}\dot{\theta}/r + g\sin(\theta)/r - 2\gamma\theta/r^2 + \tau_\theta/m, \\ \ddot{r} &= r\dot{\theta}^2 - g\cos(\theta) - \gamma(\Upsilon(r) - \Upsilon(r_0))/\sqrt{4b^2 - r^2} + \tau_r/m \end{aligned} \tag{8}$$

where $\Upsilon(r) = \arccos\left(1 - r^2/(2b^2)\right)$, γ is the stiffness of both springs divided by the mass m, τ_θ and τ_r are the control actions, g is the gravity acceleration, and r_0 is the unloaded length of the equivalent spring. The system has an equilibrium in $\theta = 0$ and $\Upsilon(r) = \Upsilon(r_0) - g/\gamma$. Its linearized dynamics is $\Delta\ddot{\theta} \simeq k_\theta \Delta\theta$, and $\Delta\ddot{r} \simeq k_r \Delta r$, with $k_\theta, k_r \in \mathbb{R}$ being two constant values. The normal modes of the linearized system are therefore two decoupled evolutions: an angular oscillation with fixed radius, and a radial oscillation with fixed angle. The nonlinear extension of the latter is trivial, since for $\theta \equiv 0$ and $\dot{\theta} \equiv 0$ the dynamics collapses into a quasi-linear one, that we studied in [12]. The other mode instead turns into a more complex oscillation, that we investigate here. In this case $c = (1, 0)$, i.e. $x_m = \theta$. We approximate $(X, \dot{X})$ as fourth order polynomials, solving in the Galerkin sense the tangency constraints introduced in [7]. To this end, we approximate the system dynamics 3rd order Taylor expansion around the equilibrium.

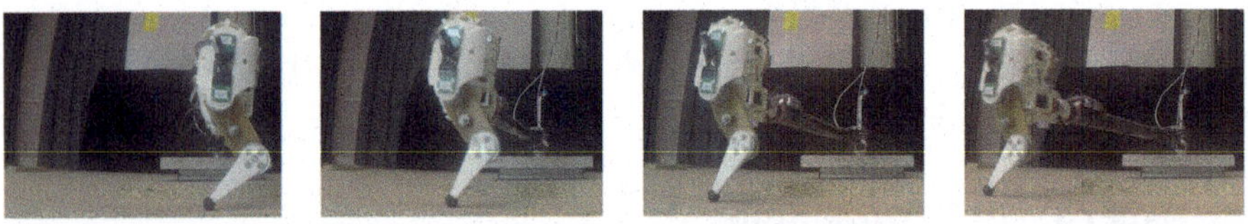

(a) Initial (b) 2nd jump (c) 4th jump (d) Final

Fig. 7. The articulated soft leg jumps forward by exploiting natural oscillations: full length covered. Panel (a) shows the initial configuration, while Panels (b, c, d) show some examples of landings. This picture is meant to give a sense of the total forward motion that it is achieved in 8 s, and 6 jumps.

3.3 Experiments with a Single Leg

We performed experiments for five different values of the orbit excitation gain α; 0.2 Nm, 0.3 Nm, 0.5 Nm, 0.7 Nm, 0.9 Nm. Target energy levels are $E^- = 21$J, $E^+ = 22$J. Due to energy losses, the desired level of energy could not be reached. Instead energy injected through τ_E is compensated by dissipation. A different steady state oscillation is reached for each value of α. Figure 3 shows oscillations resulting from three of the considered gains. Figure 4 shows the evolution of θ and r for $\alpha = 0.5$ Nm. Control action is turned on at 0s. The transient lasts for about 2 s, in which the algorithm pumps energy into the system. After that, the segmented leg starts evolving according to a stable nonlinear oscillation. Measured and ideal trajectories - i.e. $(\theta,\dot{\theta},r,\dot{r})$ and $(\theta,\dot{\theta},X(\theta,\dot{\theta}),\dot{X}(\theta,\dot{\theta}))$ respectively - are quite close to each other, as shown by the right panel in the same figure. Figure 5(a) shows the evolutions of the center of mass in Cartesian coordinates, for all the considered values of α, and for a period of 15s. The bigger is the gain, the larger are the oscillations, and the higher is the energy level reached. The resulting oscillations are slightly concave[1] and highly repeatable. Figure 5(b) illustrates the evolutions superimposed to the ideal modal manifold, i.e. to the surface $(\theta,r)=X(\theta,\dot{\theta})$. The matching is good, with larger discrepancies for high speeds and positive values of θ.

Interestingly, it is sufficient to slightly tilt the orientation of the leg's main body, to produce forward locomotion. This is shown in Fig. 6 - which depicts a single jump - and in Fig. 7 - which shows the effect of 6 jumps executed in 8 s. Note that each jump is actually produced by two oscillations. The first is only instrumental to inject energy into the system, while the second is when the actual jump happens. Only the latter is shown in figure.

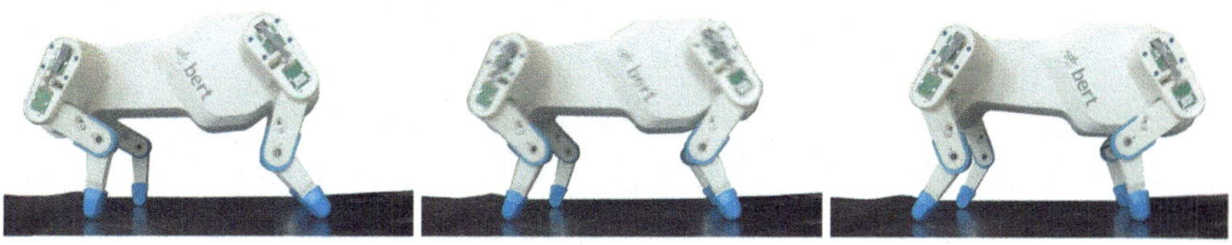

(a) (b) (c)

Fig. 8. Stills of a nonlinear oscillatory behavior induced on a soft quadruped. Each leg follows the same design of the leg used so far. The controller that we developed for the single leg testbed is applied to all four legs. The four oscillators naturally synchronize thanks to the coupling imposed by the closed kinematic chaing. The period of oscillation is about 0.8 s.

3.4 Experiments with a Quadruped

We consider here the application of proposed algorithm to the quadruped Bert [13]. This system is built using four of the above discussed soft segmented legs. We want to test if the proposed control strategy can excite stable oscillations also in this more complex system. We start by applying the algorithm separately to all the four legs. Results are shown in Fig. 8. Without any modification, the algorithm can produce and sustain a regular back-and-forth oscillation, with period 0.8 s. We then investigate if such oscillations can be used to implement locomotion patterns. This is done by employing a simple heuristic. The aim is to implement a trot gait. The two legs that at a given time are on the ground are controlled through the proposed algorithm. The other two legs are lifted enough to not interfere with the ground. They are then put forward when the two

[1] Different choices of the physical parameters (e.g. the spring stiffnesses) could have produced a convex oscillation.

stance legs are close to stopping. This event signals the beginning of the next phase. Already with such a simple algorithm, the quadruped is able to walk forward, as shown in Fig. 9. The steps are here very small, and therefore the forward speed is relatively small (around 0.15 m/s). Yet, we believe that this represents a first important proof that the theory that we are developing can be used in locomotion of complex mechanical systems.

(a) 0 seconds (b) 4 seconds

(c) 8 seconds (d) 10 seconds

Fig. 9. By implementing a simple variation on the proposed algorithm, the quadruped can move forward by exploiting modal oscillations. The distance covered in 10 s is approximately 1.5 m.

4 Experimental Insights

The experiments exhibit a quite different scenario than the one we could have expected by looking at the problem from the pure lenses of theory. For a start, they suggest that the proposed strategy can be used in practice to excite the normal modes of soft robots, generating stable and repeatable nonlinear oscillations also in the presence of many uncertainties and unmodeled dynamics in the controlled system - e.g. the actuators dynamics, the moving contact with the ground, the non zero weight of the legs, inexact identification of system parameters, neglected friction effects. Moreover, $\dot{x}_1, \dot{x}_1$ is not being directly measured, but estimated through a high pass filter. Finally, the physical system is serially actuated, and the parallel elastic behavior needs to be implemented through an opportune input mapping. First, we map τ_θ and τ_r to torques acting on x_1 and x_2 via pre-multiplication for the transpose Jacobian of the change of variables. These torques are then realized by commanding to motors a displacement equal to the torques divided by the stiffness γ. These experiments also taught us some very important lessons on where to look for improvements. First and foremost, the final energy resulted as an equilibrium between α and dissipative effects, rather than due to some stopping condition connected to the energy level E^- that could never be reached. This is because the energy regulator was built with a conservative system in mind. So experiments suggest that rather than improving the Eigenmanifold stabilization algorithm (e.g. to make it global rather than local), it is probably more practically meaningful to put some effort in developing adaptive algorithms which can dynamically adjust the value of α so to reach a desired amplitude of oscillations. For what concerns the specific leg experiment instead, we saw that mismatches from the manifold start to increase with high velocities, and that the oscillations are typically not perfectly symmetric (higher errors for positive values of δ). This suggests that a model taking into account the legs mass should be considered, therefore breaking the symmetry of $(X, \dot{X})$.

References

1. Albu-Schaeffer, A., Della Santina, C.: A review on nonlinear modes in conservative mechanical systems. Annu. Rev. Control **50**, 49–71 (2020)
2. Albu-Schaeffer, A., Eiberger, O., Grebenstein, M., Haddadin, S., Ott, C., Wimboeck, T., Wolf, S., Hirzinger, G.: Soft robotics. IEEE Robot. Autom. Mag. **15**(3), 20–30 (2008)
3. Albu-Schaeffer, A., Lakatos, D., Stramigioli, S.: One-dimensional solution families of nonlinear systems characterized by scalar functions on Riemannian manifolds. arXiv preprint arXiv:1911.01882 (2019)
4. Åström, K.J., Furuta, K.: Swinging up a pendulum by energy control. Automatica **36**(2), 287–295 (2000)
5. Buondonno, G., De Luca, A.: A recursive Newton-Euler algorithm for robots with elastic joints and its application to control. In: 2015 IEEE/RSJ International Conference on Intelligent Robots and Systems (IROS), pp. 5526–5532. IEEE (2015)
6. Calzolari, D., Della Santina, C., Albu-Schäffer, A.: PD-like regulation of mechanical systems with prescribed bounds of exponential stability: the point-to-point case. IEEE Control Syst. Lett., 1 (2021). https://doi.org/10.1109/LCSYS.2020.3046538
7. Della Santina, C., Albu Schaeffer, A.: Exciting efficient oscillations in nonlinear mechanical systems through Eigenmanifold stabilization. IEEE Control Syst. Lett. **5**(6), 1916–1921 (2021). https://doi.org/10.1109/LCSYS.2020.3048228
8. Della Santina, C., Bianchi, M., Grioli, G., Angelini, F., Catalano, M.G., Garabini, M., Bicchi, A.: Controlling soft robots: balancing feedback and feedforward elements. IEEE Robot. Autom. Mag. **24**, 75–83 (2017)
9. Della Santina, C., Catalano, M.G., Bicchi, A.: Soft Robots, pp. 1–14. Springer, Heidelberg (2020). https://doi.org/10.1007/978-3-642-41610-1_146-1
10. Garofalo, G., Ott, C.: Passive energy-based control via energy tanks and release valve for limit cycle and compliance control. IFAC-PapersOnLine **51**(22), 73–78 (2018)
11. Hutter, M., Remy, C.D., Höpflinger, M.A., Siegwart, R.: Slip running with an articulated robotic leg. In: 2010 IEEE/RSJ International Conference on Intelligent Robots and Systems (IROS), pp. 4934–4939. IEEE (2010)
12. Lakatos, D., Friedl, W., Albu-Schäffer, A.: Eigenmodes of nonlinear dynamics: definition, existence, and embodiment into legged robots with elastic elements. IEEE Robot. Autom. Lett. **2**(2), 1062–1069 (2017)
13. Seidel, D., Hermann, M., Gumpert, T., Loeffl, F.C., Albu-Schäffer, A.: Using elastically actuated legged robots in rough terrain: experiments with DLR quadruped bert. In: 2020 IEEE Aerospace Conference, pp. 1–8. IEEE (2020)
14. Zhakatayev, A., Rubagotti, M., Varol, H.A.: Time-optimal control of variable-stiffness-actuated systems. IEEE/ASME Trans. Mechatron. **22**(3), 1247–1258 (2017)

RAPID: An Algorithm for Quick Replanning Under Changed Dynamical Constraints

Sharan Nayak(✉) and Michael W. Otte

Department of Aerospace Engineering, University of Maryland, College Park, MD 20742, USA
{snayak18,otte}@umd.edu

Abstract. There are many scenarios where the dynamical constraints of a vehicle may change mid-mission either due to added payload, vehicle damage or changes to the environment. The vehicle may need to come back to the base safely without undergoing collisions with obstacles. We present RAPID (RApid Planning for Interchanged Dynamics), a sampling based motion planning algorithm for quick replanning when the dynamical constraints associated with a vehicle change. RAPID uses the planning information associated with the original dynamics as a heuristic for quick replanning. RAPID is useful when a vehicle must replan to account for changes to its dynamics but does not have enough time or resources to run a whole planning algorithm (like RRT or SST) from scratch. We compare RAPID with brute-force replan using RRT and SST using a combination of hardware and software experiments. The hardware experiments are performed using a modified Parrot Bebop 2 quadrotor fitted with an Intel UP board traversing an obstacle course. Our experimental results demonstrate RAPID provides considerable improvements over planning from scratch (RRT, SST) in solution completion time, path length, and flight time performance metrics.

Keywords: Planning and control · Mobile robots · Unmanned aerial vehicles

1 Introduction

We present **RAPID** (**RA**pid **P**lanning for **I**nterchanged **D**ynamics), a sampling-based motion planning algorithm for performing quick replan back to base when the dynamical constraints of a vehicle change. There are many situations in which the maneuvers available to a vehicle may change mid-mission, e.g., due to added payload [1] or damage to the vehicle [2]. After the change, the inbuilt controller may be unable to execute maneuvers along the original plan. In such a case it is imperative that the vehicle quickly replan to account for its new dynamical constraints so that it may return safety to base (for repair, payload delivery, etc.) without colliding with obstacles.

A naive replanning approach is to re-run the whole planning algorithm (like RRT [3] or SST [4]) with the new dynamical constraints and find a new path.

B. Siciliano et al. (Eds.): ISER 2020, SPAR 19, pp. 576–588, 2021.
https://doi.org/10.1007/978-3-030-71151-1_51

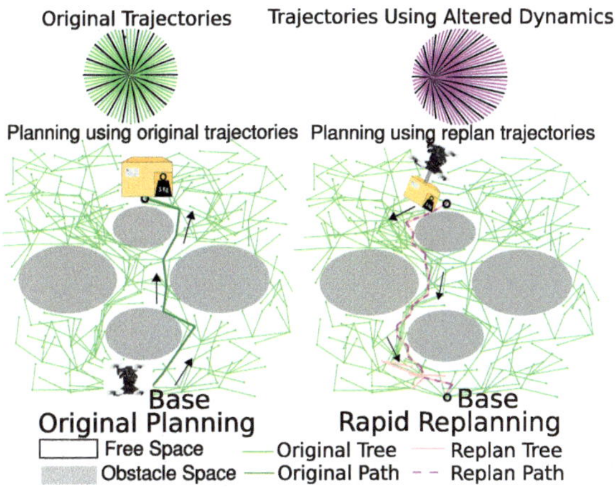

Fig. 1. Left image shows a quadrotor using its original trajectories to plan and move towards a sealed package of uneven weight. The right image shows the same quadrotor lifting the package and experiencing a dynamics change due to change in mass and shifted center of gravity and using RAPID algorithm to perform quick replanning and move back to the base location. Trajectories at 20° increments are shaded black for clarity.

However, in high stakes missions, the vehicle may not have enough time or resources to perform replanning from scratch. Instead, our method is able to quickly repair the old plan to respect the vehicle's new dynamics by leveraging the topological information gained about the environment during the initial plan. RAPID uses the original tree's cost information as a heuristic for quick replanning. Figure 1 depicts a case where a quadrotor undergoes a dynamics change due to lifting a package that causes uneven weight distribution; using RAPID, the vehicle quickly replans a valid return path to the base location.

A benefit of using RAPID is that it can be used for vehicles for which the solution to the two-point Boundary Value Problem (BVP) [5] is computationally expensive or not available. This is important because even though the two-point BVP may be solvable for the vehicle's original dynamics, there is no guarantee that the vehicle's dynamics change in a way such that two-point BVP is still solvable. We compare RAPID with brute force replan using RRT [3] and SST [4] as these are two standard algorithms that are used when two-point BVP cannot be computed.

In our work, we focus on the motion replanning aspects and assume that vehicle automatically detects changes to its dynamics. We model the vehicle's movement using trajectory libraries [6,7] and assume that these libraries exists for vehicle's common configurations and malfunctions.

2 Related Work

A survey of the sampling-based motion planning literature reveals limited work on replanning under changed dynamical constraints but numerous existing work on replanning under changed dynamic environments. For example, Bekris et al. [8] present a tree-based planner that deals with evolving configuration spaces by incrementally updating a tree-based data structure retained from previous steps and then biasing the search with a greedy and probabilistic complete exploration approach. Ferguson et al. [9] make repairs to an already existing RRT tree, removing invalid parts and then growing the changed tree until a new solution is found to account for the changed configuration space. Yoshida et al. [10] present a reactive method that uses a combination of path replanning and path deformation to account for changed environments. Otte et al. [11] present RRT-X, the first asymptotically optimal sampling based algorithm in dynamic environments that uses ideas from the D* algorithm for quick replanning. While such algorithms work for replanning in evolving configuration spaces, they are not directly applicable to replanning in response to changing vehicle dynamics. Nonetheless, our method is inspired by the same high-level idea of using an existing but outdated solution to speed up the replanning process.

Several existing works show the power of using heuristics to improve the speed and efficiency of searches. Urmson et al. [12] modify RRT to create the heuristically-guided RRT using a quality measure to balance exploration vs exploitation search leading to low cost solutions. Otte et al. [13] and Gammell et al. [14] use the ellipsoidal heuristic to perform rejection sampling and direct sampling respectively to speed up the process of generating improved solutions. Straub et al. [15] use the cost-to-go heuristic obtained from an asymmmetric bidirectional search with the reverse search tree produced using Lifelong Planning A* (LFA*) to speed up the planning process. Littlefield et al. [16] modify SST to produce the Informed-SST algorithm which uses domain-specific user defined heuristics to improve the efficiency of SST. In keeping with this trend, our replanning algorithm uses the cost-to-start heuristic obtained from the original search tree to improve the efficiency of the replan search.

Our work uses trajectory libraries to model the motion of vehicles incorporating dynamical constraints. The idea of using trajectory libraries to represent vehicle motion is not something new. Go et al. [6] presents a method using trajectory libraries and real time motion planning to synthesize animations of autonomous space, water, and land-based vehicles in interactive simulations. They also provide a method for efficient storage of trajectories in memory using only about 40 bytes per path. Frazzoli et al. [7] use trim trajectories and maneuvers to perform autonomous helicopter flight. The advantage of using trajectory libraries as noted in [17] is that no online time is wasted in calculating trajectories that a robot cannot carry out.

3 Problem Definition

In this section, we formally define the problem. Let $\mathcal{X}$ be the D-dimensional state space of a robot. Let $\mathcal{X}_{obs}$ be the obstacle space of $\mathcal{X}$. Then the free space $\mathcal{X}_{free}$ is defined as $\mathcal{X}_{obs} \setminus \mathcal{X}$. Let the home base state of the robot be denoted by $x_{base} \in \mathcal{X}_{base}$ where $\mathcal{X}_{base} \subset \mathcal{X}_{free}$. Let $\mathcal{X}_{end}$ be the end region with $\mathcal{X}_{end} \subset \mathcal{X}_{free}$. Let β be a feasible path with $\beta : [0,1] \rightarrow \mathcal{X}_{free}$ such that $\beta(0) = x_{base}$ and $\beta(1) = x_{end}$ where $x_{end} \subset \mathcal{X}_{end}$. Let $\alpha \in (0,1)$ and $\beta(\alpha) = x_{replan}$ be the state where the dynamics of the robot changes.

Problem: Find feasible path back to base incorporating dynamical changes.

Given the triple $(\mathcal{X}_{free}, x_{replan}, \mathcal{X}_{base})$, *find a feasible path* $\gamma : [0,1] \rightarrow \mathcal{X}_{free}$ *such that* $\gamma(0) = x_{replan}$ *and* $\gamma(1) \subset \mathcal{X}_{base}$ *if one exists.*

4 Technical Approach

The main idea behind RAPID is the use of topological information gained from original search tree for quick replanning. Although the edges in the original tree correspond to old dynamical maneuvers, the nodes are still valid obstacle free configurations which can be used to bias the replan towards obstacle free regions of the search space. At every iteration, RAPID greedily selects the most promising node from the original search tree to replan towards. This guides RAPID's replan away from obstacles and toward the home base.

Algorithm 1: RAPID($x_{start}, x_{end}, \mathcal{G}_{orig}, \mathcal{T}_{replan}, \delta_x, \delta_{end}$)

```
1  G_replan ← {V ← {x_start}, E ← {}}
2  Q ← {}, x_cur ← x_start, E_proc ← {}
3  for i ← 1 to M_iter do
4      if x_cur ≠ NULL then
5          V_near ← getNearestNodes(x_cur, G_orig, δ_x)
6          updatePriorityQ(Q, x_cur, V_near)
7      x_dst ← pop(Q)
8      if x_dst ≠ NULL then
9          x_cur ← NULL
10         x_src ← findClosestNode(x_dst, G_replan, E_proc)
11         if x_src ≠ NULL then
12             E_new ← Propagate(x_src, x_dst, T_replan, V)
13             if E_cur ≠ NULL then
14                 x_new ← E_new.getFinalNode()
15                 V ← V ∪ {x_new}
16                 E ← E ∪ {E_new}
17                 if endReached(x_end, x_new, δ_end) then
18                     return OutputPath(x_start, x_new)
19                 x_cur ← x_new
20             E_proc ← E_proc ∪ (x_src, x_dst)
21     else
22         break
23 return PerformBaseAlgorithm(x_start, x_cur, T_replan, δ_end)
```

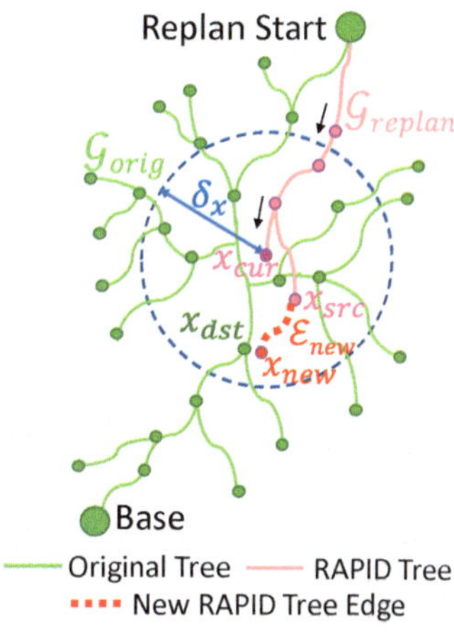

Fig. 2. New edge addition to RAPID tree

The input parameters to RAPID (Algorithm 1) are start node x_{start} (robot's current location), end node x_{end} (home base), original search tree $\mathcal{G}_{orig}$ (generated using RRT or SST), replan trajectory library $\mathcal{T}_{replan}$, neighbor radius δ_x and end radius δ_{end}. The replan search tree $\mathcal{G}_{replan}$ is initialized using a node list $\mathbf{V}$ (initialized to x_{start}) and an empty edge list $\mathbf{E}$ (line 1). We use a priority

Algorithm 2: getNearestNodes($x_{cur}, \mathcal{G}_{orig}, \delta_x$)

1 $\mathbf{V}_{near} \leftarrow \mathcal{G}_{orig}$.Nodes()
2 $\mathbf{V}_{near} \leftarrow \{x \in \mathbf{V}_{near} \mid \|x - x_{cur}\| \leq \delta_x\}$
3 **return** $\mathbf{V}_{near}$

Algorithm 3: updatePriorityQ($\mathcal{Q}, x_{cur}, \mathbf{V}_{near}$)

1 **for** x in $\mathbf{V}_{near}$ **do**
2 $\quad \mathcal{Q}$..update $(x, g(x_{cur}) + d(x_{cur}, x) + \epsilon h(x))$

Algorithm 4: findClosestNode($x_{dst}, \mathcal{G}_{replan}, \mathbf{E}_{proc}$)

1 $\mathcal{V}_{replan} \leftarrow \mathcal{G}_{replan}$.Nodes()
2 $x_{closest} \leftarrow \min_{\|x - x_{dst}\|} \{x \in \mathcal{V}_{replan}\}$
3 **if** $(x_{closest}, x_{dst}) \in \mathbf{E}_{proc}$ **then**
4 $\quad x_{closest} \leftarrow$ NULL
5 **return** $x_{closest}$

Algorithm 5: Propogate($x_{src}, x_{dst}, \mathcal{T}_{replan}, \mathbf{V}$)

1 $\mathcal{T}_{replan\_tf} \leftarrow$ Transform($x_{src}, \mathcal{T}_{replan}$)
2 $\mathcal{T}_{sort} \leftarrow \underset{\|\mathcal{E}.\text{finalNode}() - x_{dst}\|}{sort} \{\mathcal{E} \in \mathcal{T}_{replan\_tf}\}$
3 $\mathcal{T}_{sort\_best} \leftarrow \underset{N}{select\_best}\{\mathcal{E} \in \mathcal{T}_{sort}\}$
4 **for** $\mathcal{E}$ **in** $\mathcal{T}_{sort\_best}$ **do**
5 $\quad x_{fin} \leftarrow \mathcal{E}_{new}$.finalNode()
6 $\quad$ **if not** $(x_{fin} \in \mathbf{V}$ *or* Collision($\mathcal{E}$)) **then**
7 $\quad\quad$ **return** $\mathcal{E}$
8 **return** NULL

queue $\mathcal{Q}$ to store a list of potential original search trees to guide replan expansion. The current node x_{cur}, which is used for selecting nodes of $\mathcal{G}_{orig}$ to add to $\mathcal{Q}$ is initialized to x_{start} (line 2). We maintain a list $\mathbf{E}_{proc}$ to keep track of processed edges to prevent cycles in $\mathcal{G}_{replan}$ (line 1).

The execution of one iteration of RAPID occurs as follows: While $x_{cur} \neq$ NULL, (line 4), the algorithm uses getNearestNodes (Algorithm 2) to find all nodes within a radius δ_x from $\mathcal{G}_{orig}$ centered at x_{cur} and assigns it to set $\mathbf{V}_{near}$ (line 5). The algorithm uses updatePriorityQ (Algorithm 3) to update $\mathcal{Q}$ with the nearest nodes $\mathbf{V}_{near}$ (line 6). The algorithm then uses x_{dst} popped from $\mathcal{Q}$ (line 7) to find the closest node x_{src} in $\mathcal{G}_{replan}$ (line 10) using findClosestNode (Algorithm 4). We perform forward search using Propogate (Algorithm 5) to find the best edge $\mathcal{E}_{new}$ from x_{src} to x_{dst} (line 12) (Fig. 2). If $\mathcal{E}_{new} \neq$ NULL, then the new edge $\mathcal{E}_{new}$ and new node x_{new} (final node of $\mathcal{E}_{new}$) are updated in $\mathbf{V}$ (line 15) and $\mathbf{E}$ (line 16) respectively. If the end is reached (checked using endReached), the output path $\mathcal{P}_{out}$ is generated using OutputPath (line 18). If RAPID is unable to quick replan, then the brute-force replan using the algorithm from the original plan (line 23) (or any other sampling-based planning algorithm) is run to get the output path as a fallback option.

Note that the quick replanning part of RAPID is deterministic and the fallback option (assuming a sampling-based algorithm is used) is probabilistic. Hence the condition for RAPID algorithm to be probabilistic complete is that the base algorithm should be probablisitic complete.

The function getNearestNodes (Algorithm 2) performs the task of getting the nearest nodes to x_{cur} in $\mathcal{G}_{orig}$ within a ball of radius δ_x (line 2). The metric defined on the configuration space is used to calculate the distance $\|x - x_{cur}\|$ where $x \in \mathbf{V}_{near}$ (set of all nodes of $\mathcal{G}_{orig}$). All nodes $\mathbf{V}_{near}$ that lie within distance δ_x are returned.

The function updatePriorityQ (Algorithm 3) updates the priority queue $\mathcal{Q}$ with the nodes $x \in \mathbf{V}_{near}$ using the key [18] $c_{total} = g(x_{cur}) + d(x_{cur}, x) + \zeta h(x)$ where $g(x_{cur})$ is the cost-from-start, $d(x_{cur}, x)$ is the cost from x_{cur} to $x \in \mathbf{V}_{near}$, $h(x)$ is the cost-to-end heuristic, and ζ is the weight factor [19]. We choose $g(x_{cur})$

and $h(x)$ as the corresponding path costs, $d(x_{cur}, x)$ as the euclidean distance between x_{cur} and x and set ζ to 1.

The function `findClosestNode` (Algorithm 4) finds the closest node $x_{closest}$ to x_{dst} in $\mathcal{G}_{replan}$ (line 2). The node pair $(x_{closest}, x_{dst})$ is checked to make sure that is not already in the edge processed list to ensure the same pair of edge nodes is not processed again (line 3). If this condition is satisfied, then $x_{closest}$ is set to `NULL` (line 4).

The function `Propogate` (Algorithm 5) returns the best path that connects x_{src} to x_{dst}. The best path is generated by first transforming (translation and rotation) the trajectory library $\mathcal{T}_{replan}$ (line 1) using x_{src} and then sorting the trajectories $\mathcal{E}$ in $\mathcal{T}_{replan\_tf}$ using the distance metric $\|\mathcal{E}.\texttt{finalNode}() - x_{dst}\|$ (line 2). The N best sorted trajectories are selected from $\mathcal{T}_{sort}$ and assigned to $\mathcal{T}_{sort\_best}$ (line 3). For each of the sorted trajectories $\mathcal{E}$, we check if x_{fin} (last node of $\mathcal{E}$) is already in the node list $\mathbb{V}$ and if the edge $\mathcal{E}$ is collision with the obstacles (line 6). If both of these conditions are not satisfied, then $\mathcal{E}$ is returned.

4.1 Selection of Neighbor Radius δ_x

The RAPID algorithm has a tuning parameter δ_x that affects its performance. This parameter dictates the number of original tree nodes that gets pushed to $\mathcal{Q}$. A low value chosen increases the prospect of many important $\mathcal{G}_{orig}$ nodes being not considered. A high value selected increases the chances of adding $\mathcal{G}_{orig}$ nodes that are blocked by obstacles which might lead to considering more trajectories for expansion that collide with obstacles. In our experiments, we choose a value that is equal to or slightly greater than twice the length of the largest trajectory.

5 Experiments

We run multiple experiments in hardware and software to test the performance of RAPID. We compare RAPID with brute-force replan using RRT and SST, each having two different variants. The first variant is using the extend-operation [20] for node expansion while the second variant is using random propagation [4]. The extend operation (in our implementation) extends the nearest node in the existing tree by a distance ϵ to a node x_{extend} in the direction of the new sampled configuration. We then select the trajectory in corresponding $\mathcal{T}$ ($\mathcal{T}_{replan}$ for replan and $\mathcal{T}_{orig}$ for original) that minimizes $\|x_{extend} - x_i\|$ where x_i is final node of $\mathcal{E}_i \in \mathcal{T}$. The different variants are used to test the durability of RAPID in using different original tree structures for its planning. All variants have a goal bias [3] of 5% to speed up solution generation.

In our experiments, we choose the algorithm used to run the original plan as the base algorithm if RAPID is unable to quickly replan. We choose solution path length L_p, algorithm completion time T_c and flight time T_f (only hardware) as the performance metrics. We generate the trajectories in the trajectory libraries such that their start and end velocities are zero. This is done to limit the number of trajectories in the trajectory libraries and is not required to use RAPID.

Table 1. Experiment parameters

Parameter	Value (m) (Hardware)	Value (m) (Software)
Map size	5 × 4	100 × 100
Base/Replan start area radius (δ_{end})	0.3	5
Neighbor radius (δ_x)	2.2	10
Max iterations (M_{iter})	2500	2500
Propagate Trajectory Count (N)	5	5

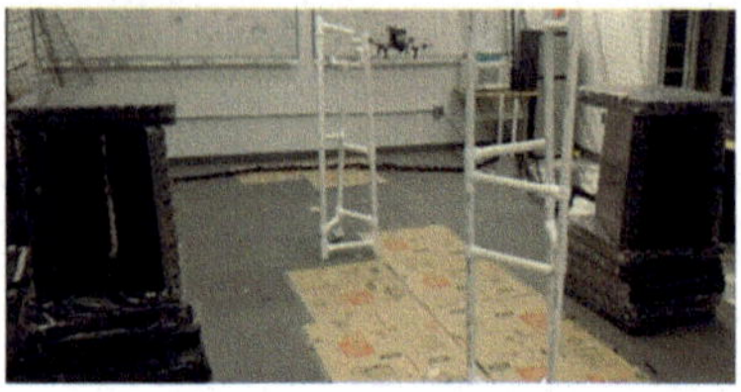

Fig. 3. Quadrotor in obstacle course

The hardware experiments are performed using a modified Bebop 2 (PRG Husky [21]) quadrotor on an obstacle course (Fig. 3) using parameters shown in Table 1. The quadrotor is fitted with Intel UP board (Fig. 6) having Intel Atom x5-z8350 4-core, 1.44 GHz CPU with 4 GB RAM to run the algorithms and navigation stack. The obstacle course has 4 obstacles (Fig. 4) and they are modeled using bounding circles for easy collision detection. The bounding circle radii are increased by 20 cm to account for non-zero vehicle size, vehicle drift and errors. The base and replan start locations are set at (0.2, 2) and (4.2, 2) respectively. The quadrotor control is limited to performing maneuvers in the horizontal plane at a fixed height. We use a Proportional-Derivative (PD) controller for position tracking and a Vicon motion capture system for state information. The dynamics change for the quadrotor is represented by changing the gains of PD controller from nominal to one-third of the nominal gains. The trajectories (Fig. 4), both original and altered, are learned by providing a reference 1 m away from origin in 10° increments over 4 s and then taking the mean of the trajectories over 10 iterations. We use these trajectories to run 10 trials and record the performance metrics. We set ϵ to 0.8 and 0.65 when using original and replanning libraries respectively.

The software experiments are performed in 2D space using parameters shown in Table 1. The obstacles are modeled as circles with uniformly sampled radii in [5, 8]. The base, replan start and obstacle center locations for each scenario are uniform randomly sampled from the parameter ranges ([1, 100], [1, 100]) such that they do not overlap each other and the obstacle surfaces are within the map area. The minimum distance between base and replan start locations is set to 70. We run 100 scenarios for each obstacle number $\{0, 5, 15, 25, 35\}$ with each scenario having 3 trials. We run experiments for a unicycle model [22] with the dynamics change represented by changed reference peak linear velocity (8 m/s to 4 m/s) and a quadrotor model [1] with a dynamics change represented by changed mass (1 kg to 1.1 kg) and shift in the center of gravity (4 cm and 1 cm parallel to the x- and y-axes). We set ϵ to 4 (for both original and replan) for quadrotor and 4 (original) and 2.5 (replan) for unicycle. The experiments are performed on a system with Intel i7-7700 4-core CPU with 32 GB RAM. The stopping criterion for both algorithms in all experiments is when a path is found between the replan start and base locations.

6 Results

The replan paths generated during comparison of RAPID vs. RRT for two trials of hardware experiments with actual path (Vicon) overlaid are shown in Fig. 5. The mean and standard deviation for 10 trials of hardware experiments are shown in Table 2. The replan paths produced during comparison of RAPID vs. RRT-Extend and RAPID vs. SST-Random for one of our software test scenarios are shown in Fig. 7. We have not shown examples of replan paths produced during comparison of RAPID vs. RRT-Random and RAPID vs. RRT-Extend due to brevity and the limit on the number of pages. The results of the software experiments are shown in Fig. 8.

Our hardware (Table 2) and software experiments (Fig. 8) demonstrate that RAPID performs better than brute-force replan using RRT and SST with respect to all performance metrics that we evaluate. The completion time for the RAPID

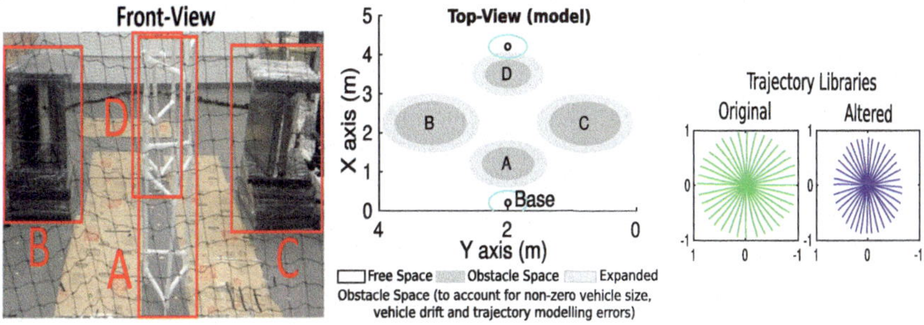

Fig. 4. For hardware experiments, left figure shows the front view of the obstacle course and middle figure shows top view (model). The original and altered (replan) trajectory libraries generated are shown to the right. The replan trajectory library is colored purple to indicate that same library is used for both RRT (blue) and RAPID (magenta).

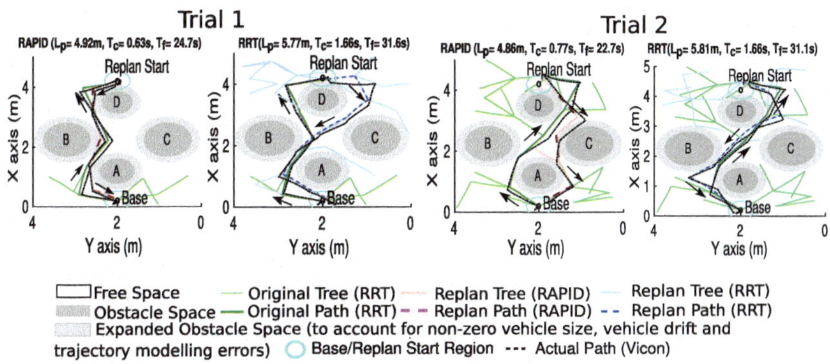

Fig. 5. Figure shows comparison of RAPID and RRT for two trials of hardware experiments. RAPID is only compared with RRT(using extend) as this is the second best performing algorithm found in the software experiments.

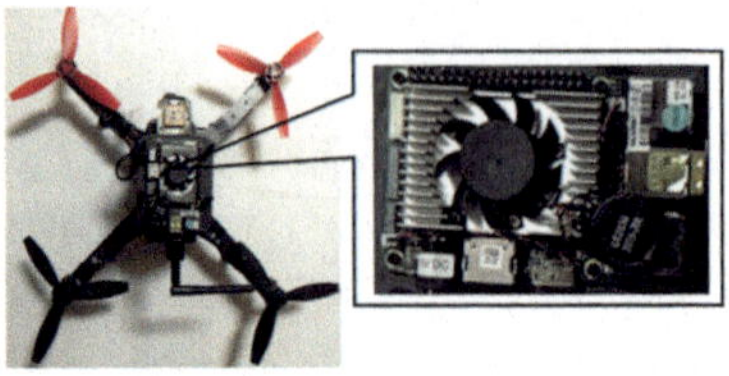

Fig. 6. Quadrotor (with Intel UP board enlarged) used in our experiments

Table 2. Mean and standard deviation of performance metrics for 10 trials of hardware experiments

Algorithm	RAPID			RRT		
Metric	T_c (s)	L_p (m)	T_f (s)	T_c (s)	L_p (m)	T_f (s)
Mean	**1.5**	**5.02**	**24.43**	2.55	6.07	33.06
Std.	2.64	0.4	3.35	1.92	1.022	3.69

algorithm is much faster than replanning from scratch. We believe this speedup can be attributed to RAPID's use of the original plan's topological information within its guiding heuristic. This is notable given that the heuristic is based on the system's original dynamics. The guiding heuristic is also responsible for causing path length to be shorter for RAPID. This is because using the heuristic guides the planning away from obstacles and towards to the goal, in comparison to planning from scratch where the path is much more random. The completion time and path length shows upward trends (Fig. 8) with increase in the number of obstacles. We believe the increase in completion time is because of increased collision checking that is necessary with more obstacles. Path length increases because paths have to wiggle and curve more to get from the start to end because of more obstacles.

While executing a maneuver in the hardware experiments, we use the same control input that is used to generate the maneuver in the trajectory library rather than perform end to end waypoint tracking. It is observed that due to the inherent error present in modelling maneuvers in the trajectory libraries, the actual maneuver executed by the quadrotor when providing the corresponding control input has a slight deviation from the modelled maneuver. This modelling error causes the quadrotor to deviate from its flight path and this deviation increases as the quadrotor progresses towards the terminal position. To decrease this deviation and to keep the quadrotor on the flight path, we select the best maneuver from the library to reach the next waypoint and apply the corresponding control input to the quadrotor.

6.1 Limitations of RAPID

We observe that RAPID is able to quick replan almost all of the time. However, in one trial of hardware experiment (out of 10 trials) and in some software experiments with percentages - RAPID-RRT Extend (2.8%), RAPID-SST Random (0.96%), RAPID-RRT Random (1.13%) and RAPID-SST Extend (3.13%), RAPID was unable to find a quick replan (and thus fell back to using a brute-force replan as specified on line 23 of Algorithm 1). Thus, in our experiments, overall RAPID is successful in quick replanning for a variety of obstacle configurations at least 96.8% (software experiments) of the time. RAPID is unable to

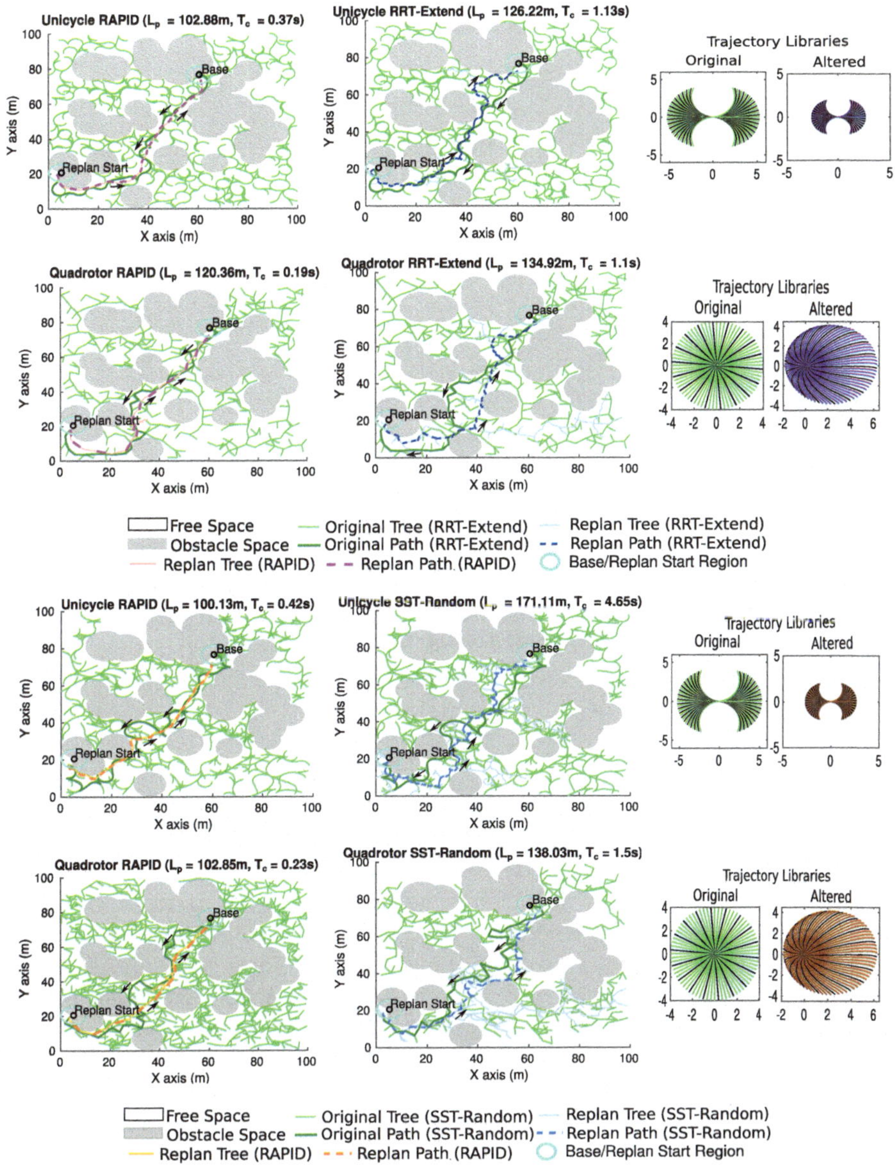

Fig. 7. Figures show replanned paths for Unicycle (top row) and Quadrotor (bottom row) using RRT-Extend (top) and SST-Random (bottom) for original plan. The original and replan maneuver libraries used in the experiments are shown in the last column. The replan trajectory libraries are colored purple and brown to show that the same trajectory library is used for multiple corresponding planning algorithms. Trajectories at 20° increments are highlighted black for clarity.

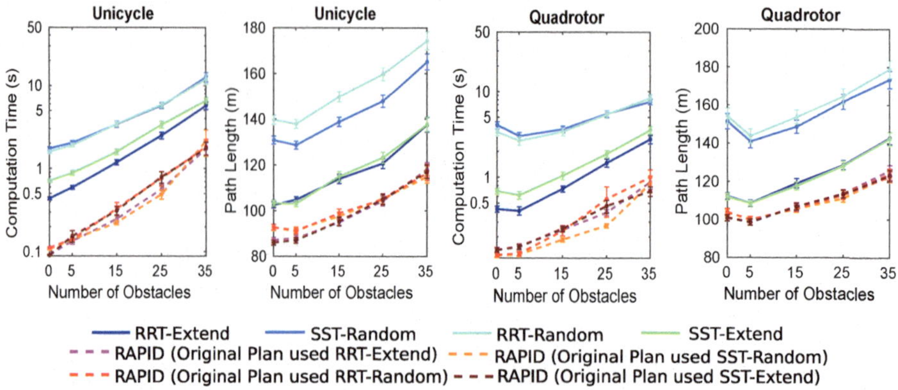

Fig. 8. Figures show completion time (semi-log scale) vs. number of obstacles (left column), path length vs. number of obstacles (right column) for Unicycle (Left) and Quadrotor (Right) respectively. Each point is a mean of values from 100 scenarios with each scenario having 3 trials.

perform a quick replan in situations when the original search tree's cost guides it through a narrow passage that the changed dynamics are incapable of maneuvering through. We also observe that RAPID performs worse when the original plan is using the extend operator (see Sect. 5) since using this operation on average causes less nodes to be added to the graph (than random propagation) and hence less nodes for utilization for RAPID.

7 Conclusion

In this paper, we present the RAPID algorithm for quick replanning under changed dynamical constraints. This algorithm uses the original tree's stored cost information to quickly replan and generate a feasible path to the start location. The advantage of RAPID is that it can be used for various types of vehicles as it does not require solving the 2-point boundary value problem. We run multiple software and hardware experiments to show RAPID performs better than existing state of the art algorithms (RRT, SST) in computation time, path length and flight time (only hardware) experiments.

Our future work involves testing RAPID in complex environments (e.g. maze) using different changed dynamics for various vehicles. We plan to study using RAPID for replanning to base or goal location whichever is closer to the location of dynamics change. We also plan to analyze the effect of changing heuristic weight ζ on the performance metrics.

Acknowledgements. This work is supported by the Minta Martin Aeronautical Research Fund, University of Maryland (UMD). Sharan Nayak is additionally supported by the Clark Doctoral Fellowship, UMD. We like to thank Chahat Deep Singh and Nitin Sanket from the Perception and Robotics Group (PRG) at UMD for allowing us to use the quadrotor system for our experiments.

References

1. Palunko, I., Fierro, R.: Adaptive control of a quadrotor with dynamic changes in the center of gravity. IFAC Proc. Vol. **44**(1), 2626–2631 (2011)
2. Mueller, M.W., D'Andrea, R.: Stability and control of a quadrocopter despite the complete loss of one, two, or three propellers. In: 2014 IEEE International Conference on Robotics and Automation (ICRA). IEEE (2014)
3. LaValle, S.M., Kuffner Jr., J.J.: Randomized kinodynamic planning. Int. J. Robot. Res. **20**(5), 378–400 (2001)
4. Li, Y., Littlefield, Z., Bekris, K.E.: Asymptotically optimal sampling-based kinodynamic planning. Int. J. Robot. Res. **35**(5), 528–564 (2016)
5. LaValle, S.M.: Planning Algorithms. Cambridge University Press, Cambridge (2006)
6. Go, J., Vu, T.D., Kuffner, J.J.: Autonomous behaviors for interactive vehicle animations. Graph. Models **68**(2), 90–112 (2006)
7. Frazzoli, E., Dahleh, M.A., Feron, E.: Real-time motion planning for agile autonomous vehicles. J. Guidance Control Dyn. **25**(1), 116–129 (2002)
8. Bekris, K.E., Kavraki, L.E.: Greedy but safe replanning under kinodynamic constraints. In: Proceedings 2007 IEEE International Conference on Robotics and Automation. IEEE (2007)
9. Ferguson, D., Kalra, N., Stentz, A.: Replanning with RRTs. In: 2006 Proceedings 2006 IEEE International Conference on Robotics and Automation, ICRA 2006. IEEE (2006)
10. Yoshida, E., Kanehiro, F.: Reactive robot motion using path replanning and deformation. In: 2011 IEEE International Conference on Robotics and Automation. IEEE (2011)
11. Otte, M., Frazzoli, E.: RRTX: asymptotically optimal single-query sampling-based motion planning with quick replanning. Int. J. Robot. Res. **35**(7), 797–822 (2016)
12. Urmson, C., Simmons, R.: Approaches for heuristically biasing RRT growth. In: Proceedings 2003 IEEE/RSJ International Conference on Intelligent Robots and Systems, IROS 2003 (Cat. No. 03CH37453), vol. 2. IEEE (2003)
13. Otte, M., Correll, N.: C-FOREST: parallel shortest path planning with superlinear speedup. IEEE Trans. Roboti. **29**(3), 798–806 (2013)
14. Gammell, J.D., Srinivasa, S.S., Barfoot, T.D.: Informed RRT*: optimal sampling-based path planning focused via direct sampling of an admissible ellipsoidal heuristic. In: 2014 IEEE/RSJ International Conference on Intelligent Robots and Systems. IEEE (2014)
15. Strub, M.P., Gammell, J.D.: Adaptively informed trees (AIT*): fast asymptotically optimal path planning through adaptive heuristics. In: 2020 IEEE International Conference on Robotics and Automation (ICRA), Paris, France, pp. 3191–3198 (2020). https://doi.org/10.1109/ICRA40945.2020.9197338.
16. Littlefield, Z., Bekris, K.E.: Informed asymptotically near-optimal planning for field robots with dynamics. In: Field and Service Robotics. Springer, Cham (2018)
17. Branicky, M.S., Knepper, R.A., Kuffner, J.J.: Path and trajectory diversity: theory and algorithms. In: 2008 IEEE International Conference on Robotics and Automation. IEEE (2008)
18. Hart, P.E., Nilsson, N.J., Raphael, B.: A formal basis for the heuristic determination of minimum cost paths. IEEE Trans. Syst. Sci. Cybern. **4**(2), 100–107 (1968)
19. Pohl, I.: Heuristic search viewed as path finding in a graph. Artif. Intell. **1**(3–4), 193–204 (1970)

20. Kuffner, J.J., LaValle, S.M.: RRT-connect: an efficient approach to single-query path planning. In: Proceedings 2000 ICRA. Millennium Conference. IEEE International Conference on Robotics and Automation. Symposia Proceedings (Cat. No. 00CH37065), vol. 2. IEEE (2000)
21. Sanket, N.J., Singh, C.D., Fermüller, C., Aloimonos, Y.: PRGFlyt: AI based indoor quadrotor autonomy framework for navigation and interaction tasks (2019). https://github.com/prgumd/PRGFlyt/wiki
22. Carona, R., Aguiar, A.P., Gaspar, J.: Control of unicycle type robots tracking, path following and point stabilization (2008)

High-Bandwidth Nonlinear Control for Soft Actuators with Recursive Network Models

Sarah Aguasvivas Manzano[1(✉)], Patricia Xu[2], Khoi Ly[3], Robert Shepherd[2], and Nikolaus Correll[1]

[1] Department of Computer Science, University of Colorado Boulder, Boulder, CO 80309, USA
nikolaus.correll@colorado.edu

[2] Sibley School of Mechanical and Aerospace Engineering, Cornell University, Ithaca, NY 14850, USA

[3] Paul M. Rady Department of Mechanical Engineering, University of Colorado Boulder, Boulder, CO 80309, USA

Abstract. We present a high-bandwidth, lightweight, and nonlinear output tracking technique for soft actuators that combines parsimonious recursive layers for forward output predictions and online optimization using Newton-Raphson. This technique allows for reduced model sizes and increased control loop frequencies when compared with conventional RNN models. Experimental results of this controller prototype on a single soft actuator with soft positional sensors indicate effective tracking of referenced spatial trajectories and rejection of mechanical and electromagnetic disturbances. These are evidenced by root mean squared path tracking errors (RMSE) of 1.8 mm using a fully connected (FC) substructure, 1.62 mm using a gated recurrent unit (GRU) and 2.11 mm using a long short term memory (LSTM) unit, all averaged over three tasks. Among these models, the highest flash memory requirement is 2.22 kB enabling co-location of controller and actuator.

1 Introduction

Model-free, learning based approaches to control soft devices have proven to outperform model-based approaches for real-life systems [5]. In most robotic designs, the placement and design of embedded sensors not only determines the sensing resolution and reliability, but also how the sensors are calibrated and modeled [1]. In the case of soft sensors, complications emerge due to the shifts in position caused by time dependent mechanical changes such as hysteresis,

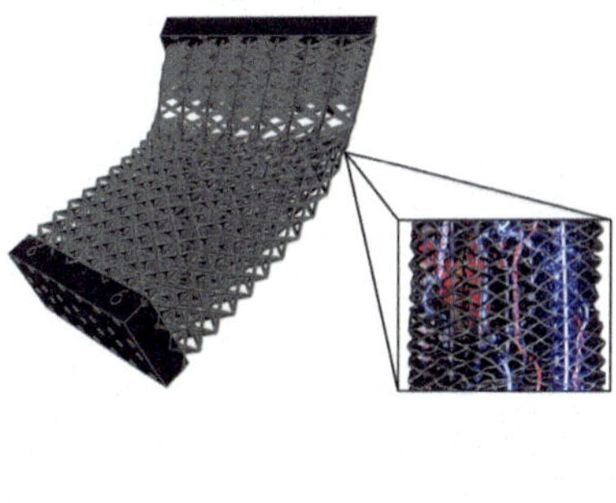

Fig. 1. Rendering of the mesh with embedded light lace sensors.

B. Siciliano et al. (Eds.): ISER 2020, SPAR 19, pp. 589–599, 2021.
https://doi.org/10.1007/978-3-030-71151-1_52

creep, and fatigue. The system used in this work is a soft artificial muscle embedded with a soft optical sensor network that we call "Light Lace" [6] described in Fig. 1.

Related Work. State-of-the-art approaches found in [1] address the challenges in machine learning (ML) for modeling and control of soft actuators. In [2], a very large (3.4 million nodes) neural network model that learns the forward dynamics of the system is used for model predictive control. This model is converted to a linear state space model where the system and input matrices are extracted from a neural network model discretized at 0.05 s, thus controlled at 20 Hz. In [3], a fully connected neural network of three hidden layers with 200 nodes each represents a discretized model at 0.033 s. For the control decision, the solver CVXGEN [11] requires a linearized model coming from automatic differentiation from a high-level deep learning package. The control loop frequency (30 Hz) is limited by the sampling rate of the sensor used in this application. Lastly, in [4] authors achieved path tracking errors in the order of centimeters (average L_2 error of 1.26 cm) and had noisy output tracking when the robot was performing tasks. Finally, [8] has shown that Long Short-Term Memory units (LSTM), are suitable for modeling the kinematic responses of soft actuators with embedded sensors in real time while being robust against sensor drift.

The aforementioned results may be limited by either the sampling rate of the system, a very large neural network model to represent the forward kinematics, or a limiting experimental mechanism that does not allow for smooth path tracking on the actuator. Complex models with large memory requirements are often difficult to embed into resource-limited microcontrollers [10] without any model reduction technique or fixed point approximation. Our approach seeks to simplify the model representation needed to control a soft actuator. This would allow to create materials that deeply embed sensing, computation and actuation, thereby leading to "materials that make robots smart" [9]. Yet, previous literature does not provide insights on techniques for automatic differentiation in platforms that cannot use high-level neural network packages enabled with automatic differentiation.

Contribution of This Paper. We present a nonlinear, predictive controller capable of functioning at high bandwidth through the use of parsimonious recursive networks and online optimization using a Newton-Raphson solver [7]. We develop this controller prototype and software architecture that are tractable in an off-the-shelf microcontroller, thus we aim for low space complexity and low memory footprint. Our approach compares favorably with similar state-of-the-art approaches [2–4] by decreased latency through reproducible numerical gradient approximations, decreased output tracking errors, increased real-time smoothness, proven repeatability and precision experimentally despite sensor input being highly nonlinear. An implementation of our framework is available open-source.[1] Finally, we demonstrate that large neural network models are not necessary for control and identification of soft robotic actuators with a low number of states to achieve an accuracy competitive with state-of-the-art approaches.

[1] https://github.com/sarahaguasvivas/nlsoft.

2 Experimental Setup

Problem Statement. We consider a segment of a fully flexible, soft mesh mounted in a solid structure, where the end effector's pose is defined as $\mathbf{y} = \{x_0, ...x_{n-1}\}$ and is actuated by m different tendon actuators $\mathbf{u} = \{u_0, ..., u_{m-1}\}$. Embedded in this mesh are w channels of light lace sensors [6] denoted as $\mathbf{l} = \{l_0, l_1, ..., l_{w-1}\}$.

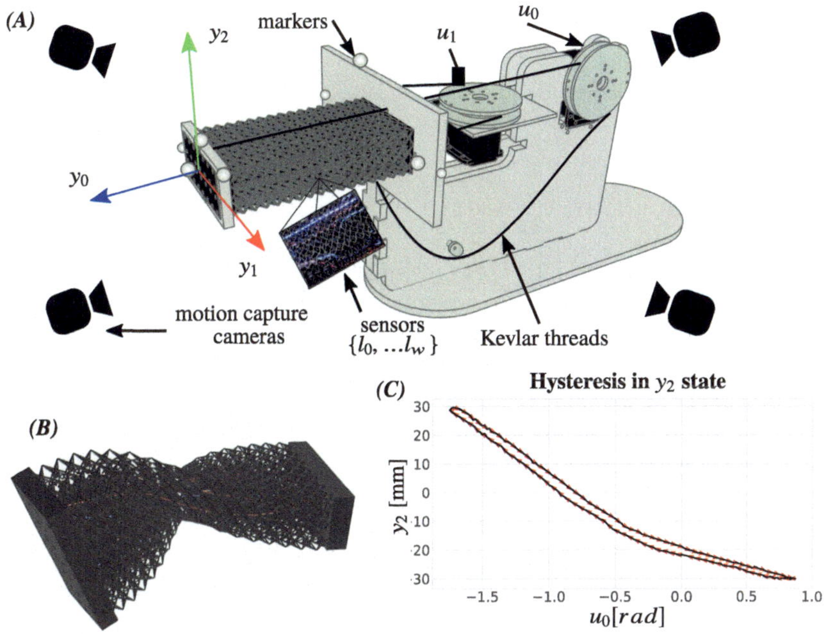

Fig. 2. Problem statement. *(A)* Experimental setup and variable definitions. *(B)* Artistic rendering to show the flexibility of the mesh and sensor materials in *(A)*. *(C)* Nonlinearity and hysteresis with respect to control input.

Materials and Methods. Figure 2 *(A)* describes the experimental test bed in this work. The undeformed mesh dimensions are $14 \times 7 \times 3\,\text{cm}^3$ in the y_0, y_1, and y_2 dimensions respectively. Two high torque servos (Dynamixel RX-64, Robotis) pull tendons that move the lattice's end effector in 3-dimensional coordinates. This end effector can move up to $\Delta\mathbf{y}_{range} = \{40.4, 28.3, 35.0\}$ mm in each direction. The tendons are connected to the end effector using high-strength Kevlar threads (Kevlar Fiber, Dupont). The inputs indicate the angular displacements of the servos with respect to the manufacturer reference datum. The optical lace sensor network consists of 11 normalized channels that are organically weaved within the soft mesh structure. We use a motion capture system (Optitrack, Natural Point, Inc.) to track the relative spatial position of the end effector with respect to its mounting base. In the real-time computation of the optimal control output, the motion capture system signals are not used towards the controls computations, but we record true positions of the end effector for evaluation in our results and computation of the errors in Sect. 4.

Light Lace Sensor Network. Our system is composed of a tendon-driven, soft, 3D printed, polyurethane mesh with stretchable lightguides distributed within the body. The 1 mm diameter, polyurethane fibers (Crystal Tec) intertwine to make up a network of optical lace sensors [6] consisting of input lines that carry light from LEDs and output lines that carry coupled light to photodiodes. The coupled light intensities vary based on the contact between input and output fibers as the system deforms to provide local strain information within the soft mesh to differentiate between twisting, bending and stretching states. This sensor network stretches and warps together with the soft mesh as seen in Fig. 2 *(B)*.

3 Nonlinear Online Controller

Similar to [7,8], we formulate a discretized output transition model that uses the feedback from the sensors and a history of past inputs and output estimates. We describe this model as $\dot{\mathbf{y}}(t) = g(\tau, \alpha, \mathbf{l})$, where τ is a queue composed by $\{\mathbf{u}(t-n_d), ..., \mathbf{u}(t)\}$; α is composed by $\{\mathbf{y}(t-d_d), ..., \mathbf{y}(t-1)\}$ and d_d and n_d, which are how far into the past the model looks at predictions and inputs respectively. We define N as the prediction horizon, that is, the number of times the nonlinear discrete model will be recursively called in order to predict future outputs with N_1 and N_2, $(N_1 < N_2 \leq N)$ being the start and end of the cost horizon defined in Eq. 1. $N_c(\leq N)$ is the control horizon, which is used towards the prediction and when $N_c < N$, we roll the τ queue until N_c and then repeat $\mathbf{u}_{N_c}$ until N.

Step 1: Data Collection. We record time series data from the sensor signals, the motion capture system, and the servo inputs $\mathbf{u}$. We run a three-stage sequence of inputs illustrated in the first row of Fig. 4, where samples of these sequences are shown along with the learned model offline predictions. Samples of this input sequence at each stage is described in the first row of Fig. 4 and is repeated 200 times. We then prepare the data depending on the values of n_d and d_d. This results in a data size of 3.8 million samples, which are then reorganized depending on n_d and d_d. The signals were collected at a discretization of an average step duration of 8.$\bar{3}$ ms or 120 Hz for a total of 8.8 h.

Step 2: Learning the Model. In this work, we compare three different types of recursive neural network (RecNN) models that were each trained using 10-fold time series split. These models are recursive because the output of the model feeds back in part of the input of the next time step in a tree-like structure that reuses the weights as seen in Fig. 3*(C)*. This requires that we define a trainable child structure for the neural network, which we call h. We test three model representations of this neural network sub-structure: A fully connected substructure h = *[Dense(5, relu), Dense(3, tanh)]*, that is similar to an RNN where the recurrent weights are the same as the feedforward ones; an LSTM-based child model trained is given by h = *[LSTM(5), Dense(5, tanh), Dense(3, tanh)]* and the GRU trained is given by h = *[GRU(5), Dense(5, relu), Dense(3, linear)]*. These lead to 243, 435, and 570 total model parameters respectively, and get recursively called N times each prediction step. These model sizes require flash memory as low as 0.97 kB, 1.7 kB and 2.22 kB using $float32$. A sample of the off-line testing set results are described in Fig. 4, and the L_2 errors over the 10 partitions are 2.52 ± 1.52 mm for the recurrent model; 0.26 ± 0.04 mm for the GRU model and 0.26 ± 0.04 mm for the LSTM model.

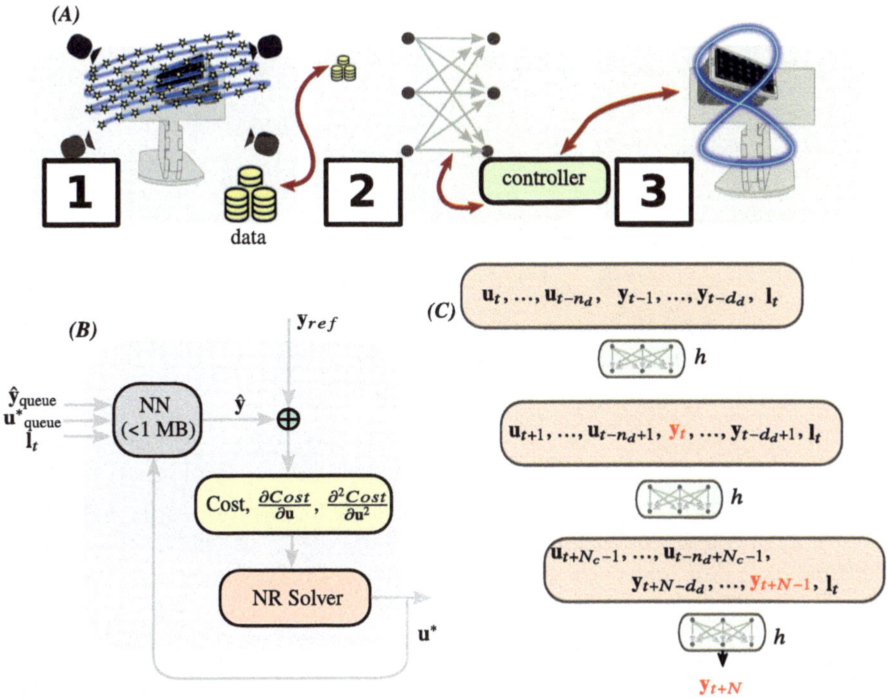

Fig. 3. *(A)* The steps from data collection to offloading the controller. *(B)* Controller description. *(C)* Recursive neural network predictions.

Cost Function. The cost function J used in this work is a combination between the standard, unconstrained, MPC cost function with additional terms that enforce smoothness in the optimal control inputs found in [7]. It is displayed in Eq. 1. $\mathbf{\Lambda} \in \mathbb{S}^{m \times m}$ and $\mathbf{Q} \in \mathbb{S}^{n \times n}$ are weighting matrices for the input changes and output square errors, respectively. The subscript ref denotes the target path to follow.

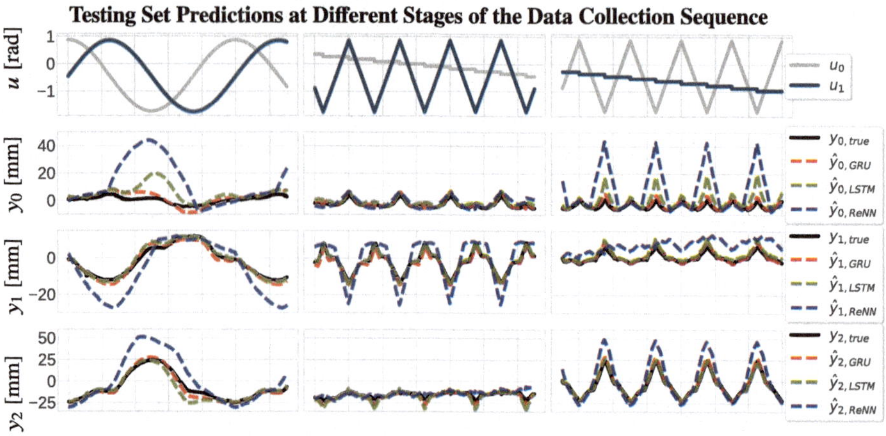

Fig. 4. Sample of the sequence swept by the block for data collection.

$$
\begin{aligned}
J = & \sum_{j=N_1}^{N_2} \|\mathbf{y}_{ref,j} - \hat{\mathbf{y}}_j\|^2_{\mathbf{Q}} + \sum_{j=0}^{N_c} \|\Delta \mathbf{u}_j\|^2_{\mathbf{\Lambda}} \\
& + \sum_{i=1}^{m} \sum_{j=1}^{N_c} \left[\frac{s}{u(n+j,i) + \frac{r}{2} - b} + \frac{s}{\frac{r}{2} + b - u(n+j,i)} - \frac{4}{r} \right]
\end{aligned} \tag{1}
$$

Newton-Raphson Solver. In order to get an optimal control input, we frame the control problem as the solution to Eq. 2, where $\frac{\partial^2 J}{\partial \mathbf{U}^2}(k)$ is the Hessian of the cost function at the current timestep k, and $\frac{\partial J}{\partial \mathbf{U}}(k)$ is the Jacobian of the cost function. This approach was first formulated in [7], where $\mathbf{U}$ as the vector composed by $\mathbf{U} = \{\mathbf{u}_k, \mathbf{u}_{k+1}, ..., \mathbf{u}_{k+N_c}\}^T \in \mathbb{R}^{N_c \times m}$.

$$
\frac{\partial^2 J}{\partial \mathbf{U}^2}(k)(\mathbf{U}(t+1) - \mathbf{U}(k)) = -\frac{\partial J}{\partial \mathbf{U}}(k) \tag{2}
$$

$$
\begin{aligned}
\frac{\partial J}{\partial \mathbf{U}}(k) \approx & -2 \sum_{j=N_1}^{N_2} (\mathbf{y}_{ref,j} - \hat{\mathbf{y}}_j)^T \mathbf{Q} \overbrace{\frac{\partial \hat{\mathbf{y}}_j}{\partial \mathbf{U}}}^{\text{Eqn. 6}} + 2 \sum_{j=0}^{N_c} \Delta \mathbf{u}_j \mathbf{\Lambda} \frac{\partial \Delta \mathbf{u}_j}{\partial \mathbf{U}} \\
& + \sum_{i=1}^{m} \sum_{j=1}^{N_c} \left[\frac{-s}{\left[u(n+j,i) + \frac{r}{2} - b\right]^2} + \frac{s}{\left[\frac{r}{2} + b - u(n+j,i)\right]^2} \right] \in \mathbb{R}^{N_c \times m}
\end{aligned} \tag{3}
$$

$$\frac{\partial^2 J}{\partial \mathbf{U}^2}(k) \approx 2 \sum_{j=N_1}^{N_2} \left[\mathbf{Q}\Big(\frac{\partial \hat{\mathbf{y}}_j}{\partial \mathbf{U}} \circ \frac{\partial \hat{\mathbf{y}}_j}{\partial \mathbf{U}}\Big) - (\mathbf{y}_{ref,j} - \hat{\mathbf{y}}_j)^T \mathbf{Q} \frac{\partial^2 \hat{\mathbf{y}}}{\partial \mathbf{U}^2} \right]$$
$$+ 2\sum_{j=0}^{N_c} \left[\mathbf{\Lambda}\Big(\frac{\partial \Delta \mathbf{u}_j}{\partial \mathbf{U}} \circ \frac{\partial \Delta \mathbf{u}_j}{\partial \mathbf{U}}\Big) + \Delta \mathbf{u}_j \mathbf{\Lambda} \frac{\partial^2 \Delta \mathbf{u}_j}{\partial \mathbf{U}^2} \right]$$
$$+ \sum_{i=1}^{m}\sum_{j=1}^{N_c} \left[\frac{2s}{\left[u(n+j,i) + \frac{r}{2} - b\right]^3} + \frac{2s}{\left[\frac{r}{2} + b - u(n+j,i)\right]^3} \right] \in \mathbb{R}^{N_c \times N_c} \tag{4}$$

Equation 3 and 4 describe the expressions used toward the Jacobian and Hessian of the cost in Eq. 1. Let $p = n_d m + d_d n + w$ be the length of the flattened input of the neural network and ε be the differentiation stencil step length such that $[\mathbf{x}_{inputs} + \varepsilon \mathbf{I}] \in \mathbb{R}^{p \times p}$.

$$\Theta = \frac{\partial \hat{\mathbf{y}}}{\partial \mathbf{x}_{inputs}} = \frac{g(\mathbf{x}_{inputs} + \varepsilon \mathbf{I})) - g(\mathbf{x}_{inputs} - \varepsilon \mathbf{I})}{2\varepsilon} + \mathcal{O}(\varepsilon^2) \in \mathbb{R}^{p \times n} \tag{5}$$

$$\frac{\partial \hat{\mathbf{y}}}{\partial \mathbf{U}} \approx \left[\sum_k^{n_d} \Theta_{mk+0,:}, \dots, \sum_k^{n_d} \Theta_{mk+(m-1),:}\right]^T \in \mathbb{R}^{n \times m} \tag{6}$$

For the second derivative ($\frac{\partial^2 \hat{\mathbf{y}}}{\partial \mathbf{U}^2}$) we do a similar treatment to obtain this matrix, the resulting matrix is in $\mathbb{R}^{m \times m}$ using a second order finite difference stencil for the second derivative. This requires that the first elements of the input vector of the neural network correspond to the control inputs as seen in Fig. 3*(C)*.

4 Results

We test our controller on the path following tasks that we named: 1) *Eight*, 2) *Pringle* and 3) *Line*, and which are defined below. We describe the *Eight* reference trajectory as $\mathbf{y_{ref}}(t) = \{Asin^2(\omega t) + y_{0,0}, Bsin(\omega t)cos(\omega t) + y_{1,0}, Csin(\omega t) + y_{2,0}\}^T$, where ω is the frequency of the wave described by this path. Parameters A and B are unitless multipliers that indicate amplitude and extension of the reference path geometry. Figure 5 shows the difference between the true location of the end effector compared with the estimated location of the end effector for three different recurrent models used in this work. Table 1 summarizes the root mean squared error statistics over 10 runs of each of the paths described. *Pringle* is defined as a hyperbolic paraboloid $\mathbf{y_{ref}}(t) = \{A(y_2^2/B^2 - y_1^2/C^2) + y_{0,0}, Bcos(2\pi\omega)t + y_{1,0}, Csin(2\pi\omega)t + y_{2,0}\}^T$, and *Line* is defined as $\mathbf{y_{ref}}(t) = \{A(y_2^2/B^2 - y_1^2/C^2) + y_{0,0}, Bsin(2\pi\omega t + 10^{-6}t^2) + y_{1,0}, Csin(2\pi\omega t + 10^{-6}t^2) + y_{2,0}\}^T$.

Table 1. Summary of the RMSE statistics in three different paths.

Model	n Parameters[a]	*Eight*	*Pringle*	*Line*
FC	243	1.36 ± 0.21 mm	1.93 ± 0.48 mm	2.02 ± 1.24 mm
GRU	435	1.29 ± 0.28 mm	1.56 ± 0.58 mm	2.01 ± 1.19 mm
LSTM	570	1.42 ± 0.39 mm	2.35 ± 0.23 mm	2.56 ± 1.44 mm

[a] Total number of parameters in the neural network, counting weights and biases.

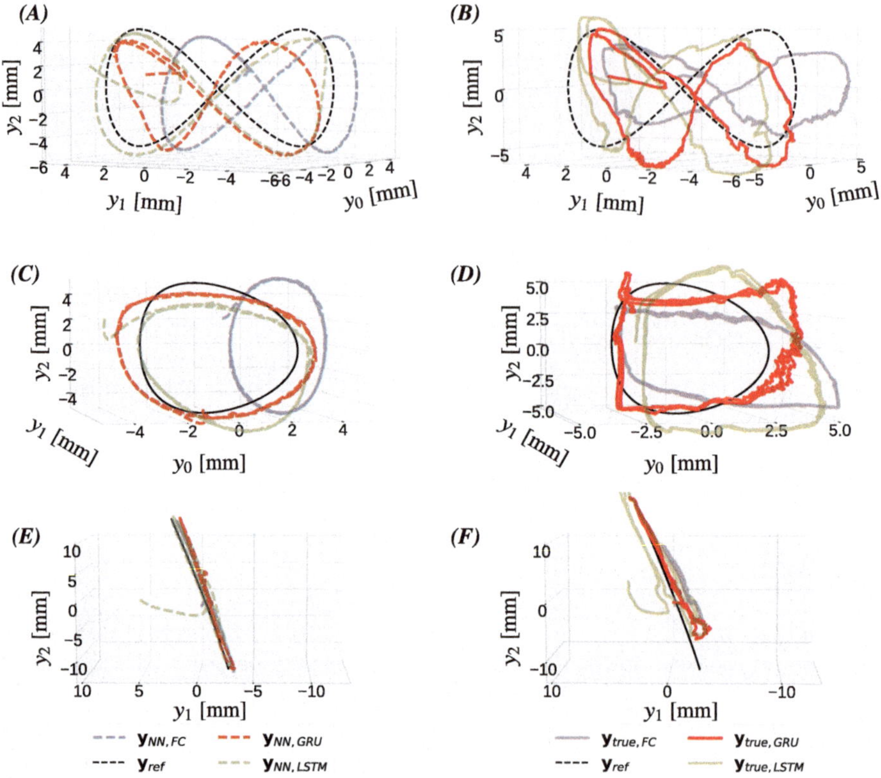

Fig. 5. 3D trajectory tracking. Left column (A, C, E) indicates the mean predicted trajectories and the right column (B, D, F) indicates the mean of the true trajectories. Best performing model from Table 1 is highlighted in each task.

Microcontroller Considerations. We timed the computation required to perform model prediction, first and second derivatives of the model in an Tensilica Xtensa LX6 microprocessor (ESP32, Espressif Systems) in the Arduino IDE for $N = 3$ and obtained around 0.25 ms, 5.86 ms for Eq. 5 and 9.07 ms for prediction, first derivative and second derivative, averaged over 5,000 steps using *nn4mc* [10] on the FC model presented in this work. When we use shared operations for the two derivatives we obtain an average time of 11.35 ms to obtain the two derivatives. These preliminary results are promising to implement the algorithm presented in commodity microcontrollers.

Mechanical and Electromagnetic Disturbance Rejection. We expose the end effector to mechanical and electromagnetic disturbances by loading magnetic balls (Nickel Set, Speks) with a total mass of 100 g at the end effector position using a 37 g crane to hold them. Figure 6 displays the response of both the sensors and the controller to the presence of such disturbances.

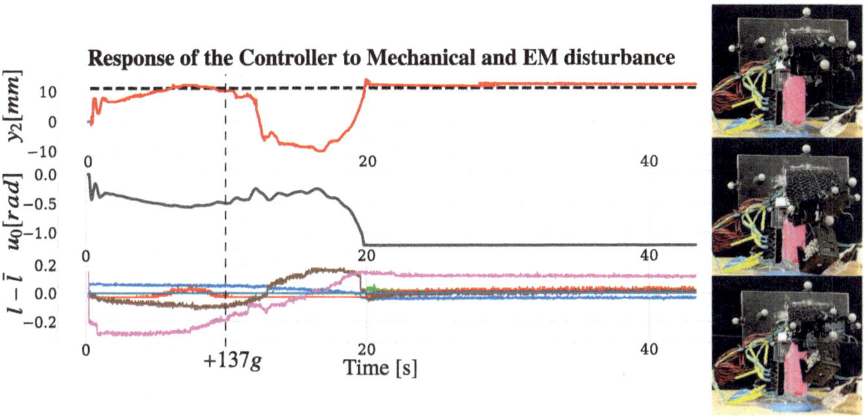

Fig. 6. Disturbance response. Top figure indicates the true position in millimeters. Second figure indicates the changes in servo command that the controller solved for. Bottom figure illustrates changes in light lace sensor signals.

Discussion. From the results displayed in Sect. 4, despite the inaccuracies in the model prediction found in Fig. 4, the optimization algorithm solves for optimal control inputs using the predicted position coming from the neural network. This means that the control performance is associated with the model prediction performance, thus the better prediction power the model has, the higher the control accuracy will be. In Fig. 5 (C) we observe that given the increased model prediction errors in the FC model in the y_0 direction, the prediction curve projects onto another plane, but we obtain a motion close to the desired motion by assigning a lower diagonal value ($1e-3$) in the first row of $\mathbf{Q}$ for the FC controller, which decreases the penalization of errors in that direction. The GRU-based model predicted the off-line testing set data closer and this reflects on the accuracy of the controller based on this model, where we evidence an increased performance relative to the other models compared. We find that even though the past history of inputs and outputs is important for the computation of the control input, when we confound this with the true measurements during the disturbance rejection experiment, models and controller proved to respond to changes in the signal coming from the sensors. This is evidenced by the changes in optimal control input signals during the disturbance rejection experiment in Fig. 6.

Limitations to This Approach. This approach is most effective when the end effector can physically reach the reference path. To prevent the servos from saturation, we clip the optimal control input, servo inputs did not reach saturation during the experiments in Fig. 5, but during the tuning process. This algorithm is most effective with knowledge of the initial state of the end effector for better initial neural network predictions. This initialization can be consistent throughout the experiments. Mechanical disturbance rejection is most effectively achieved when the sensor normalization allows for very sensitive sensor reading patterns. Therefore, the calibration used in Fig. 6 was different than that of the experiments in Fig. 5.

Conclusion. We use parsimonious neural network models in a hierarchical, recursive configuration to predict the forward kinematics of the end effector and develop a nonlinear predictive controller to optimize the control inputs. We achieve sub-centimeter accuracy that compares favorably with those in the related literature. We conclude that light lace networks [6] are a reliable sensing modality that is not affected by electromagnetic and moderate mechanical disturbances and can be used for feedback control in truly soft actuation systems. Future work includes embedding the algorithm into an off-the-shelf platform with limited resources to deploy this controller into a distributed system by approximating the Newton-Raphson solution into a closed form $\mathcal{O}(1)$ expression. We also plan to extend this controller to perform online learning of an evolving system's kinematics, further building up on our open-source nn4mc framework [10].

Acknowledgement. We would like to thank Dr. Don Soloway, Prof. Bradley Hayes and CAIRO Lab at CU Boulder and Cooper Simpson. This research has been supported by the Air Force Office of Scientific Research (Grant No. 83875–11094), we are grateful for this support.

References

1. Chin, K., Hellebrekers, T., Majidi, C.: Machine learning for soft robotic sensing and control. Adv. Intell. Syst. **2**, 1900171 (2020). https://doi.org/10.1002/aisy.201900171
2. Hyatt, P., Wingate, D., Killpack, M.D.: Model-based control of soft actuators using learned non-linear discrete-time models. Frontiers Robot. AI **6**, 2296–9144 (2019). https://doi.org/10.3389/frobt.2019.00022
3. Gillespie, M.T., Best, C.M., Townsend, E.C., Wingate, D., Killpack, M.D.: Learning nonlinear dynamic models of soft robots for model predictive control with neural networks. In: 2018 IEEE International Conference on Soft Robotics (RoboSoft), Livorno, pp. 39–45 (2018). https://doi.org/10.1109/ROBOSOFT.2018.8404894.

4. Bruder, D., Gillespie, B., Remy, C.D., Vasudevan, R.: Modeling and control of soft robots using the koopman operator and model predictive control. arXiv preprint arXiv:1902.02827 (2019)
5. Giorelli, M., Renda, F., Calisti, M., Arienti, A., Ferri, G., Laschi, C.: Neural network and Jacobian method for solving the inverse statics of a cable-driven soft arm with nonconstant curvature. IEEE Trans. Robot. **31**(4), 823–834 (2015). https://doi.org/10.1109/TRO.2015.2428511
6. Xu, P.A., Mishra, A.K., Bai, H., Aubin, C.A., Zullo, L., Shepherd, R.F.: Optical lace for synthetic afferent neural networks. Sci. Robot. **4**(34), eaaw6304 (2019)
7. Soloway, D., Haley, J.P.: Neural generalized predictive control. In: Proceedings of the 1996 IEEE International Symposium on Intelligent Control (1996)
8. Thuruthel, T.G., Shih, B., Laschi, C., Tolley, M.T.: Soft robot perception using embedded soft sensors and recurrent neural networks. Sci. Robot. **4**(26), eaav1488 (2019)
9. Hughes, D., Heckman, C., Correll, N.: Materials that make robots smart. Int. J. Robot. Res. **38**(12–13), 1338–1351 (2019)
10. Aguasvivas Manzano, S., Hughes, D.T., Simpson, C.R., Patel, R., Correll, N.: Embedded neural networks for robot autonomy. In: 2019 International Symposium on Robotics Research (ISRR), Hanoi (2019)
11. Grant, M., Boyd, S.: CVX: Matlab software for disciplined convex programming. version 2.0 beta. http://cvxr.com/cvx September 2013
12. Huber, P.J.: Robust estimation of a location parameter. Ann. Stat. **53**(1), 73–101 (1964). https://doi.org/10.1214/aoms/1177703732

Within-Hand Manipulation Planning and Control for Variable Friction Hands

Gokul Narayanan[1], Joshua Amrith Raj[1], Abhinav Gandhi[1], Aditya A. Gupte[1], Adam J. Spiers[2], and Berk Calli[1(✉)]

[1] Robotics Engineering Department, Worcester Polytechnic Institute, Worcester, MA 01609, USA
{gsathyanarayanan,jcalebchanthiraj,agandhi2,aagupte,bcalli}@wpi.edu
[2] Department Haptic Intelligence, Max Planck Institute for Intelligent Systems, Stuttgart 70372, Germany
a.spiers@is.mpg.de

Abstract. In this work, we present three vision-based within-hand manipulation methods for a variable friction hand. The system is able to plan and execute actions such as in-hand sliding and large object rotation, which are required for many within-hand manipulation tasks. We present an experimental study using objects with various geometries, and show that all the methods can reliably achieve given target object poses. We discuss the advantages and disadvantages of the proposed methods with respect to positioning accuracy and efficiency.

1 Motivation, Problem Statement, Related Work

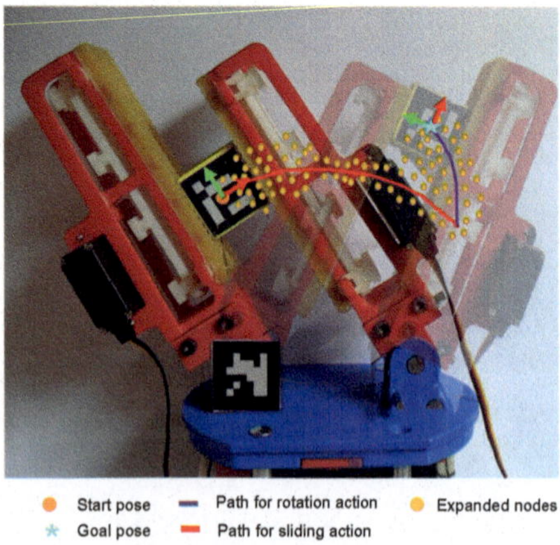

Fig. 1. WIHM from an arbitrary start pose to a desired pose via automatic planning and control using a variable friction hand.

Humans heavily rely on their within-hand manipulation (WIHM) capabilities for performing their everyday activities [1]. These actions for re-orienting and re-positioning objects within hand help people streamline manipulation operations by achieving task-oriented hand-object configurations without the need for re-grasping (e.g. re-positioning a pen or a key after picking-up to make the objects ready for use). Controlled within-hand sliding and large object reorientation are two prominent actions that are frequently needed in WIHM, but are greatly

G. Narayanan and J. A. Raj—Contributed equally to this publication.

This work was supported in part by the National Science Foundation under grant IIS-1900953.

B. Siciliano et al. (Eds.): ISER 2020, SPAR 19, pp. 600–610, 2021.
https://doi.org/10.1007/978-3-030-71151-1_53

challenging for robots even with sensorized, high-DOF, expensive hands operating in structured settings with accurate models [4,9,17].

In our previous work [14], we proposed a simple and open-source hand design that can change the effective friction of its finger surfaces to provide reliable within-hand sliding and large reorientation of a grasped object beyond the capabilities of the prior designs. These abilities were demonstrated with hard-coded actions and manual teleoperation. In this work, we expand on these abilities with the following additions (Fig. 1):

1. A generalized position and velocity-level kinematics model of the hand-object system that can work with arbitrary polygon objects,
2. Three novel approaches for conducting automatic WIHM:
   a. A feedforward control method based on a WIHM planner,
   b. A feedback control method using a visual servoing scheme,
   c. A hybrid method that combines the feedforward and feedback methods,
3. Experimental results using five objects with different shapes and sizes (in total 162 experiments),

We would like to clarify our assumptions and system properties for this current work as follows:

1. We only consider planar WIHM (object position and orientation in the grasp plane). While the variable friction mechanism can enable various 6D manipulation strategies, they are out of the scope of this paper.
2. Our system can operate with or without a support plane (i.e. when the object is lifted). All the presented experiments are without the support plane.
3. We assume that a geometric model of the object is provided. Nevertheless, this model does not need to be accurate since our feedback methods can compensate for modelling imperfections.
4. We assume that the target object has at least one flat surface to allow within-hand sliding.

Conventional WIHM methods (e.g. [12,13]) rely on accurate hand-object models, which are infeasible to obtain, whereas recent methods that relax this reliance via mechanical compliance [3] or impedance control [15] fail to achieve controlled sliding and large rotations. Our framework operates in the quasi-static level and provide reliable and precise WIHM without accurate hand-object models.

Dafle et al. [5] presented a method to perform within-hand manipulation via extrinsic dexterity by utilizing external contacts. Similarly the formulations in [13] perform controlled sliding via inertial forces and contacts with the environment. These methods, in addition to relying on the environment, also require accurate hand-object kinematics and dynamics models. In contrast, the variable friction hand does not rely on extrinsic contacts and can work with simple kinematic models with quasi-static assumptions, which greatly simplifies the planning and control approaches and boosts the system's robustness.

Finger-gaiting [15] is a method that uses three or more fingers to perform within-hand manipulation tasks without the use of extrinsic dexterity. However, these approaches require sophisticated planning methods. The methods in [2,3] take advantage of the adaptive nature of the underactuated hands and utilize vision-based control schemes for achieving accurate WIHM. Nonetheless, these methods are limited to roling contacts.

Our hand with variable friction fingers [14] performs within-hand sliding and within-hand rotation of the object by leveraging the ability to change the friction properties of the fingers. As presented in this paper, using a simple kinematics model of the gripper system and a feedback loop control, we are able to perform within-hand manipulation tasks accurately and generalize to objects of different sizes and shapes. These methods are also applicable for recently developed mechanisms that were inspired by our design [10,11], or for similar systems such as [7,16].

2 Technical Approach

In this section, we briefly discuss the functionalities of the gripper, the kinematics model and the planning and control schemes we developed. Friction plays an important role in WIHM. Effective frictional forces between the finger and object surfaces can change the outcome of a particular finger motion on the object pose. The variable friction (VF) gripper is designed to alter the friction coefficients of the robotic fingers. Each finger can be set to a high friction or a low friction state. By setting the friction configuration appropriately, we are able to achieve various WIHM behaviours. As such, moving the fingers during the following friction conditions 1) {*left finger, right finger*} = $\{high, low\}$, 2) $\{low, high\}$ and 3) $\{high, high\}$ leads to 1) sliding of the object along the right finger, 2) sliding of the object along the left finger and 3) rotating the object. For the hand's working principles, please refer to [14] and the multimedia attachment.

Kinematics Model: The WIHM kinematics provide a mapping between the actuator space, the finger space and the object pose in the Cartesian space. Our mechanism is a switching system; the motion of the object depends on the finger friction configuration, and we derive a kinematics model for each of the above-mentioned {*left finger, right finger*} configurations. The model variables d_L, d_R, θ_L and θ_R are depicted in Fig. 2 for the case of the object sliding along the right finger. Here, d_L is the distance from the left finger joint to the object's bottom corner and d_R is the distance from the right finger joint to the object's bottom corner. θ_L and θ_R are the angles of the fingers. When the object slides, one of its surfaces maintains contact with the high-friction surface while the opposite surface slides against the low-friction finger. This constraint of the object being effectively stationary against one finger is used to find the position of the object center as follows (for sliding against the right finger case):

$$\begin{bmatrix} x \\ y \end{bmatrix} = \begin{bmatrix} (d_L + b)cos(\theta_L) + (a + f_w)sin(\theta_L) \\ (d_L + b)sin(\theta_L) - (a + f_w)cos(\theta_L) \end{bmatrix} \tag{1}$$

A similar approach can be followed to derive the kinematics model for sliding along the other finger. The velocity of the object while sliding on the right finger can be obtained by differentiating Eq. (1) with respect to time. After simplifying the equation and using Eq. (1) we get,

$$\begin{bmatrix} \dot{x} \\ \dot{y} \end{bmatrix} = \begin{bmatrix} -y \\ x \end{bmatrix} \dot{\theta}_L \tag{2}$$

We observe that the Jacobians for the sliding motion on the left and the right finger are independent of the variables $d_L, d_R, \theta_L, \theta_R$ and the dimensions of the object.

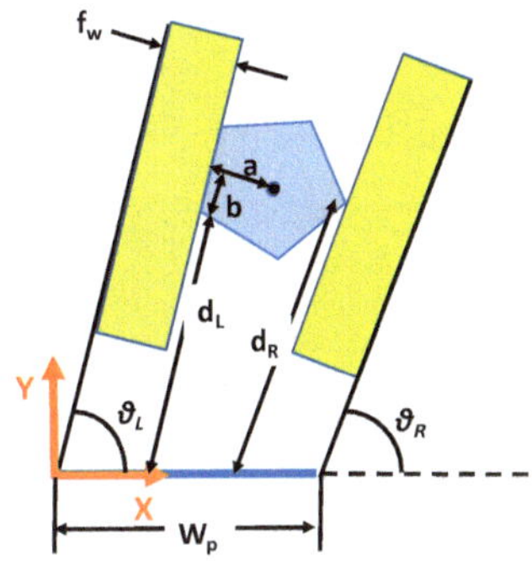

Fig. 2. Parameters of the kinematic model for the hand-object system

These equations are utilized to derive the planning and control schemes while also considering the non-holonomic and switching nature of the hand. We propose three approaches: a feedforward control method based on an WIHM planner, a feedback control method using a visual servoing scheme, and a hybrid method that combines these two methods.

WIHM Planner: For the WIHM planning problem, the state of the object is expressed in the finger space, i.e. $s = (d_L, d_R, \alpha)$, where α is the object orientation with respect to the finger. This provides an intuitive way to formulate cost functions for the actions (an alternative way would be to use the object pose in the Cartesian space, which has a one-to-one mapping to the finger space variables for the feasible object poses, and a very similar formulation would follow).

An A* planning framework is adopted to find the sequence of actions that leads the object to any target pose. The actions are defined as A = (slide_up_on_left_finger, slide_up_on_right_finger, slide_down_on_left_finger, slide_down_on_right_finger, rotate_left, rotate_right). We prefer to obtain a path with few friction state switches, since frequent switching results in non-smooth trajectories and increases the execution time. Applying the conventional A* algorithm to our problem generates a sequence of actions connecting the start state to the goal state with the lowest possible cost. The cost is given by

$$f(s') = g(s') + h(s') \tag{3}$$

where s' is the next state after an action is executed, $f(s')$ is the estimated cost from the start state to goal state, through s'. $g(s')$ is the accumulated cost, and $h(s')$ is the estimated cost to reach goal state from s' (Manhattan distance towards the goal). The trajectory generated using this formulation could (and often does) result in chattering, which means the sliding actions need to switch between the left and right fingers frequently in order to reach the goal position in the shortest possible way. Such chattering is highly undesired for our system, since it requires frequent switching between finger modes (i.e.

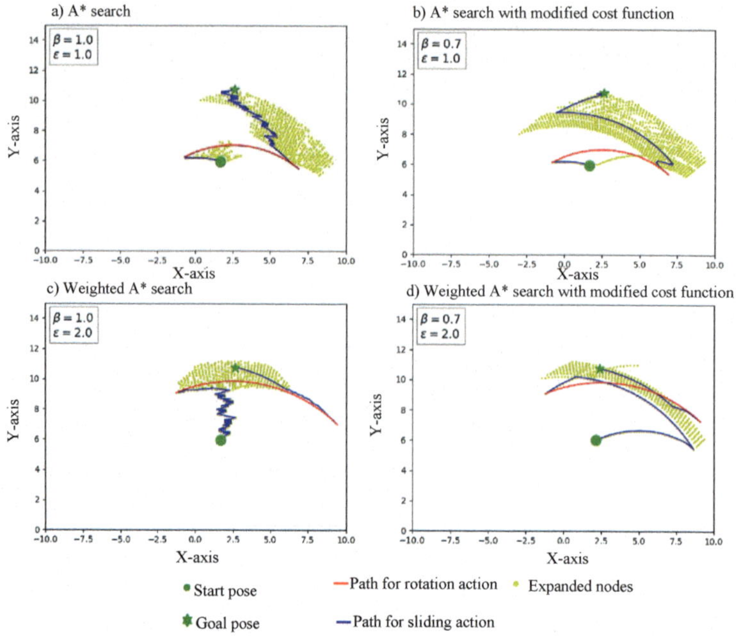

Fig. 3. Path generated by different A* searches

$\{high\_friction, low\_friction\}$ to $\{low\_friction, high\_friction\}$ and vice versa), which does not only take time, but also causes wear and tear of the hardware. Smooth paths can be achieved by relaxing the optimality constraint, and balancing between the path smoothness and the path length. This is achieved by assigning low costs to nodes which are expanded by the same action as their parent node. We define the heuristic cost $h(s)$ of a node as follows:

$$h(s) = \beta * h(s) \tag{4}$$

where

$$\beta = \begin{cases} 0 < \beta < 1, & \text{if } a == P_a \\ 1, & \text{otherwise} \end{cases} \tag{5}$$

In these equations, a is the action of the current node and P_a is the action of the parent node. The modified heuristic cost function makes the algorithm choose a smoother trajectory by favoring the parent node action, sometimes at the expense of not obtaining the shortest path. Figure 3-a presents the unmodified heuristic, and Fig. 3-b shows the trajectory with the modified heuristic. It can be observed that the path is smoother with the modified heuristic, but the number of nodes expanded are more than unmodified heuristic which makes the search algorithm significantly slower.

Therefore, we implement a weighted A* algorithm with a scaled heuristic similar to [8]:

$$f(s) = g(s) + \epsilon * h(s) \qquad where\ \epsilon > 1 \tag{6}$$

The ϵ parameter creates a bias for expanding more nodes closer to the goal, which speeds up the search (Fig. 3-c). The combination of the modified heuristic and weighted A* generates smooth paths with shorter amount of time (Fig. 3-d) compared to the classical formulation.

Visual Servoing for WIHM:
A position-based visual servoing method is designed as follows: We decompose the task into orientation correction and position correction in the finger space. We correct the positioning errors by sliding the object either along the left finger or along the right finger. We derive separate Jacobians for each of these finger friction states, and in each time step, we calculate both of the Jacobians, calculate associated object velocity vectors, and choose the one that is closest to the optimal path in the image space (i.e. a linear path).

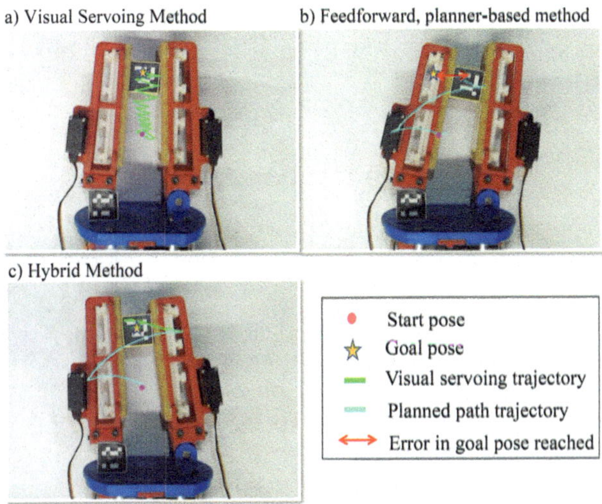

Fig. 4. Trajectory of the object from start pose to goal pose for the proposed three methods (only position correction case).

A complete rotation of the object (e.g. 90° for a rectangular prism) cannot be initiated from all points in the workspace, since this action requires large object displacements that may lead the object to the workspace limits. The motion planner finds a path that brings the object to a location where rotation is feasible, but visual servoing algorithm is essentially a local optimizer, and lacks the mechanisms to do so. In order to overcome this disadvantage, we first move the system close to the edges of the workspace, where rotations can easily be conducted (right-side edge for clockwise rotation and left-side edge for anti-clockwise rotation of the object), and then execute the rotation action. These locations are predetermined and hard-coded. Once the rotation is corrected, we continue executing position correction steps. The object trajectories obtained with this strategy may end up being far from optimum. Nevertheless, combining the offline planner and the visual servoing algorithm provides an alternative solution as follows.

The Hybrid Method: Executing the path of the offline motion planner directly on the real hardware results in significant inaccuracies (as shown in the Fig. 4-b) mainly due to the inaccuracies of the hand-object model. While utilizing vision feedback can make the system robust to these inaccuracies, using it in a standalone manner may not be preferable either since 1) we need to utilize some rules to manage the discrete rotation motions, which results in far-from-optimal trajectories, 2) high frequency chattering happen during convergence (Fig. 4-a). Therefore, we propose a hybrid controller, in which the offline planner generates 'via points', and the visual servoing algorithm tracks these intermediate goals in a close loop system. Such an approach results in both smooth and accurate WIHM Fig. 4-c. In the next sections, we analyze the performance of the motion planner, the pure visual servoing approach and the hybrid approach with experiments.

3 Experiments and Results

We conducted experiments with the setup presented in Fig. 5. The support plane below the object is removed after the initial grasp is achieved so that the WIHM is conducted in the air. To simplify the object tracking aspect and to focus on the controller performances, we used ArUco markers and the associated tracking packages [6]. We conducted experiments with all three control methods (feedforward, feedback and hybrid) using five different objects shown in Fig. 6. Four goal poses are generated: *goal-1* which requires only position correction, *goal-2* which requires position correction and one clockwise rotation, *goal-3* which requires position correction with one anti-clockwise rotation, and *goal-4* which requires position correction with two clockwise rotations. Here, we define "one rotation" as switching the contact to the adjacent edge of the object. For example, for the square prism, one rotation means a 90-degree rotation with respect to the finger surface. In our experiments, these discrete rotation references are always achieved reliably. Therefore, we do not report rotation errors since they are all zero.

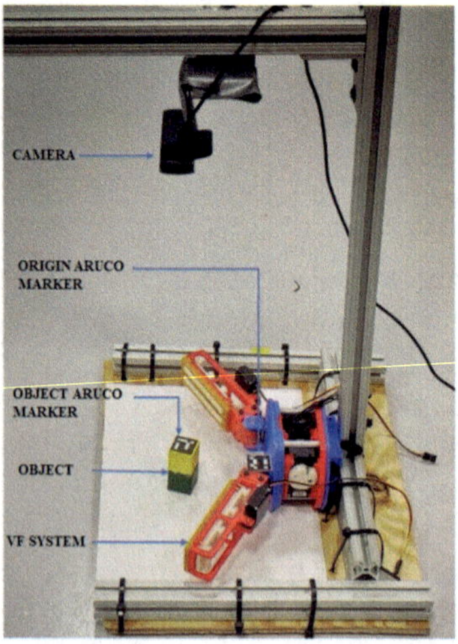

Fig. 5. The experimental setup.

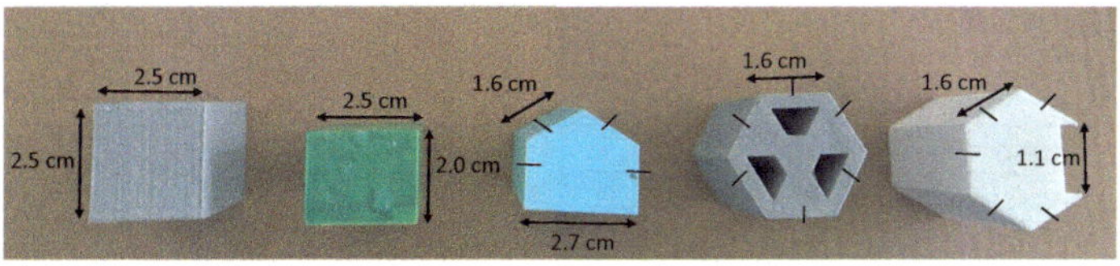

Fig. 6. Objects used in the experiments

The results with different object geometries are summarized in Table 1 (5 objects × 3 methods × 4 goals = 60 experiments total). The planner provides the most optimal path with the least number of switching, and takes the least amount of time to execute the task. However, it results in significant steady state error due to the modeling inaccuracies. Visual servoing method results in accurate positioning, but has longer execution time due to the increased switching and relatively inefficient rotation strategy. The hybrid method combines the advantages of the planner and visual servoing; it provides an accurate and efficient strategy for WIHM.

Using only the square object, we also conducted 20 experiments with each method with randomly generated goal poses (60 experiments) in order to identify any failure modes. The system was successfully able to conduct all the tasks. The results are summarized in Table 2. These results show very similar characteristics to the previous set of experiments.

In addition, we apply the *goal-4* reference 10 times repeatedly to each method using the square prism in order to study the repeatability of each method. From the results are shown in Table 3, we observe that the feedforward motion planner-based method show a high standard deviation, whereas feedback methods keep the error low consistently. The transient behavior show relatively minor fluctuations for all three methods.

Finally, in order to analyze each method's robustness to modeling inaccuracies, we also conducted experiments with the square prism object by introducing 20% error for the object size in the model. The results are presented in Table 4. The feedforward controller performance drops in terms of accuracy, path length and execution time. The visual servoing and hybrid methods generally maintained their performance.

Table 1. Averaged results for experiments with 5 different objects and 4 different goals

	Error (cm)	Switching Actions	Path length (cm)	Time Taken (s)
Motion Planner	1.21 ± 0.55	4.11 ± 0.97	24.52 ± 4.54	24.88 ± 3.65
Visual Servoing	0.41 ± 0.04	6.36 ± 1.19	28.11 ± 6.21	30.15 ± 3.70
Hybrid	0.38 ± 0.05	5.13 ± 1.08	26.76 ± 4.36	28.77 ± 3.88

Table 2. Results with square prisms with 20 randomly generated goals

	Error (cm)	Switching Actions	Path length (cm)	Time Taken (s)
Motion Planner	1.38 ± 0.57	4.35 ± 0.87	21.66 ± 5.42	25.30 ±2.49
Visual Servoing	0.39 ± 0.09	6.15 ± 0.74	27.73 ± 7.51	30.8 ± 3.27
Hybrid	0.37 ± 0.03	5.65 ± 0.98	26.51 ± 5.28	27.99 ± 4.62

Table 3. Results using square prisms with *goal*-4 as the desired pose

	Error (cm)	Switching Actions	Path length (cm)	Time Taken (s)
Motion Planner	1.29 ± 0.43	5.00 ± 0.0	23.54 ± 3.71	27.31 ± 2.03
Visual Servoing	0.41 ± 0.05	5.90 ± 0.38	27.96 ± 4.34	33.12 ± 3.01
Hybrid	0.38 ± 0.09	5.50 ± 0.40	25.92 ± 4.20	32.78 ± 3.17

Table 4. Summary of experiments with square prism focusing on modeling inaccuracies

	Error (cm)		Switching Actions		Path length (cm)		Time taken (s)	
	Correct Object Size	Wrong Object Size	Correct Object Size	Wrong Object Size	Correct Object Size	Wrong Object Size	Correct Object Size	Wrong Object Size
Motion Planner	1.16 ± 0.63	1.97 ± 0.29	4.25 ± 0.95	4.25 ± 0.95	20.54 ± 6.20	23.13 ± 6.96	24.25 ± 4.99	25.5 ± 4.57
Visual Servoing	0.40 ± 0.05	0.36 ± 0.03	5.75 ± 1.50	7.75 ± 0.5	26.52 ± 6.21	26.16 ± 9.66	32.25 ± 4.57	30.00 ± 9.12
Hybrid	0.29 ± 0.09	0.15 ± 0.08	5.75 ± 1.25	5.5 ± 1.73	26.56 ± 6.95	25.59 ± 9.56	29.00 ± 7.12	34.75 ± 4.27

4 Experimental Insights and Conclusions

It is important to note that the presented WIHM tasks achieved with our hand and the proposed controllers are beyond the capabilities of any other within-hand manipulation system in literature; to the best of our knowledge, there does not exist any other system that can automatically conduct controlled within-hand sliding and large rotations without heavily relying on accurate models. The systems in [7,10,11,16] are mechanically capable of similar actions, but no automatic WIHM schemes are proposed for those hands. Nevertheless, we believe that the controllers proposed in this work can easily be adopted to such systems.

Our experimental results show that each method has its advantages and disadvantages and can be utilized for different purposes. The feedforward method can be used in cases when smooth and fast executions are required and when the accuracy is not a major concern. Visual servoing algorithm can provide high-accuracy without requiring any planning, but it can cause chattering (since it tries to follow a straight path with the non-holonomic system), and its orientation correction strategy is inefficient. The hybrid method requires planning, but can provide smooth trajectories and accurate results. Some of these aspects can be observed in the sample trajectories in Fig. 4.

The model knowledge required by our system is simpler than the conventional methods. We do not need to know friction coefficients between the object and the fingers; the mechanism with interlaced high- and low-friction surfaces can achieve controlled sliding and rotation for a large family of objects. Our system works in the quasi-static level, therefore does not require any dynamics model. The experiments with erroneous kinematic models (Table 4) shows that the closed loop algorithms are robust to kinematic model inaccuracies as well. We believe that the proposed algorithms and the presented results strongly indicate the importance of integrating variable friction context to the planning and control scheme for reliable and robust WIHM.

References

1. Bullock, I.M., Dollar, A.M.: Classifying human manipulation behavior. In: International Conference on Rehabilitation Robotics, pp. 1–6 (2011)
2. Calli, B., Dollar, A.M.: Robust precision manipulation with simple process models using visual servoing techniques with disturbance rejection. IEEE Trans. Automat. Sci. Eng. **16**(1), 406–419 (2018)
3. Calli, B., Kimmel, A., Hang, K., Bekris, K., Dollar, A.: Path planning for within-hand manipulation over learned representations of safe states. In: International Symposium on Experimental Robotics (ISER) (2018)
4. Chong, N.Y., Choi, D., Suh, I.H.: A generalized motion/force planning strategy for multifingered hands using both rolling and sliding contacts. In: IEEE/RSJ International Conference on Intelligent Robots and Systems, pp. 113–120 (1993)
5. Dafle, N.C., Rodriguez, A., Paolini, R., Tang, B., Srinivasa, S.S., Erdmann, M., Mason, M.T., Lundberg, I., Staab, H., Fuhlbrigge, T.: Extrinsic dexterity: In-hand manipulation with external forces. In: IEEE International Conference on Robotics and Automation (ICRA), pp. 1578–1585. IEEE (2014)
6. Garrido-Jurado, S., Muñoz-Salinas, R., Madrid-Cuevas, F.J., Marín-Jiménez, M.J.: Automatic generation and detection of highly reliable fiducial markers under occlusion. Pattern Recogn. **47**(6), 2280–2292 (2014)
7. Govindan, N., Thondiyath, A.: Design and analysis of a multimodal grasper having shape conformity and within-hand manipulation with adjustable contact forces. J. Mech. Robot. **11**(5), 051012 (2019)
8. Hansen, E.A., Zhou, R.: Anytime heuristic search. J. Artif. Int. Res. **28**(1), 267–297 (2007)
9. Karayiannidis, Y., Pauwels, K., Smith, C., Kragic, D., et al.: In-hand manipulation using gravity and controlled slip. In: IEEE/RSJ International Conference on Intelligent Robots and Systems, pp. 5636–5641 (2015)
10. Lu, Q., Clark, A.B., Shen, M., Rojas, N.: An origami-inspired variable friction surface for increasing the dexterity of robotic grippers. IEEE Robot. Automat. Lett. **5**(2), 2538–2545 (2020)
11. Nojiri, S., Mizushima, K., Suzuki, Y., Tsuji, T., Watanabe, T.: Development of contact area variable surface for manipulation requiring sliding. In: IEEE International Conference on Soft Robotics, pp. 131–136 (2019)

12. Salisbury, J.K., Mason, M.: Robot Hands and the Mechanics of Manipulation. MIT Press, Cambridge (1985)
13. Shi, J., Woodruff, J.Z., Umbanhowar, P.B., Lynch, K.M.: Dynamic in-hand sliding manipulation. IEEE Trans. Rob. **33**(4), 778–795 (2017)
14. Spiers, A.J., Calli, B., Dollar, A.M.: Variable-friction finger surfaces to enable within-hand manipulation via gripping and sliding. IEEE Robot. Automat. Lett. **3**(4), 4116–4123 (2018)
15. Sundaralingam, B., Hermans, T.: Relaxed-rigidity constraints: kinematic trajectory optimization and collision avoidance for in-grasp manipulation. Auton. Robot. **43**(2), 469–483 (2019)
16. Tincani, V., Grioli, G., Catalano, M.G., Garabini, M., Grechi, S., Fantoni, G., Bicchi, A.: Implementation and control of the velvet fingers: a dexterous gripper with active surfaces. In: IEEE International Conference on Robotics and Automation, pp. 2744–2750 (2013)
17. Ueda, J., Kondo, M., Ogasawara, T.: The multifingered naist hand system for robot in-hand manipulation. Mech. Mach. Theory **45**(2), 224–238 (2010)

Nonprehensile Riemannian Motion Predictive Control

Hamid Izadinia(✉), Byron Boots, and Steven M. Seitz

University of Washington, Seattle, USA
izadinia@cs.washington.edu

Abstract. Nonprehensile manipulation involves long horizon underactuated object interactions and physical contact with different objects that can inherently introduce a high degree of uncertainty. In this work, we introduce a novel Real-to-Sim reward analysis technique, called Riemannian Motion Predictive Control (RMPC), to reliably imagine and predict the outcome of taking possible actions for a real robotic platform. Our proposed RMPC benefits from Riemannian motion policy and second order dynamic model to compute the acceleration command and control the robot at every location on the surface. Our approach creates a 3D object-level recomposed model of the real scene where we can simulate the effect of different trajectories. We produce a closed-loop controller to reactively push objects in a continuous action space. We evaluate the performance of our RMPC approach by conducting experiments on a real robot platform as well as simulation and compare against several baselines. We observe that RMPC is robust in cluttered as well as occluded environments and outperforms the baselines.

1 Introduction

Nonprehensile object manipulation is an important and yet challenging robotic task that has been studied for more than three decades [20], yet remains an area of active research [17,21]. Nonprehensile manipulation involves long horizon underactuated object interactions and physical contact with different objects that can inherently introduce a high degree of uncertainty. Suppose we want to push an object on a surface cluttered with many other objects with various shapes and properties. In such a scenario, predicting the outcome of a push action is not trivial as pushing an object not only affects its own status but also can change the status of other objects.

Traditionally, motion planning and search algorithms were adopted for nonprehensile manipulation. These algorithms traditionally have high computational cost and rely on full knowledge of a constrained workspace to produce a sequence of kinematic actions [20]. To address the computational complexity brought by uncertainty, the quasi-static planners aim to predict the motion of pushed objects by assuming a physical model of the world involving shape, friction and objects center of mass [18].

More recently, deep learning has been adopted for learning simple robotic manipulation skills using visual sensory input in a controlled scenario [2,15,16].

B. Siciliano et al. (Eds.): ISER 2020, SPAR 19, pp. 611–621, 2021.
https://doi.org/10.1007/978-3-030-71151-1_54

Central to their approach is to train a deep neural network model in a data-driven fashion to predict the cost of applying different robotic actions for accomplishing tasks. While it is essential to reliably predict the cost of taking possible actions in a real robotic task, it is challenging to obtain the data and the supervision needed for training such cost function in less constrained scenarios.

Instead of estimating the cost of different actions, can we instead simulate the real environment in an on-line closed-loop such that we can compute the outcome of taking possible actions inside the simulation? In this paper, we address this question by proposing to analyse and compute rewards through sensing and simulating the real world environment using 3D object-level geometric recomposition, a Riemannian Motion Policy (RMP), and second order dynamics. Our Real-to-Sim reward analysis technique, called Riemannian Motion Predictive Control (RMPC), simulates the effect of different trajectories and action outcomes on a real robotic platform and creates a closed-loop reactive controller to push objects in a continuous action space. To control the robot at every location on the surface, we incorporate a Riemannian Motion Policy [6] and second order dynamic model to compute the acceleration command.

Our goal is to produce an efficient second order dynamic system controller that maintains a continuous trajectory while preserving correct kinematics and smooth dynamics for nonprehensile object manipulation, in the presence of a high amount of uncertainty. To accomplish this goal, we focus on leveraging simulation to predict parameters of our second order controller. Instead of estimating the accurate physical parameters of the scene such as center of mass and surface friction we apply a reactive controller that deploys closed-loop feedback. Consequently, we are capable of correcting the motion if, due to uncertainties involved, the object deviates from the foreseen trajectory.

2 Related Work

Conventionally, planning algorithms has been deployed for solving nonprehensile manipulation by incorporating search techniques to create a trajectory of kinematic movements for a robot arm to push or rearrange objects [7,13,29]. To push various objects on simple trajectories, [9] learns to predicts suitable contact points by optimizing a scoring function trained on histogram features. To reach objects in clutter, [14] deployed Model Predictive Control (MPC) [31] to model the dynamics of a robot arm in contact with the environment. Compared to these past methods, our approach is more general and more efficient and the task we consider is more challenging and involves multiple object interactions.

More recently, deep learning has been incorporated for learning simple manipulation skills with a robot arm. [2] predict the outcome of a pushing action by training a network on pairs of RGB images of before and after push action. This work is extended in [21] to learn a forward models for manipulating a rope with a robot arm in a supervised fashion. Using a series of observations via camera images, [17] tracks a history of push interactions and learns a model for pushing in scenarios with only a single object on the table and quantizing the action

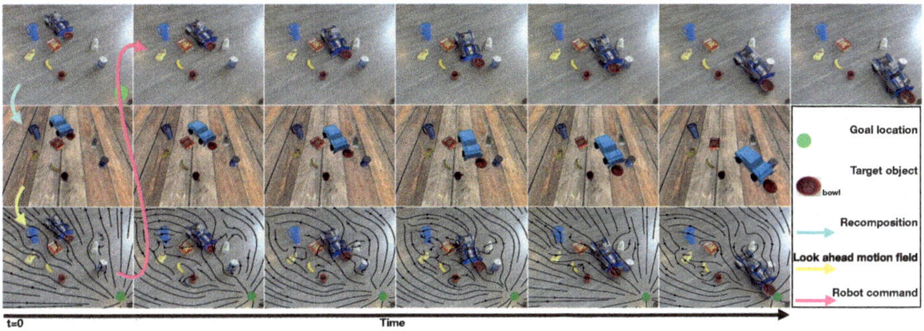

Fig. 1. Overview of our approach. **Top row** shows the rgb snapshots of the point cloud stream input to our model which we use to recompose the scene into a simulated 3D scene (**middle row**) and generate look ahead motion field (**bottom row**).

space. In contrast, our method is designed to be applicable in complex scenarios involving several objects and our controller produces continues actions.

Our work is related to recent efforts for deploying 3D and physics simulation environments for efficiently learning robotic policy. Several past works deployed domain adaptation to transfer policies learned in simulation to the real world [4,24]. Domain randomization was proposed in [26] to train highly generalizable robotic policies inside a randomized simulation environment and was shown to be successfully applicable in various robotic applications [3,11,30]. In [4] pixel-level image generation was incorporated to create auxiliary source of data. [12] learned to translate real and simulation data into a canonical representation to be used as robot observation. Several past works incorporated real images as a complementary source of data for learning control policies inside simulation that can be transferred to the real world [4,25,27]. In contrast to all these works, we generate the simulation environment by recomposing the 3D scene and incorporating simulation physics environment as a computational model for predicting the outcome of robot action trajectories in a closed loop. To our knowledge, closed loop 3D scene recomposition and simulated action look ahead has not been addressed before for efficient policy evaluation inside simulation that is applicable in the real world.

3 Approach

We assume a nonprehensile robotic manipulation setup with N objects $o_{i,i\in\{1,\ldots,N\}}$. Each object has 5 degree-of-freedom (5-DoF) defined by its rotation $R \in \mathbb{SO}(3)$ and translation $T \in \mathbb{R}^2$ and can take any arbitrary pose p in $\mathbb{SE}(2)$ on the plane. All the objects are movable on the plane and can be displaced from a location and pose to another physically feasible pose on the plane so at any time t, the state of the world s_t encapsulates the pose of the objects in the scene $s_t = (p_1, ..., p_N)$. Given a target object o_c and a goal location $l_g = (x_g, y_g)$ as inputs, we want to control the target object o_c by pushing it to l_g. To accomplish

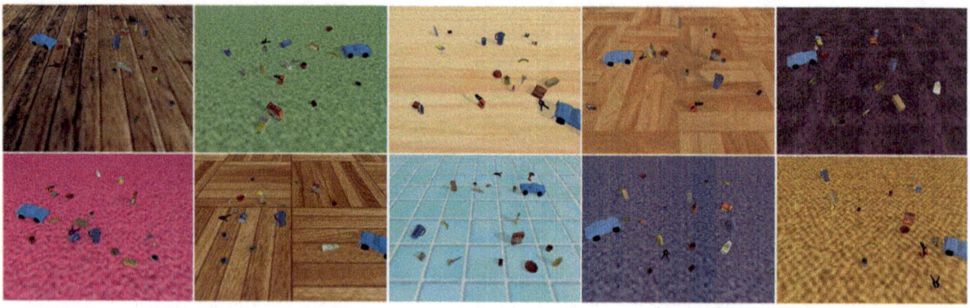

Fig. 2. Sample pushing tasks for RC-car.

that, we need to find the best sequence τ of robot actions $\tau = \{u_0, u_1, ..., u_t\}$ that pushes o_c to l_g while satisfying the constraints such as avoiding collision with other objects as well as maximizing a reward function. Each robot action u is defined by the velocity and acceleration of a push by a robot end-effector, i.e. $u_t = (v_x, v_y, a_x, a_y)$. At every iteration, our goal is to estimate a short trajectory of future actions τ that maximizes the expected sum of future rewards. By applying each action u_t at state s_t, a reward $r_t(s_t, u_t)$ will be obtained which reflects how much the object has progressed in becoming closer to the goal location. If the action u_t results in a collision between the controllable object o_c and any of the objects in the scene $o_{k,k\in\{1,...,N\},k\neq c}$ then a penalty would be considered. More formally, the reward at each step r_t is defined as:

$$r_t(s_t, u_t) = \Delta(d^t_{o_c,l_g}, d^{t-1}_{o_c,l_g}) - \sum_{i=1,i\neq p}^{N} d(p^t_i, p^{t-1}_i) \tag{1}$$

The first term in Eq. 1 reflects how much the distance of o_c to the goal location d_{o_c,l_g} has changed between time $t-1$ and t and d denotes the Euclidean distance. The second term of Eq. 1 quantifies the collision event that might occur by applying action u_t. During one step of push action, if the controllable object o_c collides with any other object o_i, it will result in an unwanted change of 5-DoF pose in o_i. To quantify the collision event, we compare the pose of each object p_i between the time t and $t-1$ by computing the Euclidean distance d between p^t_i and p^{t-1}_i. Accordingly, the reward of a trajectory τ with fixed horizon H is computed as: $R_\tau = \sum_{j=t}^{t+H} \gamma^{j-t} r_t$ with the decay factor of γ.

3.1 Closed-Loop Control

We use RMPflow [6] to compute the sequence of robot actions based on the current robot position and velocity and the configuration of scene objects. The RMPflow combines local policies in their coordinate system and computes a global policy for the robot motion. We use attractor policy for pushing to the goal and collision avoidance policy for avoiding obstacles. By combining these local motion policies the potential field of robot action is computed in continuous

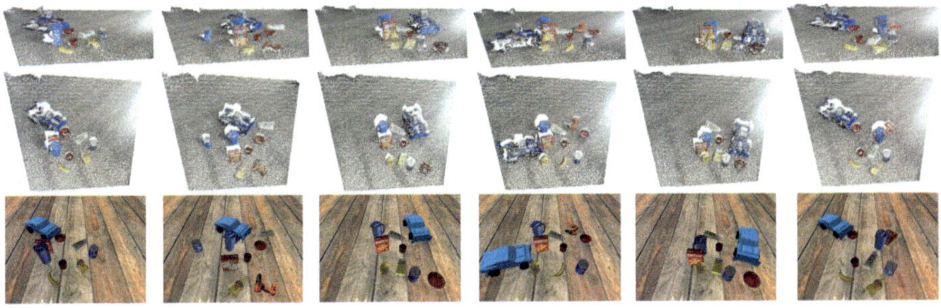

Fig. 3. Examples of fully automatic object-level scene recomposition from the point cloud. The **first** and **second** rows show the side view and the top view of the captured point cloud. The **third** row shows our object-level scene recomposition.

space which determines the robot global motion policy at every location. Suppose the average dimension o_c along its width and length is $\bar{m}_{o_c}$ and its location on the plane is $l_{o_c} = (x_{o_c}, y_{o_c})$. The local policy in the coordinate of the target object rotated by the orientation of the RMP global policy for an immediate pushing action is defined by $(v_x, v_y) = (\alpha_1 x^2 - \alpha_2 y^2 - \alpha_3(\frac{\bar{m}_{o_c}}{2})^2, \alpha_4 xy)$ and α_is are the hyperparameters of our local potential field.

3.2 Riemannian Motion Predictive Control

Using the computed RMPflow [6] we can select an action u_t at each time step t to find the optimum trajectory to push o_c to l_g. However, in dynamic systems with a high degree of uncertainty selecting the actions purely based on RMPflow may not be optimal specially in underactuated tasks [19] such as ours where the target object makes and breaks contact the with robot end-effector and other objects. Therefore, we propose Riemannian Motion Predictive Control (RMPC) where instead of fixing a single parameter for computing the RMPflow, we consider a range of weights for each node. By sampling from the different node weights, we produce different pushing trajectories at every time step t and select the next best action based on the trajectory with the highest estimated reward using Eq. 1.

Designing an accurate reward function and learning to predict the outcome of different action sequences is challenging due to inefficiencies in collecting real robot data and inherent uncertainties in under-actuated and non-prehensile object manipulation. To alleviate those challenges, our proposed method estimates the reward of different sampled action trajectories by recomposing the 3D scene and imagining the outcome of different action trajectories inside a 3D physics simulation. We leverage the power of 3D physics simulation to look ahead the outcome of pushing an object along different trajectories by proposing to recompose the 3D scene at each step during the trajectory. In our setup, we have a depth camera that captures the scene and we incorporate sensory point cloud input for object-level scene recomposition. Using the captured point

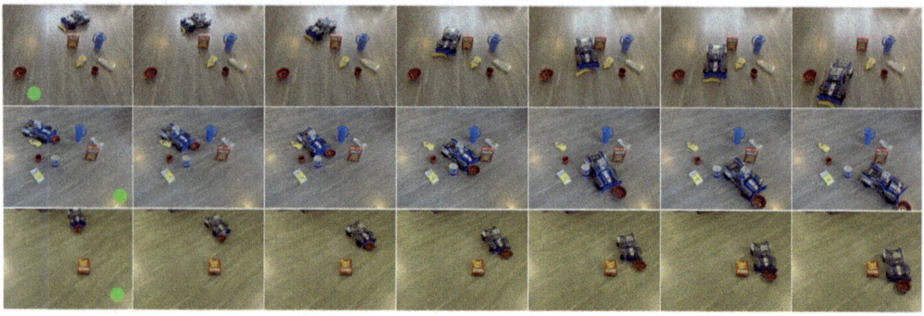

Fig. 4. Real world RC-car robot action trajectory generated by proposed RMPC. Each row shows a trial with a different initial location of the target object, other objects as obstacles. The goal location is indicated with a green dot.

cloud, we detect 3D objects in the scene. Once we recomposed the scene using object-level 3D shapes, we migrate the scene to a physics simulator to evaluate the outcome of different pushing trajectories. To train a 3D object detector that is robust to clutter and high amount of occlusion, we generate a high volume of diverse synthetic data in a simulation environment inspired from prior work of [10] and domain randomization [26]. We create a huge number of diverse scenes by randomly dropping objects in a physic simulation. Such data generation approach will capture the realistic physical arrangement of objects on a surface and results in learning a more efficient 3D object detector. Figure 1 illustrates the overview of our approach where scenes captured by point clouds are recomposed in a 3D simulation environment where we can compute the look ahead motion field.

3.3 Implementation Details

We implemented our method using PyTorch [23] and used Bullet physics engine for simulating the scenes [8]. We generate over 1 million simulated scenes diversified with random object arrangements and floor texturing to train our 3D object detector and Fig. 2 provides several examples of the generated scenes. The speed of our recomposition step is 15 frames per second. We set the hyperparameter of our local field to α_is= 1 in all our experiments.

4 Experiments

We conduct experiments both in simulation and in a real world to evaluate the performance of our proposed approach for nonprehensile manipulation. To evaluate the performance of our proposed approach for nonprehensile manipulation we conduct experiments both in simulation and in a real robot platform.

4.1 Experimental Setup

In our evaluations, We use objects of the YCB dataset [5] in simulation and in the real world experiments. For real robot experiments, we build a real RC-car hardware platform on a 1/10 chassis featuring a 4×4 suspension, and non-flat tires inspiring from the prototypes introduced in [1,28]. To control the RC-car we use move commands that define the velocity and steering direction of the RC-car through a ROS interface. In the real experiments, we localize the RC-car robot using the April tag [22] while we detect and localize other scene objects by 3D object detection as explained in Sect. 3.

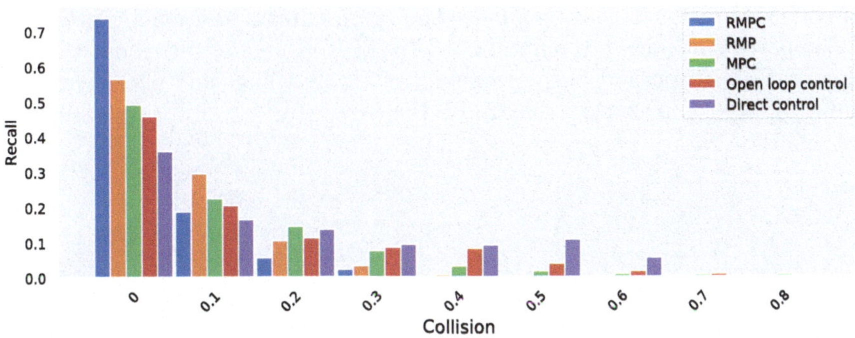

Fig. 5. Results of collision rate on held out test set using proposed Riemannian Motion Predictive Control (RMPC) on RC-car robot platform for pushing target object to goal location compare to prior works. (Higher recall in lower collision rate is better).

4.2 Results on Simulation

To evaluate our proposed motion policy we generate a held-out test set of scenes with a different combination of objects and various arrangements in simulation. To produce diverse scenarios, we randomly place objects in random locations of the workspace surface area and randomly select the target object and goal location in simulation. Figure 2 shows several examples of push tasks at test time. We use 500 held-out test scenes on our simulated RC-car robot platform and compare performance based on the collision rate. We compute the recall rate at the k%-collision ratio as evaluation criteria. For each trial, the collision ratio is computed by dividing the number of collision events by the total length of the trajectory. A higher recall rate at a lower collision ratio indicates better performance.

We compare the performance of our RMPC method against several baselines:

Riemannian Motion Predictive Control (RMPC): Our proposed RMPC recomposes scene from input point cloud and computes different trajectories using RMPflow with different weights. The next best action is selected based on future reward obtained from Eq. 1.

Riemannian Motion Policy (RMP): RMP generates the action trajectory based on the computed RMPflow with a fixed parameter similar to [6].

Model Predictive Control (MPC): MPC samples different trajectories at each step and selects the next action from the trajectory with the smallest cost in a closed-loop. The cost function is computed based on the distance of the target object and the location of the obstacles so the trajectory that moves closer to the obstacles is of higher cost.

Open-Loop Control: This baseline is similar to MPC except that it computes the whole trajectory from the initial location of the target object to the goal location and applies action sequence in an open-loop.

Direct Control: This baseline moves the target object toward goal location without considering collision avoidance and recomputes the action in closed-loop.

Comparison results are summarized in Figs. 5 and 6 and show that our proposed RMPC method outperforms all of the baselines and experiences less number of collisions during the trails. Hence RMPC obtains higher recall at lower collision ratios.

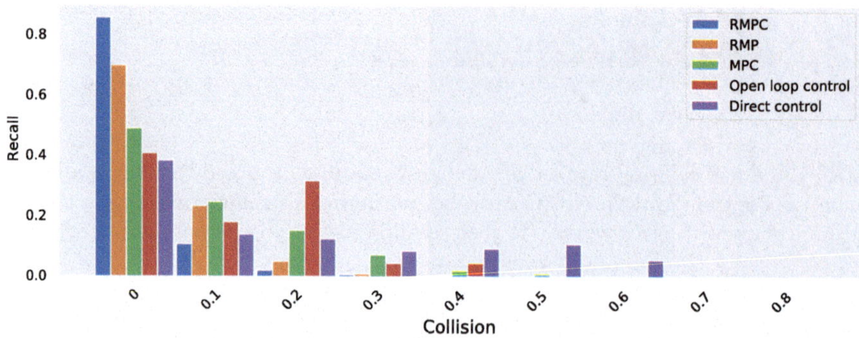

Fig. 6. Collision rate of proposed RMPC controller on RC-car robot platform for successful trajectories where target object is pushed close to the goal location on held out test set compared to prior works. (higher recall in lower collision rate is better).

4.3 Results on Real-World

In all real experiments we capture the scene point cloud using a depth camera. Given the point cloud of a scene, our 3D object detection produces the 3D bounding boxes around each object with their 3D poses as well as category labels from which we can recompose the scene in the physics simulation. Figure 3 depicts several examples of the recomposed real scene in the 3D simulator. As can be seen, our scene recomposition produces simulated scenes that are identical to the real world in terms of object scene arrangement, various object categories, and their poses. Also, our scene recomposition is robust to a high amount of clutter and occlusion in the scene.

We quantitatively compare our proposed RMPC with RMP which we observed to be the strongest baseline in Sect. 4.2. We evaluate the performance by computing the percentage of non-collision push actions in different trials. We evaluate the performance in 32 different real scene arrangements and obtained the average collision rate of 17.9% for RMPC and 35.4% average collision for RMP. Figure 4 shows the example trajectories of real RC-car pushing tasks in three different scenarios. As Fig. 4 shows, our proposed RMPC method produces a closed-loop sequence to successful actions and pushes the controllable object to the goal location (shown by green dot) without colliding to with obstacles in the scene where pushing to a goal location could be accomplished without a collision event.

5 Conclusion

Solving for the parameters of a second order dynamic system controller is challenging when the task involves objects of various properties and complex manipulation under high uncertainty specially in nonprehensile settings. In this paper, we propose a Real-to-Sim approach for analysing the reward of action trajectory in simulation and estimating the parameters of a second order dynamic system controller for nonprehensile manipulation task. With our controller, we produce a closed-loop reactive motion policy. We consider the complex task of pushing objects on a surface in a complex environment with high uncertainty and demonstrate the efficiency of our controller by comparing it to the prior works on a set of nonprehensile manipulation tasks in simulation and real-world RC-car robot platform.

Acknowledgments. This work was supported in part by the University of Washington Animation Research Labs and Google.

References

1. The MIT RACECAR (2016). https://mit-racecar.github.io
2. Agrawal, P., Nair, A.V., Abbeel, P., Malik, J., Levine, S.: Learning to poke by poking: experiential learning of intuitive physics. In: NeurIPS (2016)
3. Andrychowicz, O.M., Baker, B., Chociej, M., Jozefowicz, R., McGrew, B., Pachocki, J., Petron, A., Plappert, M., Powell, G., Ray, A., et al.: Learning dexterous in-hand manipulation. Int. J. Rob. Res. **39**, 3–20 (2020)
4. Bousmalis, K., Irpan, A., Wohlhart, P., Bai, Y., Kelcey, M., Kalakrishnan, M., Downs, L., Ibarz, J., Pastor, P., Konolige, K., et al.: Using simulation and domain adaptation to improve efficiency of deep robotic grasping. In: ICRA (2018)
5. Calli, B., Walsman, A., Singh, A., Srinivasa, S., Abbeel, P., Dollar, A.M.: Benchmarking in manipulation research: the YCB object and model set and benchmarking protocols. arXiv:1502.03143 (2015)
6. Cheng, C.A., Mukadam, M., Issac, J., Birchfield, S., Fox, D., Boots, B., Ratliff, N.: RMPflow: a computational graph for automatic motion policy generation. In: International Workshop on the Algorithmic Foundations of Robotics (2018)

7. Cosgun, A., Hermans, T., Emeli, V., Stilman, M.: Push planning for object placement on cluttered table surfaces. In: IEEE/RSJ International Conference on Intelligent Robots and Systems (2011)
8. Coumans, E.: Bullet physics simulation. In: ACM SIGGRAPH 2015 Courses (2015)
9. Hermans, T., Li, F., Rehg, J.M., Bobick, A.F.: Learning contact locations for pushing and orienting unknown objects. In: IEEE-RAS International Conference on Humanoid Robots (Humanoids) (2013)
10. Izadinia, H., Seitz, S.M.: Scene recomposition by learning-based ICP. In: CVPR (2020)
11. James, S., Davison, A.J., Johns, E.: Transferring end-to-end visuomotor control from simulation to real world for a multi-stage task. arXiv:1707.02267 (2017)
12. James, S., Wohlhart, P., Kalakrishnan, M., Kalashnikov, D., Irpan, A., Ibarz, J., Levine, S., Hadsell, R., Bousmalis, K.: Sim-to-real via sim-to-sim: data-efficient robotic grasping via randomized-to-canonical adaptation networks. In: CVPR (2019)
13. Johnson, A.M., King, J.E., Srinivasa, S.: Convergent planning. IEEE Rob. Autom. Lett. (2016)
14. Killpack, M.D., Kapusta, A., Kemp, C.C.: Model predictive control for fast reaching in clutter. Auton. Rob. **40**, 537–560 (2016)
15. Levine, S., Finn, C., Darrell, T., Abbeel, P.: End-to-end training of deep visuomotor policies. J. Mach. Learn. Res. **17**, 1334–1373 (2016)
16. Levine, S., Pastor, P., Krizhevsky, A., Ibarz, J., Quillen, D.: Learning hand-eye coordination for robotic grasping with deep learning and large-scale data collection. Int. J. Rob. Res. **37**, 421–436 (2018)
17. Li, J.K., Lee, W.S., Hsu, D.: Push-Net: deep planar pushing for objects with unknown physical properties. In: Robotics: Science and Systems (2018)
18. Lynch, K.M., Maekawa, H., Tanie, K.: Manipulation and active sensing by pushing using tactile feedback. In: IROS (1992)
19. Lynch, K.M., Mason, M.T.: Dynamic underactuated nonprehensile manipulation. In: IROS (1996)
20. Mason, M.T.: Compliance and force control for computer controlled manipulators. IEEE Trans. Syst. Man Cybern. **11**, 418–432 (1981)
21. Nair, A., Chen, D., Agrawal, P., Isola, P., Abbeel, P., Malik, J., Levine, S.: Combining self-supervised learning and imitation for vision-based rope manipulation. In: ICRA (2017)
22. Olson, E.: AprilTag: a robust and flexible visual fiducial system. In: ICRA (2011)
23. Paszke, A., Gross, S., Massa, F., Lerer, A., Bradbury, J., et al.: PyTorch: an imperative style, high-performance deep learning library. In: NeurIPS (2019)
24. Rusu, A.A., Večerík, M., Rothörl, T., Heess, N., Pascanu, R., Hadsell, R.: Sim-to-Real robot learning from pixels with progressive nets. In: CoRL (2017)
25. Sadeghi, F.: DIViS: Domain invariant visual servoing for collision-free goal reaching. In: RSS (2019)
26. Sadeghi, F., Levine, S.: CAD2RL: real single-image flight without a single real image. In: RSS (2017)
27. Sadeghi, F., Toshev, A., Jang, E., Levine, S.: Sim2real viewpoint invariant visual servoing by recurrent control. In: CVPR (2018)
28. Srinivasa, S.S., Lancaster, P., Michalove, J., Schmittle, M., Rockett, C.S.M., Smith, J.R., Choudhury, S., Mavrogiannis, C., Sadeghi, F.: MuSHR: a low-cost, open-source robotic racecar for education and research. arXiv preprint arXiv:1908.08031 (2019)

29. Stilman, M., Kuffner, J.J.: Navigation among movable obstacles: real-time reasoning in complex environments. Int. J. Humanoid Rob. **2**, 479–503 (2005)
30. Tremblay, J., Prakash, A., Acuna, D., Brophy, M., Jampani, V., Anil, C., To, T., Cameracci, E., Boochoon, S., Birchfield, S.: Training deep networks with synthetic data: bridging the reality gap by domain randomization. In: CVPRW (2018)
31. Wang, Y., Boyd, S.: Fast model predictive control using online optimization. IEEE Trans. Control Syst. Technol. **18**, 267–278 (2009)

Correction to: Experimental Robotics

Bruno Siciliano, Cecilia Laschi, and Oussama Khatib

Correction to:
B. Siciliano et al. (Eds.): *Experimental Robotics*, SPAR 19, https://doi.org/10.1007/978-3-030-71151-1

The original version of the book was inadvertently published without abstracts in chapters 19, 20. The erratum chapter and the book have been updated with the changes.

The updated version of these chapters can be found at
https://doi.org/10.1007/978-3-030-71151-1_19
https://doi.org/10.1007/978-3-030-71151-1_20

B. Siciliano et al. (Eds.): ISER 2020, SPAR 19, p. C1, 2021.
https://doi.org/10.1007/978-3-030-71151-1_55

Author Index

B. Siciliano et al. (Eds.): ISER 2020, SPAR 19, pp. 623–626, 2021.
https://doi.org/10.1007/978-3-030-71151-1

Printed by Printforce, the Netherlands